Farmland Conservation

Farmland Conservation

Evidence for the effects of interventions in northern and western Europe

Lynn V. Dicks, Joscelyne E. Ashpole, Juliana Dänhardt, Katy James, Annelie M. Jönsson, Nicola Randall, David A. Showler, Rebecca K. Smith, Susan Turpie, David Williams and William J. Sutherland

Synopses of Conservation Evidence, Volume 3

Pelagic Publishing | www.pelagicpublishing.com

Published by Pelagic Publishing
www.pelagicpublishing.com
PO Box 725, Exeter EX1 9QU

Farmland Conservation
Evidence for the effects of interventions in northern and western Europe
Synopses of Conservation Evidence, Volume 3
www.conservationevidence.com

ISBN 978-1-907-807-16-9 (Pbk)
ISBN 978-1-907-807-17-6 (Hbk)
ISBN 978-1-907-807-95-4 (ePub)
ISBN 978-1-907-807-96-1 (Mobi)
ISBN 978-1-78427-032-2 (PDF)

Series Editor: William J. Sutherland

This book should be quoted as Dicks, L.V., Ashpole, J.E., Dänhardt, J., James, K., Jönsson, A.M., Randall, N., Showler, D.A., Smith, R.K., Turpie, S., Williams, D. and Sutherland, W.J. (2014) *Farmland Conservation: Evidence for the effects of interventions in northern and western Europe*. Exeter: Pelagic Publishing.

British Library Cataloguing in Publication Data
A catalogue record for this book is available from the British Library.

Cover image: Ranscombe Farm Reserve © Phil Houghton Photography – uncropped cultivated margins with stinking chamomile and common poppy.

Contents

Advisory board

We thank the following people for advising on the scope and content of this synopsis.

About the authors

Lynn V. Dicks is a Research Fellow in the Department of Zoology, University of Cambridge.

Joscelyne E. Ashpole is a Research Assistant in the Department of Zoology, University of Cambridge.

Juliana Dänhardt has a PhD in animal ecology and is currently employed as Research Administrator at the Centre for Environmental and Climate Research, Lund University.

Katy James is a Researcher at Harper Adams University.

Annelie M. Jönsson is a PhD student in the Department of Biology at Lund University, Sweden.

Nicola Randall is a Senior Lecturer in the Department of Crop and Environmental Science and Director of the Centre for Evidence-Based Agriculture at Harper Adams University.

David A. Showler is an Ecological Consultant based in Norwich, UK.

Rebecca K. Smith is a Research Associate in the Department of Zoology, University of Cambridge.

Susan Turpie is an Agri-Environment Policy Officer in the Natural Heritage Management Team, Scottish Government.

David Williams is a Doctoral Student in the Department of Zoology, University of Cambridge.

William J. Sutherland is the Miriam Rothschild Professor of Conservation Biology at the University of Cambridge.

Acknowledgements

This synopsis was funded by the Rural Economy and Land Use Programme, Arcadia and the Natural Environment Research Council.

We would also like to thank Stephanie Prior, Rob Pople, Tara Proud and Nigel Massen for their help and advice.

About this book

The purpose of Conservation Evidence synopses

Conservation Evidence synopses **do**	Conservation Evidence synopses **do not**
Bring together scientific evidence captured by the Conservation Evidence project (over 4,000 studies so far) on the effects of interventions to conserve wildlife	Include evidence on the basic ecology of species or habitats, or threats to them
List all realistic interventions for the species group or habitat in question, regardless of how much evidence for their effects is available	Make any attempt to weight or prioritize interventions according to their importance or the size of their effects
Describe each piece of evidence, including methods, as clearly as possible, allowing readers to assess the quality of evidence	Weight or numerically evaluate the evidence according to its quality
Work in partnership with conservation practitioners, policymakers and scientists to develop the list of interventions and ensure we have covered the most important literature	Provide recommendations for conservation problems, but instead provide scientific information to help with decision-making

Who this synopsis is for

If you are reading this, we hope you are someone who has to make decisions about how best to support or conserve biodiversity. You might be a land manager, a conservationist in the public or private sector, a farmer, a campaigner, an advisor or consultant, a policymaker, a researcher or someone taking action to protect your own local wildlife. Our synopses summarize scientific evidence relevant to your conservation objectives and the actions you could take to achieve them.

We do not aim to make your decisions for you, but to support your decision-making by telling you what evidence there is (or isn't) about the effects that your planned actions could have.

When decisions have to be made with particularly important consequences, we recommend carrying out a systematic review, as the latter is likely to be more comprehensive than the summary of evidence presented here. Guidance on how to carry out systematic reviews can be found from the Centre for Evidence-Based Conservation at the University of Bangor (www.cebc.bangor.ac.uk).

The Conservation Evidence project

The Conservation Evidence project has three parts:

1. An online, **open access journal** *Conservation Evidence* publishes new pieces of research on the effects of conservation management interventions. All our papers are written by, or in conjunction with, those who carried out the conservation work and include some monitoring of its effects.
2. An ever-expanding **database of summaries** of previously published scientific papers, reports, reviews or systematic reviews that document the effects of interventions.
3. **Synopses** of the evidence captured in parts one and two on particular species groups or habitats. Synopses bring together the evidence for each possible intervention. They are freely available online and available to purchase in printed book form.

These resources currently comprise over 4,000 pieces of evidence, all available in a searchable database on the website www.conservationevidence.com.

Alongside this project, the Centre for Evidence-Based Conservation (www.cebc.bangor.ac.uk) and the Collaboration for Environmental Evidence (www.environmentalevidence.org) carry out and compile systematic reviews of evidence on the effectiveness of particular conservation interventions. These systematic reviews are included on the Conservation Evidence database.

A total of eight systematic reviews are included in this synopsis. The systematic reviews are included in 11 interventions:

*   Pay farmers to cover the cost of conservation measures (as in agri-environment schemes)
*   Provide or retain set-aside areas in farmland
*   Plant wild bird seed or cover mixture
*   Leave overwinter stubbles
*   Maintain species-rich, semi-natural grassland
*   Delay mowing or first grazing date on grasslands
*   Reduce grazing intensity on grassland (including seasonal removal of livestock)
*   Reduce fertilizer, pesticide or herbicide use generally
*   Control bracken
*   Control mink
*   Control predatory mammals and birds (foxes, crows, stoats and weasels)

Systematic reviews include:

*   The effectiveness of land-based schemes (including agri-environment schemes) at conserving farmland bird densities within the UK. http://www.environmentalevidence.org/SR11.html
*   Does sheep-grazing degrade unimproved neutral grasslands managed as pasture in lowland Britain? http://www.environmentalevidence.org/SR15.html
*   The effectiveness of current methods for the control of bracken *Pteridium aquilinum*. http://www.environmentalevidence.org/SR3.html
*   Does delaying the first mowing date benefit biodiversity in meadowland? http://www.environmentalevidencejournal.org/content/1/1/9

The following three interventions would particularly benefit from systematic reviews:

*   Plant grass buffer strips/margins around arable or pasture fields
*   Reduce tillage
*   Leave headlands in fields unsprayed (conservation headlands)

Scope of the Farmland Conservation synopsis

This synopsis covers evidence for the effects of conservation interventions for native farmland wildlife.

It is restricted to evidence captured on the website www.conservationevidence.com. It includes papers published in the journal *Conservation Evidence*, evidence summarized on our database and systematic reviews collated by the Collaboration for Environmental Evidence.

Evidence was collected from all European countries west of Russia, but not those south of France, Switzerland, Austria, Hungary and Romania.

How we decided which conservation interventions to include

A list of interventions to conserve wildlife on farmland was developed collaboratively by a team of 13 experts. An initial list of interventions based on agri-environment options available in UK countries was circulated among the group, discussed and amended at two project meetings. A number of interventions that are not currently agri-environment options were added during this process, such as 'Provide nest boxes for bees (solitary or bumblebees)' and 'Implement food labelling schemes relating to biodiversity-friendly farming'.

Interventions relating to the creation or management of habitats not considered commercial farmland (such as lowland heath, salt marsh and farm woodland) were removed, although use of such habitats for grazing commercial livestock could be included under the intervention 'Employ areas of semi-natural habitat for rough grazing'.

The list of interventions was organized into categories based on the International Union for the Conservation of Nature (IUCN) classifications of direct threats and conservation actions. Interventions that fall under the threat category 'Agriculture' are grouped by farming system, with separate sections for interventions that apply to arable or livestock farms, or across all farming types.

How we reviewed the literature

We began with a list of 1,157 references identified by collaborators at Harper Adams University College for a systematic map on the effectiveness of agri-environment schemes as interventions for conserving biodiversity in temperate Europe (Randall N.P. & James K.L. (2012) The effectiveness of integrated farm management, organic farming and agri-environment schemes for conserving biodiversity in temperate Europe – A systematic map. *Environmental Evidence*, 1 (4), 1–21. http://www.environmentalevidence.org/SR35.html). This list of references was drawn up using a peer-reviewed search protocol, with searches carried out until June 2010. It included reviews and unpublished reports. Following peer review of the systematic map, an additional search was carried out using a new search term 'Farmland or farming AND mammal or reptile or amphibian', because mammals, reptiles and amphibians were not specifically covered by the original search terms. These references were included in this synopsis, although not in the published systematic map.

Each of the references was assessed, consulting the full text where possible. Those matching the following two criteria were included in our synopsis of evidence:

- There was an intervention that conservationists would do to benefit wildlife on actively farmed land
- The effects of the intervention were monitored quantitatively.

These criteria exclude studies examining the effects of specific interventions without actually doing them. For example, predictive modelling studies and studies looking at species distributions in areas with longstanding management histories (correlative studies), were excluded. Such studies can suggest that an intervention could be effective, but do not provide direct evidence of a causal relationship between the intervention and the observed biodiversity pattern.

Studies relating to organic and integrated farming were excluded from the synopsis. These interventions were considered to be combinations of farm management techniques, rather than single interventions. Where the actual intervention (for example, reduced agri-chemical use) was clearly defined, the studies were included under that specific intervention. Studies monitoring the uptake of agri-environment schemes, but not their effects on wildlife, were also excluded.

The strategy for studies published in more than one place was to summarize the most recent, but refer to the other publications (except for PhD theses, conference proceedings and Defra reports unavailable online, which are not referred to if their main findings are published subsequently).

Altogether 588 studies were allocated to interventions they tested. Additional studies published or completed in 2010 or before were added if recommended by the expert team or identified within the literature during the summarizing process. We also searched the Conservation Evidence database using the new Sphinx-driven advanced search facility (designed and built during 2011, not available when the systematic review was conducted), for studies from Europe on arable or pasture land. This yielded an additional 38 relevant studies.

In total, 741 individual publications or reports were identified for inclusion in the synopsis.

How the evidence is summarized

Conservation interventions are grouped primarily according to the relevant direct threats, as defined in the International Union for the Conservation of Nature (IUCN)'s Unified Classification of Direct Threats (http://www.iucnredlist.org/technical-documents/classification-schemes/threats-classification-scheme). In most cases, it is clear which main threat a particular intervention is meant to alleviate or counteract. For those interventions where the main threat is Agriculture, we have divided these into All farming, Arable, Livestock and Perennial (non-timber) crops.

Not all IUCN threat types are included, only those that threaten farmland wildlife, and for which realistic conservation interventions have been suggested.

Normally, no intervention is listed in more than one place, and when there is ambiguity about where a particular intervention should fall there is clear cross-referencing.

In the text of each section, studies are presented in chronological order, so the most recent evidence is presented at the end. The summary text at the start of each section groups studies according to their findings.

At the start of each chapter, a series of **key messages** provides a rapid overview of the evidence. These messages are condensed from the summary text for each intervention.

Background information is provided where we feel recent knowledge is required to interpret the evidence. This is presented separately and relevant references are included in the reference list at the end of each background section.

Some of the references containing evidence for the effects of interventions are summarized in more detail on the Conservation Evidence website (www.conservationevidence.com). In the online synopsis, these are hyperlinked from the references within each intervention. They can also be found by searching for the reference details or species name, using the website's search facility.

The information in this synopsis is available in three ways:
- As a book, printed by Pelagic Publishing and for sale from www.pelagicpublishing.com
- As a pdf to download from www.conservationevidence.com

• As text for individual interventions on the searchable database at www.conservationevidence.com.

Terminology used to describe evidence

Unlike systematic reviews of particular conservation questions, we do not quantitatively assess the evidence, or weight it according to quality. However, to allow you to interpret evidence, we make the size and design of each trial we report clear. The table below defines the terms that we have used to do this.

The strongest evidence comes from randomized, replicated, controlled trials with paired-sites and before-and-after monitoring.

Term	Meaning
Site comparison	A study that considers the effects of interventions by comparing sites that have historically had different interventions or levels of intervention.
Replicated	The intervention was repeated on more than one individual or site. In conservation and ecology, the number of replicates is much smaller than it would be for medical trials (when thousands of individuals are often tested). If the replicates are sites, pragmatism dictates that between five and ten replicates is a reasonable amount of replication, although more would be preferable. We provide the number of replicates wherever possible, and describe a replicated trial as 'small' if the number of replicates is small relative to similar studies of its kind.
Controlled	Individuals or sites treated with the intervention are compared with control individuals or sites not treated with the intervention.
Paired sites	Sites are considered in pairs, within which one was treated with the intervention and the other was not. Pairs of sites are selected with similar environmental conditions, such as soil type or surrounding landscape. This approach aims to reduce environmental variation and make it easier to detect a true effect of the intervention.
Randomized	The intervention was allocated randomly to individuals or sites. This means that the initial condition of those given the intervention is less likely to bias the outcome.
Before-and-after trial	Monitoring of effects was carried out before and after the intervention was imposed.
Review	A conventional review of literature. Generally, these have not used an agreed search protocol or quantitative assessments of the evidence.
Systematic review	A systematic review follows an agreed set of methods for identifying studies and carrying out a formal 'meta-analysis'. It will weight or evaluate studies according to the strength of evidence they offer, based on the size of each study and the rigour of its design. All environmental systematic reviews are available at: www.environmentalevidence.org/index.html

Taxonomy

We have followed the taxonomy used in the Integrated Taxonomic Information System (www.itis.gov) and the PLANTS Database from the United States Department of Agriculture (plants.usda.gov/java/). Where possible, common names and Latin names are both given the first time each species is mentioned within each intervention.

Significant results

Throughout the synopsis we have quoted results from papers. Unless specifically stated, these results reflect statistical tests performed on the results.

Multiple interventions

Many studies investigate several interventions at once. When the effects of different interventions are separated, then the results are discussed separately in the relevant sections. However, often the effects of multiple interventions cannot be separated. When this is the case, the study is included in the section on each intervention, but the fact that several interventions were used is highlighted.

How you can help to change conservation practice

If you know of evidence relating to farmland wildlife conservation that is not included in this synopsis, we invite you to contact us, via our website www.conservationevidence.com. You can submit a previously published study by clicking 'Submit additional evidence' on the right hand side of an action page. If you have new, unpublished evidence, you can submit a paper to the *Conservation Evidence* journal. We particularly welcome papers submitted by conservation practitioners.

1 All farming systems

Support or maintain low intensity agricultural systems
We have captured no evidence for the effects of supporting or maintaining low intensity agricultural systems on farmland wildlife.

Increase the proportion of semi-natural habitat in the farmed landscape
Of five studies monitoring the effects of the Swiss Ecological Compensation Areas scheme at a landscape scale (including three replicated site comparisons), one found an increase in numbers of birds of some species, two found no effect on birds and three found some species or groups increasing and others decreasing.

Pay farmers to cover the cost of conservation measures (as in agri-environment schemes)
For birds, twenty-four studies (including one systematic review) found increases or more favourable trends in bird populations, while eleven studies (including one systematic review) found negative or no effects of agri-environment schemes. For plants, three studies found more plant species, two found fewer plant species and seven found little or no effect of agri-environment schemes. For invertebrates, five studies found increases in abundance or species richness, while six studies found little or no effect of agri-environment schemes. For mammals, one replicated study found positive effects of agri-environment schemes and three studies found mixed effects in different regions or for different species.

Apply 'cross compliance' environmental standards linked to all subsidy payments
We have captured no evidence for the effects of applying 'cross compliance' environmental standards for all subsidy payments on farmland wildlife.

Implement food labelling schemes relating to biodiversity-friendly farming (organic; LEAF Marque)
We have captured no evidence for the effects of implementing food labelling schemes relating to biodiversity-friendly farming (organic; LEAF Marque) on farmland wildlife.

Reduce field size (or maintain small fields)
We have captured no evidence for the effects of reducing field size (or maintaining small fields) on farmland wildlife.

Provide or retain set-aside areas in farmland
Thirty-seven studies (one systematic review, no randomized, replicated, controlled trials) compared use of set-aside areas with control farmed fields. Twenty-one (including the systematic review) showed benefits to or higher use by all wildlife groups considered. Thirteen studies found some species or groups used set-aside more than crops, others did not. Two found higher Eurasian skylark reproductive success and one study found lower success on set-aside rather than control fields. Four studies found set-aside had no effect on wildlife,

one found an adverse effect. Two studies found neither insects nor small mammals preferred set-aside areas.

Connect areas of natural or semi-natural habitat

All four studies (including two replicated trials) from the Czech Republic, Germany and the Netherlands investigating the effects of linking patches of natural or semi-natural habitat found some colonization by invertebrates or mammals. Colonization by invertebrates was slow or its extent varied between taxa.

Manage hedgerows to benefit wildlife (includes no spray, gap-filling and laying)

Ten studies from the UK and Switzerland (including one randomized, replicated, controlled trial) found managing hedges for wildlife increased berry yields, diversity or abundance of plants, invertebrates or birds. Five UK studies (including one randomized, replicated, controlled trial) found plants, bees and farmland birds were unaffected by hedge management. Two replicated studies found hedge management had mixed effects on invertebrates or reduced hawthorn berry yield.

Manage stone-faced hedge banks to benefit wildlife

We have captured no evidence for the effects of managing stone-faced hedge banks to benefit wildlife on farmland wildlife.

Manage ditches to benefit wildlife

Five studies (including one replicated, controlled study) from the UK and the Netherlands found ditch management had positive effects on numbers, diversity or biomass of some or all invertebrates, amphibians, birds or plants studied. Three studies from the Netherlands and the UK (including two replicated site comparisons) found negative or no clear effects on plants or some birds.

Restore or maintain dry stone walls

We have captured no evidence for the effects of restoring or maintaining dry stone walls on farmland wildlife.

Plant new hedges

Two studies (including one replicated trial) from France and the UK found new hedges had more invertebrates or plant species than fields or field margins. A review found new hedges had more ground beetles than older hedges. However, an unreplicated site comparison from Germany found only 2 out of 85 ground beetle species dispersed along new hedges. A review found lower pest outbreaks in areas with new hedges.

Protect in-field trees (includes management such as pollarding and surgery)

We have captured no evidence for the effects of protecting in-field trees on farmland wildlife.

Plant in-field trees (not farm woodland)

We have captured no evidence for the effects of planting in-field trees on farmland wildlife.

Maintain in-field elements such as field islands and rockpiles

We have captured no evidence for the effects of maintaining in-field elements such as field islands and rockpiles on farmland wildlife.

Manage woodland edges to benefit wildlife
We have captured no evidence for the effects of managing woodland edges to benefit wildlife on farmland wildlife.

Plant wild bird seed or cover mixture
Fifteen studies (including a systematic review) from the UK found fields sown with wild bird cover mix had more birds or bird species than other farmland habitats. Six studies (including two replicated trials) from the UK found birds used wild bird cover more than other habitats. Nine replicated studies from France and the UK found mixed or negative effects on birds. Eight studies (including two randomized, replicated, controlled studies) from the UK found wild bird cover had more invertebrates; four (including two replicated trials) found mixed or negative effects on invertebrate numbers. Six studies (including two replicated, controlled trials) from the UK found wild bird cover mix benefited plants; two replicated studies did not.

Plant nectar flower mixture/wildflower strips
Forty-one studies (including one randomized, replicated, controlled trial) from eight countries found flower strips increased invertebrate numbers or diversity. Ten studies (two replicated, controlled) found invertebrates visited flower strips. Fifteen studies (two randomized, replicated, controlled) found mixed or negative effects on invertebrates. Seventeen studies (one randomized, replicated, controlled) from seven countries found more plants or plant species on flower strips, four did not. Five studies (two randomized, replicated, controlled) from two countries found bird numbers, diversity or use increased in flower strips, two studies did not. Five studies (four replicated) found increases in small mammal abundance or diversity in flower strips.

Manage the agricultural landscape to enhance floral resources
A large replicated, controlled study from the UK found the number of long-tongued bumblebees on field margins was positively correlated with the number of 'pollen and nectar' agri-environment agreements in a 10 km square.

Create uncultivated margins around intensive arable or pasture fields
Twenty studies (including one randomized, replicated, controlled trial) from seven countries found uncultivated margins support more invertebrates, small mammal species or higher plant diversity than other habitats. Four studies (including two replicated studies from the UK) found positive associations between birds and uncultivated margins. Fifteen studies (including one randomized, replicated, controlled trial) from four countries found naturally regenerated margins had lower invertebrate or plant abundance or diversity than conventional fields or sown margins. Six studies (one randomized, replicated, controlled) from three countries found uncultivated margins did not have higher plant or invertebrate abundance or diversity than cropped or sown margins.

Plant grass buffer strips/margins around arable or pasture fields
Twenty studies (including two randomized, replicated, controlled studies) from four countries found grass margins benefited invertebrates, including increases in abundance or diversity. Nine studies (including two replicated, controlled trials) from the UK found grass buffer strips benefit birds, with increased numbers, diversity or use. Seven replicated studies (four controlled, two randomized) from two countries found grass buffer strips increased plant cover and species richness, a review found benefits to plants. Five studies (two

replicated, controlled) from two countries found benefits to small mammals. Six (including three replicated, controlled trials) from two countries found no clear effect on invertebrate or bird numbers.

Provide supplementary food for birds or mammals

Nine studies (two randomized, replicated, controlled) from France, Sweden and the UK found providing supplementary food increased abundance, overwinter survival or productivity of some birds. Two of the studies did not separate the effects of several interventions. Four studies (one replicated, controlled and one randomized, replicated) from Finland and the UK found some birds or mammals used supplementary food. Six replicated studies (three controlled) from Sweden and the UK found no clear effect on some birds or plants.

Make direct payments per clutch for farmland birds

Two replicated, controlled studies from the Netherlands found per clutch payments did not increase overall bird numbers. A replicated site comparison from the Netherlands found more birds bred on 12.5 ha plots under management including per-clutch payments but there were no differences at the field-scale.

Provide other resources for birds (water, sand for bathing)

A small study in France found grey partridge density was higher in areas where water, shelter, sand and food were provided.

Mark bird nests during harvest or mowing

A replicated study from the Netherlands found that marked northern lapwing nests were less likely to fail as a result of farming operations than unmarked nests.

Provide refuges during harvest or mowing

A replicated study from France found mowing refuges reduced contact between mowing machinery and unfledged quails and corncrakes. A replicated controlled study and a review from the UK found Eurasian skylark did not use nesting refuges more than other areas.

Provide foraging perches (eg. for shrikes)

We have captured no evidence for the effects of providing foraging perches on farmland wildlife.

Provide nest boxes for birds

Two studies (including one before-and-after trial) from the Netherlands and the UK found providing nest boxes increased the number of clutches or breeding adults of two bird species. A replicated study from Switzerland found nest boxes had mixed effects on the number of broods produced by two species. Eight studies (six replicated) from five countries found nest boxes were used by birds. A controlled study from the UK found one species did not use artificial nest sites. Three replicated studies (one paired) from the UK and Sweden found box location influenced use or nesting success.

Provide nest boxes for bees (solitary bees or bumblebees)

Ten studies (nine replicated) from Germany, Poland and the UK found solitary bee nest boxes were used by bees. Two replicated trials from the UK found bumblebee nest boxes had very low uptake. Two replicated studies found the local population size or number of emerging red mason bees increased when nest boxes were provided. A replicated trial in Germany found the number of occupied solitary bee nests almost doubled over three years with repeated nest box provision.

Introduce nest boxes stocked with solitary bees
We have captured no evidence for the effects of introducing nest boxes stocked with solitary bees on farmland wildlife.

Provide red squirrel feeders
We have captured no evidence for the effects of providing red squirrel feeders on farmland wildlife.

Provide otter holts
We have captured no evidence for the effects of providing otter holts on farmland wildlife.

Provide badger gates
We have captured no evidence for the effects of providing badger gates on farmland wildlife.

1.1   Support or maintain low intensity agricultural systems

- We have captured no evidence for the effects of supporting or maintaining low intensity agricultural systems on farmland wildlife.

Background
Low-intensity agricultural systems have consistently been shown to have higher biodiversity than more intensive systems. Supporting such systems may therefore benefit farmland wildlife.

1.2   Increase the proportion of semi-natural habitat in the farmed landscape

- Five studies monitored the effects of the Swiss Ecological Compensation Areas scheme at a landscape scale, including three replicated site comparisons. Of these, one found an increase in numbers of birds of some species[1]. Two found no effect on the number of bird species[5] or population densities of farmland birds[3]. Three studies[2, 4, 5] found mixed effects, with some species or groups of species increasing and others decreasing.

Background
Agricultural intensification has resulted in a loss of semi-natural habitats. These habitats include field margins, ditch banks, hedgerows, woods and ponds. Those that persist support a high proportion of the remaining farmland biodiversity.
This intervention is backed up by a body of correlative evidence, which tends to find higher biodiversity or species abundances in areas with higher proportions of semi-natural landscape. For example, in a review looking at the relationship between agricultural biodiversity and semi-natural habitats (Grashof-Bokdam & van Langevelde 2005), seven of nine studies found significantly higher spider, bird, plant or

butterfly diversity in agricultural landscapes with greater proportions of semi-natural habitat. The one study on mammal species found no such effect.

Here we summarize studies in which the proportion of semi-natural habitat in the farmed landscape has been manipulated and responses of wildlife have been monitored. Studies assessing the effects of the Swiss Ecological Compensation Areas scheme at a landscape scale are included here. This scheme obliged farmers in Switzerland to manage at least 7% of their agricultural land area as Ecological Compensation Areas, from 1998 onwards.

Grashof-Bokdam, C.J. & van Langevelde, F. (2005) Green veining: landscape determinants of biodiversity in European agricultural landscapes. *Landscape Ecology*, 20, 417–439.

A before-and-after study in 6 km^2 of mixed farmland in Switzerland (1) found that the populations of corn bunting *Miliaria calandra*, whitethroat *Sylvia communis* and common stonechat *Saxicola torquata* all increased following an increase in the proportion of land under the Ecological Compensation Areas (ECA) Scheme from 0.7% to 8.2% between 1992 and 1996 (corn buntings: 6 pairs in 1992 vs 26 in 1996; whitethroat: 15 vs 44; stonechat: 14 vs 35). In addition, across 23 study areas in Switzerland, ECA land and a 25 m buffer around it occupied only 17% of farmland but contained more 37–38% of 68 red-backed shrike *Lanius collurio* territories, 598 yellowhammer *Emberiza citrinella* territories and 35 whitethroat territories. Only 6% of Eurasian skylarks *Alauda arvensis* territories were found on ECA land.

A review on effects of the Swiss Ecological Compensation Areas (ECA) scheme on biodiversity in arable landscapes in Switzerland (2) found that effects differed between species and taxa. Bird species breeding in hedgerows (dominated by yellowhammer *Emberiza citrinella*, linnet *Carduelis cannabina*, red-backed shrike *Lanius collurio* and common whitethroat *Syliva communis*) and wetlands (mainly reed warbler *Acrocephalus scirpaceus* and marsh warbler *A. palustris*) had more territories than expected near ECAs (hedgerows: 143 territories expected vs 293 observed; wetlands: 31 territories expected vs 52 observed). For species preferring open agricultural habitats (skylark *Alauda arvensis*, common quail *Coturnix coturnix* and common kestrel *Falco tinnunculus*), fewer territories than expected were recorded near ECAs (151 expected vs 68 observed). Many compensation areas were located near vertical structures (such as hedgerows or forest edges), which may bias these results. A correlation between the proportion of ECAs in the landscape and presence of the meadow grasshopper *Chorthippus parallelus* was found but there was no such correlation for the bow-winged grasshopper *C. biguttulus*. The report reviews results from a number of studies. No details on study design, monitoring techniques or other methods were given.

A 2007 site comparison study of 23 sites in the lowlands north of the Alps, Switzerland (3) found that the percentage of farmland designated as Ecological Compensation Area (ECA) had no effect on the population density of farmland bird species or bird species with territories incorporating several habitat types. ECAs are areas managed for the primary function of providing plant and animal habitat – these include meadows farmed at a low intensity. For 37 species surveyed in 1998–1999 and again in 2003–2004, population densities in wetlands and rivers were not affected by vicinity to ECAs, although hedges and traditional orchards close to ECAs did have higher bird population densities than those further away. The 23 selected sites (covering up to 3 km^2 each) were randomly selected and surveyed 3 times each between April and June in both years of study.

A 2007 site comparison study of 516 survey points across the canton of Aargau, Switzerland (4) found no consistent effects on biodiversity across taxa. For birds, plants and butterflies, but not for snails, there were more species on Ecological Compensation Area (ECA) than non-ECA sites in the first survey (9.7 vs 7.7 bird species, 19.2 vs 14.6 plant species and 7.3 vs 5.6 butterfly species on ECA and non-ECA sites respectively). There were 4–5 snail species on both ECA and control sites. Changes over time were different on ECA plots than non-ECA plots for plants and snails, but not for birds and butterflies. Between the first survey (1996–2000) and the second survey (2001–2005) numbers of vascular plant and snail species increased on ECAs (by 5.1 and 1.4 species, respectively) but not on non-ECA fields. Across the whole landscape the number of bird species increased and the number of butterfly species decreased between the two surveys, but the changes were similar on ECA and non-ECA sites. Sampling was based on a regular 2 x 2 m grid system across the entire 1,403 km^2 of the canton of Aargau. Plants, birds, butterflies and snails were surveyed at each grid point. Whether the land or some of the 100 m radius circle plot (for the bird survey) were designated as ECA was recorded. All plots were surveyed twice, five years apart, with the first survey taking place in 1996–2000 and the second in 2001–2005. The authors note that ECAs were typically established on farmland with potential for maximum biodiversity gain, which may have affected the relative numbers of species found in the first survey.

A replicated, site comparison study of Ecological Compensation Areas (ECAs) created over 97 ha from 1993 in Switzerland (5) found that between 1988 and 2006 the number of bird species remained stable in the entire study area but increased on ECAs (high-value areas and areas of no special ecological value). Numbers declined on remaining land-use types, only slightly in nature reserves and considerably on cultivated land. More of the 22 breeding bird species recorded were within the nature reserves than ECAs or cultivated land. Population trends were calculated for 12 common species, of which 5 increased, 5 decreased and 2 remained stable. Population increases prevailed in nature reserves and high-value ECAs. Negative population trends were seen on ECAs of no special ecological value and cultivated land. ECAs included wetlands and flower-rich meadows. Breeding bird data were collected in 1988, 1989, 1999 and 2006 in different land use areas.

(1) Spiess, M., Marfurt, C. & Birrer, S. (2000) *Ecological Compensation – A Chance for Farmland Birds?* Proceedings of the IFOAM 2000: the world grows organic. Basel, Switzerland, 28–31 August. pp. 441.
(2) Herzog, F., Buholzer, S., Dreier, S., Hofer, G., Jeanneret, P., Pfiffner, L., Poiger, T., Prasuhn, V., Richner, W., Schüpbach, B., Spiess, E., Spiess, M., Walter, T. & Winzeler, M. (2006) Effects of the Swiss agri-environmental scheme on biodiversity and water quality. *Mitteilungen der Biologischen Bundesanstalt für Land-u. Forstwirtschaft* 403, 34–39.
(3) Birrer, S., Spiess, M., Herzog, F., Jenny, M., Kohli, L. & Lugrin, B. (2007) The Swiss agri-environment scheme promotes farmland birds: but only moderately. *Journal of Ornithology*, 148, S295–303.
(4) Roth, T., Amrhein, V., Peter, B. & Weber, D. (2008) A Swiss agri-environment scheme effectively enhances species richness for some taxa over time. *Agriculture, Ecosystems & Environment*, 125, 167–172.
(5) Rudin, M., Horch, P., Hugentobler, I., Weber, U. & Birrer, S. (2010) Population trends of breeding birds in the ecologically upgraded Rhine valley (canton of St. Gallen, Switzerland). *Ornithologische Beobachter*, 107, 81–100.

1.3   Pay farmers to cover the cost of conservation measures (as in agri-environment schemes)

• Twenty-six studies from four European countries (including one UK systematic review and three European reviews) looked at the effects of agri-environment schemes on birds. Twenty-four studies (including one systematic review, six site comparisons and nine reviews) found increases in population size, density or more favourable population trends of some or all birds studied on sites with agri-environment schemes compared to non-scheme sites[1, 3–6, 8, 10–12, 14, 15, 18, 19, 25, 26, 29, 31, 32, 37, 38, 40, 43–46, 48] (some of these differences were seasonal). Eleven studies (including one systematic review and four reviews) found negative or no effects[7, 10–12, 14–16, 18, 19, 25, 29, 32, 41, 44, 45]. One UK study found higher numbers of some birds where higher tier management was in place[7], another UK study found no difference between Entry Level or Higher Level Stewardship Scheme fields[47]. One study from the Netherlands found that not all agri-environment scheme agreements were sited in ideal locations for black-tailed godwit[35].

- Eleven studies from five European countries (including three replicated paired site comparisons and two reviews) looked at the effects of agri-environment schemes on plants. Seven studies (including three replicated paired site comparisons and one European review) found agri-environment schemes maintained[17, 21, 24] or had little or no effect[14, 20, 27, 34, 41] on plants, plant diversity or species richness. Three studies found increases in plant species richness in areas with agri-environment schemes[14, 21, 42]; two found decreases[14, 36]. A replicated site comparison study from Estonia found higher flower abundance on farms with agri-environment schemes in two out of four areas[30]. A review found Environmentally Sensitive Areas in England had contributed to halting the loss of semi-natural grassland habitats but were less effective at enhancing or restoring grassland biodiversity[2].

- Ten studies from three European countries (including two replicated paired site comparisons and a review) looked at the effects of agri-environment schemes on invertebrates. Six studies (including two replicated site comparisons) showed agri-environment schemes maintained[21] or had little or no effect[9, 20, 22, 30, 41] on some invertebrates in terms of diversity, abundance, species richness or bee colony growth. Five studies found increases in abundance or species richness of some invertebrates[14, 21, 30, 33, 42]. A UK study found agri-environment scheme prescriptions had a local but not a landscape-scale effect on bee numbers[39].

- Four studies (including two replicated site comparisons and a review) from the UK looked at the effects of agri-environment schemes on mammals. One study found positive effects[1]; three studies found mixed effects in different regions or for different species[11, 28, 40].

- Three of the studies above found higher numbers of wildlife on land before agri-environment schemes were introduced[15, 16, 41]. However two studies collecting baseline data found no difference in the overall number of birds[23] or earthworms and soil microorganisms[22] between areas with and without agri-environment schemes.

- A review found two out of three agri-environment schemes in Europe benefited wildlife[13].

Background

Agri-environment schemes are government or inter-governmental schemes designed to compensate farmers financially for changing agricultural practice to be more favourable to biodiversity and landscape. In Europe, agri-environment schemes are an integral part of the European Common Agricultural Policy (CAP) and Member States devise their own agri-environment prescriptions to suit their agricultural economies and environmental contexts.

Since agri-environment schemes represent many different specific interventions relevant to conservation, and where a study's results can be clearly assigned to a specific intervention, they appear in the appropriate section. This section, meanwhile, includes evidence about the success of agri-environment policies overall.

Evidence relating to the Swiss Ecological Compensation Areas with biodiversity monitoring on a landscape scale is placed under 'Increase the proportion of semi-natural habitat in the landscape'.

A replicated study from 1992 to 1994 within the South Downs Environmentally Sensitive Area, Sussex, UK (1) found that Eurasian skylark *Alauda arvensis* numbers increased but brown hare *Lepus europaeus* numbers were stable over two years on the Environmentally Sensitive Area farms. There were significantly more breeding pairs of skylark in 1993 (five males/km²) compared to 1992 (three males/km²). The number of hares remained stable over the study period. Four arable, ten mixed and three pastoral farms were studied. Hares were sampled by spotlight counting over an average of 26% of the area of each farm between November and March (1992–1993; 1993–1994). Skylarks were sampled by mapping breeding males during 2 counts along transects on 12–17 farms from April to June (1992 and 1993).

A 1997 review (2) concluded that Environmentally Sensitive Areas had made a significant contribution to halting the loss of semi-natural grasslands in England, but were less effective in enhancing and restoring grassland biodiversity, a decade after the introduction of the Environmentally Sensitive Areas scheme. The paper made a broad assessment of the effectiveness of the scheme in protecting England's lowland semi-natural grasslands. Among Environmentally Sensitive Areas of greatest significance for their lowland grassland, 6 were of 'outstanding' significance (containing >40% of the English resource of a grassland type) and 2 were of 'considerable' significance (containing 10–40% of 1 to 2 grassland types or 5–10% of 3 or more grassland types). Entry of land supporting semi-natural grassland was generally high (e.g. covering 80% of chalk grassland in the South Downs Environmentally Sensitive Area). However, there was evidence in some Environmentally Sensitive Areas that grassland habitats were declining in quality due to management being insufficiently tailored to biodiversity interest e.g. permitting use of inorganic fertilizers.

A 1998 literature review (3) found that cirl bunting *Emberiza cirlus* in the UK responded positively to Countryside Stewardship Schemes, reaching population levels of 360–388 occupied territories in 1995–1997 (Evans 1997) compared with 118 or so in the mid-1980s (Evans 1992). Some of the interventions used include reducing grassland management intensity, sowing arable field margins, managing hedgerows for wildlife, growing spring barley, reducing herbicide use and maintaining overwinter stubbles.

A 2000 literature review from the UK (4) found that the populations of four farmland birds (grey partridge *Perdix perdix*, cirl bunting *Emberiza cirlus*, corncrake *Crex crex* and Eurasian thick-knee (stone curlew) *Burhinus oedicnemus*) increased following agri-environment schemes targeted for them. The individual schemes are discussed in the relevant interventions.

A 2000 review of the effectiveness of agri-environment schemes in England (5) reported that two bird species, Eurasian thick-knee (stone curlew) *Burhinus oedicnemus* and cirl bunting *Emberiza cirlus*, had benefited from the introduction of agri-envrionment schemes. Numbers of cirl bunting increased from 118 pairs in 1989 to approximately 450 in 1998 following the introduction of measures including a 'special project' under the Countryside Stewardship Scheme. The review also stated that cirl bunting numbers showed an 82% increase in squares with Countryside Stewardship Scheme agreements between 1992 and 1998 but only a 2% increase on adjacent non-Countryside Stewardship Scheme squares. The number of Eurasian thick-knees increased from 150 pairs in 1991 to 254 by 2000 following the introduction of measures associated with agri-environment schemes including habitat management in the

Brecks Environmentally Sensitive Area and provision of nesting plots on set-asides as part of the Countryside Stewardship Scheme.

A paired site comparison study in 1992, 1998 and 1999 in south Devon, England (6) found that the number of cirl bunting *Emberiza cirlus* increased significantly more (up 72%, from 54 to 93 breeding territories) in areas participating in the Countryside Stewardship Scheme than on adjacent land not participating in the scheme (down 20%, from 124 to 96 territories) between 1992 and 1999. Countryside Stewardship Scheme land that was near to known cirl bunting breeding territories saw greater increases in cirl bunting numbers than Countryside Stewardship Scheme areas further away – of the 9 agreements further than 2 km from the nearest known breeding site in 1992, 7 remained uncolonized in 1999, 1 lost its only pair and 1 gained a pair. Forty-one 2 x 2 km^2 squares containing both land within the Countryside Stewardship Scheme and non-Countryside Stewardship Scheme land were surveyed in 1992, 1998 and 1999. In each year each tetrad was surveyed at least twice, the first time during mid-April to late May, and the second time between early June and the end of August.

A study in 1997 in two Environmentally Sensitive Areas in eastern England (7) found that higher tier options (i.e. those with more demanding prescriptions but higher financial compensation) held significantly higher densities of wading birds (northern lapwing *Vanellus vanellus*, common redshank *Tringa totanus* and common snipe *Gallinago gallinago*) than lower tiers (Tier 1: 0.02–0.04 pairs/ha; Tier 2: 0.07–0.22; Tier 3: 0.40). In addition, they held more waders for each unit of money spent on the Environmentally Sensitive Area (Tier 1: 18–46 pairs/£100,000; Tier 2: 29–114; Tier 3: 167). However, when examining 1988–1997 population trends in 4 Environmentally Sensitive Areas the authors found that all 3 species investigated declined significantly (lapwing: 1–13% decline each year; redshank: 2–19%; snipe: 7–30%).

A 2002 review of research on agri-environment schemes in England (8) summarized two reports (Wilson *et al.* 2000; ADAS 2001) evaluating the effects of the Pilot Arable Stewardship Scheme in two regions (East Anglia and the West Midlands) from 1998–2001. At the whole farm scale in winter, seed-eating songbirds, thrushes (Turdidae) and wagtails (*Motacilla* spp.) showed some benefit on agreement farms relative to control farms (numbers not given). In summer numbers of breeding northern lapwing *Vanellus vanellus*, reed bunting *Emberiza schoeniclus*, greenfinch *Carduelis chloris*, house sparrow *Passer domesticus*, starling *Sturnus vulgaris* and yellow wagtail *M. flava* were higher on agreement farms. Agreement farms had some of the following options: overwinter stubbles (sometimes preceded by reduced herbicide, followed by fallow or a spring crop), undersown spring cereals (sometimes followed by a grass or grass/clover *Trifolium* spp. ley), arable crop margins with reduced spraying (conservation headlands), grass margins or beetle banks and sown wildlife seed mixtures (pollen and nectar or wild bird seed mix). Overwinter stubble (974 and 2,200 ha in East Anglia and West Midlands respectively) and conservation headlands (605 and 1,085 ha in East Anglia and West Midlands respectively) were the most widely implemented options. The effects of the pilot scheme on birds were monitored at the farm scale over three years, relative to control areas or control farms.

A replicated site comparison study in southern England (9) found no measurable difference in experimental buff-tailed bumblebee *Bombus terrestris* colonies in terms of colony growth, worker bee traffic, number or size of worker bees, queens and males produced or diversity of pollen collected between colonies on 10 farms with substantial conservation measures and those on 10 conventional arable farms. Conservation measures included conservation headlands, set-aside and minimal use of pesticides. Experimental bumblebee colonies were placed under hedges or shrubs on each farm and every week nests were weighed and numbers of bees leaving and entering each colony counted for 10 minutes.

Colonies were analysed after four weeks. The authors suggested the lack of difference was because the buff-tailed bumblebee has a foraging range that extends beyond individual farms, which may not be true for other bumblebee species.

A replicated site comparison study of 102 sites across East Anglia and the West Midlands, UK (10) found that two years after the introduction of the Pilot Arable Stewardship Scheme (introduced in 1998) there was no difference in the number of farmland bird species observed in winter on Pilot Arable Stewardship Scheme farms and non-scheme farms. There were, however, significantly more seed-eating songbirds, wagtails and pipits (Motacillidae) on farms participating in the scheme than on farms not participating in the scheme. A further survey of 98 fields in summer found that although there were significantly more northern lapwing *Vanellus vanellus*, common starling *Sturnus vulgaris*, greenfinch *Carduelis chloris* and reed bunting *Emberiza schoeniclus* on Pilot Arable Stewardship Scheme fields, there were also fewer woodpigeon *Columba palumbus*, sedge warbler *Acrocephalus schoenobaenus* and rook *Corvus frugilegus* than on the non-Pilot Arable Stewardship Scheme farms. Fifty-four Pilot Arable Stewardship Scheme and 48 comparable non-Pilot Arable Stewardship Scheme farms were surveyed for farmland birds in both the winters of 1998–1999 and 1999–2000. Fifty Pilot Arable Stewardship Scheme and 48 non-Pilot Arable Stewardship Scheme farms were surveyed in the summer months of 1999 and 2000. The seed-eating songbirds identified included 13 species of finches (Fringillidae), buntings (Emberizidae) and sparrows (Passeridae), while the wagtails and pipits comprised 3 species. This study was part of the same monitoring project as (11, 16, 25).

A replicated site comparison study of 71–76 farms in East Anglia and the West Midlands, UK (11) found no consistent difference in the change in the number of brown hare *Lepus europaeus* and grey partridge *Perdix perdix* between 1998 and 2002 across either Arable Stewardship Pilot Scheme farmland or non-Arable Stewardship Pilot Scheme farmland. In East Anglia the density of brown hares increased on Arable Stewardship Pilot Scheme farms (from 16.2 to 20.0 hares/km^2), but not on non-Arable Stewardship Pilot Scheme farmland (12.1 hares/km^2 in both 1998 and 2002). In the West Midlands hare densities fell slightly on Arable Stewardship Pilot Scheme plots (from 4.9 to 4.3 hares/km^2) but not on non-Arable Stewardship Pilot Scheme plots (3.5 hares/km^2 in both survey years). In East Anglia grey partridge densities fell by 21% on Arable Stewardship Pilot Scheme farms (9.6 to 7.6 birds/km^2) and 68% on non-Arable Stewardship Pilot Scheme farms (5.5 to 1.8 birds/km^2) whereas in the West Midlands grey partridge densities fell by 78% on Arable Stewardship Pilot Scheme farms (3.0 to 0.8 birds/km^2) and 40% on non-Arable Stewardship Pilot Scheme farms (1.4 to 0.8 birds/km^2). Following the introduction of the Arable Stewardship Pilot Scheme in 1998 hare density data was collected after dark in the winters of 1998–1999 and 2002–2003 from 19 Arable Stewardship Pilot Scheme and 18 non-Arable Stewardship Pilot Scheme farms in East Anglia and 19 Arable Stewardship Pilot Scheme and 15 non-Arable Stewardship Pilot Scheme farms in the West Midlands. Surveys of grey partridge were made once each autumn in 1998 and 2002 on 76 farms: 20 Arable Stewardship Pilot Scheme and 19 non-Arable Stewardship Pilot Scheme farms in East Anglia and 20 Arable Stewardship Pilot Scheme and 17 non-Arable Stewardship Pilot Scheme farms in the West Midlands. This study was part of the same monitoring project as (10, 16, 25).

A study from nine areas of the UK under Environmentally Sensitive Area schemes (12) found that the impacts of Environmentally Sensitive Area designation on farmland birds were mixed. There was evidence for population increases or high numbers of some species of birds on Environmentally Sensitive Area-managed land for four Environmentally Sensitive Areas. Populations of some species were stable in six Environmentally Sensitive Areas, often in contrast to national trends, but four Environmentally Sensitive Areas saw falls in the populations of at least one target species. The authors also note that in five regions there were not adequate data for all target species.

A 2003 review of monitoring of agri-environment schemes in Europe (13) described long-term monitoring results (three years or more) for three agri-environment programmes in the UK, Ireland and the Netherlands. Some wildlife benefits were found for two of the three programmes: the Dutch Natuurbeheer (Kleijn *et al.* 2001 – study described under 'Livestock farming: Reduce management intensity on permanent grasslands') and the UK Environment-ally Sensitive Areas Scheme (17, 24) but not for the third scheme (the Irish Rural Environment Protection Scheme (20)). The benefits were not always aligned with the scheme objectives.

A 2003 review of 62 studies from 6 European countries (14) found that, overall, 54% of the species groups examined showed an increase in species richness or abundance under agri-environment schemes. Agri-environment schemes had no consistent effect on bird spe-cies. While there were individual successes, such as the 83% increase in cirl bunting *Emberiza cirlus* between 1992 and 1998 on land within the Countryside Stewardship Scheme compared with the 2% increase on adjacent land not in the scheme (6), only 13 out of 29 studies found agri-environment schemes increased bird species richness or abundance. Two studies repor-ted negative effects and nine reported both positive and negative effects. Of the 19 studies which involved statistical tests, only 4 found positive effects, 2 reported negative effects and 9 reported both positive and negative effects on species richness or abundance of birds. Half of the studies on plants that included statistical analyses (7 out of 14) found no effect, 6 studies found increased species richness/abundance and 2 found decreases. For insects and spiders (Araneae) 11 out of 17 studies that included statistical analyses found increases in species richness/abundance, none found decreases and 3 showed increases and decreases. Three out of the 62 studies included bees (Apidae). Two studies (Allen *et al.* 2001; Kleijn *et al.* 2001) found more bees (more species of bee in the case of Kleijn *et al.* 2001) on agri-environment fields compared to control fields under certain schemes. The third study (Kleijn *et al.* 1999) reported not to have found a difference in bee abundance or species richness between seven agri-environment fields and seven control fields.

A 2004 review of agri-environment scheme uptake and effectiveness in Europe (15) found that in the UK four rare bird species (grey partridge *Perdix perdix*, corncrake *Crex crex*, Eurasian thick-knee (stone curlew) *Burhinus oedicnemus* and cirl bunting *Emberiza cirlus*) benefited from agri-environment schemes (4). Although the authors note that densities of some species were higher on agri-environment scheme farms before they were designated. Similar methodological issues were found with studies in the Netherlands where studies found that, at both field and larger scales, there were no population-level benefits of agri-environment scheme designation (Kleijn *et al.* 2001) although hatching and fledging rates of some species were higher on agri-environment scheme farms (eg. Musters *et al.* 2000; Schekkerman & Müskens 2000).

A replicated, site comparison study of 74 farms in East Anglia and the West Midlands, UK (16) found few differences in the density of farmland birds on farms participating in the Arable Stewardship Pilot Scheme and non-Arable Stewardship Pilot Scheme land 5 years after the introduction of the scheme. In the West Midlands, although seed-eating songbirds, wagtails and pipits (Motacillidae), insectivores and raptors were found at higher densities on Arable Stewardship Pilot Scheme land than non-Arable Stewardship Pilot Scheme land, these higher densities were already present when measured within one year of the introduction of the scheme. Moreover, in East Anglia there were no differences in the bird densities found on Arable Stewardship Pilot Scheme and non-Arable Stewardship Pilot Scheme fields. Surveys of grey partridge *Perdix perdix* populations on 76 farms in 1998 and 2002 found that adult densities decreased uniformly on both Arable Stewardship Pilot Scheme and non-Arable Stewardship Pilot Scheme farms over the 5-year period. Bird surveys were carried out twice each winter, during the winters of 1998–1999 and 2002–2003 on 18 Arable Stewardship Pilot Scheme and 19 non-Arable Stewardship Pilot Scheme farms in East Anglia and 19 Arable Stewardship Pilot Scheme and 18 non-Arable Stewardship Pilot Scheme farms in the West Midlands. This study was part of the same monitoring project as (10, 11, 25).

A 2004 analysis of monitoring data (a replicated site comparison) in the UK (17) concluded that agri-environment schemes maintain, but do not reliably improve, plant diversity in grasslands. In 22 of 38 datasets no change was detected in the vegetation under agri-environment schemes. Nine showed some change towards the desired plant community, and seven showed further deterioration. Of 17 datasets that included non-agreement land for comparison 7 found agri-environment agreements were benefitting plant communities (deterioration or no change on non-agreement land contrasting with maintenance or restoration on agreement land). Two found more positive trends in plant communities outside agri-environment schemes than under them. In eight comparisons there was no difference between agreement and non-agreement land. Thirty-eight sets of vegetation monitoring results were analysed. They included 188 specific agri-environment schemes aimed at maintaining, enhancing or restoring grasslands or grassland landscapes in the UK. These involved repeated monitoring over up to 8 years, on between 4 and 400 locations/agri-environment schemes, using a range of sampling strategies.

A 2004 literature review of farmland bird declines in the UK (18) found that 12 of 30 declining species have shown local population density increases after the implementation of agri-environment scheme options. Five out of ten seed-eating birds responded positively to agri-environment schemes, one (cirl bunting *Emberiza cirlus*) showing large increases. Three other songbirds, corncrake *Crex crex*, grey partridge *Perdix perdix* and two wading birds responded to agri-environment scheme options. A further 7 species responded to local conservation measures and 11 species were not studied sufficiently and were found not to respond to conservation measures or were recovering following national legislation (i.e. the prohibition of organochlorine pesticides).

A 2004 literature review (19) describes how ten years of agri-environment schemes in the UK have failed to halt the decline of many formerly common farmland bird species. However, it also points out that specially-designed agri-environment scheme options have led to local-scale population increases of three rare and range-restricted species (corncrake *Crex crex*, Eurasian thick-knee (stone curlew) *Burhinus oedicnemus* and cirl bunting *Emberiza cirlus*) (4).

A replicated, paired site comparison study in 1999 and 2000 on 60 farms in 3 counties of Ireland (20) found no consistent difference between Rural Environment Protection Scheme and non-Rural Environment Protection Scheme farms in plant or ground beetle (Carabidae) diversity or abundance. Non-Rural Environment Protection Scheme farms had the greatest range in species richness and included farms with the lowest and highest numbers of plant species (23 and 50 plant species, respectively) and ground beetle species (12 and 30 ground beetle species). There were more plant species on grassland field margins on non-Rural Environment Protection Scheme farms (average 14.2 species/margin) than on Rural Environment Protection Scheme farms (12.5 species/margin). Sixty farms with Rural Environment Protection Scheme agreements at least 4-years-old were paired with 60 similar farms without agreements. The farm pairs were in three Irish counties: Laois and Offaly (largely cattle farms with pasture) and Wexford (largely mixed arable farming). On each farm, two randomly selected hedges, the adjacent field margin and one watercourse margin were surveyed for plants and ground beetles. In each field margin and watercourse margin all plant species were recorded in two 5 x 3 m quadrats and percentage cover estimated in a 1 x 3 m quadrat. All plant species in a 30 m stretch of hedge were recorded. Ground beetles were sampled in 4 pitfall traps/field margin (8 traps/farm) set at 10 m intervals in early June and late August.

A replicated, controlled trial from 1994–2004 in the five Environmentally Sensitive Areas in Northern Ireland (21) (same study as (24)) found that overall, farms without Environmentally Sensitive Area agreements showed a decrease in invertebrate diversity and a decrease in the number of plant species characteristic of infertile soils, while these decreases did not happen on Environmentally Sensitive Area farms. The number of plant species characteristic of infertile soils (stress-tolerant species) on hay meadows significantly increased from 1994–2004 on Environmentally Sensitive Area agreement farms but decreased on farms without Environmentally Sensitive Area agreements (numbers not given). Two ground beetle (Carabidae) species of conservation interest increased on farms with Environmentally Sensitive Area agreements between 1994 and 2004, each in one of the five Environmentally Sensitive Areas. The ground beetle *Cymindis vaporariorum*, characteristic of upland heaths and raised bogs, increased in the Glens and Rathlin Island Environmentally Sensitive Area; *Carabus clatratus* (a wet grassland/bog species) increased on participant farms in the West Fermanagh and Erne Lakelands Environmentally Sensitive Area. Plants, birds, spiders (Araneae) and ground beetles were monitored from 1994–2004, on farms with and without agreements, in the five Environmentally Sensitive Areas: Mourne and Slieve Croob (established 1988), Slieve Gullion (established 1994), Antrim Coast, Glens and Rathlin Island (established 1989), Sperrins (established 1994) and West Fermanagh and Erne Lakelands (established 1993). Monitoring was on permanent randomly placed quadrats in seven habitat types: wet grassland, limestone grassland, unimproved grassland, hay meadows, heather moorland, woodland and field boundaries. Quadrats were partially surveyed every three years and fully surveyed in 1994 and 2004.

A replicated, controlled trial in Estonia (22) found no difference in numbers of earthworms (Lumbricidae) or soil microbial activity between arable soils with and without agri-environment schemes, in the first two years of a pilot agri-environment scheme. There were 32–224 earthworms/m^2 of 1–5 species in the Jõgeva County area and 0–614 earthworms/m^2 of 0–5 species in the Saare County area. The grey worm *Aporrectodea caliginosa* was dominant (81–89% of all earthworm individuals) in both areas. As the scheme had been in place for one or two years only, the authors considered these results to be baseline data, showing no initial differences in soils between agri-environment and control areas. The 'Environmentally Friendly Production Scheme' required restricted nitrogen fertilizer (100 kg/ha or less) and limited field size, at least 15% of the cultivated area, to be under legumes or grass and legumes, with cereals not grown for more than 3 years in a row, uncultivated field margins and maintenance of existing landscape elements, including semi-natural habitats. The pilot scheme began in 2001. For each pilot area, earthworms were monitored in one cereal field on each of ten farms with the Environmentally Friendly Production Scheme, and five farms without it, in an adjacent reference area. Earthworms were sampled by hand-sorting from five soil blocks 50 x 50 x 40 cm. Microbial activity was sampled by estimating the activity of dehydrogenase enzymes (the fluorescein diacetate method).

A replicated study in 2005 of 2,449 1 km squares across arable and pastoral farmland in England at the start of the Entry Level Scheme (23) found that there was no difference in the total number of bird species between 1 km^2 of land participating in the scheme and areas not participating in the scheme. Eight bird species had a significantly higher occurrence on Entry

Level Scheme squares, whilst seven species (mainly non-farmland bird species) had higher occurrences on non-Entry Level Scheme squares. Three species had higher abundance on Entry Level Scheme squares (all farmland specialists: linnet *Carduelis cannabina*, tree sparrow *Passer montanus*, stock dove *Columba oenas*) and 17 species were more abundant on non-Entry Level Scheme squares. There were 975 squares on land under the Entry Level Scheme land and a further 1,474 squares on conventionally managed farmland. Each square was surveyed twice along a 2 km transect route, recording all birds seen or heard. The Entry Level Stewardship scheme was introduced in 2005 and the data from this study was used to provide a baseline against which future surveys could monitor the effectiveness of the scheme.

A replicated, before-and-after trial in Northern Ireland (24) (same study as (21)) found that the number of plant species on land managed under the Environmentally Sensitive Areas scheme over ten years was maintained but not enhanced on grasslands, and maintained in heather moorland in two of three areas for which results were reported. The number of higher plant species did not increase between 1993 and 2003 in the Environmentally Sensitive Area grassland sites, which were all in the West Fermanagh Environmentally Sensitive Area (33–41 species/transect). In heather moorland, average cover of heather increased in 1 of the 5 Environmentally Sensitive Areas (13 sites in West Fermanagh) but did not change at 2 others (43 sites in Sperrins Environmentally Sensitive Area; 6 sites in Antrim Coast Environmentally Sensitive Area). The number of plant species on heather moorland was maintained at these two Environmentally Sensitive Areas but declined between 1994 and 2004 in the Slieve Gullion Environmentally Sensitive Area (13 sites). Values are not given for heather cover or numbers of plant species on heather moorland. The study monitored plant diversity at 63 grassland sites and 93 heather moorland sites, first in 1993–1994 before the Environmentally Sensitive Area management began and again 10 years later. The sites were randomly selected from a database of farmers joining the Environmentally Sensitive Area scheme in 1993.

A replicated study in 1999 and 2003 on 84 farms in East Anglia and the West Midlands, UK (25) found that only 3 bird species (2 in East Anglia; 1 in the West Midlands) showed a significant positive response to the introduction of agri-environment schemes (Arable Stewardship Pilot Scheme), whilst one showed a significant negative effect. Meadow pipit *Anthus pratensis*, carrion crow *Corvus corone* and reed bunting *Emberiza schoeniclus* either declined less or increased on farms under agri-environment schemes, compared to conventionally managed 'control' farms. Corn bunting *Miliaria calandra* declined significantly faster on agri-environment scheme farms. Overall, only six species showed any positive response (significant or not) in both regions. Ten species showed negative responses in both regions and 12 showed a positive response in 1 region and a negative response in the other. This study was part of the same monitoring project as (10, 11, 16).

A single farm, Rawcliffe Bridge, East Yorkshire, UK (26), with a combination of conservation measures prescribed under the Entry Level Stewardship Scheme, had higher densities of some bird species than the average for UK lowland farms. Meadow pipit *Anthus pratensis*, reed bunting *Emberiza schoeniclus*, Eurasian skylark *Alauda arvensis*, grey partridge *Perdix perdix*, corn bunting *E. calandra* and yellow wagtail *Motacilla flava* occurred in higher numbers in each monitoring year than the average lowland farm density (provided by the British Trust for Ornithology). For example, there were between 12 and 22 meadow pipit pairs/100 ha at Rawbridge, compared to a national average of less than 3. Birds on the farm were monitored five times each year from 2003 to 2005 by walking the field boundaries. The number of breeding pairs/ha was estimated from clusters of sightings.

A replicated site comparison study in 2005 and 2006 on 31 farms in Seine-et-Marne, France (27) found that agri-environment measures did not benefit plant diversity. The number of plant species was higher on farms with one or two agri-environment measures than those with none at all, but farms with between three and seven different agri-environment measures had generally fewer plant species than farms with very few measures. Plant diversity (Simpson's diversity index) was unaffected by the number of agri-environment measures per farm. Twenty-six fields from 17 farms were sampled 3 times in 2005 (April; June; September). Sixty-four fields from 31 farms (including all those surveyed in 2005) were sampled twice in 2006 (April and July). Plants were recorded in 10 permanent, regularly spaced 1 m^2 (0.5 x 2 m) quadrats along the permanent margins of each field.

A replicated site comparison study in Northern Ireland (28) found that areas with agreements under the Environmentally Sensitive Areas scheme did not have more Irish hares *Lepus timidus hibernicus* than areas outside the scheme (around 0.4 Irish hares/km on average). Rabbits *Oryctolagus cuniculus* and red foxes *Vulpes vulpes* were more abundant in Environmentally Sensitive Areas (around 2 rabbits and 0.5 foxes/km on average) than in non-Environmentally Sensitive Areas (around 1 rabbit and 0.2 foxes/km on average); both species are considered pests on farmland. One-hundred-and-fifty 1 km^2 were randomly selected from within Northern Ireland's five Environmentally Sensitive Areas. A sample of 50 non-Environmentally Sensitive Area squares were matched for land use, altitude, road type and distance from the Environmentally Sensitive Area boundary. Mammals were surveyed by spotlight on night drives in mid-winter 2005, on both sides of 1 km of road bisecting each survey square. Irish hares, rabbits and red foxes were counted.

A 2007 systematic review of 29 studies incorporating data for 15 farmland bird species in the UK (29) found that there were significantly higher winter densities of farmland birds on fields under agri-environment schemes than on conventionally managed fields. Considering each scheme individually, there were greater winter densities of birds on fields within the Arable Stewardship Pilot Scheme, Countryside Stewardship Scheme, organically farmed fields, fields with set-aside, overwinter stubble and wild bird cover than on conventionally farmed fields. Overall, eight species (53%) had significantly higher winter densities on agri-environment fields compared to conventional cropping (corn bunting *Miliaria calandra*, greenfinch *Carduelis chloris*, grey partridge *Perdix perdix*, northern lapwing *Vanellus vanellus*, linnet *C. cannabina*, rook *Corvus frugilegus*, Eurasian skylark *Alauda arvensis* and song thrush *Turdus philomelos*) and no species were found to have higher densities on conventional agricultural fields compared to those fields entered under agri-environment scheme agreements. Although both organic fields and set-aside fields in summer had significantly higher densities of farmland birds, there was no difference between the number of birds on conventionally farmed fields and Arable Stewardship Pilot Scheme fields in summer. Six species (35%; grey partridge, lapwing, woodpigeon *Columba palumbus*, skylark, rook and cirl bunting *Emberiza cirlus*) of the 17 for which summer data were available were found at significantly higher densities on agri-environment scheme fields compared with fields under conventional systems. The migratory yellow wagtail *Motacilla flava* was found at lower densities on scheme fields than on conventionally managed fields. In total 29 papers describing experiments conducted between 1985 and 2005 on a total of 12,653 fields (5,381 fields under agri-environment schemes and 7,272 fields farmed conventionally) were used for the meta-analysis. The meta-analysis included 7 site comparison studies, 5 randomized controlled trials and 17 controlled trials.

A replicated site comparison study in four regions of Estonia (30) found more bumblebee *Bombus* spp. species and higher flower abundance on farms with agri-environment agreements in two of the four regions. In the central Estonian regions, with large fields and homogenous agriculture (Tartu and Jõgeva), organic and Environmentally Friendly

Production Scheme farms had more bumblebee species than conventional farms (8–9 species/farm on organic or agri-environment farms compared to around 5 species/conventional farm). There was no difference in numbers of bumblebee species or flower abundance between types of farm in the south and west Estonian regions (Võru and Saare). Bumblebees were monitored on 22 farms in each region – 10 organic farms, 6 in the Environmentally Friendly Production Scheme and 6 conventional farms with no environmental agreement. Bumblebees were counted on 500 x 2 m transects 6 times between June and August 2006 on days with temperature above 16 °C and wind speed less than 6 m/s. Flower abundance was assessed on the transects using a four point scale.

A site comparison study of 677 plots covering 38,705 ha across southern England (31) found that for 3 wading bird species population trends were more favourable (increasing or declining less rapidly) in areas under Environmentally Sensitive Area scheme options aimed at enhancing habitat than in the less expensive Environmentally Sensitive Area habitat maintenance options and in parts of the surrounding countryside not participating in the scheme. Nature reserves were shown to be most effective at maintaining wader populations. Between 1982 and 2002 common redshank *Tringa totanus* declined by 70% in the wider countryside but increased overall from 646 to 755 pairs (up 17%) on Environmentally Sensitive Area designated land, with the largest increase observed in nature reserves outside Environmentally Sensitive Areas (160%). Northern lapwing *Vanellus vanellus* showed a 48% decline in the wider countryside and increased only in nature reserves outside Environmentally Sensitive Areas (by 55%) and reserves with Environmentally Sensitive Area enhancement (121%). Common snipe *Gallinago gallinago* breeding numbers decreased everywhere (commonly with declines of 90% or more), although declines were smaller in nature reserves outside Environmentally Sensitive Areas (–66%) and reserves in Environmentally Sensitive Area enhancement (–24%). The Environmentally Sensitive Area scheme was introduced in 1987 and offered payments for either maintaining or enhancing landscape quality and biodiversity. Breeding waders were surveyed in 1982 and 2002 at lowland wet grassland sites covering ten counties in England. In both years three censuses were carried out at each site between mid-April and mid-June.

A before-and-after study examining data from 1976 to 2003 from farms across southern Sweden (32) found that four locally migrant farmland birds (northern lapwing *Vanellus vanellus*, Eurasian skylark *Alauda arvensis*, common starling *Sturnus vulgaris* and linnet *Carduelis cannabina*) showed less negative (or positive) population trends during 1987–1995, a period of agricultural extensification that included the introduction of agri-environment schemes, compared to in the preceding period of intensification (1976–1987). However, following the adoption of the Common Agricultural Policy (CAP) in 1995–2003 the species showed more negative population trends again, despite the widespread adoption of agri-environment scheme options. Three non-migrant species (house sparrow *Passer domesticus*, tree sparrow *P. montanus* and yellowhammer *Emberiza citrinella*) showed more diverse population trends and responses to agricultural changes were largely non-significant.

A replicated, site comparison study over 20 years in the UK (33) found that chalkhill blue butterflies *Polyommatus coridon* increased more on sites with agri-environment scheme agreements than sites without. Chalkhill blue numbers increased on average 3.16%/year at 66 sites with Countryside Stewardship Scheme or Environmentally Sensitive Area agreements, compared to no significant trend at non-scheme sites. Chalkhill blues were counted annually from 1981 to 2000 at 161 sites across its entire UK range. This was part of the UK Butterfly Monitoring Scheme, which takes weekly transect counts along a set route at each site and follows standardized weather conditions.

A replicated, paired site comparison in Bavaria, Germany (34) found that grasslands under the 'Agricultural Landscape Programme' (KULAP) did not have more plant species than control grasslands overall. There were 18–23 plant species/plot on sites with any KULAP agreement compared to 18–22 plant species/paired control plot (215 site pairs). When considering only sites with site-related (rather than whole farm) agreements there were more plant species under the KULAP scheme (around 22 species/site) than on paired control sites (<20 species/site; 90 site pairs). There were also more plant species/site on 58 Contract Nature Protection Scheme sites (25 species/plot) compared to paired control plots (about 17 species/plot). Nine-hundred-and-thirty-six pairs of 25 m² grassland plots were selected from 4,400 plots in the Bavarian grassland survey. All plant species within the plot were recorded between April and October (year not given). Plot pairs were in the same natural landscape, 90% within 10 km of each other. In each pair there was one with and one without an agri-environment scheme agreement.

A study of the locations of Meadow Bird Agreements in the Netherlands (35) found that 43% of the 71,982 ha area of Meadow Bird Agreements in 2004 was located on sites where meadow bird populations are constrained for reasons other than those addressed by the agri-environment management. Twenty-two percent (15,798 ha) were outside the area of known black-tailed godwit *Limosa limosa* occurrence (more than 5 breeding pairs/100 ha in a 1998–2000 survey; 90–95% of other specialist meadow bird species breed in suitable black-tailed godwit habitat). Within the black-tailed godwit area, 11% (6,166 ha) of the Meadow Bird Agreement area was on heavily drained land, 4% (2,500 ha) was in landscapes not considered open enough for meadow birds, 10% (5,400 ha) was in areas of high traffic disturbance and an estimated 8% (2,834 of the 35,000 ha for which data were available) was on sites with high predation. The authors advocated targeting Meadow Bird Agreements to the 285,000 ha of land in the Netherlands with more than 5 breeding pairs of black-tailed godwit/100 ha, but none of the other identified constraints.

A before-and-after replicated trial in the Yorkshire Dales National Park, UK (36) found that the average number of plant species in upland hay meadows fell from 19.5 species in 1980, before the introduction of agri-environment schemes, to 14.7 species in 2003, when the Countryside Stewardship Scheme and the Environmentally Sensitive Areas scheme had been employed for almost 20 years. One-hundred-and-nineteen fields surveyed in the 1980s and found to contain wood cranesbill *Geranium sylvaticum* were re-surveyed in 2003. In 47 of the fields all plant species were recorded in ten 1 m² quadrats at each site. Wood cranesbill was found in 76 of the 119 fields it had previously been found in, an extinction rate of 40%. The average nearest distance to another field containing the species increased from 121 m in 1980 to 1,072 m in 2003. Fields located over 300 m from another field containing the habitat were more likely to have lost the species.

A site comparison study of 53 2 km² plots on 14 farms in southeast Scotland (37) observed that between 2002 and 2004 the number of territorial male corn bunting *Emberiza calandra* fell by only 5% on plots that managed land according to the Farmland Bird Lifeline scheme, whereas numbers declined by 43% in non-Farmland Bird Lifeline plots in the same area. Between 2000 and 2002, before the 2002 introduction of the Farmland Bird Lifeline management practices, there was no observed change in the number of corn bunting on either group of plots, although plots destined to participate in the Farmland Bird Lifeline scheme did already have 33% higher densities of corn bunting than comparison plots. The Farmland Bird Lifeline scheme intended to reverse the declining numbers of corn bunting, a priority species in the UK Biodiversity Action Plan. Farmers were paid for a number of interventions, including delaying mowing date, providing grass margins on arable fields, farming spring cereals and turnips at low intensity, spring cropping, leaving unharvested crop and supplementary feeding. Fourteen farms, nine in Aberdeenshire and five in Fife, were surveyed every breeding season (late April to August) from 2000 to 2004.

A site comparison study of ten 3 km² plots in Austria (38) showed that, compared to conventionally managed arable land, land farmed less intensively (under agri-environment schemes) had larger numbers of ground breeding birds (16.1 vs 13.2 individuals/10 ha), Red Listed birds (2.5 vs 1.8 individuals/10 ha) and Species of European Conservation Concern (13.9 vs 10.3 individuals/10 ha). Arable land managed for the conservation of particular species had 27.6 Species of European Conservation Concern individuals/10 ha and 28.6 ground breeding individuals/10 ha compared with the 11.1 individuals/10 ha and 13.7 individuals/10 ha, respectively, on conventionally managed farmland. Reed-breeding birds on grassland benefited from similar initiatives (11.3 vs 2.8 individuals/10 ha of farmland). Habitat conservation measures appeared to benefit ground breeders on arable farmland (16.6 vs 10 individuals/10 ha). Breeding birds were surveyed during three visits between April and June 2003.

A replicated, controlled trial involving ten farms in east and central Scotland (39) found that on farms managed under the Rural Stewardship Scheme, transects covering agri-environment options (unsprayed grassy field margins, species-rich grassland uncut from March–August and hedgerows only cut every three years) attracted significantly more nest-searching and foraging queen bumblebees *Bombus* spp. than conventionally managed transects. However, on conventionally managed transects (not agri-environment scheme options), there was no significant difference between farms with and without agri-environment schemes in numbers of nest-searching queens, and conventionally managed farms had more foraging queens. Five farms that signed up to the Scottish Rural Stewardship Scheme in 2004 were paired with 5 comparison farms less than 5 km away with similar land-use but no agri-environment participation. Bumblebees were recorded on six 100 x 6 m transects/farm weekly in April–May 2009. Each farm had two arable field margin transects, two grassland transects and two hedgerow transects.

A 2009 literature review of agri-environment schemes in England (40) found that options and schemes varied in effectiveness for farmland wildlife. Breeding populations of some nationally rare birds increased after the implementation of options on arable farms (cirl bunting *Emberiza cirlus* pairs increased by 130%; Eurasian thick-knee (stone curlew) *Burhinus oedicnemus* pairs increased by 87%). A case study from a single farm found that grey partridge *Perdix perdix* numbers increased by more than 250%/year, corn bunting *Miliaria calandra* by over 100%/year and Eurasian skylark *Alauda arvensis* by 71%/year following the implementation of a number of different options. Productivity of some bird species was found to be higher on agri-environment scheme farms, which also provided key habitats. However, there was little evidence for any population-level beneficial effects of Entry Level Stewardship designations on widespread birds such as skylark or yellowhammer *E. citrinella*. Several of the studies reviewed argued that most agri-environment schemes were not well targeted to provide habitat for wading birds (Dutt 2004), although other studies argued that wader populations had declined less in regions designated as Environmentally Sensitive Areas than in the country overall (Wilson *et al.* 2005). Implementation of agri-environment schemes was also shown to benefit mammals, such as brown hare *Lepus europaeus*, with significantly higher densities on farms with agri-environment schemes than control farms in East Anglia. However, in the West Midlands, hare densities were similar between agri-environment scheme farms and control farms (11).

A replicated, paired sites study on farms across Scotland under two agri-environment scheme prescriptions (Countryside Premium Scheme and Rural Stewardship Scheme) in spring–summer 2004–2008 (41) concluded that the schemes had little impact on farmland biodiversity. Whilst 280 agri-environment scheme farms had more birds of more species than 193 non-scheme-paired farms (averages of 140 birds of 23 species on 105 Countryside Premium Scheme farms vs 108 of 20 on paired non-scheme farms; 108 birds of 19 species on 88 Rural Stewardship Scheme farms vs 86 of 17 on paired farms), trends did not vary between scheme and non-scheme farms and scheme farms had higher species richness and abundances before entering schemes. Differences held for all species and for nationally threatened species. Time since entry into the Countryside Premium Scheme did not appear to affect the number of species or bird abundance, except for a small decline in the abundance of tits *Parus* spp. In addition, no evidence was found for differing effects of schemes in different regions of Scotland, or on different farm types. There were generally more plant species and individuals and higher plant diversity on farms managed under the Countryside Premium Scheme than non-Countryside Premium Scheme farms (e.g. for one agri-environment scheme option there were 20 plant species on scheme farms vs 15 on non-scheme farms) but very limited evidence of significant differences in plant species richness, abundance or diversity between Rural Stewardship Scheme and non-Rural Stewardship Scheme farms – where there was a difference there were more plant species or higher diversity on the Rural Stewardship Scheme farms. There were no significant differences in butterfly (Lepidoptera) species richness or abundance between Countryside Premium Scheme and non-Countryside Premium Scheme farms, and no significant differences in the number of ground-active beetles (Coleoptera) between Countryside Premium Scheme or Rural Stewardship Scheme and conventionally managed farms. Plants, ground-active beetles and butterflies were already generally more abundant or more species rich on Rural Stewardship Scheme sites when they joined the scheme (during the first survey year 2004–2005).

A before-and-after study on one farm in Oxfordshire, UK (42) found that following a change to management under the Environmentally Sensitive Areas scheme (also leading to organic certification) the numbers of large moths (Lepidoptera), some species of butterfly and ground beetle (Carabidae) and the number of plant species, including butterfly larval food plant species, increased. The butterfly species that increased after Environmentally Sensitive Area management included the brown argus *Aricia agestis*, the common blue *Polyommatus icarus* and the small copper *Lycaena phlaeas*. Overall butterfly and ground beetle numbers, and numbers of pipistrelle bats *Pipistrellus pipistrellus* and *P. pygmaeus* and Daubenton's bats *Myotis daubentonii* also increased over the entire time period, but the increase did not happen after management change. Butterflies, plants, ground beetles and bats were regularly monitored on the farm from 1994 to 2006 inclusive. In 2002, the farm entered the Environmentally Sensitive Areas agri-environment scheme. The proportion of grassland increased, fertilizers, herbicides and pesticides were no longer used and the total number of livestock dropped from 180 cows and 1,000 sheep to 120 cows and 850 sheep. The land was certified organic in 2005.

A controlled study in 2002–2009 on mixed farmland in Hertfordshire, UK (43) found that the estimated population density of grey partridge *Perdix perdix* was significantly higher on land under agri-environment schemes than on conventional arable crops. This study also examined the densities found on set-aside (which were similar to those on land under other agri-environment schemes), wild bird cover (which were considerably higher than on other land uses) and the impact of predator control and supplementary food provision. Grey partridges were surveyed in March and September using dawn and dusk counts starting in 2001. Land cover within the project area was mapped and categorized as: conventional arable land, arable in agri-environment schemes, non-arable, or set-aside (which was further divided into non-rotational; wild bird cover; other rotational).

A replicated site comparison of 2,046 1 km squares of agricultural land across England in April–June 2005 and 2008 (44) (same study as (45)) found that farmland bird population responses to Entry Level Stewardship schemes varied regionally. The authors suggested that detailed, regional prescriptions may be more effective in stimulating bird population growth than uniform agri-environment schemes. Field margin management took place in 36% of squares and did not have clear impacts on 'field margin' species: 2 species responded positively in at least 1 region, 3 species showed positive and negative responses in different regions, 1 showed only a negative response and the other 6 showed no significant responses.

A large site comparison study in 2005 and 2008 of 2,046 1 km² plots of lowland farmland in England (45) (same study as (44)) found that the Countryside Stewardship Scheme and Entry Level Stewardship schemes had no consistent effect on farmland bird numbers three years after their introduction in 2005. Between 2005 and 2008, 8 Farmland Bird Index species showed significant declines on arable plots, 9 species declined significantly on pastoral plots and 6 species declined on mixed farmland squares (farmland plots covered with less than 50% arable and less than 50% pastoral farming). Only goldfinch *Carduelis carduelis*, jackdaw *Corvus monedula*, and woodpigeon *Columba palumbus* showed population increases between 2005 and 2008. Although certain farmland bird species did show landscape-specific effects, there were no consistent relationships between farmland bird numbers and whether or not the plots contained Entry Level Stewardship and Countryside Stewardship Scheme land, or the financial cost of the agri-environment interventions, or the length of hedgerows or ditches under an agri-environment scheme, or the availability of wild bird seed mix and overwinter stubbles (i.e. some species showed increases in response to a particular intervention on a particular landscape-type but not on other landscape types, and these changes were not consistent between species). The 2,046 1 km² lowland plots were surveyed in both 2005 and 2008 and classified as arable, pastoral or mixed farmland. Eighty-four percent of plots included some areas managed according to the Entry Level Stewardship or Countryside Stewardship Scheme. In both survey years 2 surveys were conducted along a 2 km pre-selected transect route through each 1 km² square.

A replicated site comparison study from 2004 to 2008 on 1,031 agricultural sites across England (46) found that in 3 out of 4 year-on-year comparisons, grey partridge *Perdix perdix* density changes and overwinter survival were higher on sites under agri-environment schemes than on sites not under schemes (partridge density changes were more positive on agri-environment scheme sites than non-agri-environment scheme sites in all comparisons except 2007–2008; overwinter survival was higher for all years except 2006–2007). However, these differences were only significant in 2005–2006 for density changes (6% increase on agri-environment scheme sites vs 11% decrease on non-agri-environment scheme sites) and 2006–2007 for overwinter survival. There were no consistent differences between agri-environment scheme and non-agri-environment scheme sites with respect to brood size. When schemes were investigated individually, only Countryside Stewardship Scheme sites and Environmentally Sensitive Area sites had significantly more positive density trends than non-scheme sites and only in 2005–2006 (6% increase on Countryside Stewardship Scheme and Environmentally Sensitive Area sites vs 12% decline on non-agri-environment scheme sites), although other years and schemes showed a similar pattern. Overwinter survival, brood size and the ratio of chicks to adults did not show consistent effects across different schemes. A higher proportion of sites under the Partridge Count Scheme implementing the options most beneficial to partridges was higher than the proportion of non-Partridge Count

Scheme sites. Various methods of succession management (rough grazing; scrub creation; scrub control; grassland creation) were negatively associated with the ratio of young to old partridges in 2008.

A small site comparison study between November 2007 and February 2008 on 75 fields in East Anglia and the West Midlands, UK (47) found no difference between the numbers of seed-eating birds in fields managed under Higher Level Stewardship of the Environmental Stewardship scheme and numbers in fields managed under Entry Level Stewardship. Entry Level Stewardship fields had overwinter stubbles, no post-harvest herbicide application and no cultivation until mid-February and were sown overwinter with wild bird seed mix. Higher Level Stewardship fields were sown with enhanced wild bird seed mix and the stubbles had the same basic Entry Level Stewardship requirements plus reduced herbicide use and cereal crop management before overwintered stubbles.

A before-and-after trial of the Entry Level Stewardship scheme (an option within the Environmental Stewardship scheme) on a 1,000 ha lowland arable farm in central England (48) observed that the number of seed-eating birds was higher on both Entry Level Steward-ship and conventionally farmed fields in the winter of 2006–2007 than during the previous winter (2005–2006) when the Entry Level Stewardship scheme was first introduced. This in-crease was greater on Entry Level Stewardship plots setting aside 5% of farmland to provide winter bird food (with an average of 70 birds/km of transect in 2007 vs five birds/km of transect in 2006) than on conventionally farmed fields (25 birds/km of transect in 2007 vs 10 birds/km of transect in 2006). Although there were also more summer breeding territories of seed-eating species, chaffinch *Fringilla coelebs*, dunnock *Prunella modularis* and robin *Erithacus rubecula* on the farm as a whole in 2007 than in the previous breeding season (2006); there was no difference in this increase between Entry Level Stewardship and conventional fields. Land managed according to the minimal environmental requirements was compared both with fields where 5% of land was removed from production and replaced with patches of winter bird food and field margins (6–8 m). Winter birds were surveyed from transects on three visits (November, December and January) in both the winters of 2005–2006 and 2006–2007 – i.e. before and after bird food patch establishment. Breeding territories were surveyed during four visits (April, May, June and July) in 2006 and 2007.

(1) Wakeham-Dawson A. (1995) Hares and skylarks as indicators of environmentally sensitive farming on the South Downs. PhD thesis. Open University.
(2) Tilzey M. (1997) Environmentally sensitive areas and grassland conservation in England. *Proceedings of the International Occasional Symposium of the European Grassland Federation (Management for Grassland Biodiversity)*. Warszawa-Łomża, Poland, 19–23 May. pp 379–390.
(3) Ovenden G., Swash A. & Smallshire D. (1998) Agri-environment schemes and their contribution to the conservation of biodiversity in England. *Journal of Applied Ecology*, 35, 955–960.
(4) Aebischer N.J., Green R.E. & Evans A.D. (2000) From science to recovery: four case studies of how research has been translated into conservation action in the UK. In N. J. Aebischer, A. D. Evans, P. V. Grice & J. A. Vickery (eds.) *Ecology and Conservation of Lowland Farmland Birds*, British Ornithologists' Union, Tring. pp. 43–54.
(5) Reid C. & Grice P. (2000) *Wildlife Gain from Agri-environment Schemes: Recommendations from English Nature's Habitat and Species Specialists English Nature*. Research Reports No. 143.
(6) Peach W., Lovett L., Wotton S. & Jeffs C. (2001) Countryside stewardship delivers cirl buntings (*Emberiza cirlus*) in Devon, UK. *Biological Conservation*, 101, 361–373.
(7) Ausden M. & Hirons G.J.M. (2002) Grassland nature reserves for breeding wading birds in England and the implications for the ESA agri-environment scheme. *Biological Conservation*, 106, 279–291.
(8) Evans A.D., Armstrong-Brown S. & Grice P.V. (2002) The role of research and development in the evolution of a 'smart' agri-environment scheme. *Aspects of Applied Biology*, 67, 253–264.
(9) Goulson D., Hughes W.O.H., Derwent L.C. & Stout J.C. (2002) Colony growth of the bumblebee, *Bombus terrestris*, in improved and conventional agricultural and suburban habitats. *Oecologia*, 130, 267–273.

(10)   Bradbury R. & Allen D. (2003) Evaluation of the impact of the pilot UK Arable Stewardship Scheme on breeding and wintering birds: Few positive responses by birds were observed in the first two years of this new agri-environment scheme. *Bird Study*, 50, 131–141.

(11)   Browne S. & Aebischer N. (2003) *Arable Stewardship: Impact of the Pilot Scheme on Grey Partridge and Brown Hare after Five Years*. Defra MAO1010.

(12)   Defra (2003) *Review of Agri-environment Scheme Monitoring Results and R&D. Ecoscope/CPM/CJC, Consulting* (RMP/1596). Final Report – Part A (V45).

(13)   Feehan J. (2003) Investing in monitoring and evaluation: an overview of practical approaches to biodiversity monitoring of agri-environment schemes. *Tearmann*, 3, 17–26.

(14)   Kleijn D. & Sutherland W.J. (2003) How effective are European agri-environment schemes in conserving and promoting biodiversity? *Journal of Applied Ecology*, 40, 947–969.

(15)   Berendse F., Chamberlain D., Kleijn D. & Schekkerman H. (2004) Declining biodiversity in agricultural landscapes and the effectiveness of agri-environment schemes. *Ambio*, 33, 499–502.

(16)   Bradbury R.B., Browne S.J., Stevens D.K. & Aebischer N.J. (2004) Five-year evaluation of the impact of the Arable Stewardship Pilot Scheme on birds. *Ibis*, 146, S171–180.

(17)   Critchley C., Burke M. & Stevens D. (2004) Conservation of lowland semi-natural grasslands in the UK: a review of botanical monitoring results from agri-environment schemes. *Biological Conservation*, 115, 263–278.

(18)   Newton I. (2004) The recent declines of farmland bird populations in Britain: an appraisal of causal factors and conservation actions. *Ibis*, 146, 579–600.

(19)   Vickery J.A., Bradbury R.B., Henderson I.G., Eaton M.A. & Grice P.V. (2004) The role of agri-environment schemes and farm management practices in reversing the decline of farmland birds in England. *Biological Conservation*, 119, 19–39.

(20)   Feehan J., Gillmor D. & Culleton N. (2005) Effects of an agri-environment scheme on farmland biodiversity in Ireland. *Agriculture, Ecosystems & Environment*, 107, 275–286.

(21)   McAdam J., McEvoy P., Flexen M. & Hoppé G. (2005) Biodiversity monitoring and policy support in Northern Ireland. *Tearmann*, 4, 15–22.

(22)   Sepp K., Ivask M., Kaasik A., Mikk M. & Peepson A. (2005) Soil biota indicators for monitoring the Estonian agri-environmental programme. *Agriculture Ecosystems & Environment*, 108, 264–273.

(23)   Chamberlain D., Noble D. & Vickery J. (2006) *Assessment of the Impacts of the Entry Level Scheme on Bird Populations: Results from the Baseline Year, 2005*. BTO Research Report 437.

(24)   McEvoy P.M., Flexen M. & McAdam J.H. (2006) The Environmentally Sensitive Area (ESA) scheme in Northern Ireland: ten years of agri-environment monitoring. *Biology and Environment: Proceedings of the Royal Irish Academy, Section B*, 106B, 413–423.

(25)   Stevens D.K. & Bradbury R.B. (2006) Effects of the Arable Stewardship Pilot Scheme on breeding birds at field and farm-scales. *Agriculture, Ecosystems & Environment*, 112, 283–290.

(26)   Bryson R.J., Hartwell G. & Gladwin R. (2007) Rawcliffe Bridge, arable production and biodiversity, hand in hand. *Aspects of Applied Biology*, 81, 155–160.

(27)   Chateil C., Abadie J.C., Gachet S., Machon N. & Porcher E. (2007) Can agri-environmental measures benefit plant biodiversity? An experimental test of the effects of agri-environmental measures on weed diversity. In (eds.) Association Nationale pour la Protection des Plantes (ANPP), Paris. pp. 356–366.

(28)   Reid N., McDonald R.A. & Montgomery W.I. (2007) Mammals and agri-environment schemes: hare haven or pest paradise? *Journal of Applied Ecology*, 44, 1200–1208.

(29)   Roberts P.D. & Pullin A.S. (2007) *The Effectiveness of Land-based Schemes (Including Agri-environment) at Conserving Farmland Bird Densities within the UK*. Systematic Review No. 11. Collaboration for Environmental Evidence/Centre for Evidence-Based Conservation, Birmingham, UK.

(30)   Viik E., Mänd M., Karise R., Koskor E., Jõgar K., Kevväi R., Martin A. & Grishakova M. (2007) The effect of agri-environmental schemes on species richness of bumble bees. In (eds.) *Agronomy*, Estonian Research Institute of Agriculture, Sacu. pp. 145–148.

(31)   Wilson A., Vickery J. & Pendlebury C. (2007) Agri-environment schemes as a tool for reversing declining populations of grassland waders: mixed benefits from Environmentally Sensitive Areas in England. *Biological Conservation*, 136, 128–135.

(32)   Wretenberg J., Lindstrom A., Svensson S. & Part T. (2007) Linking agricultural policies to population trends of Swedish farmland birds in different agricultural regions. *Journal of Applied Ecology*, 44, 933–941.

(33)   Brereton T.M., Warren M.S., Roy D.B. & Stewart K. (2008) The changing status of the Chalkhill Blue butterfly *Polyommatus coridon* in the UK: the impacts of conservation policies and environmental factors. *Journal of Insect Conservation*, 12, 629–638.

(34)   Mayer F., Heinz S. & Kuhn G. (2008) Effects of agri-environment schemes on plant diversity in Bavarian grasslands. *Community Ecology*, 9, 229–236.

(35)   Melman T.C.P., Schotman A.G.M., Hunink S. & de Snoo G.R. (2008) Evaluation of meadow bird management, especially black-tailed godwit (*Limosa limosa* L.), in the Netherlands. *Journal for Nature Conservation*, 16, 88–95.

(36) Pacha M.J. & Petit S. (2008) The effect of landscape structure and habitat quality on the occurrence of *Geranium sylvaticum* in fragmented hay meadows. *Agriculture, Ecosystems & Environment*, 123, 81–87.
(37) Perkins A., Maggs H., Wilson J., Watson A. & Smout C. (2008) Targeted management intervention reduces rate of population decline of Corn Buntings *Emberiza calandra* in eastern Scotland. *Bird Study*, 55, 52–58.
(38) Wrbka T., Schindler S., Pollheimer M., Schmitzberger I. & Peterseil J. (2008) Impact of the Austrian agri-environmental scheme on diversity of landscapes, plants and birds. *Community Ecology*, 9, 217–227.
(39) Lye G.C., Park K., Osborne J., Holland J. & Goulson D. (2009) Assessing the value of Rural Stewardship schemes for providing foraging resources and nesting habitat for bumblebee queens (Hymenoptera: Apidae). *Biological Conservation*, 142, 2023–2032.
(40) Natural England (2009) *Agri-environment Schemes in England 2009: A Review of Results and Effectiveness*. Natural England, Peterborough.
(41) Parish D., Hirst D., Dadds N., Brian S., Manley W., Smith G. & Glendinning B. (2009) *Monitoring and Evaluation of Agri-environment Schemes*. Scottish Government.
(42) Taylor M.E. & Morecroft M.D. (2009) Effects of agri-environment schemes in a long-term ecological time series. *Agriculture, Ecosystems & Environment*, 130, 9–15.
(43) Aebischer N.J. & Ewald J.A. (2010) Grey partridge *Perdix perdix* in the UK: recovery status, set-aside and shooting. *Ibis*, 152, 530–542.
(44) Davey C., Vickery J., Boatman N., Chamberlain D., Parry H. & Siriwardena G. (2010) Regional variation in the efficacy of Entry Level Stewardship in England. *Agriculture Ecosystems & Environment*, 139, 121–128.
(45) Davey C.M., Vickery J.A., Boatman N.D., Chamberlain D.E., Parry H.R. & Siriwardena G.M. (2010) Assessing the impact of Entry Level Stewardship on lowland farmland birds in England. *Ibis*, 152, 459–474.
(46) Ewald J.A., Aebischer N.J., Richardson S.M., Grice P.V. & Cooke A.I. (2010) The effect of agri-environment schemes on grey partridges at the farm level in England. *Agriculture, Ecosystems & Environment*, 138, 55–63.
(47) Field R.H., Morris A.J., Grice P.V. & Cooke A. (2010) The provision of winter bird food by the English Environmental Stewardship scheme. *Ibis*, 153, 14–26.
(48) Hinsley S.A., Redhead J.W., Bellamy P.E., Broughton R.K., Hill R.A., Heard M.S. & Pywell R.F. (2010) Testing agri-environment delivery for farmland birds at the farm scale: the Hillesden experiment. *Ibis*, 152, 500–514.

Additional references
Evans A.D. (1992) The numbers and distribution of cirl buntings *Emberiza cirlus* breeding in Britain in 1989. *Bird Study*, 39, 17–22.
Evans A.D. (1997) Cirl buntings in Britain. *British Birds*, 90, 267–282.
Kleijn D., Boekhoff M., Ottburg F., Gleichman M. & Berendse F. (1999) De effectiviteit van agrarisch natuurbeheer [The effectiveness of agri-environment schemes]. *Landschap*, 16, 227–235.
Musters J.M., Kruk M., de Graaf H.J. & ter Keurs W.J. (2000) Breeding birds as a farm product. *Conservation Biology*, 15, 363–369.
Schekkerman, H. & Müskens G. (2000) Produceren grutto's (*Limosa limosa*) in agrarisch grasland voldoende jongen voor een duurzame populatie? [Do black-tailed godwits (*Limosa limosa*) produce sufficient numbers of offspring to maintain a sustainable population?]. *Limosa*, 73, 121–134.
Wilson S., Baylis M., Sherrott A. & Howe G. (2000) *Arable Stewardship Project Officer Review*. Farming and Rural Conservation Agency report.
ADAS (2001) *Ecological Evaluation of the Arable Stewardship Pilot Scheme, 1998–2000*. ADAS.
Allen D.S., Gundrey A.L. & Gardner S.M. (2001) *Bumblebees: Technical Appendix to Ecological Evaluation of Arable Stewardship Pilot Scheme 1998–2000*. ADAS, Wolverhampton, UK.
Kleijn D., Berendse F., Smit R. & Gilissen N. (2001) Agri-environment schemes do not effectively protect biodiversity in Dutch agricultural landscapes. *Nature*, 413, 723–725.
Dutt P. (2004). *An Assessment of Habitat Condition of Coastal and Floodplain Grazing Marsh within Agri-environment Schemes*. RSPB report to Defra, London.
Wilson A. M., Vickery J. A., Brown A., Langston R. H. W., Smallshire D., Wotton S. & Vanhinsbergh D. (2005) Changes in the numbers of breeding waders on lowland wet grasslands in England and Wales between 1982 and 2002. *Bird Study*, 52, 55–69.

1.4   Apply 'cross compliance' environmental standards linked to all subsidy payments

- We have captured no evidence for the effects of applying 'cross compliance' environmental standards for all subsidy payments on farmland wildlife.

Background
Cross compliance is when farmers have to meet certain statutory standards to qualify for direct support payments such as those under the first pillar of the current Common Agricultural Policy. The standards could include, for example, keeping the land in 'good agricultural condition' or managing soil to avoid erosion. The Swiss Ecological Compensation Areas scheme, under which farmers have to manage 7% of their land to qualify for area-based payments, was made obligatory in Switzerland under cross compliance in 1998. Studies examining the effects of this scheme are included in 'Increase the proportion of semi-natural habitat in the farmed landscape'.

1.5   Implement food labelling schemes relating to biodiversity-friendly farming (organic; LEAF Marque)

- We have captured no evidence for the effects of implementing food labelling schemes relating to biodiversity-friendly farming (organic; LEAF Marque) on farmland wildlife.

Background
Some food products now carry certification labels such as the LEAF (Linking Environment and Farming) Marque (Integrated Farm Management) or are labelled as organic. These schemes are designed to allow biodiversity-friendly farming to attract a price premium, become more profitable and therefore spread, potentially benefiting biodiversity.

1.6   Reduce field size (or maintain small fields)

- We have captured no evidence for the effects of reducing field size (or maintaining small fields) on farmland wildlife.

Background
Reducing field size means having a greater number of smaller fields, with boundaries between them. One reason this approach is expected to enhance biodiversity is that field boundaries of any type provide heterogeneity, with heterogeneity thought to be a strong factor determining biodiversity on farmland.

1.7   Provide or retain set-aside areas in farmland

- We found 34 studies comparing the use of set-aside areas with control farmed fields. Two were reviews, none were randomized, replicated, controlled trials. Of these, 20 (from

Austria, Finland, Germany and the UK) showed benefits to or higher use by all wildlife groups considered[5, 10, 11, 13–15, 18, 19, 21, 23, 29, 31, 37–39, 41, 50–52, 54]. Twelve (from Finland, Germany, Ireland, Sweden and the UK) found some species or groups used set-aside more than crops, others did not[1, 3, 4, 6, 8, 22, 28, 30, 36, 40, 42–44]. Two studies (all from the UK) found no effect[12, 20, 34, 53], one found an adverse effect of set-aside[35].

- Three of the studies, all looking at Eurasian skylarks, went beyond counting animal or plant numbers and measured reproductive success. Two from the UK found higher nest survival or productivity on set-aside than control fields[14, 18]. One from the UK found lower nest survival on set-aside[35].

- Fifteen studies (from Belgium, Germany, Sweden and the UK) monitored wildlife on set-aside fields, or in landscapes with set-aside, without directly comparing with control fields or landscapes. Three looked at set-aside age and found more plants[2, 27] or insects[33] on set-aside more than a year old. Two compared use of different non-crop habitats and found neither insects[47] nor small mammals[45] preferred set-aside. Two showed increased bird numbers on a landscape scale after set-aside was introduced, amongst other interventions[26, 49]. Eight looked at the effects of set-aside management such as use of fertilizer[17], sowing or cutting regimes[7, 9, 16, 24, 25, 32, 46].

- A systematic review from the UK found significantly higher densities of farmland birds on fields removed from production and under set-aside designation than on conventionally farmed fields in both winter and summer[48].

Background

Allocation of some farmland to set-aside (fields taken out of production) was compulsory under European agricultural policy from 1992 until 2008. The idea was to reduce production. However, set-aside has also been promoted as a method of enhancing biodiversity within farmland. Set-aside can be rotational (in a different place every year or two) or non-rotational (in the same place for 5–20 years) and fields can either be sown with fallow crops or left to naturally regenerate. Unlike fallow land left for the benefit of ground-nesting birds or arable plants, set-aside is not ploughed or harrowed except for the purpose of sowing.

Set-aside is often managed by cutting and/or spraying. In some cases, set-aside land has had strips of wild flowers or grasses sown on it. Evidence for the effects of this management has been included under the following interventions: 'Plant nectar flower mixture/wildflower strips' and 'Plant grass buffer strips/margins around arable or pasture fields'.

A replicated, controlled study of 44 fields on 5 farms over 2 years in Hampshire and Wiltshire, UK (1) found that, overall, chick food was 3 times higher on fallow set-aside than on wheat. Significantly higher numbers of leafhoppers (Auchenorrhyncha) were found on first and second-year set-aside (53 vs 9/sample) and true bugs (Heteroptera) in second-year set-aside than wheat (24 vs 6). In contrast, ground beetles (Carabidae: 0.3 vs 0.8), rove beetles (Staphylinidae: 6 vs 14), leaf beetles (Chrysomelidae: 0.7 vs 1.4), aphids (Aphididae: 31 vs 74) and flies (Diptera: 38 vs 67) were all significantly less abundant on set-aside than crops (respectively). Numbers in set-aside and wheat did not differ for spiders (Araneae: 13 vs 10/sample), springtails (Collembola: 855 vs 661) or larvae of butterflies, moths and sawflies (Lepidoptera and Symphyta: 0.4 vs 0.7). Fields in the first year of the UK's five-year set-aside scheme (left fallow or drilled with grass) were sampled in June 1990. In 1991, 15 fields at 2 of the 5 farms were re-sampled to evaluate second-year fallow set-aside. Invertebrates were collected using a D-Vac suction sampler in the headlands of fields, 3 m from the field edge. Five samples of 0.5 m^2 were taken at each site.

A replicated site comparison study in 1990–1991 on 1-year-old and permanent set-aside fields in a small-scale arable region in Germany (2) found higher weed cover on permanent set-asides (89.3–94.1%) than on 1-year-old set-asides (74.2–78.5%). The number of weed species was somewhat higher along the edge of 1-year-old (average 35.1 species) than of permanent set-asides (30.7 spp.) but no such difference was found in the field centre (28.2 spp. vs 27.8 spp.). Effects of set-aside age were strongly trait and species-dependent. For example, declining, rare and threatened weed species were more common in 1-year-old (122 to 154 recordings) than on permanent set-asides (91 to 110 recordings). Most of the investigated permanent and 1-year-old set-aside fields were left uncultivated, but occasional fields were sown. In each field, two 2 m x 50 m long transects (one along the field edge and one 10 m away towards the field centre) were surveyed repeatedly. Cover estimates for each plant species and total vegetation cover were recorded. No statistical analyses were performed on the data.

A replicated, controlled site comparison study from May to October 1990 in 40 farmland sites (10 field types; 4 replicates each) near Karlsruhe, south Germany (3) found significantly more species of solitary bee in artificial reed stem nests in unsown sites with naturally developed vegetation (average 7.9 species) than in sown fields, including crops and sown grass/clover fields (average 4.6 spp.). Bee species richness increased with increasing age of the unsown set-asides and with increasing plant diversity. Wasp diversity was similar in the different field types (1 to 4 spp./field type). Smaller bee and wasp species inhabited fields with high plant diversity but were absent in fields with low plant diversity. Foraging flights took twice as long in fields with low plant diversity (35 min) than in fields with high plant diversity (15 min) for 2 investigated bee species (leaf-cutter bee *Megachile versicolor* and blue carpenter bee *Osmia caerulescens*). No such effect was found for the European potter wasp *Ancistrocerus gazella* (ca. 21 min in both sites). Unsown sites with naturally developed vegetation included one- and two-year old mown and unmown set-asides and old meadow orchards. Crops on sown fields were peas, barley, rye, clover-grass mixture and *Phacelia tanacetifolia*. Mowing of set-asides took place in late June–early July. Three artificial nests (each with two 750 ml cans filled with 180 reed stems) were located in each field centre. Female body length of bees and wasps was measured. For *Ancistrocerus gazella*, *Megachile versicolor* and *Osmia caerulescens* the time spent on foraging flights was measured on four one-year old set-asides and four old meadows. Plant surveys were conducted in May, July and October.

A replicated site comparison study of 24 one-year-old set-aside fields and 24 cereal fields in Uppland, central Sweden (4) found that 4 of 17 bird species sampled showed a significant positive association with set-aside fields: skylark *Alauda arvensis*, whinchat *Saxicola rubetra*, whitethroat *Sylvia communis* and linnet *Carduelis cannabina*. Other species showed greater association with unfarmed habitats, roads and houses, forest edges or 'open habitat'. The study plots were of similar size, edge and habitat structure. Each was sampled 7 times for 28 species of breeding bird from April–June 1992. Species with at least ten territories were examined.

A replicated, controlled site comparison study with four replicates of each treatment (5) – the same study as (3) – found that naturally regenerated set-aside fields had significantly more cavity-nesting bee and wasp nests, and more nesting species than fields sown with fallow or arable crops. The study compared bees and wasps nesting on set-aside land managed in six different ways with crop fields and old meadows in Kraichgau, southwest Germany. It used reed *Phragmites australis* stem nest boxes and recorded nesting only, not foraging activity. Set-aside fields were either sown in the year of study, with a grass-clover mix or phacelia *Phacelia tanacetifolia*, or were in their first or second year of natural regeneration, with or without mowing.

A replicated, controlled site comparison study 1989–1991 in up to 65 arable sites in the Kraichgau region, Germany (6) (same study as (3, 5)) found more plant species but fewer invertebrates on naturally developed set-aside fields than on control crop fields. There were more plant species in orchard meadows (50 species/49 m^2) and naturally developed set-asides (37–45 species/49 m^2) than in sown set-asides (10–15 species/49 m^2) and cereal fields (10–17 species/49 m^2). Plant species richness was also higher in mown than in unmown set-asides. Invertebrate numbers from suction samplers were highest in set-asides sown with clover-grass-mixes (1,500 individuals/5 m^2), intermediate in naturally developed set-asides and cereal fields (ca. 1,000 individuals/5 m^2) and lowest in *Phacelia*-sown set-asides (500 individuals/5 m^2). Invertebrate numbers caught in Malaise-traps were highest in rye fields and clover-grass-mixes (around 3,000 individuals) and lowest in naturally developed set-asides (1,000 individuals). The effect of field type and set-aside age was strongly species- or family-dependent. Up to 11 field types (4 to 5 replicates each) were investigated: 1-, 2- and 3-year-old naturally developed set-asides (mown and unmown), 1-year-old set-asides sown with either *Phacelia tanacetifolia* or a clover-grass mix, conventionally managed cereal fields (rye and barley) and low-intensity orchard meadows (>30 years old). Plant surveys (3 visits) were conducted in May to October 1990–1991 on one 49 m^2 permanent quadrat (meadows and sown fields) or on 120 m^2 (systematically changed in naturally developed fields). Insects were sampled on 4 to 5 visits in April to October using Malaise-traps (20 fields) and suction samplers (61 fields; 3 minute suctions of five 0.25 m^2 plots).

A replicated site comparison study from April to August 1993 at 21 farmland sites in Kraichgau, Germany (7) (same study as (3, 5, 6)) found that naturally regenerated set-asides and orchard meadows held more wild bee species and more individual bees than set-aside fields sown with phacelia *Phacelia tanacetifolia* (averages of 27, 28 and 10 bee species; 120, 100 and 75 bees, respectively). Also the numbers of Red Listed bee species and specialist species were higher in naturally developed than in *Phacelia*-sown set-asides. Seven field types (four replicates each) were investigated: one-, two-, three-, four- and five-year-old naturally regenerated set-asides; one-year-old set-asides sown with phacelia and orchard meadows. Wild bees were monitored for 30 minutes on each of 6 visits to each site. Along 1 100 m long transect in the field centre bees were caught using sweep nets (100 sweeps/transect). In addition, flower-visiting bees were caught.

A replicated site comparison study of four arable, ten mixed and three pastoral farms within the South Downs Environmentally Sensitive Area, UK (8) found that rotational set-aside tended to be used more than arable crops by skylarks *Alauda arvensis*, but used less or a similar amount by hares *Lepus euroaepus*. Rotational set-aside was used significantly more than arable crops during the first skylark brood period (22 vs 3–15 males/km^2). During the second brood, once set-aside had been topped or cultivated, use of set-aside by skylarks was more similar to their use of arable crops (topped: 16; cultivated: 8; arable: 9–14). Hares used winter sown cereals more than rotational set-aside in October–January (0.2–0.3 vs 0.1 hares/ha) but in February set-aside was used the same amount as crops (0.1 hares/ha). Hares were sampled by spotlight counting over an average of 26% of the area of each farm between November and March (1992–1993; 1993–1994). Skylarks were sampled by mapping breeding males during 2 counts along representative transects on 12–17 farms in April–June (1992–1993).

A small replicated, randomized study of set-aside on a farm west of Moray, Scotland (9) found that vegetation of conservation value can develop within set-aside provided species occur in the seed bank. The abundance of the dominant species false oat grass *Arrhenatherum elatius*, Yorkshire fog *Holcus lanatus* and cock's-foot *Dactylis glomerata* varied across the set-aside. Some variations were explained by sub-plot location, others by management; removing cuttings reduced false oat grass abundance. No further results are provided as the study was ongoing at time the paper was written. A 25 m-wide strip of set-aside was established in 1989 and divided into 3 plots of 25 x 28 m in 3 randomized blocks. Treatments were: cut in July, cut in September and cut in July and September to 6 cm. Each plot was divided into two sub-plots: cuttings removed or left *in situ*. Plant species composition was recorded in June–July 1993.

A small, controlled study of an arable and set-aside field on a farm near Braunschweig, Germany (10) found that arthropod numbers and species richness tended to increase with a reduction in management intensity. More species of spider were found in set-aside than arable plots with 4 levels of management intensity (set-aside: 33–36; reduced intensity: 10–22; conventional: 11–13). The effect on spider abundance was less clear. Set-aside also had a greater density of wolf spiders (Lycosidae; set-aside: 68/trap; arable 10–22) and a lower proportion of pioneer species (set-aside: 8%; reduced inputs: 49–75%; conventional: 81%). Beneficial species, such as *Carabus auratus*, were more abundant in set-aside (97–148/trap) than arable plots (1–18/trap); their activity periods were also longer in set-aside. Similar effects were seen for juvenile spider abundance (set-aside: 108/trap; reduced intensity: 50–55; conventional farming: 21). In 1992–1995 a long-term set-aside was compared with 4 plots within an arable field that differed in the input of fertilizers and pesticides (high; 30–50% reduced; none), crop rotation (3/4 course), tillage, weed control (mechanical/chemical), cultivars, drilling technique and catch crops. Six to eight emergence traps and pitfall traps sampled arthropods within each treatment. Traps were collected every 2–4 weeks throughout the year. Results for pest species are not included here.

A replicated, controlled study in summer 1991–1994 on 3 to 6 arable farms and 2 experimental sites in the Province of Bayern, Germany (11) found more plant species on rotational set-asides (17.4 species) than on control fields (10.8 species). Moreover, naturally regenerated set-asides held more plant species (range 18.3 to 32.2 spp.) than set-asides with sown clover-grass mixtures (range 16.0 to 18.9 spp.). This effect was still visible the following year, when cereal was grown on the former set-aside fields (range 13.3 to 14.2 spp. on cereal after natural regenerated set-aside vs 12.0 to 12.4 spp. on cereal after sown set-aside). Rotational set-asides were taken out of production for one year and either left to regenerate naturally or sown with a clover-grass-mix. Controls were often cereal fields. Vegetation was surveyed between June and September on total areas between 100 and 400 m². Cereal crops were surveyed yearly; cut set-asides several times a year. Note that no statistical analyses were performed on these data.

A replicated, controlled study of set-asides at four sites on two Royal Agricultural College farms, Gloucestershire, UK (12) (see also (20, 24, 25, 34)) found that small mammals showed no preference for first-year set-aside over crops. Trapping success was significantly lower in set-aside (0.6%) than the adjoining unharvested cereal crop (13%) and hedgerow (30%). Wood mice were the only species caught in set-aside. There was no significant difference in trap success between set-aside in blocks (0.6%) or strips (0.6%) or between sown (0.4%) or naturally regenerated (1.0%) set-aside, although sample sizes were very low (six captures). Following harvest, trap success in the crop decreased (4.5% to 0.5%) and significantly increased in set-aside (0.1 to 2.5%). Set-aside was either sown with a mix of wheat and rape (three sites) or left to regenerate naturally. A grid of 50 Longworth live-traps was set at each site covering a hedgerow, a 20 m strip of set-aside and a block of either set-aside or cereal crop. Trapping was undertaken for five nights/month from June–August 1995.

A replicated, paired sites before-and-after study on 7 pairs of fields in northeast Scotland in 1989–1991 (13) found that 1-year-old set-aside fields held significantly more species of birds than similar, non-set-aside fields (average of 11.9 species/10 ha for first year set-aside vs 4.8 species/10 ha for control fields). There were no differences in the years before or after set-aside. In addition, there were higher breeding densities of grey partridge *Perdix perdix*, skylark *Alauda arvensis* and Eurasian curlew *Numenius arquata* in set-aside compared with control fields. Densities of curlew, partridge, northern lapwing *Vanellus vanellus* and Eurasian oystercatcher *Haematopus ostralegus* were higher in set-aside years than before set-aside (passerine densities were not recorded before set-aside was used). Wader breeding success appeared higher on set-aside, but numbers were too small for statistical tests. The densities and number of species declined over time in set-aside fields. Set-aside fields were previously arable fields but were not cropped for at least one year.

A replicated study in the summers of 1993–1995 on 7 farms in southern England (14) found significantly higher densities of skylark *Alauda arvensis* territories on set-aside fields than on conventionally or organically-managed crop fields (0.26–0.56 territories/ha for set-aside fields vs a maximum of 0.38 territories/ha for cropped fields). Estimated nest survival was significantly higher on set-aside fields than conventionally-managed cereal fields (44% survival to fledgling on set-aside vs 11% for conventional cereals). Set-aside was both naturally regenerated from crop stubble or sown with grass.

A site comparison study of set-aside in southern Germany (15) found that numbers of plant and butterfly species were higher in naturally regenerated set-aside than cereal or set-aside sown with *Phacelia tanacetifolia*, but that the number of plant species decreased and butterfly species composition changed with set-aside age (1–4 years-old). Numbers of plant species were higher in 4 naturally regenerated set-aside fields (20–30 species) than in a cereal field (1) or 1-year-old set-aside sown with *P. tanacetifolia* (3). The number of species decreased significantly with age of naturally regenerated set-aside fields from 1 (30 plant species) to 3-years-old (20 species). Cover of annual herbs declined rapidly in the 3rd to 4th year of set-aside. The number of butterfly species was higher in naturally regenerated set-aside (11–13 species) than cereal (4) or sown set-aside (7). Butterfly species richness did not differ with set-aside age (11–13 species) but species composition changed greatly. Butterfly body size tended to decrease with set-aside age (from 24 mm to 23 mm) and mean life-span of caterpillars increased (from 61 days to 105 days). Plant species and cover were sampled in 49 m^2 plots in September 1992 and flowering plant abundance was estimated 9 times from May–October. Adult butterflies were counted along transects, nine times per field (May–October 1992) and caterpillars were sampled twice in September 1992 by sweep-netting.

A small site comparison study in Belgium (16) found 53 species of ground beetle (Carabidae) during 1 year of sampling in 3 set-aside fields, including 11 species in the Red Data Book for Flanders. The most notable species were *Amara tricuspidata* and *Harpalus froelichi*. Thirty-five of the species were considered to be breeding within the fields. The set-aside fields contained more ground beetle species than were found in previous years on cultivated arable fields (numbers not given). Two of the fields were set-aside in 1994, sown with grasses and annually mown once. These were managed without fertilizers or pesticides for two years before becoming set-aside. The other field had been left to naturally regenerate since 1992 and was partly grazed by sheep. Beetles were sampled between May 1994 and April 1995 in three pitfall traps per site, emptied fortnightly or monthly throughout the year.

A replicated, controlled study of former arable fields at 6 sites in Sweden (17) found twice as many plant species in unfertilized compared to fertilized set-aside after 10 years (30 species in the least fertile site; 10 in the most fertile). Annual cutting resulted in an increased number of species over the years. The competitive success of plant species was related to management practices but there were also interactions between management and site conditions. At each site 2 plots (10 x 20 m) were sown with a grass cover crop and 2 were left bare. Each year, one of each pair had fertilizer added (equivalent to 150 kg N/ha) and half of every plot was cut and cuttings removed (late July). Vegetation cover was assessed in the centre of each plot (8 x 1 m^2; 1975–1986).

A site comparison study from April to August 1992 on three farms in south England (18) found that skylarks *Alauda arvensis* had significantly higher productivity in set-aside fields, compared to spring-sown cereals or grass (0.5 fledglings/ha in set-aside vs 0.21 fledglings/ha in spring cereals and 0.13 fledglings/ha in silage grass). This difference was largely due to higher densities of territories (2–3 times higher in set-aside and grass, compared to cereals), more successful nests (highest on grass, but twice as high in set-aside as in cereal crops) and larger clutches in set-aside (3.9 eggs/clutch for nests in set-aside vs 3.3 eggs/clutch for spring cereals and 3.4 eggs/clutch in grass; 11 nests in each habitat type). Fledging success did not vary between habitats. No nests with chicks were found in winter-sown cereals. Set-aside consisted of 4-year-old permanent fallow sown with red fescue *Festuca rubra*, perenial rye grass *Lolium perenne* and white clover *Trifolium pratense*.

A literature review (19) looked at the effect of agricultural intensification and the role of set-aside on the conservation of farmland wildlife. It found one study that demonstrated a three-fold increase in insect density on rotational set-aside compared to conventional cereals, mainly due to increases in plant hoppers and beetle families (Carabidae, Staphylinidae and Chrysomelidae; described above (1)).

A replicated study of wood mice *Apodemus sylvaticus* in arable habitats at two Royal Agricultural College farms at Cirencester, UK (20) (see also (12, 24, 25, 34)) found that wood mice showed no preference for first-year set-aside over crops. Wood mouse *Apodemus sylvaticus* numbers were lowest on whole field set-aside (0–16), followed by hedgerow with set-aside margins (5–40) and crop (3–27). Numbers were significantly higher in woodland apart from in July (18–73). There were 2 replicate 5 ha blocks of set-aside and adjacent 20 m-wide set-aside margin strips of a similar area. A grid of 49 live traps was set in each replicate covering the 4 habitats. Trapping was undertaken monthly for a year from December 1995.

A replicated site comparison study in summer 1995 on 89 fields in the South Downs, southern England (21) found that the density of singing Eurasian skylarks *Alauda arvensis* was higher on set-aside fields than on any other field type, except undersown spring barley fields (approximately 15 birds/km^2 on 6 set-aside fields vs 22 birds/km^2 on 4 spring barley fields and 2–12 birds/km^2 on 79 other fields). Other field types were: arable fields reverted to species-rich grassland or permanent grassland, downland turf, permanent grassland, winter wheat, barley and oilseed rape.

A randomized, replicated site comparison in the winters of 1992–1993 and 1993–1994 on 40 farmland sites in Devon and East Anglia, England (22) found that only 1 taxonomic group (finches, sparrows and buntings; 7 species) showed a significant selection of set-aside habitats in both years, preferentially using sown set-aside less than 1 year old. Conversely, thrushes (four species) and hedge-dwelling species (European robin *Erithacus rubecula*, wren *Troglodytes troglodytes* and dunnock *Prunella modularis*) avoided regenerating set-aside less than one year old in Devon. At a species-level, a preference for set-aside was seen in both winters by one species in Devon (cirl bunting *Emberiza cirlus* selected sown set-aside more than one year old) and two species (plus one introduced species not considered here) in East Anglia (grey partridge *Perdix perdix* preferred older sown set-aside and yellowhammer *Emberiza citrinella* selected one year old sown cover). A further 13 species in both East Anglia and Devon preferentially selected set-aside in 1 winter. Blackbird *Turdus merula* and five other species avoided some set-aside in at least one year in Devon; no native species did so in East Anglia. The same 40 plots (50–100 ha) were surveyed each winter, although the amount of set-aside they contained varied due to rotation schemes.

A replicated, randomized site comparison study of 27 arable fields on a farm in southern Finland over 1 year (23) found that the abundance of ground beetles (Carabidae) was significantly higher in set-aside than crop fields. Set-aside contained 1,442 beetles/site compared to 334–524/site in crop fields. Of 21 species compared in set-aside and cereal fields, two were significantly more abundant in set-aside (*Trechus secalis*, *Dyschirius globosus*) and two in cereal (*Asaphidion pallipes*, *Bembidion quadrimaculatum*). The ground beetle community differed between set-aside and crop fields. Autumn breeding species dominated set-aside (70% in June), whereas spring breeders tended to use crops tilled in spring (56–80%). Twenty-seven of 150 fields were randomly selected. Six were permanent set-aside, sown with perennial grass and left for five to ten years. The others were barley, oats, sugar beet, oilseed rape and potato. Beetles were sampled using 20 pitfall traps (7 cm diameter) at each site. These were emptied every two weeks for ten consecutive weeks from June–August 1995.

A site comparison study monitoring the behaviour of individual wood mice *Apodemus sylvaticus* on two arable farms in England (24) (following on from (12)) found that set-aside established using species-rich mixes of grasses and native forbs was preferred and set-aside established using a simple grass/clover mix avoided by the mice. On average, wood mice at Jealott's Hill preferred set-aside (species-rich mixes; preference index: 0.12) and avoided crop (–0.12); at Eysey they avoided set-aside (simple mix: –0.16) and preferred other habitats (0.12). However, only at Eysey was there a significant deviation from random habitat use overall. Vegetation at Jealott's Hill contained more species but was shorter and provided less cover than that at Eysey. Set-aside was established in the 10 m next to the crop and the hedge at Jealott's Hill (1996) and on 20 m-wide margins and an adjoining 5 ha block at Eysey (1995). Nine wood mice were radio-tracked continuously for three nights at each farm (May–July 1996–1997). Vegetation data were obtained using a quadrat survey (1 m²).

A small, replicated study of set-aside on two Royal Agricultural College farms in Gloucester, England (25) (same study as (24)) found that set-aside established as margins (20 m wide; 5 ha) next to hedgerow had a more abundant and species rich small mammal community than larger (5 ha) blocks. Set-aside margins had more mammals (21 animals; 8 species caught/trap session) than larger blocks (11 animals; 5 species caught/trap session). Wood mice dominated (76% on margins; 50% on blocks). Species richness, but not diversity, was significantly greater on margins (richness: 2.4; diversity: 0.3) than blocks (richness: 2.1; diversity: 0.2). Both parameters increased from 1996 to 1997. The abundances of species changed with time and season on set-aside margins and blocks. Set-aside was established by sowing a grass/clover mix in 1995, which was cut annually in July or August. Grids of 49 traps were set in the centre of set-aside blocks and spanning the set-aside margin and adjacent hedgerow and crop. Traps were set for five nights in March, June, September and December 1996–1997 and a mark recapture technique followed.

A 2000 literature review from the UK (26) found that the populations of grey partridge *Perdix perdix*, Eurasian thick-knee *Burhinus oedicnemus* and cirl bunting *Emberiza cirlus* all increased following multiple measures including the provision of set-aside. Partridge numbers were 600% higher on farms with conservation measures aimed at partridges (including conservation headlands, planting cover crops, using set-aside and creating beetle banks) compared to farms without these measures. The UK thick-knee population increased from 150 to 233 pairs from 1991 to 1999 (interventions were set-aside provision and uncultivated plots in fields). The UK cirl bunting population increased from 118–132 pairs in 1989 to 453 pairs in 1998, with a 70% increase on fields under schemes (with overwinter stubbles, grass margins, and beneficially managed hedges and set-aside) compared to a 2% increase elsewhere.

A replicated, randomized site comparison study of non-rotational set-aside up to 9 years old at 50 farms in the eastern arable region and 50 in the western mixed farming region in the UK (27) found that plant communities differed between region, establishment method (natural regeneration or sown cover) and site age. Succession continued after 5 years, with number of plant species increasing over time (7–8 species on older sites; 4–6 on younger sites) along with proportions of perennials and plants characteristic of non-arable habitats. Species richness declined with increasing distance from the field boundary (1 m: 6–8 species; 32 m: 4–7 species). A stratified sample of farms was selected from the Integrated Arable Control System database of the Ministry of Agriculture, Fisheries and Food. One set-aside site was randomly selected per farm and one field boundary was randomly selected for vegetation sampling. Six quadrats (0.5 x 0.5 m) were sampled along 5 randomly located transects at distances of 1, 2, 4, 8, 16 and 32 m from the boundary.

A replicated site comparison study with paired sites in 1996–1997 across 92 arable farms in England (28) found 5 of 6 bird functional groups at higher densities on set-aside fields, compared to winter cereals or grassland (although thrushes only showed this preference in 1 year). On ten farms with rotational and non-rotational set-aside all groups except crows were found at higher densities on rotational fields. All groups except gamebirds (which showed no significant field preferences) were more likely to be found on set-aside than other field types. Functional groups of birds were gamebirds, pigeons, crows, skylarks *Alauda arvensis*, thrushes and seed-eating songbirds (sparrows, buntings and finches).

A replicated, paired sites comparison study in 1996–1997 on 11 farms in east and west England (29) found that set-aside fields supported more species and higher densities of birds than adjacent crop fields (1.4–7.1 birds/ha and 7–21 species for 11 set-aside fields vs 0.2–0.8 birds/ha and 2–5 species on 11 crop fields). Between 78% and 100% of species found on both field types were more abundant on set-aside. These preferences were stronger (although not significantly so) for rotational set-aside, compared to non-rotational.

A replicated, controlled, paired sites comparison study of 51 set-aside and wheat fields on 30 farms in southern and eastern England (30) found that stubble set-aside had more spiders (Araneae) and leafhoppers (Auchenorrhyncha), higher weed cover and greater plant species diversity, whilst wheat had more beetles (Coleoptera). Set-aside fields had 16 spiders, 16 leafhoppers, 0.7 leaf beetles (Chrysomelidae), 0.5 ground beetles (Carabidae) and 0.4 soldier beetles (Cantharidae)/sample on average. Wheat fields had 11 spiders, 9 leafhoppers, 0.3 leaf beetles, 0.3 ground beetles and 0.3 soldier beetles/sample on average. Numbers did not differ between set-aside and wheat for true bugs (Heteroptera; 5–6/sample), larvae of butterflies, moths (Lepidoptera) and sawflies (Tenthredinidae; 0.2–0.5 larvae/sample) or weevils (Curculionidae) (0.2 vs 0.1). Cutting set-aside (to 10–15 cm) tended to decrease invertebrate numbers compared to topping (to 25 cm) or leaving it uncut. Weed cover and diversity were significantly higher on set-aside (cover: 32%; species: 99) compared to wheat (cover: 3%; species: 41). Set-aside fields were naturally regenerated after harvest. Wheat fields received pesticides. Invertebrates were sampled using a D-Vac suction sampler in each set-aside and adjacent wheat field in June–July. Weed cover was sampled in 10 random quadrats (0.25 m^2) per field.

A replicated, controlled site comparison study in summer 1995 on 10 sites of 3 different arable habitats in the biosphere reserve Schorfheide-Chorin, northeast Germany (31) found significantly more individuals (but not families) of parasitic wasps (Hymenoptera: Parasitica) on set-aside land (>160.7 wasps/m^2) than on cereal fields (<107.5 wasps/m^2). The age of set-aside did not affect wasp numbers. There was no significant difference between numbers of parasitic wasps on set-asides and extensively managed grasslands (178.7 wasps m^2). Four winter cereal fields, 4 set-asides (1 to >10-years-old) and 2 extensively managed grasslands (1 meadow; 1 grassland grazed by sheep) were monitored. Hymenoptera were sampled from March to July 1995 using six photo-eclectors on each site. The eclectors were placed randomly and emptied every four weeks. Insects were identified to family level.

Another analysis (32) as part of the same study as in (28) found that skylark *Alauda arvensis* densities on set-aside fields ranged from zero to approximately 2.7 birds/ha. A total of 74 set-aside fields (36 rotational and 38 non-rotational) were examined, each from a different farm. Fields with approximately 30% bare earth, straw and litter had the highest densities of skylarks.

A replicated site comparison study near Karlsruhe, south Germany (33) examined the abundance and species richness of foraging bees, both solitary and social, on annually mown set-aside fields of different ages and management. The number of bee species increased with the age of set-aside fields, from 15 species on 1-year-old fields to 25 species on 5-year-old fields. Two-year-old set-aside fields had the most bee species: 29 on average, compared to 32 species for old meadows, including an average of around 5 oligolectic species (specializing on pollen from a small group of plant species). One-year-old set-aside fields sown with phacelia had an average of 13 bee species, mainly common, generalized species of bumblebee *Bombus* and *Lasioglossum*.

As part of the same study of wood mice *Apodemus sylvaticus* on an arable farm in England as that described in (24, 25) and (34) found that after harvest, mice preferred hedgerow to set-aside. Before harvest, wood mice tended to use habitats (crop, margin set-aside, block set-aside and hedgerow) at random. After harvest, set-aside was avoided. Margin and cut set-aside were avoided significantly more than block and uncut set-aside. A 3 ha block of set-aside adjoining a 20 m-wide set-aside field margin was sown (grass/clover mix) in 1995 between 2 arable fields. Twenty-four alternate 50 x 6 m-wide patches of cut and uncut set-aside were created either side of the central hedgerow. The remaining 14 m width of the margin was cut as normal. Thirty-four wood mice were radio-tracked continuously for at least three nights (June–July and September–November 1996–1997).

A replicated site comparison study in 1996–1998 on 22 farms in southern England (35) found that skylark *Alauda arvensis* nests had significantly lower survival in set-aside, compared to in cereals (22% overall survival for 525 nests in set-aside vs 38% survival for 183 nests in cereal fields). There were no differences between set-aside and other crop types (19% survival for 173 nests in grass fields, 29% survival for 60 nests in other field types) or between rotational and non-rotational set-aside. On one intensively-studied farm over 90% of 422 skylark nests were found on 10 fields of well-established, non-rotational set-aside.

A replicated site comparison study carried out in June 2000 in ten edge habitats at the arable Loddington Estate in Leicestershire, England (36) found a higher density of weevils (Curculionoidea) in edges of non-rotational set-aside than all the other habitats studied. Spider (Aranae) and rove beetle (Staphylinidae) densities were lower in set-aside than in edges of un-grazed pastures. Beetle banks, brood cover, one- and two-year-old wild bird cover, hedge bottoms, sheep-grazed pasture edges, grass/wire fence lines and winter wheat headlands were also included in the study. Invertebrates were sampled with a vacuum suction-sampler in June 2000.

A study of different set-aside crops at Loddington farm, Leicestershire (37) found that skylark *Alauda arvensis*, but not yellowhammer *Emberiza citrinella* used unmanaged set-aside more than expected compared to availability. Skylarks used unmanaged set-aside more than expected, but significantly less than set-aside sown with kale-based wild bird cover, wild bird cover strips and beetle banks. Cereal (wheat; barley) and broad-leaved crops (beans; rape) were used less than expected. Yellowhammer used unmanaged set-aside as expected compared to availability, but significantly less than cereal and set-aside with cereal-based wild bird cover or wild bird cover strips. Field margin and midfield set-aside strips were sown with kale-based and cereal-based mixtures for wild bird cover, and beetle banks. Other habitat types were: unmanaged set-aside, cereal (wheat; barley), broad-leaved crop (beans; rape) and other habitats (including permanent pasture, woodland, hedgerows, tracks and riparian areas). Thirteen skylark and 15 yellowhammer nests with chicks between 3–10-days-old were observed. Foraging habitats used by the adults were recorded for 90 minutes during 3 periods of the day.

A replicated, randomized site comparison study of 200 farms in England with set-aside (38) found that set-aside supported a range of biodiversity. Rotational set-aside supported 12 plant species/site and 1 nationally rare species (corn marigold *Chrysanthemum segetum*). On non-rotational set-aside plant species richness and cover of annuals was greater on naturally regenerated than sown grass sites (27 vs 20 species/site); cover by perennials showed the opposite trend. Older naturally regenerated sites had more perennial species but plant communities did not appear to be developing into those considered of conservation value. Twenty percent of farmers reported an increase in wild flowers and 47% reported an increase in bird numbers on rotational set-aside. Fifty-one percent of farmers reported an increase in wild flowers and 69% an increase in bird numbers on non-rotational set-aside. Bird density in set-aside was nine times higher than in crops for rotational set-aside and seven times higher for non-rotational sown grassland set-aside. Management of set-aside had minimal effect on bird abundance. Significantly more invertebrates were found in set-aside than in the adjacent crop. Vegetation was assessed on 100 rotational (spring 1996–1997) and 100 non-rotational set-aside sites (summer 1996–1997). Breeding bird territories were mapped on 63–92 farms (1996–1997). More intensive surveys were undertaken for: vegetation (8+ per year) on 6 farms, habitat use by birds and invertebrates (pitfall trapping; May–June) on 11 farms (1996–1997). Pest data are not presented here.

A replicated, controlled study in May to September 2000–2001 on six farmland sites near Vienna, Austria (39) found a higher number of ground beetle (Carabidae) species in set-aside areas than in arable fields. Sowing wildflower seed mixtures on set-aside land further increased the number of ground beetle species. The community composition of ground beetles differed between the three types of habitat. No statistical analyses were presented. Two unsown set-aside fields were >50- and 6-years-old and cut regularly. Wildflower strips (4 sites) were sown on set-aside land with the 'Voitsauer' seed mix containing 25 species of herbs and weeds between 1998 and 2000. Typical crops for the region were sown on five arable fields. One of the arable fields was under conservation contract growing a wildflower seed mix undersown in rye. Ground beetles were sampled using 4 pitfall traps 10 m apart in each habitat and site. There were five sampling periods each year, each lasting two to three days (2001) or seven days (2000).

A replicated, controlled site comparison study from November–February in 2000–2001 and 2001–2002 on 20 arable farms in eastern Scotland (40) found that, of 23 species recorded, only skylarks *Alauda arvensis* were significantly denser in fields with set-aside than fields with wild bird cover crops or conventional crops. Bird density was up to 100 times higher in wild bird cover crops than on set-aside fields. The wild bird cover crops attracted 50% more species than set-aside fields. Of eight species with sufficient data for individual analysis, seven were consistently more abundant in wild bird cover than in set-aside fields. Set-aside fields were those in which cereal stubble was left to regenerate naturally. Between 6.2 and 28.3 ha were sampled on each farm annually.

A review and meta-analysis of 127 studies comparing set-aside and conventional land (41) found that species richness and population densities of plants, birds, insects and spiders and harvestmen were significantly higher on set-aside land than on nearby conventional fields in Europe and North America. Positive effects were greatest on larger and older areas of set-aside, when the comparison conventional field contained crops rather than grasses, in countries with more arable land under agri-environment schemes and with less intensive agriculture. Overall, variation in establishment methods and types of set-aside made little difference to the positive effect on biodiversity, although species richness was increased more when set-aside was naturally regenerated rather than sown.

A replicated, controlled, paired sites comparison in summer 2003 in County Laois and County Kildare, Ireland (42) found that 18 set-aside fields had significantly higher bird species diversity and richness than 18 adjacent agricultural fields (an average of 12.8 species on set-aside vs 9.2 species on control fields). Three species – meadow pipit *Anthus pratensis*, skylark *Alauda arvensis* and woodpigeon *Columba palumbus* – were significantly more abundant on set-aside. Six species (whitethroat *Sylvia communis*, goldcrest *Regulus regulus*, blackcap *Sylvia atricapilla*, stonechat *Saxicola torquata*, tree sparrow *Passer montanus* and treecreeper *Certhia familiaris*) showed a preference for non-set-aside fields, but these were not statistically significant and were considered likely to be based on field margins rather than field management. Six species were associated with non-rotational set-aside, two with rotational set-aside, one with long-term grazed pasture set-aside and three with first year pasture set-aside.

A controlled trial in Jokioinen, southern Finland (43) from 2003 to 2004 found more spiders (Araneae) and flying insects in set-aside than in a control cereal crop, but not more plant species or ground beetles (Carabidae). Spiders were significantly more abundant in 2-year fallows, regardless of the sowing treatment (28–55 spiders/trap) than in 1-year fallows, in which spider numbers did not differ from the control cereal crop (less than 10 spiders/trap). Numbers of flying insects in the vegetation followed a similar pattern, with fewer insects in first year fallows than in stubble or two-year fallows. Numbers of ground beetles and numbers of plant species were similar across all fallow treatments and in the case of beetles also in the control cereal crop (5–25 beetles/trap; 2–14 unsown plant species/m²). Two-year fallow plots sown with red clover had fewer plant species (around 2 species/m²) than control cereal fields, which had around 16 plant species/m². Fallow treatments were established in 2003 or 2004, each on a 44 x 66 m plot. The treatments were: one- and two-year fallow sown with either grasses, or a grass-red clover *Trifolium pratense* mix; two-year rotational fallow established by undersowing spring cereal with either grasses, or a grass-red clover mix. The control was a spring barley crop. Insects were sampled using a yellow sticky trap and three pitfall traps in the centre of each plot for a week in June, July and August 2004. Unsown plant species were counted in 4 50 x 50 cm quadrats in each plot in late August 2004.

A replicated, controlled site comparison in 1999 and 2003 on 256 arable and pastoral fields across 84 farms in East Anglia and the West Midlands, England (44) found that only 2 of 12 farmland bird species analysed were positively associated with the provision of set-aside, wildlife seed mixtures or overwinter stubble. These were skylark *Alauda arvensis* (a field-nesting species) and linnet *Carduelis cannabina* (a boundary-nesting species). The study did not distinguish between set-aside, wildlife seed mixtures or overwinter stubble, classing all as interventions to provide seeds for farmland birds.

A replicated site comparison study of agri-environment scheme habitats in arable farmland in England (45) found that set-aside tended to have lower numbers of small mammals than sown grass margins. Numbers of small mammals caught in permanent set-aside (fallowed for five years or more, annual cutting of at least 90%: 1.6–2.0 mammals/plot) were lower than in 2 m grass field margins (2.9–4.4 mammals/plot) and 6 m margins (2.5–3.6 mammals/plot). In 2003, significantly fewer common shrew *Sorex araneus* and wood mouse *Apodemus sylvaticus* were captured in set-aside (shrews: 0.6; mice: 0.5) than grass margins (shrews: 0.9–1.4; mice: 0.7–1.1). The trend was similar for bank voles *Myodes glareolus* in 2004 (set-aside: 0.5 voles/plot; margins: 1.4–1.6 voles/plot). Species richness did not differ significantly between treatments (1.7–2.0 species). Twelve small mammal traps were set within 20 plots per treatment (1 m from the habitat boundary) for 4 days in November–December 2003–2004. Mammals were individually fur-clipped and released. Results from farm woodlands are not included here.

A review of the effects of agri-environment scheme options and set-aside on small mammals in the UK (46) found that results tended to depend on the management of set-aside. Studies have found that after harvest wood mice *Apodemus sylvaticus* avoided cut set-aside and crops and preferred uncut set-aside and hedge (34); that wood mice tended to avoid set-aside land relative to crop and hedgerow habitats (Tattersall & Macdonald, 2003); that wood mice used set-aside with species-rich mixes of grasses and native forbs more, and tended to avoid set-aside established using a simple grass/clover mix (24) and that set-aside established as margins next to hedgerow had a more abundant and diverse small mammal community than larger blocks (25). Although small mammal abundance did not increase as set-aside aged, one study found that species composition changed and species diversity and species richness increased (Tattersall *et al.* 2000).

A replicated site comparison study of 31 rotational set-aside fields in England (47) found that invertebrate numbers tended to be higher in uncultivated field boundaries than within set-aside fields. There were significantly lower numbers of the following groups within set-aside compared to at field edges: harvestmen (Opiliones: 0 vs 3/m^2), leafhoppers (Auchenorrhyncha: 10 vs 60), true bugs (Heteroptera: 2–10 vs 25), parasitic wasps (14 vs 20), beetles (Coleoptera: 7 vs 22), flies (Diptera: 38–42 vs 63), 'chick food items' (20–30 vs 85) and 'highly ranked predators' (1 vs 5). Aphids were more numerous in set-aside than at the field boundary (100–112 vs 10/m^2). There was no significant difference in numbers of spiders (Araneae), lacewing (Neuroptera) larvae, butterfly and moth (Lepidoptera) larvae, sawfly (Tenthredinidae) larvae and aphid predators between the margin and the field. Invertebrates were sampled in the uncultivated field boundary (0 m) and at 3 m and 50 m in to each field in mid-May. Total invertebrates (excluding springtails (Collembola) and thrips (Thysanoptera)) and those in 12 groups known to be food for farmland birds were recorded.

A 2007 systematic review (48) identified 11 papers investigating the effect of set-aside provision on farmland bird densities in the UK. In both winter and summer surveys there were significantly higher densities of farmland birds on fields removed from production and under set-aside designation than on conventionally farmed fields. The meta-analysis included experiments conducted between 1988 and 2002 from eight controlled trials and three site comparison studies.

A before-and-after study examining data from 1976 to 2003 from farms across southern Sweden (49) found that four locally migrant farmland birds showed less negative (or positive) population trends during a period of agricultural extensification, which included an increase in the area of set-aside. The authors suggested that the two could be causally linked.

A before-and-after site comparison study in 2000–2005 in Bedfordshire, England (50) found that set-aside fields sprayed in May or June supported higher densities of grey partridge *Perdix perdix*, seed-eating songbirds and skylark *Alauda arvensis*, compared to set-aside sprayed in April or crop fields (although seed-eating songbirds were equally numerous on oilseed rape *Brassica napus* fields). Early-sprayed set-aside had consistently lower densities of all species, compared to all land uses except winter-sown wheat.

A site comparison study of seven arable fields over two years in Devon, UK (51) found that in set-aside, spider abundance was higher, but number of species was similar to other arable fields. Numbers of species in set-aside (14) were similar to winter wheat (14–15) and maize (13), but higher than in winter barley, temporary grass ley (10–11) or permanent grass ley (9). Abundance was highest in set-aside (2,490 spiders), followed by wheat (2,009–2,039), maize (1,325), temporary ley (1,280), barley/temporary ley (1,087) and permanent ley (1,067). In set-aside, non-linyphiid spiders (money spiders; 1,236) accounted for a greater proportion of the total spiders sampled than in other field types. The total number of linyphiids in set-aside (1,254) was similar to numbers in ley (1,039–1,268) and maize fields (1,253). Spider numbers decreased once set-aside was cut. The set-aside field was established the year before the study and previously received low intensity management and occasional sheep grazing. Set-aside was cut once in August and was grazed over winter. A D-Vac suction sampler was used to take 6 sub-samples in each field at 2–3 week intervals from June 2001 to October 2002.

A site comparison study in 2002–2009 on mixed farmland in Hertfordshire, England (52) found that the estimated population density of grey partridge *Perdix perdix* was significantly higher on set-aside land than on conventional arable crops. The difference was strongest for rotational set-aside, with non-rotational set-aside not having a significant positive impact on partridge densities.

A site comparison study on four farms in Aberdeenshire, northeast Scotland, in summer 2005 (53) found that yellowhammers *Emberiza citrinella* from ten nests preferentially foraged on set-aside land, compared to cereal fields, but that this preference was not significant (set-aside comprising 23% of available habitat but used for 42% of foraging flights vs cereals comprising 42% of habitat and being used 25% of the time). The authors suggest that the lack of significance may be due to small sample sizes.

A replicated, controlled site comparison study in summer 2008 in northwest Scotland (54) found that croft sections in fallow had nine times more foraging bumblebees than croft sections grazed by sheep and cattle in July. In August there were more foraging bumblebees in fallow sections than sections with a silage crop, but fewer than in sections sown with a 'bird and bumblebee' conservation seed mix. Red clover *Trifolium pratense* and greater knapweed *Centaurea nigra* were two of few plant species favoured by bumblebees and were predominantly found in the fallow sections July–August. Thirty-one crofts located on Lewis, Harris, the Uists and at Durness were included in the study. In addition to the four management types mentioned, arable crops, unmanaged, sheep-grazed and winter-grazed pastures were surveyed for foraging bumblebees and bumblebee forage plants along zigzag or L-shaped transects in each croft section in June, July and August 2008. Foraging bumblebees 2 m on either side of transects were identified to species and recorded together with the plant species on which they were foraging. Inflorescences of all plant species were counted in 0.25 m^2 quadrats placed at 20 or 50 m intervals along the transects.

(1) Moreby S.J. & Aebischer N.J. (1992) Invertebrate abundance on cereal fields and set-aside land – implications for wild gamebird chicks. *British Crop Protection Council Monographs*, 50, 181–186.

(2)   van Elsen T. & Gunther H. (1992) Auswirkungen der Flächenstillegung auf die Acker-wildkraut-Vegetation von Grenzertrags-Feldern [Effects of set-aside on the wild species vegetation of marginal fields]. *Zeitschrift fur Pflanzenkrankheiten und Pflanzenschutz*, Sonderheft 13, 49–60.

(3)   Gathmann A. & Tscharntke T. (1993) Bees and wasps in trap nests on sown crop fields and self-sown fallow fields (Hymenoptera Aculeata). *Verhandlungen Gesellschaft fur Okologie*, 22, 53–56.

(4)   Berg A. & Part T. (1994) Abundance of breeding farmland birds on arable and set-aside fields at forest edges. *Ecography*, 17, 147–152.

(5)   Gathmann A., Greiler H.J. & Tscharntke T. (1994) Trap-nesting bees and wasps colonizing set-aside fields: succession and body size, management by cutting and sowing. *Oecologia*, 98, 8–14.

(6)   Greiler H.J. (1994) Insect communities in self-established and sown agricultural fallows. In W. Nentwig & H.M. Poehling (eds.) *Agrarokologie*, 11, Haupt, Bern. pp. 1–136.

(7)   Steffan-Dewenter I. & Tscharntke T. (1995) Bees on set-aside fields: impact of flower abundance, vegetation and field-age. *Mitteilungen Der Deutschen Gesellschaft Fur Allgemeine Und Angewandte Ento-mologie*, 10, 319–322.

(8)   Wakeham-Dawson A. (1995) Hares and skylarks as indicators of environmentally sensitive farming on the South Downs. PhD thesis. Open University.

(9)   Ford M.A. (1996) The transformation of surplus farmland into semi-natural habitat I. Effect of seed supply on the conservation value of Scottish set-aside exemplified by the vegetation at a site near Elgin. *Aspects of Applied Biology*, 44, 179–184.

(10) Büchs W., Harenberg A. & Zimmermann J. (1997) The invertebrate ecology of farmland as a mirror of the intensity of the impact of man? An approach to interpreting results of field experiments carried out in different crop management intensities of a sugar beet and an oil seed rape rotation including set-aside. *Biological Agriculture & Horticulture*, 15, 83–107.

(11) Hilbig W. (1997) Effects of extensification programmes in agriculture on segetal vegetation. *Tuexenia*, 295–325.

(12) Tattersall F.H., Macdonald D.W., Manley W.J., Gates S., Ferber R. & Hart B.J. (1997) Small mammals on one-year set-aside. *Acta Theriologica*, 47, 329–334.

(13) Watson A. & Rae R. (1997) Some effects of set-aside on breeding birds in northeast Scotland. *Bird Study*, 44, 245–245.

(14) Wilson J.D., Evans J., Browne S.J. & King J.R. (1997) Territory distribution and breeding success of skylarks *Alauda arvensis* on organic and intensive farmland in southern England. *Journal of Applied Ecology*, 34, 1462–1478.

(15) Steffan-Dewenter I. & Tscharntke T. (1997) Early succession of butterfly and plant communities on set-aside fields. *Oecologia*, 109, 294–302

(16) Desender K. & Bosmans R. (1998) Ground beetles (Coleoptera, Carabidae) on set-aside fields in the Campine region and their importance for nature conservation in Flanders (Belgium). *Biodiversity and Conservation*, 7, 1485–1493.

(17) Hansson M. & Fogelfors H. (1998) Management of permanent set-aside on arable land in Sweden. *Journal of Applied Ecology*, 35, 758–771.

(18) Poulsen J.G., Sotherton N.W. & Aebischer N.J. (1998) Comparative nesting and feeding ecology of skylarks *Alauda arvensis* on arable farmland in southern England with special reference to set-aside. *Journal of Applied Ecology*, 35, 131–147.

(19) Sotherton N. (1998) Land use changes and the decline of farmland wildlife: an appraisal of the set-aside approach. *Biological Conservation*, 83, 259–268.

(20) Tattersall F.H., Tew T.E. & Macdonald D.W. (1998) Habitat selection by arable wood mice: a review of work carried out by the wildlife conservation research unit. *Proceedings of the Latvian Academy of Sciences. Section B, Natural Sciences*, 52, 31–36.

(21) Wakeham-Dawson A., Szoszkiewicz K., Stern K. & Aebischer N.J. (1998) Breeding skylarks *Alauda arvensis* on Environmentally Sensitive Area arable reversion grass in southern England: survey-based and experimental determination of density. *Journal of Applied Ecology*, 35, 635–648.

(22) Buckingham D.L., Evans A.D., Morris A.J., Orsman C.J. & Yaxley R. (1999) Use of set-aside land in winter by declining farmland bird species in the UK. *Bird Study*, 46, 157–169.

(23) Kinnunen H. & Tiainen J. (1999) Carabid distribution in a farmland mosaic: the effect of patch type and location. *Annales Zoologici Fennici*, 36, 149–158.

(24) Tattersall F.H., Fagiano A.L., Bembridge J.D., Edwards P., Macdonald D.W. & Hart B.J. (1999) Does the method of set-aside establishment affect its use by wood mice? *Journal of Zoology*, 249, 472–476.

(25) Tattersall F.H., Hart B.J., Manley W.J., MacDonald D.W. & Feber R.E. (1999) Small mammals on set-aside blocks and margins. *Aspects of Applied Biology*, 54, 131–138.

(26) Aebischer N.J., Green R.E. & Evans A.D. (2000) From science to recovery: four case studies of how research has been translated into conservation action in the UK. In N.J. Aebischer, A.D. Evans, P.V. Grice & J.A. Vickery (eds.) *Ecology and Conservation of Lowland Farmland Birds*, British Ornithologists' Union, Tring. pp. 43–54.

(27) Critchley C.N.R. & Fowbert J.A. (2000) Development of vegetation on set-aside land for up to nine years from a national perspective. *Agriculture, Ecosystems & Environment*, 79, 159–174.

(28)   Henderson I.G., Cooper J., Fuller R.J. & Vickery J. (2000) The relative abundance of birds on set-aside and neighbouring fields in summer. *Journal of Applied Ecology*, 37, 335–347.

(29)   Henderson I.G., Vickery J.A. & Fuller R.J. (2000) Summer bird abundance and distribution on set-aside fields on intensive arable farms in England. *Ecography*, 23, 50–59.

(30)   Moreby S.J. & Southway S. (2000) Management of stubble-set-aside for invertebrates important in the diet of breeding farmland birds. *Aspects of Applied Biology*, 62, 39–46.

(31)   Schmitt G. & Roth M. (2000) Hymenoptera Parasitica of different agrarian habitat types in the cultural landscape of northeastern Germany. *Mitteilungen Der Deutschen Gesellschaft Fuer Allgemeine Und Angewandte Entomologie*, 12, 379–383.

(32)   Henderson I.G., Critchley N.R., Cooper J. & Fowbert J.A. (2001) Breeding season responses of Skylarks *Alauda arvensis* to vegetation structure in set-aside (fallow arable land). *Ibis*, 143, 317–321.

(33)   Steffan-Dewenter I. & Tscharntke T. (2001) Succession of bee communities on fallows. *Ecography*, 24, 83–93.

(34)   Tattersall F.H., Macdonald D.W., Hart B.J., Manley W.J. & Feber R.E. (2001) Habitat use by wood mice (*Apodemus sylvaticus*) in a changeable arable landscape. *Journal of Zoology*, 255, 487–494.

(35)   Donald P.F., Evans A.D., Muirhead L.B., Buckingham D.L., Kirby W.B. & Schmitt S.I.A. (2002) Survival rates, causes of failure and productivity of Skylark *Alauda arvensis* nests on lowland farmland. *Ibis*, 144, 652–664.

(36)   Moreby S.J. (2002) Permanent and temporary linear habitats as food sources for the young of farmland birds. In D. E. Chamberlain & A. Wilson (eds.) *Avian Landscape Ecology: Pure and Applied Issues in the Large-Scale Ecology of Birds*, International Association for Landscape Ecology (IALE(UK)), Aberdeen. pp. 327–332.

(37)   Murray K.A., Wilcox A. & Stoate C. (2002) A simultaneous assessment of farmland habitat use by breeding skylarks and yellowhammers. *Aspects of Applied Biology*, 67, 121–127.

(38)   Firbank L.G., Smart S.M., Crabb J., Critchley C.N.R., Fowbert J.W., Fuller R.J., Gladders P., Green D.B., Henderson I. & Hill M.O. (2003) Agronomic and ecological costs and benefits of set-aside in England. *Agriculture, Ecosystems & Environment*, 95, 73–85.

(39)   Kromp B., Hann P., Kraus P. & Meindl P. (2004) Viennese Programme of Contracted Nature Conservation "Biotope Farmland": monitoring of carabids in sown wildflower strips and adjacent fields. *Deutsche Gesellschaft fur allgemeine und angewandte Entomologie e.V.*, 14, 509–512.

(40)   Parish D.M.B. & Sotherton N.W. (2004) Game crops and threatened farmland songbirds in Scotland: a step towards halting population declines? *Bird Study*, 51, 107–112.

(41)   Van Buskirk J. & Willi Y. (2004) Enhancement of farmland biodiversity within set-aside land. *Conservation Biology*, 18, 987–994.

(42)   Bracken F. & Bolger T. (2006) Effects of set-aside management on birds breeding in lowland Ireland. *Agriculture, Ecosystems & Environment*, 117, 178–184.

(43)   Huusela-Veistola E. & Hyvanen T. (2006) Rotational fallows in support of functional biodiversity. *IOBC/WPRS Bulletin*, 29, 61–64.

(44)   Stevens D.K. & Bradbury R.B. (2006) Effects of the Arable Stewardship Pilot Scheme on breeding birds at field and farm-scales. *Agriculture, Ecosystems & Environment*, 112, 283–290.

(45)   Askew N.P., Searle J.B. & Moore N.P. (2007) Agri-environment schemes and foraging of barn owls *Tyto alba*. *Agriculture, Ecosystems & Environment*, 118, 109–114.

(46)   Macdonald D.W., Tattersall F.H., Service K.M., Firbank L.G. & Feber R.E. (2007) Mammals, agri-environment schemes and set-aside – what are the putative benefits? *Mammal Review*, 37, 259–277.

(47)   Moreby S.J. (2007) Invertebrate distributions between permanent field boundary habitats and temporary stubble set-aside. *Aspects of Applied Biology*, 81, 207–212.

(48)   Roberts P.D. & Pullin A.S. (2007) *The Effectiveness of Land-based Schemes (Including Agri-environment) at Conserving Darmland Bird Densities within the UK.* Systematic Review No. 11. Collaboration for Environmental Evidence/Centre for Evidence-Based Conservation, Birmingham, UK.

(49)   Wretenberg J., Lindstrom A., Svensson S. & Part T. (2007) Linking agricultural policies to population trends of Swedish farmland birds in different agricultural regions. *Journal of Applied Ecology*, 44, 933–941.

(50)   Henderson I.G., Ravenscroft N., Smith G. & Holloway S. (2009) Effects of crop diversification and low pesticide inputs on bird populations on arable land. *Agriculture, Ecosystems & Environment*, 129, 149–156.

(51)   Woolley C. (2009) Estimating population change and dispersal activity of spiders in an agricultural landscape. PhD thesis. University of Plymouth.

(52)   Aebischer N.J. & Ewald J.A. (2010) Grey Partridge *Perdix perdix* in the UK: recovery status, set-aside and shooting. *Ibis*, 152, 530–542.

(53)   Douglas D.J.T., Benton T.G. & Vickery J.A. (2010) Contrasting patch selection of breeding Yellowhammers *Emberiza citrinella* in set-aside and cereal crops. *Bird Study*, 57, 69–74.

(54)   Redpath N., Osgathorpe L.M., Park K. & Goulson D. (2010) Crofting and bumblebee conservation: the impact of land management practices on bumblebee populations in northwest Scotland. *Biological Conservation*, 143, 492–500.

Additional references
Tattersall, F.H., Avundo, A.E., Manley, W.J., Hart, B.J. & Macdonald, D.W. (2000) Managing set-aside for field voles (*Microtus agrestis*). *Biological Conservation*, 96, 123–128.
Tattersall, F.H. & Macdonald, D.W. (2003) Wood mice in the arable ecosystem. In F.H. Tattersall & W.J. Manley (eds.) *Conservation and Conflict: Mammals and Farming in Britain.* Linnean Society Occasional Publications, Westbury Publishing, West Yorkshire. pp. 82–96.

1.8   Connect areas of natural or semi-natural habitat

- All four studies (including one site comparison and two replicated trials) from the Czech Republic, Germany and the Netherlands investigating the effects of habitat corridors or restoring areas of natural or semi-natural habitat between existing patches found some degree of colonization of these areas by invertebrates or mammals. However for invertebrates one unreplicated site comparison study reported that the colonization process was slow[2], and three studies found that the extent of colonization varied between invertebrate taxa[1, 2, 4, 5].

- One small, replicated study from the Czech Republic investigated colonization of two bio-corridors by small mammal species[3]. It found more small mammal species in the bio-corridors than in an adjacent forest or arable fields.

- All three studies from Germany and the Netherlands looking at the effects on invertebrates found mixed results. One replicated study found more species of some wasps (cavity-nesting wasps and caterpillar-hunting wasps) in grass strips connected to forest edges than in isolated strips[5]. An unreplicated study found that the abundance of three ground beetle species substantially increased in an arable field undergoing restoration to heathland but that typical heathland species failed to colonize over the 12 year period[1]. One study found that 2 out of 85 ground beetle species used a meadow and hedge-island strip extending from semi-natural habitats into arable farmland[2]. In the same study the habitat strip did not function well for ground beetles and harvestmen but was colonized by snails and spiders[4].

Background
This intervention involves the creation of habitat corridors between currently isolated natural/semi-natural habitats or the restoration of natural/semi-natural habitats between existing patches.

Habitat fragmentation, as well as destruction, may be an important driver of population declines. Small areas hold fewer species than large ones and if individuals are unable to cross areas of converted habitat then populations in separate habitat patches will become isolated. This potentially makes them more vulnerable to extinction, from natural variations in birth and death rates or sex ratios, from inbreeding depression and from outside pressures; both natural (such as storms or wildfires) and man-made (such as hunting or continued habitat loss). However the precise effects of habitat fragmentation, as opposed to loss, are debated (e.g. Fahrig 1997).

Theoretically, the number of species surviving in a habitat fragment is determined by its size and its effective distance to other habitat patches (MacArthur & Wilson 1967). Connecting remaining areas of natural or semi-natural habitat is therefore often seen as a way to increase the viability of populations, but there is considerable debate as to the effectiveness of such 'wildlife corridors' (e.g. Beier & Noss 1998).

MacArthur R.H. and Wilson E.O. (1967) *The Theory of Island Biogeography*. Princeton University Press, Princeton, New Jersey.

Fahrig L. (1997) Relative effects of habitat loss and fragmentation on population extinction. *The Journal of Wildlife Management*, 61, 603–610.

Beier P. & Noss R.F. (1998) Do habitat corridors provide connectivity? *Conservation Biology*, 12, 1241–1252.

An unreplicated habitat restoration study from 1973 to 1984 on a heathland reserve in the Netherlands (1) found a substantial loss of ground beetle (Carabidae) species in an ex-arable field undergoing restoration to heathland over the 12-year period. Many of the ground beetle species that disappeared or decreased were able to disperse and capable of flight. The adjacent heathland and a young coppiced oak forest did not lose any species characteristic of their respective habitats over the same period. The numbers of several ground beetle species (*Amara communis, Pterostichus versicolor, A. lunicollis*) increased substantially in the field over the 12-year period, and the authors attribute this increase to the restoration process, which involved management to promote nutrient impoverishment of the soil. A small group of species that favour dense heather (*Calluna* spp.; *Erica* spp.) vegetation and that were found in the adjacent heathland had not colonized the restoration field by the end of the study. Cultivation of the ca. 5 ha field ceased in 1972; prior to which it had mainly been used for growing wheat. The vegetation was thereafter mown annually and the cuttings removed in order to impoverish the soil. Sets of 3 pitfall traps (25 x 25 cm, 10 m apart) were established in the restoration field, the heath, the forest and a 3–4 m-wide sand bank running between the field and the forest. Ground beetles were sampled weekly throughout the year for 12 years.

An unreplicated site comparison study from 1982 to 1991 in western Germany (2) (same study as (4)) found that out of 85 ground beetle (Carabidae) species sampled, only 2 used a young habitat strip as a dispersal corridor. The 2 ground beetle species (*Carabus nemoralis* and *Notiophilus palustris*) which appeared to use a meadow and hedge strip as a dispersal corridor were initially present in the semi-natural source habitat and gradually appeared along the strip over the 9 years following planting (1982 to 1990). Although three other ground beetle species also immigrated to the corridor, they were able to fly, so the linear shape of the habitat was unlikely to be important to them and it could not be confirmed that they originated from the studied source habitat. Twenty-five ground beetle species present in the source habitat showed no tendency to disperse to the corridor. The corridor was established in 1982, consisting of a 1.6 km long, 10 m-wide meadow strip, along which 9 400 m^2 hedge islands were planted as stepping stones. It was attached at one end to an area of old mixed semi-natural habitat (woods, hedge fragments, ponds surrounded by small reeds and wet and dry meadows) and extended into intensive arable farmland. Ground beetles were sampled along the corridor using six pitfall traps in hedge islands and meadow strips from 1982 to 1990. Semi-natural habitats and adjacent arable fields were sampled from 1990 to 1991.

A small replicated study from 1992 to 1996 in an arable area in the Czech Republic (3) found that from the third year after planting, 2 bio-corridors (10 m wide, planted with trees and shrubs) had more small mammal species and individuals than 2 adjacent fields or a forest. The bio-corridors had 8 small mammal species (supporting both field and forest species) and 128–143 captures, compared to 5 species and 47–68 captures in fields (maize and wheat) and 66 captures in the forest. The mammal community in the forest differed from that of the bio-corridors and fields, where wood mouse *Apodemus sylvaticus* and common vole *Microtus arvalis* tended to dominate. During the autumn (from 1994), the wood mouse population peaked in bio-corridors, but few were caught in (bare) fields. The two bio-corridors were planted in 1991, one extended perpendicular to a forested area into an arable field and the second extended from the end of the first bio-corridor further into the crop. They were fenced and ploughed in the first years after planting to allow short-lived weeds to grow in the herb layer. Fifty snap traps were set in a 150 m line in each habitat and left for 3 nights twice in spring and autumn from 1992 to 1996 and in summer 1994.

The same unreplicated site comparison study as (2), between 1982 and 1998 (4) found marked differences in the effectiveness of the meadow and hedge-island habitat strip as a dispersal corridor for four invertebrate taxa: ground beetles (Carabidae), harvestmen (Opiliones), spiders (Araneae) and snails (Gastropoda). Nine years after planting, the strip did not (or not yet) function well as a dispersal corridor for ground beetles or harvestmen. Snails were the best colonizers, with the highest proportion of species migrating to the strip, including target woodland species. The authors suggest that passive travel by small snails on mammals or birds may have contributed to this. Spiders also had a high proportion of immigrating species, but many of them were not present in the source habitat and may have passively 'ballooned' in from the surrounding area, rather than using the strip as a dispersal corridor. The authors concluded that while the hedge islands appeared to be working as stepping stones for species able to travel passively, this was not true for actively moving invertebrates such as ground beetles or harvestmen, perhaps because of the age, size or connectedness of hedge islands at the time of study. In addition to the sampling regime described in (2) invertebrates were sampled from the surrounding area in 1992–1994 and 1997–1998. Spiders, harvestmen and ground beetles were sampled using pitfall traps and snails were sampled by flotation (in 1984, 1987 and 1990).

A replicated study in 2004 in Lower Saxony, Germany (5) found that the numbers of cavity-nesting wasp (Hymenoptera) species, brood cells and caterpillar-hunting wasp (Eumenidae) brood cells in trap nests were higher in grass strips connected to forest edges than in trap nests in isolated grass strips. The number of wasp species was significantly higher in connected (2.3 species) than in highly isolated grass strips (0.8); differences were not significant between connected and slightly isolated (1.2) or between slightly and highly isolated strips. Numbers of wasp brood cells were significantly higher in connected (30 brood cells) than slightly (7) and highly isolated grass strips (4), caterpillar-hunting wasps showed the same pattern. Numbers did not differ between strip types for spider-hunting wasps (Sphecidae), species richness of parasitoids or numbers of parasitized brood cells. At each of 12 arable sites, 9–12 traps were placed in 3 types of 3 m-wide grass strip: 'connected' strips connected via a corridor to a forest edge (traps set 200 m from forest), 'slightly isolated' strips separated from forest by a cereal field (traps 200 m from forest, no connecting corridor) and 'highly isolated' strips 600 m from the nearest forest edge (no connecting corridor). Distances between trap nests were at least 600 m. Trap nests consisted of 4 plastic tubes filled with common reed *Phragmites australis* sections (2–10 mm diameter) and were installed at a height of 1.0–1.2 m from April–September 2004.

(1) van Dijk T.S. (1986) Changes in the carabid fauna of a previously agricultural field during the first twelve years of impoverishing treatments. *Netherlands Journal of Zoology*, 36, 413–437.
(2) Gruttke H. (1994) Dispersal of carabid species along a linear sequence of young hedge plantations. In K. Desender, M. Dufrene, M. Loreau, M. L. Luff & J. P. Maelfait (eds.) *Carabid Beetles: Ecology and Evolution*, Kluwer Academic Publishers, The Netherlands. pp. 299–303.
(3) Bryja J. & Zukal J. (2000) Small mammal communities in newly planted biocorridors and their surroundings in southern Moravia (Czech Republic). *Folia Zoologica*, 49, 191–197.
(4) Gruttke H. & Willecke S. (2000) Effectiveness of a newly created habitat strip as dispersal corridor for invertebrates in an agricultural landscape. *Proceedings of the Environmental Encounters Series: Workshop on ecological corridors for invertebrates: strategies of dispersal and recolonisation in today's agricultural and forestry landscapes*. Neuchatel, May, 45. pp. 67–80.
(5) Holzschuh A., Steffan-Dewenter I. & Tscharntke T. (2009) Grass strip corridors in agricultural landscapes enhance nest-site colonization by solitary wasps. *Ecological Applications*, 19, 123–132.

1.9   Manage hedgerows to benefit wildlife (includes no spray, gap-filling and laying)

- Ten studies from Switzerland and the UK (three replicated and controlled studies of which one was randomized) found that managing hedges for wildlife resulted in increased berry yields[10], species diversity or richness of plants[2, 8, 9] and invertebrates[2] and diversity[15] or abundance[5, 11, 14, 19, 20] of farmland birds.

- Five studies from the UK (including one replicated, controlled and randomized study) found that hedge management did not affect plant species richness[3, 6, 12], numbers of bumblebee queens[16] or farmland birds[17, 18]. Two replicated studies have shown mixed or adverse effects, with hedge management having mixed effects on invertebrates[6, 12] or leading to reduced hawthorn berry yield[4].

- A replicated site comparison in the UK[1] found hedges cut every two years had more suitable nesting habitat for grey partridge than other management regimes. A replicated study from the UK found that hawthorn berry yield was reduced when management involved removing fruit-bearing wood[7].

> **Background**
> Hedges can be key habitats for farmland biodiversity, but they may need managing to maximize their value. Managing hedges to benefit wildlife involves one or more of the following management changes: reduce cutting frequency; reduce or avoid spraying; mow vegetation beneath hedgerows; fill gaps in hedges; coppice or lay to restore traditional hedge structure.
> See also 'Agri-Chemicals – Reduce fertilizer, pesticide or herbicide use generally', which includes studies monitoring hedgerow biodiversity in response to management outside hedgerows.

A study of nine farms in England and one in Scotland from 1979 to 1981 (1) found that the most suitable nesting habitat for grey partridge *Perdix perdix* was in hedges trimmed biennially compared to those unmanaged, occasionally managed, cut annually, sides cut annually, boundaries with verges cut or regularly grazed. Grey partridge breeding density and recruitment increased with the length of field boundary, amount of dead grass and height of earth bank at the hedge base. Nests were sited where dead grass, bramble *Rubus fruticosus* agg. and leaf litter were significantly more abundant and bank height was higher. Field boundaries and hedges were surveyed in late winter. Breeding density was surveyed in March. Four farms were searched for nests and nest success was recorded. Forty-two grey partridge nests were recorded. Hedge characteristics were recorded around each nest and at a randomly chosen 'non-nest' site within 100 m.

A replicated, controlled, randomized study of 14 hedges at 10 sites throughout Northern Ireland in 1991 (2) found that hard coppicing increased numbers of plant species and laying increased numbers of invertebrate species. Treatments had higher numbers of plant species/plot compared to the control (26) but this was only significant for coppicing (with planting in gaps; 31–34), not pollarding (28) or laying (27). There were significantly more invertebrate orders in laid hedges (4.1) than the control (2.6). Numbers in coppiced (3.6–3.8) and pollarded (3.2) hedges did not differ significantly from the control. There were similar numbers of plant species and invertebrate orders in Environmentally Sensitive Area and non- Environmentally Sensitive Area hedges. The 14 hedges were dominated by hawthorn *Crataegus monogyna*, had 150 m of uniform height and density of trees, had permanent pasture on both sides and were largely overgrown and unmanaged. The 5 treatments were applied to 25 m lengths of each hedge. Hedges were fenced to exclude grazing and were cut to 1.5 m each third year where appropriate. Plants were listed within each plot during the summer. Invertebrates were sampled using shelter traps (20 x 5 cm) in the hedgerow canopy during May. Additional plant data are available for 1992 and 1994 (3).

In the same study as (2), (3) found that although hard coppicing, pollarding (to 1.5 m) and laying initially increased plant species diversity, after four years it had no effect on species richness. In 1991, 2 treatments had more plant species than the control (25 species; coppicing: 31–33; pollarding 28); the exception was laying (27). By 1994, although more species were recorded from all treatments than the control (23), none were significantly different (coppicing: 25–26; laying: 25; pollarding: 25). In 1991 the 6 Environmentally Sensitive Area hedges and 8 non-Environmentally Sensitive Area hedges both had a mean of 20 plant species. Environmentally Sensitive Area hedges coppiced with mixed planting had significantly more species than control Environmentally Sensitive Area hedges (23 vs 19). By 1994 Environmentally Sensitive Area sites had slightly higher plant diversity than non- Environmentally Sensitive Area sites (1992: 21 vs 20; 1994: 22 vs 20).

A replicated study of hedgerows in Cambridgeshire, UK (4) found that hawthorn *Crataegus monogyna* berry yield was significantly higher in unmanaged hedgerows than those that were laid (and trimmed after 5 years) or coppiced (and trimmed after 5 years) or pollarded to 1.5 m. Yield decreased with more extreme management treatments (unmanaged: 148–161 g/2.5 m^2; laid: 48–85 g/2.5 m^2; coppiced: 3 g/2.5 m^2; pollarded: 2 g/2.5 m^2). There was some compensation for reduced hawthorn yields in laid and coppiced hedgerows through increased rose hip yields, although rose hip samples were too small for analysis. There were no significant differences between laid (84 g/2.5 m^2) and laid and trimmed hedgerows (86 g/2.5 m^2); the two hedgerow treatments were visually difficult to separate. The weights of 50 berries were lighter in coppiced plots (11 g) than all other treatments (15–16 g). There were no significant differences in berry dry matter content between treatments (45–51%). Hawthorn and rose *Rosa canina* agg. berries were harvested from three to eight replicates in October 1997. Hedgerows were laid or coppiced in 1990–1991 and were trimmed or pollarded in 1995–1996. Berries were harvested within each plot (20–40 m long sections) from five 50 × 50 cm quadrats on the side of hedges, 1 m above ground.

A 2000 literature review (5) found that the UK population of cirl bunting *Emberiza cirlus* increased from between 118 and 132 pairs in 1989 to 453 pairs in 1998 following a series of schemes designed to provide overwinter stubbles, grass margins, beneficially managed hedges and set-aside. Numbers on fields under these schemes increased by 70%, compared with a 2% increase elsewhere.

A replicated study of hedgerows within seven arable and pastoral farms in England and Wales (6) found that cutting frequency and timing affected invertebrate numbers but not plant diversity. Abundance of individual invertebrate groups tended to decline with regular hedge cutting. However, although numbers of some taxa such as jumping plant lice (Psyllids) were higher in uncut sections, cutting increased others, notably herbivores and detritivores such as true bugs (Heteroptera; uncut: 4/plot; annual: 22–28; biennial: 15–23), beetles (Coleoptera; uncut: 4/plot; annual: 5–9; biennial: 8–13), springtails (Collembola) and thrips (Thysanoptera). Cutting in February rather than September reduced numbers of butterflies and moths (Lepidoptera; 33 vs 65/plot) and flies (Diptera; 82 vs 118/plot), but increased beetles (Coleoptera; 9–13 vs 5–8) the following summer. Cutting frequency (uncut, annual, biennial and triennial) and timing did not affect numbers of plant species in the hedge or hedge base. Hedge dimensions were greatest on annually cut hedges and smallest on those uncut. The longer the hedge was left between cuts, the more berries were produced (uncut, biennial, annual). Berry numbers were reduced with triennial cuts. Each hedgerow received replicated treatments (15–21) of each cutting frequency and timing. Data were obtained on the abundance of berries (autumn), shrubs, hedge-base flora and invertebrates (May and July) within each hedgerow plot. The same study is presented in Marshall *et al.* 2001.

A replicated study of hedgerows in Cambridgeshire and Warwickshire, England (7) (same site as (4)) found that hawthorn *Crataegus monogyna* berry yield was significantly reduced when management involved removing fruit-bearing wood. Yield was significantly higher in sections that had been laid (282 g/2.5 m^2) or uncut (219–421) than those that had been cut (4–10), coppiced (3–26) pollarded (70) or grubbed out (0). Yield differences were due to greater numbers of berries rather than increased berry size. At Monks Wood, the dry matter content was significantly higher in uncut sections; this was not the case at Drayton. At Drayton there were 5 randomized replicate plots (12 m long) of the following 5 treatments: unfenced or fenced control cut annually; fenced uncut; coppiced; grubbed out and replanted with blackthorn. At Monks Wood there were 2 randomized trials each of 10–12 (20 m long) plots that received 3–5 replicates of 3 (uncut, coppiced or laid) or 2 (uncut or pollarded to 1.5 m) treatments. Berries were harvested within each plot from 5 50 × 50 cm quadrats on the side of hedges, 1 m above ground.

A replicated, controlled study of three hedgerows in farmland at Long Ashton Research Station, Somerset (8) found that hedgerow management, particularly sowing perennial seed mix, increased botanical diversity in the hedge base. Plant species diversity in sown plots was significantly higher than in plots where the hedge was cut, in 2 arable fields (17–38 vs 13–27 species in sown and unsown plots respectively) and 1 grassland field (23–27 vs 16–23). In the grassland field, there was little difference between treatments (unmanaged: 15–21; autumn cut: 16–23; selective herbicide: 17–23; no fertilizer: 19–24) and the initial increase in number of plant species in the sown plot did not persist. In the cereal fields plots sown without selective herbicides tended to have more plant species than plots sown with selective herbicides (18–27 vs 16–23). There was no overall difference between the number of plant species in autumn- (14–26 species) and spring-cut hedges (13–27). Excluding fertilizer (13–31 plant species) and applying selective herbicide (17–29 plant species) tended to increase the number of plant species, although the initial increase due to fertilizer exclusion only persisted in 1 of the 2 arable fields. Total herbicide application initially reduced the number of plant species to 4, but species rapidly recovered (15–24). The number of true bug (Heteroptera) species was higher in plots treated with selective herbicide than other treatments (grassland: 10 vs 7 true bug species with and without selective herbicide respectively; arable: 5 vs 1–3). Three hedgerows with low botanical diversity and high annual weed densities were selected. Treatments were applied to consecutive 1 m-wide plots along each hedge bottom. Vascular plants were recorded in May and July–August 1997–1999. Invertebrates were sampled using a D-Vac suction sampler (four 5-second samples).

A site comparison study of 60 hedgerows on 2 neighbouring arable farms in Wiltshire, UK (9) found that coppiced and gapped-up hedges had the greatest number of plant species (23 species on average) followed by those with adjacent sown grass and grass/wildflower strips (2, 4 or 20 m wide; Manor Farm: 17 species) and those with a 0.5 m sterile strip created with a broad-spectrum herbicide (Noland's Farm: 15 species). Hedges with adjacent sown strips had a lower abundance of pernicious weed species. The composition of woody species within hedges did not differ between the two farms (Manor Farm: 22 woody species, Noland's Farm: 16 woody species). All 23 sampled hedges on Noland's Farm were trimmed annually and had the vegetation at the hedge base cut. The 37 sampled hedges on Manor Farm were trimmed in alternate years, and 9 were coppiced and gapped-up. Hedge vegetation was assessed in 25 m long plots in the middle of a field edge, on both sides of each hedge, in June 1996.

A replicated site comparison study of hedgerows at two arable and one mixed farm in England (10) found that berry yield was significantly higher in hedges managed but uncut for at least two years (143–175 g/2.5 m^2) than those cut annually (4–11 g/2.5 m^2), but both had significantly lower yields than those uncut for many years (305–530 g/2.5 m^2). There was no significant difference in the percentage dry matter content between treatments (uncut: 36–42% dry matter; uncut ≥ 2 years: 34–44%; annual cut: 35–41%). The farms were in Yorkshire, Cambridgeshire and Buckinghamshire (mixed). Five hedges of each cutting regime were identified per site. Hawthorn berries were harvested (September–October 2001) from 10 quadrats (50 × 50 cm) on the side of hedges, 1 m above ground and at 10 m intervals (or the next nearest hawthorn to 10 m).

A small replicated, controlled study from May–June in 1992–1998 in 1 experimental area with managed hedges (3 km^2) and 4 conventionally managed arable farms in Leicestershire, England (11) found that the abundance of nationally declining songbird species and species of conservation concern significantly increased through time in the site with managed hedges. Although there was no overall difference in bird abundance, species richness or diversity between the experimental and control sites, numbers of nationally declining species rose by 102% (except for Eurasian skylark *Alauda arvensis* and yellowhammer *Emberiza citrinella*). Nationally stable species rose (insignificantly) by 47% (with 8 species exhibiting net increases, especially greenfinch *Carduelis chloris* 68%, and 4 species exhibiting net decreases). The author concluded that managing hedges to increase shrubby vegetation, as part of an integrated management package, provides the greatest benefits to species of conservation concern but does not affect species diversity at the farm scale.

A review of the literature on the impacts of agricultural management on bats, their habitats and invertebrate prey in Europe (12) found one study that reported complex impacts on invertebrates from management that affects hedgerow structure (see (6)).

A replicated study of 751 hedges restored under 100 Countryside Stewardship Scheme agreements and 774 hedges restored under 100 Environmentally Sensitive Area agreements in England investigated the effects on the hedgerow network over 5 years (13). Limited data were presented comparing biodiversity pre- and post-works. Overall, the majority of hedges under the agreements were less than 2 m wide at the base (Countryside Stewardship Scheme: 65%; Environmentally Sensitive Area: 81%) and under 2 m tall (Countryside Stewardship Scheme: 48%; Environmentally Sensitive Area: 57%). Trees were present in 53–56% of hedges. Overall, 21% of Countryside Stewardship Scheme and 38% of Environmentally Sensitive Area hedges were classified as species-rich (compared to an average of 26% of hedges in England). The average number of basal flora species per hedge was six species under Countryside Stewardship Scheme and eight under Environmentally Sensitive Area agreements. Significantly more pre-works hedges were over 2 m in height (Countryside Stewardship Scheme: 53%; Environmentally Sensitive Area: 62%) compared to post-works hedges (Countryside Stewardship Scheme: 28%; Countryside Stewardship Scheme: 34%). Under Environmentally Sensitive Area agreements, significantly more pre-works hedges were over 2 m wide (20%) than post-work hedges (11%). Countryside Stewardship Scheme hedges with a high structural variability tended to be pre-works (16 vs 12%) including hedges of a gappy nature and of various heights. Only 11% of the pre-works Countryside Stewardship Scheme hedges were stock-proof, compared to 18% post-works. Hedges pre- and post-restoration works (coppicing, laying or planting) were sampled using the national local hedgerow procedure. A maximum of eight hedges were sampled from any one agreement.

A replicated site comparison study (14) found that on average 50% of hedgerows in Ecological Compensation Areas on farmland in the Swiss plateau were of 'good ecological quality' (based on national guidelines for Ecological Compensation Area target vegetation). Ecological quality was higher for Ecological Compensation Area hedges in the 'pre-alpine hills' zone than in the more intensively farmed 'lowland' zone, due to more old trees and fewer invading plants. The centres of territories of hedgerow birds were significantly more frequent in or near Ecological Compensation Area hedges (293 territories), suggesting that hedgerow birds were attracted to or favoured these areas. Plant species and hedgerow characteristics were recorded for 317 Ecological Compensation Area hedgerows (total length 44 km) in 11 study areas between 1998 and 2001. Territories of breeding birds were mapped in 23 study areas, based on 3 visits between mid-April and mid-June.

A 2007 site comparison study on 23 sites in the lowlands north of the Alps, Switzerland (15) found 23 out of 100 hedges managed as Ecological Compensation Areas had at least 1 of the 37 surveyed bird species present, compared to 13/100 hedges outside the agri-environment scheme. The 23 sites (covering up to 3 km² each) were randomly selected and surveyed 3 times each between April and June in both years of study.

A replicated, controlled trial of the Rural Stewardship agri-environment scheme on five farms in Scotland (16) found that hedgerows dominated by hawthorn *Crataegus monogyna* or blackthorn *Prunus spinosa* were less attractive than field margins or grasslands to nest-searching queen bumblebees *Bombus* spp. in April and May. There was no significant difference in numbers of foraging or nesting queens between hedgerows managed under the agri-environment scheme (winter-cut every three years; gaps filled; vegetation below un-mown and unsprayed) and conventionally managed hedgerows. The study took place before the woody species comprising the hedgerow came into flower. Nest-searching and foraging queen bumblebees were recorded on 6 100 x 6 m transects on each farm, once a week from 14 April to 16 May 2009, on dry days with temperatures of 5–25°C. Each farm had two arable field margin transects, two grassland (non-boundary) transects, and two hawthorn- or blackthorn-dominated hedgerow transects. On farms with the Rural Stewardship Scheme, one of each transect type was under the agri-environment scheme.

A replicated site comparison of 2,046, 1 km² plots of lowland farmland in England in 2005 and 2008 (17) (same study as (18)) found that management of hedges and ditches under Entry Level Stewardship did not have clear impacts on farmland bird species. Management had significant positive impacts on five species in at least one region of England, but these effects were often very weak and four of the same species showed negative responses in other regions. The other five 'hedgerow' species investigated were never positively associated with boundary management. Generally, effects appeared to be more positive in the north of England.

A site comparison study of 2,046, 1 km² plots of lowland farmland in England (18) (same study as (17)) found that 3 years after the 2005 introduction of the Countryside Steward-ship and Entry Level Stewardship schemes there was no association between the length of hedgerow managed according to the agri-environment scheme and farmland bird numbers. Hedgerow specialist species, including the yellowhammer *Emberiza citrinella* and common whitethroat *Sylvia communis*, showed no significant population response, whereas there were greater numbers of starling on arable, pastoral and mixed farmland with hedgerow manage-ment. For example, in mixed farmland plots starling populations increased by 0.2 individuals for each 1 km of hedgerow. On the other hand, grey partridge *Perdix perdix* appeared to be detrimentally affected, with an apparent decline of 0.3 individuals for every 1.1 km of hedgerow managed according to the agri-environment schemes. The 2,046, 1 km^2 lowland plots were surveyed in both 2005 and 2008 and classified as arable, pastoral or mixed farmland. Eighty-four percent of plots included some area managed according to the Entry Level Stewardship or Countryside Stewardship Scheme. In both survey years, 2 surveys were conducted along a 2 km pre-selected transect route through each 1 km^2 square.

A replicated study in February 2008 across 97, 1 km^2 plots in East Anglia, England (19) (part of the same study as (17)) found that 4 farmland bird species showed strong positive responses to field boundaries managed under agri-environment schemes. These were blue tit *Parus (Cyanistes) caeruleus*, dunnock *Prunella modularis*, common whitethroat *Sylvia communis* and yellowhammer *Emberiza citrinella*. A further five (Eurasian blackbird *Turdus merula*, song thrush *T. philomelos*, Eurasian bullfinch *Pyrrhula pyrrhula*, long-tailed tit *Aegithalos caudatus* and winter wren *Troglodytes troglodytes*) showed weak positive responses and Eurasian reed buntings *Emberiza schoeniclus* showed a weak negative response. The boundaries were classed as either hedges, ditches or hedges and ditches and most were managed under the Entry Level Stewardship scheme.

A replicated site comparison study on farms in two English regions (20) found that summer yellowhammer *Emberiza citrinella* numbers were significantly higher in hedges under environmental stewardship management than in conventionally managed hedges. On East Anglian farms, this was true for both Entry Level Stewardship and Higher Level Stewardship hedge management options (estimated >1.5 yellowhammers/m in Higher Level Stewardship hedges compared to <0.5 yellowhammers/m in conventional hedges). On Cotswolds farms it was only true for hedges managed as 'high environmental value hedges' under Higher Level Stewardship (estimated 0.5 yellowhammers/m), while hedges managed under Entry Level Stewardship did not have more yellowhammers than conventional hedges (estimated <0.2 yellowhammers/m). Hedgerows managed under Entry Level Stewardship are cut every two or three years in winter only. Surveys were carried out in the summers of 2008 and 2009 on up to 30 Higher Level Stewardship farms and 15 non-stewardship farms in East Anglia, and up to 19 Higher Level Stewardship and 8 non-stewardship farms in the Cotswolds.

(1) Rands M.R.W. (1987) Hedgerow management for the conservation of partridges *Perdix perdix* and *Alectoris rufa*. *Biological Conservation*, 40, 127–139.

(2) McAdam J.H., Bell A.C. & Henry T. (1994) The effect of restoration techniques on flora and microfauna of hawthorn-dominated hedges. *Proceedings of the Hedgerow Management and Nature Conservation: British Ecological Society Conservation Ecology Group*. Wye College, University of London. pp. 25–32.

(3) McAdam J.H., Bell A.C., Gilmore C., Mulholland F. & Henry T. (1996) The effects of different hedge restoration strategies on biodiversity. *Aspects of Applied Biology*, 44, 363–367.

(4) Sparks T.H. & Martin T. (1999) Yields of hawthorn *Crataegus monogyna* berries under different hedgerow management. *Agriculture, Ecosystems and Environment*, 72, 107–110.

(5) Aebischer N.J., Green R.E. & Evans A.D. (2000) From science to recovery: four case studies of how research has been translated into conservation action in the UK. In N. J. Aebischer, A. D. Evans, P. V. Grice & J. A. Vickery (eds.) *Ecology and Conservation of Lowland Farmland Birds*, British Ornithologists' Union, Tring. pp. 43–54.

(6) Maudsley M.J., Marshall E.J.P. & West T.M. (2000) *Guidelines for Hedge Management to Improve the Conservation Value of Different Types of Hedge*. Defra BD2102.

(7) Sparks T.H., Robinson K.A. & Downing S.L. (2000) Hedgerow management and the yield of hawthorn *Crataegus monogyna* berries. *Aspects of Applied Biology*, 58, 421–424.

(8) Marshall E.J.P., West T.M. & Maudsley M.J. (2001) Treatments to restore the diversity of herbaceous flora of hedgerows. In C. Barr & S. Petit (eds.) *Hedgerows of the World: Their Ecological Functions in Different Landscapes*, International Association for Landscape Ecology, 10th Annual Conference of the International Association for Landscape Ecology. September, Birmingham, UK. pp. 319–328.

(9) Moonen A.C. & Marshall E.J.P. (2001) The influence of sown margin strips, management and boundary structure on herbaceous field margin vegetation in two neighbouring farms in southern England. *Agriculture, Ecosystems & Environment*, 86, 187–202.
(10) Croxton P.J. & Sparks T.H. (2002) A farm-scale evaluation of the influence of hedgerow cutting frequency on hawthorn (*Crataegus monogyna*) berry yields. *Agriculture, Ecosystems & Environment*, 93, 437–439.
(11) Stoate C. (2002) Multifunctional use of a natural resource on farmland: wild pheasant (*Phasianus colchicus*) management and the conservation of farmland passerines. *Biodiversity and Conservation*, 11, 561–573.
(12) Bat Conservation Trust (2003) *Agricultural Practice and Bats: A review of Current Research Literature and Management Recommendations*. Defra BD2005.
(13) Catherine Bickmore Associates (2004) *Hedgerow Maintenance and Restoration under the ESA and Countryside Stewardship Schemes*. Defra MA01008.
(14) Herzog F., Dreier S., Hofer G., Marfurt C., Schüpbach B., Spiess M. & Walter T. (2005) Effect of ecological compensation areas on floristic and breeding bird diversity in Swiss agricultural landscapes. *Agriculture, Ecosystems & Environment*, 108, 189–204.
(15) Birrer S., Spiess M., Herzog F., Jenny M., Kohli L. & Lugrin B. (2007) The Swiss agri-environment scheme promotes farmland birds: but only moderately. *Journal of Ornithology*, 148, S295–303.
(16) Lye G.C., Park K., Osborne J., Holland J. & Goulson D. (2009) Assessing the value of Rural Stewardship schemes for providing foraging resources and nesting habitat for bumblebee queens (Hymenoptera: Apidae). *Biological Conservation*, 142, 2023–2032.
(17) Davey C., Vickery J., Boatman N., Chamberlain D., Parry H. & Siriwardena G. (2010) Regional variation in the efficacy of Entry Level Stewardship in England. *Agriculture Ecosystems & Environment*, 139, 121–128.
(18) Davey C.M., Vickery J.A., Boatman N.D., Chamberlain D.E., Parry H.R. & Siriwardena G.M. (2010) Assessing the impact of Entry Level Stewardship on lowland farmland birds in England. *Ibis*, 152, 459–474.
(19) Davey C.M., Vickery J.A., Boatman N.D., Chamberlain D.E. & Siriwardena G.M. (2010) Entry Level Stewardship may enhance bird numbers in boundary habitats. *Bird Study*, 57, 415–420.
(20) Field R.H., Morris A.J., Grice P.V. & Cooke A.I. (2010) Evaluating the English Higher Level Stewardship scheme for farmland birds. *Aspects of Applied Biology*, 100, 59–68.

Additional reference
Marshall E.J.P., Maudsley M.J., West T.M. & Rowcliffe H.R. (2001) Effects of management on the biodiversity of English hedgerows. Pages 361–365 in: C. Barr & S. Petit (eds.) *Hedgerows of the World: Their Ecological Functions in Different Landscapes*. International Association for Landscape Ecology, 10th Annual Conference of the International Association for Landscape Ecology. September, Birmingham.

1.10   Manage stone-faced hedge banks to benefit wildlife

- We have captured no evidence for the effects of managing stone-faced hedge banks to benefit farmland wildlife on farmland wildlife.

Background
Stone-faced hedge banks are traditional boundary features in some agricultural landscapes, such as in the southwest of England. Management for biodiversity involves maintaining the wall with traditional materials.

1.11   Manage ditches to benefit wildlife

- Five out of a total of eight studies from the Netherlands and the UK (including one replicated, controlled paired study and three replicated site comparisons) looking at the effects of managing ditches on biodiversity found that this intervention resulted in increased invertebrate biomass or abundance[3, 4], plant species richness[5, 11], emergent plant cover[6], amphibian diversity and abundance[6], bird visit rates[4] and higher numbers of some bird species or positive impacts on some birds in plots with ditches managed under agri-environment schemes[8–10].

- One replicated, controlled and paired study from the Netherlands found higher plant diversity on ditch banks along unsprayed edges of winter wheat compared to those sprayed with pesticides[1].

- Three studies from the Netherlands and the UK (including two replicated site comparisons) found that ditch management had negative or no clear effects on some farmland bird species[8-10] or plants[2,7].

Background

Ditch bank biodiversity is declining in agricultural landscapes and so management is required to maintain and increase species diversity. Ditch wildlife has been shown to be affected by agricultural management practices such as mowing and grazing regimes (e.g. van Strien *et al.* 1989) and by ditch management practices including cleaning/dredging technique, season and frequency (e.g. van Strien *et al.* 1991; Twisk *et al.* 2000, 2003).

In the Netherlands, botanical agri-environment schemes to enhance biodiversity are most commonly applied to ditch banks. Farmers are encouraged to follow the recommended management, i.e. low stocking rate, first mowing at the end of June or beginning of July, no fertilization and deposition of dredging material at the high end of the ditch bank.

van Strien A.J, van Der Linden, J., Melman, T.C.P & Noordervliet, M.A.W. (1989) Factors affecting the vegetation of ditch banks in peat areas in the western Netherlands. *Journal of Applied Ecology*, 26, 989–1004.
van Strien A.J., van der Burg T., Rip W.J. & Strucker R.C.W. (1991) Effects of mechanical ditch management on the vegetation of ditch banks in Dutch peat areas. *Journal of Applied Ecology*, 28, 501–513.
Twisk W., Noordervliet M.A.W. & Ter Keurs W.J. (2000) Effects of ditch management on caddisfly, dragonfly and amphibian larvae in intensively farmed peat areas. *Aquatic Ecology*, 34, 397–411.
Twisk W., Noordervliet M.A.W. & Ter Keurs W.J. (2003) The nature value of the ditch vegetation in peat areas in relation to farm management. *Aquatic Ecology*, 37, 191–209.

A replicated, controlled paired study in 1991–1992 of ditch banks on arable farms in the Netherlands (1) found higher plant diversity and more important/rare plant species on ditch banks along unsprayed edges of winter wheat compared to those sprayed with pesticides. Ditch banks next to unsprayed edges of winter wheat had 65 plant species and a floristic value of 2,201 (scoring system based on the importance of different plant species in terms of rarity) compared to those sprayed with pesticides (50 species; floristic value 1,181). There was no significant difference on banks along unsprayed and sprayed edges of sugar beet *Beta vulgaris* (species: 48 and 41; floristic values: 3,616 and 3,029 respectively) and potato crops (species: 46 and 41; floristic values: 1,961 and 1,864 respectively). Frequency and cover of species and floristic value of vegetation was recorded in two plots on each ditch, one along a sprayed and one an unsprayed edge of sugar beet (7), potato (8) and winter wheat (20) fields in June–July.

A replicated, controlled study of species sown on ditch banks on six farms in the western peat district of the Netherlands (2) found that, overall, there was no significant difference between overall species or species-level germination and establishment, plant survival or reproduction (flowering/seed-setting) under three cutting regimes. However, on high-productivity ditch banks, germination (7% vs 4–5%), establishment (20% vs 7–15%) and reproduction (21–39% vs 15–27%) of many species were higher under 'conventional management' than the 3 cutting treatments. On low-productivity ditch banks, plants tended to have lower survival under 'conventional management' (60% vs 70–80%) and higher reproduction under 'conventional management' and with the first cut in May (33–40% vs 23–26%). One ditch bank was selected on each farm and was divided into four treatments, each with five replicates: two cuts (July and September); three cuts (June; August; September); two cuts (May; September); 'conventional management' (standard cutting and grazing – varied between farms). A mown/artificial gap (15 x 15 cm) was created for each of the 9 species in each plot. Approximately 100 seeds were sown of each species in October 2001. Numbers of seedlings and established plants (≥4 cm) were monitored each month until September 2003. Biomass samples were collected from plots (20 x 50 cm) in July 2001 (pre-treatment biomass) and before each mowing event in 2002–2003; dry weights were recorded.

A replicated, controlled study of 32 ditches in arable and pastoral land in 2005 in Leicestershire, UK (3) (same study as (4)) found that bunded ditches, which dammed water, had significantly greater invertebrate biomass than controls (dry weight: 10 g/m^2 vs 4 g/m^2). Invertebrate families other than flies (Diptera) showed a more mixed response to bunding. Ditches were bunded (small dams placed across ditches) and slightly widened in 5–20 m lengths, with equal length control sections approximately 50 m upstream. Five insect emergence traps (0.5 mm mesh; surface area 0.1 m^2) were spaced along each section. Samples were collected every two weeks (April–August 2005); invertebrates identified to family and recorded as biomass estimates.

A replicated, controlled (paired) study of drainage ditches in arable and pastoral areas of Leicestershire, UK (4) (same study as (3)) found that that wetting-up ditches resulted in higher invertebrate and bird numbers. The following were significantly greater in bunded (dammed ditches) compared to non-bunded ditches: bird visit rates (1.0 vs 0.5 visits/month), emergent aquatic insect biomass (1,400 vs 900 individuals/m^2), surface-active fly (Diptera) adults (in arable ditches in 2005; 85–100 vs 60–65/sample) and fly larvae and butterfly/moth (Lepidoptera) larvae (in pastoral ditches in 2006). There was no difference for invertebrates active in the grass layer. Vascular plant species richness was lower and bare ground cover higher in bunded ditches than controls in 2005 due to disturbance during creation. Sampling involved bird observations (45 minutes; 1–2/month), fixed/floating traps for emerging aquatic insects, pitfall traps and sweep-netting for terrestrial invertebrates and a botanical quadrat (0.25–0.5 m^2) survey. Data was obtained between April 2005–March 2007; all year for birds and spring–summer for other groups.

A replicated site comparison study from 1999 to 2004 in the Netherlands (5) found that ditch management affected plant diversity. Diversity was significantly higher on farms with ecologically managed ditches (mown once in September, cuttings removed to reduce nutrient input) buffered with ≥ 3 m-wide field margin strips (36–65 plant species/400 m^2) and organic farms (converted to organic less than 5 years ago: 32 plant species/400 m^2; converted more than 5 years ago: 36–52 plant species/400 m^2) than conventional farms (26–34 species/400 m^2). On ecologically managed farms plant diversity increased significantly over 6 years (up to 27%) there was a small shift to less common plant species and a decrease in the number of nitrogen rich species and Ellenberg nitrogen-values. There tended to be more nitrogen poor species on ecologically-managed and organic farms compared to conventional farms. Four ecologically managed farms, 18 conventional and 20 organic arable farms were studied. Cutting date varied on conventional and organic farms, but cuttings were never removed, and ditches on both farm-types did not have buffer field margin strips. On ecologically managed farms, plant species surveys of 100 m of ditch bank spread over the whole farm were undertaken once a year during 1999–2004. On organic (in 2001) and conventional (2003) farms, plant species presence was recorded on 10 x 25 m of ditch bank along a transect (May–June).

A replicated site comparison study of 18 agri-environment scheme-managed and 24 conventionally managed ditches within pasture and perpendicular to 8 nature reserve borders in the western peat district of the Netherlands (6) found that amphibian diversity and abundance (and emergent plant cover) was significantly higher in agri-environment than conventional ditches. Adult green frog *Rana esculenta* numbers in conventional ditches declined with distance from reserves; this was not the case in agri-environment scheme ditches. Farmers managing ditches under agri-environment schemes are encouraged to reduce grazing/mowing intensity, reduce fertilizer inputs, and not to deposit mowing cuttings or sediments from ditch cleaning on the ditch banks. Relative amphibian abundance was measured in ditches in April–May and/or May–July 2008 just inside reserves and at 4 distances (0–700 m) from reserve borders. Three methods were used during each sampling period: 5 minute counts, 20 dip-net samples and 2 overnight funnel traps. Habitat variables including percentage cover of aquatic plants were also estimated.

A replicated site comparison (paired) study of ditch banks on six dairy farms in the western peat district of the Netherlands (7) found that agri-environment scheme ditch management did not result in increased plant diversity or decreased productivity over ten years. The total number of plant species on ditch banks under agri-environment scheme management decreased significantly between the periods 1993–1995 (31 species) and 2000–2003 (29 species); numbers of target plant species did not differ (7 species). Productivity on agri-environment scheme ditch banks measured as grass/broadleaved plant ratio increased significantly (1993–1995: 0.37; 2000–2003: 0.44) and Ellenberg nitrogen values increased in 4 and decreased in 2 farms (1993–1995: 5.82; 2000–2003: 5.92). Differences between agri-environment scheme and surrounding ditch banks tended to decrease over the study period. Plant diversity data were obtained from agri-environment scheme ditch banks in July–August 1993–1995 and May 2000–2003 (42 repeatedly sampled plots) and non-agri-environment scheme ditch banks surrounding 5 of the farms (78 plots/year). Five replicate biomass samples were taken from agri-environment scheme ditch banks in mid-May 2000–2002 (9–72 plots) before grazing and mowing. Two productivity measures were also derived from botanical data: grass/broadleaved plant ratios and Ellenberg nitrogen-values.

A replicated site comparison of 2,046, 1 km squares of agricultural land across England in 2005 and 2008 (8) (same study as (9, 10)) found that management of hedges and ditches under Entry Level Stewardship did not have clear impacts on farmland bird species. Management had significant positive impacts on five species in at least one region of England, but these effects were often very weak and four of the same species showed negative responses in other regions. The other five 'hedgerow' species investigated were never positively associated with boundary management. Generally, effects appeared to be more positive in the north of England.

A large site comparison study of 2,046, 1 km^2 plots of lowland farmland in England (9) (same study as (8, 10)) found that 3 years after the 2005 introduction of the Countryside Stewardship Scheme and Entry Level Stewardship Scheme, there was no consistent association between the length of ditches managed according to the agri-environment scheme on a plot and farmland bird numbers. There were higher numbers of linnet *Carduelis cannabina* and reed bunting *Emberiza schoeniclus*, which are known to nest in ditch bank vegetation, in plots with ditches managed according to the Countryside Stewardship Scheme and Entry Level Stewardship compared to other plots. However, this difference was not observed for other species also expected to benefit from ditch management, including yellowhammer *Emberiza citrinella* and yellow wagtail *Motacilla flava*. Between 2005 and 2008, Eurasian skylark *Alauda arvensis* and grey partridge *Perdix perdix* declines were greater in plots with lengths of ditch management than other plots. For example, grey partridges showed decreases of 1.3 birds for each 0.08 km of ditch on pastoral farmland. The 2,046, 1 km^2 lowland plots were surveyed in both 2005 and 2008 and classified as arable, pastoral or mixed farmland. Eighty-four percent of plots included some area managed according to Entry Level Stewardship or the Countryside Stewardship Scheme. In both survey years 2 surveys were conducted along a 2 km pre-selected transect route through each 1 km^2 square.

A replicated study in February 2008 across 97, 1 km^2 plots in East Anglia, England (10) (part of the same study as (8, 9)) found that 4 farmland birds showed strong positive responses to field boundaries (hedges and ditches) managed under agri-environment schemes. These species were blue tit *Parus (Cyanistes) caeruleus*, dunnock *Prunella modularis*, common whitethroat *Sylvia communis* and yellowhammer *Emberiza citrinella*. Six other species showed weak or negative responses: Eurasian blackbird *Turdus merula*, song thrush *T. philomelos*, Eurasian bullfinch *Pyrrhula pyrrhula*, long-tailed tit *Aegithalos caudatus*, winter wren *Troglodytes troglodytes*, and Eurasian reed bunting *Emberiza schoeniclus*. The boundaries were classed as either hedges, ditches or hedges and ditches and most were managed under the Entry Level Stewardship scheme.

A replicated study of 24 pastoral ditches in 2008 in the Netherlands (11) found that delaying twice yearly mowing dates resulted in higher plant diversity. The highest number of seed-setting species was recorded following mowing on 1 July and 1 September, which was 126% higher than under the conventional regime of mowing on 1 June and 1 August. The effect of the mowing date differed between plant species. Species richness was significantly higher and biomass significantly lower on ditches in nature reserves compared to those under long-term agri-environment schemes (>16 years), short-term agri-environment schemes (<6 years) and conventional management. Plots were mown twice on a unique combination of an early (15 May; 1 June; 15 June; 1 July) and late date (1 August; 15 August; 1 September; 15 September). Before mowing, presence of species, target species with ripe seeds and biomass was recorded in 16 plots under different biannual mowing treatments within 6 randomly selected ditches under each of the 4 management systems: nature reserves, long-term agri-environment schemes, short-term agri-environment schemes and conventional farms.

(1) de Snoo G.R. & van der Poll R.J. (1999) Effect of herbicide drift on adjacent boundary vegetation. *Agriculture, Ecosystems and Environment*, 73, 1–6.
(2) Blomqvist M.M., Tamis W.L.M., Bakker J.P. & van der Meijden E. (2006) Seed and (micro) site limitation in ditch banks: Germination, establishment and survival under different management regimes. *Journal for Nature Conservation*, 14, 16–33
(3) Aquilina R., Williams P., Nicolet P., Stoate C. & Bradbury R. (2007) Effect of wetting-up ditches on emergent insect numbers. *Aspects of Applied Biology*, 81, 261–262.
(4) Defra (2007) *Wetting up Farmland for Birds and Other Biodiversity*. Defra BD1323.
(5) Manhoudt A.G.E., Visser A.J. & de Snoo G.R. (2007) Management regimes and farming practices enhancing plant species richness on ditch banks. *Agriculture, Ecosystems & Environment*, 119, 353–358.
(6) Maes J., Musters C.J.M. & de Snoo G.R. (2008) The effect of agri-environment schemes on amphibian diversity and abundance. *Biological Conservation*, 141, 635–645.
(7) Blomqvist M.M., Tamis W.L.M. & de Snoo G.R. (2009) No improvement of plant biodiversity in ditch banks after a decade of agri-environment schemes. *Basic and Applied Ecology*, 10, 368–378.
(8) Davey C., Vickery J., Boatman N., Chamberlain D., Parry H. & Siriwardena G. (2010) Regional variation in the efficacy of Entry Level Stewardship in England. *Agriculture Ecosystems & Environment*, 139, 121–128.
(9) Davey C.M., Vickery J.A., Boatman N.D., Chamberlain D.E., Parry H.R. & Siriwardena G.M. (2010) Assessing the impact of Entry Level Stewardship on lowland farmland birds in England. *Ibis*, 152, 459–474.
(10) Davey C.M., Vickery J.A., Boatman N.D., Chamberlain D.E. & Siriwardena G.M. (2010) Entry Level Stewardship may enhance bird numbers in boundary habitats. *Bird Study*, 57, 415–420.
(11) Leng X., Musters C.J.M. & de Snoo G.R. (2011) Effects of mowing date on the opportunities of seed dispersal of ditch bank plant species under different management regimes. *Journal for Nature Conservation*, 19, 166–174.

1.12   Restore or maintain dry stone walls

- We have captured no evidence for the effects of restoring or maintaining dry stone walls on farmland wildlife.

Background
Dry stone walls are constructed without the use of cement or mortar. They may provide an important habitat for plants and other farmland wildlife.

1.13   Plant new hedges

- Two studies from France and the UK compared newly planted hedges with control areas. Both (including one replicated trial) found newly planted hedges had higher abundance, species richness or diversity of beetles[2, 8] or spiders[8] than crop fields or field margins. The replicated study also found vascular plant species diversity and grass species richness were higher in newly planted hedges than recently established grass field margins[8]. A review found newly established hedges supported more ground beetles than older hedges[3].

- A small-scale study from the UK found that local hawthorn plants exhibited better growth and were more stock proof than those of eight other provenances[7]. A literature review found lower pest outbreaks in areas with new hedges[5]. A replicated study in the UK[6] found that the diversity of arthropods supported by newly planted hedges varied between seven different plant species.

- An unreplicated site comparison study in Germany found that 2 out of 85 ground beetle species used newly planted hedges as stepping stones for dispersal[1]. Results from the same study found that invertebrates that moved passively (attached to mammals and birds), such as snails, benefited most from the hedge-islands compared to actively moving ground beetles and harvestmen[4].

Background
Agricultural intensification, which has included increasing field size and pesticide use, has resulted in a loss of field margin habitats, such as hedgerows. These features can provide a relatively undisturbed habitat for wildlife in intensively managed agricultural landscapes. Hedge planting has therefore been investigated to determine whether it can enhance biodiversity; evidence to date focuses mainly on invertebrates.

An unreplicated site comparison study from 1982 to 1991 in western Germany (1) (same study as (4)) found that only 2 ground beetle (Carabidae) species (out of 85 sampled) used a sequence of young hedge plantations as stepping stones for their dispersal. Two forest or forest-edge ground beetle species, present in nearby semi-natural habitat, gradually appeared along a meadow and hedge strip over the nine years following hedge planting (1982 to 1990). Twenty-five ground beetle species from the semi-natural habitat showed no tendency to use the hedge plantations as stepping stones. In 1982, 9 small hedge islands (each 400 m^2) were planted at intervals along a 10 m-wide meadow strip, attached at one end to mixed wooded and open semi-natural habitats (woods, hedge fragments, ponds surrounded by small reeds, wet and dry meadows) and extending 1.6 km into arable fields. Ground beetles were sampled using six pitfall traps/section in hedge islands and meadow strips from 1982 to 1990. Semi-natural habitats and adjacent arable fields were sampled from 1990 to 1991.

A small-scale study in 1996 in France (2) found that ground beetle (Carabidae) diversity declined with distance from a newly planted hedge in intensive arable farmland. Rare ground beetle species decreased and the most abundant species *Pterostichus melanarius* became more dominant with distance from the hedge. The hedge was planted in 1995 and comprised two 200 m sections of shrubs divided by a 100 m section of mixed fodder crop (oats and cabbages). It was separated from the adjacent barley crop by a 9 m-wide zone planted with oats and sorghum. Ground beetles were sampled using pitfall traps in the hedge (15 traps) and at 10–110 m from the centre of the hedge (4 traps at each of 5 distances). Traps were emptied every 2–4 weeks (April to mid-October 1996). Fenced pitfall traps (12 in the hedge and three 110 m into the crop) were used to estimate absolute densities and were emptied every day for 8 days in June 1996.

A 1999 review of literature (3) found two unpublished studies showing that newly planted hedges supported field species of ground beetle (Carabidae). In one study the youngest hedge, 3 years old, had more ground beetles than 5-, 9- or 40-year-old hedges. Another study in Germany showed that newly planted hedges linking patches of semi-natural habitat were not used as corridors by forest or openland ground beetle species (1).

The same unreplicated site comparison study as (1), between 1982 and 1998, (4) found marked differences in the effectiveness of the hedge-island and meadow habitat strip as a dispersal corridor for four invertebrate taxa: ground beetles (Carabidae), harvestmen (Opiliones), spiders (Araneae) and snails (Gastropoda). Nine years after planting, the hedge-island and meadow strip did not (or not yet) function well as a dispersal corridor for ground beetles or harvestmen. Snails were the best colonizers, with the highest proportion of species migrating to the strip, including target woodland species. The authors suggest that passive travel by small snails on mammals or birds may have contributed to this. Spiders also had a high proportion of immigrating species, but many of them were not present in the source habitat and may have passively 'ballooned' in from the surrounding area, rather than using the strip as a dispersal corridor. The authors concluded that while the hedge islands appeared to be working as stepping stones for species able to travel passively, this was not true for actively moving invertebrates, such as ground beetles or harvestmen, perhaps because of the age, size or connectedness of hedge islands at the time of study. In addition to the sampling regime described in (1), invertebrates were sampled from the surrounding area in 1992–1994 and 1997–1998. Spiders, harvestmen and ground beetles were sampled using pitfall traps and snails were sampled by flotation (in 1984, 1987 and 1990).

A 2000 literature review (5) looked at which agricultural practices can be altered to benefit ground beetles (Carabidae). It included one study (2), which is outlined above, that found a greater diversity of ground beetles near newly planted hedges. Another study (El Titi 1991) of whole farming systems found lower pest outbreaks in areas with new hedges on farms managed under integrated farming.

A replicated study in 1998 and 1999 in mid-Wales (6) found that 7 species planted in 2 hedgerows in semi-upland farmland supported significantly different numbers of arthropods: common gorse *Ulex europaeus* (1,007 arthropods), sessile oak *Quercus petraea* (436), blackthorn *Prunus spinosa* (381), hawthorn *Crataegus monogyna* (258), silver birch *Betula pendula* (180), rowan *Sorbus aucuparia* (110) and ling heather *Calluna vulgaris* (53). Sessile oak was the most diverse in terms of arthropod orders, with 13 out of 15 orders recorded, 2 of which were not found on any other host. Hawthorn and common gorse *Ulex europaeus* were the next most diverse, each with one unique arthropod order. Common gorse *Ulex europaeus*, sessile oak, blackthorn and rowan between them had representatives of all 27 families of beetles (Coleoptera), true bugs (Hemiptera) and moths and butterflies (Lepidoptera) recorded in the study. All planted species had a similar or better growth rate than the commonly planted hawthorn, apart from sessile oak and ling heather. Planting was undertaken in 1996 within the fenced (2 m wide) margins of 2 fields. Margins were divided into eight 6 m plots, which were planted with a double row of 30–40 plants of each species, replicated across 3 blocks. Invertebrates were sampled by tree beating at five points/plot in June, August and September (1998–1999).

A randomized, replicated small-scale study from 1995 to 1997 in mid-Wales (7) found that hawthorn *Crataegus monogyna* plants, propagated from seeds sourced from a local hawthorn population (local provenance), exhibited better growth and had a more stock proof growth form than those of eight other provenances. Local plants had the latest bud-burst, least severe mildew symptoms and more thorns compared to those of other provenances (four British, four continental European). Hawthorn of local provenance grew tallest at the upland site, but was relatively slow-growing at the lowland site. In terms of establishment, fenced plots had lower hawthorn mortality than unfenced, sheep-grazed plots (4% vs 100% mortality respectively) at the upland site. Mortality was low at the lowland site (fenced: 1%, unfenced: 3%). Fenced plots with mulching had approximately 320% greater growth than unmulched sections. One experimental hedge was established at one upland and one lowland site, both sites were grazed by sheep. The 2 hedges had 3 replicate blocks of 4 10 m strips that were either fenced, mulched or fenced and mulched. Within each strip, nine plants of each provenance were planted in a random order in 1995. Each plant was measured (February 1995–1997), scored for powdery mildew *Podosphaera clandestina* (July and August) and the date of bud burst was recorded (1995–1996).

A replicated study in winter 2002 in Oxfordshire, UK (8) found that the total abundance, species richness and diversity of beetles (Coleoptera) and spiders (Araneae), as well as abundance and species richness of rove beetles (Staphylinidae), was higher in hedge bases than in field margins, but there was no difference between recently planted (2–5-years-old) and mature hedgerows (40–60-years-old). Grass cover was lower, but the number of grass species higher, in the bases of recently established hedgerows compared with recently sown grass margins (3–4-years-old). The diversity of vascular plant species was greater in recently established and mature hedgerows, as well as mature field margins (ca. 50-years-old) compared with recently sown grass margins. The bases of recently planted hedgerows had fewer vascular plant species and lower cover of tall perennial wild flowers and mosses compared with mature field margins. Five geographically separate replicates of each of the four habitats were sampled for beetles and spiders in February 2002 by taking 12 soil core samples in a 70 m-long sampling section. Percentage cover of vascular plant species, moss and bare ground was estimated, and biomass (dry matter) and organic carbon content were measured.

(1) Gruttke H. (1994) Dispersal of carabid species along a linear sequence of young hedge plantations. In K. Desender, M. Dufrene, M. Loreau, M. L. Luff & J. P. Maelfait (eds.) *Carabid Beetles: Ecology and Evolution*, Kluwer Academic Publishers, the Netherlands. pp. 299–303.
(2) Fournier E. & Loreau M. (1999) Effects of newly planted hedges on ground-beetle diversity (Coleoptera, Carabidae) in an agricultural landscape. *Ecography*, 22, 87–97.
(3) Kromp B. (1999) Carabid beetles in sustainable agriculture: a review on pest control efficacy, cultivation impacts and enhancement. *Agriculture, Ecosystems & Environment*, 74, 187–228.
(4) Gruttke H. & Willecke S. (2000) Effectiveness of a newly created habitat strip as dispersal corridor for invertebrates in an agricultural landscape. *Proceedings of the Environmental Encounters Series: Workshop on ecological corridors for invertebrates: strategies of dispersal and recolonisation in today's agricultural and forestry landscapes*. Neuchatel, May, 45. pp. 67–80.
(5) Holland J.M. & Luff M.L. (2000) The effects of agricultural practices on Carabidae in temperate agroecosystems. *Integrated Pest Management Reviews*, 5, 109–129.

(6) Hayes M.J., Jones A.T., Sackville Hamilton N.R., Wildig J. & Buse A. (2001) Studies on the restoration of Welsh hedges. *Proceedings of the 10th Annual Conference of the International Association for Landscape Ecology, Hedgerows of the World: Their ecological functions in different landscapes.* Birmingham, UK, 5–8 September. pp. 339–348.
(7) Jones A.T., Hayes M.J. & Hamilton N.R.S. (2001) The effect of provenance on the performance of *Crataegus monogyna* in hedges. *Journal of Applied Ecology*, 38, 952–962.
(8) Pywell R.F., James K.L., Herbert I., Meek W.R., Carvell C., Bell D. & Sparks T.H. (2005) Determinants of overwintering habitat quality for beetles and spiders on arable farmland. *Biological Conservation*, 123, 79–90.

Additional reference
El Titi A. (1991) The Lautenbach project 1978–89: integrated wheat production on a commercial arable farm, south-west Germany. In L.G. Firbank, N. Carter & G.R. Potts (eds.) *The Ecology of Temperate Cereal Fields.* Blackwell, Oxford.

1.14   Protect in-field trees (includes management such as pollarding and surgery)

- We have captured no evidence for the effects of protecting in-field trees on farmland wildlife.

Background
This intervention may involve managing in-field trees, using techniques such as pollarding, or tree surgery.

1.15   Plant in-field trees (not farm woodland)

- We have captured no evidence for the effects of planting in-field trees on farmland wildlife.

Background
This intervention may involve planting trees within fields; it does not involve planting farm woodland.

1.16   Maintain in-field elements such as field islands and rockpiles

- We have captured no evidence for the effects of maintaining in-field elements such as field islands and rockpiles on farmland wildlife.

Background
This intervention may involve maintaining in-field features, such as field islands which can include small patches of woodland or grass, and within field rockpiles.

1.17   Manage woodland edges to benefit wildlife

- We have captured no evidence for the effects of managing woodland edges to benefit wildlife on farmland wildlife.

1.18   Plant wild bird seed or cover mixture

- Thirty individual studies investigated the effects on birds of sowing wild bird seed or cover mix; 21 studies found positive effects. Fourteen studies from the UK (including one systematic review and nine replicated controlled trials, of which four were randomized, and three reviews) found that fields sown with wild bird cover mix had higher abundance[10, 13, 19–21, 25, 27, 28, 35, 38], density[11, 18, 21, 22, 30, 32, 36, 39], species diversity[18, 20, 21] and species richness[22, 27, 28] of birds than other farmland habitats. Six studies from the UK (including one review and two replicated studies) found that birds showed a preference for wild bird cover[33, 40] and used it significantly more than other habitats[1, 4, 9, 48]. One review found the grey partridge population increased substantially on farms where conservation measures including cover crops were in place[3]. Nine replicated studies from France and the UK reported mixed or negative effects of wild bird cover on birds compared to other farmland habitats[12, 14, 17, 23, 29, 41–46]. Six studies found that mixtures including kale[11, 13, 18] or a mixture of kale and/or other species[5, 21, 27] attracted the largest number of bird species or highest bird abundance[22].

- Twelve studies from the UK looked at the effects of wild bird cover strips on invertebrates. Seven studies from the UK (including one review and four replicated controlled studies, of which two were also randomized) found positive effects. Farmland habitats sown with wild bird cover mix were used more by butterflies[4], and had a higher abundance or species richness of butterflies and/or bees[28, 29, 31, 35, 37, 49] than other farmland habitats. One review found wild bird cover benefited invertebrates[26]. Four studies (including one review and two replicated trials) reported mixed or negative effects of wild bird cover on invertebrate numbers compared with other farmland habitats[2, 6–8, 39]. One study found that bees and butterflies showed preferences for particular plant species[34, 47].

- Eight studies from the UK looked at plants and wild bird cover. Six studies (including two reviews and two replicated controlled trials) found that planting wild bird cover mix was one of the three best options for conservation of annual herbaceous plant communities[16], benefited plants[6, 26] and resulted in increased plant diversity[31] and species richness[28, 29, 35]. However two replicated studies (of which one was a site comparison) found mixed/negative effects for plant species richness[2, 24].

- One replicated trial from the UK found that small mammal activity was higher in wild bird cover than in the crop in winter but not in summer[29].

A study of habitat use by yellowhammers *Emberiza citrinella* in 1993, 1995 and 1997 on a mixed farm in Leicestershire, UK (1) found that in summer, yellowhammers used both cropped and uncropped habitats, including wild bird cover, and in winter wild bird cover was used more than all other habitats relative to its availability. In summer, wild bird cover strips (8 m wide) were used significantly more than wheat or field boundaries (2 m wide) but less than barley. In winter, cereal-based wild bird cover was used significantly more than all other habitats and kale-based *Brassica* spp. bird cover was used significantly more than cereal and rape crops. A 15% area of arable land was managed for game birds. Yellowhammer nests were observed for 1.5–2 hours when nestlings were 4–10-days-old and 5–15 foraging trips per nest were plotted in May–June 1993 and 1995. A 60 ha area of the farm was also walked 7 times in November–December and February–March 1997 and habitat use was recorded.

A replicated trial from 1995 to 1998 in Hampshire, UK (2) recorded fewer flowering plant species, bee (Apidae), fly (Diptera) and butterfly (Lepidoptera) species on a single field margin strip sown with wild bird cover seed mix established for three years compared to three strips sown with a diverse wildflower seed mix. There were 20 flowering plant species, 8 bee (Apidae), 3 fly (Diptera) and 3 butterfly (Lepidoptera) species on the single field margin strip sown with wild bird cover seed mix established for 3 years in 1998, and 24, 9, 7 and 8 plant, bee, butterfly and fly species respectively on 3 wildflower seed mix strips in the same study. The wild bird mix strip had more plant species but fewer bee, fly and butterfly species than a single naturally regenerated field margin strip (16, 9, 4 and 6 plant, bee, butterfly and fly species respectively on the naturally regenerated strip). The field margins were established or sown in 1995. Numbers of inflorescences or flowers and flower-visiting bees, wasps (Hymenoptera), flies and butterflies were counted on a 200 x 2 m transect in each strip, once a month from May to August 1998.

A 2000 literature review from the UK (3) found that populations of grey partridge *Perdix perdix* were 600% higher on farms where conservation measures aimed at partridges were in place, compared to farms without these measures (Aebischer 1997). Measures included the provision of conservation headlands, planting cover crops, using set-aside and creating beetle banks.

A small study of set-aside strips from 1995 to 1999 at Loddington, Leicestershire, UK (4) found that set-aside sown with wild bird cover was used by nesting Eurasian skylark *Alauda arvensis* and butterflies (Lepidoptera) significantly more than other habitats. The majority of skylark territories found were within set-aside strips (margins or midfield) sown with wild bird cover (1995: 76%; 1996: 65%; 1997: 71%; 1999: 55%), although the habitat covered only 8–10% of the area. The habitat was also used more for foraging than all habitats, except linseed *Linum usitatissimum*. Transects along wild bird cover set-aside strips also had more butterfly records than any other habitat in 1997 and 1998 (28–40% vs 1–18%). Wild bird cover was sown with either cereal-based or kale-based *Brassica* spp. mixtures. Skylark territories were recorded in 1995–1997 and 1999 and nests were located in 1999 and foraging trips observed for two 1.5 hour periods. Two butterfly transects were walked weekly from April–September.

A replicated, randomized study from 1998 to 2000 of annual and biennial crops in Norfolk, Hertfordshire and Leicestershire, UK (5) found that bird species tended to use a variety of crops. Yellowhammers *Emberiza citrinella* used mainly cereals. Greenfinch *Carduelis chloris* tended to use borage *Borago officinalis*, sunflowers *Helianthus* spp. and mustard *Brassica juncea*. Crops used by several bird species included kale *Brassica oleracea*, quinoa *Chenopodium quinoa*, fat hen *Chenopodium album* and linseed *Linum usitatissimum*. Buckwheat *Fagopyron esculentum* was used a small amount and, apart from greenfinch, few others used sunflower or borage. Crops were sown in a randomized block design with three replicates at each of the three farms. Plots were 20 or 50 m x either 12 or 16 m. Numbers of birds feeding in, or flushed from, each plot were recorded before 11:00 at weekly intervals from October–March 1998–2000.

A review (6) of two reports (Wilson *et al.* 2000; ADAS 2001) evaluating the effects of the Pilot Arable Stewardship Scheme in two regions in the UK (East Anglia and the West Midlands) from 1998 to 2001 found that 'wildlife seed mix' benefited plants, bumblebees *Bombus* spp., bugs (Hemiptera) and sawflies (Symphyta), but not ground beetles (Carabidae). The wildlife seed mix option could be wild bird seed mix or nectar and pollen mix for pollinators, and the review does not distinguish between these mixes. The effects of the pilot scheme on plants, invertebrates and birds were monitored over three years, relative to control areas, or control farms. Only plants and invertebrates were measured within individual options. Wildlife seed mix was the least widely implemented option, with total areas of 106 and 152 ha in East Anglia and the West Midlands respectively.

A replicated study in June 2000 in ten edge habitats on an arable farm in Leicestershire, England (7) found that first-year wild bird cover had the highest density (not significant) of caterpillars (Lepidoptera). Weevil (Curculionidae) densities were similar in first- and second-year wild bird cover but lower than in edges of non-rotational set-aside. Spider (Araneae) and rove beetle (Staphylinidae) densities were lower in wild bird cover than in ungrazed pasture edges. Type of neighbouring crop did not affect invertebrate densities in the different habitats. Apart from the four habitats mentioned above, beetle banks, brood cover, hedge bottoms, sheep-grazed pasture edges, grass/wire fence lines and winter wheat headlands were included in the study. Invertebrates were sampled with a vacuum suction sampler in June 2000. This study was part of the same experimental set-up as (8, 9).

A replicated study from 1995 to 1999 of arable habitats on a farm in Leicestershire, UK (8) found that the abundance of some invertebrate groups was higher in non-crop strips (wild bird cover or grass beetle banks), whereas other groups were more abundant in crops. Four invertebrate groups tended to have significantly higher densities in non-crop strips than crops in all years: spiders (Araneae) 7 vs 1–5 individuals/sample, true bugs (Homoptera) 29 vs 1–4, typical bugs (Heteroptera) 10–58 vs 0–9, and key 'chick food insects' 65 vs 2–10. In three of the years, true weevils (Curculionidae) were found at significantly higher densities in non-crop strips and beans (0–11) than other crops (0–2). In contrast, in 3 or 4 of the years, densities in crops were significantly higher than non-crops for true flies (Diptera) 20–230 vs 25–100 individuals and aphids (Aphididae). Moth and butterfly larvae (Lepidoptera) and ground beetles (Carabidae) differed significantly in only one or two years, when density was higher in crops than non-crops. Total beetles (Coleoptera) varied between years and habitats. Sawfly larvae (Symphyta), leaf beetles (Chrysomelidae) and soldier beetles (Cantharidae) showed no significant differences. Wild bird cover was sown as 2–5 m-wide strips along field boundaries and re-sown every few years with a cereal or kale-based *Brassica* spp. mixture. Grass strips (1 m wide) were sown onto a raised bank along edges or across the centre of fields. Invertebrates were sampled each year in the centre of 5–11 grass/wild bird cover strips and 3 m into 3–4 pasture, 8–12 wheat, 6–8 barley, 3–6 oilseed rape and 4 field bean fields. Two samples of 0.5 m^2 were taken in each habitat using a D-Vac suction sampler in June 1995–1999. This study was part of the same experimental set-up as (7, 9).

A study of different set-aside crops on a farm in Leicestershire, UK (9) found that Eurasian skylark *Alauda arvensis* and yellowhammer *Emberiza citrinella* used wild bird cover set-aside (kale *Brassica napus* set-aside, cereal set-aside, annual/biennial crop strips) more than expected compared to availability. Skylarks also used wild bird cover more than unmanaged set-aside, broad-leaved crops and other habitats. Yellowhammer used wild bird cover strips more than expected. Cereal set-aside wild bird cover was used significantly more than beetle banks, kale set-aside wild bird cover, unmanaged set-aside and other habitats. Wild bird cover strips were used significantly more than kale set-aside, unmanaged set-aside and other habitats. Field margin and midfield set-aside strips were sown with kale-based and cereal-based mixtures for wild bird cover and beetle banks. Other habitat types were: unmanaged set-aside, cereal (wheat; barley), broad-leaved crop (beans; rape) and other habitats. Thirteen skylark and 15 yellowhammer nests with chicks between 3–10-days-old were observed. Foraging habitat used by the adults was recorded for 90 minutes during 3 periods of the day. This study was part of the same experimental set-up as (7, 8).

A small, replicated controlled study from May–June 1992–1998 in Leicestershire, UK (10) found that the abundance of nationally declining songbirds and species of conservation concern significantly increased on a 3 km^2 site where 20 m-wide mid-field and field-edge strips were planted with game cover crops (alongside several other interventions). However, there was no overall difference in bird abundance, species richness or diversity between the experimental and three control sites. Numbers of nationally declining species rose by 102% (except for Eurasian skylark *Alauda arvensis* and yellowhammer *Emberiza citrinella*). Nationally stable species rose (insignificantly) by 47% (8 species increased; 4 decreased). The other interventions employed at the same site were managing hedges, beetle banks, supplementary feeding, predator control and reducing chemical inputs generally.

A replicated, randomized, controlled study over the winters of 1998–2001 on 161 arable farms across England (11) (same study as (18)) found that, overall, all bird species analysed exhibited higher densities on wild bird cover crops than on conventional crops except Eurasian skylark *Alauda arvensis*, which preferred cereal stubbles. Although all species showed non-random and different wild bird cover crop preferences, kale *Brassica* spp. was preferred by the greatest number of species. Additionally, bird abundance was significantly greater on wild bird cover crops located adjacent to hedgerows than those located midfield. Ten annual crops and 4 biennial crops were planted each year at each of 192 sites with 3 replicates/crop. At 11 and 13 sites in 1999–2000 and 2000–2001 respectively, strips containing the same crop were grown in pairs, 1 against a hedgerow and 1 infield, to determine location preference.

A replicated site comparison study of 88 farms in East Anglia and the West Midlands, UK (12) found that between 1998 and 2002 there was no difference in the decrease in autumn densities of grey partridge *Perdix perdix* on farms that planted wild bird cover mixtures and farms that did not. Surveys for grey partridge were made once each autumn in 1998 and 2002 on 88 farms: 38 farms that planted wild bird cover and 50 farms that did not.

A replicated, controlled study over the winters of 1997–1998, 1998–1999 and 2000–2001 on one arable, autumn-sown crop farm in County Durham, England (13) found that farmland bird abundance was significantly higher in wild bird cover crops than commercial crops (420 birds/km^2 in wild bird cover vs 30–40/km^2 for commercial crops). Of 11 species with sufficient data for analysis, all species-year combinations exhibited significant preferences for wild bird cover crops. Of the wild bird cover crops, kale *Brassica napus* crops were preferred by nine species and quinoa *Chenopodium quinoa* crops by six species; cereals and linseed *Linum usitatissimum* were also used. The wild bird cover crops were planted in c. 20 m-wide strips along one edge of arable wheat, barley or oilseed rape fields. There were approximately 15 experimental and 15 control fields. Bird counts were conducted twice monthly from October–March in 1997–1998 and three times per month from October–December as well as twice monthly from January–March in 1998–1999 and 2000–2001.

A replicated, controlled, before-and-after study from 1998 to 2003 (3 years habitat manipulation and 3 years monitoring) in 4 cereal farms (12–20 km^2) in the Beauce, Grande Beauce and Champagne Berrichonne regions, France (14) found that grey partridge *Perdix*

perdix populations were unaffected by cover strips. Neither breeding density nor the reproductive success of breeding pairs increased in managed compared to control areas. The survival rate was significantly lower in managed areas for all winters except for one winter in one site. Observations suggested that cover strips attracted predators, such as foxes *Vulpes vulpes* and hen harriers *Circus cyaneus*, causing the managed land to become 'ecological traps'. Cover strips (500–1,000 ha/farm) were either set-asides or, typically, a maize-sorghum mixture. Partridges were surveyed in March and mid-December to early-January to assess overwinter mortality, and in August to assess reproductive success.

A 2004 review of experiments on the effects of agri-environment measures on livestock farms in the UK (15) found that in one experiment in southwest England (the Potential to Enhance Biodiversity in Intensive Livestock farms (PEBIL) project, also reported in (25)), birds preferred grass margins sown with plants providing seed food and cover, over plots of grassland subject to various management treatments. The review assessed results from seven experiments (some incomplete at the time of the review) in the UK and Europe.

A replicated study in the summers of 1999–2000 comparing ten different conservation measures on arable farms in the UK (16) found that wildlife seed mixtures (site-specific mixture, but largely planted for birds) appeared to be one of the three best options for the conservation of annual herbaceous plant communities. Uncropped cultivated margins and no-fertilizer conservation headlands were the other two options. The average numbers of plant species in different conservation habitats were wildlife seed mixtures: 6.7; uncropped cultivated margins: 6.3; undersown cereals: 5.9; naturally regenerated grass margins: 5.5; no-fertilizer conservation headlands: 4.8; spring fallows: 4.5; sown grass margins: 4.4; overwinter stubbles: 4.2; conservation headlands: 3.5; grass leys: 3.1. Plant species richness was highest in wildlife seed mixtures due to the range of sown species and a high number of annual arable species. Plants were surveyed on a total of 294 conservation measure sites (each a single field, block of field or field margin strip) on 37 farms in East Anglia (dominated by arable farming) and 38 farms in the West Midlands (dominated by more mixed farming). The ten habitats were created according to agri-environment scheme guidelines. Vegetation was surveyed once in each site in June–August in 1999 or 2000 in thirty 0.25 m^2 quadrats randomly placed in 50–100 m randomly located sampling zones in each habitat site. All vascular plant species rooted in each quadrat, bare ground, or litter and plant cover were recorded.

A replicated, randomized study from November 2003 to March 2004 in 205 cereal stubble fields in arable farmland in south Devon, UK (17) found no clear changes in habitat use by seed-eating birds after the establishment of wild bird cover crops on some stubble fields. The target species, cirl bunting *Emberiza cirlus*, made insignificant use of wild bird cover crops (average of two individuals/plot). Only two plots contained more than five individuals and use of the habitat dropped drastically in March, which the authors suggest makes the habitat a poor alternative to stubbles. High numbers of other seed-eating species including chaffinch *Fringilla coelebs* and yellowhammer *Emberiza citrinella* were recorded on the wild bird cover crops, especially those containing a mixture of rape, millet, linseed *Linum usitatissimum*, kale *Brassica* spp. and quinoa *Chenopodium quinoa* (maximum seed-eating bird count 491 on wild bird cover vs 191 on barley fields). Only song thrush *Turdus philomelos* abundance was significantly positively related to wild bird cover presence. However, few stubble fields contained wild bird cover crops (13 fields with 24 wild bird cover strips) and the results may have been confounded by low sample size.

A replicated, randomized, controlled study over the winters of 1998–2001 in 192 plots of arable fields in lowland England (18) (same study as (11)) found significantly higher density and diversity of farmland birds on wild bird cover crops than conventional crops. Although there were no significant differences between wild bird covers containing a single plant species and conventional crops, bird density was 50 times higher on 'preferred' wild bird covers. Kale *Brassica oleracae viridus*-dominated wild bird covers supported the widest range of bird species (especially insectivores and seed-eaters), quinoa *Chenopodium quinoa*-dominated wild bird covers were mainly used by finches and tree sparrows *Passer montanus* and (unharvested) seeding cereals were mainly used by buntings. Sunflowers *Helianthus* spp., phacelia *Phacelia* spp. and buckwheat *Fagopyron esculentum* were the least preferred wild bird covers. All species, except Eurasian skylark *Alauda arvensis*, corn bunting *Miliaria calandra* and rook *Corvus frugilegus*, were significantly denser on wild bird cover. The differences between wild bird covers were more marked in late-winter as kale and quinoa *Chenopodium quinoa* retained seeds for longer periods. Within each plot one wild bird cover and up to four conventional crops were surveyed at least once.

A replicated, randomized, controlled study from June–September 2001–2002 of 21 cereal farms in eastern Scotland (19) found that farmland birds were significantly more abundant on fields containing wild bird cover crops than on fields with conventional crops. A total of 25 species were recorded with up to 80 times more birds seen in wild bird cover than conventional crops. Over all month-crop combinations bird density was significantly higher on wild bird cover crops for all groups except finches in July. Bird density increased steadily over all months of the study on wild bird cover crops but remained relatively constant on conventional crops. Wild bird cover crops contained up to 90% more weed species and 280% more important bird-food weeds than conventional crops. The wild bird cover crops were composed mainly of kale *Brassica* spp., quinoa *Chenopodium quinoa* and triticale *Triticosecale* spp. and were sown in 20 x 650 m strips. A random sample of 4.9 ha of conventional crops was made on each farm.

A replicated, randomized, controlled study from November–February in 2000–2001 and 2001–2002 on 20 arable farms in eastern Scotland (20) found that farmland bird abundance and diversity were significantly higher in fields containing wild bird cover crops (0.6–4.2 ha sampled annually) than fields with set-aside, fields with overwinter stubble or fields with conventional crops. Bird density was up to 100 times higher/ha in wild bird cover crops than on control fields. Wild bird cover crops attracted 50% more species than set-aside and stubble fields and 91% more than conventional fields. Of eight species with sufficient data for individual analysis, seven were consistently significantly more abundant in wild bird cover than in control crops. However, Eurasian skylarks *Alauda arvensis* were significantly more abundant in set-aside and stubble fields. The authors point out that many of the species that favour wild bird cover crops are those currently causing concern because of their declining populations.

A review of the results of four projects conducted from 1998 to 2004 on wild bird cover crops planted in arable farms in England (21) found that the density and diversity of bird species increased significantly when wild bird cover crops were included in the farm. Four studies reported greater use of wild bird cover crops than of commercial crops during winter (October–March). One study reported an increase in bird abundance when wild bird cover crops were introduced into areas that previously lacked them. Kale *Brassica napus* and quinoa *Chenopodium quinoa* were used by the most species. Buckwheat *Fagopyron esculentum* was rarely used by species in any of the studies. Millet was used by more species than any other cereal. Three other studies also found that the location of wild bird covers within the whole-farm configuration had an effect on bird densities. Wild bird covers located close to hedges were favoured. Four studies found that a mixture of wild bird cover crops will produce the highest bird density and diversity.

A replicated, controlled, paired sites study over winter 1997–1998 and summer 1999–2000 in arable farmlands in southern England and the Scottish lowlands (22) found that songbird density and species richness were higher in wild bird cover crops in both seasons. In total, more species were recorded in wild bird cover winter crops than control plots (26 vs 10 species). Similarly, summer wild bird cover crops contained more species than control plots (14 vs 10 species). Songbird abundance was significantly higher on wild bird cover winter (10–50 individuals/ha vs 1) and summer (3 individuals/ha vs 0.4) crops. There was a significantly higher abundance of declining songbird species in the kale *Brassica oleracea* and quinoa *Chenopodium quinoa*, but not cereal wild bird cover crops. Winter wild bird cover plots were sown with kale, quinoa or cereal, while summer wild bird cover plots were predominantly triticale. Thirty experimental and 30 control plots were used in winter, with 6 experimental and 6 control plots in summer.

A replicated study in 1999 and 2003 on 256 arable and pastoral fields across 84 farms in East Anglia and the West Midlands, England (23) found that only 2 of 12 farmland bird species analysed were positively associated with the provision of wildlife seed mixtures, overwinter stubble or set-aside. These were Eurasian skylark *Alauda arvensis* (a field-nesting species) and Eurasian linnet *Carduelis cannabina* (a boundary-nesting species). The study did not distinguish between set-aside, wildlife seed mixtures or overwinter stubble, classing all as interventions to provide seeds for farmland birds.

A replicated site comparison study in 1999 and 2003 in the UK (24) found that 33 field margins sown with a locally specific 'wildlife seed mixture' had greater numbers of perennial plants and pernicious weeds after 4 years, but the total number of plant species did not increase (7–8 plant species/margin). This option was not considered the best option for the conservation of arable plants. The most commonly sown plant species were brassicas (sown at 14 sites). Cereals, maize *Zea mays*, buckwheat *Fagopyron esculentum*, borage *Borago officinalis*, grasses, legumes, teasel *Dipsacus fullonum* and phacelia *Phacelia tanacetifolia* were also sown at some sites. Plants were surveyed in thirty 0.025 m^2 quadrats within a 100 m sampling zone. Percentage cover and plant species were recorded.

A randomized, replicated, controlled trial from 2003 to 2006 in southwest England (25) found that plots of permanent pasture sown with a wild bird seed mix attracted more foraging songbirds (dunnock *Prunella modularis*, wren *Troglodytes troglodytes*, European robin *Erithacus rubecula*, seed-eating finches (Fringillidae) and buntings (Emberizidae)) than 12 control plots, managed as silage (cut twice in May and July, and grazed in autumn/winter). Dunnocks, but not chaffinches *Fringella coelebs* or blackbirds *Turdus merula*, nested in hedgerows next to the sown plots more than expected, with 2.5 nests/km compared to less than 0.5 nests/km in hedges next to experimental grass plots. Twelve experimental plots (50 x 10 m) were sown on 4 farms with a mix of crops including linseed *Linum usitatissimum* and legumes. There were 12 replicates of each management type, monitored over 4 years. This study was part of the same experimental set-up as (27, 37, 46).

A 2007 review of published and unpublished literature (26) found experimental evidence of benefits of wild bird seed or cover mix to plants (one study (16)) and invertebrates (true bugs (Hemiptera) (Gardner *et al*. 2001) and bumblebees *Bombus* spp. (Allen *et al*. 2001)).

A randomized, replicated, controlled trial from 2003 to 2006 in southwest England (27) found that plots of permanent pasture sown with a mix of crops including linseed *Linum usitatissimum* and legumes attracted more birds, and more bird species, than control treatments, in both summer and winter. Three plots (50 x 10 m) were established on each of four farms in 2002, re-sown in new plots each year and monitored annually from 2003 to 2006. Legumes sown included white clover *Trifolium repens*, red clover *T. pratense*, common vetch *Vicia sativa* and bird's-foot trefoil *Lotus corniculatus*. There were 12 replicates of each treatment. This study was part of the same experimental set-up as (25, 37, 46).

A replicated, controlled trial in 2005–2006 in Warwickshire, UK (28) found that field corners or margins sown with a wild bird seed mix had more birds and bird species in winter than all other treatments, and more plant species, bumblebees *Bombus* spp. and butterflies (Lepidoptera) (individuals and species) than control plots sown with winter oats. Fifty-five birds/plot from 4 species on average were recorded on the wild bird seed plots compared to 0.1–1 bird/plot and 0.1–0.7 species on average on control crop plots, plots sown with wild-flower seed mix and plots left to naturally regenerate. There were 11 plant species/m^2, 25 bumblebees and 4 bumblebee species/plot, 25 butterflies and 6 butterfly species/plot on wild bird seed plots, compared to 2 plant species/m^2, no bumblebees, 1 butterfly and 0.9 butterfly species/plot in control cereal crop plots. Each treatment was tested in one section of margin and one corner in each of four fields. The wild bird seed mix (five species) was sown in April 2006 and fertilized in late May 2006. The crop (oats) was sown in October 2005. Plants were monitored in three 1 m^2 quadrats/plot in July 2006. Butterflies, bumblebees and flowering plants were recorded on a 6 m-wide transect 5 times between July and September 2006. Farmland birds were counted on each plot on seven counts between December 2006 and March 2007. The second monitoring year of the same study is presented in (35).

A replicated trial in 2004 and 2005 on four farms in England (29) found that plants, insects, mammals and birds all used sown wild bird seed mix plots more than wheat crop at some times of year. The number of flowers and flowering species, the abundance and number of species of butterflies (Lepidoptera) and the number of bumblebee species *Bombus* spp., were all higher in the wild bird mix than in the crop. Small mammal activity was higher in the wild bird mix in winter (around 25 mammals/100 trap nights in wild bird mix, compared to around 8 in the crop), and higher in the crop in summer (around 10 mammals caught in the crop, compared to less than 1 on average in the wild bird mix). The number of birds and bird species were higher in the wild bird mix than the crop in December and January (around 100 birds of over 3 species per count on average in the wild bird mix, compared to less than 10 birds or <1 species in the crop), but not in February and March. Eurasian linnet *Acanthis cannabina* (at three sites) and reed bunting *Emberiza schoeniclus* (at one site) were the most abundant bird species recorded in the wild bird mix. A seed mix containing white millet *Echinochloa esculenta*, linseed *Linum usitatissimum*, radish *Raphanus sativus* and quinoa *Chenopodium quinoa* was sown in a 150 x 30 m patch in the centre of an arable field (winter wheat) on each of four farms in Cambridgeshire, Bedfordshire, Oxfordshire and Buckinghamshire, in April 2004 and 2005. Plants, bees and butterflies were counted in summer 2005. Small mammals were trapped in November–December 2005 and May–June 2005. Birds were counted once a month between December 2004 and March 2005.

A 2007 systematic review identified five papers investigating the effect of winter bird cover on farmland bird densities in the UK (30). There were significantly higher densities of farmland birds in winter on fields with winter bird cover than on adjacent conventionally managed fields. The meta-analysis included experiments conducted between 1998 and 2001 from two controlled trials and one randomized control trial.

A replicated, controlled, randomized study on 28 arable farms in East Anglia and southern England (31) found that as the area sown with cover crops increased, plant diversity in both regions, numbers of butterflies (Lepidoptera) in East Anglia and bees (Apidae) in southern England increased. Results also suggested that cover crops sown in strips have greater butterfly diversity than those sown in blocks, this did not appear to be the case for bees, but numbers recorded were low in the wet cool summer. One of six treatments was randomly allocated to each farm (two replicates per region): 1.5 ha or 6 ha of project-managed un-cropped land in either strips or blocks, or 1.5 ha or 6 ha of farm-managed uncropped land. Two organic farms were also selected per region. Uncropped land was split into four equal areas comprising a floristically-enhanced grass mix, a plant mix to provide summer cover and foraging (e.g. mustard, legume, cereal mixture), a mix to provide winter cover and foraging (e.g. cereal/kale *Brassica* spp./quinoa *Chenopodium quinoa* mixture) and annual cultivation to encourage annual arable plants. Plants (April and June) and insects were assessed within and at the edge of three fields (cereal crop, non-cereal crop and uncropped field in 2006–2009). Butterfly, bee and hoverfly (Syrphidae) diversity and abundance were recorded during transect walks in July.

A replicated, randomized, controlled study in September, November, December and February in 2004–2005 in 7 grassland farms (87–96% grass) in western Scotland (32) found that songbirds responded significantly more positively to wild bird cover crops in grassland compared to arable regions. Average songbird densities were two orders of magnitude greater in wild bird cover crops than conventional crops (average 51 birds/ha vs 0.2). The average density of songbirds in wild bird cover in the grassland region was more than double that in wild bird cover in the arable region at the same time of year (average 61 and 29 birds/ha respectively). Average bird densities in grassland conventional crops were just 14% of that in the arable region. On each site an average of 1.2 ha of wild bird cover and 10.3 ha of conventional crops was randomly sampled. Arable farm data from a previous study was used for comparison.

A replicated experiment in northeast Scotland over three winters 2002–2005 (33), found that unharvested seed-bearing crops were most frequently selected by birds (28% of all birds despite these patches occupying less than 5% of the area surveyed). For nine species seed-bearing crops were used more than expected (based on available crop area) in at least one winter. Outside agri-environment schemes (the Rural Stewardship Scheme and Farmland Bird Lifeline) cereal stubble was the most selected habitat. In total, 53 lowland farms (23 in Rural Stewardship Scheme, 14 in Farmland Bird Lifeline and 16 not in a scheme) were assessed. Over 36,000 birds of 10 species were recorded.

A randomized, replicated study in 2006 and 2007 in Warwickshire, UK (34) (same study as (47)) found that butterflies (Lepidoptera) and bumblebees *Bombus* spp. displayed different preferences for 13 annual and perennial plant species, 10 of which were typical components of wild bird seed mixtures. In 2006, more butterflies were found in plots sown with lucerne *Medicago sativa* (6.3 butterflies/plot) than plots sown with borage *Borago officinalis* (0.3), chicory *Cichorium intybus* (0.8) and sainfoin *Onobrychis viciifolia* (0.8). More butterfly species were found in lucerne plots (3.5 species/plot) than in borage, chicory, sainfoin and fodder radish *Raphanus sativus* (0.3–0.5). In 2007, red clover *Trifolium pratense* plots had the largest number of butterflies, significantly more than chicory (3.3 vs 0.0 butterflies/plot), whilst all other plant species ranged between 0.3–2.3. In both years, bumblebees were most abundant in phacelia *Phacelia tanacetifolia* plots (134 and 38.5 bumblebees/plot in 2006 and 2007), followed by borage (100 and 32). Crimson clover *T. incarnatum* and sunflower *Helianthus annuus* (37 and 26 respectively) had more bumblebees than other plant species (0–6) in 2006. Red clover plots had more bumblebees (21) than buckwheat *Fagopyrum esculentum*, chicory, linseed *Linum usitatissimum*, lucerne, mustard *Brassica juncea* or sweet clover *Melilotus officinalis* in 2007. The number of bumblebee species recorded in crimson clover, phacelia, borage and sunflower was significantly higher than all other plant species (2.8–4.0 vs 0–1.3 species/plot) in 2006. In 2007 red clover in addition to the 4 species from 2006 had significantly more bumblebee species than mustard (3.0–3.3 vs 0.5 species/plot). Short-tongued bees showed a significant preference for phacelia and borage compared with all other treatments in both years. Long-tongued bees showed a significant preference for crimson clover over all other species apart from borage and phacelia in 2006, and red clover in 2007 (although they also showed a strong preference for crimson clover and sainfoin in 2007). Peak flowering of many important bee forage species was in late July, including phacelia, borage, red clover and sweet clover. Thirteen species were sown in single species stands in 6 x 4 m plots with 4 replicates in May 2006. Annual species were re-established in the same plots in May 2007. Abundance and diversity of butterflies and bumblebees were recorded on transects in each plot six times between July and September 2006 and May and September 2007. On each visit the percentage cover of flowers of all dicot species/plot was estimated.

The second monitoring year of the same study as (28) in the UK (35) found that wild bird seed mix plots had more birds in winter (86 birds/plot, of 6 species on average) than control cereal plots, plots sown with wildflower seed mix or plots left to naturally regenerate (2 birds/plot or less and 0.4–1.6 species/plot on average). Wild bird seed plots also had more bumblebee *Bombus* spp. and butterfly (Lepidoptera) individuals and species than naturally regenerated or control cereal plots and more vacuum-sampled invertebrates than control plots. Wild bird seed plots had 8 plant species/m^2, 40 bumblebees and 4 bumblebee species/plot, 18 butterflies and 6 butterfly species/plot, compared to 3 plant species/m^2, no bumblebees and 1 butterfly/plot on control cereal plots. Control plots had 254 vacuum-sampled canopy-dwelling invertebrates/m^2 on average, compared to 840–1,197/m^2 on other treatments. Plants were monitored in three 1 m^2 quadrats/plot in June 2007. Butterflies, bumblebees and flowering plants were recorded in a 6 m-wide transect 6 times between July and September 2006 and 2007. Invertebrates in the vegetation were vacuum sampled in early July 2007. Farmland birds were counted on each plot on four counts between December 2007 and March 2008. The crop control in year two was winter wheat.

A 2009 literature review of agri-environment schemes in England (36) found that high densities of seed-eating songbirds and Eurasian skylark *Alauda arvensis* were found on land planted with wild bird seed or cover mix and on stubble fields. A survey in 2007–2008 found that densities of seed-eating songbirds were highest on wild bird seed or cover mix, compared to other agri-environment scheme options.

A randomized, replicated, controlled trial from 2003 to 2006 in southwest England (37) found plots on permanent pasture annually sown with a mix of legumes, or grass and legumes, supported more common bumblebees *Bombus* spp. (individuals and species) than seven grass management options. In the first two years, the numbers of common butterflies (Lepidoptera) and common butterfly species were higher in plots sown with legumes than in five intensively managed grassland treatments. No more than 2.2 bumblebees/transect were recorded on average on any grass-only plot in any year, compared to over 15 bumblebees/transect in both sown treatments in 2003. The plots sown with legumes generally had fewer butterfly larvae than all grass-only treatments, including conventional silage and six different management treatments. Experimental plots 50 x 10 m were established on permanent pastures (more than 5-years-old) on 4 farms. There were nine different management types, with three replicates/farm, monitored over four years. Seven management types involved different management options for grass-only plots, including mowing and fertilizer addition. The two legume-sown treatments comprised either a mix of crops sown partly for wild birds, including linseed *Linum usitatissimum* and legumes, uncut, or spring barley *Hordeum vulgare* undersown with a grass and legume mix (white clover *Trifolium repens*, red clover *T. pratense*, common vetch *Vicia sativa*, bird's-foot trefoil *Lotus corniculatus* and black medick *Medicago lupulina*) cut once in July. Bumblebees and butterflies were surveyed along a 50 m transect line in the centre of each experimental plot, once a month from June to September annually. Butterfly larvae were sampled on two 10 m transects using a sweep net in April and June–September annually. This study was part of the same experimental set-up as (25, 27, 46).

A 2009 literature review of European farmland conservation practices (38) found that margins sown with wild bird cover had high numbers of some invertebrates which are important bird food, but lower numbers than on margins sown with a wildflower mix. Cover crops such as quinoa *Chenopodium quinoa* and kale *Brassica oleracea* provided more food for seed-eating birds in late winter than other field margin types and supported large numbers of some songbird species.

A controlled study in 2002–2009 on mixed farmland in Hertfordshire, England (39), found that the estimated population density of grey partridges *Perdix perdix* was significantly higher on land sown with wild bird cover than on conventional arable crops. This study also examined the densities found on land under various agri-environment schemes and set aside (which were higher than those on wild bird cover) and the impact of predator control and supplementary food provision. Grey partridges were surveyed in March and September using dawn and dusk counts starting in 2001. Land cover within the project area was mapped and categorized as: conventional arable land, arable in agri-environment schemes, non-arable, or set-aside (which was further divided into non-rotational; wild bird cover; other rotational).

A 2010 follow-up review of experiments on the effects of agri-environment measures on livestock farms in the UK (40), found that in one experiment in southwest England (the Potential for Enhancing Biodiversity on Intensive Livestock Farms PEBIL project BD1444, also reported in (25)) found small insect-eating birds preferred field margins sown with a diverse mixture of plants that provided seed food, compared to grass margins subject to different management techniques, despite there being no difference in the number of insects between the two sets of treatments. The preference for wild bird cover was attributed to easier accessibility (less dense ground cover). The review assessed results from four experimental projects (one incomplete at the time of the review) in the UK.

A replicated site comparison study in 2005 and 2008 of 2,046, 1 km squares of agricultural land across England (41) (same study as (42)) found that 4 of 8 regions had at least 2 farmland birds that showed positive responses to wild bird cover and overwinter stubble fields. Across all 15 bird species thought to benefit from these interventions, only one region (the northwest) showed significantly more positive responses than would be expected by chance. Some species responded positively in some regions and negatively in others.

A replicated site comparison study in 2005 and 2008 of 2,046, 1 km^2 plots of lowland farmland in England (42) (same study as (41)) found that 3 years after the 2005 introduction of two agri-environment schemes, Countryside Stewardship Scheme and Entry Level Stewardship, there was no consistent association between the provision of wild bird cover and farmland bird numbers. European greenfinch *Carduelis chloris*, stock dove *Columba oenas*, starling *Sturnus vulgaris* and woodpigeon *Columba palumbus* showed more positive population change (population increases or smaller decreases relative to other plots) in the 9 km^2 and 25 km^2 areas immediately surrounding plots planted with wild bird cover mix than in the area surrounding plots not planted with wildlife seed mixture. Although Eurasian linnet *Carduelis cannabina* and rook *Corvus frugilegus* also showed positive associations with wild bird cover mix at the 25 km^2 scale, plots with wild bird cover were associated with a greater decline in grey partridge *Perdix perdix* populations at both scales between 2005 and 2008. The 2,046, 1 km^2 lowland plots were surveyed in both 2005 and 2008 and classified as arable, pastoral or mixed farmland. Eighty-four percent of plots included some area managed according to Entry Level Stewardship or the Countryside Stewardship Scheme. In both survey years, 2 surveys were conducted along a 2 km pre-selected transect route through each 1 km^2 square.

A replicated site comparison study from 2004 to 2008 in England (43) found that the ratio of young-to-old grey partridges *Perdix perdix* was higher in 2007 and 2008 on sites with higher proportions of wild bird cover. Brood sizes were also related to wild bird cover in 2008 only. Overwinter survival was positively related to wild bird cover in 2004–2005, but negatively in 2007–2008. There were no relationships between wild bird cover and year-on-year density trends. Spring and autumn counts of grey partridge were made at 1,031 sites across England as part of the Partridge Count Scheme.

A replicated site comparison study between November 2007 and February 2008 of 52 fields in East Anglia and the West Midlands (44) (same study as (45)) found no difference between the number of seed-eating birds in fields managed under Higher Level Stewardship of the Environmental Stewardship scheme (fields sown with enhanced wild bird seed mix) than in fields managed under Entry Level Stewardship of the Environmental Stewardship scheme (fields sown with wild bird cover mix). In East Anglia, but not the West Midlands, there were significantly more seed-eating birds on fields planted with wild bird cover under the Environmental Stewardship scheme (59.3 birds/ha) than non-Environmental Stewardship fields planted with a game cover (2.1 birds/ha). Seed-eating birds were surveyed on two visits to each site between 1 November 2007 and 29 February 2008.

A replicated site comparison study in winter 2007–2008 on farms in East Anglia and the West Midlands, England (45) (same study as (44)) found that more seed-eating farmland songbirds (including tree sparrow *Passer montanus* and corn bunting *Emberiza calandra*) were found on Higher Level Stewardship wild bird seed mix sites (6–11 birds/ha) than on non-stewardship game cover crops (<0.5 birds/ha) in East Anglia, but not in the West Midlands (2–4 birds/ha on both types). The survey was carried out on 27 farms with Higher Level Stewardship, 13 farms with Entry Level Stewardship and 14 with no environmental stewardship.

A replicated study from April–July in 2006 on four livestock farms in southwest England (46) found that dunnock *Prunella modularis*, but not Eurasian blackbird *Turdus merula* or chaffinch *Fringella coelebs*, nested at higher densities in hedges alongside field margins sown with wild bird seed crops, or barley undersown with grass and clover, compared to those next to grassy field edges under various management options (dunnock: approximately 2.5 nests/km for seed crops vs 0.3/km for grass margins; blackbirds: 1.0 vs 1.3; chaffinch: 1.5 vs 1.4). Margins were 10 x 50 m and located adjacent to existing hedgerows. Seed crop margins were sown with barley (undersown with grass/legumes) or a kale *Brassica* spp./quinoa *Chenopodium quinoa* mix. There were 12 replicates of each treatment; 3 replicates on each farm. This study was part of the same experimental set-up as (25, 27, 37).

A replicated, randomized study in 2006 and 2007 in Warwickshire, UK (47) (same study as (34)) found bee (Apidae) and butterfly (Lepidoptera) abundance and species richness were higher in stands of specific sown plant species. Bumblebee *Bombus* spp. abundance and species richness were significantly higher on plots sown with phacelia *Phacelia tanacetifolia* and borage *Borago officinalis* (32–85 bees/plot) compared to other treatments (1–22 bees/plot). Crimson clover *Trifolium incarnatum* (10–21 bees/plot), sunflower *Helianthus annuus* (10–22) and in 2007 red clover *Trifolium pratense* (20) also tended to have high bee abundances (other plant species: 1–11 bees/plot). Short- and long-tongued bees showed differences in preferences. In 2006, butterfly abundance and species richness were significantly higher in plots with lucerne *Medicago sativa* compared to borage, chicory *Cichorium intybus* and sainfoin *Onobrychis viciifolia*. In 2007 butterfly abundance was higher in red clover compared with chicory, but the number of species did not differ between treatments. Mobile and immobile butterfly species showed differences in preferences. Flowers of buckwheat *Fagopyrum esculentum* were the most abundant followed by phacelia, borage and sunflower in 2006. In 2007 fodder radish, red clover and sweet clover *Melilotus officinalis* also had high flower abundance. Mustard *Brassica juncea* and linseed *Linum usitatissimum* had the least abundant flowers in both years, along with other species each year. Thirteen species were sown in single species stands: nine small-seeded crop species typically sown in wild bird seed mixes and four wild flower species typically sown in pollen and nectar seed mixes. The species were sown in May each year in adjacent 6 x 4 m plots in a randomized block experiment with 4 replicates. Butterflies and bumblebees were sampled by walking transects through each plot on six occasions from May–September. Flower cover was estimated at the same time.

A replicated study on four farms in Gloucestershire and Oxfordshire, England, in 2007 (48) found that grey partridge *Perdix perdix* released in coveys in the autumn used cover crops more frequently than birds released in pairs in the spring. Four farms were studied. Birds were radio-tagged and their positions marked on a 1:5,000 map.

A replicated, controlled study in summer 2008 in northwest Scotland (49) found that croft sections (an agricultural system specific to Scotland, consisting of small agricultural units with rotational cropping regimes and livestock production) sown with a brassica-rich 'bird and bumblebee' conservation seed mix had 47 times more foraging bumblebees than sheep-grazed sections and 16 times more bumblebees *Bombus* spp. than winter-grazed pastures in June. In July the 'bird and bumblebee' mix sections had 248 and 65 times more bumblebees than sections grazed by sheep or both sheep and cattle respectively. The number of bumblebees in July was also significantly higher (4–16 times) in 'bird and bumblebee' sections than in arable, fallow, silage, and winter-grazed pasture sections. The availability of bumblebee forage plant flowers was lower in 'bird and bumblebee' sections than in silage sections in June, but no other significant differences involving the conservation mix were detected. Plant species in the legume (Fabaceae) family were the most frequently visited by foraging bumblebees. Tufted vetch *Vicia cracca* was one of a few plant species favoured by bumblebees and was predominantly found in 'bird and bumblebee' sections in July–August, although it was not part of the seed mixture. Thirty-one crofts located on Lewis, Harris, the Uists and at Durness were included in the study. The 'bird and bumblebee' conservation mix was sown for several bird species and foraging bumblebees, species sown included kale *Brassica oleracea*, mustard *Brassica* spp., phacelia *Phacelia* spp., fodder radish *Raphanus sativus*, linseed *Linum usitatissimum* and red clover *Trifolium pratense*. In addition to the seven management types mentioned, unmanaged pastures were surveyed for foraging bumblebees and bumblebee forage plants along zigzag or L-shaped transects in each croft section once in June, July and August 2008. Foraging bumblebees 2 m either side of transects were identified to species and recorded together with the plant species on which they were foraging. Flowers of all plant species were counted in 0.25 m^2 quadrats at 20 or 50 m intervals along the transects.

(1)   Stoate C. & Szczur J. (1997) Seasonal changes in habitat use by yellowhammers (*Emberiza citrinella*). *Proceedings of the 1997 Brighton Crop Protection Conference – Weeds*. Farnham. pp. 1167–1172.
(2)   Carreck N.L., Williams I.H. & Oakley J.N. (1999) Enhancing farmland for insect pollinators using flower mixtures. *Aspects of Applied Biology*, 54, 101–108.
(3)   Aebischer N.J., Green R.E. & Evans A.D. (2000) From science to recovery: four case studies of how research has been translated into conservation action in the UK. In N.J. Aebischer, A.D. Evans, P.V. Grice & J.A. Vickery (eds.) *Ecology and Conservation of Lowland Farmland Birds*, British Ornithologists' Union, Tring. pp. 43–54.
(4)   Boatman N.D. & Bence S.L. (2000) Management of set-aside to enhance biodiversity: the wild bird cover option. *Aspects of Applied Biology*, 62, 73–78.
(5)   Boatman N.D. & Stoate C. (2002) Growing crops to provide food for seed-eating farmland birds in winter. *Aspects of Applied Biology*, 67, 229–236.

(6) Evans A.D., Armstrong-Brown S. & Grice P.V. (2002) The role of research and development in the evolution of a 'smart' agri-environment scheme. *Aspects of Applied Biology*, 67, 253–264.

(7) Moreby S.J. (2002) Permanent and temporary linear habitats as food sources for the young of farmland birds. In D.E. Chamberlain & A. Wilson (eds.) *Avian Landscape Ecology: Pure and Applied Issues in the Large-Scale Ecology of Birds*, International Association for Landscape Ecology (IALE(UK)), Aberdeen. pp. 327–332.

(8) Moreby S.J. & Southway S. (2002) Cropping and year effects on the availability of invertebrate groups important in the diet of nestling farmland birds. *Aspects of Applied Biology*, 67, 107–112.

(9) Murray K.A., Wilcox A. & Stoate C. (2002) A simultaneous assessment of farmland habitat use by breeding skylarks and yellowhammers. *Aspects of Applied Biology*, 67, 121–127.

(10) Stoate C. (2002) Multifunctional use of a natural resource on farmland: wild pheasant (*Phasianus colchicus*) management and the conservation of farmland passerines. *Biodiversity and Conservation*, 11, 561–573.

(11) Boatman N.D., Stoate C., Henderson I.G., Vickery J.A., Thompson P.G.L. & Bence S.L. (2003) *Designing Crop/Plant Mixtures to Provide Food for Seed-eating Farmland Birds in Winter*. Defra 339.

(12) Browne S. & Aebischer N. (2003) *Arable Stewardship: Impact of the Pilot Scheme on Grey Partridge and Brown Hare after Five Years*. Defra MAO1010.

(13) Stoate C., Szczur J. & Aebischer N. (2003) Winter use of wild bird cover crops by passerines on farmland in northeast England: Declining farmland species were more abundant in these crops which can be matched to the birds' requirements. *Bird Study*, 50, 15–21.

(14) Bro E., Mayot P., Corda É.V.E. & Reitz F. (2004) Impact of habitat management on grey partridge populations: assessing wildlife cover using a multisite BACI experiment. *Journal of Applied Ecology*, 41, 846–857.

(15) Buckingham D.L., Atkinson P.W. & Rook A.J. (2004) Testing solutions in grass-dominated landscapes: a review of current research. *Ibis*, 146, S2 163–170.

(16) Critchley C., Allen D., Fowbert J., Mole A. & Gundrey A. (2004) Habitat establishment on arable land: assessment of an agri-environment scheme in England, UK. *Biological Conservation*, 119, 429–442.

(17) Defra (2004) *Comparative Quality of Winter Food Sources for Cirl Bunting Delivered through Countryside Stewardship Special Project and CS Arable Options*. Defra BD1626.

(18) Henderson I.G., Vickery J.A. & Carter N. (2004) The use of winter bird crops by farmland birds in lowland England. *Biological Conservation*, 118, 21–32.

(19) Parish D.M.B. & Sotherton N.W. (2004) Game crops as summer habitat for farmland songbirds in Scotland. *Agriculture, Ecosystems & Environment*, 104, 429–438.

(20) Parish D.M.B. & Sotherton N.W. (2004) Game crops and threatened farmland songbirds in Scotland: a step towards halting population declines? *Bird Study*, 51, 107–112.

(21) Stoate C., Henderson I.G. & Parish D.M.B. (2004) Development of an agri-environment scheme option: seed-bearing crops for farmland birds. *Ibis*, 146, S203–209.

(22) Sage R.B., Parish D.M.B., Woodburn M.I.A. & Thompson P.G.L. (2005) Songbirds using crops planted on farmland as cover for game birds. *European Journal of Wildlife Research*, 51, 248–253.

(23) Stevens D.K. & Bradbury R.B. (2006) Effects of the Arable Stewardship Pilot Scheme on breeding birds at field and farm-scales. *Agriculture, Ecosystems & Environment*, 112, 283–290.

(24) Critchley C.N.R., Walker K.J., Pywell R.F. & Stevenson M.J. (2007) The contribution of English agri-environment schemes to botanical diversity in arable field margins. *Aspects of Applied Biology*, 81, 293–300.

(25) Defra (2007) *Potential for Enhancing Biodiversity on Intensive Livestock Farms (PEBIL)*. Defra BD1444.

(26) Fisher G.P., MacDonald M.A. & Anderson G.Q.A. (2007) Do agri-environment measures for birds on arable land deliver for other taxa? *Aspects of Applied Biology*, 81, 213–219.

(27) Pilgrim E.S., Potts S.G., Vickery J., Parkinson A.E., Woodcock B.A., Holt C., Gundrey A.L., Ramsay A.J., Atkinson P., Fuller R. & Tallowin J.R.B. (2007) Enhancing wildlife in the margins of intensively managed grass fields. In J. J. Hopkins, A. J. Duncan, D. I. McCracken, S. Peel & J. R. B. Tallowin (eds.) *High Value Grassland: Providing Biodiversity, a Clean Environment and Premium Products. British Grassland Society Occasional Symposium No.38*, British Grassland Society (BGS), Reading. pp. 293–296.

(28) Pywell R. & Nowakowski M. (2007) *Farming for Wildlife Project: Annual Report 2006/7*. NERC report 6441.

(29) Pywell R.F., Shaw L., Meek W., Turk A., Shore R.F. & Nowakowski M. (2007) Do wild bird seed mixtures benefit other taxa? *Aspects of Applied Biology*, 81, 69–76.

(30) Roberts P.D. & Pullin A.S. (2007) *The Effectiveness of Land-based Schemes (Including Agri-environment) at Conserving Farmland Bird Densities Within the UK*. Systematic Review No. 11. Collaboration for Environmental Evidence/Centre for Evidence-Based Conservation, Birmingham, UK.

(31) Anon (2008) *Managing Uncropped Land in Order to Enhance Biodiversity Benefits of the Arable Farmed Landscape*. Home-Grown Cereals Authority Project RD-2004–3137.

(32) Parish D.M.B. & Sotherton N.W. (2008) Landscape-dependent use of a seed-rich habitat by farmland passerines: relative importance of game cover crops in a grassland versus an arable region of Scotland. *Bird Study*, 55, 118–118.

(33) Perkins A.J., Maggs H.E. & Wilson J.D. (2008) Winter bird use of seed-rich habitats in agri-environ-ment schemes. *Agriculture, Ecosystems & Environment*, 126, 189–194.

(34) Pywell R., Hulmes L., Meek W. & Nowakowski M. (2008) *Creation and Management of Pollen and Nectar Habitats on Farmland: Annual Report 2007/8*. NERC report 6443.

(35) Pywell R. & Nowakowski M. (2008) *Farming for Wildlife Project: Annual Report 2007/8*. NERC report 6366.

(36) Natural England (2009) *Agri-environment Schemes in England 2009: A Review of Results and Effectiveness*. Natural England, Peterborough.

(37) Potts S.G., Woodcock B.A., Roberts S.P.M., Tscheulin T., Pilgrim E.S., Brown V.K. & Tallowin J.R. (2009) Enhancing pollinator biodiversity in intensive grasslands. *Journal of Applied Ecology*, 46, 369–379.

(38) Vickery J.A., Feber R.E. & Fuller R.J. (2009) Arable field margins managed for biodiversity conservation: a review of food resource provision for farmland birds. *Agriculture, Ecosystems & Environment*, 133, 1–13.

(39) Aebischer N.J. & Ewald J.A. (2010) Grey Partridge *Perdix perdix* in the UK: recovery status, set-aside and shooting. *Ibis*, 152, 530–542.

(40) Buckingham D.L., Atkinson P.W., Peel S. & Peach W. (2010) New conservation measures for birds on grassland and livestock farms. *Proceedings of the BOU – Lowland Farmland Birds III: Delivering Solutions in an Uncertain World*. British Ornithologists' Union. pp. 1–13.

(41) Davey C., Vickery J., Boatman N., Chamberlain D., Parry H. & Siriwardena G. (2010) Regional variation in the efficacy of Entry Level Stewardship in England. *Agriculture Ecosystems & Environment*, 139, 121–128.

(42) Davey C.M., Vickery J.A., Boatman N.D., Chamberlain D.E., Parry H.R. & Siriwardena G.M. (2010) Assessing the impact of Entry Level Stewardship on lowland farmland birds in England. *Ibis*, 152, 459–474.

(43) Ewald J.A., Aebischer N.J., Richardson S.M., Grice P.V. & Cooke A.I. (2010) The effect of agri-environment schemes on grey partridges at the farm level in England. *Agriculture, Ecosystems & Environment*, 138, 55–63.

(44) Field R.H., Morris A.J., Grice P.V. & Cooke A. (2010) The provision of winter bird food by the English Environmental Stewardship scheme. *Ibis*, 153, 14–26.

(45) Field R.H., Morris A.J., Grice P.V. & Cooke A.I. (2010) Evaluating the English Higher Level Stewardship scheme for farmland birds. *Aspects of Applied Biology*, 100, 59–68.

(46) Holt C.A., Atkinson P.W., Vickery J.A. & Fuller R.J. (2010) Do field margin characteristics influence songbird nest-site selection in adjacent hedgerows? *Bird Study*, 57, 392–395.

(47) Pywell R.F., Meek W.R., Hulmes L. & Nowakowski M. (2010) Designing multi-purpose habitats: utilisation of wild bird seed species by pollinating insects. *Aspects of Applied Biology*, 100, 421–426.

(48) Rantanen E.M., Buner F., Riordan P., Sotherton N. & Macdonald D.W. (2010) Habitat preferences and survival in wildlife reintroductions: an ecological trap in reintroduced grey partridges. *Journal of Applied Ecology*, 47, 1357–1364.

(49) Redpath N., Osgathorpe L.M., Park K. & Goulson D. (2010) Crofting and bumblebee conservation: the impact of land management practices on bumblebee populations in northwest Scotland. *Biological Conservation*, 143, 492–500.

Additional references

Aebischer N.J. (1997) Gamebirds: management of the grey partridge in Britain. In M. Bolton (ed.) *Conservation and the Use of Wildlife Resources*. Chapman & Hall, London. pp. 131–151.

Wilson S., Baylis M., Sherrott A. & Howe, G. (2000) *Arable Stewardship Project Officer Review*. Farming and Rural Conservation Agency report.

ADAS (2001) *Ecological Evaluation of the Arable Stewardship Pilot Scheme, 1998–2000*. ADAS report.

Allen D.S., Gundrey A.L. & Gardner S.M. (2001) *Bumblebees. Technical Appendix to Ecological Evaluation of Arable Stewardship Pilot Scheme 1998–2000*. ADAS, Wolverhampton.

Gardner S.M., Allen D.S., Woodward J., Mole A.C. & Gundrey A.L. (2001) *True Bugs. Technical Appendix to Ecological Evaluation of Arable Stewardship Pilot Scheme 1998–2000*. ADAS, Wolverhampton.

Walker K., Pywell R.F., Carvell C, Meek W.R. (2001). *Arable Weed Survey of the Salisbury Plain Training Area*. Centre for Ecology and Hydrology, Abbots Ripton.

1.19   Plant nectar flower mixture/wildflower strips

- A total of 80 individual studies have in some way investigated the effects of flowering strips on biodiversity. Sixty-four individual studies show some benefits to one or more wildlife groups.

- Sixty-five individual studies reported the effects of flower strips on invertebrates. Of these, fifty reported positive effects. Forty-one studies from eight European countries (including 5 reviews and 23 replicated controlled studies, of which 1 was randomized and 2 site comparisons) found evidence that flower strips had a positive influence on invertebrate numbers with increased abundance[3, 5, 7, 13, 18, 30, 31, 34, 35, 40, 42, 52, 58, 60, 65, 75, 76, 84, 90, 91, 104], species richness/diversity[23, 24, 33, 44, 54, 59, 63, 92, 99], or both[11, 12, 16, 25, 29, 36, 43, 64, 72, 74, 78 ,81, 87, 89, 94, 95, 97, 101, 103]. Ten studies (nine replicated of which two were controlled) found invertebrates visited[8, 21, 48, 53, 88] or foraged on[2, 6, 10, 51, 67] flower strips but did not specify increases/decreases in numbers. Two studies found effects on ground beetles other than changes in numbers. One replicated controlled study showed that ground beetles were more active[4] or had enhanced feeding/reproductive conditions[14] in flower strips. A review found flower strips supported ground beetle species that were rarely found in crops[69]. Fifteen studies reported mixed or negative effects of flower strips on invertebrates[9, 15, 17, 26, 27, 38, 41, 46, 50, 71, 79, 82, 83, 96, 98, 102]. Six studies found no significant effects[1, 32, 39, 47, 49, 55, 56].

- Twenty-one studies looked at the effects of flower strips on plants. Sixteen studies from seven European countries (including ten replicated controlled studies of which one was randomized) found evidence that flower strips had higher plant cover[20], number of flowers[87], diversity[52, 86], and species richness[24, 25, 28, 33, 37, 49, 55, 57, 63, 81, 85, 86, 89, 90, 92, 105]. One review found flower strips benefited plants but did not specify how[50]. Four studies found negative or no effects of flower strips on the number or diversity of plant species[1, 9, 45, 68]. Five studies described the effects of different margin establishment or management techniques on plants[19, 62, 66, 78, 88].

- Seven studies investigated birds and wildflower strips. Four replicated, controlled studies from Switzerland and the UK (two of which were randomized) and one review of European studies found evidence that plots sown with a wildflower or legume seed mix had a positive influence on birds. Flower strips attracted more birds or bird species[22, 80, 99] and the number of birds using flower strips increased over time[77]. Eurasian skylarks preferentially foraged in, and nested in or near, sown weed patches and were less likely to abandon their territories when they included sown weed patches[93]. However one replicated trial in Switzerland[100] found barn owls avoided sown wildflower areas. Two winter recording periods of the same replicated, controlled study in the UK[81, 89] found there were not more bird species or individuals on wildflower plots compared to control margins.

- All five studies investigating the effects of wildflower strips on small mammals (four replicated studies from Switzerland and one review of studies from northwestern Europe) found evidence that small mammals benefit from strips sown with wild flowers or flowers rich in pollen and nectar, with increases in abundance[34, 70, 73], density and species richness[100]. One replicated study from Switzerland reported that most common vole home ranges and core regions of their territories were found within a wildflower strip[61].

- Nineteen studies (of which eight were replicated, controlled) reported positive effects of sowing phacelia[2, 8, 10, 18, 21, 30, 33, 51, 53, 88, 103] and/or other plant species such as borage and red clover[3, 13, 30, 33, 48, 58, 60, 63, 67, 85, 88, 101, 103, 104]. Three replicated studies found that sowing phacelia had negative impacts or no effects on biodiversity[9, 15, 46].

Background

Flowering plants are sown in strips or blocks, providing forage resources for bees and other flower-visiting insects. Increased insect numbers may then provide food for more birds. Nectar flower mixture may include agricultural varieties of flowering plants such as clovers.

See 'Agri-chemicals – Restrict certain pesticides' for a study looking at control of slugs with restricted use of molluscicide in oilseed fields with wildflower strips (Friedli & Frank 1998).

Friedli J. & Frank T. (1998) Reduced applications of metaldehyde pellets for reliable control of the slug pests *Arion lusitanicus* and *Deroceras reticulatum* in oilseed rape adjacent to sown wildflower strips. *Journal of Applied Ecology*, 35, 504–513.

A replicated, controlled study in June to October 1985–1986 and February to August 1987 in a winter wheat field in north Germany (1) could not detect consistent differences in arthropod numbers between sown flower strips (1 m wide), winter wheat strips (12 m wide) and the winter wheat control field. The number of arthropod pest species, weed species cover and crop yield did not differ significantly between the strip types. Arthropod abundance varied greatly over the season, between years and between trapping methods. Five flower strips (1 m wide, each separated by a 12 m wheat strip) were sown in 1.2 ha of a 1.8 ha arable field with a seed mix containing Crimson clover *Trifolium incarnatum*, red clover *T. pratense*, lupin *Lupinus angustifolius* and winter rape *Brassica napa* in May 1985 and 1986. The remaining area of the same arable field was used as a control. Arthropods were sampled using pitfall and yellow bowl traps in all three treatments throughout the season in all three years. Plant biomass, vegetation cover, weed species composition and frequency were monitored monthly.

A replicated study in 1989 in Hertfordshire, UK (2) found that seven species of bumblebee *Bombus* spp., including the long-tongued common carder bee *B. pascuorum*, and one cuckoo bumblebee *B. [Psithyrus] vestalis* foraged on plots sown with phacelia *Phacelia tanacetifolia*. Of observed worker bumblebee visits, 97% were for nectar, not pollen. The plots each flowered for 6 to 8 weeks, with a maximum flower density of more than 4,000 flowers/m^2 on the plot sown in late May. The plot sown in July flowered until early December. Three 9 m^2 plots of phacelia were sown at Rothamsted Research experimental farm in May and July 1989. Bee and flower densities were recorded weekly. Flowers were counted in a 0.25 m^2 area of each plot. Bees were counted at 09:00, 11.00, 13.00 and 15.00 in each plot; their behaviour, species and gender were recorded.

A replicated, controlled study in late June to August 1989 in central Sweden (3) found that margins sown with different mixtures of legumes attracted significantly more bumblebees *Bombus* spp., butterflies (Lepidoptera), flies (Diptera, excluding hoverflies Syrphidae) and honey bees *Apis mellifera* than other habitats. Margins dominated by red clover *Trifolium pratense* were most attractive to bumblebees (299 individuals in red clover margins out of a total of 413 individuals recorded on all margin-types and the control) and butterflies (75 of 242 individuals). Honey bees (2,374 of 2,422 individuals) and flies (excluding hoverflies) (679 of 984 individuals) preferred margins dominated by white melilot *Melilota albus*. Hoverflies did not show significant preferences for any treatment. Turnip rape (*Brassica napa, B. rapa*), white melilot and red clover dominated the honey bee pollen loads in a hive 1 km away. There were 20 experimental plots (2 x 10 m) with 4 replicates of 5 treatments: field margin sown with legume mix dominated by white melilot, field margin sown with legume mix dominated by red clover, naturally regenerated field margin, field margin along ditch containing wild herbs and grasses, and species-rich semi-natural pasture. Flower visiting insects were counted three times a week by slowly walking transects.

A replicated, controlled study in the summers of 1990–1991 in a cereal field in Switzerland (4) found higher recapture rates of three ground beetle species (Carabidae: *Carabus granulatus*, *Poecilus cupreus*, *Pterostichus melanarius*) in wildflower strips (57.7%, 41.8% and 19.8% recaptured in a different trap to the trap of first capture) than in the cereal control area (20.0%, 26.7% and 8.8% recaptured) indicating these species were more active in the wildflower strips. The activity density of four ground beetle species (*P. cupreus*, *Pterostichus anthracinus*, *Pt. melanarius*, *C. granulatus*) was significantly higher in wildflower margins than in the crop. In 1991, two species moved significantly more from the cereal crop to the wildflower margins than vice versa (*P. cupreus* and *Pt. melanarius*). After harvest, only two species *Harpalus rufipes* and *Pterostichus niger* showed a strong association with wildflower strips, with most individuals being recaptured in wildflower strips, irrespective of the habitat they were initially caught in (crop or wildflower strip). Four wildflower strips (1.5 m wide) were studied. Strips were sown in 1989 at 12, 24 and 36 m apart in one part of a winter cereal field. The remaining area of cereal field was used as a control. After establishment, strips were left untreated for three years. Ground beetles were sampled from May–September 1990 and April–July 1991 using a network of numbered pitfall traps (diameter 7 cm) placed in rows in the strips and the cereal field. Captured beetles were individually marked and released about 10 cm from the trap they were caught in. This study was performed within the same experimental site as (5, 11, 12, 14).

A controlled study in winter 1990–1991 in one cereal field in central Switzerland (5) found generally more overwintering arthropods in wildflower strips than in the adjacent cereal crop. Five times more beetles (Coleoptera) were recorded in soil samples from the wildflower strips than from the crop (1,032 vs 209 individuals/m^2 respectively). Similar patterns were found for samples from photo-eclector traps. Rove beetles (Staphylinidae) and ground beetles (Carabidae) were more abundant in wildflower strips than in the crop, although the greatest abundance of both groups was found in conventional field margins. Other arthropod groups such as spiders (Aranae) and mites (Acari) also had higher densities in wildflower strips than in the crop. More arthropods overwintered in wild plants than in cereal stubbles. Of all arthropods found in cereal stubbles, 48% were found in cereal taken from the wildflower strips, 41% in cereals from conventional field margins and 11% from samples within the crop. Five 1.5 m-wide wildflower margins were established around one cereal field in 1989. The margins were sown with a mixture of wild flower species including clover *Trifolium* spp. and species from the Brassicaceae family. Overwintering arthropods were sampled from soil cores and photo-eclectors. Vegetation samples of 22 plant species and the cereal stubbles were taken twice a month from November 1990 to April 1991. Arthropods overwintering in the plants were hatched in the laboratory. Note that no statistical analyses were performed on the data presented in this paper. This paper summarizes a large study which is partly published elsewhere. It was performed within the same experimental site as (4, 11, 12, 14).

A replicated study in 1992–1993 in one arable field in Baden-Württemberg, Germany (6) recorded 58 species of wild bee (Apidae) either nesting or foraging on wildflower plots (sown with 'Tübingen' nectar and pollen mixture), including 11 species of true bumblebee *Bombus* spp. and 5 species of cuckoo bumblebee *Bombus* [Psithyrus] spp. Thirty-five bee species foraged on flowers from the Tübingen wildflower mixture. In total, over 50 herbaceous plant species were recorded in the Tübingen wildflower plots in 1992, and over 60 in Tübingen plots grown over 2 years in 1993. Ladybirds (Coccinellidae), hoverflies (Syrphidae), green lacewings (Chrysopidae) and butterflies (Lepidoptera) were also observed on the sown strips, including the swallowtail butterfly *Papilio machaon*. Three strips of the commercially available 'Tübingen nectar and pollen mixture' (40% phacelia *Phacelia tancetifolia*; 25% buckwheat *Fagopyron esculentum*) were sown at the edge of an arable field. Two strips were sown only in the first year; one strip was sown in both years.

A randomized, replicated trial from 1987 to 1991 in Oxfordshire, UK (7) found that field margins sown with wildflower seed mix had more adult meadow brown butterflies *Maniola jurtina* but not more larvae than unsown margins in two of the three study years. In 1990 and 1991, sown plots had 4–52 meadow browns/50 m and unsown plots 4–10 meadow browns/50 m. In all 3 years, there were more meadow brown butterflies on uncut margins, or margins cut in spring or autumn than in margins cut in summer (sown margins: 4–22 meadow browns/50 m with summer cut; 14–52 meadow browns/50 m without summer cut). There was no difference in the abundance of meadow brown larvae (three larvae/plot on average) between treatments. Two-metre-wide field margins were established around arable fields in October 1987. In 1988 margins were either left to naturally regenerate or sown with a wildflower seed mix (17 wild flower species, 6 grass species, with a wild flower:grass weight ratio of 1:4). Both treatments were rotavated before sowing. Fifty-metre-long plots were managed in one of the following ways: uncut, cut once in June with hay collected, cut in April and June with hay collected, cut in April and September with hay collected, cut in April and June with hay left lying (unsown margins only) or sprayed once a year in summer (unsown margins only). There were six replicates of each treatment. Adult meadow brown butterflies were monitored weekly along walked transects in the experimental plots from June to September 1989 and from April to September 1990 and 1991. Meadow brown larvae were sampled in spring 1991, by sweep netting and visual searching. This study is part of the same experimental set-up as (16, 31, 32, 35, 37, 47, 105).

A replicated, controlled, site comparison study in 1990 in the Kraichgau region, Germany (8) found set-aside fields newly sown with phacelia *Phacelia tanacetifolia* attracted many honey bees *Apis mellifera* (foraging bees not quantified), but no cavity-nesting solitary bees (Apidae) made nests in bundles of reed *Phragmites australis* stems placed in the phacelia-sown fields. In contrast 12 bee species nested in reed stems placed in 2-year-old naturally regenerated set-aside fields mown in late June in the same study. Four set-aside fields were sown with phacelia. Bundles of reed stems for cavity-nesting bees (and wasps Sphecidae, Eumenidae) were placed in the four newly sown phacelia set-aside fields in April 1990 and sampled in October 1990. This trial was part of a larger study (9).

A replicated, controlled study in 1989–1991 in up to 65 arable sites in the Kraichgau region, Germany (9) (same study as (8)) found lower plant species richness and invertebrate abundance on phacelia *Phacelia tanacetifolia* sown set-aside fields than on naturally developed set-asides. Plant species richness was lowest in sown set-asides (10–15 species/49 m^2) and cereal fields (10–17 spp./49 m^2) and highest in orchard meadows (50 spp./49 m^2) and naturally developed set-asides (37–45 spp./49 m^2). Invertebrate numbers from suction samplers were lowest in phacelia-sown set-asides (500 individuals/5 m^2), intermediate in naturally developed set-asides and cereal fields (ca. 1,000 ind./5 m^2) and highest in set-asides sown with clover-grass-mixes (1,500 ind./5 m^2). Invertebrate numbers caught in Malaise traps were highest in rye fields and clover-grass mixes (around 3,000 ind.) and lowest in naturally developed set-asides (1,000 ind.). Further studies and single species comparisons showed that the effect of field type and set-aside age was strongly species/family-dependent. Up to 11 field types (4–5 replicates each) were investigated: 1, 2 and 3-year-old naturally developed set-asides (mown and unmown), 1-year-old set-asides sown with either phacelia or a clover-grass mix, conventionally managed cereal fields (rye and barley), and low-intensity orchard meadows (<30 years old). Plant surveys (3 visits) were conducted in May to October 1990–1991 on one 49 m² permanent quadrat (meadows and sown fields) or on 120 m² (systematically changed in naturally developed fields). Insects were sampled on 4–5 visits in April to October using Malaise traps (20 fields) and suction samplers (61 fields; 3 minute suctions in five 0.25 m² plots).

A replicated, controlled study in June–July 1993 in four pairs of winter wheat plots in Hampshire, UK (10) found a higher proportion of hoverflies (Syrphidae) with phacelia *Phacelia tanacetifolia* pollen in their stomachs in plots with sown phacelia strips than in plots without strips. There was no difference in egg production between female hoverflies in plots with and without *Phacelia* strips. There were non-significant trends of more *Aphidius* spp. parasitoid wasps, other wasps (Braconidae and Proctotrupoidea) and hoverflies in phacelia strips than in the crop. Four pairs of winter wheat plots (minimum size 5 ha; minimum width 100 m) were either managed according to the Integrated Farming System (IFS) or conventionally. At IFS plots, strips of phacelia were sown along the longest edge (300–400 m) in April 1993. Conventional control plots did not have strips. Invertebrates were sampled either using fluorescent-yellow water traps (19 cm diameter) located at different distances from the phacelia strip/field edge or using a D-Vac. Traps were emptied weekly. Five D-Vac samples were taken once in two plots. Hoverflies were dissected and pollen content in the stomach as well as number of eggs in females recorded. This study system was extended and further studied by (18).

A replicated, controlled study in the winter of 1990–1991 in four within-field wildflower strips in a cereal field in the Bernese Seeland, Switzerland (11) found more than four times higher densities and more than twice as many overwintering species of ground beetles (Carabidae), rove beetles (Staphylinidae) and spiders (Araneae) in sown wildflower strips than in the winter cereal areas between them. The proportion of ground beetle and rove beetle larvae was significantly higher in cereal fields than in the wildflower strips. Four wildflower strips (1.5 m wide) were sown in 1989, 12, 24 and 36 m apart in one part of a winter cereal field. The remaining area of cereal field was used as a control. Strips were sown with a variety of wild flowering plants and left untreated for three years. Soil samples (diameter 10 cm, depth 20 cm) were taken 8 times from December to March in the 4 strips and the 3 cereal areas between the wildflower strips. After hand-sorting all samples for arthropods, the samples were extracted in a Berlese apparatus for five days and then hand-sorted again. Beetles were determined to species level, spiders to family level. This study was performed within the same experimental site as (4, 5, 12, 14).

A replicated, controlled study in Switzerland in the summers of 1990 and 1991 in one cereal field (12) found more ground beetle (Carabidae) species in wildflower strips than in the cereal area between these strips. The number of ground beetles was also higher in the wildflower strips, but only during the first year. Both ground beetle abundance and diversity was higher in the cereal area between the wildflower strips than in the control area of the same field. Ground beetle numbers decreased with increasing distance from the wildflower strips. Many of the ground beetle species that were only found in wildflower strips in 1990 dispersed into the cereal areas in 1991. Four wildflower strips (1.5 m wide) were studied. Strips were sown in 1989, 12, 24 and 36 m apart in one part of a winter cereal field. The remaining area of cereal field without strips was used as a control. Strips were sown with a variety of wild flowering plants and left untreated for three years. Ground beetles were sampled weekly throughout the summer using pitfall traps (diameter 7 cm) placed in rows in the strips and the cereal field. This study was performed within the same experimental site as (4, 5, 11, 14).

A series of four studies from 1991 to 1993 in Hampshire, UK (13) found the abundance of some hoverfly (Syrphidae) species was higher in areas with sown flower strips than in control sections and hoverflies preferred foraging on certain plant species. Two trials showed hoverflies foraged on sweet alyssum *Lobularia maritima*, buckwheat *Fagopyrum esculentum*, coriander *Coriandrum sativum*, borage *Borago officinalis*, sunflower *Helianthus annuus* and dwarf marigold *Calendula officinalis* compared to other plant species or field margins. A field-scale trial found no difference in the total number of hoverflies, but more marmalade flies *Episyrphus balteatus* (39 vs 4 individuals/field boundary) in a winter wheat field with a 2 m-wide strip of coriander than a control field. More aphids (Aphidoidea) were found on marked wheat ears in the control field. In 1993 an unreplicated, controlled study found more males and females of 3 hoverfly species/genera (*E. balteatus*, *Metasyrphus corollae* and *Eristalis* spp.) in a 2 m-wide strip (240 m long, divided into 0.75 x 10 m plots) sown with 13 plant species (including amaranthus *Amaranthus* spp., safflower *Carthamus tinctorius* and quinoa *Chenopodium quinoa*) in a spring barley field than in a control strip on the same field between 7 and 14 July. Plant species used for foraging had small (less than 4 mm diameter) white or yellow flowers and easily accessible anthers and pollen (buckwheat, coriander, gold-of-pleasure *Camelina sativa* and texsel *Brassica carinata*). Hoverflies were recorded using transect walks and fluorescent yellow water traps. Ten wheat plants around each yellow water trap were used to count the number of aphids.

A replicated, controlled study from April to July 1991 in Switzerland in one winter rye field (14) found enhanced feeding and reproductive conditions for the ground beetle *Poecilus cupreus* (Carabidae) in sown wildflower strips and cereal strips adjacent to wildflower strips, than in a cereal control area in the same field. Male *P. cupreus* (and females in early season only) had generally higher crop fullness and satiation in the wildflower strip-managed area than in the control area, indicating higher food availability. Females in the wildflower and cereal strips were generally larger and heavier, and had more ripe eggs in their ovaries (except in May), than females in the control area. Ground beetles were sampled weekly using five pitfall traps (diameter 7 cm) in each of the three habitats (wildflower strip, cereal strip and cereal control area). Captured beetles were dissected to analyse different size and reproductive measures and gut contents. This study was performed within the same experimental site as (4, 5, 11, 12).

A replicated, controlled study from April to August 1993 in the Kraichgau region, Germany (15) found that wild bee (Apidae) species richness was lower on set-asides sown with phacelia *Phacelia tanacetifolia* (around 10 species) than on 2-year-old naturally developed set-asides (around 27 spp.) and orchard meadows (around 28 spp.). The number of Red Listed bee species and specialist species were lower on phacelia-sown set-asides than on orchard meadows and naturally developed set-asides. Wild bee abundance in phacelia-sown set-asides (around 75 individuals) was lower than in orchard meadows (around 120 ind.) and 2-year-old naturally developed set-asides (around 100 ind.), but higher than in naturally developed set-asides of different ages. Seven field types (4 replicates each) were investigated in 21 farmland sites: 1-, 2-, 3-, 4- and 5-year-old naturally developed set-asides; 1-year-old set-asides sown with phacelia, and orchard meadows. Wild bees were monitored on six 30 minute visits at each site. Bees were caught using sweep nets (100 sweeps/transect) along one 100 m transect in the field centre. Plant species richness and the abundance of flowering plants was recorded at each visit. Additional plant surveys on a 49 m^2 quadrat were performed in July and August.

A randomized, replicated study from 1989 to 1991 on the Oxford University Farm, Oxfordshire, UK (16) found that butterfly (Lepidoptera) abundance and species richness were higher in sown wildflower margins (21–91 individuals; 7–10 species) than in unsown, naturally generated margins (14–39 ind.; 6–9 spp.) from the second year after establishment. Cutting during summer reduced butterfly diversity and density in the margins, but there were no such effects of cutting in spring and autumn. Cutting in summer also led to an immediate decline in the number of flowering plants directly after the treatment. However, the number of flowers in cut margins had increased by September when it was higher than in uncut margins. Existing field margins (0.5 m wide) were extended by 1.5 m in October 1987. The extended margins were rotavated and either left to naturally regenerate or sown with a wildflower seed mix in March 1988. Fifty-metre-long plots were managed in one of the following ways: uncut, cut once in summer hay collected, cut spring and summer hay collected, cut spring and autumn hay collected, cut spring and summer hay left lying (unsown margins only), sprayed once a year in summer (unsown margins only). There were eight replicates of each treatment. Butterflies were monitored weekly along transects from June to September 1989 and from April to September 1990 and 1991. Transects were divided into 50 m sections corresponding to the experimental plots. This study was part of the same experimental set-up as (7, 31, 32, 35, 37, 47, 105).

A replicated, controlled study in April–June 1993 in one winter rape field near Bern, Switzerland (17) found lower numbers of pest species (mainly pollen beetles *Meligethes* spp. and cabbage weevils *Ceutorhynchus* spp.) near a sown weed strip than near a field boundary, at least early in the season. There was no difference in the abundance of predators and parasitoids between transects near the weed strip and the boundary. A 1.5 m-wide weed strip was sown with a seed mixture containing 25 varieties of annual, biennial and perennial plant species in the middle of a 3.8 ha winter rape field in spring 1992. The composition of the seed mixture was designed to provide flowering plants over the whole growing season. The strip was not cut or treated for three years. Adult and larval arthropods in the rape field were sampled weekly along transects at 3, 10, 20 and 50 m from the weed strips and the opposite field boundary from April–June 1993 using several different methods (visual counting, sweep netting, dissecting of rape pods and using water traps).

A replicated, controlled study in June–July 1993 and 1994 using 4 pairs of adjacent winter wheat plots in Hampshire, UK (18) found higher numbers of total cereal parasitoids (average 33 vs 5/0.5 m²), gamebird chick food insects (25 vs 2) and parasitic wasps *Aphidius* spp. (13 vs 12) in phacelia *Phacelia tanacetifolia* sown strips than in the adjacent crop in one of the study years, 1994. In the same year, the abundance of braconid wasps (Braconidae) was higher in plots with a sown phacelia strip (but not in the phacelia strip itself) than in plots without a strip. No significant differences in numbers of any other arthropod group considered were found between phacelia strips and the other habitats. Four pairs of winter wheat plots (minimum size 5 ha; minimum width 100 m) were either managed according to the integrated farming system (IFS) or conventionally managed. At IFS plots 1 m-wide strips of phacelia were sown (1 g/m²) along the longest edge of the plot (300–400 m) in April each year. Conventional control plots did not have phacelia strips. Invertebrates were sampled either using fluorescent yellow water traps (19 cm diameter) located at different distances into the phacelia strip/field edge or using a D-Vac. Traps were emptied weekly; D-Vac samples (two set-ups) were taken once a year in each plot. Tiller counts were made to assess aphid numbers, species, life-stage and aphid mummies five times yearly in each plot. This study used an extended version of the experimental set-up in (10).

A replicated, controlled, randomized study of a sown wildflower margin at a farm in Oxfordshire, UK (19) found that margin management affected plant species richness. Seventy plant species were recorded in the sown wildflower margin, including 28 of 36 sown species and 42 unsown species. A single cut in June resulted in a significant reduction of sown (2 vs 4 species/quadrat) and unsown species diversity (5 vs 6–8). Grass-specific herbicide did not affect overall species diversity, however sown and unsown grass diversity was reduced and sown and unsown herbaceous plant diversity significantly increased in herbicide-sprayed plots. Unsprayed plots were dominated by black grass *Alopecurus myosuroides*, however the species was eliminated by a December application of herbicide. Sown crested dog's-tail *Cynosurus cristatus* was eliminated by a second treatment of herbicide in April; late mowing in June also decreased this species. The wildflower/grass seed mix was sown on 21 contiguous margin plots (3 m wide by 12 m). Plots were grouped into three blocks, within which they randomly received one of seven treatments: unmanaged, cut in April, cut in April and May, cut in May, cut in May and June, cut in June or grass-specific herbicide (fluazifop-P-butyl) application in April. Cuttings were removed. Half of each plot received grass-specific herbicide application in December. Vegetation in sub-plots was sampled in five 0.1 m² quadrats in July 1995.

A replicated, controlled, randomized study of four field margins in southern and eastern England (20) found that plant cover was higher in margins sown with grass or grass/wildflower mixtures than naturally regenerated margins, and diversity tended to be higher with more complex seed mixtures. Percentage plant cover was significantly higher on spring-sown and Breckland autumn-sown grass or grass/wildflower plots than naturally regenerating plots. Plant cover did not differ with seed mixture diversity or management treatment (unmanaged; cut; grass herbicide), although cover tended to be lower on cut plots in the first year. In 1994 plant diversity was higher in plots sown with more complex seed mixtures (32–37 species) than those sown with grass-only (22–27) or naturally regenerated (21–25). In 1995 grass-seed-only plots tended to be the least diverse (15–21 species), but naturally regenerated plots (18–28) were as diverse as some complex seed mixtures (23–31). Species diversity did not differ between management treatments. Margins were created in each field and divided into 6 plots (4 x 30 m). Each was (randomly) sown with a seed mixture: grass, low cost mix (3 grass: 7 wild flower species), alkaline soil mix (6: 16), neutral soil mix (5: 15), acid soil mix (6: 16) and 1 naturally regenerated treatment. Plots were divided into 10 m sub-plots, which were either unmanaged, cut once, or treated with grass-specific herbicide. Plants were sampled in each sub-plot in summer 1994–1995.

A replicated study in 1994 and 1995 in Hertfordshire, UK (21) found that plots sown with 2 commercial nectar and pollen seed mixtures, Tübingen mixture (40% phacelia *Phacelia tanacetifolia*) and Ascot Linde mixture (25% phacelia) attracted 14 species of bees/wasp (Hymenoptera), including all 6 common UK bumblebee (*Bombus* spp.) species and 3 cuckoo bumblebee species *Bombus* [Psithyrus] spp. across 2 years. A small number of solitary bees (Andrenidae; Megachilidae) of three species (no more than two individuals on any plot) were recorded. The plots also attracted 14 hoverfly (Syrphidae) species and 6 butterfly (Lepidoptera) species. Phacelia attracted 87-99% of all bee visits and 31–98% of all hoverfly visits over the 2 years. Buckwheat *Fagopyrum esculentum*, a nectar source that comprised 20% of both seed mixtures by weight, attracted 1% or less of all bee visits, but up to 36% of hoverfly visits. Phacelia flowered for a long period on all plots. The main flowering period lasted four weeks, but some flowering continued for several months afterwards. The sown species successfully competed with previously existing weeds. In April 1994 Tübingen mixture was sown on a 25 x 25 m plot. In 1995 both mixtures were sown on 19 x 14 m plots: Tübingen mixture sown in April and May; Ascot Linde mixture sown in May and June. In each plot plants and flowers were counted in four 1 m² quadrats. Insect density and diversity were recorded at least three times a week/plot.

A replicated, controlled study in 1995–1996 in Cambridgeshire, UK (22) found that a set-aside strip sown with a mix of 11 wildflower species ('Tübinger Mischung' or 'bee mixture') attracted more birds (average 45–131 individuals) than strips sown with 3 different grass mixtures (18–121 individuals) or a grass and wildflower mixture (33–100 individuals). However the 'bee mixture' attracted the lowest number of bird species (8–15 species). Strips sown with a grass and wildflower mixture attracted more bird species (16–25 species) than the bee mixture, but fewer species than strips sown with a diverse grass mixture (23–33 species). Most of the yellowhammers *Emberiza citrinella* recorded in the study were found on the bee mixture strip. No statistical analyses were performed on these data. Five seed mixtures were sown on 15 set-aside areas (minimum 20 x 100 m) on 1 farm in autumn 1993 and 1994. Only one strip was sown with the bee mixture, three to four strips were sown for all other set-aside strips. Seed mixtures contained: only grass species (3 mixes of 3 to 6 species), mix of grasses and wild flowers (6 grass and 8 wild flower species) or only wild flowers (11 species). Birds were recorded on ten 15 minute point counts between June and September 1995 and July and October 1996. Individual bird locations were recorded in three categories: field boundary, set-aside strip or crop. After each count the strips were walked to flush any birds present but not visible during the count.

A replicated, controlled study in summer 1995 near Göttingen, Germany (23) found higher arthropod species richness on potted mugwort *Artemisia vulgaris* plants placed in sown wildflower strips compared to the cereal field, but not compared to other margin types. The predator-prey ratio in wildflower strips did not differ from the control winter wheat field but was significantly lower than in a 6-year-old uncultivated field margin. The effect of wildflower strips on numbers of individual arthropod species varied between species, with some species (e.g. the aphid *Macrosiphoniella oblonga* and the fruit fly *Oxyna parietina*), but not all found in higher numbers in wildflower strips than in the control. Two types of wildflower strip were sown with either a wildflower seed mixture or a phacelia *Phacelia tanacetifolia* mix. Other margin types were one-year-old and six-year-old uncultivated margins and cereal strips. There were four replicates of each margin type. Potted mugwort plants (four pots) were placed in all margin types and the control. All herbivores and their predators on the plants were recorded during six visits in June and July. In September, all mugwort plants were dissected to assess numbers of arthropods feeding inside the plants. Results from the same study are also presented in (24, 49).

A replicated, controlled study in summer 1995 near Göttingen, Germany (24) found higher species richness of plants and arthropods colonizing potted mugwort *Artemisia vulgaris* and red clover *Trifolium pratense* plants in wildflower strips than in unsprayed cereal control edges. However, the number of arthropod species on mugwort did not differ between any of the other established margin types. The number of arthropod species colonizing red clover flower heads decreased significantly with increasing distance from wildflower strips into adjacent cereal fields, but no such decline was found for individual numbers. Two types of wildflower strips were sown either with a wildflower seed mixture (19 species) or phacelia mixture (*Phacelia tanacetifolia* plus 3 species). Other margin types were one-year-old and six-year-old naturally regenerated margins and cereal strips. Potted mugwort (four pots) and red clover (three pots) plants were placed in all margin types and the controls. Mugwort plants were visited six times in June and July to count all herbivores and their predators on the plants before being taken to the lab in September to assess all arthropods feeding inside the plants. Red clover flower heads were collected five times in June–July and dissected for arthropods living inside the plants. Vegetation of all margins was surveyed in June. Results from the same study are also presented in (23, 49).

A replicated study in 1994–1996 in Gloucestershire, UK (25) found higher plant species richness (23 vs 19 species) as well as higher abundance and diversity of butterflies (Lepidoptera) in sown wildflower margins than in naturally regenerated margins. Cutting and subsequent grazing of the sown margins significantly decreased butterfly diversity (5.6 vs 6.8 species) but not abundance (14.6 vs 16.3 individuals). Margins were established around two organically-managed arable fields by either sowing a seed mix (containing five grasses; six wild flowers) or by natural regeneration in 1994. In 1996 part of the margins were cut in June and grazed in July. The rest was left untreated. Butterflies were monitored along transects weekly from May to September 1996. Plant species and flower abundance were recorded in May and September 1996.

A small replicated, controlled study in Switzerland (26) found that ground beetle (Carabidae) species richness was not significantly higher in sown weed strips than in adjacent crops, but ground beetle species richness decreased with distance from the strips. The oldest weed strip (2 years old) contained the highest number of ground beetle species (10 species/trap), followed by the adjacent rape field (9/trap) and one of the 1-year-old weed strips (9–10/trap), although the differences were not significant. The other one-year-old weed strip had 8 species/trap and other crops 6–8/trap. In 1992, numbers of ground beetle species in rape and wheat plots decreased with distance from weed strips (15% and 35% decreases respectively). Weed strips contained similar numbers of species in their first and second year. Three to five species were found only in the strips. Strips were sown with 25 weed species and were 1 (2 strips) or 2 years old (1 strip); they were not mown. Ground beetles were sampled using 4 pitfall traps/site, emptied every 14 days from April–September 1992 and 1993.

A replicated trial from 1994 to 1996 in central Germany (27) found that few solitary bee and wasp (Hymenoptera: Aculeata) species occupied reed *Phragmites australis* stem nest boxes in set-aside fields sown with a clover-grass mix relative to nest boxes placed in semi-natural grasslands (quantitative details are lacking from the report of this trial). Three replicates in each of five habitat types were studied: set-aside fields (sown with clover-grass mixture), sown field margin strips, extensively-managed grassland, chalk grassland, orchard meadows. Ten reed stem nest boxes were placed in each site. In autumn, nests were dissected and occupants identified. This study is part of the same study set-up as (39).

A replicated trial in 1995 near Wageningen in the Netherlands (28) found that 4 m-wide field margins planted with wild flowers had more plant species than margins left to naturally regenerate or sown with rye grass *Lolium perenne* 2 years after establishment. On average there were 13.7 plant species/0.25 m^2 in wildflower margins, 8.6 in naturally regenerated margins and 5.9 in grass-sown margins. There were fewer plant species in the 1 m of wildflower margin closest to the arable field (11–12 species/0.25 m^2) than in plots more than 1.5 m away (14–17 species); this pattern was not observed in other treatments. Two prominent arable weeds, creeping thistle *Cirsium arvense* and couch grass *Elymus repens*, both had lower biomass in wildflower-sown margins than in naturally regenerated margins (0.1 g/m^2 and 6 g/m^2 respectively in wildflower-sown margins; 33 g/m^2 and 28 g/m^2 in naturally regenerated plots). Wildflower-sown margins had similar couch grass biomass to the grass-sown plots, but much lower creeping thistle biomass (8 g/m^2 of creeping thistle in grass margins). In 1993, 27 experimental plots (8 x 4 m) were established on the boundaries of 3 arable fields. Wildflower plots were sown with 30 broadleaved (non-grass) wildflower species. All plots were mown once a year, without removing cuttings. There were three replicates of each treatment on each field. Plant biomass and species richness were measured in eight 0.5 x 0.5 m plots along a single transect across each margin in August 1995.

A replicated, controlled study in 1994–1996 near Hannover, Germany (29) found higher numbers of spider (Araneae) species and individuals (peak 435 individuals/m^2) in sown wild-flower strips than in cereal fields. Spider abundance varied throughout the year. Spider species richness increased from the first to the third year following margin establishment. Abundance and population dynamics of aphids (Aphididae) on wheat tillers differed between the years (peak 7.5 aphids/tiller), but abundance generally increased with increasing distance from the wildflower strips. Note that no statistical tests were presented in this study. Two wildflower strips (1.5 x 30 m) were sown in 2 different winter wheat fields in 1994, with a seed mixture containing 19 non-grass plant species. The strips were cut annually after harvest. Spiders were sampled with a D-Vac both in the strips and in the fields at defined distances from the strip. Aphid numbers were assessed using tiller counts at the same sample sites.

A replicated study in summer 1996 in central Germany (30) found that both species richness and abundance of spiders (Araneae) caught in sown wildflower strips depended greatly on the species composition of the seed mixtures used. Highest species richness was reported in plots containing phacelia *Phacelia tanacetifolia* and Egyptian clover *Trifolium alexandrinum* (40 spider species) and lowest diversity (30 species) in plots with phacelia, buckwheat *Fagopyrum esculentum*, common sunflower *Helianthus annuus* and common mallow *Malva sylvestris*. Spider abundance was highest in plots containing sundial lupin *Lupinus perenne* and common vetch *Vicia sativa* in both pitfall traps and photoeclectors (155/124 individuals), significantly higher than in both naturally regenerated plots (97/56 individuals) and plots with fodder radish *Raphanus sativus oleiferus* (104/49 individuals). Note that most results in this study are not statistically tested. Eight different types of strip with 3 replicates each were tested: 6 seed mix-tures contained mainly flowering plants (1–12 species), 1 mixture contained mainly grass seeds (2 species plus white clover *T. repens*) and 1 naturally regenerated treatment. Spiders were sampled using two pitfall traps and two photoeclectors in each plot.

A randomized, replicated trial from 1987 to 1991 in Oxfordshire, UK (31) found that field margins sown with a wildflower seed mix had more spiders (Araneae), but not more spider species, than naturally regenerated margins on all dates. Cutting, especially summer cutting, significantly reduced the abundance of spiders. Two-metre-wide field margins were estab-lished around arable fields in October 1987. They were either left to naturally regenerate or sown with a wildflower seed mix (17 wildflower species, 6 grass species, with a wild-flower:grass weight ratio of 1:4) in March 1988. Both treatments were rotavated before sowing. Fifty-metre-long plots were managed in one of the following ways: uncut; cut once in June with hay collected; cut in April and June with hay collected; cut in April and September with hay collected. There were six replicates of each treatment. Spiders were sampled using a suction trap (D-Vac) in September 1987 and 1988 and in May, July and September in 1989, 1990 and 1991. This study was part of the same study set-up as (7, 16, 32, 35, 37, 47, 105).

A randomized, replicated trial from 1987 to 1996 in Oxfordshire, UK (32) found no difference in the number of pseudoscorpions (Pseudoscorpionida) between naturally regenerated field margins and those sown with a wildflower mix. More pseudoscorpions (*Chthonius ischnocheles* and *C. orthodactylus*) were found in unmanaged field margin plots (95 pseudoscorpions in total on sown and unsown plots) than in cut treatments (19–53 pseudo-scorpions). Plots cut in spring and summer had fewer pseudoscorpions than other margins (19 pseudoscorpions on sown and unsown plots). Plots cut just once in June or cut twice but not in June had intermediate numbers of pseudoscorpions (29 and 53 pseudoscorpi-ons respectively). Pseudoscorpions were sampled from the litter layer (not the soil) using a suction trap (D-Vac) in May, July and September 1995 and 1996. This study was part of the same study set-up as (7, 16, 31, 35, 37, 47, 105).

Two replicated trials from 1995 to 1998 in Hertfordshire and Hampshire, UK (33) monitored flower-visiting insects on sown flower strips. One trial (Hampshire 1995–1998) found more flower-visiting insect species and plant species on strips sown with a wildflower mix than on a naturally regenerated margin or a margin sown with wild bird cover mix in 1998. One trial (Hertfordshire 1996–1997) found plots sown with 6 annual plant species were visited by 39 invertebrate species (including bees Apidae, flies Diptera and butterflies Lepidoptera) in the summers after sowing. Wildflower strips attracted 24 invertebrate species, compared to 14 and 19 species on the wild bird strip and naturally regenerated strip respectively. There were 24 flowering plant species on the wildflower strips compared to 20 and 16 on the wild bird strip and naturally regenerated strip (Hampshire). Five plant species attracted many insects or species: wild carrot *Daucus carota*, black knapweed *Centaurea nigra*, oxeye daisy *Leucanthemum vulgare*, bird's-foot trefoil *Lotus corniculatus* and black medick *Medicago lupulina*. Butterflies only visited phacelia *Phacelia tanacetifolia*, borage *Borago officinalis* and marigold *Calendula officinalis* out of six plant species sown in the Hertfordshire study. Short-tongued bumblebees, buff-tailed *Bombus terrestris/lucorum* and red-tailed bumblebees *B. lapidarius/ruderarius* were the most abundant wild bee visitors, and bees were most numerous on phacelia, borage and (second year only) cornflower *Centaurea cyanus*. Five field margin strips were established in the Hampshire study in 1995, three sown with perennial grass and wildflower mix, one with wild bird mix, one naturally regenerated. In the Hertfordshire study, four plots were sown with six annual plant species in 1996 and 1997. In both studies, the number of flowers, flower-visiting bees, wasps (Hymenoptera), flies and butterflies were counted (monthly from May–August 1998 in the Hampshire study, several times a week in the Hertfordshire study). The Hertfordshire study was part of the same study as (48).

A 1999 review of research into field margins in northwest Europe (34) found that numbers of invertebrates and small mammals increased with the establishment of wildflower margins. Three studies reported that 1–1.5 m-wide flower strips resulted in higher numbers of invertebrates within the strips and field as a whole (Klinger 1987; (4); Nentwig 1992). One study in Switzerland found that 3 m strips were used intensively by small mammals and resulted in a population increase of common shrew *Sorex araneus* in spring and summer (Baumann 1996).

A replicated, randomized study in Oxfordshire, UK (35) found that from 1995 to 1996 total numbers of invertebrates and leafhoppers (Auchenorrhyncha) were significantly higher in sown wildflower margins than in unsown, naturally regenerated margins. Cut plots (cut in summer alone, spring and summer or spring and autumn) had significantly lower numbers of all invertebrates, spiders (Araneae), true bugs (Heteroptera) and leafhoppers than uncut plots in all seasons, apart from spiders and true bugs in May. Numbers of all invertebrates were significantly higher in treatments cut twice a year than annually. Cutting in spring and autumn resulted in higher numbers of invertebrates. The abundance of spiders was significantly higher in plots cut bi-annually in spring and autumn than in spring and summer (in July and September samples). Existing field margins (0.5 m wide) were extended by 1.5 m in October 1987. These were rotavated and left to naturally regenerate or sown with a wildflower seed mix. Six management treatments were applied with 6 replicates in a randomized block design on 50 m-long plots: uncut, cut once in summer; cut in spring and summer; cut in spring and autumn; cut in spring and summer (hay left lying); sprayed once a year in summer. Invertebrates were sampled using a D-Vac suction sampler at 10 m intervals along each plot in May, July and September in 1995–1996. This study is part of the same study design as (7, 16, 31, 32, 37, 47, 105).

A 1999 review of literature (36) found that four experimental studies (5, 11, 14, 26) found higher numbers or species diversity of ground beetles (Carabidae) in sown wildflower strips in cereal fields.

A replicated, randomized study from 1987 to 1992 in Oxfordshire, UK (37) found that species richness and abundance of sown plant species were higher in 1.5 m-wide extensions to margins than the original margin sections (0.5 m-wide). Species richness of sown wild-flowers was significantly higher in new compared to old sections (3.6–6.3 vs 0.1–0.9/species quadrat), frequencies of species showed the same pattern. After three years, the original margin sections only had 20% of the species found in the new margins. Cutting in spring and autumn increased the number of species (6–7/quadrat), whereas under other treatments numbers declined sharply in the first year after sowing (from 6 to 4 species) and remained significantly lower (uncut: 4; summer cut: 3–4; spring/summer cut, 4–5). There was no significant difference between numbers in margins cut once or uncut. Individual species showed a range of responses to cutting regimes. Plants were sampled in 3 permanent quadrats (50 x 100 m) at 10 m intervals in existing and new sections of margins. Relative frequencies were recorded as presence/absence in eight sub-sections of the quadrat four times/year from July–September. This study is part of the same study design as (7, 16, 31, 32, 35, 47, 105).

A randomized, replicated controlled trial from 1993 to 1996 near Bristol, UK (38) found that 4 m-wide field margins sown with a nectar flower mixture had more suction-sampled invertebrates, but not more ground beetles (Carabidae), than control cropped margins or margins sown with grass. There were around 200 invertebrates/sample on margins sown with a wildflower/grass mix and naturally regenerated margins, compared to 110–130 invertebrates/sample on control or grass-sown plots. Wolf spiders (Lycosidae) were more abundant on grass and wildflower-sown margins than on control or naturally regenerated margins (numbers not given). There was no difference in the number of ground beetle species (average 8 species/plot), nor in the numbers of the four most commonly caught ground beetle species, between margin types. In a 2 m-wide margin there were more over-wintering invertebrates in the soil of the wildflower-sown half than the naturally regenerating half, but this difference was not found in 4 m-wide replicated experimental plots. Three field margins were established in spring 1993. Experimental plots 10 x 4 m were either sown with arable crop (control), rye grass *Lolium perenne* or a wildflower and grass seed mix, or left to naturally regenerate. There were three replicate plots in each margin. All plots were cut annually after harvest, and cuttings left in place. Another 100 x 2 m-wide field margin, 50 m sown with a wildflower mix and 50 m unsown, was used to monitor wintering invertebrates. Ground beetles were sampled in eight pitfall traps in or near each margin for one week in June for four years. Invertebrates were sampled using a vacuum sampler on plots in two of the three margins in June 1994. Arthropods were extracted from soil samples taken from plots in two margins in December 1993 and February 1994.

A replicated study in 1994–1996 near Göttingen, Germany (39) found no significant differences in the body mass and sex ratio of red mason bees *Osmia rufa* in sown wildflower strips on set-aside land compared to field margins (mostly naturally regenerated) and three types of grassland. Overall, female body mass was correlated with flower availability. Sex ratio was correlated with female body mass; relatively more female larvae were found in habitats with large females. Ten artificial nesting aids for solitary bees were placed in five arable habitats (set-asides sown with wildflower seed mixes, mostly naturally regenerated field margins, extensive bio-dynamic grasslands, chalky grasslands and orchard meadows). There were three replicates in each habitat type. Unparasitized cocoons of *Osmia rufa* were weighed and sex determined in the lab. This study is part of the same study set-up as (27).

A 2000 literature review (40) looked at which agricultural practices can be altered to benefit ground beetles (Carabidae). It found four studies (1, 4, 11, 14) showing that wildflower strips increased ground beetle numbers in adjacent cereal fields.

A replicated, controlled study in summer 1997 and 1998 in Switzerland (41) found that the number of spider (Araneae) and butterfly (Lepidoptera) species in wildflower strips sown on set-aside areas did not differ significantly from winter wheat fields. The number of spider species in wildflower strips was among the lowest reported (average 20 species), significantly lower than in low-intensity meadows (more than 25 species) and forest edges (more than 35 species). Butterfly species richness in the wildflower strips (more than 6 species) differed significantly only from forest edges (less than 4 species). However, for both taxa, wildflower strips attracted some species that were never or only rarely found in other habitats. The investigated habitat types were forest edges, arable fields (winter wheat and intensively managed meadow) and ecological compensation areas including hedgerows, extensively managed and low-intensity meadows, wildflower strips on set-aside land and orchard meadows. There were 109 sites in two arable regions. Spiders were collected in pitfall traps in May and June 1997. Butterflies were observed during 6 visits (10 minutes; covering 0.25 ha) in each site in 1998. This was part of the same study as (54).

A replicated, controlled study in winter 1995–1996 in northwest Switzerland (42) found significantly higher abundances of arthropods in sown wildflower strips than in adjacent arable habitats in two of three paired sites on two arable farms. Species numbers were generally higher in the wildflower strips but this was not statistically tested. Many of the most frequent arthropod species were pest predators, e.g. rove beetles (Staphylinidae), ground beetles (Carabidae) and spiders (Araneae). Two of the wildflower strips (5 years old; 4–5 m wide) were paired to winter wheat fields on an integrated farm. Both strips were sown with grass-clover mixtures and an additional 14 wildflower species and cut 2–3 times a year. The third wildflower strip (2 years old; 3 m wide) was on an organic farm and paired to a ploughed strip (formerly wildflower strip; 6 months old). In each habitat, 24 soil samples were taken 3 times during the study period using a soil borer (25 cm depth; 8 cm diameter). In arable habitats soil samples were taken 30 m parallel to the field margins.

The initial findings of a controlled, replicated site comparison study of the Swiss Ecological Compensation Areas scheme in 1999–2005 in Switzerland (43) found more ground beetles (Carabidae) and ground beetle species in wildflower strips than in adjacent arable crops. The same was true for ground beetle species with specific habitat requirements. Ground beetles were sampled using funnel pitfall traps on 11 wildflower strips and comparison crop strips. Plants, ground beetles, spiders (Araneae), butterflies (Lepidoptera), grasshoppers (Orthoptera) and breeding birds were monitored on grasslands in 3 case study areas of around 5 km².

A replicated, controlled study in April to July 1999 in Witzwil, Switzerland (44) found higher species richness of ground beetles (Carabidae) and spiders (Araneae) in three sown wildflower strips than in adjacent crop fields. However, the number of individuals was generally lower in strips than in the crop. Abundance and species richness of specialist species was clearly higher in strips than crop fields for both ground beetles and spiders. On some of the sites, species richness (especially for spiders) appeared higher in the edge samples of the crop fields than in the centre. Note that no statistical analyses were performed on the data in this study. Wildflower strips (2-, 5- and 6-years-old) were sown with a mixture of 38 native wild and cultivated plant species and were not managed during the sample period. The crop fields (winter wheat; summer wheat and rye) were treated with pesticides in autumn 1998 and spring 1999. Ground beetles and spiders were caught in pitfall traps. Four traps each were placed in the wildflower strip. In the crop field, traps were placed 15 and 70–100 m from the wildflower strip.

A site comparison study in 1996 in Wiltshire, UK (45) found that coppiced and gapped-up hedges had higher plant diversity than those with adjacent sown grass and grass/wildflower strips. Hedges with adjacent sown strips had lower abundances of pernicious weed species. Sixty hedgerows on two neighbouring arable farms were studied. All 23 sampled hedges on Noland's Farm were trimmed annually and had the vegetation at the hedge base cut. The 37 sampled hedges on Manor Farm were trimmed in alternate years, and nine were coppiced and gapped-up. Hedge vegetation was assessed in 25 m-long plots in the middle of a field edge on both sides of each hedge in June.

A replicated study in 1993 in Germany (46) found that 1-year-old set-aside fields sown with phacelia *Phacelia tanacetifolia* had similar numbers of bees (Apidae) but fewer bee species (13 species/field on average) than 1- to 5-year-old naturally regenerated set-aside fields (15–29 species/field). Bees found on phacelia were mainly common species of bumblebee *Bombus* spp. and the solitary bee genus *Lasioglossum* spp., whereas several endangered and specialized bees were found foraging on naturally regenerated set-aside. The percentage cover of flowers did not differ between ages of set-aside but was higher (more than 25% cover of flowers) in phacelia-sown set-aside fields and old meadows than on naturally regenerated set-aside (around 10%). The following field types were studied: 1- to 5-year-old naturally regenerated set-aside fields; 1-year-old phacelia-sown set-aside fields; >30-year-old orchard meadows. There were 4 replicates of each field type, with a total of 28 sites. Fields were set-aside after harvest in autumn and mown in July. Orchard meadows were mown once or twice in June–August. Between May and August percentage of flower cover was estimated on five occasions. Bee species and the flowering plant species visited by the bees were surveyed 6 times between April and August on 30 minute transects in each field.

A randomized, replicated trial from 1987 to 1996 in Oxfordshire, UK (47) found no difference in numbers of predatory sheet web spider *Lepthyphantes tenuis* between field margins sown with a wildflower mix and naturally regenerated margins. In September, when most of the spiders were caught, there were significantly fewer *L. tenuis* individuals in margins (sown and unsown) that were cut in June (around 10 individuals/m^2) compared to more than 15/m^2 in plots cut in spring and autumn, or not cut. In May and July, plots with a recent cut (April- or June-cut treatments respectively) also had lower numbers of *L. tenuis* than other plots. *L. tenuis* individuals were counted in invertebrate samples collected using a suction trap (D-Vac) in May, July and September 1990, 1991, 1995 and 1996. This was part of the same study design as (7, 16, 31, 32, 35, 37, 105).

A replicated study in 1996 and 1997 in Hertfordshire, UK (48) (same study as (33)) found that plots sown sequentially from mid-April to mid-July with a mix of six annual flowering species (cornflower *Centaurea cyanus*, common mallow *Malva sylvestris* (both native), borage *Borago officinalis*, buckwheat *Fagopyrum esculentum*, marigold *Calendula officinalis* and phacelia *Phacelia tanacetifolia*) provided continuous forage for pollinators from mid-June to mid-November. The mix attracted 15 bee (Apidae), 17 fly (Diptera) and 6 butterfly (Lepidoptera) species and the common wasp *Vespula vulgaris*. The most numerous insects were the honey bee *Apis mellifera* and red-tailed bumblebee *Bombus lapidarius/B. ruderarius* (not distinguished in the study). Abundance of flies varied over the season, while abundance of butterflies was low. Butterflies and bumblebees *Bombus* spp. preferred borage and phacelia, while solitary bees and flies preferred marigold. Mallow and buckwheat did not contribute much to flower density or pollinator diversity. Four plots (22 x 14 m or 20 x 13 m) were sown each year (91 or 22 kg/ha) at monthly intervals, then harrowed and irrigated as necessary.

Flower density was recorded weekly in four random 1 m^2 quadrats in each plot. Pollinators were recorded in the outer 3 m of each plot on 21–34 days from the mid/end of June to the end of October/beginning of November.

A replicated study from April to September 1995 near Göttingen, Germany (49) (same study as (23, 24)) found that sown wildflower strips had higher plant species richness and could suppress the abundance of aggressive arable weeds. However, arthropod species richness and abundance in wildflower strips did not differ from the other margin types. Both the abundance and species richness of arthropods found on red clover *Trifolium pratense* plants in wheat fields decreased with increasing distance from the margins, however the decrease in abundance was less pronounced in fields with sown wildflower strips where dispersal from the margin into the field was higher than for control margins. Five margin types (3 m wide; 100–150 m long) around 4 cereal fields were studied: sown with a mixture of 19 wild flower species; sown with a phacelia *Phacelia tanacetifolia* mixture; 1-year-old naturally developed; 6-year-old naturally developed; control strips sown with winter wheat or oats. Potted plants of mugwort *Artemisia vulgaris* (four pots/margin) and red clover (three pots/margin) were used to study plant-arthropod communities. Red clover pots were also arranged in cereal fields 4, 8 and 12 m from wildflower strips to assess dispersal. Mugwort pots were set out in May and visited weekly to count all arthropods feeding inside the plants, leaf miners and galls. In September the plants were dissected and all larvae and pupae found inside the plants were individually reared in the lab to estimate parasitization rates. Red clover pots were set out in April. At five visits in June and July, flower heads were sampled and dissected, and larvae and pupae found inside the plants were reared in the lab for species determination.

A 2002 review (50) of two reports (Wilson *et al.* 2000; ADAS 2001) evaluating the effects of the Pilot Arable Stewardship Scheme in two regions (East Anglia and the West Midlands) from 1998 to 2001 found that 'wildlife seed mix' benefited plants, bumblebees *Bombus* spp., bugs (Hemiptera) and sawflies (Symphyta) but not ground beetles (Coleoptera). The wildlife seed mix option could be nectar and pollen mix for pollinators or wild bird seed mix, and the review does not distinguish between these. The effects of the pilot scheme on plants, invertebrates (bumblebees; true bugs; ground beetles; sawflies) and birds were monitored over three years, relative to control areas, or control farms. Only plants and invertebrates were measured within individual options. Wildlife seed mix was the least widely implemented option with total areas of 106 and 152 ha in East Anglia and the West Midlands respectively.

A replicated trial in Switzerland from 1996 to 1999 (51) found 26 honey bees *Apis mellifera*/m^2 and 0.2 bumblebees *Bombus* spp./m^2 foraging on a plot sown with phacelia *Phacelia tanacetifolia* in 1996. Two plots sown with 50% white clover *Trifolium repens*, one in each of 1998 and 1999, had 1.7 and 3.9 foraging honey bees/m^2 respectively. All 3 plots were located on one trial farm; plots measured approximately 0.3 ha. Five to six honey bee colonies were established adjacent to the plots several days before surveying.

A small-scale replicated, controlled trial in summer 2000 in North Yorkshire, UK (52) found significantly more bumblebees *Bombus* spp. and butterflies (Lepidoptera) on four 6 m-wide margins sown with a grass and wildflower seed mix than on four naturally regenerated, grass-sown or control cropped margins. Spring numbers of ground beetles (Carabidae) and ground-dwelling spiders (Araneae) were higher in all treatments compared with cropped margins. Margins sown with a grass and wildflower mix harboured more pollen beetles *Meligethes* spp. than naturally regenerated margins. Plant diversity was higher in margins sown with a grass and wildflower mix. Four margins of winter cereal fields (all adjacent to hedges) on 2 farms were split into 72 m-long plots and sown in September 1999 with either a grass mix, a grass and wildflower mix, cereal crop or left to regenerate naturally. Ground and canopy-dwelling invertebrates, bumblebees, butterflies and plants were surveyed from late April to late September 2000 using pitfall traps, sweep netting, transects and quadrats.

A small-scale study in 1994 in Vestby, Norway (53) found the number of bumblebees *Bombus* spp. visiting a single 2 x 210 m strip sown with phacelia *Phacelia tanacetifolia* peaked at 237 individuals (0.6 bumblebees/m^2) on 17 July, and gradually declined to 93 bumblebees (0.2/m^2) on 28 July. Maximum numbers of honey bees *Apis* spp. foraging on the phacelia strip were recorded on 14 July with 3,739 honey bees (9.0/m^2); honey bee abundance declined steadily after 18 July with the lowest numbers recorded on 28 July (22 honey bees). The strip was sown in May 1994 along the boundary of a cereal field and a 'habitat island' (area of semi-natural habitat within the farmed landscape). Bees were surveyed over a three week period (5–28 July).

A site comparison study in 1997 and 1998 in the region of Rafz, Switzerland (54) (part of the same study as (41)) found that butterfly (Lepidoptera) species richness was significantly higher in wildflower strips planted as Ecological Compensation Areas than in intensively managed wheat fields. Eleven wildflower strips and 20 wheat fields were sampled. Butterflies were observed for 10 minute periods on 0.25 ha of each site on 5 occasions from May to August 1998, between 10:00 and 17:30 on sunny days with temperatures of at least 18°C.

A replicated trial in 2001–2002 at three sites in the UK (55) found that margins of sugar beet *Beta vulgaris* fields sown with wildflowers had more plant species, but not more invertebrates (individuals or species) than margins sown with grasses, crops, or margins left to naturally regenerate. Wildflower margins had 35 plant species/m, compared to around 17 spp./m in naturally regenerated margins, 15 spp./m in grass margins and 6–11 spp./m in barley *Hordeum vulgare* or beet margins. Fertilized wildflower-sown margins, tested at 2 sites, had fewer species than those without fertilizer, around 30 plant species/m. Naturally regenerated margins had more invertebrate individuals (>1,700 caught) and invertebrate groups (45 groups) than other margin types. However, the difference in invertebrate numbers between different treatments was fairly small (>900–>1,700 individuals; 35–45 groups caught). In autumn 2001, 50 x 6 m margins at the edges of beet fields were planted in 3 beet growing regions with either sugar beet, spring barley, grasses (8 species), wildflowers (20% of seeds by weight from 20 species) or allowed to naturally regenerate. At two sites a sixth margin of wildflowers with nitrogen fertilizer applied was established. There were two replicates of each treatment at each site. In summer 2002, plants (including crop plants) were counted in the margins and invertebrates sampled using pitfall traps, set for two weeks.

A replicated, controlled, before-and-after study from 2000 to 2002 in winter cereal fields on four UK farms (56) found no significant effects of flower-rich margins or an aphid sex-pheromone treatment (nepetalactone, designed to enhance Aphidiinae parasitoid wasps that use aphids as their hosts) on aphid populations due to cold and wet weather at the beginning of summer 2001, resulting in small parasitoid populations. Invertebrates were sampled by *in situ* counting, suction-sampling, pan-traps and pitfall traps along four transects in three fields (with or without wildflower-rich/tussocky grass margins) on each farm. Sex-pheromone was released from formulated strips in the grassy field margin in autumn and in the crop in spring, starting autumn 2000. This study was part of the same experimental set-up as (60).

A replicated, controlled study from 1988 to 1997 in south-central Sweden (57) found higher plant species richness in two plots sown with wildflowers (32 wildflower and grass species) than in plots planted with rose bushes *Rosa canina*, sown with a clover *Trifolium pratense* and grass seed mix or in adjacent untreated field boundaries (control), 9 years after establishment. At the third site (organic), plots sown with wildflowers and/or planted with rose bushes had lower weed and couch grass *Elytrigia repens* cover compared with untreated field boundaries, naturally regenerated plots and plots sown with a clover and grass mix, seven years after establishment. In two of the field boundaries, total weed cover decreased in all treatments except 'clover and grass' where it remained stable or increased. Couch grass cover increased in all treatments in two of the boundaries. Plant species richness tended to decline in most treatments over time, however in experimental plots sown with wildflowers and/or planted with rose bushes, 14–20 of the original 32 wildflower species were still present 7 or 9 years after establishment. In 'clover and grass' plots the clover component decreased or totally disappeared, while sown and unsown grasses and weeds increased. At the organic site, wildflower sown plots and naturally regenerated plots had similar species richness but different species compositions due to a high cover of annual weeds in the naturally regenerated plots. Four replicates of three–four treatments were established in experimental plots on each field boundary in 1988 or 1990, either by widening an existing boundary or re-establishing a previously removed dirt road (organic site). All plots were cut annually in late summer and the cuttings removed. Vegetation surveys were carried out twice in experimental plots (1991–1993 and 1997) and once in untreated field boundaries (1997) in three to five 0.25 m^2 quadrats. It is not clear whether the results for clover and grass plots were a direct result of planting nectar flowers or grass.

A replicated, controlled trial in 2000–2002 in North Yorkshire, UK (58) found that 6 m-wide field margin plots sown, or half-sown with a native 'grass and wildflower' seed mix supported significantly more bumblebees *Bombus* spp. than margins sown with a 'tussocky grass' mix, or control cropped field margins. Wildflower-sown margins supported significantly more bumblebees than naturally regenerated margins, but only in the first year of the three-year study and this difference was not significant when data were averaged across all three years. Wildflower sown margins supported consistently high numbers of bumblebees, whereas naturally regenerated margins had one bumper year for bumblebees and were poor in the other two years. The three most popular forage plant species were cornflower *Centaurea cyanus*, bird's-foot trefoil *Lotus corniculatus* and spear thistle *Cirsium vulgare*. The study was carried out on three arable field margins of one farm. Each margin was split into five 72 x 6 m plots and each plot subjected to 1 of 5 treatments: naturally regenerated, sown tussocky grass mix, sown grass and wildflower mix, split treatment of 3 m tussocky grass and 3 m grass and wildflower mix, or cropped to the edge. Bumblebee activity was surveyed using a standard 'bee walk' methodology.

A replicated, controlled study in May to September 2000 and 2001 in the suburban area of Vienna, Austria (59) found more ground beetle (Carabidae) species in sown wildflower strips on set-aside land than in arable fields and on unsown set-asides. Ground beetle community composition differed between the three habitat types. No statistical analyses were presented in this paper. There were six farmland sites. Wildflower strips (four sites) were sown on set-aside land with the 'Voitsauer' seed mix containing 25 species of wildflowers and weeds between 1998 and 2000. The two unsown set-asides aged 6 and >50-years-old were cut regularly. Typical crops for the region were sown on five arable fields. One of the arable fields was under conservation contract, growing a wildflower seed mix undersown in rye. Ground beetles were sampled using 4 pitfall traps 10 m apart in each habitat and site. There were five sampling periods each year, each lasting two to three days (2001) or seven days (2000).

A series of three replicated, controlled studies from 2000 to 2003 in the UK (60) monitored beneficial invertebrates and aphids (Aphidoidea) in crop fields with and without wildflower margins. All three studies found some beneficial invertebrates were more abundant in fields with flower-rich margins. One study (2000–2003; four farms) found more ground beetles

(Carabidae) and *Pterostichus* spp. and smaller aphid populations in fields with flower-rich margins than those with tussocky grass margins, or no margins. Numbers of adult aphid-eating hoverflies were similar in winter wheat fields with and without flower margins. In the same study in 2003 (four new locations with different crop types; same study as (56)) more aphids were parasitized and there were more parasitoids in a broccoli field with a flower-rich margin than in one with control (no field margins) and areas treated with a pheromone to enhance numbers of aphid-specific parasitoid wasps (Aphidiinae). In a 2002 study on four fields more predatory invertebrates were found next to flower-rich set-aside strips than conventional field margins in mid-July (margins 24 m wide, including phacelia *Phacelia tanacetifolia*, sunflowers *Helianthus* spp. and yellow sweet clover *Melilotus officinalis*). More cereal aphids, but fewer rove beetles (Staphylinidae) were found next to flower-rich margins in wheat or pea fields respectively. Cereal aphid numbers were unaffected by field margins. The study was continued in four winter wheat and four pea fields. It found that abundance of ground beetles, *Harpalus* spp., field overwintering and predatory invertebrates and ground beetle species richness was higher in pea fields with flower-rich set-aside strips than fields with control margins. However numbers of field overwintering invertebrates, predatory invertebrates, rove beetles, *Pterostichus* spp. and rove beetle species richness were lower in wheat fields with wildflower strips. Pea aphid *Acyrthosiphon pisum* abundance did not differ between pea fields with or without set-aside strips. Numbers of some hoverflies or aphids were unaffected by the presence of sown margins in all studies. Invertebrates were sampled using a range of methods, including suction sampling, pan traps and pitfall traps. The 2000–2003 study consisted of four transects in three winter wheat fields: one field with wildflower-rich margin, one with tussocky grass margin, one with neither.

A replicated study in 2000 and 2001 south of Bern, Switzerland (61) found that most home ranges and core regions of common vole *Microtus arvalis* territories lay within a 6-year-old wildflower strip (91% total home range, 100% core region found within wildflower strip). Thus, vole activity in the adjacent crop fields (maize and wheat in 2000, maize and sugar beet in 2001) was very low. Vole home ranges in the wildflower strip were small (median size 125 m^2) compared to findings from other studies and habitats. The authors suggest that an abundance of food in the wildflower strip may account for the small range size. Daily home range sizes were stable between days (overlap of 61–99%). The wildflower strip (130 x 6 m) was dominated by tansy *Tanacetum vulgare*, Fuller's teasel *Dipsacus fullonum*, wild parsnip *Pastinaca sativa* and grasses, and had not been mown. In total 118 voles were captured using Longworth traps. Radio-transmitters (2 g) were attached with a nylon cable around the animals' neck. Each vole was tracked every 60 seconds using automatic telemetry for 1 to 5 days. Data from 40 voles tracked for at least 24 h with >100 bearings per day were analysed. Individual home ranges were analysed in 2000 and 2001; in 2002 above ground activity patterns of 20 voles were analysed.

A replicated study from 2001 to 2004 in Switzerland (62) showed that both seed composition of wildflower mixtures and micro-climatic conditions influenced the emerging plant community in sown wildflower strips. The number of plant species established from sown mixtures was relatively high (around 25 species/25 m^2) in dry to moist sites, whereas fewer species (around 15 species/25 m^2) established on wet or shady sites. Seed mixtures containing high proportions (>75%) of grasses often resulted in grass-dominated margins. Problematic weed species established but did not dominate any strip community. No clear effects of cutting could be shown during the four study years. Eighty wildflower strips on 35 farms were studied. The strips were hand-sown in April and May 2001 with 4 types of seed mixture, each mixture adapted to the micro-climatic conditions at the site: mixture with legumes; mixture without legumes; grass component of 75%; grass component of 90%. Additional strips were established in 2003 using seed mixtures with 20% or 40% grasses. Strips were cut once, twice or not at all in their first year and annually in August from their second year onwards. Cuttings were removed. Half the strip was left uncut. Vegetation was surveyed annually in June from the second year after establishment.

A replicated, controlled, paired sites comparison study in 2003 in central and eastern England (63) found that bumblebee *Bombus* spp. foraging activity and species richness were significantly enhanced at 28 uncropped field margins sown with a 'wildlife seed mixture', compared to paired control sites of conventionally managed cereal or 16 'conservation head-lands'. Wildlife seed mixture margins contained significantly more grass, non-grass and perennial plant species than control sites, with over double the total number of species. Flowering herbaceous plants were more abundant and diverse in wildlife seed margins, and these margins provided the widest range of forage species. This result was dependent upon key forage species being included in the seed mixture, including red clover *Trifolium pratense*, bird's-foot trefoil *Lotus corniculatus* and borage *Borago officinalis*, the latter being of particular importance to short-tongued bumblebee species such as *Bombus terrestris* and *B. lucorum*. The seed mixture contained grasses and annual and perennial broad-leaved herbs. Nineteen farms were surveyed in East Anglia and 17 farms in the West Midlands. Three agri-environment scheme (Arable Stewardship Pilot Scheme) options were studied: field margins sown with a wildlife seed mixture (28 sites); conservation headlands with no fertil-izer (16 sites); naturally regenerated field margins (18 sites). Fifty-eight conventional cereal field margins were used as a control and paired with Arable Stewardship Pilot Scheme sites. Bumblebees were surveyed along 100 x 6 m or 50 x 6 m transects twice, in July and August. Vegetation was surveyed in twenty 0.5 x 0.5 m quadrats.

A replicated study in June and September 2002 in Yorkshire, UK (64) found that beetle (Coleoptera) abundance and species richness were strongly influenced by the type of seed mixture sown on experimental plots. A mix containing mainly flowering plants but no tus-sock grass species ('Fine grass and forbs mix') had fewer beetles and beetle species than a mix containing wildflowers, fine and tussock grass species ('Tussock grass and forbs mix') and a mix with tussock and fine grass species but no flowering plants ('Countryside Steward-ship mix'). Beetle diversity (Shannon-Weiner index) did not differ between the different seed mixes. Plant communities in the grass-only and tussock grass and wildflower mixes were more similar to each other than to the fine grass and wildflower mix. The three seed mixtures were each randomly sown on three of nine experimental plots in each of five blocks on one farm in autumn 2001. Plots measured 25 x 5 m. Seed mixes contained 3–7 grass and 0–19 wild-flower species. The strips were cut once in July with cuttings left *in situ*. Plant diversity and cover and vegetation structure were surveyed in June and September 2002 using 0.5 x 0.5 m quadrats and a 'drop disk'. Beetles were sampled using a Vortis (Burkland Ltd., UK) suction sampler. Five samples (15 suctions for 10 seconds) were taken in each plot (total area sampled 1.32 m^2) on each visit. This study was extended in (91).

A replicated study in 2005 in the province of South Holland, the Netherlands (65) found that natural enemies (parasitoids, hoverflies (Syrphidae) and predatory bugs (Hemiptera)) were generally more abundant near sown flower strips than further into the crop. There were no clear effects of sown field margins and flower strips on pest populations. No figures were presented. A total of 15 km of perennial field margins and flower strips were sown along field edges and across fields on five neighbouring farms in a 400 ha area. Flower availability, natural enemy and key pest densities were measured in 2005.

A replicated study from 2000 to 2006 in England (66) found red clover *Trifolium pratense* and other legumes tended to establish well after hay, rich in red clover, was spread over former arable fields, however, seeds of these species may already have been present in the seedbank. Leguminous species tended to increase in abundance in three fields between 2003 and 2006 (red clover present in 18–55% quadrats in 2003; 44–90% of quadrats in 2006). However, in the 2 other fields, both red and white clover *T. repens* decreased (red clover present in 25–34% quadrats in 2003 to 2–11% in 2006; white clover 25–97% in 2003 to 22–48% in 2006). Of the undesirable weeds, creeping thistle *Cirsium arvense* (87–98% in 2003 to 37–94% in 2006) and spear thistle *C. vulgare* (10–54% in 2003 to 0–38% in 2006) tended to decrease in abundance between 2003 and 2006. Common ragwort *Senecio jacobaea* increased in 3 fields (0–15% in 2003 to 6–36% in 2006) and declined in 1 (19% in 2003 to 0% in 2006). Hay and cuttings were obtained from nearby farms and the study site and were spread over fields once between April and August 2000–2003. Fields were summer grazed by livestock. The presence of species was recorded in 100 random nested quadrats (1 x 1 m and 2 x 2 m) in each field until 2006.

A replicated trial in 2001–2003 on a farm in North Yorkshire, UK (67) found highest bumblebee *Bombus* spp. abundance on plots sown annually with a cover crop mix of five herbaceous species (including borage *Borago officinalis*, fodder radish *Raphanus sativus* and common melilot *Melilotus officinalis*). Short-tongued bumblebees (*B. terrestris*, *B. lucorum* and *B. pratorum*) strongly preferred this annual seed mixture over two perennial grass and wildflower seed mixes. Long-tongued bumblebee species *B. hortorum* and *B. pascuorum* preferred the perennial grass and wildflower seed mixtures, but were not more abundant on the 'diverse' than the 'basic' mix. On average 70% of pollen collected by buff-tailed bumblebee workers *B. terrestris* was from borage and 76% of pollen collected by common carder bee *B. pascuorum* workers came from red clover *Trifolium pratense*. Five 6 x 30 m plots of each seed mixture were established in April 2001 (the annual seed mixture plots re-sown each year after ploughing). Bumblebees were monitored May–August 2002–2003 in 4 x 30 m transects down the centre of each plot. Both perennial grass and wildflower mixes contained 5 grass species, the basic mix contained 3 herbaceous species (black knapweed *Centaurea nigra*, bird's-foot trefoil *Lotus corniculatus* and red clover) and the diverse mix contained 18 herbaceous species.

A replicated trial from 2001 to 2004 in Belgium (68) found that field margins sown with wildflower mix had similar numbers of plant species to naturally regenerated margins after three years. The number of plant species decreased over time in sown plots (from 22–23 species/plot in July 2002 to 13–16 species/plot in July 2004) and the proportion of legumes also decreased. The relative abundance of perennial plants increased and the relative abundance of annuals decreased over time on all the field margin plots, regardless of treatment. In September 2001, 10 m lengths of two 10 x 180 m arable field margins were either sown with 77 commercially available wildflower/grass species (mix 1), sown with 63 native, locally sourced wildflower and grass species (mix 2) or left to naturally regenerate. One margin was in a sunny location; the other shaded by trees. The margins were mown twice each year in late June and September, from 2002 to 2004. Each combination of treatments was replicated three times. Plants were recorded in July and October from 2002 to 2004.

A 2006 review on the effects of the Swiss Ecological Compensation Areas scheme in Switzerland (69) found wildflower strips sown with native flowers on set-aside land attracted ground beetle species (Carabidae) that were never or only rarely found in wheat fields. No details on study design, monitoring techniques or other methods were given.

A replicated, controlled study in September 2004 in four regions north of the Alps in Switzerland (70) found slightly more montane water vole *Arvicola terrestris scherman* hills in sown wildflower field margins than in conventional field margins and sown wildflower strips on set-aside land. Significantly more vole hills and holes were recorded in the different types of field margin than in the crops or on the edge of crop fields. Montane water voles were the most commonly recorded species (98% of observations); common voles *Microtus arvalis* made up 2% of observations. The European mole *Talpa europaea* was not recorded at any site. Three types of field margin were compared: 17 sown field margins (5 x 120 m) established in 2001–2003 with seed mixtures containing native wildflowers, grasses and legumes, 11 conventional field margins generally species-poor and cut several times yearly (0.5–2 x 100–200 m) and 7 wildflower strips sown on set-aside land (at least 5 m wide and 120 m long). On each site 9 plots (5 x 5 m) were investigated: 3 plots on the field margin, 3 on the edge of the crop field and 3 plots 10 m into the crop. Vole and mole hills and holes were counted on three visits. Shape and distribution of the hills and holes were used to distinguish between the three species.

A replicated, controlled study in 2002 and 2004 (April–July) in central Switzerland (71) found more spiders (Araneae) and ground beetles (Carabidae) in wildflower strips and extended field margins than in permanent road margins. There were more ground beetle species in wildflower strips and extended field margins than in road margins, whereas there were fewer spider species in wildflower strips than in road margins. No statistical analyses were performed on the data. Four extended field margins (1-year-old in 2002), 4 sown wildflower strips (1- to 4-years-old) and 4 permanent meadow strips (road margins, less than 10-years-old) in 2 different regions on 12 farmland sites were compared. All sites were 100–250 m long and 0.5–5.0 m wide, except for one 50 m-wide wildflower strip. No information was provided about seed mixtures used for margin establishment, however the existing vegetation on the sites was either grass dominated or a species-poor to species-rich flora dominated by flowering herbs. Arthropods were sampled using pitfall (funnel) traps placed in groups of 4 at least 10 m apart in each site. The traps were emptied weekly for three weeks in April–May and two weeks in June–July.

A replicated, controlled trial in 2004 in thirty-two 10 km grid squares across England (72) found significantly more bumblebee species *Bombus* spp. in field margins sown with wildflower or 'pollen and nectar' seed mixes (more than 3 bumblebee species/transect) than in grassy margins (1.3–1.4 spp./transect) and control cropped margins (0.1 spp./transect). Pollen and nectar margins had more individuals (86 bees/transect) than any other treatment. Wildflower margins had more individuals (43/transect) than grassy (6–8/transect) and control cropped margins (0.2/transect). Field margins were 6 m wide and part of agri-environment scheme agreements. Five field margin types were investigated: grass and wildflower mix (sown between 1999 and 2003); 'pollen and nectar'-rich margin (sown between 2002 and 2003); grass mix (sown between 1993 and 2000); grass mix (sown between 2002 and 2003); control cropped margins. Wildflower mixes were variable in species composition but typically consisted of perennial wildflowers, fine-leaved and tussock-grass species, sown in an 80% grass:20% wildflower ratio by weight. 'Pollen and nectar' mixes typically consisted of at least four nectar-rich wildflower species (such as clover *Trifolium* spp. and bird's-foot trefoil *Lotus corniculatus*) and 4 grass species sown in an 80% grass:20% wildflower ratio by weight. All 5 margin types were surveyed within each 10 km grid square (excluding the grass and wildflower mix which was not present in all squares), giving a total of 151 margins. Bumblebees were counted on a 100 x 6 m transect in each field margin, once in July and once in August.

A replicated, controlled study in the summer of 2003 in central Switzerland (73) found higher densities of small mammals (mainly common voles *Microtus arvalis*) in wildflower strips than in low-intensity meadows, conventionally farmed artificial grasslands and autumn-sown wheat fields. Small mammal species richness in wildflower strips was equal to that in conventionally farmed habitats and low-intensity meadows, but lower than in herbaceous strips. Over the summer, small mammal density increased most in the wildflower strips and herbaceous strips than in low-intensity meadows, conventionally farmed artificial grassland and autumn-sown wheat fields. Wildflower strips (fallows sown with seed mixtures of native plants) and herbaceous strips (consisting mainly of herbaceous plants such as thistles *Cirsium* spp., common teasel *Dipsacus sylvestris*, St John's wort *Hypericum perforatum*, common mallow *Malva sylvestris* and mulleins *Verbascum* spp.) were not cut regularly during the growing season, whereas the other grassland habitats were cut at least twice. A capture-recapture method was used to estimate small mammal densities. Small mammals were trapped and individually marked during 60-hour trapping sessions in March, May and July.

A replicated, controlled trial from 2001 to 2004 at 6 sites across central and eastern England (74) found 6 m-wide margins of cereal fields sown with pollen and nectar flower mixture supported significantly more foraging bumblebee *Bombus* spp. species and individuals than cropped, grassy or naturally regenerated field margins. Bumblebees included the long-tongued species *B. ruderatus* and *B. muscorum*. Wildflower mixture supported significantly more foraging bumblebee species and individuals than cropped field margins, including conservation headlands, in all three years of monitoring, and grassy or naturally regenerated unsown field margins in years two and three. In the third study year (2004) wildflower and pollen and nectar mixtures supported similar numbers of bumblebee species and individuals. Wildflower margins had more flowers in May–June than July–August (approximately 4,200 vs 2,000 forage flowers/plot). Pollen and nectar margins had few flowers in May and June when bumblebee queens of late-emerging species are foraging but a large number of flowers in July–August (2,000 vs 9,000 forage flowers/plot). The number of flowers declined in the nectar and pollen mix in year three. Flower numbers remained similar between years two and three in the wildflower mix. Native varieties of red clover *Trifolium pratense* and bird's-foot trefoil *Lotus corniculatus* flowered earlier than agricultural varieties. Two experimental plots (6 x 50 m) were established in each field along 2 margins. There were 6 treatments: wildflower mixture (21 native wildflower; 4 fine grass species); pollen and nectar mixture (4 agricultural legume; 4 fine grass species); tussocky grass mixture; conservation headland; natural regeneration; crop (control). Foraging bumblebees were counted in May–late August on 6 m-wide transects 6–11 times/margin. Flower abundance was estimated along bumblebee transects in 2002, 2003 and 2004.

A 2007 review of two studies in England (75) found that margins sown with a nectar and pollen mix consistently attracted more foraging bumblebees *Bombus* spp. than other field margin options. Two replicated controlled trials (72, 74) monitored the use of arable field margins sown with grass, wildflower and pollen and nectar seed mixes. One used 6 sites (74), the other 32 sites (72). Both studies found higher numbers of bumblebees on margins sown with pollen and nectar mix, although the number of bumblebee individuals and species increased over time in the wildflower mix in one study (74) and supported higher numbers of some species in the other (72). The review recommends pollen and nectar mix for a rapid and positive impact on the number of foraging bumblebees but suggests that wildflower mix is important in catering for a wider range of bumblebee species across the whole season.

A replicated, controlled trial from 2003 to 2005 in eastern and central England (76) found that forage patches sown with a 20% legume seed (clovers *Trifolium* spp. and bird's-foot trefoil *Lotus corniculatus*): 80% grass mix attracted significantly higher densities of bumble-bees *Bombus* spp. than control patches of non-crop vegetation typical of the site (average 26 bumblebees/200 m^2 on sown forage patches compared to 2 bumblebees/200 m^2 on control patches). Honey bees *Apis mellifera* and cuckoo bumblebees (*Bombus* [Psithyrus] spp.) were not found in greater densities on forage patches. The study also showed that bumblebee densities on sown forage patches were higher in areas with a greater proportion of arable land in a surrounding 1 km radius than in landscapes with less arable and more grassland, woodland and urban habitats. Eight areas with varying proportions of arable, grassland, woodland and urban areas in the surrounding landscape were studied. Four treatments were established in each area from autumn 2003 to spring 2004: sown forage patches of 0.25, 0.5 and 1 ha and 1 control patch of non-crop vegetation typical of the area. Bumblebees and honey bees were surveyed monthly from May to September 2005 on two 2 x 100 m transects in each forage patch.

A randomized, replicated, controlled trial from 2002 to 2006 in eastern England (77) (same study as (82, 83)) found that the number of birds using sown wildflower margins in summer increased by 29% between 2003 and 2006. The management of sown wildflower field margins affected bird use more than the seed mix used. Bird densities were higher on disturbed and grass-specific herbicide-treated plots than on cut plots (no actual bird densities given, only model results). Bird densities were linked to densities of diurnal ground beetles (Carabidae), especially in disturbed and grass-specific herbicide-treated plots. Field margin plots (6 x 30 m) were established using one of three seed mixes: Countryside Stewardship grass mix, tussock grass mix and a mixture of grasses and wildflowers designed for pollinating insects. The margins were managed in spring from 2003 to 2005 with 1 of 3 treatments: cut to 15 cm; soil disturbed by scarification until 60% of the area was bare ground; treated with grass-specific herbicide at half the recommended rate. There were five replicates of each treatment combination at two farms. Birds were surveyed five to eight times between April and July from 2002 to 2006.

A replicated trial from 2001 to 2005 across 9 regions in Switzerland (78) found that wildflower strips sown with a 'locally adapted' mix of grass and flower species (species local to the area) had between 10 and 27 target plant species/20 m^2, and more butterflies (Lepidoptera), grasshoppers (Orthoptera) and ground beetles (Carabidae) than conventionally cropped margins. The sown 'locally adapted' field margins had more butterfly and grasshopper species and individuals than standard wildflower strips and 4 to 40 times more grasshopper and butterfly species and individuals than conventional cropped margins. There were on average 3.0 unusual (not ubiquitous) butterfly species and 5.4 unusual grasshopper species in the sown margins. Wildflower strips and 'locally adapted' sown field margins consistently had more ground beetle species and individuals than conventional margins. Conventional margins tended to have more spider (Araneae) species (statistically significant only in one region in 2002). The total abundance of spiders and ground beetles was highest in wildflower strips (2,500–4,800 individuals/margin/year in total) followed by locally adapted sown margins (2,000–4,000) compared with cropped margins (1,000–2,000). Seventy field margins (5 x 120 m) were sown with seeds of up to 38 grass and wildflower species ('locally adapted' mix) in 2001 and 2003. Butterflies and grasshoppers were counted five times between May and August 2003 and 2005 on seven locally adapted sown margins and compared with standard wildflower strips, conventional cropped margins, extensively managed sown hay meadows and biodiversity-rich meadows. Ground beetles and spiders were sampled using pitfall traps for five weeks from April to July in 2002 and 2004 on four locally adapted sown margins, four standard wildflower margins and four conventional margins. Plants were monitored on all locally adapted sown margins in June 2002–2005.

A small replicated site comparison study in 2005 in Oxfordshire, UK (79) found that field margins sown with a wildflower mix had fewer grasshopper and cricket (Orthoptera) species and individuals than margins sown with a grass and flower mix (floristically enhanced grassy margins). Wildflower margins had less than four individual insects from one species/margin on average, compared to ten individuals from four species on narrow grass and flower margins (2 m wide). They did not have more grasshopper/crickets than sown grassy margins or existing grassy tracks. The wildflower margins had the lowest grass cover (less than 60%) compared to 100% for sown grass and flower margins. Three replicates of five field margin types were monitored on a large mixed farm: grass and wildflower mix (2 m); grass and wildflower mix (6 m); grass only mix; wildflower mix; grassy track. Grasshoppers and crickets were surveyed using a sweep net over two 20 minute periods in a 50 m section of each margin in late July or August 2005.

A randomized, replicated, controlled trial from 2003 to 2006 in southwest England (80) found that plots of permanent pasture sown with a grass and legume seed mix attracted more birds and more bird species than control treatments, in both summer and winter. Three plots (50 x 10 m) were established on each of four farms in 2002, re-sown in new plots each year and monitored annually from 2003 to 2006. Sown legumes included white clover *Trifolium repens*, red clover *T. pratense*, common vetch *Vicia sativa* and bird's-foot trefoil *Lotus corniculatus*. This study was part of the same experimental set up as (97).

A replicated, controlled trial in 2005–2006 in Warwickshire, UK (81) found that field corners or margins sown with a wildflower mix had more plant species, more bumblebees *Bombus* spp. (species and individuals) and more butterfly (Lepidoptera) species than control plots sown with winter oats. There were 17 plant species/m^2, 7 bumblebees and 2 bumblebee species/plot, and 5 butterfly species/plot on average in wildflower plots, compared to 2 plant species/m^2, no bumblebees and 1 butterfly species/plot in cereal crop plots. Two declining butterfly species, small copper *Lycaena phlaeas* and common blue *Polyommatus icarus* were only found in wildflower plots. Wildflower plots did not have more butterfly individuals, or more birds in winter (species or individuals) than control crop plots. The wildflower mix (25 broadleaved non-grass species; 4 grass species 10%:90% wildflowers to grasses by weight) was sown in August 2005 and treated with grass-specific herbicide in November 2005. Plots were cut three times in 2006, and cuttings removed. Each treatment was tested in one section of margin and one corner in each of four fields on one farm. Plants were monitored in three 1 m^2 quadrats/plot in July 2006. Butterflies, bumblebees and flowering plants were recorded in a 6 m-wide transect 5 times between July and September. Farmland birds were counted on each plot on seven counts between December 2006 and March 2007. Results from the second year of monitoring are presented in (89).

A randomized, replicated, controlled trial from 2002 to 2006 in eastern England (82) (same study as (77, 83)) found that field margins sown with a flower mix designed for pollinating insects did not support more butterflies (Lepidoptera) or bumblebees *Bombus* spp. than a floristically enhanced tussocky grass seed mixture. There were 35–47 bumblebees of 4 species and 18–20 butterflies of 6 species/125 m^2 plot on average in margins sown with some non-grass species in the mix, compared to 10 bumblebees of 2 species and 12 butterflies of 5 species on grass-only margins. Different types of management did not affect the abundance of bees and butterflies or the number of butterfly species but there were more bumblebee species on plots treated with grass-specific herbicide in spring (average 4 species/125 m^2 compared to 3 species on cut or disturbed plots). Field margin plots (6 x 30 m) were established in 2000–2001 using 1 of 3 seed mixes: Countryside Stewardship mix (7 grass species, sown at 20 kg/ha), tussock grass mix (7 grass species, 11 wildflowers, sown at 35 kg/ha) and a mixture of grasses and wildflowers designed for pollinating insects (4 grass species, 16–20 wildflowers, sown at 35 kg/ha). Margins were managed in spring from 2003 to 2005 with 1 of 3 treatments: cut to 15 cm; soil disturbed by scarification until 60% of the area was bare ground; treated with grass-specific herbicide in spring at half the recommended rate. There were five replicates of each treatment combination on three farms.

A randomized, replicated, controlled trial from 2002 to 2006 in eastern England (83) (same study as (77, 82)) found that field margins sown with a flower mix designed for pollinating insects supported fewer planthoppers (Auchenorrhyncha) than those sown with a grass-only seed mixture. Flower-rich sown margins had 25–40 planthoppers/plot on average (depending on management) while grass-only margins had 30–70 planthoppers/plot. Field margin plots (6 x 30 m) were established in 2000–2001 using 1 of 3 seed mixes: Countryside Stewardship mix (7 grass species, sown at 20 kg/ha), tussock grass mix (7 grass, 11 wildflower species, sown at 35 kg/ha) and a mixture of grasses and wildflowers designed for pollinating insects (4 grass species, 16–20 wildflowers, sown at 35 kg/ha). The margins were managed in spring from 2003 to 2005 with 1 of 3 treatments: cut to 15 cm; soil disturbed by scarification until 60% of the area was bare ground; treated with grass-specific herbicide in spring at half the recommended rate. There were five replicates of each treatment combination on three farms.

A replicated, controlled study from 2005 to 2008 in England and Scotland (84) found the average number of worker bumblebees *Bombus* spp. was greater on margins where legume-rich seed mix was established than on other field margins (grassy margins or track edges). There was an observed decline in the relative number of foraging worker bumblebees on legume-sown margins after they had been established for more than three years (data from five farms). No formal statistical analyses were performed on these data. On each of 41 farms, four 100 x 2 m field margins were surveyed for bumblebees. Two of the margins were sown with a legume-rich seed mix in either April 2005, 2007 or 2008. The other two margins were track edges or grassy margins. Bumblebee surveys were made on all four margins on each farm from 2005 to 2008, twice in June/early July and late July/early August.

A replicated, controlled study from 2001 to 2004 in the UK (85) found arable margins sown with a legume-grass seed mix had more bumblebee *Bombus* spp. forage plant species (almost 100% cover of Alsike clover *Trifolium hybridum* and red clover *T. pratense* one year after establishment) over 4 years compared to naturally regenerated margins. The cover of Alsike clover declined from a peak of approximately 33% in 2002 to 2.5% in 2004, whilst red clover cover peaked at around 85% in 2003 and declined to 20% in 2004. Clover-sown plots were invaded by perennial grasses including false oat grass *Arrhenatherum elatius* in the third and fourth years of the study when clover cover decreased substantially. Bee visits were not reported in this study, however the results of fixed-time transect walks in the clover margins are reported in (Edwards & Williams 2004), which found a 300-fold increase in bumblebee forager numbers in the margins planted with clover, however no control count was carried out for comparison. Two 6 m-wide margins were established on one farm and subdivided into 3 plots. There were two margin types: naturally regenerated, or sown with a mixture of grasses and leguminous species including two species of clover. Three different management treatments were applied to the sub-plots in the first year (2001): cut three times with cuttings left; cut three times with cuttings removed; cut six times with cuttings left. From 2002 to 2004 all plots were cut in late summer and the cuttings removed. Forage plants were monitored in 0.25 m^2 quadrats every 1 m along a 30 m transect in early August 2001, 2002, 2003 and 2004.

A replicated, controlled study in 2000–2003 in Norway (86) found that plant species diversity was higher in strips sown with a grass/wildflower mixture than strips left to regenerate naturally or in the grass crop. The number of plant species was significantly higher in sown grass/wildflower strips (17–18 species/quadrat) than naturally regenerated strips (10–12 spp.) or the grass crop (7–9 spp.). The same was true for the number of meadow herbs (sown: 7–10 species; unsown: 1–3 spp.; crop: 1 spp./quadrat). Plant diversity (Shannon diversity index) was also significantly higher in sown grass/wildflower strips than in either naturally regenerated margins or in the crop (sown: 1.8–1.9, unsown: 1.3–1.6, crop: 1.0–1.3). Four of the 22 sown meadow wildflower species did not establish. Naturally regenerated strips and the grass crop had many species in common by the fourth study year, and grasses and perennial weeds dominated in the crop and unsown strips. By the fourth year some sown species were recorded in the unsown strips or grass crop and woody species from an existing semi-natural margin were recorded in the sown strips. The total number of plant species did not vary with distance from the existing margin. Four strips (2 m wide) were ploughed perpendicular to an existing semi-natural margin in May 2000. One half of each was left to regenerate naturally; the other half was sown with a grass/meadow wildflower (22 species) seed mixture (5 g/m^2). Wildflower seeds were local to the area; grass seeds were cultivated varieties. Sown strips did not receive fertilizer and were cut once (late September). Permanent quadrats (0.5 x 0.5 m) were sampled in the grass crop and strips in June 2000–2003.

A replicated, controlled site-comparison study in 2005 in the Netherlands (87) found that the number of flowers was 10 times higher in plots sown (or planted) with 17 insect-pollinated plant species than outside the plots (approximately 4,650 vs 480 flowers/plot within/outside flower plots respectively). The number and diversity of bees (Apidae) and hoverflies (Syrphidae) was significantly higher (60–80% higher) in flower plots than on control transects. Outside the flower plots, hoverfly abundance was significantly higher 50 m away from the flower plots but not at any other distance. The lowest numbers of bees and bee species were recorded 50 m away from the flower plots. Seventeen species of annual and perennial plants were either transplanted or sown in fenced 10 x 10 m plots at 5 locations in intensive farmland. Hoverflies and bees were surveyed at 10 sampling locations along a 1,500 m transect running away from each flower plot and along 5 1,500 m control transects. All transects ran alongside ditches. Bees and hoverflies were sampled using window traps, yellow water pans and nets four times between June and September 2004.

Two randomized, replicated studies from 2005 to 2007 in Yorkshire and Warwickshire, UK (88) studied different 'pollen and nectar' seed mixes. The Yorkshire study (2005–2007) looked at six different seed mixes and found two agricultural varieties of red clover *Trifolium pratense* had the highest cover. There were more flowers of sown plant species in an agricultural clover mix in 2005 and in wild Somerset red clover in 2006. There were more red clover flowers in an agricultural mix in all three years. Bird's-foot trefoil *Lotus corniculatus* flowers were more abundant in the first and third years in wild Somerset red clover mix than in agricultural mixes. Plots measured 48 x 6 m and were replicated twice. The Warwickshire study (2006–2007) (same study as (103)) found butterfly (Lepidoptera) abundance and species richness were highest in plots sown with lucerne *Medicago sativa* or red clover (3–6 butterflies/plot, 2–3.5 species vs 0–3 butterflies and 0–3 species for all other plots). Phacelia *Phacelia tanacetifolia* plots had the highest number of bumblebees *Bombus* spp. (39–134 bumblebees/plot), followed by borage *Borago officinalis* (32–100). Phacelia, crimson clover *Trifolium incarnatum*, borage, sunflower *Helianthus annuus* and red clover (2007 only) plots had significantly more bumblebee species than all other plots in 2006 (3–4 vs 0–1 species/plot). Short-tongued bees preferred phacelia and borage in both years. Long-tongued bees preferred crimson clover, borage, phacelia or red clover. In the first year there were more annual than perennial flowers. Flowering of many important bee forage species peaked in late July. Thirteen annual and perennial plant species were sown individually in 6 x 4 m plots, replicated 4 times in May 2006; annual species were re-sown in May 2007. Butterflies and bumblebees were surveyed six times in each plot (July–September 2006 and May–September 2007). On each visit the percentage cover of all flowers was estimated.

A replicated, controlled trial from 2005 to 2007 in Warwickshire, UK (89) (the second monitoring year of the same study as (81)) found that wildflower plots had more plant species, bumblebees *Bombus* spp. and butterflies (Lepidoptera) (individuals and species) than naturally regenerated or control cereal plots, and more vacuum-sampled invertebrates than control plots. Wildflower plots did not have more birds in winter than control plots. On wildflower plots there were 10 plant species/m^2, 63 bumblebees and 5 bumblebee species/plot, 18 butterflies and 6 butterfly species/plot, compared to 3 plant species/m^2, 0 bumblebees, and 1 butterfly/plot on control cereal plots. Control cereal plots had 254 vacuum-sampled canopy-dwelling invertebrates/m^2 on average, compared to 840–1,197/m^2 on other treatments. Plants were monitored in three 1 m^2 quadrats in each plot in June 2007. Butterflies, bumblebees and flowering plants were recorded in a 6 m-wide transect 6 times between July and September 2006 and 2007. Invertebrates in the vegetation were vacuum-sampled in early July 2007. Farmland birds were counted on each plot on four counts between December 2007 and March 2008. The crop control in year two was winter wheat.

A replicated, controlled trial from 2002–2004 in County Wexford, Ireland (90) found that 1.5–3.5 m-wide margins of permanent pasture fields fenced, rotavated and sown with a wildflower seed mix had more springtails (Collembola: Anthropleona), spiders (Araneae), flies (Diptera) and plant species than control margins. Wildflower margins had 18 plant species/plot in July 2002, decreasing to 11 plant species/plot in 2004, compared to less than 5 plant species/plot throughout the study in control plots and unrotavated fenced margins. Some undesirable weeds, such as broad-leaved dock *Rumex obtusifolius*, established in rotavated plots but were less abundant in plots sown with wildflower mix (broad-leaved dock cover was 6–25% in wildflower sown plots and 26–50% in naturally regenerated plots in 2002 and 2004). Grazing on half of each plot from 2003 to 2004 (one year after margin establishment) did not affect the number of plant species the following year. Plots were 30 m long (1.5–3.5 m wide) with 3 replicates of each treatment combination. The wildflower mix had 10 grass species and 31 non-grass species. Plants were monitored in permanent quadrats in July 2002, May and July 2003 and May 2004. Invertebrates were sampled in six emergence traps/plot between May and September 2003. Ground areas under the emergence traps were sampled with a vacuum sampler.

A replicated study in summer 2002–2004 and 2006 on three farms in England (91) (study extended from (64)) found a greater abundance (but not species richness) of herbivorous beetle species (Coleoptera) in seed mixtures including wildflowers than in grass-only mixtures. However there were more predatory beetle species and individuals in margins containing tussock grass species regardless of whether the mixture also included wildflowers. Margin management (i.e. soil scarification) also had a positive effect on species richness of predatory beetles. Three different seed mixtures were sown: grass only; tussock grass and wildflowers; fine grass and wildflowers. Each of the seed mixtures was randomly sown on three of 9 experimental plots (25 x 5 m) in each of 5 blocks on 3 farms in autumn 2001. From 2003 three different management practices were applied in each replicate block in May each year: cutting the vegetation to 10–15 cm, application of grass-specific herbicide (Fuazifop-*p*-butyl) at 0.8 l/ha and scarification of 60% of the soil surface. Plant diversity and cover and vegetation structure were surveyed yearly in June using 0.5 x 0.5 m quadrats and vertical drop pins. Beetles were sampled using a Vortis suction sampler (75 suctions of 10 seconds each) over a fixed area (equivalent to 1.45 m^2) in each plot on each sampling date. Rove beetles (Staphylinidae), ground beetles (Carabidae), ladybirds (Coccinellidae), leaf-beetles (Chrysomelidae) and weevils (Curculionoidea) were determined to species level and categorized as herbivorous or predatory.

A site comparison study between 1997 and 2004 in central Switzerland (92) found wild-flower strips sown with 20–40 species contained significantly more (8–60% more) plant, butterfly (Lepidoptera), ground beetle (Carabidae) and spider (Araneae) species than crop fields in the same region. Estimated total numbers of species were 149 (plant), 19 (butterfly), 85 (ground beetle) and 134 (spider) on Ecological Compensation Area wildflower strips and 50, 19, 78 and 104 species on conventional crop fields respectively. Rare or threatened species were not found more frequently on Ecological Compensation Area sites. The increased num-ber of species was a response of common species. The study sampled 78 wildflower strips and 72 crop fields in a predominantly arable region.

A replicated, controlled study from March–July 2006 in mixed farmland near Bern, Switzerland (93) found that Eurasian skylarks *Alauda arvensis* with territories that included undrilled patches sown with six annual weed species were significantly less likely to abandon the territory and more likely to use undrilled patches as nesting and foraging sites. Nests were significantly more likely to be built within or close to undrilled patches (60% of skylark nests were within 5 m of an undrilled patch). Skylarks preferentially foraged in undrilled patches over all other crop types; undrilled patches covered 0.17–0.63% of the foraging area but were accessed on 12.6% of observed foraging flights. Plant cover ranged from 35 to 50%, and plant height ranged from 5 to 80 cm in the plots. Undrilled patches were composed of either four 3 x 12 m patches/ha (in 7 fields) or a single strip 2.5 x 80 m (in 14 fields). In autumn 2005 undrilled patches were sown with six annual weed species including common corn-cockle *Agrostemma githago* in winter wheat fields. Skylark territories were surveyed over one breeding season (2006) in 21 experimental sites and 16 control wheat fields.

A replicated, controlled study in summer 2001 in intensively managed farmland around Bern, Switzerland (94) found that the number of species and individuals, biomass and individual weights of most sampled arthropod predators increased with the age of sown wildflower sites. Conversely the number of rove beetles (Staphylinidae) and rove beetle bio-mass was highest in newly created wildflower sites, but the weight of individual rove beetles increased with age of wildflower sites. Control wheat fields had among the lowest species richness, density and biomass of predators, but these values were only significantly lower than in the oldest wildflower strips for spider (Araneae) and ground beetle (Carabidae) bio-mass and true bug (Heteroptera) density. Vegetation cover had a significant influence on spider assemblages. Ground beetle species assemblages were strongly correlated with veget-ation cover, field size and soil water content in wildflower sites. Five different habitats with 4 replicates were surveyed at 20 sites (average 0.8 ha). The 4 sown wildflower habitats had been established for 1, 2, 3 and 4 years (1-year-old sites sown in May 2001) and were sown with a seed mixture containing 25 native plant species, not treated with fertilizer, pesticides or cut. Winter wheat fields were used as controls. Spiders, ground beetles and rove beetles were sampled using three photo-eclectors/site for two consecutive months. True bugs were sampled 4 times along 80 m transects using sweep-nets (100 sweeps/transect). Vegetation cover, volume of soil pores and sand content were determined.

A replicated, controlled study in June and July 2006 in north Germany (95) found more hoverflies (Syrphidae) and hoverfly species in broad (12–25 m wide) and narrow (3–6 m) sown wildflower strips (7 sites each) than in grassy margins (3 m-wide, 7 sites), wheat-wheat boundaries (7 sites) and within wheat fields adjacent to the margins (7 sites). Hoverfly density and species richness (total hoverflies and aphid-eating hoverflies) also increased with increasing amount of arable land around the site at smaller scales (0.5 and 1 km) but not at larger scales (2 and 4 km). Margins were located along a gradient of habitat complexities in the surrounding landscape, ranging from 30% to 100% arable land. Hoverflies were sampled by sweep netting (one sweep per footstep) along 100 m transects.

A replicated, controlled study in the summers of 2004–2005 in northwest Switzerland (96) found wildflower strips had a variable effect on parasitism and predation of eggs and larvae of two common butterfly/moth (Lepidoptera) cabbage pests. Parasitization rates of cabbage moth *Mamestra brassicae* eggs and larvae as well as small white butterfly *Pieris rapae* larvae on one farm did not differ between plots with and without wildflower strips. However on a second farm, parasitization rates of small white butterfly larvae and predation rates of cabbage moth eggs were significantly higher in plots with adjacent wildflower strips. Wildflower strips did not affect the spatial pattern of parasitization in the fields. Six cabbage *Brassica oleracea* fields were studied on two organic farms. Two blocks (45 x 25 m) were studied on each field, one with and one without a wildflower strip (3 x 35 m). Wildflower strips were sown with seed mixtures containing 24 native plant species and were not treated with pesticides or mown. Egg parasitization rates were assessed by placing laboratory eggs pinned to paper cards on the ground underneath labelled plants in a 3 x 3 m grid for three days. Eggs were incubated for four weeks at 22°C to rear any parasitoids. Missing and damaged eggs were counted to estimate the predation rate. Butterfly larvae were sampled on randomly selected plants and parasitization rates were determined using DNA-based techniques.

A randomized, replicated, controlled trial from 2003 to 2006 in southwest England (97) (same experimental set up as (80)) found plots on permanent pasture annually sown with a mix of legumes, or grass and legumes, supported more common bumblebees *Bombus* spp. (individuals and species) than seven grass management options. In the first two years, there were more common butterflies (Lepidoptera) and common butterfly species in plots sown with legumes than in five intensively managed grassland treatments. No more than 2.2 bumblebees/transect were recorded on average on any grass-only plot in any year, compared to over 15 bumblebees/transect in both sown treatments in 2003. Plots sown with legumes generally had fewer butterfly larvae than all grass-only treatments, including conventional silage and six different management treatments. Experimental plots 50 x 10 m were established on permanent pastures (more than 5-years-old) on 4 farms. There were nine different management types, with three replicates/farm, monitored over four years. Seven management types involved different management options for grass-only plots, including mowing and fertilizer addition. The two legume-sown treatments comprised either barley *Hordeum vulgare* undersown with a grass and legume mix (including white clover *Trifolium repens*, red clover *T. pratense*, and common vetch *Vicia sativa*) cut once in July, or a mix of crops (including linseed *Linum usitatissimum*) and legumes, uncut. Bumblebees and butterflies were surveyed along a 50 m transect line in the centre of each experimental plot, once a month from June to September annually. Butterfly larvae were sampled on two 10 m transects using a sweep net in April and June–September annually.

A replicated, controlled, paired sites study in summer 2005 in northwest Switzerland (98) found densities of several spider (Araneae) families were higher in wheat fields with adjoining sown wildflower areas than in fields with grassy margins. Crab spiders (Thomisidae), ground spiders (Gnaphosidae) and wolf spiders (Lycosidae) as well as young orb weaver spiders (Araneidae) had higher densities in fields with adjacent sown wildflower areas. However, spider diversity and the total number of spider species were not significantly different in wheat fields adjoined by sown wildflower areas than fields with grassy margins. Twenty winter wheat fields were studied (0.5–4.1 ha in size); 10 fields had adjoining sown wildflower areas; 10 were adjoined by grassy margins. Wheat fields were treated with herbicides, fungicides and mainly mineral fertilizers but no insecticides. Sown wildflower areas (a Swiss agri-environment scheme) were sown with a mixture of 25 wildflower species, and were not treated with pesticides, fertilizers or mown. Sown wildflower areas were 0.4–2.3 ha in size (minimum 25 m wide) and between 2- and 6-years-old. Grassy margins were ca. 0.7 m wide and mown several times/year. Spiders were sampled from May to June using pitfall traps (0.2 l, 6.5 cm diameter) and a suction sampler (0.1 m diameter).

A 2009 literature review of European farmland conservation practices (99) found that field margins sown with a wildflower mix had higher arthropod diversities than adjacent crops, or margins sown with grass seed only. Several bird species were also found to use wildflower strips more than margins sown with grass seed only.

A replicated study in 2005 in western Switzerland (100) found that small mammal density and species richness were higher in wildflower areas than crops, but wildflower areas were avoided by barn owls *Tyto alba*. Wildflower areas (2-years-old, 1 ha) had more small mammal species and individuals (6 species, 458–1,285 individuals/ha) compared to crops or meadows (2–5 species, 0–680/ha). In May and July, small mammal densities were significantly higher in wildflower areas (458–1,030 individuals/ha) and winter wheat *Triticum aestivum* (90–680/ha) than in tobacco *Nicotiana tabacum*, permanent and intensive meadows and in May maize *Zea mays* (0–10/ha) (in July the density in maize was 200/ha). In September density was significantly higher in wildflower areas (1,285/ha) than in winter wheat (0); other habitats had intermediate densities (5–60/ha). Barn owls significantly preferred cereal crops relative to availability and avoided wildflower areas and all other crop types. The estimated index of habitat selection by barn owls in order of decreasing preference was wheat, meadows, other crops and lastly wildflower areas. Four arable sites were studied. Small mammal population size was estimated using capture-mark-recapture. Mammal traps were placed at 20 points along 2 parallel 45 m transects in each habitat and set over 3 nights and days in May, July and September 2005. Seven breeding male barn owls were radio-tagged from June to September 2005 and hunting or resting locations recorded.

A replicated study in summer 2007 in south Sweden (101) found higher densities and species richness of butterflies (Lepidoptera) and bumblebees *Bombus* spp. in sown wildflower strips than in strips consisting mainly of grass species (greenways or '*beträdor*'). Eighty-six percent of the recorded butterflies and 83% of the bumblebees were found in wildflower strips. Butterfly density was nearly 20 times higher in wildflower strips than in the grass strips. The most common flowers visited were field scabious *Knautia arvensis* and knapweeds *Centaurea* spp. for butterflies, and knapweeds for bumblebees (72% of all recordings). The presence of bushes adjacent to the strip positively affected the number of butterfly species and individual numbers of both butterflies and bumblebees. Butterflies and bumblebees were recorded on 1 wildflower strip (6 transects) and 3 grass strips (14 transects) on 5 occasions on 4 arable farms. Butterflies and bumblebees were counted within 2 m either side of the observer and the flower species visited by the insects noted.

A replicated, randomized study from 2005 to 2007 in Warwickshire, UK (102) found no difference in the number of bumblebees *Bombus* spp. or bumblebee species between plots sown with ten different flowering plant and grass seed mixtures, but recorded a significant increase in the number of bumblebee individuals and species in sub-plots treated with the grass-specific herbicide propyzamide in 2007. These sub-plots also showed a significant decrease in grass cover (from 45 to 2%) and an increase in the cover of sown wildflowers (from 24 to 56%), bare ground (from 4 to 16%) and undesirable weeds (from 4 to 14%). The number and cover of sown wildflowers decreased over the years in favour of competitive grass species. Ten different seed mixes (3 replicates each) were sown in plots (6 x 10 m) in April 2005. The seed mixes contained four to six flowering plant species and one to four grass species sown in different proportions. Plots were cut three times in 2005 and twice in 2006 with cuttings left in place. The grass-specific herbicide fluazifop-P-butyl was sprayed in plots with rye grass nurse crop in April 2006. In November 2006 all plots were split into 2 sub-plots (3 x 10 m), of which 1 was sprayed with propyzamide. The percentage cover of vascular plants was recorded in two randomly placed 1 x 1 m quadrats in each plot (2005–2006) or sub-plot (2007) respectively. Bumblebee abundance and diversity were monitored twice each year in late summer 2006 and 2007. These results were also presented in (88) but are only reported here.

A replicated, randomized study in 2006 and 2007 in Warwickshire, UK (103) (same study as (88)) found that bee (Apidae) and butterfly (Lepidoptera) abundance and species richness were higher in plots sown with specific wildflower species. Bumblebee *Bombus* spp. abundance and species richness were significantly higher on plots sown with phacelia *Phacelia tanacetifolia* and borage *Borago officinalis* (32–85 bees/plot) compared to other treatments (1–22/plot). Crimson clover *Trifolium incarnatum* (10–21/plot), sunflower *Helianthus annuus* (10–22/plot) and red clover *T. pratense* (20/plot) also tended to have high bumblebee abundances (other species: 1–11/plot). Short- and long-tongued bees had different preferences. In 2006, butterfly abundance and species richness were significantly higher in plots with lucerne *Medicago sativa* compared to borage, chicory *Cichorium intybus* and sainfoin *Onobrychis viciifolia*. In 2007 butterfly abundance was higher in red clover compared with chicory, but the number of species did not differ between treatments. Mobile and immobile butterfly species had different preferences. Flowers of buckwheat *Fagopyrum esculentum* were the most abundant followed by phacelia, borage and sunflower in 2006. In 2007 fodder radish *Raphanus sativus*, red clover and sweet clover *Melilotus officinalis* also had high flower abundance. Mustard *Brassica juncea* and linseed *Linum usitatissimum* had the fewest flowers in both years, along with other species each year. Thirteen species were sown in single species stands: four wildflower species typically sown in pollen and nectar seed mixes and nine small-seeded crop species typically sown in wild bird seed mixes. The species were sown in May each year in adjacent 6 x 4 m plots in a randomized block experiment with 4 replicates. Butterflies and bumblebees were sampled by walking transects through each plot on six occasions from May–September. Flower cover was estimated at the same time.

A replicated, controlled site comparison study in summer 2008 in northwest Scotland (104) found that croft sections (an agricultural system specific to Scotland, consisting of small agricultural units with rotational cropping regimes and livestock production) sown with a brassica-rich 'bird and bumblebee' conservation seed mix had 47 times more foraging bumblebees *Bombus* spp. than sheep-grazed sections and 16 times more bumblebees than winter-grazed pastures in June. In July the 'bird and bumblebee' mix sections had 248 and 65 times more bumblebees than sections grazed by sheep or both sheep and cattle respectively. The number of bumblebees in July was also significantly higher (4–16 times) in 'bird and bumblebee' sections than in arable, fallow, silage and winter-grazed pasture sections. The availability of bumblebee forage plant flowers was lower in 'bird and bumblebee' sections than in silage sections in June but no other significant differences involving the conservation mix were detected. Foraging bumblebees most frequently visited plant species in the legume (Fabaceae) family. Tufted vetch *Vicia cracca* was one of a few plant species favoured by bumblebees and was predominantly found in 'bird and bumblebee' sections in July–August, although it was not part of the seed mixture. Thirty-one crofts located on Lewis, Harris, the Uists and at Durness were studied. Species sown in the bird and bumblebee mix included kale *Brassica oleracea*, mustard *Brassica* spp., phacelia *Phacelia* spp. and red clover *Trifolium pratense*. In addition to the seven management types mentioned, unmanaged pastures were surveyed for foraging bumblebees and bumblebee forage plants along zigzag or L-shaped transects in each croft section once in June, July and August 2008. Foraging bumblebees 2 m either side of transects were identified to species level and recorded together with the plant species on which they were foraging. Flowers of all plant species were counted in 0.25 m^2 quadrats at 20 or 50 m intervals along the transects.

A randomized, replicated trial from 1987 to 2000 in Oxfordshire, UK (105) found that the number of plant species on 2 m-wide margins sown with a wildflower seed mix in 1988 declined by about half over 13 years. There were 23–24 plant species/quadrat in 1988 and 9–12 plant species/quadrat in 2000. The most rapid decline was in the first two years, when many annual species were lost. Sown plots retained more perennial plant species than naturally regenerated plots throughout the 13 years (around 10 vs 8 perennial species/quadrat respectively in 2000). After 13 years sown plots tended to have more species than naturally regenerated plots (9–12 vs 7–9 species/plot respectively in 2000) but this difference was not statistically significant. There was no effect of different mowing regimes on the numbers of plant species, although in the early years mown plots had more plant species than uncut plots. Sown plots that were cut twice retained a greater proportion of sown species (50–60%) than plots cut once or uncut (<40%). Sowing reduced the colonization of margins by unsown perennial species at first, but by 2000 many perennial species, including couch grass *Elymus repens*, were similarly abundant in sown and unsown plots. Plant species were monitored three times a year from 1988 to 1990 and once in July 2000 in three 0.5 x 1 m quadrats/plot. This was part of the same study set-up as (7, 16, 31, 32, 35, 37, 47).

(1)   Nentwig W. (1989) Augmentation of beneficial arthropods by strip management. 2. Successional strips in a winter-wheat field. *Journal of Plant Diseases and Protection*, 96, 89–99.
(2)   Williams I.H. & Christian D.G. (1991) Observations on *Phacelia Tanacetifolia* Bentham (Hydrophyllaceae) as a food plant for honey bees and bumble bees. *Journal of Apicultural Research*, 30, 3–12.
(3)   Lagerlöf J., Stark J. & Svensson B. (1992) Margins of agricultural fields as habitats for pollinating insects. *Agriculture Ecosystems & Environment*, 40, 117–124.
(4)   Lys J.-A. & Nentwig W. (1992) Augmentation of beneficial arthropods by strip-management. 4. Surface activity, movements and activity density of abundant carabid beetles in a cereal field. *Oecologia*, 92, 373–382.
(5)   Bürki H.M. & Hausammann A. (1993) Überwinterung von Arthropoden im Boden und an Ackerkräutern künstlich angelegter Ackerkrautstreifen. *Agrarökologie*, 7, 1–144.
(6)   Engels W., Schulz U. & Rädle M. (1994) Use of the Tübingen mix for bee pasture in Germany. In A. Matheson (eds.) *Forage for Bees in an Agricultural Landscape*, International Bee Research Association, Cardiff. pp. 57–65.
(7)   Feber R.E., Smith H. & MacDonald D.W. (1994) The effects of field margin restoration on the meadow brown butterfly (*Maniola jurtina*). *British Crop Protection Council Monographs*, 58, 295–300.
(8)   Gathmann A., Greiler H.J. & Tscharntke T. (1994) Trap-nesting bees and wasps colonizing set-aside fields: succession and body size, management by cutting and sowing. *Oecologia*, 98, 8–14.
(9)   Greiler H.J. (1994) Insect communities in self-established and sown agricultural fallows. In W. Nentwig & H. M. Poehling (eds.) *Agrarokologie*, 11, Haupt, Bern. pp. 1–136.
(10) Holland J.M., Thomas S.R. & Courts S. (1994) *Phacelia tanacetifolia* flower strips as a component of integrated farming. *British Crop Protection Council Monographs*, 58, 215–220.
(11) Lys J.-A. & Nentwig W. (1994) Improvement of the overwintering sites for Carabidae, Staphylinidae and Araneae by strip-management in a cereal field. *Pedobiologia*, 38, 238–242.
(12) Lys J.-A., Zimmermann M. & Nentwig W. (1994) Increase in activity density and species number of carabid beetles in cereals as a result of strip-management. *Entomologia experimentalis et applicata*, 73, 1–9.
(13) MacLeod A. (1994) Provision of plant resources for beneficial arthropods in arable ecosystems. PhD thesis. University of Southampton.

(14) Zangger A., Lys J.-A. & Nentwig W. (1994) Increasing the availability of food and the reproduction of *Poecilus cupreus* in a cereal field by strip-management. *Entomologia experimentalis et applicata*, 71, 111–120.

(15) Steffan-Dewenter I. & Tscharntke T. (1995) Bees on set-aside fields: impact of flower abundance, vegetation and field-age. *Mitteilungen Der Deutschen Gesellschaft Fur Allgemeine Und Angewandte Entomologie*, 10, 319–322.

(16) Feber R.E., Smith H. & Macdonald D.W. (1996) The effects on butterfly abundance of the management of uncropped edges of arable fields. *Journal of Applied Ecology*, 33, 1191–1205.

(17) Hausammann A. (1996) Strip-management in rape crop: is winter rape endangered by negative impacts of sown weed strips? *Journal of Applied Entomology*, 120, 505–512.

(18) Holland J.M. & Thomas S.R. (1996) *Phacelia tanacetifolia* flower strips: their effect on beneficial invertebrates and gamebird chick food in an integrated farming system. *Acta Jutlandica*, 71, 171–182.

(19) Marshall E.J.P. & Nowakowski M. (1996) Interactions between cutting and a graminicide on a newly-sown grass and wild flower field margin strip. *Aspects of Applied Biology*, 44, 307–312.

(20) West T.M. & Marshall E.J.P. (1996) Managing sown field margin strips on contrasted soil types in three environmentally sensitive areas. *Aspects of Applied Biology*, 44, 269–276.

(21) Carreck N.L. & Williams I.H. (1997) Observations on two commercial flower mixtures as food sources for beneficial insects in the UK. *Journal of Agricultural Science*, 128, 397–403.

(22) Clarke J.H., Jones N.E., Hill D.A. & Tucker G.M. (1997) The management of set-aside within a farm and its impact on birds. *Proceedings of the 1997 Brighton Crop Protection Conference*. 1–3, pp. 1179–1184.

(23) Denys C. (1997) Do field margins contribute to enhancement of species diversity in a cleared arable landscape? Investigations on the insect community of mugwort (*Artemisia vulgaris* L). *Mitteilungen Der Deutschen Gesellschaft Fur Allgemeine Und Agewandte Entomologie, Band 11, Heft 1–6, Dezember – Entomologists Conference*, 11, 69–72.

(24) Denys C., Tscharntke T. & Fischer R. (1997) Colonization of wild herbs by insects in sown and naturally developed field margin strips and in cereal fields. *Verhandlungen der Gesellschaft fur Okologie*, 27, 411–418.

(25) Feber R.E. & Hopkins A. (1997) Diversity of plant and butterfly species on organic farmland field margins in relation to management. *Proceedings of the British Grassland Society Fifth Research Conference*. University of Plymouth, Devon, UK, 8–10 September. pp. 63–64.

(26) Frank T. (1997) Species diversity of ground beetles (Carabidae) in sown weed strips and adjacent fields. *Biological Agriculture & Horticulture*, 15, 297–307.

(27) Gathmann A. & Tscharntke T. (1997) Bienen und Wespen in der Agrarlandschaft (Hymenoptera Aculeata): Ansiedlung und Vermehrung in Nisthilfen [Bees and wasps in the agricultural landscape (Hymenoptera Aculeata): colonization and augmentation in trap nests]. *Mitteilungen Der Deutschen Gesellschaft Fur Allgemeine Und Angewandte Entomologie*, 11, 91–94.

(28) Kleijn D., Joenje W. & Kropff M.J. (1997) Patterns in species composition of arable field boundary vegetation. *Acta Botanica Neerlandica*, 46, 175–192.

(29) Lemke A. & Poehling H.M. (1997) Effects of sown weed strips in winter wheat on the abundance of cereal aphids and spiders. *Mitteilungen Der Deutschen Gesellschaft Fur Allgemeine Und Angewandte Entomologie*, 11, 237–240.

(30) Weiss B. & Buchs W. (1997) Reaction of the spider coenoesis on different kinds of rotational set aside in agricultural fields. *Mitteilungen Der Deutschen Gesellschaft Fur Allgemeine Und Angewandte Entomologie*, 11, 147–151.

(31) Baines M., Hambler C., Johnson P.J., Macdonald D.W. & Smith H. (1998) The effects of arable field margin management on the abundance and species richness of Araneae (spiders). *Ecography*, 21, 74–86.

(32) Bell J.R., Gates S., Haughton A.J., Macdonald D.W., Smith H., Wheater C.P. & Cullen W.R. (1999) Pseudoscorpions in field margins: Effects of margin age, management and boundary habitats. *Journal of Arachnology*, 27, 236–240.

(33) Carreck N.L., Williams I.H. & Oakley J.N. (1999) Enhancing farmland for insect pollinators using flower mixtures. *Aspects of Applied Biology*, 54, 101–108.

(34) de Snoo G.R. & Chaney K. (1999) Unsprayed field margins – what are we trying to achieve? *Aspects of Applied Biology*, 54, 1–12.

(35) Haughton A.J., Bell J.R., Gates S., Johnson P.J., Macdonald D.W., Tattersall F.H. & Hart B.H. (1999) Methods of increasing invertebrate abundance within field margins. *Aspects of Applied Biology*, 54, 163–170.

(36) Kromp B. (1999) Carabid beetles in sustainable agriculture: a review on pest control efficacy, cultivation impacts and enhancement. *Agriculture, Ecosystems & Environment*, 74, 187–228.

(37) Smith H., Feber R. & MacDonald D. (1999) Sown field margins: why stop at grass? *Aspects of Applied Biology*, 54, 275–282.

(38) Thomas C.F.G. & Marshall E.J.P. (1999) Arthropod abundance and diversity in differently vegetated margins of arable fields. *Agriculture, Ecosystems & Environment*, 72, 131–144.

(39) Gathmann A. & Tscharntke T. (2000) Habitatbewertung mit einer häufigen Wildbienenart – Körpergrösse und Geschlechterverhältnis bei *Osmia rufa* (L.) (Hymenoptera: Megachilidae) [Habitat

evaluation using an abundant wild bee species – body size and sex ratio in *Osmia rufa* (L.) (Hymenoptera: Megachilidae).]. *Mitteilungen Der Deutschen Gesellschaft Fur Allgemeine Und Angewandte Entomologie*, 12, 607–610.

(40) Holland J.M. & Luff M.L. (2000) The effects of agricultural practices on Carabidae in temperate agroecosystems. *Integrated Pest Management Reviews*, 5, 109–129.

(41) Jeanneret P., Schüpbach B., Steiger J., Waldburger M. & Bigler F. (2000) Evaluation of ecological measures: biodiversity. Spiders and butterflies. *Agrarforschung*, 7, 112–116.

(42) Pfiffner L. & Luka H. (2000) Overwintering of arthropods in soils of arable fields and adjacent semi-natural habitats. *Agriculture, Ecosystems & Environment*, 78, 215–222.

(43) Herzog F., Gunter M., Hofer G., Jeanneret P., Pfiffner L., Schlapfer F., Schupbach B. & Walter T. (2001) Restoration of agro-biodiversity in Switzerland. In Y. Villacampa, C.A. Brebbia & J.L. Uso (eds.) *Ecosystems and Sustainable Development III*, 10, WIT Press, Southampton. pp. 397–406.

(44) Luka H., Lutz M., Blick T. & Pfiffner L. (2001) The influence of sown wildflower strips on ground beetles and spiders (Carabidae & Araneae) in an intensively cultivated agricultural area (Grosses Moos, Switzerland). *Schweiz. Peckiana*, 1, 45–60.

(45) Moonen A.C. & Marshall E.J.P. (2001) The influence of sown margin strips, management and boundary structure on herbaceous field margin vegetation in two neighbouring farms in southern England. *Agriculture, Ecosystems & Environment*, 86, 187–202.

(46) Steffan-Dewenter I. & Tscharntke T. (2001) Succession of bee communities on fallows. *Ecography*, 24, 83–93.

(47) Bell J.R., Johnson P.J., Hambler C., Haughton A.J., Smith H., Feber R.E., Tattersall F.H., Hart B.H., Manley W. & Macdonald D.W. (2002) Manipulating the abundance of *Lepthyphantes tenuis* (Araneae: Linyphiidae) by field margin management. *Agriculture, Ecosystems & Environment*, 93, 295–304.

(48) Carreck N. & Williams I. (2002) Food for insect pollinators on farmland: insect visits to flowers of annual seed mixtures. *Journal of Insect Conservation*, 6, 13–23.

(49) Denys C. & Tscharntke T. (2002) Plant-insect communities and predator-prey ratios in field margin strips, adjacent crop fields, and fallows. *Oecologia*, 130, 315–324.

(50) Evans A.D., Armstrong-Brown S. & Grice P.V. (2002) The role of research and development in the evolution of a 'smart' agri-environment scheme. *Aspects of Applied Biology*, 67, 253–264.

(51) Fluri P. & Frick R. (2002) Honey bee losses during mowing of flowering fields. *Bee World*, 83, 109–118.

(52) Meek B., Loxton D., Sparks T., Pywell R., Pickett H. & Nowakowski M. (2002) The effect of arable field margin composition on invertebrate biodiversity. *Biological Conservation*, 106, 259–271.

(53) Dramstad W.E., Fry G.L.A. & Schaffer M.J. (2003) Bumblebee foraging – is closer really better? *Agriculture Ecosystems & Environment*, 95, 349–357.

(54) Jeanneret P., Schüpbach B., Pfiffner L. & Walter T. (2003) Arthropod reaction to landscape and habitat features in agricultural landscapes. *Landscape Ecology*, 18, 253–263.

(55) May M. & Nowakowski M. (2003) Using headland margins to boost environmental benefits of sugar beet. *British Sugar Beet Review*, 71, 48–51.

(56) Powell W., Walters K., A'Hara S., Ashby J., Stevenson H. & Northing P. (2003) Using field margin diversification in agri-environment schemes to enhance aphid natural enemies. *IOBC/WPRS Bulletin*, 26, 123–128.

(57) Bokenstrand A., Lagerlöf J. & Torstensson P.R. (2004) Establishment of vegetation in broadened field boundaries in agricultural landscapes. *Agriculture, Ecosystems & Environment*, 101, 21–29.

(58) Carvell C., Meek W.R., Pywell R.F. & Nowakowski M. (2004) The response of foraging bumblebees to successional change in newly created arable field margins. *Biological Conservation*, 118, 327–339.

(59) Kromp B., Hann P., Kraus P. & Meindl P. (2004) Viennese Programme of Contracted Nature Conservation 'Biotope Farmland': monitoring of carabids in sown wildflower strips and adjacent fields. *Deutsche Gesellschaft fur allgemeine und angewandte Entomologie e.V.*, 14, 509–512.

(60) Powell W., A'Hara S., Harling R., Holland J.M., Northing P., Thomas C.F.G. & Walters K.F.A. (2004) *Managing Biodiversity in Field Margins to Enhance Integrated Pest Control in Arable Crops ('3-D farming' project)*. HGCA, 356.

(61) Briner T., Nentwig W. & Airoldi J.P. (2005) Habitat quality of wildflower strips for common voles (*Microtus arvalis*) and its relevance for agriculture. *Agriculture Ecosystems & Environment*, 105, 173–179.

(62) Jacot K., Eggenschwiler L. & Bosshard A. (2005) Botanical development of restored species rich field margins. *Agrarforschung*, 12, 10–15.

(63) Pywell R.F., Warman E.A., Carvell C., Sparks T.H., Dicks L.V., Bennett D., Wright A., Critchley C.N.R. & Sherwodd A. (2005) Providing foraging resources for bumblebees in intensively farmed landscapes. *Biological Conservation*, 121, 479–494.

(64) Woodcock B.A., Westbury D.B., Potts S.G., Harris S.J. & Brown V.K. (2005) Establishing field margins to promote beetle conservation in arable farms. *Agriculture, Ecosystems & Environment*, 107, 255–266.

(65) Alebeek F.V., Wiersema M., Rijn P.V., Wäckers F., Belder E.d., Willemse J. & Gurp H.V. (2006) A region-wide experiment with functional agrobiodiversity (FAB) in arable farming in the Netherlands.

Proceedings of the Landscape Management for Functional Biodiversity 2nd Working Group Meeting 16–19 May. Zürich-Reckenholz, Switzerland, 29. pp. 141–144.

(66)   Allcorn R.I., Akers P. & Lyons G. (2006) Introducing red clover *Trifolium pratense* to former arable fields to provide a foraging resource for bumblebees *Bombus* spp. at Dungeness RSPB reserve, Kent, England. *Conservation Evidence*, 3, 88–91.

(67)   Carvell C., Westrich P., Meek W.R., Pywell R.F. & Nowakowski M. (2006) Assessing the value of annual and perennial forage mixtures for bumblebees by direct observation and pollen analysis. *Apidologie*, 37, 326–340.

(68)   Cauwer B.D., Reheul D., D'Hooghe K., Nijs I. & Milbau A. (2006) Disturbance effects on early succession of field margins along the shaded and unshaded side of a tree lane. *Agriculture, Ecosystems & Environment*, 112, 78–86.

(69)   Herzog F., Buholzer S., Dreier S., Hofer G., Jeanneret P., Pfiffner L., Poiger T., Prasuhn V., Richner W., Schüpbach B., Spiess E., Spiess M., Walter T. & Winzeler M. (2006) Effects of the Swiss agri-environmental scheme on biodiversity and water quality. *Mitteilungen der Biologischen Bundesanstalt für Land-u. Forstwirtschaft*, 403, 34–39.

(70)   Jacot K., Beerli C. & Eggenschwiler L. (2006) Improved field margins and their effects on voles and moles. *Agrarforschung*, 14, 212–217.

(71)   Luka H., Uehlinger G., Pfiffner L., Mühlethaler R. & Blick T. (2006) Extended field margins – a new element of ecological compensation in farmed landscapes – deliver positive impacts for Articulata. *Agrarforschung*, 13, 386–391.

(72)   Pywell R.F., Warman E.A., Hulmes L., Hulmes S., Nuttall P., Sparks T.H., Critchley C.N.R. & Sherwood A. (2006) Effectiveness of new agri-environment schemes in providing foraging resources for bumblebees in intensively farmed landscapes. *Biological Conservation*, 129, 192–206.

(73)   Aschwanden J., Holzgang O. & Jenni L. (2007) Importance of ecological compensation areas for small mammals in intensively farmed areas. *Wildlife Biology*, 13, 150–158.

(74)   Carvell C., Meek W.R., Pywell R.F., Goulson D. & Nowakowski M. (2007) Comparing the efficacy of agri-environment schemes to enhance bumble bee abundance and diversity on arable field margins. *Journal of Applied Ecology*, 44, 29–40.

(75)   Carvell C., Pywell R. & Meek W. (2007) The conservation and enhancement of bumblebees in intensively farmed landscapes. *Aspects of Applied Biology*, 81, 247–254.

(76)   Heard M.S., Carvell C., Carreck N.L., Rothery P., Osborne J.L. & Bourke A.F.G. (2007) Landscape context not patch size determines bumble-bee density on flower mixtures sown for agri-environment schemes. *Biology Letters*, 3, 638–641.

(77)   Henderson I.G., Morris A.J., Westbury D.B., Woodcock B.A., Potts S.G., Ramsay A. & Coombes R. (2007) Effects of field margin management on bird distributions around cereal fields. *Aspects of Applied Biology*, 81, 53–60.

(78)   Jacot K., Eggenschwiler L., Junge X., Luka H. & Bosshard A. (2007) Improved field margins for a higher biodiversity in agricultural landscapes. *Aspects of Applied Biology*, 81, 277–283.

(79)   Marshall G.M. (2007) The effect of arable field margin structure and composition on Orthoptera assemblages. *Aspects of Applied Biology*, 81, 231–238.

(80)   Pilgrim E.S., Potts S.G., Vickery J., Parkinson A.E., Woodcock B.A., Holt C., Gundrey A.L., Ramsay A.J., Atkinson P., Fuller R. & Tallowin J.R.B. (2007) Enhancing wildlife in the margins of intensively managed grass fields. In J.J. Hopkins, A.J. Duncan, D.I. McCracken, S. Peel & J.R.B. Tallowin (eds.) *High Value Grassland: Providing Biodiversity, a Clean Environment and Premium Products. British Grassland Society Occasional Symposium No.38*, British Grassland Society (BGS), Reading. pp. 293–296.

(81)   Pywell R. & Nowakowski M. (2007) *Farming for Wildlife Project: Annual Report 2006/7*. NERC report 6441.

(82)   Pywell R.F., Meek W.M., Carvell C. & Hulmes L. (2007) The SAFFIE project: enhancing the value of arable field margins for pollinating insects. *Aspects of Applied Biology*, 81, 239–245.

(83)   Ramsay A.J., Potts S.G., Westbury D.B., Woodcock B.A., Tscheulin T.R., Harris S.J. & Brown V.K. (2007) Response of planthoppers to novel margin management in arable systems. *Aspects of Applied Biology*, 81, 47–52.

(84)   Edwards M. (2008) *Syngenta Operation Bumblebee Monitoring Report 2005–2008*. Syngenta.

(85)   Gardiner T., Edwards M. & Hill J. (2008) Establishment of clover-rich field margins as a forage resource for bumblebees *Bombus* spp. on Romney Marsh, Kent, England. *Conservation Evidence*, 5, 51–57.

(86)   Hovd H. (2008) Occurrence of meadow herbs in sown and unsown ploughed strips in cultivated grassland. *Acta Agriculturae Scandinavica. Section B, Plant Soil Science*, 58, 208–215.

(87)   Kohler F., Verhulst J., van Klink R. & Kleijn D. (2008) At what spatial scale do high-quality habitats enhance the diversity of forbs and pollinators in intensively farmed landscapes? *Journal of Applied Ecology*, 45, 753–762.

(88)   Pywell R., Hulmes L., Meek W. & Nowakowski M. (2008) *Creation and Management of Pollen and Nectar Habitats on Farmland: Annual report 2007/8*. NERC report 6443.

(89)   Pywell R. & Nowakowski M. (2008) *Farming for Wildlife Project: Annual Report 2007/8*. NERC report 6366.

(90) Sheridan H., Finn J.A., Culleton N. & O'Donovanc G. (2008) Plant and invertebrate diversity in grassland field margins. *Agriculture Ecosystems & Environment*, 123, 225–232.

(91) Woodcock B.A., Westbury D.B., Tscheulin T., Harrison-Cripps J., Harris S.J., Ramsey A.J., Brown V.K. & Potts S.G. (2008) Effects of seed mixture and management on beetle assemblages of arable field margins. *Agriculture, Ecosystems & Environment*, 125, 246–254.

(92) Aviron S., Nitsch H., Jeanneret P., Buholzer S., Luka H., Pfiffner L., Pozzi S., Schüpbach B., Walter T. & Herzog F. (2009) Ecological cross compliance promotes farmland biodiversity in Switzerland. *Frontiers in Ecology and the Environment*, 7, 247–252.

(93) Fischer J., Jenny M. & Jenni L. (2009) Suitability of patches and in-field strips for Skylarks *Alauda arvensis* in a small-parcelled mixed farming area. *Bird Study*, 56, 34–42.

(94) Frank T., Aeschbacher S., Barone M., Kunzle I., Lethmayer C. & Mosimann C. (2009) Beneficial arthropods respond differentially to wildflower areas of different age. *Annales Zoologici Fennici*, 46, 465–480.

(95) Haenke S., Scheid B., Schaefer M., Tscharntke T. & Thies C. (2009) Increasing syrphid fly diversity and density in sown flower strips within simple vs. complex landscapes. *Journal of Applied Ecology*, 46, 1106–1114.

(96) Pfiffner L., Luka H., Schlatter C., Juen A. & Traugott M. (2009) Impact of wildflower strips on biological control of cabbage lepidopterans. *Agriculture Ecosystems & Environment*, 129, 310–314.

(97) Potts S.G., Woodcock B.A., Roberts S.P.M., Tscheulin T., Pilgrim E.S., Brown V.K. & Tallowin J.R. (2009) Enhancing pollinator biodiversity in intensive grasslands. *Journal of Applied Ecology*, 46, 369–379.

(98) Schmidt-Entling M.H. & Dobeli J. (2009) Sown wildflower areas to enhance spiders in arable fields. *Agriculture Ecosystems & Environment*, 133, 19–22.

(99) Vickery J.A., Feber R.E. & Fuller R.J. (2009) Arable field margins managed for biodiversity conservation: a review of food resource provision for farmland birds. *Agriculture, Ecosystems & Environment*, 133, 1–13.

(100) Arlettaz R., Krahenbuhl M., Almasi B., Roulin A. & Schaub M. (2010) Wildflower areas within revitalized agricultural matrices boost small mammal populations but not breeding Barn Owls. *Journal of Ornithology*, 151, 553–564.

(101) Haaland C. & Gyllin M. (2010) Butterflies and bumblebees in greenways and sown wildflower strips in southern Sweden. *Journal of Insect Conservation*, 14, 125–132.

(102) Pywell R.F., Meek W.R., Hulmes L. & Nowakowski M. (2010) Creation and management of pollen and nectar habitats on farmland. *Aspects of Applied Biology*, 100, 369–374.

(103) Pywell R.F., Meek W.R., Hulmes L. & Nowakowski M. (2010) Designing multi-purpose habitats: utilisation of wild bird seed species by pollinating insects. *Aspects of Applied Biology*, 100, 421–426.

(104) Redpath N., Osgathorpe L.M., Park K. & Goulson D. (2010) Crofting and bumblebee conservation: the impact of land management practices on bumblebee populations in northwest Scotland. *Biological Conservation*, 143, 492–500.

(105) Smith H., Feber R.E., Morecroft M.D., Taylor M.E. & Macdonald D.W. (2010) Short-term successional change does not predict long-term conservation value of managed arable field margins. *Biological Conservation*, 143, 813–822.

Additional references

Klinger K. (1987) Auswirkungen eingesäter Randstreifen an einem Winterweizen-Feld auf Raubarthropodenfauna und den Getreideblattlausbefall [Effects of field margin strips along a winter wheat field on predatory arthropods and cereal aphids]. *Journal of Applied Entomology*, 104, 47–58.

Nentwig W. (1992) Die nühtzlingsfördernde Wirkung von Unkrautern in angesäten Unkrautstreifen [Augmentation of beneficial arthropods by sown weed strips in agricultural areas]. *Z. Pflanzenkrankheiten and Pflanzenschuzt, Sonderheft*, 33, 33–40.

Baumann L. (1996) The influence of field margins on populations of small mammals – a study of the population ecology of the common vole (*Microtus arvalis*) in sown weed strips. *Field Margin Newsletter*, 6/7, 26–33.

Wilson S., Baylis M., Sherrott A. & Howe G. (2000) *Arable Stewardship Project Officer Review*. Farming and Rural Conservation Agency report.

ADAS (2001) *Ecological Evaluation of the Arable Stewardship Pilot Scheme, 1998–2000*. ADAS report.

Edwards M. & Williams P.H. (2004) Where have all the bumblebees gone, and could they ever return? *British Wildlife*, 15, 305–312.

1.20   Manage the agricultural landscape to enhance floral resources

- One large replicated controlled trial showed that the average abundance of long-tongued bumblebees on field margins was positively correlated with the number of 'pollen and nectar' agri-environment agreements in a 10 km grid square [1].

Background

Managing landscapes to enhance nectar and pollen resources for flower-visiting insects is increasingly recognized as an important strategy to enhance the agricultural pollination service and to conserve pollinator populations. It could involve increasing the diversity or area of flowering crops or conserving aspects of the landscape, such as flower-rich meadows, woodlands or river banks, which provide important floral resources.

In Europe, recent research has shown that higher coverage of the mass-flowering crop oilseed rape *Brassica napus* in the landscape is associated with higher numbers of foraging worker bumblebees *Bombus* spp. at focal sampling points, but not with enhanced bumblebee reproductive success or colony densities (Westphal *et al.* 2003; Herrmann *et al.* 2007; Westphal *et al.* 2009). This work is not summarized by Conservation Evidence because the evidence is correlative: the area of flowering crops was not increased as a conservation measure.

Westphal C., Steffan-Dewenter I. & Tscharntke T. (2003) Mass flowering crops enhance pollinator densities at a landscape scale. *Ecology Letters*, 6, 961–965.

Herrmann F., Westphal C., Moritz R.F.A. & Steffan-Dewenter I. (2007) Genetic diversity and mass resources promote colony size and forager densities of a social bee (*Bombus pascuorum*) in agricultural landscapes. *Molecular Ecology*, 16, 1167–1178.

Westphal C., Steffan-Dewenter I. & Tscharntke T. (2009) Mass flowering oilseed rape improves early colony growth but not sexual reproduction of bumblebees. *Journal of Applied Ecology*, 46, 187–193.

A replicated, controlled trial in 2004 in thirty-two 10 km grid squares across England (1) found the abundance of long-tongued bumblebees *Bombus* spp., mostly common carder bee *B. pascuorum* and garden bumblebee *B. hortorum* recorded on trial field margins (various planting treatments, including sown grass and wildflower margins) was positively correlated with the total number of pollen and nectar mix agri-environment agreements in each 10 km square. There is no record of the numbers of long-tongued bumblebees in these grid squares before the agreements were implemented. Bumblebees were counted on a 100 x 6 m transect in each of 151 field margins, once in July and once in August.

(1)   Pywell R.F., Warman E.A., Hulmes L., Hulmes S., Nuttall P., Sparks T.H., Critchley C.N.R. & Sherwood A. (2006) Effectiveness of new agri-environment schemes in providing foraging resources for bumblebees in intensively farmed landscapes. *Biological Conservation*, 129, 192–206.

1.21   Create uncultivated margins around intensive arable or pasture fields

- Thirty-nine studies (including 13 replicated controlled trials, of which 3 were also randomized, and 4 reviews) from 8 European countries compared wildlife on uncultivated margins with other margin options. Twenty-four found benefits to some wildlife groups (including 11 replicated controlled trials, of which one was also randomized, and 4 reviews). Nineteen studies (including one randomized, replicated controlled trial) from

Germany, Ireland, Lithuania, Norway, the Netherlands and the UK found that uncultivated margins support more invertebrates (including bees) and/or higher plant diversity or species richness than conventionally managed field margins[1, 2, 4, 7, 8, 13, 17, 18, 20, 24, 38, 39, 41–43, 45] or other field margin options[1, 2, 11, 18, 25, 26, 29, 37, 46]. One replicated, controlled study[35] showed that uncultivated margins supported more small mammal species than meadows and farmed grasslands. Four studies (two replicated UK studies; two reviews) reported positive associations between birds and field margins[19, 34 ,46, 48] including food provision. A review from the UK found grass margins (including naturally regenerated margins) benefited plants and some invertebrates[23].

- Fifteen studies (including one randomized, replicated, controlled trial) from Germany, the Netherlands, Norway and the UK found that invertebrate and/or plant species richness or abundance were lower in naturally regenerated than conventionally managed fields[18, 24, 31, 32, 42] or sown margins[2, 3, 5, 7, 9, 10, 12, 15, 16, 18, 24, 25, 29, 41, 45, 49]. Six studies (including one randomized, replicated, controlled trial) from Belgium, Germany and the UK found that uncultivated margins did not have more plant or invertebrate species or individuals than cropped[24] or sown margins[6, 12, 14, 15, 33]. A review found grass margins (including naturally regenerated margins) did not benefit ground beetles[23].

- Five studies (including three replicated controlled trials) from Ireland and the UK reported declines in plant species richness[43, 49] and invertebrate numbers[5, 28, 31, 36] in naturally regenerated margins over time. One replicated trial[22] found that older naturally regenerated margins (6-years-old) had more invertebrate predators (mainly spiders) than newly established (1-year-old) naturally regenerated margins.

- Five studies (including one replicated randomized trial) from the Netherlands and the UK found that cutting margins had a negative impact on invertebrates[3, 5, 9, 10, 13, 14, 16, 21, 24, 31, 32] or no impact on plant species[49]. One replicated controlled study found cut margins were used more frequently by yellowhammers when surrounding vegetation was >60 cm tall[44].

- Seven studies (including four replicated controlled trials and a review) from Ireland, the Netherlands, Norway and the UK reported increased abundance or biomass of weed species in naturally regenerated margins[11, 24, 26, 28, 41, 43, 46].

Background

This intervention allows the field margin vegetation to regenerate naturally, without planting, although it can involve subsequent mowing. The field margins are not fertilized and only spot-treated with herbicides if injurious weeds occur.

See also 'Plant grass buffer strips/margins around arable or pasture fields'; 'Leave uncropped, cultivated margins or plots (includes 'lapwing plots')'.

A replicated study in 1988 in East Anglia, UK (1) found that ground beetles (Carabidae), true bugs (Heteroptera) and spiders (Araneae) were more abundant in uncropped headlands than cropped conservation (restricted pesticides) and conventional headlands. For each group, significantly more individuals were found in uncropped headlands (average number of individuals per site: spiders: 210; ground beetles: 260; true bugs: 50) than in conservation or sprayed headlands or in crops (spiders: 100–110; ground beetles: 70–160; true bugs: 15–30). Ground beetles were twice as abundant in crops adjacent to uncropped and conservation headlands than adjacent to sprayed headlands. Numbers of species were higher in un-cropped headlands (ground beetles: 21 species; true bugs: 5; spiders: 26) than conservation and sprayed headlands (ground beetles: 15–18 species; true bugs: 2–3; spiders: 15–17). Spider diversity was significantly higher in uncropped (Simpson's index: 6) than conservation and sprayed headlands and in the crop (2–3); ground beetles (4–8) and true bug (1–3) diversity did not differ. True bug nymphs *Nabis ferus* penetrated further into crops adjacent to uncropped and conservation headlands than sprayed headlands. Headlands represented the outer 6 m of 8 barley fields at 3 locations. Parallel grids of 6 x 50 m were set up in the headlands and the crop (8–14 m from the headland) and sub-divided into fifteen 10 x 2 m sections. One pit-fall trap was placed in each section (15 traps/grid). A further 2 x 50 m grid was set up in the verge parallel to the field margin and divided into five 2 x 10 m sections (5 traps/grid). Traps were emptied after 14 days in June–July 1988. A Dietrick Vacuum sampler was used along 5 transect lines (0–15 m into the crop); 2 samples consisting of 5 subsamples (each 0.4 m^2) were taken 3 weeks apart.

Further results for ground beetles (Carabidae) from the same study (1) are presented in a second paper (2) which found that ground beetles were more abundant in uncropped head-lands than conservation headlands (restricted pesticides), fully sprayed headlands and crops. There were significantly more ground beetles on uncropped headlands (3–21/trap) than fully sprayed headlands (3–6/trap) or the main crop (3–9/trap). Conservation headlands ten-ded to have lower numbers than uncropped headlands (3–14/trap). There also tended to be more ground beetles in the crop adjacent to uncropped headlands than conservation or fully sprayed headlands, but the difference was not significant. There were significantly more ground beetle species (total number of species across all sites) on uncropped headlands (36 species across 3 sites) compared to sprayed headlands (conservation: 32 across 4 sites; fully sprayed: 24 across 2 sites) or the crop (31 in 8 sites). There was no significant difference between the vegetation cover under different treatments. Plant cover was also measured in five 25 x 25 cm quadrats in each grid.

A randomized, replicated trial from 1987 to 1991 in Oxfordshire, UK (3) found more adult meadow brown butterflies *Maniola jurtina* on 2 m-wide naturally regenerated field mar-gins left uncut, or cut in spring or autumn than on margins cut in summer (4–10 meadow browns/50 m with summer cut; 4–15 meadow browns/50 m without). Unsown margins had 4–10 meadow browns/50 m in 1991 and 1992, fewer than margins sown with a wildflower mix (4–52 meadow browns/50 m). There was no difference between treatments in abund-ance of meadow brown larvae (3 larvae/plot on average). There were more meadow browns on all of the experimental field margins than on narrow, unmanaged field boundaries of a neighbouring farm (numbers not given). Two metre-wide field margins were established around arable fields in October 1987. They were either left to naturally regenerate or sown with a wildflower seed mix in March 1988. Both treatments were rotavated before sowing. Fifty metre-long plots were managed in one of the following ways: uncut; cut once in June hay collected; cut April and June hay collected; cut in April and September hay collected; cut April and June hay left lying (unsown margins only); sprayed once a year in summer (unsown margins only). There were six replicates of each treatment. Adult meadow brown butterflies were monitored weekly along walked transects in the experimental plots from June to September 1989 and from April to September 1990 and 1991. Meadow brown larvae were sampled in spring 1991 by sweep netting and visual searching. This study was part of the same experimental set-up as (5, 13, 14, 16, 21, 49).

A small replicated, controlled study from 1990 to 1992 in East Anglia, UK (4) found that ground beetles (Carabidae) were more abundant in uncropped headlands than conservation headlands (cropped, no herbicides or insecticides) and sprayed headlands (as main wheat field). Uncropped strips had a significantly greater abundance of ground beetles (2,487) compared to conservation (1,474) and sprayed headlands (938). Species diversity tended to be higher in uncropped headlands (43) compared to conservation (41) and sprayed headlands (35). Different species reacted differently to treatments. There were a number of species that were restricted to uncropped or conservation headlands and one restricted to sprayed headlands. Numbers of species and overall abundance varied with season. Two 120 m strips of each treatment were established in a randomized block design along one headland of a 19 ha wheat field. Ground beetles were sampled using 3–5 pitfall traps in the middle of each plot, 3 m from the field boundary. Catches were collected every 1–2 weeks from February to August. Aphid numbers were also sampled but are not presented here.

A randomized, replicated study from 1989 to 1991 in Oxfordshire, UK (5) found that butterfly (Lepidoptera) abundance and species richness was lower in unsown, naturally generated margins (14–39 individuals; 6–9 species) than in sown wildflower margins (21–91 individuals; 7–10 species) from the second year after establishment. Spraying with herbicides (Roundup[TM]) and cutting during summer reduced butterfly diversity and density in the margins, but there were no such effects of cutting in spring and autumn. Both cutting in summer and spraying led to an immediate decline in the number of flowering plants directly after the treatment. In the cut margins, however, the number of flowers had increased by September when it was higher than in uncut margins. Butterflies were monitored weekly along transects from June to September 1989 and from April to September 1990 and 1991. Transects were divided into 50 m sections corresponding to the experimental plots. Monitoring was done according to standard methods and only under suitable conditions. This study was part of the same experimental set-up as (3, 13, 14, 16, 21, 49).

A replicated, controlled, randomized study of four field margins in three Environmentally Sensitive Areas in England (6) found that plant cover was higher in margins sown with grass or grass/wildflower mixtures than those naturally regenerated, but plant diversity within naturally regenerated margins was similar to some margins sown with diverse seed mixtures. In 1994 plant diversity was higher in plots sown with more complex seed mixtures (32–37) than those sown with grass only (22–27) or regenerated naturally (21–25). In 1995 grass seed only plots tended to be the least diverse (15–21), but naturally regenerated plots (18–28) were as diverse as some complex seed mixtures (23–31). Species diversity did not differ between management treatments. Margins were created in each field and divided into 6 plots (4 x 30 m). Each was (randomly) sown with a seed mixture: grass, low cost mix (3 grass: 7 wildflower), alkaline soil (6:16), neutral soil (5:15), acid soil (6:16) and one natural regeneration. Plots were divided into 10 m sub-plots, which were either unmanaged, cut once, or received grass herbicide. Plants were sampled in each sub-sub-plot in summer 1994–1995. The three Environmentally Sensitive Areas studied were the Breckland Environmentally Sensitive Area in Suffolk, the Somerset Levels and Moors Environmentally Sensitive Area and the South Wessex Downs Environmentally Sensitive Area in Wiltshire and Dorset. The same study is presented in Marshall et al. 1994.

A replicated, controlled study in summer 1995 in an intensively farmed landscape near Göttingen, Germany (7) found higher arthropod species richness on potted mugwort Artemisia vulgaris plants placed in uncultivated margins (one- and six-years-old) compared to a cereal field but not compared to other margin types. The predator-prey ratio was significantly higher in the six-year-old margin than in all other margin types and the control. The effect of uncultivated margins on individual arthropod numbers was species-dependent but slightly more individuals were found in the one-year-old than in the six-year-old uncultivated margins. Investigated margin types besides the two types of uncultivated margin were wildflower strips (wildflower seed mixture or *Phacelia* spp. only) and cereal strips/headlands. Potted mugwort plants (four pots) were placed in all margin types and the control (one winter wheat field). All herbivores and their predators on the plants were recorded during six visits in June and July. In September all mugwort plants were dissected in the lab to assess numbers of arthropods feeding inside the plants. Results from the same study are also presented in (8, 22).

A replicated, controlled study in summer 1995 near Göttingen, Germany (8) found higher species richness of arthropods colonizing potted mugwort *Artemisia vulgaris* plants in naturally regenerated margins than in unsprayed cereal control edges. However arthropod species numbers on mugwort did not differ between any of the established margin types or between mugwort plants placed in one- or six-year-old regenerated margins and mugwort plants in larger set-aside areas of the same vegetation and age. Effects of the margins on individual abundance was not clear, but polyphagous spiders of the genus *Theridion* were recorded in significantly higher numbers on mugwort in 6-year-old regenerated margins than in 1-year-old margins, wildflower strips and sprayed cereal edges. Besides the 1- and 6-year-old naturally regenerated margins, wildflower strips (19 species sown), *Phacelia* strips (*P. tanacetifolia* plus 3 species), sown cereal strips and cereal control edges were investigated. Potted mugwort plants (four pots) were placed in all margin types and the controls. Mugwort plants were visited six times in June and July to count all herbivores and their predators on the plants before being taken to the lab in September to assess all arthropods feeding inside the plants. Vegetation of all margins was surveyed in June. Results from the same study are also presented in (7, 22).

A replicated study in 1994–1996 in Gloucestershire, UK (9) (same study as (10)) found that plant species richness, as well as abundance and diversity of butterflies (Lepidoptera), was lower in naturally regenerated margins than in sown wildflower margins (for plants: 19 vs 23 species). Cutting and subsequent grazing of naturally regenerated margins significantly decreased butterfly diversity (3 vs 6 species) but not abundance (5 vs 10 individuals). Margins were established around two organically-managed arable fields by either sowing a seed mix (containing five grasses; six forbs) or by natural regeneration in 1994. In 1996 part of the margins were cut in June and grazed in July. The rest was left untreated. Butterflies were monitored along transects weekly from May to September 1996. Abundance of all plants present as well as flower abundance at the time of the survey was recorded in May and in September 1996.

A replicated study in summer 1996 in Gloucestershire, UK (10) (same study as (9)) found lower overall butterfly (Lepidoptera) abundance and species richness in ten naturally regenerated experimental plots than in ten plots sown with a wild grass/flower seed mixture. Vegetation removal (plots cut for silage in June; grazed by cattle in July) had no effect on butterfly abundance, however butterfly species richness was lower in cut/grazed plots. Plant species richness was on average lower in naturally regenerated plots than in sown plots (19 vs 23 species). Vegetation removal had no effect on plant species richness but non-defoliated plots had more wildflower species in flower in July. In September 1994 20 contiguous 50 m-long experimental plots were created in the margins of 2 adjoining organic fields on one farm by widening the existing 0.5 m margin to 2 m width. Presence of all sown and unsown plant species were recorded as well as wildflowers in flower (May and July 1996). Butterflies were monitored weekly in June–September along a transect route.

A replicated trial from 1993 to 1996 on farmland near Wageningen, the Netherlands (11) found that 4 m-wide field margins left to naturally regenerate had more plant species than margins sown with rye grass *Lolium perenne* 2 years after establishment. On average there were 9 plant species/0.25 m^2 in naturally regenerated margins, compared to 6 in grass-sown margins and 14 in margins sown with 30 non-grass wildflower species. Two prominent arable weeds, creeping thistle *Cirsium arvense* and couch grass *Elymus repens*, both had higher biomass in the naturally regenerated margins than in wildflower or grass-sown margins (33 g/m^2 and 28 g/m^2 respectively in naturally regenerated plots, compared to 0–8 g/m^2 and 6–9 g/m^2 in sown margins). In 1993 experimental plots (8 x 4 m) were established on boundaries of 3 arable fields. All plots were mown once a year without removing cuttings. There were three replicates of each treatment on each field. Plant biomass and number of species were measured in eight 0.5 m x 0.5 m plots on a single transect line across each margin in August 1995.

A replicated study in summer 1996 in central Germany (12) found that spider (Araneae) abundance in naturally regenerated plots (97 and 56 individuals in pitfall traps and photoeclectors respectively) was significantly lower than plots containing sundial lupin *Lupinus perenne* and common vetch *Vicia sativa* (155 and 124 individuals in pitfall traps and photoeclectors) and similar to plots with fodder radish *Raphanus sativus oleiferus* (104 and 49 individuals). Note that most results in this study were not statistically tested. Eight different types of strip with 3 replicates each were tested: 6 seed mixtures contained mainly flowering plants (1–12 species), 1 mixture contained mainly grass seeds (2 species plus white clover *Trifolium repens*) and 1 naturally regenerated treatment. Spiders were sampled using two pitfall traps and two photoeclectors in each plot.

A randomized, replicated, before-and-after trial from 1987 to 1991 in Oxfordshire, UK (13) found that spider (Araneae) abundance and species richness were higher after field margins were established on unmanaged plots (from <5 species at the start of the study to between 5 and 12 species following field margin establishment). Naturally regenerated field margins had fewer spiders but not fewer spider species, than field margins sown with a wildflower seed mix on all dates. Cutting, especially summer cutting, significantly reduced the abundance of spiders. Spraying with herbicide reduced the numbers of spiders, but not the number of spider species, relative to control plots in two of the three years. Spiders were sampled using a suction trap (D-Vac) in September 1987 and 1988, and in May, July and September in 1989, 1990 and 1991. This study was part of the same experimental set-up as (3, 5, 14, 16, 21, 49).

A randomized, replicated trial from 1987 to 1996 in Oxfordshire, UK (14) found that pseudoscorpions (Pseudoscorpionida) favoured unmanaged field margins (not cut or sprayed), but there was no difference in numbers between naturally regenerated margins and those sown with a wildflower mix. More pseudoscorpions (*Chthonius ischnocheles* and *C. orthodactylus*) were found in unmanaged field margin plots (95 pseudoscorpions in total on sown and unsown plots) than in cut or sprayed treatments (19–53 pseudoscorpions). Plots cut in April and June with hay removed, or sprayed with herbicide in summer, had fewer pseudoscorpions than other margins (19 and 21 pseudoscorpions respectively). Plots cut just once in June, cut twice but not in June or cut in April and June but with hay left lying, had intermediate numbers of pseudoscorpions (29, 53 and 30 pseudoscorpions). Pseudoscorpions were sampled from the litter layer (not the soil) using a suction trap (D-Vac) in May, July and September 1995 and 1996. This study was part of the same experimental set-up as (3, 5, 13, 16, 21, 49).

A trial from 1995 to 1998 in Hampshire, UK (15) found the same number of bee (Apidae) species (9), but fewer flowering plant, fly (Diptera) and butterfly (Lepidoptera) species, on a single naturally regenerated field margin strip established for 3 years than on 3 strips sown with a diverse wildflower seed mix in the same study (16, 4 and 6 species respectively on the naturally regenerated margin vs 24, 7 and 8 species on the sown margins). The field margins were established (or sown) in 1995. The number of flowers and flower-visiting bees, wasps, flies and butterflies were counted on a 200 x 2 m transect in each strip, once a month from May to August 1998.

A randomized, replicated study from 1995 to 1996 in Oxfordshire, UK (16) found that total numbers of invertebrates and leafhoppers (Auchenorrhyncha) were significantly lower in unsown, naturally generated margins than in sown wildflower margins. Cut plots (in summer alone, spring and summer or spring and autumn) had significantly lower numbers of all invertebrates, spiders (Araneae), true bugs (Heteroptera) and leafhoppers than uncut plots in all seasons, apart from spiders and true bugs in May. There was no effect of cutting frequency or timing or leaving/removing hay on invertebrate numbers. Invertebrates were sampled using a D-Vac suction sampler at 10 m intervals along each plot in May, July and September in 1995–1996. This study was part of the same experimental set-up as (3, 5, 13, 14, 21, 49).

A 1999 review of literature (17) found that uncropped field margins, left to naturally regenerate, were shown to increase ground beetle (Carabidae) numbers by three studies (Müller 1991, (2), plus one unpublished study). In one case (2) there were more beetles than in conventional crop margins.

A randomized, replicated, controlled trial from 1993 to 1996 in Bristol, UK (18) found that 4 m-wide field margins left to naturally regenerate had more suction-sampled invertebrates but not more ground beetles (Carabidae) than control cropped margins or margins sown with grass. There were around 180 invertebrates per sample on naturally regenerated margins, compared to 110–130 invertebrates/sample on control or grass-sown plots. There was no difference in the number of ground beetle species (average of eight species/plot), nor in the numbers of the four most commonly caught ground beetle species, between margin types. In a 2 m-wide margin there were more over-wintering invertebrates in the soil of the wildflower sown half than the naturally regenerating half, but this difference was not found in 4 m-wide replicated experimental plots. Three field margins were established in spring 1993 at one site. Experimental plots 10 x 4 m were either sown with arable crop (control), rye grass *Lolium perenne* or a wildflower and grass seed mix, or left to naturally regenerate. There were three replicate plots in each margin. All plots were cut annually after harvest, and cuttings left in place. Another 100 x 2 m-wide field margin, with 50 m sown with a wildflower mix and 50 m unsown, was used to monitor wintering invertebrates. Ground beetles were sampled in eight pitfall traps in or near each margin for a week in June for four years, 1993–1996. Invertebrates were sampled using a vacuum sampler on plots of two of the three margins in June 1994. Arthropods were extracted from soil samples taken from plots of two margins in December 1993 and February 1994.

A 2000 literature review (19) found that the UK population of Eurasian thick-knees *Burhinus oedicnemus* increased from 150 pairs in 1991 to 233 in 1999, following an agri-environment scheme designed to provide uncultivated plots in fields and set-aside.

A replicated, controlled trial in 1999 of arable field margins in the UK (20) found that margins allowed to regenerate naturally for one year supported significantly more honey bees *Apis* spp. and bumblebees *Bombus* spp. than unsprayed cropped margins managed as conservation headlands (average 10–50 bees/transect on naturally regenerated margins compared to <3 bees/transect in conservation headlands). The trial was replicated once on each of five farms, with two uncropped field margins and one control conservation headland margin per farm. Margins were 4–6 m wide and located on the boundary of spring-sown cereal fields. Transects (0.5 x 50 m^2) parallel to the field edge were walked at 8–10 day intervals over a 40-day period in each margin to record bee numbers, species and flower preferences.

A randomized, replicated trial from 1987 to 1996 in Oxfordshire, UK (21) found greater numbers of predatory sheet web spiders *Lepthyphantes tenuis* on field margins left uncut and unsprayed with herbicide. In September, when most of the spiders were caught, there were significantly fewer *L. tenuis* spiders in margins (sown and unsown) that were cut in June (around 10 spiders/m², compared to >15 spiders/m² in plots cut in spring and autumn, or not cut). In May and July plots with a recent cut (April or June-cut treatments respectively) also had lower numbers of *L. tenuis* than other plots. Spraying unsown plots with herbicide reduced the numbers of *L. tenuis* later in the same year (average 4 and 10 spiders/m² in sprayed plots in July and September respectively, compared to 8 and 20 spiders/m² on un-sprayed plots in July and September). Plots where the vegetation was cut but not removed did not have more spiders than plots where cut vegetation was removed. *L. tenuis* individuals were counted in invertebrate samples collected using a suction trap (D-Vac) in May, July and September 1990, 1991, 1995 and 1996. This study was part of the same experimental set-up as (3, 5, 13, 14, 16, 49).

A replicated study from April to September 1995 near Göttingen, Germany (22) found higher predator abundance (mainly spiders Araneae) and higher predator-prey ratios in 6-year-old than in 1-year-old naturally developed field margins. In addition, predator-prey ratios were higher in large, naturally developed fallows than in the field margins. These results emphasize the importance of habitat age and area for the establishment of natural enemy populations. However, arthropod species richness in naturally developed margins did not differ from other margin types. Potted plants of mugwort *Artemisia vulgaris* (four pots per margin) and red clover *Trifolium pratense* (three pots per margin) were used to study plant-arthropod communities. Red clover pots were also set out in winter wheat fields at 4, 8 and 12 m distances adjacent to strips sown with cereal and wildflower mix. Red clover pots were set out in April 1995. On five visits in June and July 1995, flower heads of red clover were sampled, dissected and the larvae and pupae of arthropods feeding inside the plants reared in the lab for species determination. Results from the same study are also presented in (7, 8).

A review (23) of two reports (Wilson *et al.* 2000; ADAS 2001) evaluating the effects of the Pilot Arable Stewardship Scheme in two regions (East Anglia and the West Midlands) from 1998 to 2001 found that grass margins benefited plants, bumblebees *Bombus* spp., true bugs (Hemiptera) and sawflies (Symphyta) but not ground beetles (Carabidae). The grass margins set of options included sown grass margins, naturally regenerated margins, beetle banks and uncropped cultivated wildlife strips. The review does not distinguish between these, although the beneficial effects were particularly pronounced on sown or naturally regenerated grassy margins for true bugs. The effects of the pilot scheme on plants and invertebrates (bumblebees; true bugs; ground beetles; sawflies) were monitored over three years, relative to control areas. Grass margins were implemented on total areas of 361 and 294 ha in East Anglia and West Midlands respectively.

A small replicated, controlled trial in the summer of 2000 in North Yorkshire, UK (24) found that four naturally regenerated field margins had higher plant diversity, but not more bumblebees *Bombus* spp. or butterflies (Lepidoptera) (species or individuals) than four cropped margins. A number of rare or uncommon arable weeds were recorded in naturally regenerated margins, but also a much higher abundance of barren brome *Anisantha sterilis* than in any other treatment. Spring numbers of ground beetles (Carabidae) and ground-dwelling spiders (Araneae) were higher in naturally regenerated margins than cropped margins. Harvestmen (Opiliones) avoided naturally regenerated margins in favour of any sown habitat in autumn. Four margins of winter cereal fields, all adjacent to hedges, were split into 72 m long plots and sown in September 1999 with either grass, grass and wild flowers, cereal crop or left to regenerate naturally on two farms. Ground and canopy-dwelling invertebrates, butterflies and plants were surveyed from late April to late September 2000 using pitfall traps, sweep netting, transects and quadrats.

A replicated trial in 2001–2002 in the UK (25) found that margins of sugar beet *Beta vulgaris* fields left to naturally regenerate had more invertebrates (individuals and species), but not more plant species, than margins sown with wildflowers, crops or grasses. Naturally regenerated margins had over 1,700 invertebrates in total, from 45 groups. However the difference in invertebrate numbers between different treatments was fairly small (over 900 to over 1,700 individuals; 35–45 groups caught). Naturally regenerated margins had around 17 plant species/m, compared to 35 plant species/m on wildflower margins, 15 species/m for grass margins and 6–11 species/m for barley or beet margins. In autumn 2001 50 m x 6 m margins at the edges of beet fields were planted with either sugar beet, spring barley, grasses (8 species), nothing (natural regeneration) or wildflowers. There were two replicates of each treatment at each of three sites. In summer 2002 plants (including crop plants) were counted in the margins and invertebrates sampled using pitfall traps, set for two weeks.

A replicated trial from 1998 to 2000 in Wiltshire, UK (26) found that naturally regenerated field margins had more undesirable weed species than sown field margin plots, but more predatory beetles (Coleoptera) in the second year. There was no difference in the total abundance of invertebrates between field margin treatments. Eleven 100 x 2 m field margin plots were left to regenerate naturally on one farm. Thirty-eight were sown with a grass seed mix of either 3 grass species (12 plots), 6 grass species (13 plots) or 6 grass and 4 herb species (13 plots) in autumn 1998. The plots were around four fields under a Countryside Stewardship Agreement on the Harnhill Manor Farm. Invertebrates were sampled using pitfall traps (five traps/plot) in spring and autumn and suction traps in summer. Plants were recorded in four 1 m^2 quadrats/plot in summer.

A replicated, controlled study between 1988 and 1997 in Sweden (27) found higher plant species richness in experimental field margin plots allowed to regenerate naturally than in plots sown with a clover and grass seed mixture after one year. Seven years after establishment naturally regenerated plots, clover and grass plots and control boundaries had higher cover of weeds in total and of couch grass *Elytrigia repens* than plots planted with rose bushes *Rosa canina* and/or meadow plants. Couch grass increased in all treatments but significantly so in naturally regenerated plots and plots with clover and grass. Plots with meadow plants and naturally regenerated plots had similar species richness but quite different species compositions due to a high cover of annual weeds in the latter. In 1990 four replicates of each treatment (naturally regenerated, planted with rose bushes and/or sown with meadow plants, sown with clover *Trifolium* spp. and grass mixture) were established randomly along the stretch of a previously removed dirt road. All plots were cut annually in late summer and the cuttings removed. Vegetation surveys were carried out twice in experimental plots (1991 and 1997) and once in control boundaries (1997) in three to five 0.25 m^2 quadrats. It is not clear whether the results for clover and grass plots were a direct result of planting nectar flowers or grass.

A replicated, controlled trial from 1999 to 2002 on arable field margins in North Yorkshire, UK (28) found 6 m-wide naturally regenerated, uncultivated field margin plots supported significantly more foraging bumblebees *Bombus* spp. than margins sown with tussocky grass, or control cropped field margins, but only in one year (2001) of this three year study. In 2001 the bumblebees were mostly foraging on spear thistle *Cirsium vulgare*, a pernicious agricultural weed that had to be controlled by cutting at the end of that summer. In the other two years (2000 and 2002) the naturally regenerated field margins did not support significantly more bumblebees than the control or grass-sown sites. Naturally regenerated margins were the only treatment that did not support consistent numbers of bumblebees in all three years. The naturally regenerated field margins supported fewer bumblebees (18 individuals and 2.7 species/100 m on average) than margins sown with a wildflower seed mixture (29 individuals; 3.0 species/100 m) but the two treatments were not directly compared in the analysis. Three cereal field margins on one farm were divided into five 72 m x 6 m long plots and subjected to 5 different treatments: natural regeneration (6 m wide), sown 'tussocky' grass mixture (6 m wide), sown 'grass and wildflower' mixture (6 m wide), split treatment of 3 m-wide 'tussocky' grass mixture adjacent to hedge and 3 m-wide sown 'grass and wildflower' mixture adjacent to crop, and margin cropped to the edge. Plots were cut and herbage removed following establishment of the seed mixtures. Wildflower plots were cut in August 2001 and 2002 and the herbage removed. Transects were walked along the central line of each plot recording bumblebee activity and identifying foraging bumblebees to species level.

A replicated study in the summers of 1999–2000 on arable farms in the UK (29) found that naturally regenerated grassy margins had more plant species than sown grassy margins, but were not considered one of the best options for the conservation of annual herbaceous plant communities. The naturally regenerated margins were dominated by three grasses (different species from the sown margins) and thistles. Average numbers of plant species in the different conservation habitats were wildlife seed mixtures 6.7; uncropped cultivated margins 6.3; undersown cereals 5.9; naturally regenerated grass margins 5.5; no-fertilizer conservation headlands 4.8; spring fallows 4.5; sown grass margins 4.4; overwinter stubbles 4.2; conservation headlands 3.5; grass leys 3.1. Plants were surveyed on a total of 294 conservation measure sites (each a single field, block of field or field margin strip) on 37 farms in East Anglia (dominated by arable farming) and 38 farms in the West Midlands (dominated by more mixed farming). The ten habitats were created according to agri-environment scheme guidelines. Vegetation was surveyed once in each site in June–August in 1999 or 2000. The vegetation was examined in thirty 0.25 m^2 quadrats randomly placed in 50–100 m randomly located sampling zones in each habitat site. Top cover and plant cover was estimated with 1–30 pin hits.

A replicated, paired site comparison study in 2000 in Ireland (30) found that wider, uncultivated margins (average 181 cm wide) with reduced agrochemical inputs on Rural Environment Protection Scheme (REPS) farms did not have higher plant or ground beetle (Carabidae) diversity or abundance than margins on non-REPS farms (average 145 cm). There were around 11 plant species and 21–22 ground beetle species/margin on both types of farm. Fourteen arable farms with Rural Environment Protection Scheme agreements at least 4-years-old were paired with 14 similar farms without agreements. On each farm two randomly selected field margins were surveyed for plants and ground beetles.

A small-scale controlled study in 2000–2001 in Essex, UK (31) found densities of lesser marsh grasshoppers *Chorthippus albomarginatus* (69% of all grasshoppers found) and meadow grasshoppers *C. parallelus* (31% of all grasshoppers found) in two Countryside Stewardship Scheme field margins (one margin naturally regenerated, one margin created from existing grass ley) were not statistically different than in intensively managed habitats (arable field; heavily grazed cattle and sheep pastures). Adult density of both grasshopper species was higher on lightly grazed pasture and a disused farm track than in either field margin. Grasshopper density was initially higher in the sown grass margin than the naturally regenerated margin or control grazed pasture 3 years after establishment (0.4, 0.1 and 0.3 grasshoppers/m^2 respectively). Seven years into the 10-year agreement, grasshopper density had decreased in the sown and naturally regenerated margins (0.05 grasshoppers/m^2) but increased substantially in the control grazed pasture (1.2 grasshoppers/m^2). The authors suggested that annual cutting for hay was the reason for the reduced grasshopper populations in the margins. In each of 9 study sites (2 field margins; 1 arable field; 1 lightly grazed pasture; 1 heavily grazed cattle pasture; 1 heavily grazed sheep pasture; 1 hay meadow; 1 set-aside grassland; 1 disused farm track) 10 quadrats (2 x 2 m^2) were randomly positioned in a 100 m^2 plot. Grasshoppers were counted in quadrats once in July and once in August (2000 and 2001).

A site comparison study from 2001 to 2005 of organic arable fields in the Netherlands (32) found that greater numbers of overwintering generalist predators were recorded in unmown perennial field margins compared to mown grass strips and bare fields. Higher numbers of generalist predators (ground beetles Carabidae; spiders Araneae; rove beetles Staphylinidae) were found in unmown margins (202 individuals/m^2) than mown strips (124/m^2) and bare fields (152/m^2). Over twice as many overwintering ground beetles were found within margins (101/m^2) than mown strips and fields (33–48/m^2). The same was true for other beetles (margins: 112/m^2; mown grass strips: 45/m^2; bare fields: 36/m^2). One farm system sampled had numerous field margins (21% of area), whilst the other had few (5% of area). To catch overwintering arthropods, pitfall traps were set within enclosures (1 x 1 m^2) in March–May 2004, three within unmown field margins, three within short-mown grass strips and six in bare soil plots in fields. Pests and pest predation were also sampled, but results are not presented here.

A replicated trial from 2001 to 2004 in Belgium (33) found that naturally regenerated margins had similar numbers of plant species to margins sown with wildflower mix after three years. In naturally regenerating plots the number of plant species increased (unshaded margin only) or remained similar from 8–15 species/plot in July 2002 to 12–15 species/plot in July 2004. The relative abundance of perennial plants increased and the relative abundance of annuals decreased over time on all the field margin plots, regardless of treatment. In naturally regenerated margins the proportion of legumes increased over time whilst in sown margins the proportion of legumes decreased significantly. In September 2001, 10 m lengths of two 10 x 180 m arable field margins were either left to naturally regenerate or sown with one of two wildflower/grass species mixtures containing 63 or 77 plant species. One margin was in a sunny location; the other shaded by trees. The margins were mown twice, in late June and September, each year from 2002 to 2004. Each combination of treatments was replicated three times. Plants were recorded in July and October from 2002 to 2004.

A replicated study in 1999 and 2003 on arable and pastoral fields in the UK (34) found that a combination of creating uncultivated and planted margins around fields was strongly positively associated with the presence of 4 out of 12 farmland bird species analysed. These species were Eurasian skylark *Alauda arvensis* (a field-nesting species), chaffinch *Fringilla coelebs*, whitethroat *Sylvia communis* and yellowhammer *Emberiza citrinella* (all boundary-nesting species). The other species analysed were corn bunting *Miliaria calandra*, lapwing *Vanellus vanellus*, yellow wagtail *Motacilla flava*, dunnock *Prunella modularis*, greenfinch *Carduelis chloris*, linnet *C. cannabina*, reed bunting *E. schoeniclus* and tree sparrow *Passer montanus*. The study did not distinguish between uncultivated and planted margins. On the 256 study fields birds were recorded using territory-mapping techniques between 1 April and 31 July 2003. Sites were visited eight times and all registrations plotted on a farm map. Territories were assigned a habitat unit based on their location.

A replicated, controlled study in the summer of 2003 in central Switzerland (35) found that small mammal density was higher in herbaceous strips than in low-intensity meadows, conventionally farmed artificial grasslands and autumn-sown wheat fields. Small mammal species richness in herbaceous strips (six species) was higher than in any other studied habitat (two species each). The increase in small mammal density over the summer was higher in herbaceous strips and wildflower strips than in the other three habitats. Herbaceous strips consisted mainly of herbaceous plants, such as thistles *Cirsium* spp., common teasel *Dipsacus sylvestris*, St John's wort *Hypericum perforatum*, common mallow *Malva sylvestris* and mulleins *Verbascum* spp. On the 15 study sites, herbaceous strips and wildflower strips were not regularly cut during the growing season, whereas other grassland habitats were cut at least twice. Small mammals were trapped and individually marked during 60-hour trapping sessions in March, May and July. Traps were checked every eight hours. A capture-recapture method was used to estimate small mammal densities.

A replicated, controlled trial in central and eastern England (36) found that naturally regenerated field margins supported a greater number and diversity of foraging bumblebees *Bombus* spp. than cropped margins (including conservation headlands), but only in the first year of the study. In subsequent second and third years, bumblebee numbers were not significantly different from cropped treatments, but this may be due to the presence of more attractive floral resources planted on the same field margins for the experiment. Six sites were studied and two experimental plots (50 x 6 m) established in each cereal field along two margins. Six treatments were assigned to plots: conservation headland; natural regeneration; tussocky grass mixture; wildflower mixture; pollen and nectar mixture; crop (control treatment). Foraging bumblebees were counted from May to late August on 6 m-wide transects between 6 and 11 times in each margin.

A replicated, controlled study in the summers of 1997–2000 and 2003 in Essex, UK (37) found that naturally regenerated 6 m margins had higher plant species richness (35 species) than grass-sown 6 m margins (20 species), 7 years after margin establishment under the Countryside Stewardship Scheme. Butterfly (Lepidoptera) abundance was higher in 6 m-wide Countryside Stewardship Scheme margins (naturally regenerated and grass-sown margins not distinguished) than in control margins. Comparisons between 6 m margins (naturally regenerated and grass-sown margins were not distinguished) and control sections showed 54 vs 19 butterflies/km/visit. The meadow brown butterfly *Maniola jurtina* also occurred in higher numbers in Countryside Stewardship Scheme field margins: 6 m margins (naturally regenerated and grass-sown margins not distinguished) and their control sections had 22 vs 5/km/visit. Butterfly abundance and species richness did not change over the study period in either 6 m margins or in a transect across farmland. Six metre margins were established on three farms either through natural regeneration or by sowing with a grass-seed mixture, and all were cut annually after 15 July.

A replicated, controlled trial in 2005–2006 in Warwickshire, UK (38) found that field corners or margins left to naturally regenerate for one year had more bumblebees *Bombus* spp. (species and individuals) than control crop plots, or plots sown with wild bird seed or wildflower seed mix. There were 55 bumblebees and 5 bumblebee species/plot on average on naturally regenerated plots, compared to no bumblebees on control crop plots and 7 bumblebees of 2 species/plot on sown plots. Naturally regenerated plots also had more butterfly and plant species than control cereal plots (5–6 butterfly species/plot and 7 plant species/m^2, compared to 1 butterfly species/plot and 2 plant species/m^2 in cereal crop plots). Naturally regenerated plots did not have more butterfly individuals, or more birds in winter (species or individuals) than control crop plots. Plots were located on one farm and were left as unmanaged wheat stubble for all of 2006. Each treatment was tested in one section of margin and one corner in each of four fields. Plants were monitored in three 1 m^2 quadrats/plot in July 2006. Butterflies, bumblebees and flowering plants were recorded in a 6 m-wide transect

5 times between July and September. Farmland birds were counted on each plot on seven counts between December 2006 and March 2007.

A replicated trial in Lithuania in 2006–2007 (39) found that uncropped field margins had significantly higher plant diversity than margins within wheat crops, on both organic and intensive farms. Fifteen field margin areas were left to regenerate naturally (uncropped) and compared with fifteen margins of a winter wheat crop, across three farms in Lithuania. One farm was managed organically; the other two conventionally. Plants in the margins were monitored in June and July 2006 and 2007 in 0.5 m^2 sample plots.

A replicated, paired, site comparison study in 2003 in 3 regions of Germany (40) found that 21 uncultivated fallow strips adjacent to organic wheat fields had an average of 6.3 bee (Apidae) species, 2.6 solitary bee individuals/100 m^2 and 8.5 bumblebee *Bombus* spp. individuals/100 m^2. Uncultivated fallow strips adjacent to conventional wheat fields had an average of 3.9 bee species, 1.1 solitary bee individuals/100 m^2 and 3.7 bumblebees/100 m^2. Bee species richness was 60% higher on uncultivated strips adjacent to organic wheat fields than those adjacent to conventional wheat fields, and had 136% more solitary bees and 130% more bumblebees. Strips adjacent to organic wheat fields also had more flowering plant species and higher flower cover. Species richness and abundance of bees in fallow strips appeared to be limited by foraging resources, which were more abundant when adjacent fields were organic. However, only bees that gather pollen from a range of plants were found on fallow strips during surveys. Specialist bees did not appear to benefit from fallow strips, suggesting that they do not completely compensate for missing semi-natural habitats. Bees were surveyed along 100 m transects 4 times in May–June 2003 in 42 paired fallow strips adjacent to organic/conventional fields. Flowering plants were surveyed in bee transects and in two transects along the centre and edge of the adjacent field. All fallow strips were mown once a year, with an average width of 2.6 m.

A replicated, controlled study in 2000–2003 of a grass crop field in Norway (41) found that plant species diversity was higher in strips sown with a grass/wildflower mixture than strips left to regenerate naturally or in the grass crop. There were 10–12 plant species/quadrat in four 2 m-wide naturally regenerated strips on average, compared to 17–18 species/quadrat in a strip sown with grass and flower mixture, and 7–9 species/quadrat in a control strip of the main grass crop. Naturally regenerated strips were dominated by grasses and perennial weeds. Four strips (2 m wide) were ploughed, perpendicular to an existing semi-natural margin, in May 2000. One half of each was left to regenerate naturally; the other half was sown with a grass/meadow flower (22 species) seed mixture. Sown strips did not receive fertilizer and were cut once (late September). Permanent quadrats (0.5 x 0.5 m) were sampled in the grass crop and strips in June 2000–2003.

The second monitoring year of the same replicated, controlled study as (38) in the UK from 2005–2007 (42) found that naturally regenerated plots had more plant species and more vacuum-sampled invertebrates (individuals and groups) than control plots, but not more butterflies (Lepidoptera) or birds in winter. Naturally regenerated plots had 6 plant species/m^2; 7 bumblebee *Bombus* spp. individuals/plot; 5 butterfly individuals and 2 butterfly species/plot, compared to 3 plant species/m^2; 0 bumblebee individuals; and 1 butterfly individual/plot on control cereal plots. Control plots had 254 vacuum-sampled canopy-dwelling invertebrates/m^2 on average, compared to 840–1,197/m^2 on other treatments. Plants were monitored in three 1 m^2 quadrats per plot in June 2007. Butterflies, bumblebees and flowering plants were recorded in a 6 m-wide transect 6 times between July and September in 2006 and 2007. Invertebrates in the vegetation were vacuum-sampled in early July 2007. Farmland birds were counted on each plot on four counts between December 2007 and March 2008. The crop control in year two was winter wheat.

A replicated, controlled trial from 2002–2004 in County Wexford, Ireland (43) found that 1.5–3.5 m-wide margins of permanent pasture fields fenced to exclude livestock had more springtails (Collembola: Anthropleona) and spiders (Araneae) than control margins. Margins that were rotavated and left to naturally regenerate also had more flies (Diptera) and plant species than control plots. These margins had around 12 plant species/plot in 2002, degrading to just over 5 plant species/plot in 2004, compared to less than 5 plant species/plot throughout the study in control plots and unrotavated fenced margins. Some undesirable weeds, such as broad-leaved dock *Rumex obtusifolius*, established in rotavated plots. Margin width made no difference to plant species richness. Allowing grazing on half of each plot from 2003–2004 (one year after margin establishment) did not affect the number of plant species the following year. Plots were 30 m long, with 3 replicates of each treatment combination. Plants were monitored in permanent quadrats in July 2002, May and July 2003 and May 2004. Invertebrates were sampled in six emergence traps per plot, between May and September 2003. Ground areas under the emergence traps were sampled with a vacuum sampler.

A replicated, controlled study in 2005–2006 on mixed lowland farms in Scotland (44), found that a larger proportion of early-summer yellowhammer *Emberiza citrinella* foraging flights were in field margins (32% of 233 flights from 10 nests), compared to cereal crops (8%). However, in late summer, cereal fields were used more (up to 56% of 506 flights) and field margins less (down to 15%). Field margins supported higher total invertebrate abundance than spring or winter barley across the summer period (average total invertebrate abundance was 45 in margins compared to 28 and 23 in spring and winter barley respectively). In 2006, sections of margins around some nests were cut down to the soil. These patches measured 15 x 1 m and comprised 2% of margin area. They were used for 3% of 172 foraging flights in early summer and 34% of 77 foraging flights in late summer. Cut patches were used more frequently in margins with swards >60 cm tall. The authors suggest that yellowhammers used cut patches disproportionately as the uncut sections grew taller and so reduced access to invertebrates. The study was carried out on five farms. Yellowhammer foraging flights were recorded from May–August 2005. Thirty yellowhammer nests with nestlings were observed, each for a 3 hour period between 07.00 h and 11.00 h. Foraging locations of adult birds from the nest site were recorded on sketch maps, and following the observation period each foraging site was visited and the distance from the nest measured.

A replicated, controlled study in summer 2006 in north Germany (45) found that species richness and abundance of hoverflies (Syrphidae) during the wheat peak-ripening stage was higher in naturally developed grass strips (3 m wide, 7 sites) than in wheat-wheat boundaries (seven sites) and within the wheat fields adjacent to the margins (seven sites), but lower than in sown flower strips (seven sites each). Hoverfly density and species richness increased with an increasing amount of arable land at smaller scales (0.5 and 1 km around site) but not at larger scales (2 and 4 km). This was true for all hoverflies and all aphid-eating hoverfly species. Margins were located along a gradient of different habitat complexities in the surrounding landscape, ranging from 30% to 100% arable land. Hoverflies were captured by sweep netting (one sweep per footstep) along 100 m transects.

A literature review in 2009 of European farmland conservation practices (46) found that sown uncropped field margins were used by foraging bumblebees *Bombus* spp. more than other margin types, including naturally regenerated margins. Naturally regenerated margins were found to hold many important food species for birds (both invertebrate and plant). In addition rare plants, such as rough poppy *Papaver hybridum*, may be found in naturally regenerating margins. The authors argued that on poor soils with a diverse seed bank, naturally regenerating margins may have a greater diversity of plants and be of greater conservation value than seeded grass margins, but if soils are rich then they can become dominated by a few species.

A replicated, site comparison study in 2010 on lowland farmland in England (47) found no consistent association between the provision of uncultivated field margins on arable or pastoral farmland and farmland bird numbers three years after the 2005 introduction of the Countryside Stewardship Scheme and Entry Level Stewardship agri-environment schemes. Although plots with field margins did see more positive population changes (increases or smaller decreases relative to other plots) of rook *Corvus frugilegus*, starling *Sturnus vulgaris* and woodpigeon *Columba palumbus* the effect was small. For example, starlings showed increases of only 0.0002 individuals for every 0.001 km² of margin in mixed farmland plots. Other species expected to benefit from margin provision including corn bunting *Emberiza calandra*, grey partridge *Perdix perdix*, kestrel *Falco tinnunculus*, jackdaw *Corvus monedula*, reed bunting *E. schoeniclus*, and common whitethroat *Sylvia communis* showed no effect of margin management. Yellowhammers *Emberiza citrinella*, which were also expected to benefit from margin creation, showed a positive association in mixed landscapes, but a negative association on grassland plots. The 2,046, 1 km² lowland plots were surveyed in both 2005 and 2008 and classified as arable, pastoral or mixed farmland. Eighty-four percent of plots included some area managed according to Entry Level or Countryside Stewardship agri-environment schemes. In both survey years, 2 surveys were conducted along a 2 km pre-selected transect route through each 1 km² square, with all birds seen or heard recorded in distance bands.

A replicated study in 2007 in Gloucestershire and Oxfordshire, England (48) found that grey partridge *Perdix perdix* released in pairs in the spring used field margins more frequently than birds released in coveys in the autumn. Four farms were studied. Birds were radio-tagged and their positions marked on a 1:5,000 map.

A randomized, replicated trial from 1987 to 2000 in Oxfordshire, UK (49) found that the number of plant species on naturally regenerated 2 m-wide margins declined by about half over 13 years. There were 13–15 plant species/quadrat in 1988 and 7–9 plant species/quadrat in 2000. The most rapid decline was in the first two years, when many annual species were lost. Herbicide-sprayed plots had fewer perennial plant species than other management treatments from 1989 onwards (<6 perennial species/quadrat in 2000, compared to 6–8 for other treatments). After 13 years, naturally regenerated plots tended to have fewer species than plots sown with a wildflower seed mix (9–12 species/plot in 2000), but this difference was not statistically significant. There was no effect of different mowing regimes on the numbers of plant species, although in the early years mown plots had more plant species than uncut plots. Plant species were monitored three times a year from 1988 to 1990, and once in July 2000 in three 0.5 x 1 m quadrats/plot. This study was part of the same experimental set-up as (3, 5, 13, 14, 16, 21).

(1) Hassall M., Hawthorne A., Maudsley M., White P. & Cardwell C. (1992) Effects of headland management on invertebrate communities in cereal fields. *Agriculture Ecosystems & Environment*, 40, 155–178.
(2) Cardwell C., Hassall M. & White P. (1994) Effects of headland management on Carabid beetle communities in Breckland cereal fields. *Pedobiologia*, 38, 50–62.
(3) Feber R.E., Smith H. & MacDonald D.W. (1994) The effects of field margin restoration on the meadow brown butterfly (*Maniola jurtina*). *British Crop Protection Council Monographs*, 58, 295–300.
(4) Hawthorne A. & Hassall M. (1995) The effect of cereal headland treatments on carabid communities. Arthropod natural enemies in arable land I – density, spatial heterogeneity and dispersal, *Acta Jutlandica*, 70, 185–198.
(5) Feber R.E., Smith H. & Macdonald D.W. (1996) The effects on butterfly abundance of the management of uncropped edges of arable fields. *Journal of Applied Ecology*, 33, 1191–1205.

(6) West T.M. & Marshall E.J.P. (1996) Managing sown field margin strips on contrasted soil types in three environmentally sensitive areas. *Aspects of Applied Biology*, 44, 269–276.

(7) Denys C. (1997) Do field margins contribute to enhancement of species diversity in a cleared arable landscape? Investigations on the insect community of mugwort (*Artemisia vulgaris* L). *Mitteilungen Der Deutschen Gesellschaft Fur Allgemeine Und Agewandte Entomologie, Band 11, Heft 1–6, Dezember – Entomologists Conference*, 11, 69–72.

(8) Denys C., Tscharntke T. & Fischer R. (1997) Colonization of wild herbs by insects in sown and naturally developed field margin strips and in cereal fields. *Verhandlungen der Gesellschaft fur Okologie*, 27, 411–418.

(9) Feber R.E. & Hopkins A. (1997) Diversity of plant and butterfly species on organic farmland field margins in relation to management. *Proceedings of the British Grassland Society Fifth Research Conference*. University of Plymouth, Devon, UK, 8–10 September. pp. 63–64.

(10) Hopkins A. & Feber R.E. (1997) Management for plant and butterfly species diversity on organically farmed grassland field margins. *Proceedings of the International Occasional Symposium of the European Grassland Federation: Management for grassland biodiversity*. Warszawa-Lomza, Poland, 19–23 May. pp. 69–73.

(11) Kleijn D., Joenje W. & Kropff M.J. (1997) Patterns in species composition of arable field boundary vegetation. *Acta Botanica Neerlandica*, 46, 175–192.

(12) Weiss B. & Buchs W. (1997) Reaction of the spider coenoesis on different kinds of rotational set aside in agricultural fields. *Mitteilungen Der Deutschen Gesellschaft Fur Allgemeine Und Angewandte Entomologie*, 11, 147–151.

(13) Baines M., Hambler C., Johnson P.J., Macdonald D.W. & Smith H. (1998) The effects of arable field margin management on the abundance and species richness of Araneae (spiders). *Ecography*, 21, 74–86.

(14) Bell J.R., Gates S., Haughton A.J., Macdonald D.W., Smith H., Wheater C.P. & Cullen W.R. (1999) Pseudoscorpions in field margins: Effects of margin age, management and boundary habitats. *Journal of Arachnology*, 27, 236–240.

(15) Carreck N.L., Williams I.H. & Oakley J.N. (1999) Enhancing farmland for insect pollinators using flower mixtures. *Aspects of Applied Biology*, 54, 101–108.

(16) Haughton A.J., Bell J.R., Gates S., Johnson P.J., Macdonald D.W., Tattersall F.H. & Hart B.H. (1999) Methods of increasing invertebrate abundance within field margins. *Aspects of Applied Biology*, 54, 163–170.

(17) Kromp B. (1999) Carabid beetles in sustainable agriculture: a review on pest control efficacy, cultivation impacts and enhancement. *Agriculture, Ecosystems & Environment*, 74, 187–228.

(18) Thomas C.F.G. & Marshall E.J.P. (1999) Arthropod abundance and diversity in differently vegetated margins of arable fields. *Agriculture, Ecosystems & Environment*, 72, 131–144.

(19) Aebischer N.J., Green R.E. & Evans A.D. (2000) From science to recovery: four case studies of how research has been translated into conservation action in the UK. In N. J. Aebischer, A.D. Evans, P.V. Grice & J.A. Vickery (eds.) *Ecology and Conservation of Lowland Farmland Birds*, British Ornithologists' Union, Tring. pp. 43–54.

(20) Kells A.R., Holland J.M. & Goulson D. (2001) The value of uncropped field margins for foraging bumblebees. *Journal of Insect Conservation*, 5, 283–291.

(21) Bell J.R., Johnson P.J., Hambler C., Haughton A.J., Smith H., Feber R.E., Tattersall F.H., Hart B.H., Manley W. & Macdonald D.W. (2002) Manipulating the abundance of *Lepthyphantes tenuis* (Araneae: Linyphiidae) by field margin management. *Agriculture, Ecosystems & Environment*, 93, 295–304.

(22) Denys C. & Tscharntke T. (2002) Plant-insect communities and predator-prey ratios in field margin strips, adjacent crop fields, and fallows. *Oecologia*, 130, 315–324.

(23) Evans A.D., Armstrong-Brown S. & Grice P.V. (2002) The role of research and development in the evolution of a 'smart' agri-environment scheme. *Aspects of Applied Biology*, 67, 253–264.

(24) Meek B., Loxton D., Sparks T., Pywell R., Pickett H. & Nowakowski M. (2002) The effect of arable field margin composition on invertebrate biodiversity. *Biological Conservation*, 106, 259–271.

(25) May M. & Nowakowski M. (2003) Using headland margins to boost environmental benefits of sugar beet. *British Sugar Beet Review*, 71, 48–51.

(26) Asteraki E.J., Hart B.J., Ings T.C. & Manley W.J. (2004) Factors influencing the plant and invertebrate diversity of arable field margins. *Agriculture, Ecosystems & Environment*, 102, 219–231.

(27) Bokenstrand A., Lagerlöf J. & Torstensson P.R. (2004) Establishment of vegetation in broadened field boundaries in agricultural landscapes. *Agriculture, Ecosystems & Environment*, 101, 21–29.

(28) Carvell C., Meek W.R., Pywell R.F. & Nowakowski M. (2004) The response of foraging bumblebees to successional change in newly created arable field margins. *Biological Conservation*, 118, 327–339.

(29) Critchley C., Allen D., Fowbert J., Mole A. & Gundrey A. (2004) Habitat establishment on arable land: assessment of an agri-environment scheme in England, UK. *Biological Conservation*, 119, 429–442.

(30) Feehan J., Gillmor D. & Culleton N. (2005) Effects of an agri-environment scheme on farmland biodiversity in Ireland. *Agriculture, Ecosystems & Environment*, 107, 275–286.

(31) Gardiner T. & Hill J. (2005) A study of grasshopper populations in Countryside Stewardship Scheme field margins in Essex. *British Journal of Entomology and Natural History*, 18, 73–80.

(32) Alebeek F.v., Kamstra J.H., Kruistum G.v. & Visser A. (2006) Improving natural pest suppression in arable farming: field margins and the importance of ground dwelling predators. *Proceedings of the Landscape Management for Functional Biodiversity*, 2nd Working Group Meeting, 16–19 May, Zürich-Reckenholz, Switzerland, 29. pp. 137–140.

(33) Cauwer B.D., Reheul D., D'Hooghe K., Nijs I. & Milbau A. (2006) Disturbance effects on early succession of field margins along the shaded and unshaded side of a tree lane. *Agriculture, Ecosystems & Environment*, 112, 78–86.

(34) Stevens D.K. & Bradbury R.B. (2006) Effects of the Arable Stewardship Pilot Scheme on breeding birds at field and farm-scales. *Agriculture, Ecosystems & Environment*, 112, 283–290.

(35) Aschwanden J., Holzgang O. & Jenni L. (2007) Importance of ecological compensation areas for small mammals in intensively farmed areas. *Wildlife Biology*, 13, 150–158.

(36) Carvell C., Meek W.R., Pywell R.F., Goulson D. & Nowakowski M. (2007) Comparing the efficacy of agri-environment schemes to enhance bumble bee abundance and diversity on arable field margins. *Journal of Applied Ecology*, 44, 29–40.

(37) Field R.G., Gardiner T. & Watkins G. (2007) The use of farmland by butterflies: a study on mixed farmland and field margins. *Entomologist's Gazette*, 58, 3–15.

(38) Pywell R. & Nowakowski M. (2007) *Farming for Wildlife Project: Annual Report 2006/7*. NERC report 6441.

(39) Balezentiene L. (2008) Organic and intensive farming impact on phytodiversity. *Vagos*, 79, 30–36.

(40) Holzschuh A., Steffan-Dewenter I. & Tscharntke T. (2008) Agricultural landscapes with organic crops support higher pollinator diversity. *Oikos*, 117, 354–361.

(41) Hovd H. (2008) Occurrence of meadow herbs in sown and unsown ploughed strips in cultivated grassland. *Acta Agriculturae Scandinavica. Section B, Plant Soil Science*, 58, 208–215.

(42) Pywell R. & Nowakowski M. (2008) *Farming for Wildlife Project: Annual Report 2007/8*. NERC report 6366.

(43) Sheridan H., Finn J.A., Culleton N. & O'Donovanc G. (2008) Plant and invertebrate diversity in grassland field margins. *Agriculture Ecosystems & Environment*, 123, 225–232.

(44) Douglas D.J.T., Vickery J.A. & Benton T.G. (2009) Improving the value of field margins as foraging habitat for farmland birds. *Journal of Applied Ecology*, 46, 353–362.

(45) Haenke S., Scheid B., Schaefer M., Tscharntke T. & Thies C. (2009) Increasing syrphid fly diversity and density in sown flower strips within simple vs. complex landscapes. *Journal of Applied Ecology*, 46, 1106–1114.

(46) Vickery J.A., Feber R.E. & Fuller R.J. (2009) Arable field margins managed for biodiversity conservation: a review of food resource provision for farmland birds. *Agriculture, Ecosystems & Environment*, 133, 1–13.

(47) Davey C.M., Vickery J.A., Boatman N.D., Chamberlain D.E., Parry H.R. & Siriwardena G.M. (2010) Assessing the impact of Entry Level Stewardship on lowland farmland birds in England. *Ibis*, 152, 459–474.

(48) Rantanen E.M., Buner F., Riordan P., Sotherton N. & Macdonald D.W. (2010) Habitat preferences and survival in wildlife reintroductions: an ecological trap in reintroduced grey partridges. *Journal of Applied Ecology*, 47, 1357–1364.

(49) Smith H., Feber R.E., Morecroft M.D., Taylor M.E. & Macdonald D.W. (2010) Short-term successional change does not predict long-term conservation value of managed arable field margins. *Biological Conservation*, 143, 813–822.

Additional references
Müller L. (1991) Auswirkungen der extensivierungsförderung auf Wirbellose [The effects of extensification on invertebrates]. *Faunistisch-Ökologische Mitteilungen*, S10, 41–70.

Marshall E.J.P., West T.M. & Winstone L. (1994) Extending field boundary habitats to enhance farmland wildlife and improve crop and environmental protection. *Aspects of Applied Biology*, 40, 387–391.

Wilson S., Baylis M., Sherrott A. & Howe G. (2000) *Arable Stewardship Project Officer Review*. Farming and Rural Conservation Agency report.

ADAS (2001) *Ecological Evaluation of the Arable Stewardship Pilot Scheme, 1998–2000*. ADAS report.

1.22   Plant grass buffer strips/margins around arable or pasture fields

• Nineteen studies from Finland, the Netherlands, Sweden and the UK (including seven replicated controlled studies, of which two were randomized, and three reviews), found that planting grass buffer strips (some margins floristically-enhanced) increased arthropod abundance[3, 7, 8, 28, 38, 41, 45, 46, 49–53, 55–57, 60, 61, 68, 69], species richness[3, 36, 49, 57, 69] and diversity[36, 38, 63, 64, 69]. A review found grass margins benefited bumblebees and some other invertebrates but did not distinguish between the effects of several different margin types[19].

- Nine studies from the UK (including seven replicated studies, of which two were controlled and two reviews) found that planting grass buffer strips (some margins floristically-enhanced) benefits birds, resulting in increased numbers[13, 17, 42, 52, 65, 67], densities[18, 65], species richness[2] and foraging time[21].

- Seven studies from the Netherlands and the UK (all replicated, of which four were controlled and two randomized), found that planting grass buffer strips (some margins floristically-enhanced) increased the cover[1] and species richness of plants[5, 20, 22, 35, 38, 63]. A review found grass margins benefited plants but did not distinguish between the effects of several different margin types[19].

- Five studies from Finland and the UK (including two replicated, controlled trials and a review), found that planting grass buffer strips benefits small mammals[58]: including increased activity[9] and numbers[32, 43, 48].

- Six studies from the Netherlands and the UK (including three replicated, controlled trials) found that planting grass buffer strips had no clear effect on insect numbers[12, 15, 20, 25, 62], bird numbers[37] or invertebrate pest populations[34]. A replicated site comparison found sown grassy margins were not the best option for conservation of rare arable plants[44].

Background

This intervention involves planting field margins with a grass-rich seed mixture. It includes 'floristically-enhanced' grass margins available under the English Higher Level Stewardship scheme. The margins are not fertilized and only spot-treated with herbicides if necessary.

A replicated, controlled, randomized study in the UK (1) found that plant cover was higher in margins sown with grass or grass/wildflower mixtures than those naturally regenerated, but plant diversity tended to be lower with grass-only seed mixtures. Margins were created in each field and divided into six plots (4 x 30 m). Each was (randomly) sown with a seed mixture: grass, low cost mix (3 grass: 7 herb species), alkaline soil (6: 16), neutral soil (5: 15), acid soil (6: 16) and one natural regeneration. Plots were divided into 10 m sub-plots, which were either unmanaged, cut once or received grass herbicide. The same study is reported in West & Marshall (1996).

A replicated, controlled study in 1995–1996 in Cambridgeshire, UK (2) found that more bird individuals (average 20% of all individual birds recorded) and more bird species (average 56% of all bird species counted in 1995–1996) used the sown set-aside strips than the adjacent crop area (average 7% individuals and 33% species) in both years. Across all habitats 44 species were recorded in 1995 and 31 spp. in 1996. However, the highest proportions of both individuals and species were recorded in field boundaries (average 68% of all individuals and 80% of all spp.). The highest species richness was found in the most species rich grass mix. The seed mixture 'Tübinger Mischung', designed to provide nectar for bees (Apidae) and containing only wildflowers, attracted the largest number of birds but the lowest number of bird species. Yellowhammer *Emberiza citrinella*, red-legged partridge *Alectoris rufa* and pheasant *Phasianus colchicus* were the most recorded species in set-aside strips. Note that no statistical analyses were performed on these data. Five seed mixtures were sown on 15 set-aside areas (minimum 20 m x 100 m) in autumn 1993 and 1994. Seed mixtures contained only grass species (3 mixes of 3 to 6 species), a mix of grasses and wildflowers (6 grass and 8 wildflower species) or only wildflowers (11 species). Birds were recorded during 15 minute point counts on 10 occasions between June and September 1995 and July and October 1996. Individual bird locations were recorded in three categories: field boundary, set-aside strip or crop. After each count, the strips were walked to flush any birds present but not visible during the count.

A replicated study in summer 1996 in Gloucestershire, UK (3), found higher overall butterfly abundance and species richness in plots sown with a wild grass/flower seed mixture (four grasses, five wildflowers) than in naturally regenerated plots. Vegetation removal had no effect on butterfly abundance, but overall species richness was lower in plots cut for silage in June and grazed by cattle in July. Plant species richness was on average higher in sown than naturally regenerated plots (23 vs 19 species). Vegetation removal had no effect on plant species richness but uncut/ungrazed plots had more wildflower species in flower in July. In September 1994, twenty 50 m long experimental plots were created in the margins of 2 adjoining organic fields. Ten plots were sown; ten were allowed to naturally regenerate. Presence of all sown and unsown plant species were recorded as well as wildflower species in flower (May and July 1996). Butterflies were monitored weekly June–September along a transect route.

A replicated trial in the Netherlands (4) found that 4 m-wide field margins sown with rye grass *Lolium perenne* had fewer plant species than margins left to naturally regenerate or sown with wildflowers two years after establishment. On average there were 5.9 plant species/0.25 m^2 in grass-sown margins, compared with 8.6 in naturally regenerated margins and 13.7 in margins sown with 30 non-grass wildflower species. Two prominent arable weeds, creeping thistle *Cirsium arvense* and couch grass *Elymus repens* both had lower biomass in the grass-sown margins than in naturally regenerated margins (8 g/m^2 and 9 g/m^2 respectively in grass-sown margins; 33 g/m^2 and 28 g/m^2 in naturally regenerated plots). Wildflower-sown margins had similar couch grass biomass to the grass-sown plots, but much lower creeping thistle biomass (0.1 g/m^2 in wildflower margins). In 1993, 27 experimental plots (8 x 4 m) were established on the boundaries of 3 arable fields. There were three replicates of each treatment on each field. All plots were mown once a year, without removing cuttings. Plant biomass and number of species were measured in eight 0.5 m x 0.5 m plots on a single transect line across each margin, in August 1995.

A replicated, controlled, randomized study from 1987 to 1991 in Oxfordshire, England (5) found that a grass ley sown with a species-rich mix of grasses and wildflowers retained more sown plant species and had more naturally regenerating species than a conventionally sown ley. Loss of sown species increased at high fertilizer levels. Numbers of sown and naturally regenerating plant species were lower under a silage than a hay cutting regime. The more species-rich ley was less productive and so easier to manage by infrequent mowing. Field margins (7–9 m wide) were established in 1987 around three arable fields. In 1988 they were divided into 50 m-long plots and half (randomly assigned) in each field were sown with each grass ley mix: conventional (two grass and one clover *Trifolium* species) or a more species-rich mix, comprising six indigenous grasses and three forbs (excluding rye grass *Lolium perenne* and white clover *T. repens*).

A replicated study in summer 1996 in central Germany (6) found that spider (Araneae) species richness in plots sown with a grass mixture (36 spider species) was lower than plots sown with lacy phacelia *Phacelia tanacetifolia* and Egyptian clover *Trifolium alexandrinum*

(40 spider spp.), but higher than in plots sown with lacy phacelia, buckwheat *Fagopyrum esculentum*, common sunflower *Helianthus annuus* and common mallow *Malva sylvestris* (30 spider spp.). Spider abundance in grass mix plots (126 and 65 individuals in pitfall traps and photoeclectors respectively) was lower than in plots containing sundial lupine *Lupinus perennis* and common vetch *Vicia sativa* (155 and 124 individuals in pitfall traps/photoeclectors), but only significantly lower in the eclector samples. Note that most results in this study are not statistically tested. Eight different types of strips with 3 replicates each were tested: 6 seed mixtures containing mainly flowering plants (1–12 species), 1 mixture containing mainly grasses (red fescue *Festuca rubra* (64%), perennial rye grass *Lolium perenne* (25%) plus white clover *T. repens* (10%)) and 1 naturally regenerated treatment. Spiders were sampled using two pitfall traps and two photoeclectors in each plot.

A replicated, controlled, randomized study in Finland (7) found that spider (Araneae) abundance was higher in perennial grass/clover *Trifolium* spp. strips than in the crop (approximately 1,200–3,000 vs 400–900 spiders respectively). Wolf spiders (Lycosidae) dominated in grass strips (57–77% of total catches). Money spiders (Linyphiidae) were also more common in grass strips than in the crop. Total spider catches including wolf spiders decreased with distance into the crop. Perennial grass/clover strips 12 m wide were sown with a mixture of timothy *Phleum pratense*, meadow fescue *Festuca pratensis*, red clover *T. pratense* and white clover *T. repens* at the ends of 24 plots in 1991. Spiders were sampled with a pitfall trap in the centre of the grass strip and at 12, 66 and 120 m into each plot from the grass strip 8–10 times (each trapping period lasting one week) between sowing and harvest (1992–1994).

A replicated study in the summers of 1997–1998 in three regions across the UK (8) found that the total percentage of grass cover in planted grass strips affected the abundance of sawfly (Symphyta) larvae positively. Sawfly larvae numbers were found to increase with strip age and to decrease with the amount of cock's-foot *Dactylis glomerata* (both trends non-significant). There was no difference in the average catch of sawfly larvae between beetle banks and strips planted along existing field margins. Numbers of insects used by gamebirds as chick food increased with strip age and area, but there was also a significant difference between farms. There was a non-significant trend for chick-food insect numbers to increase with the proportion of red fescue *Festuca rubra*. Cocksfoot, red fescue and perennial rye grass *Lolium perenne* were the predominant grasses in most strips, being most common in 35, 25 and 17 strips respectively. A total of 116 strips were sampled on 32 farms. Grass strips had been established 0.5–12 years previously, both along pre-existing field margins and across cropped fields (beetle banks). Invertebrates were sampled by sweep-netting at the base of the vegetation in mid-June to mid-July. Percentage cover of all plant species and vegetation height was measured in 0.25 m^2 quadrats.

A replicated study from 1992 to 1998 in England (9) found that small mammal activity was significantly greater in field margins than in open fields in both organic (142 vs 86 mammals respectively) and conventional systems (139 vs 78). There was no difference between systems. The same trend was seen in both systems for wood mouse *Apodemus sylvaticus* (margins: 40–80%; field: 20–60% of population activity), bank vole *Myodes glareolus* (75–95% vs 5–25%) and common shrew *Sorex arenaeus* (40–90% vs 10–60%). The difference between activity in the margin and field was greater during winter than summer. Seeded margins showed a rapid increase in activity over 4 years for wood mouse (year 1: 15–16 trapped; year 4: 28–30), bank vole (year 1: 2–8; year 4: 16–36) and common shrew (year 1: 6–7; year 4: 18–19). Two to four new field margins were sampled within organic and conventional fields at two farms, in Essex and Leicestershire. Mark-recapture programmes were undertaken using Longworth traps over 10 nights each season from 1992 to 1998. Traps were set at 0, 1, 2, 3, 4, 5, 10, 20, 40 m into the field, replicated 5 times at each site. Additional, 'one-off' trapping sessions were under-taken over one year at five pairs of organic/conventional farms.

A replicated study in 1996 in the Netherlands (10) found that different arthropod populations responded differently to mowing. After mowing, populations of bugs (Heter-optera), aphids (Aphidoidea), parasitic wasps (Ichneumonidae), hoverflies (Syrphidae) and rove beetles (Staphylinidae) increased to between >1.5 and nearly 2.5 times their population size prior to mowing. Mowing had the opposite effect on populations of spiders (Araneae), harvestmen (Opiliones), and moths and butterflies (Lepidoptera), reducing these populations by half or more. Ten grass margins (3 m x 900 m) on five farms were sown with grasses, including giant fescue *Festuca gigantea*, timothy *Phleum pratense* and cock's-foot *Dactylis glom-erata*. Grassy margins were mown on approximately half of the farms at the beginning of July. Arthropods were sampled using two pyramid traps/margin, installed for a three-week period five times during the 1996 growing season.

A 1999 review of research into field margins in northwest Europe (11) found that biod-iversity was enhanced by establishing grass margins. Three studies found that establishing grass margins increased beneficial predatory invertebrates. Another study found no increase in invertebrate predators, but a higher abundance of field mice *Apodemus* spp., skylark *Alauda arvensis*, meadow pipit *Anthus pratensis*, blue-headed wagtail *Motacilla flava flava* and linnet *Carduelis cannabina* in grass margins compared to normal crop edges, particularly when the grass was tall. However, one study in the Netherlands reported that in the short-term most newly created grass margins are less species-rich than existing verges.

A randomized, replicated, controlled trial between 1993 to 1996 in Bristol, UK (12) found that 4 m-wide field margins sown with rye grass *Lolium perenne* did not have more suction-sampled invertebrates, over-wintering invertebrates in the soil or ground beetles (Carabidae) than control cropped margins. There were 110–130 invertebrates/sample on control (cropped) and grass-sown plots. There was no difference in the number of ground beetle spe-cies (average of 8 species/plot), nor in the numbers of the 4 most commonly caught ground beetle species between margin types. Wolf spiders (Lycosidae) were more abundant on grass and wildflower-sown margins than on control or naturally regenerated margins (numbers not given). Three field margins were established in spring 1993. Experimental plots 10 x 4 m were either sown with arable crop (control), rye grass or a wildflower and grass seed mix, or left to naturally regenerate. There were three replicate plots in each margin. All plots were cut annually after harvest, and cuttings left in place. Ground beetles were sampled in eight pitfall traps in or near each margin, for one week in June for four years, 1993–1996. Invertebrates were sampled using a vacuum sampler on plots of two of the three margins in June 1994. Arthropods were extracted from soil samples taken from plots of one margin in December 1993 and February 1994.

A 2000 literature review (13) found that the UK population of cirl buntings *Emberiza cirlus* increased from 118–132 pairs in 1989 to 453 pairs in 1998 following a series of agri-environment schemes designed to provide overwinter stubbles, grass margins, and beneficially managed hedges and set-aside. Numbers on fields under the specific agri-environment scheme increased by 70%, compared with a 2% increase elsewhere.

A 2000 literature review (14) looked at which agricultural practices can be altered to benefit ground beetles (Carabidae). It found four European studies demonstrating that ground beetles use grassy strips (including one experimental study (12)). The other three studies considered habitat use, rather than directly testing the intervention.

A small, controlled study in 1999 in North Yorkshire, UK (15) found no significant dif-ference in numbers of ground beetles (Carabidae) in 3 m and 6 m sown grass margins and cropped edges (numbers are not presented). Five margins 3 m wide, four 6 m wide and 4 cropped field edges were sampled on one arable farm. A line of 8 pitfall traps, 5 m apart, were placed 1.5 m from the hedge base in each margin. Traps were set for 29 days in April–May and 7 days in September 1999. This study was carried out at the same experimental site as (25, 32).

A site comparison study in 1996 in Wiltshire, UK (16) found that coppiced and gapped-up hedges (hedges cut to the ground and gaps planted with hedging plants) had higher plant diversity than those with adjacent sown grass and grass/wildflower strips. Hedges with adjacent sown strips had a lower abundance of pernicious weed species. Sixty hedgerows on two neighbouring arable farms were studied. All 23 sampled hedges on Noland's Farm were trimmed annually and had the vegetation at the hedge base cut. The 37 sampled hedges on Manor Farm were trimmed in alternate years, and 9 were coppiced and gapped-up. Hedge vegetation was assessed in 25 m long plots in the middle of a field edge, on both sides of each hedge, in June 1996.

A 2001 paired site comparison study in south Devon (17) found that fields with 6 m grass margins were associated with increases in cirl bunting *Emberiza cirlus* numbers. Six of seven Countryside Stewardship Scheme plots, that had 6 m grass margins and were within 2.5 km of former cirl bunting territories gained birds, whereas there were declines of 20% in cirl bunting numbers on land not participating within the Countryside Stewardship Scheme. Forty-one 2 x 2 km^2 squares containing both land within the Countryside Steward-ship Scheme and non-Countryside Stewardship Scheme land were surveyed in 1992, 1998 and 1999. Each tetrad was surveyed at least twice each year, the first time during mid-April to late May and the second time between early June and the end of August.

A replicated, controlled study in winter 1999–2000 and summer 2000 in the West Mid-lands, UK (18) found 16 times higher winter densities of seed-eating birds (larks Alaudidae, finches Fringillidae, buntings Emberizidae and sparrows Passeridae) within 6 m of boundar-ies of fields with Countryside Stewardship Scheme grass margins than on fields without (1.1 vs 0.1 birds/ha). Twice as many blackbirds *Turdus merula* were found near the boundaries of fields without Countryside Stewardship Scheme grass margins than those with grass margins (1.8 vs 0.9 birds/ha). A total of 388 grass fields on 23 pastoral farms were surveyed 4 times each in winter and in summer. No statistical analysis was performed.

A 2002 review (19) of two reports (Wilson *et al.* 2000; ADAS 2001) evaluating the effects of the Pilot Arable Stewardship Scheme in two regions (East Anglia and the West Midlands) from 1998 to 2001 found that grass margins benefited plants, bumblebees *Bombus* spp., true bugs (Hemiptera) and sawflies (Symphyta), but not ground beetles (Carabidae). The grass margins set of options included sown grass margins, naturally regenerated margins, beetle banks and uncropped cultivated wildlife strips. The review does not distinguish between these, although the beneficial effects were particularly pronounced on sown or naturally regenerated grassy margins for true bugs. The effects of the pilot scheme on plants and inver-tebrates (bumblebees; true bugs; ground beetles; sawflies) were monitored over three years, relative to control areas. Grass margins were implemented on total areas of 361 and 294 ha in East Anglia and West Midlands respectively.

A replicated, controlled study during the summer of 2000 in North Yorkshire, UK (20) found grass margins contained more plant species than cropped margins but fewer species than margins sown with a grass and wildflower mix. Bumblebee *Bombus* spp. abundance and butterfly (Lepidoptera) diversity did not differ between treatments. However there were more meadow brown butterflies *Maniola jurtina* in grass margins and grass and wildflower margins than in naturally regenerated or control cropped margins. Spring numbers of ground beetles (Carabidae) and ground-dwelling spiders (Araneae) were higher in all treatments compared with the crop. Harvestmen (Opiliones) preferred grass margins to the crop in autumn. Four margins of winter cereal fields, all adjacent to hedges, on two farms, were split into 72 m-long plots and sown in September 1999 with a grass mix, grass and wildflower mix, cereal crop or left to regenerate naturally. Ground and canopy-dwelling invertebrates, butterflies and plants were surveyed from late April to late September 2000 using pitfall traps, sweep netting, transects and quadrats.

A controlled study from 1995 to 1997 and 1999 in Oxfordshire, UK (21) found that yellowhammers *Emberiza citrinella* spent significantly more time foraging in grass margins and field boundaries than other habitats. A significantly greater number of foraging visits per unit area of available habitat were made to grass margins and field boundaries than to all other habitat types. There was no significant difference between use of grass margins and field boundary habitats or between cut and uncut grass margins. However, greater use was made of both cut and uncut grass margins combined than field boundaries. Total area surveyed was 142.8 ha in 1995–1997 and 107.0 ha in 1999. Five habitat types were studied on one mixed arable and pastoral farm: cut or uncut grass margins (2 or 10 m wide, at edge of arable field), field boundaries, arable fields (winter-sown cereals) and grass fields (pasture, silage and hay).

A replicated trial in 2001–2002 in the UK (22) found that margins of sugar beet *Beta vulgaris* fields sown with grasses had fewer plant species, and slightly fewer invertebrates (individuals or species) than margins sown with wildflowers, or left to regenerate naturally. Grass-sown margins had 15 plant species/m, compared to 35 and 17 plant species/m for wildflower and naturally regenerated margins respectively, and 6–11 species/m for barley *Hordeum vulgare* or beet margins. The difference in invertebrate numbers between different treatments was fairly small (over 900 to over 1,700 individuals; 35–45 groups caught). In autumn 2001, 50 x 6 m margins at the edges of beet fields were sown with either sugar beet, spring barley, grasses (8 species), wildflowers (20% of seeds by weight, from 20 species) or allowed to naturally regenerate. There were two replicates of each treatment at each of three sites. In summer 2002, plants (including crop plants) were counted in the margins, and invertebrates sampled using pitfall traps, set for two weeks.

A replicated trial in Wiltshire, UK from 1998 to 2000 (23) found that sown grassy field margins suppressed undesirable weed species, but did not enhance the abundance of invertebrates, relative to naturally regenerated uncultivated margins. Sown plots had significantly lower cover of undesirable weeds (nettle *Urtica dioica*, creeping thistle *Cirsium arvense* and black grass *Alopecurus myosuroides*) than naturally regenerated plots. There was no difference in the total abundance of invertebrates between field margin treatments. In 2000, there were more predatory beetles (Coleoptera) in naturally regenerated plots than in sown plots. Thirty-eight 100 x 2 m field margin plots were sown with a grass seed mix consisting of either 3 grass species (12 plots), 6 grass species (13 plots) or 6 grass and 4 wildflower species (13 plots) in autumn 1998. Eleven plots were left to regenerate naturally. The plots surrounded four fields under a Countryside Stewardship Agreement on the Harnhill Manor Farm, Wiltshire, UK. Invertebrates were sampled using pitfall traps (five traps/plot) in spring and autumn and suction traps in summer. Plants were recorded in four 1 m^2 quadrats/plot in summer.

A replicated, controlled study in 1988–1997 in south-central Sweden (24), found that experimental plots sown with a clover *Trifolium* spp. and grass mix in a re-established field boundary on an organic farm had lower plant species richness than plots planted with rose bushes *Rosa canina* and/or sown with meadow plants or allowed to regenerate naturally one year after establishment. In two other (widened) field boundaries in a conventional system, clover and grass plots had fewer plant species nine years after establishment than plots with meadow plants. Seven years after establishment, total weed cover at the organic farm was higher in plots sown with a clover and grass mix, in natural regeneration plots and in reference boundary sections compared with plots with rose bushes and/or meadow plants. In two field boundaries total weed cover decreased in all treatments, except the clover and grass mix where it remained stable or increased. Four replicates of three or four treatments were established in experimental plots at the site of each field boundary in 1988 or 1990, either by widening an existing boundary or re-establishing a previously removed dirt road (organic site). All plots were cut annually in late summer and the cuttings removed. Vegetation surveys were carried out twice in experimental plots (1991–1993 and 1997) and once in reference boundaries (1997) in three to five 0.25 m² quadrats. It is not clear whether the results for clover and grass plots were a direct result of planting nectar flowers or grass.

A replicated, controlled trial in 2000–2002 in North Yorkshire (25) found 6 m-wide field margin plots sown with a 'tussocky grass' seed mix supported no more bumblebees *Bombus* spp. than conventionally cropped field margins. The study was carried out on three arable field margins of one farm. Each margin was split into five 72 m x 6 m plots and each plot subjected to one of five treatments: naturally regenerated, sown tussocky grass mix, sown grass and wildflower mix, split treatment of 3 m tussocky grass and 3 m grass and wildflower mix, or cropped to the edge. Bumblebee activity was surveyed using a standard 'bee walk' methodology. This study was carried out at the same experimental site as (15, 32).

A replicated study in the summers of 1999–2000 in the UK (26) found that sown grass margins had fewer plant species than six other conservation measures, including naturally regenerated margins. Sown grass margins were not considered one of the best options for conservation of annual herbaceous plant communities. Average numbers of plant species in the different conservation habitats were: sown grass margins 4.4; wildlife seed mixtures 6.7; uncropped cultivated margins 6.3; undersown cereals 5.9; naturally regenerated margins 5.5; no-fertilizer conservation headlands 4.8; spring fallows 4.5; overwinter stubbles 4.2; conservation headlands 3.5; grass leys 3.1. Plants were surveyed on a total of 294 conservation measure sites (each a single field, block of field or field margin strip), on 37 farms in East Anglia (dominated by arable farming) and 38 farms in the West Midlands (dominated by more mixed farming). The ten habitats were created according to agri-environment scheme guidelines. Vegetation was surveyed once in each site in June–August in 1999 or 2000. The vegetation was examined in thirty 0.25 m² quadrats randomly placed in 50–100 m randomly located sampling zones in each habitat site. All vascular plant species rooted in each quadrat as well as bare ground or litter were recorded.

A replicated study in 2000 and 2002 in Dorset, UK (27) found that field margins dominated by grass species supported different invertebrate communities in the adjacent crop to margins dominated by wildflower species. In 2000, margins dominated by grass species were associated with ground beetles (Carabidae); *Bembidion* spp. and *Carabus* spp., as well as rove beetles (Staphylinidae); *Tachinus* spp. Margins dominated by wildflowers were associated with ladybirds (Coccinellidae) and weevils (Curculionidae). In 2002, grassy margins were associated with ground beetles, *Bembidion* spp. and click beetles (Elateridae). Thirty field boundary lengths from six study fields on one farm were assessed for plant species cover. Invertebrates were sampled in eight pitfall traps adjacent to the field boundary.

A replicated, controlled study in 1997–2000 in Essex, UK (28) found that total butterfly (Lepidoptera) abundance, but not species richness, was higher in 6 m-wide grass margins (average 45.8 butterflies/km/visit) (the study did not distinguish between sown and naturally regenerated grass margins) than in control cropped sections (average 20.9). Of the 'key' grassland butterfly species, only the meadow brown *Maniola jurtina* had greater abundance in grass margins (average 18.9 butterflies/km/visit) than in controls (average 8.9). Significantly more butterflies, including *M. jurtina*, were found in a sown grass margin established adjacent to a permanent set-aside field than on all other margin types. Sown grass margins (not adjacent to permanent set-aside fields) had the lowest abundance of gatekeeper *Pyronia tithonus*, skipper *Thymelicus* spp. and large skipper *Ochlodes venata* butterflies. Five grass margins were established on two farms according to the requirements of the Countryside Stewardship Scheme in 1996 and sown with grass seed mixtures (6 or 9 species). In addition, three margins were established by natural regeneration on one farm, and on both farms one arable field edge without margins was used as a control. Butterfly abundance was monitored weekly along transects from late June to early August 1997–2000. All butterflies were recorded, but special note was taken of 'key' grassland species: meadow brown, gatekeeper, small skipper *Thymelicus sylvestris*, Essex skipper *T. lineola*, and large skipper. This study is part of the same experimental set-up as (29, 36, 45, 46).

A replicated, controlled study in 1997–2000 in Essex, UK (29) found that numbers of the gatekeeper butterfly *Pyronia tithonus* increased in sown grass margins one year after establishment and were significantly higher in 2000 than in 1997. Although more gatekeepers were recorded in grass margins than in control sites (without margins) during most visits (except for one farm in 1998), abundance was significantly higher at only one farm. More gatekeepers were observed on grass margins with adjacent hedgerows and on control sites with hedgerows than on the grass margins without hedgerows. Grass margins (2 m wide, 141–762 m long) were established in October 1996–2000 by sowing 3 different grass seed mixtures (8 margins: 4–5 species, mainly cock's-foot *Dactylis glomerata*; 1 margin: 6 spp., mainly red fescue *Festuca rubra*). Three field edges without margins (1 on each of 3 farms, 133–343 m long, 100–300 m hedgerow) were used as controls. Gatekeeper abundance was monitored weekly along transects in July and August. This study is part of the same experimental set-up as (28, 36, 45, 46).

A small-scale controlled study in 2000–2001 in Essex, UK (30) found densities of lesser marsh grasshoppers *Chorthippus albomarginatus* (69% of all grasshoppers found) and meadow grasshoppers *C. parallelus* (31% of all grasshoppers found) in two Countryside Stewardship Scheme field margins (one margin naturally regenerated, one margin created from existing grass ley) were not statistically different than in intensively managed habitats (arable field, heavily grazed cattle and sheep pastures). Adult density of both grasshopper species was higher on lightly grazed pasture and a disused farm track than in either field margin. Grasshopper density was initially higher in the sown grass margin than the naturally regenerated margin or control grazed pasture three years after establishment (0.4, 0.1 and 0.3 grasshoppers/m² respectively). Seven years into the 10-year agreement, grasshopper density had decreased in the sown and naturally regenerated margins (0.05 grasshoppers/m²) but increased substantially in the control grazed pasture (1.2 grasshoppers/m²). The authors suggested that annual cutting for hay was the reason for the reduced grasshopper populations in the margins. In each of nine study sites (two field margins, one arable field, one lightly grazed pasture, one heavily grazed cattle pasture, one heavily grazed sheep pasture, one hay meadow, one set-aside grassland, one disused farm track) 10 quadrats (2 x 2 m²) were randomly positioned in a 100 m² plot. Grasshoppers were counted in quadrats once in July and once in August (2000 and 2001).

A replicated study in winter 2002 in Oxfordshire, UK (31) found that the total abundance, species richness and diversity of beetles (Coleoptera) and spiders (Araneae), as well as abundance and species richness of rove beetles (Staphylinidae) were lower in field margins than in hedge bases, but there was no difference between recently sown (3–4-years-old) and mature field margins (about 50-years-old). The important aphid predator *Bembidion lampros* (ground beetle Carabidae) occurred in higher densities in both recently sown grass margins (8 individuals/m^2) and mature (40–60-years-old) hedge bases (12/m^2) compared with mature field margins and recently planted hedges (2–5-years-old). Another important aphid predator the rove beetle *Tachyporus hypnorum* was also found at highest densities in recently sown grass margins, although this finding was non-significant. Recently sown grass margins had higher grass cover, but lower grass species abundance and vascular plant species diversity than in the other habitats. Recently sown grass margins also had lower vascular plant species richness and lower wildflower and moss cover than mature field margins. Four overwintering habitats for beetles and spiders were surveyed at one site: recently sown grass margin, mature field margin, recently planted hedge base and mature hedge base. Five geographically separate replicates of each of the 4 habitats were sampled for beetles and spiders in February 2002 by taking 12 soil core samples in a 70 m long sampling section. Percentage cover of vascular plant species, moss and bare ground was estimated, and biomass (dry matter) and organic carbon content were measured.

A replicated, controlled trial in North Yorkshire, UK (32) found more bank voles *Clethrionomys glareolus* and common shrews *Sorex araneus* on sown grass field margins in autumn than on control cropped margins, but no such differences in spring. There were 13–14 and 26–38 bank voles/autumn trapping period on 3 m and 6 m margins respectively, compared to 0–1 voles on control margins. There were 14–15 and 10–13 common shrews/autumn trapping period on 3 m and 6 m margins respectively, compared to 1–4 common shrews on control margins. Wood mice *Apodemus sylvaticus* were found in similar numbers on all margin types in autumn and spring (0–29 mice/trapping period). Four 3 m-wide and four 6 m-wide field margins were established in autumn 1997 and sown with a mix of grasses on one arable farm. Small mammals were trapped in spring (April–May) and autumn (September–October) 1999 and 2000 on sown field margins and four conventional cropped field edges (controls). On four separate nights in each trapping period, 20 Longworth traps were set 10 m apart on each margin: 10 along the edge furthest from the crop, and 10 placed 2 m into the crop. This study was carried out at the same experimental site as (15, 25).

A replicated study in June and September 2002 in Yorkshire, UK (33) found that beetle (Coleoptera) abundance and species richness on experimental plots was strongly influenced by the type of seed mixture. A mix containing only grass species ('Countryside Stewardship Scheme mix') had a higher abundance and species richness of beetles than a mix containing mainly flowering plants and no tussock grass species ('fine grass and forbs mix'). A third mix containing wildflowers, fine and tussock grass species ('tussock grass and forbs mix') had similarly high beetle numbers and richness to the grass-only Countryside Stewardship Scheme mix. Beetle diversity (Shannon-Weiner index) did not differ between the different seed mixes. The plant communities in the grass-only and tussock grass and forbs mixes were more similar to each other than to the fine grass and forbs mix. Each of the seed mixtures was randomly sown on three of nine experimental plots in each of five blocks on one farm in autumn 2001. Plots measured 25 x 5 m. Seed mixes contained 3–7 grass and 0–19 forb species. The strips were cut once in July with cuttings left in place. Plant diversity and cover and vegetation structure were surveyed in June and September 2002 using 0.5 x 0.5 m quadrats and a 'drop disk'. Beetles were sampled using a Vortis© (Burkland Ltd., UK) suction sampler. Five samples (15 suctions for 10 seconds) were taken in each plot (total area sampled 1.32 m^2) on each visit. This study was extended in (57).

A replicated study in 2005 in the Netherlands (34) found no clear effects of sown field margins and flower strips on invertebrate pest populations. No figures were presented. A total of 15 km of perennial field margins and flower strips were sown along field edges and across fields in a 400 ha area on 5 arable fields. Flower availability, natural enemy and key pest densities were measured in 2005.

A replicated, controlled study in summer 2004 in lowland England (35) found sown margins contained more species of grasses and wildflowers (including perennials) as well as more foodplants for birds, butterfly (Lepidoptera) larvae and bumblebees *Bombus* spp. foodplants than cereal field headlands. Margins sown with a mixture of grasses and wild-flowers had fewer weed species than unsown sites and compared to grass-only sown margins they had a greater number of plant species and up to 60% more perennial wildflowers. Annual plants were more prevalent in grass-sown margins up to two years old, but species composition was not related to age in older margins. One hundred and sixteen margins were studied in eight regions. Five types of margin (minimum length 120 m) were monitored: sown with grass mix (less than two years old), sown with grass mix (more than two years old), sown with grass and wildflower mix, naturally regenerated and normal cereal field margins (control). This study was part of the same experimental set-up as (44).

A replicated, controlled study in 1997–2000 in Essex, UK (36) found significantly greater butterfly (Lepidoptera) species richness on 2 m sown grass margins, but not on 6 m (study does not distinguish sown from naturally regenerated 6 m margins) grass margins compared to control sites (field edges without grass margins). Butterfly species richness was also higher on 2 m grass margins sown with a more diverse seed mixture. Significantly higher butterfly diversity was found on 2 m grass-sown margins adjacent to hedgerows than on those without hedgerows. No significant differences were found in butterfly species richness between 6 m margins and controls. Plant species richness was higher on both 2 and 6 m margins run-ning alongside hedgerows than on those without hedgerows. The 6 m margins established by natural regeneration held the highest plant species richness. Twenty-six margins were estab-lished on three farms in October 1996–1998: grass-sown (2 m wide); grass-sown (6 m wide); naturally regenerated (6 m wide); control crop (2 and 6 m wide). Grass sown margins were established using a range of grass seed mixtures containing common grass species. All plant species in the margins and adjacent hedgerows were recorded in July–August 1998–2000 and abundance measured using the DAFOR scale. Butterflies in margins and control sites were monitored weekly along transects between 1997 and 2000 in suitable weather. This study is part of the same experimental set-up as (28, 29, 45, 46).

A replicated, paired site comparison study in 2006 in the UK (37) found that installing 6 m-wide grass field margin strips along arable fields had no effect on the number of birds or bird species found to breed or forage on farmland. Under the Countryside Stewardship Scheme, these 6 m-wide grass field margin strips were either created through natural regeneration, sowing grass species, or sowing a grass/wildflower mixture. The study surveyed 7 pairs of fields (one with field margins managed under the Countryside Stewardship Scheme; one con-ventionally farmed) and the 12.5 ha area surrounding each field, from each of 3 different parts of the UK 4 times during the breeding season.

A replicated, paired sites comparison study in mid-summer 2003 in southern England (38) found that plants, bees (Apidae) and grasshoppers (Orthoptera) were all more abundant or had higher diversity on fields with 6 m-wide sown grassy margins, compared to control fields without margins. For example, an average of 5.2 grasshopper and cricket individuals and 1.8 species were found in fields with grass margins, compared to 0.9 individuals and 0.6 species in control fields without grass margins. However spiders (Araneae), ground beetles (Carabidae) and farmland birds did not respond positively to grass margins, with 11–18 bird species/site for fields with grass margins, compared to 11–15 species/site for fields without margins. Forty-two arable field sites in 21 pairs of fields with and without grass margins were studied. Vegetation was assessed in 1 x 5 m quadrats and plant cover assessed visually. Numbers of nesting birds were assessed using territory mapping. Bees were surveyed from June to mid-July using butterfly and sweep nets along a transect for 15 minutes. Spiders were sampled in pitfall traps in the crop and margin. Grasshopper numbers and activity were measured through sweep netting and visual/audial assessment.

A replicated, controlled study in 2001–2005 on ten farms in England (39) found that of three margin types, birds favoured margins sown with a tussock grass and wildflower mix in 2003 and a fine grass and wildflower mix in 2004 to margins sown with grass species only. Flower abundance and species richness was highest in margins sown with a fine grass and wildflower mix and lowest in the standard grass mix margins. Bumblebee *Bombus* spp. abundance and species richness were highest on the tussock grass and wildflower mix and lowest on the standard grass margin mix. More birds were associated with scarified than cut grass margins in July 2004. Scarified margins had greater plant diversity and more unsown plant species. Cutting maintained plant species diversity in the grass and wildflower mixes. Grass-specific herbicide application benefited fine grass and flower species. The effects of management treatments on invertebrate abundance were habitat and group specific. Scarified margins had greater beetle (Coleoptera) and true bug (Hemiptera) diversity at some sites. Cutting and grass-specific herbicide application in both tussock grass and fine grass wild-flower mixes increased abundance of true bugs and planthoppers (Fulgoroidea). Butterfly (Lepidoptera) diversity in the standard grass margins was enhanced by scarification, and increased in the tussock grass and wildflower mix through grass-specific herbicide application. In autumn 2001, three grass mixes (standard Countryside Stewardship Scheme mix, tussock grass and wildflower mix, fine-leaved grass and wildflower mix) were sown in 6 m-wide margins at three sites, with five replicates. From 2003 each margin type was subjected to three different management treatments: cutting, scarification or selective grass-specific herbicide application.

A replicated, controlled trial in 2004 in thirty-two 10 km grid squares across England (40) found that 6 m-wide sown grass margins had more bumblebee *Bombus* spp. species, and a higher abundance of foraging bumblebees, than conventionally cultivated and cropped field margins (on average 6–8 bumblebees of 1.3–1.4 species/transect on grassy margins, compared to 0.2 bumblebees of 0.1 species/transect for cropped margins). Older grassy margins, sown more than three years previously, did not attract more foraging bumblebees than those sown in the previous two years. Field margins were 6 m-wide and part of agri-environment scheme agreements. Five field margin types were investigated: grass mix (sown between 1993 and 2000); grass mix (sown between 2002 and 2003); grass and wildflower mix (sown between 1999 and 2003); 'pollen and nectar'-rich margin (sown between 2002 and 2003); control cropped margins. Grass mixes typically included species such as cock's-foot *Dactylis glomerata* and timothy *Phleum pratense*. All five margin types were surveyed within each 10 km grid square (excluding the grass and wildflower mix which was not present in all squares), giving a total of 151 margins. Bumblebees were counted on a 100 x 6 m transect in each field margin, once in July and once in August.

A replicated study in 2003–2004 in Devon, UK (41) found that the density of the meadow grasshopper *Chorthippus parallelus* was significantly higher in 6 m margins than 2 m-wide margins (study does not distinguish between sown and naturally regenerated margins), grazed pasture or long-term set-aside. Two-metre-wide margins supported higher meadow grasshopper densities than intensively grazed pastures, but not Countryside Stewardship Scheme P1 pastures (lightly grazed). Within the 6 m-margins, grasshoppers were more abundant on the outer edge (adjacent to the crop) than on the inner edge (adjacent to a hedge). Meadow grasshoppers were found at highest densities in swards measuring between 30 and 50 cm tall. At low abundances, dock species *Rumex* spp. and cock's-foot *Dactylis glomerata* had a positive impact on meadow grasshopper density, but a negative impact at high abundances. Meadow grasshopper density was negatively impacted by bare ground and rye grasses *Lolium* spp. Fifteen farms were surveyed; 12 were subject to a Countryside Stewardship Scheme prescription and of the Countryside Stewardship Scheme farms; 3 were organic production systems. Five habitat types were surveyed: intensive dairy pasture; long-term set-aside; Countryside Stewardship Scheme P1 grazed pasture; Countryside Stewardship Scheme 2 m-wide field margin; Countryside Stewardship Scheme 6 m-wide field margin. Forty-one arable field margins were studied. Thirty-six of the margins were established through sowing (the study does not specify seed mixture); the remaining five were naturally regenerated. Margins were all cut, but at different frequencies and to different extents in July–August. Grasshoppers were surveyed between July and September in 2003, and July–August in 2004. Two sample blocks (2 x 30 m) were set up in long-term set-aside, grazed pasture and intensive dairy pasture, one at the field edge and one 30 m from the field edge. In fields with margins, one sample block (2 x 30 m for 2 m margins; 6 x 30 m for 6 m margins) was established at the field edge. Grasshoppers were sampled in 0.5 m^2 box quadrats. Vegetation height and cover were measured in five 0.5 m^2 quadrats.

A replicated study in 1999 and 2003 in East Anglia and the West Midlands, England (42) found that a combination of creating uncultivated and planted margins around fields was strongly positively associated with 4 out of 12 farmland bird species analysed. These were skylark *Alauda arvensis* (a field-nesting species) and chaffinch *Fringilla coelebs*, whitethroat *Sylvia communis* and yellowhammer *Emberiza citrinella* (all boundary-nesting species). The study did not distinguish between uncultivated and planted margins. The study was carried out on 256 arable and pastoral fields on 84 farms.

A replicated study in 2003 and 2004 in England (43) found that sown grass field margins tended to have higher numbers of small mammals than set-aside. Numbers of captured small mammals were highest in 2 m margins (2.9–4.4 individuals), followed by 6 m margins (2.5–3.6) and set-aside (1.6–2.0). Numbers of small mammals captured were correlated with sward height in 2 m margins. In 2003, significantly more common shrews *Sorex arenaeus* were captured in 2 m margins (1.4 individuals) than set-aside (0.6) and more wood mice *Apodemus sylvaticus* were found in 6 m margins (1.1) than set-aside (0.5). The trend was similar for bank voles *Clethrionomys glareolus* in 2004: 6 m margins (1.6), 2 m margins (1.4) and set-aside (0.5). Species richness did not differ significantly (1.7–2.0). Species richness, total number of small mammals captured, and the number of bank voles and common shrews captured was higher in 6 m margins cut every 2–3 years compared to those cut annually, although this was only significant for common shrews in 2003. Following establishment, 2 m margins were cut at 2–3 year intervals. For 6 m margins, eight 2 m strips at the edges of margins were cut annually and 12 were cut every 2–3 years. Twelve small mammal traps were set within 20 plots per treatment (1 m from the habitat boundary) for 4 days in November–December 2003–2004. Mammals were individually fur-clipped and released.

A replicated site comparison study in the UK (44) found that sown grassy margins more than two years old had 87–95% cover with grasses. Those sown with wildflowers had on average 28% cover with non-grass broadleaved plants, compared to 14% cover in margins sown with a simple grass seed mix. This option was not considered the best option for conservation of arable plants. A total of 75 sown grass margins were surveyed in 2004. Twenty-two of them were sown with some non-grass flowering species, as well as grasses. Margins were randomly selected from eight UK regions. Plants were surveyed in thirty 0.025 m^2 quadrats within a 100 m sampling zone of each margin and percentage cover across all quadrats estimated. This study was part of the same experimental set-up as (35).

A replicated, controlled study in 1997–2000 in Essex, UK (45) found that total butterfly (Lepidoptera) abundance was higher in grass margins (average 66.6 butterflies) than in control sections (field edges without margins) (average 25.6). Of the 'key' grassland butterfly species, both meadow brown *Maniola jurtina* and skipper butterflies *Thymelicus* spp. had higher abundance in sown grass margins (average 15.5 and 13.9 individuals respectively) than in controls (average 3.6 and 1.2 respectively). Between 1997 and 2000 there was a significant reduction in the abundance of total butterflies (from an average of 100.6 to 47.0), *Thymelicus* spp. (from 32.4 to 3.9) and large skipper *Ochlodes venata* (from 15.3 to 0.6) in the margins. During the same time, the average abundance of gatekeeper *Pyronia tithonus* increased from 2.2 to 12.9 in the margins. Grass margins were established as described in (29). Butterfly abundance was monitored weekly along transects from late June to early August 1997–2000. All butterflies were recorded, but special note was taken of 'key' grassland species: meadow brown, gatekeeper, small skipper *Thymelicus sylvestris*, Essex skipper *T. lineola*, large skipper. This study is part of the same experimental set-up as (28, 29, 36, 46).

A replicated, controlled study in the summers of 1997–2000 and 2003 in Essex, UK (46) found that butterfly (Lepidoptera) abundance was higher in 2 m and 6 m-wide Countryside Stewardship Scheme margins than in control margins (field edges without established grass margins). Comparisons between grass-sown 2 m margins and control sections showed 64 vs 24 butterflies/km/visit and 54 vs 19 for 6 m margins (study does not distinguish between the effects of sown and naturally regenerated 6 m margins). The meadow brown butterfly *Maniola jurtina* also occurred in higher numbers in Countryside Stewardship Scheme field margins, 2 m margins and their control sections had 15 vs 4 individuals/km/visit, 6 m margins had 22 vs 5/km/visit. Butterfly abundance and species richness did not change over the study period in either 2 or 6 m margins or in a transect across farmland. Plant species richness declined significantly within sown field margins of both widths from 1998 to 2003 as the sown grass species became dominant. Sown 6 m margins had lower plant species richness in 2003 (20 species) compared with naturally regenerated 6 m margins (35 species). Eleven margins were studied on three farms. Two metre margins were sown with a grass-only seed mixture and the vegetation left uncut after the first year. Six metre margins were established through natural regeneration or by sowing, all cut annually after 15 July. This study was part of the same experimental set-up as (28, 29, 36, 45).

A randomized, replicated, controlled trial from 2002 to 2006 in eastern England (47) (same study as (50, 51)) found that the management of sown grass field margins affected bird use more than the seed mix used. Bird densities were higher on disturbed and grass-specific herbicide-treated plots than on cut plots (no actual bird densities given, only model results). Bird densities were linked to densities of diurnal ground beetles (Carabidae), especially in disturbed and grass-specific herbicide-treated plots. The number of birds using the margins in summer increased by 29% between 2003 and 2006. In winter, there were twice as many birds on cut margins than uncut margins, and twice as many birds in the second year than the first. Field margin plots (6 x 30 m) were established using one of three seed mixes: Countryside Stewardship mix (seven grass species), tussock grass mix and a mixture of grasses and wildflowers designed for pollinating insects. The margins were managed in spring from 2003 to 2005 with one of three treatments: cut to 15 cm, soil disturbed by scarification until 60% of the area was bare ground, treated with grass-specific herbicide at half the recommended rate. There were five replicates of each treatment combination, at two farms. Birds were surveyed five to eight times between April and July from 2002 to 2006. In the winters of 2004–2005 and 2005–2006, birds were surveyed on 6 m margins on 10 farms with 2 seed mixes (tussocky grass and fine grass). Margins were either cut in autumn or uncut. There were four replicates of each treatment combination per farm.

A 2007 literature review of the effects of agri-environment scheme options on small mammals in the UK (48) identified three studies that found small mammal abundance tended to be higher in grass margins compared to cropped fields (9, 32) (Macdonald *et al.*, 2000). One study (32) also found that wider grass margins had the highest numbers of bank voles *Myodes glareolus*.

A small replicated site comparison study in 2005 in Oxfordshire, UK (49) found that field margins sown with a grass and wild flower mix had more species and individual grasshoppers and crickets (Orthoptera) than margins sown with grass only, wildflower margins, or grassy tracks. Narrow grass and wildflower margins (2 m wide) had an average of 10 insects from 4 species, compared to 1–4 individuals from less than 2 species for the other margin types. Wide grass and wildflower margins (6 m) also had more species and individuals than others (8 individuals; 2.5 species/margin on average), but this was not always statistically significant. Seventy-three percent of all crickets and grasshoppers caught were in margins sown with a grass and flower mix. The grass and flower margins had intermediate vegetation height (30–40 cm), low cover of bare ground and intermediate grass and flower cover compared to other margins. Three replicates of five field margin types were monitored on a large mixed farm: grass and wildflower mix (2 m), grass and wildflower mix (6 m), grass only mix, wildflower mix, grassy track. Grasshoppers and crickets were surveyed using a sweep net over two 20 minute periods in a 50 m section of each margin, in late July or August 2005.

A randomized, replicated, controlled trial from 2002 to 2006 in eastern England (50) (same study as (47, 51)) found that floristically enhanced grassy margins supported more bumblebees *Bombus* spp. and butterflies (Lepidoptera) than grass-only margins. For bees and butterflies, there was no difference in abundance or number of species between the grass and wildflower mix and the pollinating insect mix (35–47 bumblebees of 4 species and 18–20 butterflies of 6 species/125 m^2 plot on average on the grass and wildflower mix and the pollinating insect mix, compared to 10 bumblebees of 2 species and 12 butterflies of 5 species on grass-only margins). Different types of management did not affect the abundance of bees and butterflies or the number of butterfly species, but there were more bumblebee species on plots treated with grass-specific herbicide in spring (average 4 species/125 m^2, compared to 3 species on cut or disturbed plots). Field margin plots (6 x 30 m) were established in 2000–2001 using 1 of 3 seed mixes: Countryside Stewardship mix (7 grass species, sown at 20 kg/ha), tussock grass mix (7 grass species, 11 wildflowers, sown at 35 kg/ha) and a mixture of grasses and wildflowers designed for pollinating insects (4 grass species, 16–20 wildflowers, sown at 35 kg/ha). The margins were managed in spring from 2003 to 2005 with 1 of 3 treatments: cut to 15 cm; soil disturbed by scarification until 60% of the area was bare ground; treated with grass-specific herbicide in spring at half the recommended rate. There were five replicates of each treatment combination, at three farms.

A randomized, replicated, controlled trial from 2002 to 2006 in eastern England (51) (same study as (47, 50)) found that sown grass margins had a greater abundance of planthoppers (Auchenorrhyncha) than margins sown with a grass and wildflower mix. Grass-only margins had 30–70 planthoppers/plot on average (depending on management), while other margins had 25–45 planthoppers/plot. There were fewer planthoppers in disturbed (scarified) plots (20–30 planthoppers/plot on average, for all seed mix treatments) than in those cut or treated with grass-specific herbicide in spring (35–70 planthoppers/plot on average). Field margin plots (6 x 30 m) were established in 2000–2001 using 1 of 3 seed mixes: Countryside Stewardship mix (7 grass species, sown at 20 kg/ha), tussock grass mix (7 grass species, 11 wildflowers, sown at 35 kg/ha) and a mixture of grasses and wildflowers designed for pollinating insects (4 grass species, 16–20 wildflowers, sown at 35 kg/ha). The margins were managed in spring from 2003 to 2005 with 1 of 3 treatments: cut to 15 cm, soil disturbed by scarification until 60% of the area was bare ground, treated with grass-specific herbicide in spring at half the recommended rate. There were five replicates of each treatment combination at three farms. Planthoppers were sampled in June and September 2003 and 2004 by suction sampling (75 10-second sucks/plot in total, sampled from 40.5 m^2/plot).

A 2007 literature review in Leicestershire, UK (52) found that grass margins contained large numbers of overwintering invertebrates such as rove beetles (Staphylinidae) and ground beetles (Carabidae) as well as high numbers of yellowhammer *Emberiza citrinella* and whitethroat *Sylvia communis* nests; yellowhammer had higher survival than in adjacent hedgerows.

A 2008 literature review of grasshoppers (Acrididae) and bush-crickets (Tettigoniidae) (order: Orthoptera) in 6 m-wide grass margins around arable fields in eastern England, UK (53) found that grass margins appear to increase grasshopper and bush-cricket diversity and abundance in landscapes with small to intermediate field sizes (38), but not in landscapes with large and intensively-farmed fields (30). The review suggested that landscape context at different scales, management routines such as cutting and species composition within the margins affect grasshopper/cricket populations in field margins.

A replicated, controlled study in summer 2005 in south Wiltshire, UK (54) found that a higher proportion of grain aphids *Sitobion avenae* were parasitized in winter wheat fields with wide margins (5–6 m) compared to standard margins (<1 m) ten days after inoculation with aphids. One month after inoculation more aphids were parasitized at 20 m from the wide margin compared with the standard margins. Flying predators reduced aphid numbers by 90% and 93% in fields with standard and wide field margins respectively one month after inoculation whereas ground-dwelling predators achieved reductions of only 40% and 18%. Ground-dwelling predators had no additional effect on aphid abundance compared to when only flying predators were present in fields with either wide or standard width field margins. Spiders (Araneae) were more abundant in suction samples collected in fields with wide margins. Flying predators and balloon flies (Empididae) were more abundant in fields with standard margins. Exclusion cages were used to investigate the effect of ground-dwelling and flying predators in isolation or together on aphid abundance, as well as in the absence of predators in fields with different margin widths (standard (<1 m) or wide (5–6 m)) in ten winter wheat fields. Two transects with exclusion cages were established in each field, at 20 and 80 m from margin. Aphid abundance was monitored 4 days before and 10, 20 and 32 days after inoculation with aphids on 10 June.

A paired, replicated, controlled study in spring 2006 in Berkshire, UK (55) found that earthworms (Lumbricidae), woodlice (Isopoda), and rove beetles (Staphylinidae), as well as three main feeding groups (litter consumers, soil ingesters and predators) had higher abundance and species density in sown grass strips compared with the field bean crop. However the presence of grass strips did not increase soil macrofaunal diversity outside the field margin in either the adjacent crop, or under the adjacent hedgerow. The species composition of soil macrofaunal communities in grass strips was different compared with other habitats on a within-field and a within-farm scale. Six-metre-wide grass strips were established and managed according to the Countryside Stewardship Scheme guidelines in 2000–2001. Soil core samples were collected in April–May along transects perpendicular to paired hedgerow boundaries at 0 m (under hedge), 3 m (in grass strip/crop), and 9 and 27 m into the crop. Five other habitats were sampled for the within-farm analysis (winter wheat fields, pasture, set-aside, coniferous Scot's pine *Pinus sylvestris* plantation, and broadleaf plantation) in May.

A replicated, controlled study in 2005 in Cambridgeshire, UK (56) found that the species density and abundance of woodlice (Isopoda) and beetles (Coleoptera), as well as the species density of earthworms (Lumbricidae) was higher in sown grass margins than in the winter wheat crop. Species densities and abundances/m^2 (grass margin average vs crop average) were woodlouse density 2.8 vs 0 and woodlouse abundance 74 vs 0; beetles 17.3 vs 10.0 and 80 vs 41; earthworms 5.1 vs 3.8 and 281 vs 244; millipedes (Diplodopa) 3.2 vs 3.5 and 36 vs 38; centipedes (Chilopoda) 2.1 vs 1.8 and 14 vs 18. Scarified plots had lower abundance and fewer species of woodlice compared with spring cut and herbicide treated plots while species composition was similar to that of the crop. Scarified plots also had fewer soil-feeders and litter-feeders, and predatory species densities were lower, compared with the other plots. Field margins were created in 2001 with four replicated blocks of nine treatments (three seed mixtures x three management regimes) in one arable field. Soil macrofauna was sampled through soil cores in April and October 2005.

A replicated study in summer 2002–2004 and 2006 in England (57) (extension of (33)) found higher abundance and species richness of predatory beetles (Coleoptera) in margins containing tussock grass species than in margins with fine grasses only (independent of the presence of wildflowers). However the abundance of herbivorous beetle species was lower in margins containing only grass species and higher in grass mixtures with a wildflower component. Soil scarification had a positive effect on species richness of predatory beetles. Three different seed mixtures were sown: grass only; tussock grass and wildflowers; fine grass and wildflowers. Each of the seed mixtures was randomly sown on three of nine experimental plots in each of five blocks on three farms in autumn 2001. Plots measured 25 x 5 m. From 2003, three different management practices were applied in each replicate block in May each year: cutting the vegetation to 10–15 cm; application of grass-specific herbicide (fuazifop-*p*-butyl) at 0.8 l/ha; and scarification of 60% of the soil surface. Plant diversity and cover and vegetation structure were surveyed yearly in June using 0.5 x 0.5 m quadrats and vertical drop pins. A Vortis suction sampler (75 suctions of 10 seconds each) was used over a fixed area (equivalent to 1.45 m^2) in each plot on each sampling date to collect beetles. Rove beetles (Staphylinidae), ground beetles (Carabidae), ladybirds (Coccinellidae), leaf-beetles (Chrysomelidae) and weevils (Curculionoidea) were determined to species level and categorized as either herbivorous or predatory.

A replicated, controlled habitat selection study in 2003–2005 in southwestern Finland (58) found that field voles *Microtus agrestis* in riparian field margins moved on average longer distances in narrow (≤5 m) filter strips than in wide (>15 m) buffer zones. Home range sizes tended to be larger in narrow than in wide margins, although these differences were not significant. Field voles were most frequently found in control plots where vegetation was left uncut with no supplementary food or cover added, in both narrow and wide riparian field margins. Crop fields and all mown habitat types were used significantly less by field voles in wide buffer zones than in narrow filter strips. Overall, mown plots were used less than unmown plots. In wide buffer zones, voles used mown habitats proportionally significantly less than other available habitats, whereas in narrow filter strips there was no difference in use between mown and unmown plots. Supplementary food appeared to attract voles in unmown plots in both wide and narrow riparian field margins, but not in mown plots. Mown plots with supplementary food provided were avoided by voles in wide margins. Riparian field margin width did not affect the proportional use of crop fields and field margin habitats from late autumn to spring (summer use not tested). Field margins were created under an agri-environment scheme prior to the study. In mid-June 2005, one 210 m-long section in each of 4 riparian field margins was divided into fourteen 15 m-long experimental plots, half of which were mown to <20 cm. Food and/or cover was added to mown/unmown plots (total of eight treatments). Trapping and radio-tracking field voles started two weeks after habitat manipulation. Radio-tracking for the seasonal habitat-use analysis was done in summer (June 2003), late autumn (December 2003), winter (January 2004) and spring (April–May 2005).

A replicated study in May–August 2004–2006 in Aberdeenshire, Scotland (59) found that yellowhammers *Emberiza citrinella* appeared to use cut field margins (sown or naturally regenerated) significantly more in late than early summer for foraging. Cut patches were used more frequently in margins with swards >60 cm tall. The authors suggest that yellowhammers used cut patches disproportionately as the uncut sections grew taller and so reduced access to invertebrates.

A replicated, controlled trial in spring 2008 in Scotland (60) found that on farms under the Rural Stewardship agri-environment scheme, 1.5 to 6 m-wide grassy field margins attracted higher densities of foraging queen bumblebees *Bombus* spp. in spring than conventionally managed field margins (more than 3 queens/100 m on grassy margins, compared to 1 queen/100 m on conventional margins). However, when counts on conventionally managed field margins were compared on farms with and without agri-environment schemes, farms without the agri-environment agreement had more foraging queens. This raises the possibility that farms with the Rural Stewardship Scheme agreement supported similar numbers of queens overall, but they were preferentially distributed on the agri-environment field margins. Margins on 10 arable farms were studied, 5 of which participated in the Rural Stewardship Scheme. Six habitat types were studied using 100 m transects on each farm: Rural Stewardship Scheme grass margin, conventionally managed arable field margin, species rich grassland, unfarmed grassland, Rural Stewardship Scheme hedgerow, conventionally managed hedgerow. The number of bumblebee queens within 3 m of the transect were recorded, once a week over a five-week period.

A replicated mark-release-recapture study in summer 2007 in Oxfordshire, UK (61) found overall higher abundance of nine common larger farmland moth (Lepidoptera) species in the margins and centres of arable fields with 6 m-wide perennial grass margins than in fields with standard 1 m margins, but this varied highly between species. Six moth species which contributed to the higher abundance of moths in wide field margins were less mobile; moving a shorter distance between captures and being more frequently recaptured at the site of first capture. Nectar availability (number of flowerheads) was higher in wide margins, both for overall nectar plant species and plant species known to be moth favourites. Plant species richness and diversity was similar in hedgerows surrounding fields bordered with wide margins and with standard margins. Five Heath pattern actinic light traps (6 W) were positioned in each of four arable fields: one in the centre and one in each field margin (1 m from hedgerow). All traps were >100 m apart and >50 m from hedgerow intersections. Traps were operated on the 32 nights (dusk till dawn) with suitable weather between 5 June and 14 July. Nectar availability was assessed at each trap site on 25 June by counting the number of flowerheads present on field margins 10 m either side of the trap locations. Percentage cover and species richness of woody plant species (excluding trees) was estimated in hedges bordering the fields.

A replicated, controlled study in 2006 in central Oxfordshire, UK (62) found no difference in moth (Lepidoptera) abundance or diversity between 6 m (agri-environment scheme option) and 1 m (standard option) -wide field margins in four lowland farmland areas. In each area, one farm with standard margins and one with 6 m-wide margins were sampled.

Three Heath pattern actinic light traps (6 W) were set up on each farm ≥100 m apart and >50 m from hedgerow intersections. Traps were placed on 2 m^2 white cotton sheets 1 m from hedgerows bordering fields with no banks or ditches. All farms were sampled once during each of 11 discrete fortnightly periods from mid-May to mid-October 2006. Sampling was carried out from dusk till dawn during nights with suitable weather conditions. At dawn, all individuals were identified to species, species-pair or genus, marked with a unique number and released where caught.

A series of three replicated trials in the Netherlands (63) found that the number of plant species in field margins and adjacent ditch banks increased in the 4 years following establishment of 2–3 m-wide sown grass and wildflower field margins. More field margins and ditch banks showed a decline in cover of agricultural weeds following margin establishment. For both butterflies (Lepidoptera) and dragonflies (Odonata), more than half the transects showed increased species diversity in field margins, in the two to eight years following the establishment of margins. Ninety field margins at least 2 m wide were established on 21 farms across the Netherlands and monitored for 2 to 6 years. On 20 of the farms, 107 ditch banks alongside 3 m-wide field margins were also monitored. Most margins were planted with grasses. All margins and ditch banks were mown at least once a year and cuttings removed. Plant species richness was measured in permanent quadrats or sections. Butterflies were counted in 50 m transect counts along field margins on six farms, and dragonflies on five farms. Transect counts were either every week, or two to five times during summer.

A 2009 literature review of European farmland conservation practices (64) found that sown grass margins had higher arthropod diversity than adjacent crops, and also held higher abundances of soil invertebrates. The availability of bird food-species was also higher than in crops, although use of grass-only strips by several bird species (yellowhammer *Emberiza citrinella*, red-legged partridge *Alectoris rufa*, greenfinch *Carduelis chloris*, linnet *C. cannabina*) was lower than for margins planted with wildflower mixes.

A replicated study in February 2008 in East Anglia, England (65) found that field margins managed under agri-environment schemes had a positive influence on 19 out of 24 farmland bird species. However, only yellowhammer *Emberiza citrinella* and possibly blackcap *Sylvia altricapilla* showed a strong positive response to agri-environment scheme margins affecting species densities. Great tit *Parus major* and common starling *Sturnus vulgaris* showed weak positive responses. Field margins were categorized as grassy/weedy, bare/fallow or wild-bird cover (although very few fields had wild bird cover) and most were managed under the Entry Level Stewardship scheme. Ninety-seven 1 km^2 plots were included in the study. All field boundaries within each square were walked and all birds present mapped. Squares were visited twice; once in April to mid-May, and once in mid-May to June.

A replicated site comparison study in 2004–2008 across England (66) found that grey partridge *Perdix perdix* brood size was negatively associated with the proportion of a site under planted grass buffer strips (association significant in 2008). The ratio of young:old partridges was negatively related to the proportion of grass strips in 2005 and 2008. However, year-on-year changes in partridge density and overwinter survival were positively correlated with the proportion of grass buffer strips on a site, this relationship was significant from 2006 to 2007 (year-on-year changes) and 2005–2006 (overwinter survival). Spring and autumn counts of grey partridge were made at 1,031 sites across England as part of the Partridge Count Scheme.

A replicated site comparison study in three regions in England (67) found that hedges alongside wildflower-rich grass field margins ('floristically enhanced' margins) under Higher Level Stewardship had more yellowhammers *Emberiza citrinella* (estimated 0.4 birds/m) compared to hedges without a grass margin (estimated 0.2 birds/m). Hedges alongside unenhanced grass margins, either conventionally managed or managed under Entry Level Stewardship, did not have more yellowhammers. Surveys were carried out on 69 farms with Higher Level Stewardship in East Anglia, the West Midlands and the Cotswolds and on 31 farms across all 3 regions with no environmental stewardship.

A replicated study in summer 2007 in south Sweden (68) found lower densities and species richness of butterflies (Lepidoptera) and bumblebees *Bombus* spp. in margins mainly sown with a mix of grass species, 4 m-wide (greenways or 'beträdor') than in sown wild-flower strips. Fourteen percent of the recorded butterflies, and 17% of the bumblebees, were found in grass strips, and butterfly density was nearly 20 times lower in grass strips than in wildflower strips. Bumblebees were almost absent in the sown grass strips. However, the presence of bushes adjacent to grass strips positively influenced butterfly species richness and abundance of both butterflies and bumblebees. Butterflies and bumblebees were recorded on 3 grass strips (14 transects) and 1 wildflower strip (6 transects) on 5 occasions on 4 arable farms. Butterflies and bumblebees were counted within 2 m either side of the observer and the flower species visited by the insects noted.

A replicated, controlled study in the summers of 2008–2009 in Berkshire, UK (69) found that butterfly (Lepidoptera) abundance, species richness and diversity were positively associated with the number of sown wildflower species in existing grass buffer strips. Butterfly species richness was higher in plots that had received a combined treatment of scar-ification and grass-specific herbicide application compared with single treatment and control plots. Butterfly abundance and diversity were higher in plots that were both scarified and treated with grass-specific herbicide than single treatment, but not control plots. Sown wild-flower cover and species richness was higher in the combined treatment plots in both years, and there was a significant increase in wild flower cover from 2008 to 2009. In both years, species richness of unsown wildflowers (annuals, perennials and in total) was higher in the combined scarification/grass-specific herbicide treatments. It was also higher in scarification-only than in grass-specific herbicide-only and control plots, but it decreased in scarified plots from 2008 to 2009. Six metre-wide grass buffer strips were created on two arable farms in 2004 and managed under an Entry Level Stewardship agreement from 2005. Four treatments were randomly established within each of three replicate blocks/site in early spring 2008: scarific-ation; selective grass-specific herbicide application; scarification and selective grass-specific herbicide; control. Scarification was always followed by sowing a wildflower seed mixture. All plots were cut in autumn and cuttings left in place. In both years vegetation was assessed once in June, in ten randomly placed 0.25 m^2 quadrats within each treatment plot avoid-ing the edges. Percentage cover of all plant species was estimated on an eight point scale. Abundance, diversity and species richness of adult butterflies was recorded during standard transect walks along the centre of each treatment plot (25 x 4 m). Each plot was sampled eight times/year between May and September.

(1) Marshall E.J.P., West T.M. & Winstone L. (1994) Extending field boundary habitats to enhance farm-land wildlife and improve crop and environmental protection. *Aspects of Applied Biology*, 40, 387–391.
(2) Clarke J.H., Jones N.E., Hill D.A. & Tucker G.M. (1997) The management of set-aside within a farm and its impact on birds. *Proceedings of the 1997 Brighton Crop Protection Conference*. 1–3. pp. 1179–1184.
(3) Hopkins A. & Feber R.E. (1997) Management for plant and butterfly species diversity on organically farmed grassland field margins. *Proceedings of the International Occasional Symposium of the European Grassland Federation: Management for grassland biodiversity*. Warszawa-Lomza, Poland, 19–23 May. pp. 69–73.
(4) Kleijn D., Joenje W. & Kropff M.J. (1997) Patterns in species composition of arable field boundary vegetation. *Acta Botanica Neerlandica*, 46, 175–192.
(5) Smith H., McCallum K. & Macdonald D.W. (1997) Experimental comparison of the nature conserva-tion value, productivity and ease of management of a conventional and a more species-rich grass ley. *Journal of Applied Ecology*, 34, 53–64.

(6)   Weiss B. & Buchs W. (1997) Reaction of the spider coenoesis on different kinds of rotational set aside in agricultural fields. *Mitteilungen Der Deutschen Gesellschaft Fur Allgemeine Und Angewandte Entomologie*, 11, 147–151.

(7)   Huusela-Veistola E. (1998) Effects of perennial grass strips on spiders (Araneae) in cereal fields and impact on pesticide side-effects. *Journal of Applied Entomology*, 122, 575–583.

(8)   Barker A.M. & Reynolds C.J.M. (1999) The value of planted grass field margins as a habitat for sawflies and other chick-food insects. *Aspects of Applied Biology*, 54, 109–116.

(9)   Brown R.W. (1999) Margin/field interfaces and small mammals. *Aspects of Applied Biology*, 54, 203–206.

(10)   Canters K.J. & Tamis W.L.M. (1999) Arthropods in grassy field margins in the Wieringermeer: scope, population development and possible consequences for farm practice. *Landscape and Urban Planning*, 46, 63–69.

(11)   de Snoo G.R. & Chaney K. (1999) Unsprayed field margins – what are we trying to achieve? *Aspects of Applied Biology*, 54, 1–12.

(12)   Thomas C.F.G. & Marshall E.J.P. (1999) Arthropod abundance and diversity in differently vegetated margins of arable fields. *Agriculture, Ecosystems & Environment*, 72, 131–144.

(13)   Aebischer N.J., Green R.E. & Evans A.D. (2000) From science to recovery: four case studies of how research has been translated into conservation action in the UK. In N.J. Aebischer, A.D. Evans, P.V. Grice & J.A. Vickery (eds.) *Ecology and Conservation of Lowland Farmland Birds*, British Ornithologists' Union, Tring. pp. 43–54.

(14)   Holland J.M. & Luff M.L. (2000) The effects of agricultural practices on Carabidae in temperate agroecosystems. *Integrated Pest Management Reviews*, 5, 109–129.

(15)   Telfer M.G., Meek W.R., Lambdon P., Pywell R.F., Sparks T.H. & Nowakowski M. (2000) The carabids of conventional and widened field margins. *Aspects of Applied Biology*, 58, 411–416.

(16)   Moonen A.C. & Marshall E.J.P. (2001) The influence of sown margin strips, management and boundary structure on herbaceous field margin vegetation in two neighbouring farms in southern England. *Agriculture, Ecosystems & Environment*, 86, 187–202.

(17)   Peach W., Lovett L., Wotton S. & Jeffs C. (2001) Countryside stewardship delivers cirl buntings (*Emberiza cirlus*) in Devon, UK. *Biological Conservation*, 101, 361–373.

(18)   Buckingham D.L., Peach W.J. & Fox D. (2002) Factors influencing bird use in different pastoral systems. *Proceedings of the British Grassland Society/British Ecological Society Conference*. University of Lancaster, 15–17 April. pp. 55–58.

(19)   Evans A.D., Armstrong-Brown S. & Grice P.V. (2002) The role of research and development in the evolution of a 'smart' agri-environment scheme. *Aspects of Applied Biology*, 67, 253–264.

(20)   Meek B., Loxton D., Sparks T., Pywell R., Pickett H. & Nowakowski M. (2002) The effect of arable field margin composition on invertebrate biodiversity. *Biological Conservation*, 106, 259–271.

(21)   Perkins A.J., Whittingham M.J., Morris A.J. & Bradbury R.B. (2002) Use of field margins by foraging yellowhammers *Emberiza citrinella*. *Agriculture, Ecosystems & Environment*, 93, 413–420.

(22)   May M. & Nowakowski M. (2003) Using headland margins to boost environmental benefits of sugar beet. *British Sugar Beet Review*, 71, 48–51.

(23)   Asteraki E.J., Hart B.J., Ings T.C. & Manley W.J. (2004) Factors affecting the plant and invertebrate diversity of arable field margins. *Agriculture, Ecosystems & Environment*, 102, 219–231.

(24)   Bokenstrand A., Lagerlöf J. & Torstensson P.R. (2004) Establishment of vegetation in broadened field boundaries in agricultural landscapes. *Agriculture, Ecosystems & Environment*, 101, 21–29.

(25)   Carvell C., Meek W.R., Pywell R.F. & Nowakowski M. (2004) The response of foraging bumblebees to successional change in newly created arable field margins. *Biological Conservation*, 118, 327–339.

(26)   Critchley C., Allen D., Fowbert J., Mole A. & Gundrey A. (2004) Habitat establishment on arable land: assessment of an agri-environment scheme in England, UK. *Biological Conservation*, 119, 429–442.

(27)   Powell W., A'Hara S., Harling R., Holland J.M., Northing P., Thomas C.F.G. & Walters K.F.A. (2004) *Managing Biodiversity in Field Margins to Enhance Integrated Pest Control in Arable Crops ('3-D farming' Project)*. HGCA 356.

(28)   Field R.G., Gardiner T., Mason C.F. & Hill J. (2005) Agri-environment schemes and butterflies: the utilisation of 6 m grass margins. *Biodiversity and Conservation*, 14, 1969–1976.

(29)   Field R.G. & Mason C.F. (2005) The utilization of two-metre Countryside Stewardship Scheme grass margins by the gatekeeper *Pyronia tithonus* (L). *Journal of Natural History*, 39, 1533–1538.

(30)   Gardiner T. & Hill J. (2005) A study of grasshopper populations in Countryside Stewardship Scheme field margins in Essex. *British Journal of Entomology and Natural History*, 18, 73–80.

(31)   Pywell R.F., James K.L., Herbert I., Meek W.R., Carvell C., Bell D. & Sparks T.H. (2005) Determinants of overwintering habitat quality for beetles and spiders on arable farmland. *Biological Conservation*, 123, 79–90.

(32)   Shore R.F., Meek W.R., Sparks T.H., Pywell R.F. & Nowakowski M. (2005) Will Environmental Stewardship enhance small mammal abundance on intensively managed farmland? *Mammal Review*, 35, 277–284.

(33)   Woodcock B.A., Westbury D.B., Potts S.G., Harris S.J. & Brown V.K. (2005) Establishing field margins to promote beetle conservation in arable farms. *Agriculture, Ecosystems & Environment*, 107, 255–266.

(34)   Alebeek F.V., Wiersema M., Rijn P.V., Wäckers F., Belder E.D., Willemse J. & Gurp H.V. (2006) A region-wide experiment with functional agrobiodiversity (FAB) in arable farming in the Netherlands. *Proceedings of the Landscape Management for Functional Biodiversity 2nd Working Group Meeting 16–19 May 2006*. Zürich-Reckenholz, Switzerland, 29. pp. 141–144.

(35)   Critchley C.N.R., Fowbert J.A., Sherwood A.J. & Pywell R.F. (2006) Vegetation development of sown grass margins in arable fields under a countrywide agri-environment scheme. *Biological Conservation*, 132, 1–11.

(36)   Field R.G., Gardiner T., Mason C.F. & Hill J. (2006) Countryside Stewardship Scheme and butterflies: a study of plant and butterfly species richness. *Biodiversity and Conservation*, 15, 443–452.

(37)   Kleijn D., Baquero R.A., Clough Y., Díaz M., Esteban J.d., Fernández F., Gabriel D., Herzog F., Holzschuh A., Jöhl R., Knop E., Kruess A., Marshall E.J.P., Steffan-Dewenter I., Tscharntke T., Verhulst J., West T.M. & Yela J.L. (2006) Mixed biodiversity benefits of agri-environment schemes in five European countries. *Ecology Letters*, 9, 243–254.

(38)   Marshall E.J.P., West T.M. & Kleijn D. (2006) Impacts of an agri-environment field margin prescription on the flora and fauna of arable farmland in different landscapes. *Agriculture, Ecosystems & Environment*, 113, 36–44.

(39)   Ogilvy S.E., Clarke J.H., Wiltshire J.J.J., Harris D., Morris A., Jones N., Smith B., Henderson I., Westbury D.B., Potts S.G., Woodcock B.A. & Pywell R.F. (2006) SAFFIE - research into practice and policy. *Proceedings of the HGCA Conference, Arable Crop Protection in the Balance: Profit and the Environment*, 14, 1–12.

(40)   Pywell R.F., Warman E.A., Hulmes L., Hulmes S., Nuttall P., Sparks T.H., Critchley C.N.R. & Sherwood A. (2006) Effectiveness of new agri-environment schemes in providing foraging resources for bumblebees in intensively farmed landscapes. *Biological Conservation*, 129, 192–206.

(41)   Smith D.W. (2006) Managing agri-environment grass fields and margins for Orthoptera and farmland birds. PhD thesis. University of Reading.

(42)   Stevens D.K. & Bradbury R.B. (2006) Effects of the Arable Stewardship Pilot Scheme on breeding birds at field and farm-scales. *Agriculture, Ecosystems & Environment*, 112, 283–290.

(43)   Askew N.P., Searle J.B. & Moore N.P. (2007) Agri-environment schemes and foraging of barn owls *Tyto alba*. *Agriculture, Ecosystems & Environment*, 118, 109–114.

(44)   Critchley C.N.R., Walker K.J., Pywell R.F. & Stevenson M.J. (2007) The contribution of English agri-environment schemes to botanical diversity in arable field margins. *Aspects of Applied Biology*, 81, 293–300.

(45)   Field R.G., Gardiner T., Mason C.F. & Hill J. (2007) Agri-environment schemes and butterflies: the utilisation of two metre arable field margins. *Biodiversity and Conservation*, 16, 465–474.

(46)   Field R.G., Gardiner T. & Watkins G. (2007) The use of farmland by butterflies: a study on mixed farmland and field margins. *Entomologist's Gazette*, 58, 3–15.

(47)   Henderson I.G., Morris A.J., Westbury D.B., Woodcock B.A., Potts S.G., Ramsay A. & Coombes R. (2007) Effects of field margin management on bird distributions around cereal fields. *Aspects of Applied Biology*, 81, 53–60.

(48)   Macdonald D.W., Tattersall F.H., Service K.M., Firbank L.G. & Feber R.E. (2007) Mammals, agri-environment schemes and set-aside - what are the putative benefits? *Mammal Review*, 37, 259–277.

(49)   Marshall G.M. (2007) The effect of arable field margin structure and composition on Orthoptera assemblages. *Aspects of Applied Biology*, 81, 231–238.

(50)   Pywell R.F., Meek W.M., Carvell C. & Hulmes L. (2007) The SAFFIE project: enhancing the value of arable field margins for pollinating insects. *Aspects of Applied Biology*, 81, 239–245.

(51)   Ramsay A.J., Potts S.G., Westbury D.B., Woodcock B.A., Tscheulin T.R., Harris S.J. & Brown V.K. (2007) Response of planthoppers to novel margin management in arable systems. *Aspects of Applied Biology*, 81, 47–52.

(52)   Stoate C. & Moorcroft D. (2007) Research-based conservation at the farm scale: development and assessment of agri-environment scheme options. *Aspects of Applied Biology*, 81, 161–168.

(53)   Gardiner T., Hill J. & Marshall E.J.P. (2008) Grass field margins and Orthoptera in eastern England. *Entomologist's Gazette*, 59, 251–257.

(54)   Holland J.M., Oaten H., Southway S. & Moreby S. (2008) The effectiveness of field margin enhancement for cereal aphid control by different natural enemy guilds. *Biological Control*, 47, 71–76.

(55)   Smith J., Potts S. & Eggleton P. (2008) The value of sown grass margins for enhancing soil macro-faunal biodiversity in arable systems. *Agriculture, Ecosystems & Environment*, 127, 119–125.

(56)   Smith J., Potts S.G., Woodcock B.A. & Eggleton P. (2008) Can arable field margins be managed to enhance their biodiversity, conservation and functional value for soil macrofauna? *Journal of Applied Ecology*, 45, 269–278.

(57)   Woodcock B.A., Westbury D.B., Tscheulin T., Harrison-Cripps J., Harris S.J., Ramsey A.J., Brown V.K. & Potts S.G. (2008) Effects of seed mixture and management on beetle assemblages of arable field margins. *Agriculture, Ecosystems & Environment*, 125, 246–254.

(58) Yletyinen S. & Norrdahl K. (2008) Habitat use of field voles (*Microtus agrestis*) in wide and narrow buffer zones. *Agriculture, Ecosystems & Environment*, 123, 194–200.

(59) Douglas D.J.T., Vickery J.A. & Benton T.G. (2009) Improving the value of field margins as foraging habitat for farmland birds. *Journal of Applied Ecology*, 46, 353–362.

(60) Lye G.C., Park K., Osborne J., Holland J. & Goulson D. (2009) Assessing the value of Rural Stewardship schemes for providing foraging resources and nesting habitat for bumblebee queens (Hymenoptera: Apidae). *Biological Conservation*, 142, 2023–2032.

(61) Merckx T., Feber R.E., Dulieu R.L., Townsend M.C., Parsons M.S., Bourn N.A.D., Riordan P. & Macdonald D.W. (2009) Effect of field margins on moths depends on species mobility: field-based evidence for landscape-scale conservation. *Agriculture, Ecosystems & Environment*, 129, 302–309.

(62) Merckx T., Feber R.E., Riordan P., Townsend M.C., Bourn N.A.D., Parsons M.S. & Macdonald D.W. (2009) Optimizing the biodiversity gain from agri-environment schemes. *Agriculture, Ecosystems & Environment*, 130, 177–182.

(63) Musters C.J.M., Alebeek F.V., Geers R.H.E.M., Korevaar H., Visser A. & Snoo G.R.D. (2009) Development of biodiversity in field margins recently taken out of production and adjacent ditch banks in arable areas. *Agriculture, Ecosystems & Environment*, 129, 131–139.

(64) Vickery J.A., Feber R.E. & Fuller R.J. (2009) Arable field margins managed for biodiversity conservation: a review of food resource provision for farmland birds. *Agriculture, Ecosystems & Environment*, 133, 1–13.

(65) Davey C.M., Vickery J.A., Boatman N.D., Chamberlain D.E. & Siriwardena G.M. (2010) Entry Level Stewardship may enhance bird numbers in boundary habitats. *Bird Study*, 57, 415–420.

(66) Ewald J.A., Aebischer N.J., Richardson S.M., Grice P.V. & Cooke A.I. (2010) The effect of agri-environment schemes on grey partridges at the farm level in England. *Agriculture, Ecosystems & Environment*, 138, 55–63.

(67) Field R.H., Morris A.J., Grice P.V. & Cooke A.I. (2010) Evaluating the English Higher Level Stewardship scheme for farmland birds. *Aspects of Applied Biology*, 100, 59–68.

(68) Haaland C. & Gyllin M. (2010) Butterflies and bumblebees in greenways and sown wildflower strips in southern Sweden. *Journal of Insect Conservation*, 14, 125–132.

(69) Blake R., Woodcock B., Westbury D., Sutton P. & Potts S. (2011) New tools to boost butterfly habitat quality in existing grass buffer strips. *Journal of Insect Conservation*, 15, 221–232.

Additional references

West T.M. & Marshall E.J.P. (1996) Managing sown field margin strips on contrasted soil types in three environmentally sensitive areas. *Aspects of Applied Biology*, 44, 269–276.

Macdonald, D.W., Tew, T.E., Todd, I.A., Garner, J.P. & Johnson, P.J. (2000) Arable habitat use by wood mice (*Apodemus sylvaticus*) 3. A farm-scale experiment on the effects of crop rotation. *Journal of Zoology*, 250, 313–320.

Wilson S., Baylis M., Sherrott A. & Howe G. (2000) *Arable Stewardship Project Officer Review*. Farming and Rural Conservation Agency report.

ADAS (2001) *Ecological Evaluation of the Arable Stewardship Pilot Scheme, 1998–2000*. ADAS report.

1.23   Provide supplementary food for birds or mammals

- A total of 18 individual studies investigated the effects of providing supplementary food. Nine studies from France, Sweden and the UK (seven replicated studies, of which six were controlled and two also randomized) found that the provision of supplementary food increased farmland bird abundance[9, 16, 25], breeding population size[19], density[2], body mass[3,19], hatching, nestling growth and fledging rates[1, 17], increased overwinter survival of a declining house sparrow population[8] and that fed male hen harriers bred with more females than control birds[6]. Two studies[9, 25] did not separate the effects of several other interventions carried out on the same study site. Four studies from the UK and Finland (three replicated studies, of which one was controlled and one randomized) found that farmland songbirds[11, 12, 20] and field voles (field voles on unmown plots only)[24] used supplementary food when provided, including the majority of targeted species such as tree sparrow, yellowhammer and corn bunting[15].

- Five replicated studies from the UK (of which two were also controlled) found that the provision of supplementary food had no clear effect on farmland bird breeding abundance[21], European turtle dove reproductive success, territory size or territory density[7], overwinter survival of three stable house sparrow populations[8], tree sparrow nest box use[10], or the abundance of weed seeds on the soil surface[22]. One replicated, controlled study from Sweden found no effect of supplementary food provision on common starling clutch size or nestling weight, and lower fledging rates in nests which received supplementary food compared to nests without supplementary food in one year[4].

- Four studies from the UK (two replicated of which one was also randomized and controlled) found that the use of supplementary food by farmland birds varied between species and region[14, 15], depended upon the time of year[5, 23] and proximity to other feeding stations[18] and natural feeding areas[13].

Background

This intervention may involve the provision of supplementary food for birds or mammals in farmland habitats; such food typically includes seeds. Providing supplementary food for farmland wildlife may be particularly important when food resources in the wider farmed environment are scarce.

Food supply is one of the key factors determining mortality and reproductive rates. Providing supplementary food is therefore often used as a technique to support small populations. However, feeding is only likely to have a positive effect on a population if the food supply is limiting either reproduction or survival.

As with all interventions in this synopsis, studies that investigate population-level effects are most useful for conservationists. This is especially true for supplementary feeding, as many birds have large foraging ranges and the appearance of increased numbers at a feeding station, or even in the habitat surrounding feeders may not represent an increase in numbers but a redistribution of the same birds, and could even hide a population decline. It is also important to note that the effect of providing food can be confounded by many factors. For example, variations in natural food supplies due to population cycles or irregular fruiting, whilst droughts or other extreme weather and pollution levels can also affect how populations respond to food.

See also 'Provide other resources for birds (water, sand for bathing)' for studies which investigate the impact of providing resources other than food.

A replicated, controlled study in mixed farmland in northeast Scotland between 1971 and 1973 (1) found that carrion crow *Corvus corone* nestlings in nests provided with supplementary food had significantly higher hatching, survival and fledging rates than those in control (unfed) nests. With human 'predation' included: 79% of fed nests hatching at least one chick (n = 11), 71% having at least one chick surviving for 10 days and 71% fledging at least one chick (n = 10) compared to 61% (n = 28), 54% (n = 15) and 43% (n = 12) for controls. With human 'predation' excluded: 92% of fed nests hatching at least one chick (n = 12), 83% having at least one chick surviving for 10 days and 83% fledging at least one chick (n = 10) compared to 55% (n = 22), 55% (n = 12) and 45% (n = 10) for controls. Nestlings from fed nests, however, were no heavier than those from controls, when comparing first-hatched with first-hatched etc. Supplementary food consisted of one domestic hen's egg and five dead hen chicks provided every day from when laying began and increasing to one egg and ten chicks from the seventh day after hatching until fledging. Other experiments in this study found that winter feeding (a hen's egg and five chicks provided between January and April 1973) led to crows laying clutches earlier but did not affect clutch size (average laying date of 13 April and 4.4 eggs/clutch for fed territories, n = 10 vs 18 April and 4.3 eggs/clutch for controls, n = 21). Further experiments examined the effect of moving supplementary food further from nests, but the author argued that these results were confounded by supplementary food being taken by non-target birds. Finally, additional experiments in the study found that crow nesting density did not increase following the provision of supplementary food and additional nesting sites in breeding territories.

A small study of feeding as a management option for grey partridges *Perdix perdix* at an arable farm in France (2), found that partridge density was higher in an area with 'partridge cafeterias', than the area without. In spring 1973, the population on the 424 ha farm was 71 pairs (1 pair/6 ha) and 4 single birds. In spring 1974, a total of 48 pairs (1 pair/4.7 ha) and 4 single birds were recorded in the southern section (224 ha), where 27 partridge cafeterias had been constructed. The northern section (200 ha), with no cafeterias, had 24 pairs (1 pair/8.3 ha). Cafeterias comprised a barrel with a feed mixture (grain and weed seeds), a mini-midden to provide maggots and insects and a sand-bath, sheltered by a leaning roof that collected rainwater in a drinking trough. Stoats *Mustela erminea* and mice *Mus* spp. were also controlled with traps at the 'cafeterias'. Small shrubs were planted next to cafeterias to provide shelter. Where possible they were placed one/territory.

A replicated, controlled study in the breeding seasons of 1985–1987 in grasslands on Öland, southern Sweden (3), found that female northern wheatear *Oenanthe oenanthe*, but not males, that were provided with supplementary food were significantly heavier than unfed controls (average of 26.9 g for 53 fed females and 24.4 g for 42 fed males vs 24.3 g and 23.7 g for 48 and 32 unfed controls). However, there was no effect when females were feeding older chicks, which were able to regulate their body temperature. A few days after hatching, most food was delivered to chicks, not consumed by adults. Food consisted of 7 g of mealworms provided either during incubation, or for the entire breeding season.

A replicated, controlled study in grasslands in southern Sweden between 1982 and 1990 (4) found that common starling *Sturnus vulgaris* supplied with supplementary food showed only occasional differences in egg weight, no differences in clutch size or nestling weights and fledging rates were actually lower in fed nests in 1990 (4.3 young/nest for fed nests vs 5.6 for controls). However common starlings supplied with food began laying significantly earlier than controls (first laying date 21 April–5 May for fed nests vs 22 April–10 May for controls). There were no such differences between nests in the years when supplementary food was not supplied. Supplementary food consisted of approximately 100 g of mealworms placed in small feeders either on the outside or inside of nest boxes, supplied to different colonies in 1982 and 1985 and a subset of nests at a third colony in 1990. Feeding began approximately one month before laying started and stopped once all females began laying. Feeding represented more than the daily energetic needs of a pair of starlings.

A study of habitat use by yellowhammer *Emberiza citrinella* in 1997 on a mixed farm in Leicestershire UK (5), found that in winter supplementary feeding sites were used more than cereal and rape crops and in late winter more than wild bird cover. In early winter, cereal-based wild bird cover was used significantly more than all other habitats and supplementary feeding sites, kale-based wild bird cover and field boundaries were used more than cereal and rape crops. In late winter, supplementary feeding sites were used significantly more than all other habitats including wild bird cover. A 15% area of the arable land was managed for game birds, and in winter grain was distributed along some hedges and supplied in hoppers at permanent feeding sites. A 60 ha area of the farm was walked 7 times in November–December and February–March 1997 and habitat use was recorded.

A randomized, replicated, controlled study in heathland on Orkney Mainland, Scotland, in 1999–2000 (6) found that male hen harriers *Circus cyaneus* provided with supplementary food (chicken *Gallus domesticus* chicks and quarter pieces of European rabbit *Oryctolagus cuniculus* or brown hare *Lepus europaeus*) bred with significantly more females than control (unfed) males (100% of 11 fed males mated and 36% mated with more than one female vs 80% of 9 unfed males mated; 11% mated with more than one female). There was no effect of feeding on clutch size or hatching success (average of approximately 5.1 eggs/clutch for 13 fed clutches vs 4.7 eggs/clutch for 4 unfed clutches), but productivity still increased. Hooded crows *Corvus cornix* were also removed from all territories.

A replicated cross-over study in 1999–2000 in ten mixed agricultural and natural habitat sites in Norfolk and Suffolk, England (7) found that European turtle dove *Streptopelia turtur* reproductive success, territory size or territory density did not differ between years when supplementary food was provided and control (unfed) years (24 nests studied; daily survival rates of 79–97% for fed nests vs 85–98% for unfed). However, doves were frequently observed eating the food. The authors argued that the experimental sites were too small (mostly 200–400 ha) to affect the wide-ranging doves.

A replicated study of four house sparrow *Passer domesticus* populations in mixed farmland in Oxfordshire, UK (8) found that supplementary feeding in winter increased overwinter survival of a declining population, but not three populations understood to be stable. Monthly overwinter survival rate varied between populations, with that of the declining population lower than the other three populations (0.8 vs 0.9). The apparent survival rate of the declining population over the November–March period increased from 0.39 in 1998–1999 to 0.65 in 1999–2000. There was no effect of supplementary feeding on the other three populations. Three populations were selected at random and the fourth was selected for the availability of historical records, which indicated an 80% decline over the last 30 years. Populations were 6–24 km apart. Supplementary seed food was freely provided to the populations during winter 1999–2000. Nest recording, mark re-sighting and microsatellite-based molecular genetics were used for sampling.

A small replicated, controlled study from May–June 1992–1998 in Leicestershire, UK (9), found that the abundance of nationally declining songbirds and species of conservation concern significantly increased on a 3 km² site where supplementary food was provided from hoppers and by hand (alongside several other interventions). However there was no overall difference in bird abundance, species richness or diversity between the experimental and three control sites. Numbers of nationally declining species rose by 102% (except for Eurasian skylark *Alauda arvensis* and yellowhammer *Emberiza citrinella*). Nationally stable species rose (insignificantly) by 47% (eight species increased; four decreased). The other interventions employed at the same site were managing hedges, beetle banks, wild bird seed cover strips, predator control and reducing chemical inputs generally.

A replicated, controlled paired site study from March–August in 2000–2003 in 20 paired nest box groups (10 placed along wetland edges and 10 in farmlands) in Rutland, England (10) found that tree sparrow *Passer montanus* showed no preference for nest boxes supplied with supplementary food (4 fed boxes colonized vs 4 unfed). There was no difference in the number of nesting attempts made by birds with or without supplementary food although the mean clutch size was significantly higher in nests closer to supplementary food (5.6 compared to 5.0 eggs/clutch). The authors point out that the small spatial scale of the study (1 km between pairs) may have confounded any effect of supplementary feeding. Nest box groups consisted of 5 nest boxes placed 2–20 m apart. Sunflower seeds were randomly provided to one nest box group within each pair.

A study at a farmland site in northwest England between January 2003 and February 2004 (11) found that twite *Carduelis flavirostris* used a supplementary feeding station (established in spring 2002) frequently outside the breeding season, with up to 250 birds seen at once. However, twite used the station far less during the breeding season, when they relied more on wild seeds. Birds from another feeding station (see (12)) and other breeding colonies up to 20 km away used the feeding station, as well as individuals from a nearby colony of 20–30 birds. Supplementary food consisted of nyjer *Guizotia abyssinica* seed spread in a thick 2 m x 5 cm line on a 2 m x 2 m patch of bare earth and replenished every week. This study was part of the same experimental set-up as (12, 13).

A study at a farmland site in northwest England between January 2003 and February 2004 (12) used an identical procedure to (11) at a site 12.6 km away and found that twite *Carduelis flavirostris* used the supplementary feeding station frequently outside the breeding season, with up to 150 birds being seen at once. However, twite used the station far less during the breeding season, when they relied more on wild seeds. A large number of birds from near the feeding station in (11) and colonies up to 20 km away used the feeding station, as well as birds from the 2 nearby colonies (each approximately 1.5 km away and 20–30 birds). This study was part of the same experimental set-up as (11, 13).

A study at a farmland site in northwest England between January 2003 and February 2004 (13) used an identical procedure to (11) to establish a feeding station approximately 1 km from a colony of six pairs of twite *Carduelis flavirostris*. This station was only used occasionally and only by one or two birds at a time. The author suggests that the lack of use could have been due to the small size of the colony and the fact that it was not positioned close to natural feeding areas for twite. This study was part of the same experimental set-up as (11, 12).

A randomized, replicated, controlled study at three farmland sites in England in the winters of 1999–2000 until 2001–2002 (14) found that farmland birds showed mixed responses to supplementary food. Chaffinch *Fringella coelebs*, linnet *Carduelis cannabina* and yellowhammer *Emberiza citrinella* all showed significant short-term increases on at least one plot provided with food (chaffinch densities increased by 80–200% on 3 of 6 fed plots; yellowhammer densities increased by 230–400% on 4 of 6 fed plots; data for linnets not provided). There were no corresponding short-term changes on nearby control plots. Skylark *Alauda arvensis* did not show any consistent response to food at any of the sites and there was no longer term impact of feeding on bird densities. Supplementary food consisted of 36 kg/ha of mixed grains broadcast over fields. The authors suggest that the lack of effect of feeding at some sites may be due to a very low natural seed density in the soil, meaning that even with supplementary food, the level of food was too low to attract birds.

The results from two replicated studies from the UK found that the factors affecting the use of supplementary food by a range of farmland songbirds were not consistent across species or regions (15). The 'Bird Aid' programme (run between October and March in the winters of 2000–2001 until 2002–2003 across the UK) found that all three target species (tree sparrow *Passer montanus*, yellowhammer *Emberiza citrinella* and corn bunting *Miliaria calandra*) used supplementary food, consisting of 25 kg of seeds supplied each week. Tree sparrow and yellowhammer tended to use feeding stations more if they were closer to cover and in mixed landscapes; the opposite was true for corn bunting. The Winter Food for Birds project, run from October 2002 to March 2003 at ten replicates of seven sites across eastern England, found that six of eight target species used supplementary food, consisting of 5 kg each of millet and sunflower seeds supplied each week, sufficiently often for analysis. At both the local and landscape scale, only human habitats and woodlands had uniform effects, increasing and decreasing the use for three and four species respectively. All other habitats had different impacts on different species. Results from the same experimental set-up are also presented in (16, 18, 19, 21, 23).

A replicated, controlled study from November–July 2002–2004 in ten sites each containing seven feeding stations (placed at the centre of a 2 x 2 km tetrad) separated at set distances from each other (100 m, 500 m, 1 km, 2 km, 5 km and 10 km) in East Anglia, UK (16) found that supplementary provision of seeds increased local seed-eating bird abundance, especially species of conservation concern. Yellowhammer *Emberiza citrinella* and chaffinch *Fringilla coelebs* used the feeding stations most extensively (93–100% of all stations). Although genuine population trends were difficult to infer from the experimental setup, the authors argued that food provisioning increased the local abundance of several otherwise declining species (yellowhammer, reed bunting *Emberiza schoeniclus*, house sparrow *Passer domesticus* and chaffinch) over two winters. Colour-ring re-sighting and radio-tracking revealed that target seed-eaters move small distances between food resources (500 m–1 km) and the authors suggest placing food resources (overwinter stubbles and wild bird cover crops) at a minimum of 1 km apart in order to be cost effective in reaching the largest number of populations. Supplementary seed (10 kg of equally distributed sunflower hearts and millet) was replenished weekly. Bird use of the feeding stations was monitored twice weekly (20 min observation sessions). Results from the same experimental set-up are also presented in (15, 18, 19, 21, 23).

A replicated, controlled, randomized paired study of 15–16 nest-box colonies of starling *Sturnus vulgaris* in 1998–1999 in southern Sweden (17) found that food supplementation increased growth and survival of nestlings. Greater availability of pasture also increased survival, but tended to have a smaller effect. Fledging success increased with supplemental feeding and local availability of pasture (<10% pasture: fed 0.92, unfed 0.80; >10% pasture: fed 1.0, unfed 0.95). Nestling growth (tarsus length) was significantly higher following supplemental feeding, but was not affected by habitat (<10% pasture: fed 32.2 mm, unfed 32.8 mm; >10% pasture: fed 32.6 mm, unfed 32.9 mm). Feather growth rate showed the same pattern. There was no effect on chick condition and no interactions between effects of habitat and feeding. Colonies were over 1 km apart and each comprised 8 boxes. Nest boxes were visited to determine clutch size, hatching date, fledging success and to band (day 1) and measure (day 10 and 14) nestlings. Agricultural land-use was classified in a radius of 500 m around each colony. On the day of hatching, two broods with similar clutch size were matched from different habitat classes and were randomly selected to receive supplemental food or not. From the fourth day after hatching, half of the breeding pairs were given a bowl containing 84 g of mealworm larvae twice daily (1998–1999).

The Winter Food for Birds project (see (15)) was continued in the winter of 2003–2004 and this study discusses the data from both winters (18). For four songbird species (blue tit *Parus caeruleus*, chaffinch *Fringilla coelebs*, great tit *P. major* and robin *Erithacus rubecula*), feeding stations were used more frequently and by more birds if they were more than 500 m from other stations, compared with stations less than 500 m from neighbours. The same pattern was seen (but not significant) in blackbird *Turdus merula* and house sparrow *Passer domesticus*. Yellowhammer *Emberiza citrinella* and reed bunting *E. schoeniclus*, however, used clustered sites more. There was no significant impact of distance on feeder use by greenfinch *Carduelis chloris*, goldfinch *C. carduelis* or dunnock *Prunella modularis*. All species used multiple stations if they were closer than 500 m apart, but used only single stations if they were more widely spaced. The authors use this information to recommend that stations are placed at least 1 km apart to maximize cost-effectiveness (i.e. to ensure the maximum number of birds have access to supplementary food). Results from the same experimental set-up are also presented in (15, 16, 19, 21, 23).

A replicated, controlled study from November–March 2004–2007 in ten experimental and ten control tetrads (composed of four 1 km² sites) of arable farmland in East Anglia, UK (19) found that provision of seeds during winter significantly increased body mass and breeding population sizes of seed-eating bird species. Supplementary food was most used in early to mid-winter for generalist species and late winter for specialist species (such as chaffinch *Fringilla coelebs* and yellowhammer *Emberiza citrinella*). Radio-tracking and mark-recapture techniques revealed that resource patches (such as wild bird cover crops and overwinter stubbles) should be separated by 1.1–1.3 km to be both cost and conservation effective for priority species (like yellowhammer). The authors suggest that year-round resource delivery could be achieved by placing breeding habitat 2.7–3.6 km from winter food patches. They caution that specific inter-patch distances may vary according to species and habitat but should be based on species of conservation concern. Experimental sites contained one central feeding station provided freely with seed (10 kg of equally distributed millet, rape, wheat and sunflower seeds; replenished twice a week) and were fenced (50 cm in height) using 50 mm mesh wire and bamboo canes to exclude gamebirds. This study was an extension of and used partly the same experimental set-up as (15, 16, 18, 21, 23).

A series of randomized, replicated trials at two sites in England in the winters of 2000–2001 and 2001–2002 (20) found that five songbird species took supplementary food when provided and preferentially took wheat over oats and oats over barley. Tree sparrow *Passer montanus* and reed bunting *Emberiza schoeniclus* also fed on maize, preferring it to all cereals except wheat, while house sparrow *P. domesticus* preferred maize to all cereals. Corn bunting *E. calandra* and yellowhammer *E. citrinella* preferred all cereals to maize. Tree sparrow selected both cereals and oily seeds (e.g. sunflower seeds, oilseed rape), but avoided rye grass seed. All species preferred cereals to sunflower seeds and none showed any distinction between wheat and a 'weed seed mix'. At one site, food was provided in tubular feeders, in the other it was heaped on the ground. Survival rates of birds were not monitored.

A replicated study using the same data as (15) and combining it with data from control areas between 2000 and 2003 (21) did not find robust evidence for supplementary winter feeding increasing breeding abundances of farmland songbirds. There were no effects of the Bird Aid programme on target species, although sites used more frequently had increased populations of yellowhammer *Emberiza citrinella* and corn bunting *Miliaria calandra*, but decreased populations of tree sparrow *Passer montanus*. Four of five insect-eating/generalist species declined faster in Winter Food for Birds programme sites than in controls. There was no such effect for six seed-eating species. Declines in dunnock *Prunella modularis*, robin *Erithacus rubecula* and yellowhammer *Emberiza citrinella* were lower in Winter Food for Birds sites provided with more food and centrally-placed Winter Food for Birds sites, compared to those provided with less food or those around the periphery of Winter Food for Birds clusters. Results from the same experimental set-up are also presented in (15, 16, 18, 19, 23).

A replicated, controlled trial in 1999–2002 on arable fields on three farms in Hampshire, Lincolnshire and Yorkshire, UK (22) investigated whether addition of supplementary bird seed affected the abundance of weed seeds on the soil surface by attracting seed-eating birds. The study found no difference in mid-winter weed seed densities in sites with and without addition of bird seed, even though a separate study at the same site recorded more birds on areas with added seed. This may have been due to low winter seed predation by birds. Post-treatment winter seed densities varied from 151 to 398 seeds/m² on control plots and 92 to 365 seeds/m² on treatment blocks. There was some evidence that birds changed the species composition of weed seeds, as there were fewer larger-sized weed seeds in treatment blocks. Bird seed (36 kg/ha, 444 seeds/m², including cracked maize and linseed/soya/barley or sorghum/millet) was applied 3 times from November–March on two 100 ha blocks, with a further 2 control blocks, on each farm. Seed densities were estimated from 10 soil scrapes (0.2 x 0.2 m), from at least 10 field edge and 10 mid-field locations in each block, before the first bird seed application and 2–3 weeks after each application.

A further study, using the same data as (15) investigated how use of supplementary food by farmland songbirds varied over winter months (23). Supplementary food-use peaked in or before January for five generalists and 'human-associated' seed-eating species (blackbird *Turdus merula*, goldfinch *Carduelis carduelis*, greenfinch *C. chloris*, house sparrow *Passer domesticus* and robin *Erithacus rubecula*), whilst yellowhammer *Emberiza citrinella*, reed bunting *E. schoeniclus*, chaffinch *Fringilla coelebs* and dunnock *Prunella modularis* all used supplementary food most in February or later. Use by great tits *Parus major* and blue tits *P. caeruleus* declined overwinter. The authors suggest the first group use food when temperatures are lowest and daylight hours shortest, whilst the second group (which are heavily dependent on farmland seed) use food when naturally-occurring food sources are at their lowest. They caution that these results are likely to be dependent on the mix of farming types across the landscape, with eastern England being dominated by arable fields. Results from the same experimental set-up are also presented in (15, 16, 18, 19, 21).

A replicated, controlled habitat selection study in four riparian field margins in the municipality of Jokioinen, southwestern Finland (24) found that supplementary food appeared to attract field voles *Microtus agrestis* in uncut plots in both wide buffer zones and narrow filter strips but not in mowed plots. Mowed food plots in wide buffer zones were avoided by voles. In mid-June 2005, one 210 m-long section in each margin was divided into fourteen 15 m-long experimental plots, half of which were mown to <20 cm. Food and/or cover was added to mowed/unmowed plots to create eight treatments. The remaining plots were interspersed between experimental plots. Trapping and radio-tracking field voles started two weeks after habitat manipulation.

A controlled study in 2002–2009 on mixed farmland in Hertfordshire, England (25) found that the number of grey partridges *Perdix perdix* increased significantly on an experimental site where supplementary food was provided (along with several other interventions), but only slightly on a control site without supplementary food. This increase was apparent in spring (from fewer than 3 pairs/km² in 2002 to 12 in 2009, with a high of 18 pairs/km² on the experimental site, compared to approximately 1 pair/km² on the control site in 2002, increasing to approximately 4 pairs/km² in 2009) and autumn (from fewer than 10 birds/km² in 2002 to approximately 65 in 2009, with a high of 85 birds/km² compared to approximately 4 birds/km² on the control site in 2002, increasing to approximately 15 in 2009). Food consisted of wheat from a hopper, provided from October to March. The experimental site also had predator control and habitat creation.

(1) Yom-Tov Y. (1974) The effect of food and predation on breeding density and success, clutch size and laying date of the crow (*Corvus corone* L.). *Journal of Animal Ecology*, 43, 479–498.
(2) Westerskov K.E. (1977) Covey-oriented partridge management in France. *Biological Conservation*, 11, 185–191.
(3) Moreno J. (1989) Body-mass variation in breeding northern wheatears – a field experiment with supplementary food. *Condor*, 91, 178–186.
(4) Källander H. & Karlsson J. (1993) Supplemental food and laying date in the European starling. *Condor*, 95, 1031–1034.
(5) Stoate C. & Szczur J. (1997) Seasonal changes in habitat use by yellowhammers (*Emberiza citrinella*). *Proceedings of the 1997 Brighton Crop Protection Conference – Weeds*. Farnham, pp. 1167–1172.
(6) Amar A. & Redpath S.M. (2002) Determining the cause of the hen harrier decline on the Orkney Islands: an experimental test of two hypotheses. *Animal Conservation*, 5, 21–28.

(7)   Browne S.J. & Aebischer N.J. (2002) The effect of supplementary feeding on territory size, territory density and breeding success of the turtle dove *Streptopelia turtur*: a field experiment. *Aspects of Applied Biology*, 67, 21–26.

(8)   Hole D.G., Whittingham M.J., Bradbury R.B., Anderson G.Q.A., Lee P.L.M., Wilson J.D. & Krebs J.R. (2002) Widespread local house-sparrow extinctions – agricultural intensification is blamed for the plummeting populations of these birds. *Nature*, 418, 931–932.

(9)   Stoate C. (2002) Multifunctional use of a natural resource on farmland: wild pheasant (*Phasianus colchicus*) management and the conservation of farmland passerines. *Biodiversity and Conservation*, 11, 561–573.

(10) Field R.H. & Anderson G.Q.A. (2004) Habitat use by breeding tree sparrows *Passer montanus*. *Ibis*, 146, 60–68.

(11) Raine A. (2004) Providing supplementary food as a conservation initiative for twite *Carduelis flavirostris* breeding in the South Pennines near Worsthorne, Lancashire, England. *Conservation Evidence*, 1, 23–25.

(12) Raine A. (2004) Providing supplementary food as a conservation initiative for twite *Carduelis flavirostris* breeding in the South Pennines near Littleborough, West Yorkshire, England. *Conservation Evidence*, 1, 26–28.

(13) Raine A. (2004) Providing supplementary food as a conservation initiative for twite *Carduelis flavirostris* breeding in the South Pennines near Midgley, West Yorkshire, England. *Conservation Evidence*, 1, 29–30.

(14) Robinson R.A., Hart J.D., Holland J.M. & Parrott D. (2004) Habitat use by seed-eating birds: a scale-dependent approach. *Ibis*, 146, 87–98.

(15) Siriwardena G.M. & Stevens D.K. (2004) Effects of habitat on the use of supplementary food by farmland birds in winter. *Ibis*, 146, 144–154.

(16) Defra (2005) *The Consequences of Spatial Scale for Agri-environment Schemes Designed to Provide Winter Food Resources for Birds*. Defra BD1616.

(17) Granbom M. & Smith H.G. (2006) Food limitation during breeding in a heterogeneous landscape. *Auk* 123, 97–107.

(18) Siriwardena G.M., Calbrade N.A., Vickery J.A. & Sutherland W.J. (2006) The effect of the spatial distribution of winter seed food resources on their use by farmland birds. *Journal of Applied Ecology*, 43, 628–639.

(19) Defra (2007) *Understanding the Demographic Mechanisms Underlying Effective Deployment of Winter Prescriptions for Farmland Bird Recovery*. Defra BD1628.

(20) Perkins A.J., Anderson G. & Wilson J.D. (2007) Seed food preferences of granivorous farmland passerines. *Bird Study*, 54, 46–53.

(21) Siriwardena G.M., Stevens D.K., Anderson G.Q.A., Vickery J.A., Calbrade N.A. & Dodd S. (2007) The effect of supplementary winter seed food on breeding populations of farmland birds: evidence from two large-scale experiments. *Journal of Applied Ecology*, 44, 920–932.

(22) Holland J.M., Smith B.M., Southway S.E., Birkett T.C. & Aebischer N.J. (2008) The effect of crop, cultivation and seed addition for birds on surface weed seed densities in arable crops during winter. *Weed Research*, 48, 503–511.

(23) Siriwardena G.M., Calbrade N.A. & Vickery J.A. (2008) Farmland birds and late winter food: does seed supply fail to meet demand? *Ibis*, 150, 585–595.

(24) Yletyinen S. & Norrdahl K. (2008) Habitat use of field voles (*Microtus agrestis*) in wide and narrow buffer zones. *Agriculture, Ecosystems & Environment*, 123, 194–200.

(25) Aebischer N.J. & Ewald J.A. (2010) Grey Partridge *Perdix perdix* in the UK: recovery status, set-aside and shooting. *Ibis*, 152, 530–542.

1.24   Make direct payments per clutch for farmland birds

- Two replicated and controlled studies from the Netherlands[1, 3] found limited evidence for increased wading bird populations on farms with per-clutch payments. One study found no population effects over three years[1]. The second[3] found slightly higher breeding densities of wading birds, but not higher overall numbers.

- A replicated and controlled study[1] found higher hatching success of northern lapwing and black-tailed godwit on farms with payment schemes than control farms.

- A replicated site comparison from the Netherlands that looked at the effects of per-clutch payments in combination with postponed agricultural activities found more birds bred on

12.5 ha plots under the per-clutch payment and postponed agricultural activities scheme but found no differences at the field-scale[2].

Background

Most agri-environment schemes aim to compensate farmers for the cost of conservation management on their land, irrespective of the outcomes. The Netherlands, however, also has a scheme where farmers are paid directly, based on the number of breeding bird pairs on their land.

A replicated, controlled study on intensive dairy grassland in the western Netherlands between 1993 and 1996 (1) found that northern lapwing *Vanellus vanellus* and black-tailed godwit *Limosa limosa* showed higher hatching success on 15 farms offered per-clutch payments for farmland birds than on 9 control farms (65% vs 48% for lapwing, 63% vs 39% for black-tailed godwit). A non-significant difference was also seen for common redshank *Tringa totanus* (39% vs 21%). There were no differences in treatment during 1993–1994, before payments. The number of control farms was reduced to three in 1995–1996, because the farmers on other farms had become too involved in conservation for their farms still to be considered true controls. No other bird conservation measures were in place and the cost was estimated at €40/clutch. Population-level impacts were not observed, possibly due to the relatively short time-scale and small number of farms.

A replicated site comparison study of 42 fields in the Netherlands (2) found that more birds bred on 12.5 ha scheme plots consisting of a mixture of fields with postponed agricultural activities and fields with a per-clutch payment scheme than on conventionally farmed plots. A survey of individual fields found there was no difference in bird abundance and breeding on those fields with postponed agricultural activities only and on conventionally farmed fields. The number of bird species on each type of farmland also did not differ between agri-environment scheme and non-agri-environment scheme plots. The agri-environment scheme, which intended to promote the conservation of Dutch meadow birds, prohibited changes in field drainage, pesticide application (except for patch-wise control of problem weeds) and any agricultural activity between 1 April and early June. Additionally, farmers of surrounding fields were paid for each meadow bird clutch laid on their land (though no agricultural restrictions were in place on these fields). The study surveyed 7 pairs of fields (one within the agri-environment scheme, one conventionally farmed) and the 12.5 ha area surrounding each field, from each of 3 different parts of the Netherlands 4 times during the breeding season.

A replicated, controlled, paired sites study in the western Netherlands in 2003 (3) found slightly higher breeding densities of birds on 19 grassland plots with per-clutch payments for wading bird clutches, compared to 19 paired, control plots, both when delayed mowing was also used and when per-clutch payment was the only scheme used (13 territories/plot for combined schemes, 13 territories/plot for per-clutch payment and 11 territories/plot for controls). However, birds were not more abundant under either scheme, compared with controls (approximately 125 birds/plot for combined schemes, 125 birds/plot for per-clutch payment and 110 birds/plot for controls). Wader breeding densities were higher (but not significantly so) on combined and per-clutch payment plots (approximately 7 territories/plot for combined schemes, 7 territories/plot for per-clutch payment and 5 territories/plot for controls). When individual wader species were analysed, there were higher numbers of redshank *Tringa totanus* on combined or per-clutch payment plots (approximately 5 birds/plot for combined schemes, 5 birds/plot for per-clutch payment and 3 birds/plot for controls), but there were no significant differences in breeding densities for redshank, northern lapwing *Vanellus vanellus*, Eurasian oystercatcher *Haematopus ostralegus* or black-tailed godwit *Limosa limosa*. The authors suggest that groundwater depth, soil hardness and prey density were what drove these patterns. All farms had been operating the schemes for at least three (and an average of four) years before the study.

(1) Musters C.J.M., Kruk M., De Graaf H.J. & Keurs W.J.T. (2001) Breeding birds as a farm product. *Conservation Biology*, 15, 363–369.
(2) Kleijn D., Baquero R.A., Clough Y., Díaz M., Esteban J.D., Fernández F., Gabriel D., Herzog F., Holzschuh A., Jöhl R., Knop E., Kruess A., Marshall E.J.P., Steffan-Dewenter I., Tscharntke T., Verhulst J., West T.M. & Yela J.L. (2006) Mixed biodiversity benefits of agri-environment schemes in five European countries. *Ecology Letters*, 9, 243–254.
(3) Verhulst J., Kleijn D. & Berendse F. (2007) Direct and indirect effects of the most widely implemented Dutch agri-environment schemes on breeding waders. *Journal of Applied Ecology*, 44, 70–80.

1.25   Provide other resources for birds (water, sand for bathing)

- A small study in France found that grey partridge density was higher in areas where a combination of supplementary food, water, shelter and sand for bathing were provided[1].

Background

This intervention involves providing supplementary resources, other than food, for birds on farmland. These resources can include water, or sand for bathing.

See also 'Provide supplementary food for birds or mammals' for studies that investigate the effects of providing supplementary food.

A small study of feeding as a management option for grey partridges *Perdix perdix* at an arable farm in France (1) found that partridge density was higher in an area with 'partridge cafeterias', than the area without. In spring 1973 the population on the 424 ha farm was 71 pairs (1 pair/6 ha) and 4 single birds. In spring 1974, a total of 48 pairs (1 pair/4.7 ha) and 4 single birds were recorded in the southern section (224 ha), where 27 partridge cafeterias had been constructed. The northern section (200 ha), with no cafeterias, had 24 pairs (1 pair/8.3 ha). Cafeterias comprised a barrel with a feed mixture (grain and weed seeds), a mini-midden to provide maggots and insects and a sand-bath, sheltered by a leaning roof that collected rainwater in a drinking trough. Stoats *Mustela erminea* and mice *Mus* spp. were also controlled with traps at the 'cafeterias'. Small shrubs were planted next to cafeterias to provide shelter. Where possible they were placed one/territory.

(1) Westerskov K.E. (1977) Covey-oriented partridge management in France. *Biological Conservation*, 11, 185–191.

1.26   Mark bird nests during harvest or mowing

- One replicated study from the Netherlands[1] found that marked northern lapwing nests were less likely to fail as a result of farming operations than unmarked nests.

Background

Marking the nests of ground-nesting birds may reduce the accidental destruction by farmers during harvest or mowing.

A replicated study in 2005–2006 on arable farms in Noordoostpolder and Oostelijk Flevoland, the Netherlands (1) found that marked northern lapwing *Vanellus vanellus* nests were significantly less likely to fail as a result of farming operations than unmarked nests (0–9% of 1,644 marked nests destroyed vs 15–42% of 229 unmarked nests). However, overall survival rates did not differ significantly (37–73% success for marked nests vs 38–66% for unmarked), with some evidence that marked nests were deserted or predated more often. Nests on the marked farms (121 in 2005, 113 in 2006) were marked with 2 bamboo poles (1 m high) by 151–171 volunteers, and farmers told of their presence. On the control farms, no markers were put in place and farmers were not informed of the nests.

(1)   Kragten S., Nagel J.A.N.C. & De Snoo G.R. (2008) The effectiveness of volunteer nest protection on the nest success of northern lapwings *Vanellus vanellus* on Dutch arable farms. *Ibis*, 150, 667–673.

1.27   Provide refuges during harvest or mowing

- Three studies examined the effect of providing refuges for birds during harvest or mowing in France and the UK. One replicated study in France[1] found evidence that providing refuges during mowing reduced contact between mowing machinery and unfledged quail and corncrakes. However one replicated controlled study and a review from the UK found that Eurasian skylark did not use nesting refuges more than other areas[2, 3].

Background

During mowing and harvesting operations, ground-nesting birds frequently remain in long grass or crops for as long as possible. If mowing/harvest occurs from the outside of the field inwards, this behaviour can leave the birds trapped in the centre of the field and killed as the last patch is harvested. However, if unharvested refuges are left in fields then it is possible that chicks and adults will remain in them and survive.

A replicated study in 1996–1997 in 62 hay fields in Bourgogne, France (1) found that contact between mowing machinery and unfledged quail *Cortunix cortunix* and corncrakes *Crex crex* was reduced by approximately 50% and 33% respectively by leaving 10 m-wide uncut strips in the centre of fields. In addition, unmowed strips had the highest concentrations of corncrakes, quails and passerines (7.7 birds/ha, 3.8 birds/ha and 10.8 birds/ha respectively in 1996). All refuge areas were mown within the first 10 days of August using the 'outside-in' method. During mowing, observers with binoculars recorded birds in the refuge areas.

A 2010 review of four experiments on the effects of agri-environment measures on livestock farms in the UK (2) found one trial from 2006 to 2008 demonstrating that uncut nesting refuges for Eurasian skylark *Alauda arvensis* in silage fields were not used more than other areas. Refuge plots of 1 ha were cut with a raised mowing height in the first silage cut, then left uncut for the rest of the season. The plots were preferred for re-nesting for two weeks following the first cut, but subsequently did not have higher nest densities than other areas. Skylarks continually re-nest rather than re-nesting in a batch after each cut. After the second cut, safe areas were completely avoided by skylarks. This study formed part of the same Defra-funded project (BD1454) as in (3) for which no reference is given in the review.

A replicated, controlled study in 2007 on seven fields in Dorset, UK (3) found that after the first cut, Eurasian skylark *Alauda arvensis* did not nest in half hectare 'safe nesting plots' which were cut 10 cm higher than the rest of the field. Following the second cut of the main field, grass on safe nesting plots was avoided with new nesting attempts taking place on the surrounding cut grass. In 2007, 13 safe nesting plots were established on 7 fields with skylark territories. Safe nesting plots were mown 10 cm higher during the first field cut (approximately 27 cm grass height on safe nesting plots vs 9 cm on normal cut areas), and were not mown at all during the second cut. They were left unmown/ungrazed until the end of August when skylark breeding had ceased.

(1) Broyer J. (2003) Unmown refuge areas and their influence on the survival of grassland birds in the Saône valley (France). *Biodiversity and Conservation*, 12, 1219–1237.
(2) Buckingham D.L., Atkinson P.W., Peel S. & Peach W. (2010) New conservation measures for birds on grassland and livestock farms. *Proceedings of the British Ornithologists' Union – Lowland Farmland Birds III: Delivering solutions in an uncertain world*. British Ornithologists' Union. pp. 1–13.
(3) Defra (2010) *Modified Management of Agricultural Grassland to Promote In-field Structural Heterogeneity, Invertebrates and Bird Populations in Pastoral Landscapes*. Defra BD1454.

1.28 Provide foraging perches (eg. for shrikes)

• We have captured no evidence for the effects of providing foraging perches (eg. for shrikes) on farmland wildlife.

Background
If prey are plentiful but birds have low hunting success then it may be possible to increase population sizes by making hunting more effective, for example by providing perches for birds to use.

1.29 Provide nest boxes for birds

• Two studies (including one before-and-after study) from the Netherlands and the UK found that following the provision of nest boxes there was an increase in the number of Eurasian kestrel clutches[1] and breeding tree sparrows[8]. One replicated study from Switzerland found the number of Eurasian wryneck broods in nest boxes declined over five years whilst the number of Eurasian hoopoe broods increased[10].

• Eight studies from Finland, the Netherlands, Sweden, Switzerland and the UK (six were replicated) found that nest boxes in agricultural habitats were occupied by Eurasian kestrel[1, 4], long-eared owl[1], common starling[3, 5], tits *Parus* spp.[7], tree sparrow[8], stock dove and jackdaw[9], and Eurasian wryneck and Eurasian hoopoe[10]. Whilst two studies from the UK (a replicated, paired site study and a controlled study) found that carrion crows did not nest in artificial trees[2] and tree sparrows showed a preference for nest boxes in wetland habitat, compared to those in farmland sites[6].

• Two replicated studies from Sweden found that nest success within boxes was related to the amount of pasture available[5] and nest boxes positioned higher above the ground had higher occupancy, numbers of eggs and numbers of hatched young[3].

Background

This intervention involves providing nest boxes for birds on farmland. Nest boxes can provide suitable habitats for hole-nesting birds where more natural nesting habitats such as tree cavities are scarce (Newton 1994). Nest boxes may consist of a wooden box with a circular (or other shaped) entrance hole and can be installed on poles or attached to trees.

See also 'Provide owl nest boxes (Tawny owl, Barn owl)' for studies looking at the effects of providing nest boxes for owls that commonly nest in farm buildings, such as barn owls or tawny owls.

Newton I. (1994) The role of nest sites in limiting the numbers of hole-nesting birds: a review. *Biological Conservation*, 70, 265–276.

A before-and-after study of Eurasian kestrel *Falco tinnunculus* from 1959 to 1965 in the Oostelijk Flevoland in the Netherlands (1) found that there was an increase from 20 breeding pairs to 109 clutches following nest box installation in the study area (natural vegetation, plantation and 20% crops). In 1959, there were 25 nest boxes and approximately 20 breeding kestrel pairs, of which 11 used a nest box. In 1960, once the 243 boxes had been installed, 109 clutches were found in the 3 blocks (2 natural vegetation; 1 cultivated); only a few were outside the boxes. In 1960–1965, there were 16–62 kestrel clutches in the cultivated block (80% plantation; 20% common crops; 117 boxes), there were also 0–12 long-eared owl *Asio otus* clutches. In each block, 81 kestrel nest boxes (50 x 30 x 30 cm) on 2 m poles were placed in 9 rows of 9 (330 m apart). An additional 36 nest boxes were placed in one half of the cultivated block in the winter of 1961–1962.

A controlled study in mixed farmland in northeast Scotland in 1971 (2) found that carrion crows *Corvus corone* did not nest in artificial trees, irrespective of whether they were provided with supplementary food or not. In one experiment, a line of 15 artificial trees (3–6 m branches tied to fence posts and provided with an old crow's nest) were set up, approximately 70 m apart. Two pairs of crows established territories, but neither attempted to breed. A second experiment provided a single artificial tree in 2 occupied territories, 70 m from the tree used by the resident pair. Neither artificial tree was used, as the resident pairs successfully defended their territories.

A replicated study of 48 common starling *Sturnus vulgaris* nest boxes at two pasture sites in Sweden (3) found that compared to boxes at 1.5 m or 3 m above ground, those at 4.5 m had significantly higher occupancy (100% vs 75%), numbers of eggs (94 vs 55–57) and numbers of hatched young (86 vs 46–53). Mean date for first egg was also 2.6 days earlier in the highest boxes. Although there was also a greater number of fledged young in the highest boxes (27 vs 11–19), the average number fledged did not differ significantly between heights (3.4 vs 2.3). Wooden nest boxes (50 mm diameter entrance) were put up at the three heights on eight trees at each site. Boxes were inspected throughout the breeding season.

A replicated study in agricultural sites in southern Finland (4) found that Eurasian kestrel *Falco tinnunculus* occupied 18–22% of 161 nest boxes between 1985 and 1995, with no differences between small, intermediate and large boxes. Boxes sheltered from prevailing weather were more likely to be occupied than exposed boxes (25% of 80 sheltered boxes used vs 17% of 81 exposed boxes). There were no significant differences in clutch size or number of fledglings produced between nest box types and orientations, with success related to laying date and vole *Microtus* spp. abundance. Occupied boxes were, on average, further from forest edges, roads and inhabited houses, and closer to grassy ditches than unoccupied boxes. Boxes were 25 × 27.5 × 25 cm, with a 12.5 x 25 cm entrance (small); 34 × 35 × 20 cm, with a 12 x 34 cm entrance (intermediate); or 33.5 × 45 × 30 cm, with a 12 x 33.5 cm hole (large).

A replicated study of common starling *Sturnus vulgaris* nest boxes in southern Sweden (5) found that nest success within boxes was high and was related to the amount of pasture available. There were between 1 and 8 (average 4) breeding attempts initiated in each colony of 8 boxes. Only 8% of 609 nests failed during laying or incubation and an additional 5% during nestling rearing. Breeding attempts and the proportion of hatchlings that fledged increased and nest failures decreased with an increase of pasture in the surrounding area. In 1994, 19 breeding colonies of a row of 8 nest boxes on trees (5–10 m apart, 1.8 m above ground) were established. An additional 13 colonies were installed in 1996–1998. Nest boxes were visited every 1–2 days to record egg-laying, hatching and fledging.

A replicated, paired site study from March–August in 2000–2003 in 20 paired nest box groups (10 placed along wetland edges and 10 in farmland) in Rutland, England (6) found that tree sparrow *Passer montanus* showed a significant preference for nest boxes in wetland habitat, compared to those in farmland sites (8 wetland nest boxes colonized vs no farmland sites). Nest box groups consisted of 5 nest boxes placed 2–20 apart. Sunflower seeds were randomly provided to one nest box group within each pair.

A replicated trial in arable farming landscapes in Norfolk, England, in the summers of 1997–2001 (7) found that tits *Parus* spp. nested in a higher proportion of hanging woodcrete boxes (38% of 48 boxes occupied), compared to tree-mounted woodcrete boxes (25% of 48) or thick and thin wooden boxes (20% and 16% of 48 boxes respectively). Patterns were the same for great tit *Parus major*, blue tit *P. (Cyanistes) caeruleus* and all species combined (also including coal tit *P. (Periparus) ater* and marsh tit *P. (Poecile) palustris*), although a higher proportion of great tits used woodcrete boxes (91% of great tits vs 47% of blue tits). Clutch size, brood size and number of young fledged by blue tits and great tits did not differ significantly between box types. Woodcrete boxes were either attached to a tree trunk (18 cm high, base 18 cm diameter) or free-hanging (19 cm high, base 11 cm diameter). Wooden boxes were 16.5 x 15 x 19.5 cm, and of either 1.9 cm or 2.4 cm thick wood. All designs had a 3.2 cm diameter entrance. Another trial found that a higher proportion of tit *Parus* spp. nests were in 50 green nest boxes (72% of 41 nests) than in 50 brown boxes (28%), and in 50 boxes with circular entrances (68%) compared to those with a wedge-shaped entrance (32%).

A trial from 2003 to 2005 on a single farm, Rawcliffe Bridge, East Yorkshire, UK (8) found that nest boxes were 54%, 50% and 68% occupied in 2003, 2004 and 2005 respectively. In 2003, all five boxes designed for tree sparrow *Passer montanus* were occupied. In 2005, 20 tree sparrow boxes (70% of the 28 provided) were occupied. The number of breeding tree sparrows on the farm increased from 6 to 20 pairs between 2003 and 2005. In the years 2003, 2004 and 2005, 32, 60 and 84 bird nest boxes were put up, including some designed for tree sparrows. They were inspected in February each year. Birds on the farm were monitored five times each year from 2003 to 2005 by walking the field boundaries. The number of breeding pairs/ha was estimated from clusters of sightings.

A replicated study in 1988–2000 in Noord-Brabant, the Netherlands (9) found that stock dove *Columba oenas* used nest boxes provided in mixed agricultural habitats, laying in total an average of 118 eggs laid/year with 52% hatching and 84% of chicks fledging (an average of 52 chicks/year). Boxes were 20 x 20 x 50 cm, with an 8 x 8 cm entrance hole and placed 3–5 m above the ground in trees, 20–30 m apart. Jackdaw *Corvus monedula* also used the nest boxes, but were removed from 1995 onwards.

A replicated study from 2002 to 2008 of 625 nest boxes inside agricultural shacks and buildings in Valais, Switzerland (10) found that 5% were occupied by Eurasian wryneck *Jynx torquilla* in 2008. Of the 269 monitored locations (2–3 boxes/location), 32 (12%) were occupied by a wryneck in 2008; 23 of those locations had a wryneck nest box. Within the occupied locations, 19 wryneck broods occurred in 1 of the 56 available Eurasian hoopoe *Upupa epops* nest boxes and 14 occurred in 1 of the 22 available wryneck nest boxes. Locations that had been occupied in the past had a higher probability of occupancy. The presence of hoopoes had no influence on the nest box choice. Wryneck nest boxes had no effect on reproductive output, however, in general, nestlings from broods in wryneck nest boxes had a higher body mass than those in hoopoe boxes (27 vs 25 g). The wryneck population inhabiting the hoopoe nest boxes declined from 72 broods in 2002 to 34 broods in 2007, potentially due to competition with the hoopoe population (1998: 20 broods, 2007: 160 broods). The study site was largely of fruit plantations, vineyards and vegetable cultures. A pair of hoopoe boxes were installed at each location from 1998 to 2003 and a further 135 wryneck boxes were installed at half of the locations in 2008.

(1) Cavé A.J. (1968) The breeding of the kestrel *Falco tinnunculus*, in the reclaimed area Ootelijk Flevoland. *Netherlands Journal of Zoology*, 18, 313–407.
(2) Yom-Tov Y. (1974) The effect of food and predation on breeding density and success, clutch size and laying date of the crow (*Corvus corone* L.). *Journal of Animal Ecology*, 43, 479–498.
(3) Svensson S. (1991) Preferences for nest site height in the starling *Sturnus vulgaris* – an experiment with nest-boxes. *Ornis Svecica*, 1, 59–62.
(4) Valkama J. & Korpimaki E. (1999) Nestbox characteristics, habitat quality and reproductive success of eurasian kestrels. *Bird Study*, 46, 81–88.
(5) Smith H.G. & Bruun M. (2002) The effect of pasture on starling (*Sturnus vulgaris*) breeding success and population density in a heterogeneous agricultural landscape in southern Sweden. *Agriculture, Ecosystems & Environment*, 92, 107–114.
(6) Field R.H. & Anderson G.Q.A. (2004) Habitat use by breeding tree sparrows *Passer montanus. Ibis*, 146, 60–68.
(7) Browne S.J. (2006) Effect of nestbox construction and colour on the occupancy and breeding success of nesting tits Parus spp. *Bird Study*, 53, 187–192.
(8) Bryson R.J., Hartwell G. & Gladwin R. (2007) Rawcliffe Bridge, arable production and biodiversity, hand in hand. *Aspects of Applied Biology*, 81, 155–160.
(9) Potters H. (2009) Broedbiologie van een kleine populatie nestkastbewonende Holenduiven in westelijk Noord-Brabant (Breeding biology of a small population of stock pigeon *Columba oenas* in North-Brabant). *Limosa*, 82, 1–12.
(10) Zingg S., Alletaz R. & Schaub M. (2010) Nestbox design influences territory occupancy and reproduction in a declining, secondary cavity-breeding bird. *Ardea*, 98, 67–75.

1.30 Provide nest boxes for bees (solitary bees or bumblebees)

• Ten studies (nine replicated trials and a review of studies) from Germany, Poland and the UK of solitary bee nest boxes all showed the nest boxes were readily used by bees[1, 3–13]. Two replicated studies found the local population size[13] or number of emerging red mason bees increased when nest boxes were provided[5]. One replicated trial in Germany[6] showed that the number of occupied solitary bee nests almost doubled over three years with repeated nest box provision at a given site.

• Two replicated trials tested bumblebee nest boxes and both found very low uptake, 2% or less[2, 14].

• Occupancy rates of solitary bee nest boxes, where reported (two replicated studies), were between 1 and 26% of available cavities[1, 13]. Five studies (four replicated trials and a review of studies) reported the number of bee species found in the nest boxes – between 4.6 and 33 species[3, 4, 6, 7, 9, 10].

- One replicated study from Germany found nest boxes should be placed 150–600 m from forage resources[8]. A replicated study from Poland found the highest production of red mason bees per nest was from nesting materials of reed stems or wood[11].

Background

This intervention involves providing nest boxes for solitary bees or bumblebees. The majority of the studies summarized here tested the effect of providing nest sites on solitary bees. Solitary bee species nest either in cavities such as hollow stems or bored holes in wood or masonry, or in the ground. Many of the studies used solitary bee nest boxes consisting of common reed *Phragmites australis* stems; other options include wood or paper tubes. The success of nest boxes in providing suitable alternative nesting habitats for bumblebees may depend on a number of factors including nest box design, siting and availability of foraging resources (Lye *et al.* 2011).

Lye G.C., Park K.J., Holland J.M. & Goulson D. (2011) Assessing the efficacy of artificial domiciles for bumblebees. *Journal for Nature Conservation*, 19, 154–160.

A replicated study in 1966–1969 from 20 sites in southern England found that (1) red mason bees *Osmia rufa* readily occupied artificial nest boxes comprising of metal food cans filled with drinking straws (straw diameter 5–7 mm). In the first year of the trial, 349 cans were recovered, of these 44 (13%) had one or more straws occupied by a red mason bee nest. Over the following two years, there was a tendency by this species to reoccupy cans. *Osmia caerulescens* and species of *Megachile* also occupied the cans. In March 1966, 398 cans were distributed across sites in 8 counties; 349 cans were recovered in September. In 1967, 1968 and 1969 cans were again placed at several of the sites. Cans were attached to tree branches and fence posts 1–2 m above the ground. The open end of each can faced east or south and was tilted slightly downwards from the horizontal to prevent rain entering.

A trial (unequally replicated) of 654 bumblebee *Bombus* spp. nest boxes over 3 years (1989–1991) in farmland, gardens and fenland in Cambridgeshire, UK (2) found only 10 boxes were occupied (1.5%). The nest boxes tested were wooden boxes raised 10 cm or 1 m above the ground, or nest sites constructed with bricks and concrete tiles on the ground. Dry moss, felt or shredded textiles were added as bedding. Two common and widespread bumblebee species used boxes of both types: the early bumblebee *Bombus pratorum* and the common carder bee *B. pascuorum*.

A replicated, controlled study in May to October 1990 in 40 farmland sites (10 field types, 4 replicates each) near Karlsruhe, south Germany (3) (same study as (4)) found a significantly higher species richness of solitary bees (Apidae) in artificial reed *Phragmites australis* stem nests in unsown sites with naturally developed vegetation (approximately 7.9 species) than in sown fields (approximately 4.6 spp.). Unsown sites with naturally developed vegetation included one- and two-year-old mown and unmown set-asides and old meadow orchards. Crops on sown fields were peas, barley, rye, clover *Trifolium* spp.-grass mixture and phacelia *Phacelia tanacetifolia*. Mowing of set-asides took place in late June–early July. Three artificial nests (each with two 750 ml cans filled with approximately 180 reed stems) were located in each field centre.

A replicated study in April 1990 of 240 bundles of reed *Phragmites australis* stems in 40 fields of 10 management types, in Kraichgau, southwest Germany (4) (same study as (3)) found that of 43,200 available reed stems, 292 were occupied by a total of 14 bee (Apidae) species and 9 wasp species (Hymenoptera). Five species of bee considered to be endangered in Germany occupied the reed stem nests: *Anthidium lituratum, Heriades crenulatus, Megachile alpicola, Osmia gallarum* and *Osmia leaiana*. The two endangered *Osmia* species were exclusively found in nests in old meadows (more than 30 years old with several old fruit trees). The other three also nested in stems provided in two-year-old mown set-aside, and two species (*A. lituratum* and *M. alpicola*) used reed stems in a variety of field types, including cereal crops. The proportion of larvae in the nests that died from disease or failed parasitism was 13%; 2% were successfully parasitized. Two-hundred-and-forty bundles of reed stems in tins were put out, 6 in each of 40 fields of 10 management types, including various types of set-aside, crop fields and old meadows. This study is also referred to by (7).

A replicated six-year trial at two experimental farms near Poznan, western Poland (5) demonstrated that the red mason bee *Osmia rufa* readily nests in bundles of reed *Phragmites australis* stems 7–8 mm in diameter. Bundles of reed stems in roofed containers were set out in March from 1989 to 2004. In winter each year, occupied reed stems were collected and kept in refrigerators over winter. The following spring, overwintered reed stems were placed out in incubators along with new nest boxes. In the first year (1989), 1,750 red mason bee cocoons were introduced with the nest boxes at each site. The behaviour of emerging bees was observed. At one site the total number of emerging red mason bees increased from 1,453 in 1989 to 108,973 in 1994 (a 75-fold increase). At the other site the number of emerging red mason bees increased from 1,519 in 1989 to 13,413 in 1992, after which the population was resettled for other experiments.

A replicated trial from 1994 to 1996 in central Germany (6) found that reed *Phragmites australis* stem nest boxes were occupied by 13 species of bee (Apidae), 19 species of wasp and 17 species of parasite and parasitoid (Hymenoptera). In total, 8,303 nests were made. The number of occupied stems almost doubled over 3 years from 1,761 in 1994 to 3,326 in 1996. One-hundred-and-fifty reed stem nest boxes (plastic tubes filled with 150 lengths of 20 cm reed stem) were placed at 15 different sites. Three replicates in each of five habitat types were studied: sown field margin strips, set aside fields (sown with clover *Trifolium* spp.-grass mixture), extensively used grassland, chalk grassland, orchard meadows. Ten reed stem nest boxes were placed in each site. In autumn, nests were dissected and occupants identified. This study is also referred to by (7).

A review of a series of 4 trials (two (4, 6) already described above) between 1990 and 1996 in Germany (7) found 33 bee species (Apidae) (not including parasitic bees) used reed *Phragmites australis* bundles placed in tins or plastic tubes attached to wooden posts across a variety of agricultural and semi-natural habitats including orchard meadows, old hay meadows, set-aside fields, field margins and chalk grasslands. Two studies documented predation and parasitism rates in reed bundles in tins or plastic tubes attached to wooden posts in various semi-natural and agricultural habitats. The average percentage killed by predators or parasites was 21% for bees and 28% for wasps.

A replicated trial in 1997 of reed *Phragmites australis* stem nest boxes at 15 different agricultural sites near Göttingen in Lower Saxony, Germany (8) (same study as (9)) found nest boxes had a 50% chance of being occupied by 2 specialized (oligolectic) species of bee (Apidae) – *Chelostoma rapunculi* and *Megachile lapponica* – at a distance of 256–260 m from a patch of their required forage plants. The study also found that female solitary bees of 4 medium to large European species *Andrena barbilabris, A. flavipes, A. vaga* and the red mason bee *Osmia rufa* have a maximum foraging range between 150 to 600 m, so nest boxes have to be placed within this distance of forage resources. There was no colonization of nest boxes by *C. rapunculi* more than 300 m from a patch of its food plant, bellflowers *Campanula* spp.. Nest boxes consisted of

150–180 stem sections of common reed, with diameters of 2–10 mm, 15–20 cm-long and put in 10–13 cm diameter plastic tubes or tins. Reed-filled tubes were attached to 1.5 m-long wooden posts above the ground, with 4 nest boxes to a post. Two posts (8 nest boxes) were placed at 15 different sites from April to October. Patches of bellflower and willowherb *Epilobium* spp. were recorded in a 1 km radius from the nest boxes at each site. Nesting females (141 individuals) of the 4 other solitary bee species were marked, then moved in darkened boxes 50 to 2000 m away from their nests. Returning individuals were recorded. This experiment was run on sandy grasslands near Mannheim in 1995, and on chalk grasslands near Göttingen in 1997, between 10:00 h and 18:00 h on sunny days during the main flight period for each bee species.

A replicated trial in 1997 of 120 reed *Phragmites australis* stem nest boxes at 15 different agricultural sites near Göttingen in Lower Saxony, Germany (9) (same study as (8)) found the boxes were occupied by 11 species of bee (Apidae). The red mason bee *Osmia rufa* and the common yellow face bee *Hylaeus communis* were the most widespread and common nest box occupants in this study. Fourteen percent of bee brood cells were attacked by natural enemies (brood parasites, parasitoids or predators). Nest boxes consisted of 150–180 stem sections of common reed, with diameters of 2–10 mm, cut 20 cm-long and put in 10.5 cm diameter plastic tubes. Reed-filled tubes were attached to wooden posts 1–1.2 m above the ground, with 4 nest boxes to a post. Two posts (8 nest boxes) were placed at 15 different sites from April to October. In October, occupied reeds were cut open and the number of brood cells in each stem counted. Occupants were reared in the laboratory and identified to species where possible.

A replicated study from 1998 to 1999 in 45 orchard meadows in central Germany (10) recorded 17,278 cells from 13 species of solitary bee (Apidae) using 540 reed stem nest boxes. Orchards were either mown once or twice a year, grazed (usually by sheep) or had no management for at least 5 years. In each orchard, 3 wooden posts (1.5 m height, 5–7 cm diameter), each with 4 'nesting traps' (total of 540 traps), were set up at regular distances from April to September in 1998 and 1999. The traps comprised 150–180, 20 cm-long common reed *Phragmites australis* stem sections, packed into 10.5 cm diameter plastic tubes. Traps were collected at the end of September and bees and wasps identified to genus, or species, where possible, as were parasitoids and parasites of stored food.

A replicated study in 2000 and 2001 at an agricultural experimental station in Poznan County, Poland (11) tested 6 different nesting materials for the red mason bee *Osmia rufa* and found all materials were used by female bees, but the highest production of bees per nest was from reed *Phragmites australis* stems (3.5 bees/nest in 1999) or wood (7.2 bees/nest in 2000). Nests in paper tubes were all parasitized. Nests in plastic were well occupied (80–100%) but had a low success rate (0.2–1.8 bees/nest), partly due to mould. For each trial, 150 nests of each of the following materials were tested: reed stems, plastic tubes, paper tubes (bundles), wood, cork (grooved boards joined together in blocks), and holes drilled into wood, lined with printer acetate.

A replicated study in 1998 of 12 trap nests in each of 5 orchard meadows near Göttingen, Germany (12) found trap nests (bundles of common reed *Phragmites australis* stems) were used as nest sites by the red mason bee *Osmia rufa*. Three years later, in autumn 2001, a total of 974 newly developed females were counted in 60 such nests and 222 of them were observed re-stocking nests. Bundles of common reed stems (approximately 153 stems, cut 15–20 cm-long) in 10–13 cm diameter plastic tubes, attached to 1.5 m-long wooden posts in groups of 4 were placed in 5 orchard meadows. In autumn 2001, all female adults inside the nests were marked with a plastic bee marker. The stems were closed again and stored until spring 2002, when they were placed in emergence boxes on the posts they came from. Trap nests were observed for two or three 30–60 minute periods from 16 to 22 May 2002.

A replicated study of 30 orchard meadows in Lower Saxony, Germany (13) found that increasing nest site availability resulted in an increase of red mason bee *Osmia rufa* local population size from 80 to 2,740 brood cells/site from 1998 to 2002. Each trap nest contained an average of four red mason bee brood cells in common reed *Phragmites australis* stems. The mean proportion of suitable stems used by the red mason bee increased from 1% in 1998 to 26% in 2002 (highest 96%). The proportion of orchard meadows occupied by the red mason bee also increased, from 84% in 1998 to 100% in 2001 and 2002. Following removal of all brood cells in 2003, figures returned to those of 1998. Habitat connectivity did not affect the number of red mason bee brood cells/site. Population size and rates of parasitism (1992: 93%, 2002: 100% of populations) significantly affected population growth rates. The proportion of brood cells of other bee and wasp (Hymenoptera) species in traps decreased with increasing red mason bee occupancy. At each site, 12 trap nests (of 153 common reed segments) were installed each year. Nests were collected in September and then returned to the same posts in the spring (along with new traps).

A replicated trial in 2008 (14) of 150 underground bumblebee *Bombus* spp. nest boxes on Scottish farmland found very low uptake rates. Just 2% of 150 were used. The boxes were made with two pairs of flower pots placed mouth to mouth, buried in the ground. Fifteen underground boxes were placed on each of ten farms in March and April.

(1)   Free J.B. & Williams I.H. (1970) Preliminary investigations on the occupation of artificial nests by *Osmia rufa* L. (Hymenoptera, Megachilidae). *Journal of Applied Ecology*, 73, 559–566.
(2)   Fussell M. & Corbet S. (1992) The nesting places of some British bumblebees. *Journal of Apicultural Research*, 31, 32–41.
(3)   Gathmann A. & Tscharntke T. (1993) Bees and wasps in trap nests on sown crop fields and self-sown fallow fields (Hymenoptera Aculeata). *Verhandlungen Gesellschaft fur Okologie*, 22, 53–56.
(4)   Gathmann A., Greiler H.J. & Tscharntke T. (1994) Trap-nesting bees and wasps colonizing set-aside fields: succession and body size, management by cutting and sowing. *Oecologia*, 98, 8–14.
(5)   Wójtowski F., Wilkaniec Z. & Szymas B. (1995) Increasing the total number of *Osmia rufa* (L.) (Megachilidae) in selected biotopes by controlled introduction method. In J. Banaszak (ed.) *Changes in the Fauna of Wild Bees in Europe*, Pedagogical University, Bydgoszcz, Poland. pp. 177–180.
(6)   Gathmann A. & Tscharntke T. (1997) Bienen und Wespen in der Agrarlandschaft (Hymenoptera Aculeata): Ansiedlung und Vermehrung in Nisthilfen [Bees and wasps in the agricultural landscape (Hymenoptera Aculeata): colonization and augmentation in trap nests]. *Mitteilungen Der Deutschen Gesellschaft Fur Allgemeine Und Angewandte Entomologie*, 11, 91–94.
(7)   Tscharntke T., Gathmann A. & Steffan-Dewenter I. (1998) Bioindication using trap-nesting bees and wasps and their natural enemies: community structure and interactions. *Journal of Applied Ecology*, 35, 708–719.
(8)   Gathmann A. & Tscharntke T. (2002) Foraging ranges of solitary bees. *Journal of Animal Ecology*, 71, 757–764.
(9)   Steffan-Dewenter I. (2002) Landscape context affects trap-nesting bees, wasps, and their natural enemies. *Ecological Entomology*, 27, 631–637.
(10) Steffan-Dewenter I. & Leschke K. (2003) Effects of habitat management on vegetation and above-ground nesting bees and wasps of orchard meadows in Central Europe. *Biodiversity and Conservation*, 12, 1953–1968.
(11) Wilkaniec Z. & Gieidasz K. (2003) Suitability of nesting substrates for the cavity-nesting bee *Osmia rufa*. *Journal of Apicultural Research*, 42, 29–31.
(12) Steffan-Dewenter I. & Schiele S. (2004) Nest site fidelity, body weight and population size of the red mason bee, *Osmia rufa* (Hymenoptera: Megachilidae), evaluated by mark-recapture experiments. *Entomologia Generalis*, 27, 123–131.
(13) Steffan-Dewenter I. & Schiele S. (2008) Do resources or natural enemies drive bee population dynamics in fragmented habitats? *Ecology* 89, 1375–1387.
(14) Lye G. (2009) Nesting ecology, management and population genetics of bumblebees: an integrated approach to the conservation of an endangered pollinator taxon. PhD thesis. Stirling University.

1.31   Introduce nest boxes stocked with solitary bees

• We have captured no evidence for the effects of introducing nest boxes stocked with solitary bees on farmland wildlife.

Background
This intervention may involve introducing nest boxes stocked with solitary bees on farmland. Captive rearing of solitary bees may be used to augment or re-establish wild populations on farmland.
See also 'Provide nest boxes for bees (solitary bees or bumblebees)'.

1.32   Provide red squirrel feeders

• We have captured no evidence for the effects of providing red squirrel feeders on farmland wildlife.

Background
This intervention may involve providing supplementary food for red squirrels *Sciurus vulgaris* in specially-designed feeders. The red squirrel has experienced population declines in the UK and Ireland but is widespread in most areas of Europe (Shar *et al.* 2008).

Shar S., Lkhagvasuren D., Bertolino S., Henttonen H., Kryštufek B. & Meinig H. (2008) *Sciurus vulgaris*. IUCN Red List of Threatened Species. Version 2012.2. Available at www.iucnredlist.org/details/ 20025/0. Accessed 9 January 2013.

1.33   Provide otter holts

• We have captured no evidence for the effects of providing otter holts on farmland wildlife.

Background
This intervention may involve constructing or installing artificial holts to provide shelter for otters along waterways.

1.34   Provide badger gates

• We have captured no evidence for the effects of providing badger gates on farmland wildlife.

Background
This intervention may involve installing gates where badger paths cross netting or fences, to allow badgers to pass through without damaging the fence. The badger gate may consist of a frame with a wooden flap that opens both ways (Natural England 2011).

Natural England (2011) *Badger Gates in Rabbit-proof Fencing*. Natural England Technical Information Note TIN026. Second Edition 1 January 2011.

2 Arable farming

Key messages

Increase crop diversity
Four studies (including one replicated, controlled trial) from Belgium, Germany and Hungary found more ground beetle or plant species or individuals in fields with crop rotations or on farms with more crops in rotation than monoculture fields.

Implement 'mosaic management', a Dutch agri-environment option
A replicated, controlled before-and-after study from the Netherlands found mosaic management had mixed effects on population trends of wading bird species. A replicated, paired sites study from the Netherlands found one bird species had higher productivity under mosaic management.

Take field corners out of management
A replicated site comparison from the UK found a positive correlation between grey partridge overwinter survival and taking field corners out of management. Brood size, ratio of young to old birds and density changes were unaffected.

Leave overwinter stubbles
Eighteen studies investigated the effects of overwinter stubbles. Thirteen studies (including two replicated site comparisons and a systematic review) from Finland, Switzerland and the UK found leaving overwinter stubbles benefited some plants, invertebrates, mammals or birds. Three UK studies (one randomized, replicated, controlled) found only certain birds were positively associated with overwinter stubbles.

Create beetle banks
Five reports from two replicated studies (one controlled) and a review from Denmark and the UK found beetle banks had positive effects on invertebrate numbers, diversity or distributions. Five replicated studies (two controlled) found lower or no difference in invertebrate numbers. Three studies (including a replicated, controlled trial) from the UK found that beetle banks, alongside other management, had positive effects on bird numbers or usage. Three studies (one replicated site comparison) from the UK found mixed or no effects on birds; two found negative or no clear effects on plants. Two studies (one controlled) from the UK found harvest mice nested on beetle banks.

Plant nettle strips
A small study from Belgium found nettle strips in field margins had more predatory invertebrate species than the crop, but fewer individuals than the crop or natural nettle stands.

Leave unharvested cereal headlands in arable fields
We have captured no evidence for the effects of leaving unharvested cereal headlands on farmland wildlife.

Leave cultivated, uncropped margins or plots (includes 'lapwing plots')
Seventeen of nineteen individual studies looking at uncropped, cultivated margins or plots (including one replicated, randomized, controlled trial) primarily from the UK found benefits to some or all target farmland bird species, plants, invertebrates or mammals. Two studies (one replicated) from the UK found no effect on ground beetles or most farmland birds. Two replicated site comparisons from the UK found cultivated, uncropped margins were associated with lower numbers of some bird species or age groups in some areas.

Plant crops in spring rather than autumn
Seven studies (including two replicated, controlled trials) from Denmark, Sweden and the UK found sowing crops in spring had positive effects on farmland bird numbers or nesting rates, invertebrate numbers or weed diversity or density. Three of the studies found the effects were seasonal. A review of European studies found fewer invertebrates in spring wheat than winter wheat.

Undersow spring cereals, with clover for example
Eleven studies (including three randomized, replicated, controlled trials) from Denmark, Finland, Switzerland and the UK found undersowing spring cereals benefited some birds, plants or invertebrates, including increases in numbers or species richness. Five studies (including one replicated, randomized, controlled trial) from Austria, Finland and the UK found no benefits to invertebrates, plants or some birds.

Create rotational grass or clover leys
A controlled study from Finland found more spiders and fewer pest insects in clover leys than the crop. A replicated study from the UK found grass leys had fewer plant species than other conservation habitats. A UK study found newer leys had lower earthworm abundance and species richness than older leys.

Convert or revert arable land to permanent grassland
All seven individual studies (including two replicated, controlled trials) from the Czech Republic, Denmark and the UK looking at the effects of reverting arable land to grassland found no clear benefits to birds, mammals or plants.

Reduce tillage
Thirty-four studies (including seven randomized, replicated, controlled trials) from nine countries found reducing tillage had some positive effects on invertebrates, weeds or birds. Twenty-seven studies (including three randomized, replicated, controlled trials) from nine countries found reducing tillage had negative or no clear effects on some invertebrates, plants, mammals or birds. Three of the studies did not distinguish between the effects of reducing tillage and reducing chemical inputs.

Add 1% barley into wheat crop for corn buntings
We have captured no evidence for the effects of adding 1% barley into wheat crop for corn buntings on farmland wildlife.

Create corn bunting plots
We have captured no evidence for the effects of creating corn bunting plots on farmland wildlife.

Create skylark plots

All four studies (two replicated, controlled trials) from Switzerland and the UK investigating the effect of skylark plots on Eurasian skylarks found positive effects, including increases in population size. A replicated study from Denmark found skylarks used undrilled patches in cereal fields. Three studies (one replicated, controlled) from the UK found benefits to plants and invertebrates. Two replicated studies (one controlled) from the UK found no significant differences in numbers of invertebrates or seed-eating songbirds.

Plant cereals in wide-spaced rows

Two studies (one randomized, replicated, controlled) from the UK found planting cereals in wide-spaced rows had inconsistent, negative or no effects on plant and invertebrate abundance or species richness.

Sow rare or declining arable weeds

Two randomized, replicated, controlled studies from the UK identified factors important in establishing rare or declining arable weeds, including type of cover crop, cultivation and herbicide treatment.

Use new crop types to benefit wildlife (such as perennial cereal crops)

We have captured no evidence for the effects of using new crop types to benefit wildlife (such as perennial cereal crops).

Plant more than one crop per field (intercropping)

All five studies (including three randomized, replicated, controlled trials) from the Netherlands, Poland, Switzerland and the UK looking at the effects of planting more than one crop per field found increases in the number of earthworms or ground beetles.

2.1   Increase crop diversity

- All four studies (including one replicated, controlled study and one review) from Belgium, Germany, Hungary and unspecified European countries reported a positive effect of crop rotations on ground beetles or plants. Three studies found higher ground beetle species richness and/or abundance[1, 2, 4] and one study found higher plant species richness[3] in rotation fields or on farms with more crops in rotation compared to monoculture fields.

- A study from Hungary found that fields in monoculture had a more stable and abundant ground beetle community than fields within a rotation[1, 4].

Background
Farmland heterogeneity is thought to be key in determining on-farm biodiversity (Benton *et al.* 2003). Therefore, increasing the range of different crops grown in a given year may increase the biological value of a farm.

Benton T.G., Vickery J.A. & Wilson J.D. (2003) Farmland biodiversity: is habitat heterogeneity the key? *Trends in Ecology & Evolution*, 18, 182–188.

A site comparison study from 1977 to 1979 in Hungary (1) found that monoculture fields had a more stable and abundant ground beetle (Carabidae) community than fields within a rotation. Significantly more beetles were caught in a monoculture maize *Zea mays* field than in rotation (maize/wheat) fields each year (1,203–2,511 vs 368–1,057 beetles) and the activity density was almost 3 times higher in the monoculture (5.1 vs 1.6 beetles/trap/week). Ground beetle species diversity, however, was higher on rotation fields (18–26 species/year) than the monoculture (18–21). The activity periods of the most common species were longer in the monoculture. The ground beetle community also appeared to be more stable in the mono-culture, with 65% of species being caught in the spring and autumn compared to just 31% caught in both seasons in the rotation fields. Each year two types of maize field were stud-ied: the monoculture had been planted with maize for almost 20 years (400 ha); rotation fields previously sown with winter wheat were planted with maize in the first year of the study (28–78 ha). Monoculture and rotation fields were chosen to be as near to each other as possible (200–30,000 m apart). Invertebrates were sampled using a line of 10 pitfall traps extending from the field margin towards the centre of each field. Traps were checked weekly from May to August (monoculture) or October (rotations) each year.

A site comparison study in 1988 near Ghent, Belgium (2) found more ground beetles (Carabidae) and more ground beetle species in a maize *Zea mays* field in crop rotation than in maize or perennial rye grass *Lolium perenne* fields grown in monoculture. On the crop rotated maize field, 3,500 ground beetles of 18 species were caught, compared to 3,000 beetles of 15 species on the maize monoculture field. Fourteen species were caught on the rye grass mono-culture in a different year. Ground beetles were collected every fortnight from six pitfall traps in each field, from May until September 1988 (1980 for the rye grass monoculture). Fields were small, just a few hectares in size.

A replicated, controlled study in summer 1991–1994 in the Province of Bayern, Germany (3) found that farms in the Bavarian Cultural Landscape Programme (Bayerisches Kultur-landschaftsprogramm), an extensification programme, had more plant species than control farms (15.6 vs 13.8 plant species). There are no restrictions on fertilizer or pesticide use but some less common crops (e.g. flax and grass seeds) can be included in the crop rotation in this extensification programme. The study did not measure the actual number of crops in rotation on these farms. Vegetation was surveyed between June and September on total areas between 100 and 400 m^2. Cereal crops were surveyed yearly; cut set-asides several times a year. Note that no statistical analyses were performed on the data.

A 1999 literature review (4) found two studies that compared ground beetles (Carabidae) in maize *Zea mays* fields in monoculture with those in crop rotation. One found more ground beetles in monocultures in large fields (28–400 ha; (1)), the other (with smaller fields measur-ing just a few hectares) is reported to have found no major differences, except for the presence of 3 additional species in crop-rotated maize (2).

(1) Lovei G.L. (1984) Ground beetles (Coleoptera, Carabidae) in 2 types of maize fields in Hungary. *Pedobiologia*, 26, 57–64.
(2) Desender K. & Alderweireldt M. (1990) The carabid fauna of maize fields under different rotation regimes. *Mededelingen van de Faculteit Landbouwwetenschappen Universiteit Gent*, 55, 493–500.
(3) Hilbig W. (1997) Effects of extensification programmes in agriculture on segetal vegetation. *Tuexenia*, 295–325.
(4) Kromp B. (1999) Carabid beetles in sustainable agriculture: a review on pest control efficacy, cultivation impacts and enhancement. *Agriculture, Ecosystems & Environment*, 74, 187–228.

2.2   Implement 'mosaic management', a Dutch agri-environment option

- A replicated, controlled before-and-after study from the Netherlands[2] found that northern lapwing population trends changed from decreases to increases following the introduction of mosaic management. Three other species of wading bird did not show such a response and Eurasian oystercatcher populations did less well under mosaic management than other management types.

• A replicated, paired sites study in the Netherlands[1] found that black-tailed godwit had higher productivity under mosaic management than other management types due to higher nest survival, and nests were less likely to be trampled by livestock or destroyed by mowing under mosaic management.

Background
Mosaic management is a Dutch agri-environment scheme that, rather than concentrating on individual farms, attempts to coordinate management across groups of farms. Interventions include delayed and staggered mowing, refuge strips and nest protection and aim to provide suitable foraging habitat for wader chicks throughout the year.

A replicated, paired sites comparison in 2004–2005 on six wet grassland sites in the Netherlands (1) found that the reproductive productivity of black-tailed godwit *Limosa limosa* was significantly higher on sites managed under a 'mosaic management' agri-environment scheme, compared to non-scheme sites (average of 0.28 chicks fledged/breeding pair for scheme sites vs 0.16 chicks/pair on non-scheme sites). Differences were due to higher nest survival on mosaic management sites (50% vs 33%), as there were no differences in the number of chicks hatching in successful nests (3.4 chicks/successful nest vs 3.2), or the fledging rate of chicks (11% fledging success on all sites). Nests were equally likely to be predated on scheme and non-scheme sites (32% predated vs 37%), but were more likely to be trampled or destroyed by mowing on non-scheme sites (6% vs 29%). Most fields in five scheme sites and about 50% in the sixth, had nests marked (to reduce losses due to farming activities); at non-scheme sites almost 100% of nests were marked in three, some in two, and none in one. The number of nests on different sites was not provided.

A replicated, controlled, before-and-after study in 1996–2008 in eight wet grassland areas in Friesland and Groningen, the Netherlands (2) found that northern lapwing *Vanellus vanellus* population trends moved from a 7% annual decrease to a 4% annual increase following the introduction of mosaic management in 2000–2001. Three other species (black-tailed godwit *Limosa limosa*, common redshank *Tringa totanus* and Eurasian oystercatcher *Haematopus ostralegus*) did not show any change in trend after the introduction. When comparing trends on the mosaic management sites with 29 farms using individual conservation management, 46 farms with standard management and 42 nature reserves, only northern lapwing populations increased significantly more on mosaic management sites, compared to the others. Oystercatcher populations did significantly less well on mosaic management sites, compared to nature reserves.

(1) Schekkerman H., Teunissen W. & Oosterveld E. (2008) The effect of 'mosaic management' on the demography of black-tailed godwit *Limosa limosa* on farmland. *Journal of Applied Ecology*, 45, 1067–1075.
(2) Oosterveld E.B., Nijland F., Musters C.J.M. & Snoo G.R. (2010) Effectiveness of spatial mosaic management for grassland breeding shorebirds. *Journal of Ornithology*, 152, 161–170.

2.3   Take field corners out of management

- A replicated site comparison study in the UK[1] found that taking field corners out of management was positively correlated with grey partridge overwinter survival. However it had no effect on grey partridge brood size, the ratio of young to old birds or year-on-year density changes.

Background

Field corners can be taken out of management on both arable and livestock farms. This can involve either not managing or planting corners with grass.
See also 'All farming systems: Plant grass buffer strips/margins around arable or pasture fields'.

A replicated site comparison study in 2004 to 2008 in England (1) found that grey partridge *Perdix perdix* overwinter survival was positively correlated with taking field corners out of management, significantly so in 2007–2008. There were no relationships between taking field corners out of management and brood size, the ratio of young to old birds or year-on-year density changes. Spring and autumn counts of grey partridge were made at 1,031 sites across England as part of the Partridge Count Scheme.

(1) Ewald J.A., Aebischer N.J., Richardson S.M., Grice P.V. & Cooke A.I. (2010) The effect of agri-environment schemes on grey partridges at the farm level in England. *Agriculture, Ecosystems & Environment*, 138, 55–63.

2.4   Leave overwinter stubbles

- Eighteen studies (including four reviews and one systematic review) investigated the effects of overwinter stubbles on farmland wildlife. Thirteen studies from Finland, Switzerland and the UK (six replicated trials, including two site comparisons, four reviews and a systematic review) found evidence that leaving overwinter stubbles provides some benefits to plants[3, 5, 6, 12, 14], insects[3], spiders[9], mammals[12, 15] and farmland birds[1, 2, 4, 6, 13, 15, 16]. These benefits include higher densities of farmland birds in winter[13, 15], increased grey partridge productivity[18], and increased cirl bunting population size[1, 2] (in combination with several other conservation measures) and territory density[15].

- One replicated site comparison study from the UK found evidence that leaving overwinter stubbles had inconsistent or no effects on farmland bird numbers[16, 17]. Three studies found only certain bird species showed positive associations with overwinter stubbles. Two replicated studies (of which one also randomized and controlled) found that only Eurasian skylark[7] or both Eurasian skylark and Eurasian linnet[10] benefited, out of a total 23 and 12 farmland bird species tested respectively. One study found that only grey partridge and tree sparrow showed positive population responses to areas with overwinter stubbles[17].

- Two studies from the UK (one randomized; one replicated and controlled) found that different farmland bird species benefited from different stubble heights[8, 11]. One replicated site comparison study found mixed effects between different stubble management options on seed-eating bird abundance[19, 20].

Background

This intervention involves leaving crop stubbles in fields until at least February–March. These stubbles may provide an important food source for seed-eating birds over the winter (Campaign for the Farmed Environment 2011).

The availability and extent of overwinter stubbles may have an important influence on bird populations. One study from the UK (Gillings *et al.* 2005) showed that national population trends of 16 out of 26 bird species were positively influenced by the presence of overwinter stubbles. The same study also predicted that yellowhammer *Emberiza citrinella* and Eurasian skylark *Alauda arvensis* population declines would be lessened in areas with a high proportion of stubbles and that 10-year population stability/growth could be achieved by increasing the coverage of stubbles within 1 km squares to 15 ha or 20 ha respectively. Additionally, a 2008 literature review and analysis of the Environmental Stewardship scheme, particularly Entry Level Stewardship in the UK (Vickery *et al.* 2008), suggested that, for Eurasian skylark, approximately 0.1 km^2 of stubble/km^2 would be needed to prevent population declines. The authors also suggest that having these patches over 1 km apart would maximize winter use.

Gillings S., Newson S.E., Noble D.G. & Vickery J.A. (2005) Winter availability of cereal stubbles attracts farmland birds and positively influences breeding population trends. *Proceedings of the Royal Society B*, 272, 733–739.

Vickery J., Chamberlain D., Evans A., Ewing S., Boatman N., Pietravalle S., Norris K. & Butler S. (2008) *Predicting the Impact of Future Agricultural Change and Uptake of Entry Level Stewardship on Farmland Birds*. British Trust for Ornithology, The Nunnery, Thetford.

Campaign for the Farmed Environment (2011) *Guide to Voluntary Measures 2011 Edition*. Campaign for the Farmed Environment, Warwickshire.

A 2000 literature review (1) found that the UK population of cirl bunting *Emberiza cirlus* increased from between 118 and 132 pairs in 1989 (Evans 1997) to 453 pairs in 1998 (Wotton *et al.* 2000) following a series of agri-environment schemes designed to provide overwinter stubbles, grass margins, and beneficially managed hedges and set-aside. Numbers on fields under specific agri-environment schemes increased by 70%, compared with a 2% increase on neighbouring land not under the scheme.

A 2001 paired site comparison study in south Devon, UK (2) found that the presence of areas of spring sown barley followed by overwinter stubbles was associated with an increase in the number of cirl bunting *Emberiza cirlus*. Along with a number of other Countryside Stewardship Scheme management options found to be important for cirl bunting, an increase in the area of spring sown barley followed by overwinter stubbles coincided with an increase in the number of cirl bunting pairs from 1997 to 1998. Six of seven Countryside Stewardship Scheme plots that had 6 m grass margins and were within 2.5 km of former cirl bunting territories gained birds, and there was a tendency for farms providing grass margins to also include spring sown barley (followed by overwinter stubbles). The association between grass margin uptake and overwinter stubble uptake leads the authors to suggest that overwinter stubbles (and spring sown barley) may have a positive influence on cirl bunting, although these results are not definitive. More generally, there were declines of 20% in cirl bunting numbers on land not participating in the Countryside Stewardship Scheme. Forty-one 2 x 2 km^2 squares containing both land within the Countryside Stewardship Scheme and non-Countryside Stewardship Scheme land were surveyed in 1992, 1998 and 1999. In each year, squares were surveyed for cirl bunting at least twice, the first time during mid-April to late May, and the second time between early June and the end of August.

A review (3) of two reports (Wilson *et al.* 2000; ADAS 2001) evaluating the effects of the Pilot Arable Stewardship Scheme in two regions (East Anglia and the West Midlands) in the UK from 1998 to 2001 found that overwinter stubbles benefited plants, bumblebees *Bombus* spp. and true bugs (Hemiptera), especially when followed by spring fallow. Stubbles also benefited ground beetles (Carabidae) and sawflies (Symphyta). The effects of the pilot scheme on plants, invertebrates and birds were monitored over three years, relative to control areas, or control farms. Only plants and invertebrates were measured within individual options. Overwinter stubbles were the most widely implemented options, with total areas of 974 and 2,200 ha in East Anglia and West Midlands respectively.

A replicated study in the winters of 1997–1998 and 1998–1999 on 122 stubble fields on 32 farms in central England (4) found that 13 bird species were found using stubble fields. Four species (Eurasian linnet *Carduelis cannabina*, Eurasian skylark *Alauda arvensis*, reed bunting *Emberiza schoeniclus* and corn bunting *Miliaria calandria*) were found more frequently on intensively-farmed barley *Hordeum* spp. stubbles than intensive or organic wheat *Triticum* spp., whilst woodpigeons *Columba palumbus* were found most frequently on organic wheat. Intensive barley stubbles had the highest cover of weeds (51% cover on intensive barley; 40% on intensive wheat; 28% on organic wheat). Weed seed densities in March were highest on undersown organic wheat stubble fields compared to intensive barley or wheat stubbles. Weed seed density decreased the least on undersown organic wheat stubbles between October and March compared to intensive barley or wheat stubbles (11% decline on undersown organic wheat stubbles; 23% decline on intensive wheat stubbles; 35% decline on intensive barley). Seventeen stubble fields contained organic wheat with the previous crop undersown with rye grass *Lolium* spp. and white clover *Trifolium repens*. Sixty-seven fields were managed for intensive wheat and 38 fields for intensive barley; both intensively-managed crops received inorganic fertilizer and pesticide applications. Each study field was either overwintering as stubble or entered into the first year of a set-aside scheme. Plants were surveyed in forty 20 x 20 cm quadrats in each field in October. Seed densities were recorded in 27 fields from 10 soil cores/field in October 1997 and March 1998. Birds were surveyed monthly on parallel transects.

A replicated study in the summers of 1999–2000 comparing ten different conservation measures on arable farms in the UK (5) found that overwinter stubbles had high total plant cover, but not as many plant species as some other measures. Overwinter stubbles and spring fallows had relatively high total plant cover, and over 50% cover of grasses. Litter cover was higher while richness of annual plant species was lower in overwinter stubbles compared with spring fallows, probably due to cultivation in spring fallows. The average numbers of plant species in the different conservation habitats were: overwinter stubbles 4.2; wildlife seed mixtures 6.7; uncropped cultivated margins 6.3; undersown cereals 5.9; uncultivated margins 5.5; no-fertilizer conservation headlands 4.8; spring fallows 4.5; sown grass margins 4.4; conservation headlands 3.5; grass leys 3.1. Uncropped cultivated margins, wildlife seed mixtures and no-fertilizer conservation headlands appeared to be the best options for conservation of annual broadleaf plant communities. Plants were surveyed on 37 farms in East Anglia (dominated by arable farming) and 38 farms in the West Midlands (dominated by more mixed farming). The study included 294 habitat sites (defined as a single field, block of field or field margin strip). Vegetation was surveyed once in each site in June–August in 1999 or 2000, in thirty 0.25 m^2 quadrats randomly placed in 50–100 m randomly located sampling zones in each habitat site. All vascular plant species rooted in each quadrat, bare ground or litter and plant cover were recorded.

A replicated, randomized study from November 2003 to March 2004 in 205 cereal stubble fields under a range of management intensities in arable farmland in south Devon, UK (6) found that barley stubbles following low-input herbicide were more beneficial for cirl bunting *Emberiza cirlus* than wheat or conventionally managed stubbles. The number of breeding cirl bunting territories the previous season and small field size (probably, as the authors point out, because cirl buntings prefer to forage near hedgerows and because smaller fields are less intensively managed) also correlated positively with population size. Overall, barley fields were generally preferred by seed-eating species. Low-input barley stubbles had significantly higher seed abundance and broadleaved weed cover (approximately four times greater). Fields where stubbles were grazed over winter led to significantly lower densities of seed-eating birds in general. The authors point out that seed-eating bird species' preference for barley stubbles was independent of the positive correlation with broadleaved weed density and should be taken into account when planning prescriptions.

A replicated, randomized, controlled study from November–February in 2000–2001 and 2001–2002 in 20 arable farms in eastern Scotland (7) found that, of 23 bird species recorded, only Eurasian skylark *Alauda arvensis* was significantly more abundant in fields with stubble left over winter than fields with wild bird cover crops or conventional crops. Stubble fields were those in which cereal and oilseed rape stubbles were left over winter. Six to 28 ha were sampled on each farm annually.

A replicated, controlled study in winter 2003–2004 on 20 wheat fields on 12 lowland farms in central England (8) found that seed-eating songbirds and invertebrate-feeding birds were more abundant on stubble fields cut to 6 cm, whereas Eurasian skylark *Alauda arvensis* and partridges (Phasianidae) were more abundant on fields with uncut stubble, approximately 14 cm tall (seed-eaters: 343 individuals seen on approximately 25 of 120 visits to cut fields vs 89 individuals on 15 visits to control fields; invertebrate-eaters: 623 birds on 17 visits vs 34 on 5 visits; skylark: 557 on 50 visits vs 814 on 80 visits; partridges: 5 on 2 visits vs 235 on 27 visits). Crows (Corvidae) and pigeons (Columbidae) showed no response to stubble cutting. Each field was split so that half was cut (late October 2003) to approximately 6 cm tall, with the other half left as a control.

A controlled trial from 2003 to 2004 in Jokioinen, southern Finland (9) found that uncultivated barley stubble had significantly more spiders (Araneae) than a control spring barley crop, but similar numbers of ground beetles (Carabidae) and unsown plant species. The stubble field had around 20 spiders/trap, compared to around 5 spiders/trap in the control plot. A 44 x 66 m plot of uncultivated spring barley stubble was established in 2004 (the barley sown and harvested in 2003), and compared with an equivalent plot of spring barley crop sown in 2004. Insects were sampled using a yellow sticky trap and three pitfall traps in the centre of each plot for a week in June, July and August 2004. Unsown plant species were counted in four 50 x 50 cm quadrats in each plot in late August 2004.

A replicated study in 1999 and 2003 on 256 arable and pastoral fields across 84 farms in East Anglia and the West Midlands, England (10) found that only 2 of 12 farmland bird species analysed were positively associated with the provision of overwinter stubble, set-aside or wildlife seed mixtures. The two species were Eurasian skylark *Alauda arvensis* (a field-nesting species) and Eurasian linnet *Carduelis cannabina* (a boundary-nesting species). The study did not distinguish between set-aside, wildlife seed mixtures or overwinter stubble, classing all as interventions to provide seeds for farmland birds.

A small randomized site comparison study in winter 2004–2005 in central England (11) found that seed-eating songbirds and invertebrate-feeding birds were found at higher densities on sections of fields where stubble had been cut short (404 seed-eating birds and

244 invertebrate-feeding birds recorded on uncut stubble plots vs 77 and seven on cut stubble). Eurasian skylark *Alauda arvenis*, partridges (Phasianidae), pigeons *Columba* spp., and meadow pipit *Anthus pratensis* were found at higher densities in areas of uncut stubble (241 skylark, 100 partridges, 37 pigeons and 81 meadow pipit on uncut plots vs 27, 7, 12 and 9 on cut plots). In addition, skylarks and invertebrate-feeders were found at higher densities on scarified (i.e. lightly tilled) sections of fields than control (unscarified) sections (339 skylarks and 1,371 invertebrate feeders on scarified plots vs 241 and 251 on controls). The stubble on one half of each field was cut in the winter of 2004–2005 (late December–early February) before the fields were surveyed between December 2004 and March 2005.

A 2007 review of published and unpublished literature (12) found experimental evidence of benefits of overwinter stubble to plants (one study: (5)) and use of overwinter stubbles by brown hares *Lepus europaeus* (one correlative study not included here). This review assessed the evidence for wider benefits of UK agri-environment prescriptions aimed at conserving wild birds on arable land.

A 2007 systematic review identified five papers investigating the effect of overwinter stubble provision on farmland bird densities in the UK (13). There were significantly higher densities of farmland birds in winter on fields with stubbles than on conventionally managed fields. In particular, there were greater densities of seed-eating songbirds and crows (Corvidae) on fields with stubbles than on control fields. The meta-analysis included experiments conducted between 1992 and 2002 from three controlled trials, one time series, and one site comparison study.

A replicated before-and-after study in 1989–2005 on 28 selected arable fields in the western Swiss Plateau (14) found that populations of two hornwort species (*Anthoceros agrestis* and *Phaeoceros carolinianus*, the latter rare) declined between surveys carried out before and after introduction of the Swiss agri-environment scheme in 1999. An index of hornwort abundance was greater during an initial survey in 1989–1995 than in a repeat survey of the same sites in 2005–2007. Hornwort abundance was strongly affected by the availability of stubble fields. The proportion of stubble fields left unmanaged after harvest was found to decrease between the survey periods. The scheme appeared suboptimal for conserving hornwort taxa because it did not support the maintenance of autumn or winter stubble fields (which, in turn, declined as a result of soil conservation measures introduced in 2005). Selected fields (on average 1–2 ha) were surveyed every September–October, and observations of hornwort occurrences (of *A. agrestis* and *P. carolinianus* gametophytes, based on 20 minutes search by 2 people) were used to calculate an abundance index. Crop type and management were also recorded.

A 2009 literature review of agri-environment schemes in England (15) found a 146% increase in cirl bunting *Emberiza cirlus* territory density on land under a Countryside Steward-ship Scheme 'special project', which (amongst other interventions) increased the amount of weedy overwinter stubbles in the target area between 1992 and 2003. In addition, the national population increased from 319 to nearly 700 pairs over the same period (Wotton *et al.* 2000, (2), Wotton *et al.* 2004). Generally, the review found high densities of seed-eating songbirds and Eurasian skylark *Alauda arvensis* on stubbles and wild bird seed or cover mix, compared to other land uses, and a survey in the winter of 2007–2008 found the highest densities of skylark on stubble fields, compared with other agri-environment scheme options (Field *et al.* in press). The review also stated that overwinter stubbles deliver benefits for brown hare *Lepus europaeus*, but did not provide further details.

A replicated site comparison in 2005 and 2008 of agricultural land across England in 2005 and 2008 (16) (same study as (17)) found that four of eight regions had at least two farmland bird species that showed positive responses to wild bird cover and overwinter stubble fields.

Across all 15 species thought to benefit from these interventions, only one region (the north-west) showed significantly more positive responses than would be expected by chance. Some species responded positively in some regions and negatively in others. There were 2,046, 1 km^2 lowland study plots, surveyed in 2005 and 2008.

A large replicated site comparison study in 2005 and 2008 of lowland farmland in England (17) (same study as (16)) found that three years after the 2005 introduction of the Countryside Stewardship schemes and Entry Level Stewardship schemes, there was no consistent association between the provision of stubbles and farmland bird numbers. Grey partridge *Perdix perdix* and tree sparrow *Passer montanus* were the only two species that showed more positive population change (population increases or smaller decreases relative to other plots) from 2005 to 2008 in the 9 km^2 and 25 km^2 areas immediately surrounding plots planted with stubble than in the area surrounding plots without stubbles. The effect of stubbles was small, however, with tree sparrow numbers increasing by 0.05 at the 9 km^2 scale for every 0.07 km^2 of stubble and by 0.07 at the 25 km scale for every 0.14 km^2 of stubble. The 2,046, 1 km^2 low-land plots were surveyed in both 2005 and 2008 and classified as arable, pastoral or mixed farmland. Eighty-four percent of plots included some area managed according to Entry Level Stewardship or the Countryside Stewardship Scheme. In both survey years, 2 surveys were conducted along a 2 km pre-selected transect route through each 1 km^2 square.

A replicated site comparison study from 2004 to 2008 in England (18) found that the proportion of young grey partridges *Perdix perdix* in the population (ratio of young to old grey partridge) was positively associated with the amount of sites left as overwinter stubble. However, when stubbles were used in conjunction with other interventions, the results were mixed. In conjunction with small field sizes and reduced chemical inputs, stubbles were weakly positively correlated with year-on-year changes in partridge density but negatively related to brood size. In conjunction with undersowing spring cereals, stubbles were negat-ively associated with year on year changes (in 2006–2007), overwinter survival (2004–2005, 2005–2006 and generally).

A replicated site comparison study between November 2007 and February 2008 of 75 fields in East Anglia and the West Midlands, England (19) (same study as (20)) found no differences in the number of seed-eating birds or Eurasian skylark *Alauda arvensis* on Environmental Stewardship stubbles and non-Environmental Stewardship stubbles. There was also no signi-ficant difference in the number of seed-eating birds on stubbles managed under Higher Level Stewardship of the Environmental Stewardship scheme (18 birds/ha) than in fields managed under Entry Level Stewardship (8.5 birds/ha). Skylarks, however, were found to be more numerous on Higher Level Stewardship fields (9.3 birds/ha) than Entry Level Stewardship fields (1.2 birds/ha). Entry Level Stewardship stubbles had no post-harvest herbicide and no cultivation until mid-February, Higher Level Stewardship stubbles had the basic Entry Level Stewardship requirements plus reduced herbicide use and cereal crop management prior to the overwinter stubbles. Non-Environmental Stewardship stubbles were rotational stubbles without restrictions on herbicide or cultivation practices. Seed-eating birds were surveyed on 2 visits to each site between 1 November 2007 and 29 February 2008.

A replicated site comparison study in winter 2007–2008 on farms in East Anglia and the West Midlands, England (20) (same study as (19)) found more seed-eating farmland songbirds on overwinter stubbles managed under Entry Level Stewardship than on non-stewardship stubbles in the West Midlands (average 6.0 birds/ha on Entry Level Stewardship vs 2.5 birds/ha on conventionally managed stubble). This difference was not significant for farms in East Anglia (3.5 birds/ha on Entry Level Stewardship stubble vs 0.7 birds/ha on con-ventionally managed stubble fields). Overwinter stubble fields in stewardship schemes have restrictions on herbicide use and cultivation times. The group of birds analysed included tree sparrow *Passer montanus* and corn bunting *Emberiza calandra*, but not grey partridge *Perdix perdix*. More of these birds used overwinter stubbles on Higher Level Stewardship farms than on Entry Level Stewardship farms. There were 5 birds/ha compared to 2 birds/ha on average, on stubble fields on Higher Level Stewardship and Entry Level Stewardship farms respectively. The survey was carried out in winter 2007–2008 on 27 farms with Higher Level Stewardship, 13 farms with Entry Level Stewardship and 14 with no environmental stewardship.

(1)   Aebischer N.J., Green R.E. & Evans A.D. (2000) From science to recovery: four case studies of how research has been translated into conservation action in the UK. In N. J. Aebischer, A. D. Evans, P. V. Grice & J. A. Vickery (eds.) *Ecology and Conservation of Lowland Farmland Birds*, British Ornithologists' Union, Tring. pp. 43–54.
(2)   Peach W., Lovett L., Wotton S. & Jeffs C. (2001) Countryside stewardship delivers cirl buntings (*Emberiza cirlus*) in Devon, UK. *Biological Conservation*, 101, 361–373.
(3)   Evans A.D., Armstrong-Brown S. & Grice P.V. (2002) The role of research and development in the evolution of a 'smart' agri-environment scheme. *Aspects of Applied Biology*, 67, 253–264.
(4)   Moorcroft D., Whittingham M.J., Bradbury R.B. & Wilson J.D. (2002) The selection of stubble fields by wintering granivorous birds reflects vegetation cover and food abundance. *Journal of Applied Ecology*, 39, 535–547.
(5)   Critchley C., Allen D., Fowbert J., Mole A. & Gundrey A. (2004) Habitat establishment on arable land: assessment of an agri-environment scheme in England, UK. *Biological Conservation*, 119, 429–442.
(6)   Defra (2004) *Comparative Quality of Winter Food Sources for Cirl Bunting Delivered through Countryside Stewardship Special Project and CS Arable Options*. Defra BD1626.
(7)   Parish D.M.B. & Sotherton N.W. (2004) Game crops and threatened farmland songbirds in Scotland: a step towards halting population declines? *Bird Study*, 51, 107–112.
(8)   Butler S.J., Bradbury R.B. & Whittingham M.J. (2005) Stubble height affects the use of stubble fields by farmland birds. *Journal of Applied Ecology*, 42, 469–476.
(9)   Huusela-Veistola E. & Hyvanen T. (2006) Rotational fallows in support of functional biodiversity. *IOBC/WPRS Bulletin*, 29, 61–64.
(10)   Stevens D.K. & Bradbury R.B. (2006) Effects of the Arable Stewardship Pilot Scheme on breeding birds at field and farm-scales. *Agriculture, Ecosystems & Environment*, 112, 283–290.
(11)   Whittingham M.J., Devereux C.L., Evans A.D. & Bradbury R.B. (2006) Altering perceived predation risk and food availability: management prescriptions to benefit farmland birds on stubble fields. *Journal of Applied Ecology*, 43, 640–650.
(12)   Fisher G.P., MacDonald M.A. & Anderson G.Q.A. (2007) Do agri-environment measures for birds on arable land deliver for other taxa? *Aspects of Applied Biology*, 81, 213–219.
(13)   Roberts P.D. & Pullin A.S. (2007) *The Effectiveness of Land-based Schemes (including Agri-environment) at Conserving Farmland Bird Densities within the UK*. Systematic Review No. 11. Collaboration for Environmental Evidence/Centre for Evidence-Based Conservation, Birmingham, UK.
(14)   Bisang I., Bergamini A. & Lienhard L. (2009) Environmental-friendly farming in Switzerland is not hornwort-friendly. *Biological Conservation*, 142, 2104–2113.
(15)   Natural England (2009) *Agri-environment Schemes in England 2009: A Review of Results and Effectiveness*. Natural England, Peterborough.
(16)   Davey C., Vickery J., Boatman N., Chamberlain D., Parry H. & Siriwardena G. (2010) Regional variation in the efficacy of Entry Level Stewardship in England. *Agriculture Ecosystems & Environment*, 139, 121–128.
(17)   Davey C.M., Vickery J.A., Boatman N.D., Chamberlain D.E., Parry H.R. & Siriwardena G.M. (2010) Assessing the impact of Entry Level Stewardship on lowland farmland birds in England. *Ibis*, 152, 459–474.
(18)   Ewald J.A., Aebischer N.J., Richardson S.M., Grice P.V. & Cooke A.I. (2010) The effect of agri-environment schemes on grey partridges at the farm level in England. *Agriculture, Ecosystems & Environment*, 138, 55–63.
(19)   Field R.H., Morris A.J., Grice P.V. & Cooke A. (2010) The provision of winter bird food by the English Environmental Stewardship scheme. *Ibis*, 153, 14–26.
(20)   Field R.H., Morris A.J., Grice P.V. & Cooke A.I. (2010) Evaluating the English Higher Level Stewardship scheme for farmland birds. *Aspects of Applied Biology*, 100, 59–68.

Additional references
Evans A.D. (1997) Cirl buntings in Britain. *British Birds*, 90, 267–282.
Wilson S., Baylis M., Sherrott A. & Howe G. (2000) *Arable Stewardship Project Officer Review*. Farming and Rural Conservation Agency report.

Wotton S.R., Langston R.H.W., Gibbons D.W. & Pierce A.J. (2000) The status of the cirl bunting *Emberiza cirlus* in the UK and the Channel Islands in 1998. *Bird Study*, 47, 138–146.

ADAS (2001) *Ecological Evaluation of the Arable Stewardship Pilot Scheme, 1998–2000*. ADAS report.

Wotton S., Rylands K., Grice P., Smallshire S. & Gregory R. (2004) The status of the cirl bunting in Britain and the Channel Islands in 2003. *British Birds*, 97, 376–384.

Field R.H., Morris A.J., Grice P.V. & Cooke A. (In press) Winter use of seed-bearing crops by birds within the English Environmental Stewardship Scheme. *Ibis*.

2.5   Create beetle banks

- Fourteen reports from eight studies out of a total 24 reports from 12 individual studies (including 8 replicated studies, of which 3 were controlled and 4 literature reviews) from Denmark and the UK found that beetle banks provide some benefits to farmland biodiversity.

- Sixteen reports from eight individual studies looked at invertebrates and beetle banks. Five reports from two replicated studies (of which one was controlled) and a review found positive effects on invertebrate densities/numbers[1, 2, 8], distribution[10], or higher ground beetle density and species diversity in spring and summer but not winter[17]. Six reports from three replicated studies (of which one was randomized and controlled) found that invertebrate numbers varied between specific grass species sown on beetle banks[2, 4-6, 20, 21]. Two replicated studies (one paired and controlled) found that the effect of beetle banks varied between invertebrate groups or families[9, 14]. Five replicated studies (of which two were controlled) found lower or no difference in invertebrate densities or numbers on beetle banks relative to other habitats[3, 5, 6, 11, 13]. One review found lesser marsh grasshopper did not forage on two plant species commonly sown in beetle banks[23].

- Six studies looked at birds and beetle banks. Two reviews and one replicated controlled trial found positive effects on bird numbers (in combination with other farmland conservation measures[7, 16]) or evidence that birds used beetle banks[23]. Two studies (one replicated site comparison) found mixed effects on birds[15, 24]. One replicated study found no farmland bird species were associated with beetle banks[22].

- One replicated, paired, controlled study and a review looked at the effects of beetle banks on plants and found either lower plant species richness on beetle banks in summer[18], or that grass margins including beetle banks were generally beneficial to plants but these effects were not pronounced on beetle banks[12].

- One controlled study and a review found beetle banks acted as nest sites for harvest mice[19, 23].

Background

Beetle banks are grassy mounds, about 2 m wide, that run across the middle of large arable fields. They may be created using two-directional ploughing and sown with a mix of grass species (HGCA 2008). They are intended to provide habitat, especially during winter, for predatory insects such as beetles and spiders. They may also provide foraging habitats for birds and habitat for small mammals.

HGCA (2008) *Beneficials on Farmland: Identification and Management Guidelines*. ADHB-HGCA, London.

A replicated, randomized study in spring 1988–1990 on one beetle bank on a mixed farm in north Hampshire, UK (1) found weak evidence for a shift of predatory invertebrate activity from the beetle bank into the wheat crop over time. Individuals of the ground beetle (Carabidae) *Demetrias atricapillus* were more abundant on or very near the beetle bank in the first half of the study period (on average 12.2 individuals/m^2 at 0–3 m distance from the beetle bank 14 April–3 May 1989) after which they were more evenly distributed (on average 0.4/m^2 at 0–60 m from the beetle bank 8–22 May 1989) in 1989–1990. There was no consistent pattern in the distribution of the rove beetle (Staphylinidae) *Tachyporus hypnorum* in 1989–1990, although lower numbers were found on the beetle bank than in the crop by the end of the study in 1989. Money spiders (Linyphiidae) were more abundant on the beetle bank than in the crop, and significantly so for all but one sample date in 1989. There was evidence of some dispersal of money spiders and limited crop invasion by wolf spiders (Lycosidae) in spring 1990. The beetle bank was created in autumn 1986 through two-directional ploughing, it was 0.4 m high, 1.5 m wide and 290 m long. The bank crossed a 7 ha field with chalky-flint soil. The crop during the study was winter wheat in all years. The beetle bank was treated with a broad-spectrum herbicide in spring 1987 to remove broad-leaved herbs before the different treatments were hand-sown. Six replicates of each treatment (four single grass species, two mixes of three or four of the grass species, and bare ground) were created. Predation pressure was studied by placing dishes of prey at different distances along transects running from the beetle bank out into the crop (in 1988: 0, 1, 5 and 15 m, in 1989: 0, 3, 10, 30 and 60 m). The number of prey items remaining after 24 h was recorded. Dishes were active over one 24-h-period/week for 7 weeks. Dispersal into the crop was studied by taking weekly vacuum-net samples along five of the same cock's-foot *Dactylis glomerata* transects as above in 1989 (this part of the study is also reported in (2)). In 1990, 10 perpendicular transects of barrier pitfall traps were placed at regular intervals along the beetle bank, avoiding non-grass treatments. Traps were placed 1, 4, 20 and 50 m into the crop, and set for one 3-day-period each week and then emptied weekly throughout April and May. Vacuum-net samples were also collected in 1990 from five transects adjacent to five cock's-foot plots but at the same distances as the barrier pitfall traps. This study was part of the same experimental set-up as (2–4, 21).

A replicated, randomized, controlled study in the two winters of 1987–1989 at a mixed/arable farm in north Hampshire, UK (2) found that two beetle banks sown with four different grass species (creeping bent *Agrostis stolonifera*, cock's-foot *Dactylis glomerata*, Yorkshire fog *Holcus lanatus* and perennial rye grass *Lolium perenne*) produced densities of polyphagous invertebrate predators (invertebrates that feed on many different food sources) of up to 150 individuals/m^2 in the first winter and over 1,500/m^2 in the second winter. In the first winter, on bank 1 (in a 7 ha field) creeping bent held fewer predators (39/m^2) than the other 3 grass species (66–102/m^2), and similar numbers to bare ground (30/m^2). In the second winter, Yorkshire fog held more predatory invertebrates (648–1,398/m^2) than creeping bent (273–488/m^2) and perennial rye grass (276–394/m^2) on both banks as well as cock's-foot (218/m^2) on bank 2 (in a 20 ha field) but not on bank 1 (cock's-foot: 1,488/m^2). In comparison, densities in the field were much lower (26–29/m^2). In the second winter the two most abundant species were the ground beetle (Carabidae) *Demetrias atriacapillus* and the rove beetle (Staphylinidae) *Tachyporus hypnorum*. In spring 1989 *D. atriacapillus* occurred in higher numbers on or immediately adjacent to the banks up until 3 May (average density 12/m^2 at 0–3 m). After this date the distribution of this beetle throughout the field was more even (0.4/m^2 at 0–60 m). Significantly higher abundances of *T. hypnorum* occurred at 0 and 60 m into the field up until 18 April 1989 after which there were no consistent spatial patterns for this species, although there were lower numbers on the banks than in the field at the end of the study (22 May). The beetle banks were created in cereal fields on chalky-flint soil in autumn 1986 and treated with a broad-spectrum herbicide prior to hand-sowing in spring 1987. Six replicates of each treatment (four single grass species, two mixes of three or four of the grass species, and bare ground) were created. Predator communities were studied (November–February) through ground-zone searching in quadrats and destructive sampling (digging up turf samples) in the banks as well as mid-field. Crop penetration by emigrating predators was studied (once a week April–late May 1989) through transects of vacuum-net sampling at 0, 3 10, 30 and 60 m distance perpendicular to the cock's-foot treatments on bank 1. This study was part of the same experimental set-up as (1, 3, 4, 21).

A replicated, randomized, controlled study in the three winters from 1987 to 1990 on two farms in Hampshire, UK (3) (part of the same study as (2) but extended with a third winter and a third beetle bank in a 51 ha field, also on chalky-flint soil, on a second farm) found that three years after beetle bank establishment, total predator densities and both ground beetle (Carabidae) and spider (Araneae) community compositions were not different to those in natural field boundaries. The tussock-forming grass cock's-foot *Dactylis glomerata* supported highest densities of ground beetles on all three beetle banks in the third winter. Community composition of both ground beetles and spiders changed significantly throughout the study in favour of species that prefer boundary or more permanent habitats. See (2) for methods of beetle bank creation, experimental design and methods of predator sampling. This study was part of the same experimental set-up as (1, 2, 4, 21).

A replicated study on one beetle bank on a mixed farm in north Hampshire, UK (4) found the densities of both ground beetles (Carabidae) and rove beetles (Staphylinidae) in four grass treatments showed two peaks in density over the study period (the seven winters of 1987–1988 to 1993–1994) in the second and sixth winter after establishment. The pattern was the same for spiders (Araneae) in cock's-foot *Dactylis glomerata*, but in Yorkshire fog *Holcus lanatus*, creeping bent *Agrostis stolonifera* and perennial rye grass *Lolium perenne* the densities steadily increased and peaked in the fifth winter. Ground beetle densities over the seven year period in the different grass plots were as follows: cock's-foot 11–110 individuals/m^2; creeping bent 3–15/m^2; perennial rye grass 2–11/m^2 (only 5 winters); Yorkshire fog 1–76/m^2. The respective rove beetle densities over the 7 (or 5) winters were: cock's-foot 1–125 individuals/m^2; creeping bent 0–67/m^2; perennial rye grass 2–79/m^2; Yorkshire fog 2–113/m^2. Cock's-foot and Yorkshire fog generally had the highest densities of predators but not always significantly so. The grass species composition in plots sown with cock's-foot, Yorkshire fog and creeping bent remained relatively similar (min. 85% of original grass species left) during the study. Plots of false oat-grass *Arrhenatherum elatius* and red fescue *Festuca rubra* were created and added to the study in 1991. The 290 m-long beetle bank was created in spring 1987 and split into six blocks, each further sub-divided into eight plots with one treatment/plot. The eight treatments were sown cock's-foot (3 g/m^2), sown Yorkshire fog (4 g/m^2), sown perennial rye grass (3 g/m^2), sown creeping bent (8 g/m^2), mix of 3 grass species (cock's-foot, Yorkshire fog, perennial rye grass), mix of 4 grass species (previous 3 species plus creeping bent), bare ground, and sown flowering plants to provide pollen and nectar resources. Predatory invertebrates were sampled by taking two 20 x 20 x 10 cm turf samples/plot/winter. Percentage cover of grasses was measured in the four original grass treatments in October 1992 by placing six 25 x 25 cm quadrats in each grass plot, and in winter 1993–1994 it was measured in the collected turf samples. For methods in the first three winters see (1). This study was part of the same experimental set-up as (1–3, 21).

A replicated study in the winters of 1993–1994 to 1995–1996 on a lowland arable estate in Leicestershire, UK (5) (this study was continued in (20)) found that the average total predator, ground beetle (Carabidae) and rove beetle (Staphylinidae) (excluding aphid-specific species) density was higher in one hedge than one beetle bank over three winters. Out of five different grass species and areas of naturally regenerated vegetation, false oat-grass *Arrhenatherum ela-tius*, cock's-foot *Dactylis glomerata* and timothy *Phleum pratense* held the highest densities of total predators, ground beetles and rove beetles on two other beetle banks. Beetle banks were 360–400 m long, 2–2.5 m wide and sown in 1992–1993. Invertebrates were collected from soil samples using a cylindrical borer. This study was part of the same experimental set-up as (14, 15, 19, 20).

A replicated study in the summers of 1997–1998 in three regions (southern England, East Anglia and the Midlands) across the UK (6) found no difference in the average catch of saw-fly (Hymenoptera: Symphyta) larvae between beetle banks and grass strips planted along existing field margins. The total percentage cover of grass in planted grass strips affected the abundance of sawfly larvae positively. There were non-significant trends for sawfly larvae numbers to increase with strip age and to decrease with the amount of cock's-foot *Dactylis glomerata*. Numbers of gamebird chick-food insects increased with strip age and area, but there was also a significant difference between farms. There was a non-significant trend for chick-food insect numbers to increase with the proportion of red fescue *Festuca rubra*. Cock's-foot, red fescue and perennial rye grass *Lolium perenne* were the predominant grasses in most strips, being most common in 35, 25 and 17 strips respectively. A total of 116 grass strips (83 along pre-existing field margins and 33 beetle banks) on 32 farms were surveyed. For the overall analysis, the 11 strips on 3 of the farms were excluded. Grass strips had been established 0.5–12 years previously, both along pre-existing field margins and across cropped fields (beetle banks). Invertebrates were sampled by sweep-netting at the base of the veget-ation in mid-June to mid-July. Percentage cover of all plant species and vegetation height was measured in 0.25 m^2 quadrats. Apart from where stated, this study does not distinguish between the effects of creating beetle banks and planting grass buffer strips/margins around arable or pasture fields.

A 2000 literature review from the UK (7) found that populations of grey partridge *Perdix perdix* were 600% higher on farms with conservation measures aimed at partridges in place, compared to farms without these measures (Aebischer 1997). Measures included the provision of conservation headlands, planting cover crops, using set-aside and creating beetle banks.

A 2000 literature review (8) looked at which agricultural practices can be altered to benefit ground beetles (Carabidae). It found three studies, two in the UK (2, 5) and one in Denmark (a PhD thesis), showing higher ground beetle numbers in arable fields close to beetle banks.

A replicated, paired, controlled study in the two winters of 1997–1999 and summer 1999 on five farm estates in the UK (9) found different patterns of density and diversity for ground beetles (Carabidae), rove beetles (Staphylinidae) and spiders (Araneae) between five pairs of beetle banks and field margins in two consecutive winters. Rove beetle diversity was lower in beetle banks than in field margins in both winters, but density in beetle banks increased sig-nificantly between winters. There were no significant effects on ground beetles. The overall catch of chick-food invertebrates was lower in 22 beetle banks than in paired field margins on 5 farm estates, but the abundance of key prey groups was similar. There was no differ-ence in grasshopper and bushcricket (Orthoptera) species richness between the two habitats (on average 1.4 species in beetle banks, 1.8 in field margins), but older beetle banks held higher abundances of grasshoppers and bushcrickets. Both abundance and species richness of butterflies and moths (Lepidoptera) was significantly lower in beetle banks than in field margins in June, July and August, but both habitats peaked in July. Destructive turf samples were collected randomly from the two habitats to assess predatory invertebrates. Chick-food invertebrates and grasshoppers and bushcrickets were sampled through sweep-netting and butterflies and moths through standard transect walks. This study was part of the same experimental set-up as (10, 11, 17, 18).

A replicated, controlled study in 1998 in two sites with autumn-sown crops on an estate in Hampshire, UK (10) found that boundary-overwintering ground beetle (Carabidae) species (species that migrate into fields in spring) were clustered near two beetle banks and a hedgerow in the early part of the season (March), after which activity-densities were more evenly spread until they clustered again later in the summer (July). The distribution of field-inhabiting species (species resident in fields year-round) was fairly uniform or more associated with the centre of the fields through the early part of the season. The two sites differed in the latter part of the season with one displaying a gappy distribution near the beetle bank, and the other clustering near the hedgerow and the beetle bank. The distribution of overwintering ground beetles in January was irregular within the beetle banks and the hedgerow, but there was no apparent pattern in distribution of active beetles from February to July. Two sets of ten transects (connected, paired pitfall traps at 5, 25, 50, 75, 100 and 150 m into the crop) were set up at each site. At site A, transects extended at 10 m intervals into the winter barley crop at right angles from both sides of a beetle bank sown with cock's-foot *Dactylis glomerata*. At site B, transects extended into the crop from one side of a beetle bank sown with cock's-foot and from a hedgerow at the opposite side of the field, parallel to the bank, leaving a 50 m gap between traps at the furthest distance. Transects of pairs of un-connected pitfall traps were established within the beetle banks and the hedgerow. Pairs of traps were set at 10 m intervals and opened concurrently with the within-field traps for 72 h-periods March–July (A) or February–June (B). Fifteen 20 x 20 x 20 cm turf samples were removed from the beetle banks and the hedgerow in early January. This study was part of the same experimental set-up as (9, 11, 17, 18).

A replicated, controlled study in 1998–1999 (winter–summer) on 5 farm estates in Hampshire and Wiltshire, UK (11) found that diversity and average total abundance of chick-food invertebrates in sweep-net samples was higher in permanent field margins (65 individuals from 15 samples) than beetle banks (47 individuals from 15 samples) in 1999 and this was consistent between farms. In winter, the amount of plant litter, dead grass and tussocks that form important nesting material for game birds was higher in beetle banks (61%) than in field margins (27%) but overall vegetation cover in the two habitats was not different, and similar to that in summer (62–97%). Older beetle banks had higher diversity but not abundance of invertebrates. Invertebrate diversity also increased with plant diversity in both beetle banks and field margins. Invertebrate abundance and diversity was measured by vacuum suction sampling and sweep-netting. Vegetation cover and composition was assessed with 0.71 m² quadrats. Four to 22 banks/margins on 1 to 5 estates were included in the study in the 2 years. This study was part of the same experimental set-up as (9, 10, 17, 18).

A review (12) of two reports (Wilson *et al.* 2000, ADAS 2001) evaluating the effects of the Pilot Arable Stewardship Scheme in two regions (East Anglia and the West Midlands) in the UK, from 1998 to 2001 found that grass margins benefited plants, bumblebees *Bombus* spp., bugs (Hemiptera) and sawflies (Symphyta) but not ground beetles (Coleoptera). The grass margins set of options included sown grass margins, naturally regenerated margins, beetle banks and uncropped cultivated wildlife strips, but the review does not distinguish between these different options. None of the beneficial effects were pronounced on beetle banks. The effects of the pilot scheme on plants and invertebrates (bumblebees, true bugs, ground beetles, sawflies) were monitored over three years, relative to control areas. Grass margins were implemented on total areas of 361 and 294 ha in East Anglia and the West Midlands respectively.

A replicated study in June 2000 in ten edge habitats on a lowland arable farm in Leicester-shire, England (13) found that beetle banks contained the highest density of sawfly (Symphyta) larvae, significantly higher compared to hedge bottoms and winter wheat head-lands, but not compared to grass/wire fence lines or edges of un-grazed pasture. Spider (Araneae) and rove beetle (Staphylinidae) densities were lower in beetle banks than in un-grazed pastures. Set-aside contained a higher density of weevils (Curculionidae) than beetle banks. There was no difference in ground beetle (Carabidae) or caterpillar (Lepidoptera) densities between habitats. Type of neighbouring crop did not affect invertebrate densities in the different habitats. Apart from the six habitats mentioned above, brood cover, one- and two-year-old wild bird cover and sheep-grazed pasture edges were included in the study. Invertebrates were sampled with a vacuum suction sampler in June 2000.

A replicated study from 1995 to 1999 of arable habitats on a farm in Leicestershire, UK (14) found that the abundance of some invertebrate groups was higher in non-crop strips (grass beetle banks or wild bird cover), whereas other groups were more abundant in crops. Four in-vertebrate groups tended to have significantly higher densities in non-crop strips than crops in all years: spiders (Araneae) 7 vs 1–5 individuals/sample; true bugs (Homoptera) 29 vs 1–4; typical bugs (Heteroptera) 10–58 vs 0–9; and key 'chick food insects' 65 vs 2–10. In 3 of the years, true weevils (Curculionidae) were found at significantly higher densities in non-crop strips and beans (0–11) than other crops (0–2). In contrast, in 3 or 4 of the years, densities in crops were significantly higher than non-crops for: true flies (Diptera) 20–230 vs 25–100 individuals and aphids (Aphididae). Moth and butterfly larvae (Lepidoptera) and ground beetles (Carabidae) differed significantly in only one or two years, when density was higher in crops than non-crops. Total beetles (Coleoptera) varied between years and habitats. Saw-fly larvae (Symphyta), leaf beetles (Chrysomelidae) and soldier beetles (Cantharidae) showed no significant differences. Grass strips (1 m wide) planted as beetle banks were sown onto a raised bank along edges or across the centre of fields. Wild bird cover was sown as 2–5 m-wide strips along field boundaries and re-sown every few years with a cereal or kale-based mixture. Invertebrates were sampled each year in the centre of 5–11 grass/wild bird cover strips and 3 m into 3–4 pasture, 8–12 wheat, 6–8 barley, 3–6 oilseed rape and 4 field bean fields. Two samples of 0.5 m² were taken in each habitat using a D-Vac suction sampler in June 1995–1999. This study was part of the same experimental set-up as (5, 15, 19, 20).

A study of different set-aside crops on an arable farm in Leicestershire, UK (15) found that Eurasian skylark *Alauda arvensis*, but not yellowhammer *Emberiza citrinella*, used beetle banks more than expected compared to availability. Skylarks used beetle banks (planted tussocky perennial grasses) more than expected compared to availability and significantly more than unmanaged set-aside, broad-leaved crops and other habitats. Yellowhammer used beetle banks as expected compared to availability but significantly less than cereal and wild bird cover cereal set-aside. Field margin and midfield set-aside strips were sown with kale-based and cereal-based mixtures for beetle banks and wild bird cover. Other habitat types were: unmanaged set-aside, cereal (wheat, barley), broad-leaved crop (beans, rape) and other habitats. Thirteen skylark and 15 yellowhammer nests with chicks between 3–10 days old were observed. Foraging habitat used by the adults was recorded for 90 minutes during 3 periods of the day. This study was part of the same experimental set-up as (5, 14, 19, 20).

A small replicated, controlled study from May–June 1992–1998 in Leicestershire, UK (16) found that the abundance of nationally declining songbirds and bird species of conservation concern significantly increased on a 3 km² site where beetle banks were created (alongside several other interventions), although there was no overall difference in bird abundance, species richness or diversity between the experimental and 3 control sites. Numbers of nationally declining species rose by 102% (except for Eurasian skylark *Alauda arvensis* and yellowhammer *Emberiza citrinella*). Nationally stable species rose (insignificantly) by 47% (8 species increased; 4 decreased). The other interventions employed at the same site were managing hedges, wild bird cover strips, supplementary feeding, predator control and reducing chemical inputs generally.

A replicated, paired, controlled study on five conventional arable estates in Hampshire and Wiltshire, UK (17) found that ground beetle (Carabidae) population patterns and vegetation composition in beetle banks and field margins changed across seasons. In winter there was no difference in ground beetle density (range: about 200–300/m^2), species richness (range: 15–22 species in total) or diversity between beetle banks and field margins, but species richness increased with age in beetle banks. Ground beetle density and species diversity was higher in beetle banks than field margins in both spring and summer (beetle banks had on average about 75 individuals/m^2 in spring and about 90/m^2 in summer whilst field margins had about 45 and 60/m^2 in each season respectively). Only eight sites were included in the spring analysis. Ground beetle species composition was similar in the two habitats during winter and summer. The winter catches contained especially large proportions of *Bembidion lampros*. In spring the species composition was different with far fewer *B. lampros* and more larvae (not identified to species). Total plant cover was high in both habitats in both seasons but significantly higher in field margins during summer. However cover of tussocky grasses was higher in beetle banks in both seasons and did not decline with bank age. Field margins had higher species diversity in summer and higher species richness in both seasons compared with beetle banks. Both measures however increased with beetle bank age so that older banks had a similar number of species to margins. A total of 22 beetle banks were included in this study, ranging from <1 to 14 years old, each paired with a conventional permanent margin in the adjacent field. Ground beetle populations were sampled in four periods: winter (January–February), spring (May), summer (August) and winter (February) through destructive sampling (vacuum suction-sampling and digging up turf samples). Vegetation composition was investigated in winter (January–February) and summer (July) through quadrats placed on the ground. This study was part of the same experimental set-up as (9, 10, 11, 18).

A replicated, paired, controlled study in the summers of 1998–1999 and late winter 1998–1999 on five farm estates in southern UK (18) found lower summer plant species richness and diversity in beetle banks compared with conventional arable field margins. Both measures increased with age of beetle banks in summer. Beetle banks had higher cover of tussock-forming and other grasses, but lower cover of herbaceous, woody and nectar-providing plants in the second summer. In winter there was no difference in overall plant cover between beetle banks and field margins but average species richness was lower in beetle banks. Species richness increased with age in beetle banks. There was no relationship between tussock cover and age of bank in winter. Beetle banks, aged <1–14 years, were sown mainly with cock's-foot *Dactylis glomerata* and had received no, or little, active management since establishment. Percentage cover of all plant species was assessed in twenty 0.5 m^2 quadrats along each bank or margin in July 1998, January–February and July–August 1999. This study was part of the same experimental set-up as (9–11, 17).

A controlled study in autumn 1998 on a predominantly arable farm in Leicestershire, UK (19) found overall more harvest mouse *Micromys minutus* nests in beetle banks (117 nests/ha) than field margins (14 nests/ha) although this difference was not statistically tested. Beetle banks were created and sown with grasses such as cock's-foot *Dactylis glomerata* between September 1992 and 1994 and cut regularly in the year of establishment. Field margins were often adjacent to a hedgerow and normally left uncut. The two habitats were hand searched for harvest mouse nests in September to November, in a total of 1.8 km of beetle banks and 9.8 km of field margins. This study was part of the same experimental set-up as (5, 14, 15, 20).

A replicated study in 1994–1998 including two beetle banks on an arable estate in Leicestershire, UK (20) (a continuation of (5)) found higher densities of invertebrate predators in false oat grass *Arrhenatherum elatius* ($2,045/m^2$) than in red fescue *Festuca rubra* ($1,492/m^2$), crested dog's-tail *Cynosurus cristatus* ($1,380/m^2$) and naturally regenerated vegetation ($1,060/m^2$). Rove beetles (Staphylinidae), were the dominant family in the predatory invertebrate catch, and showed the same significant pattern ($1,716/m^2$ in false oat grass, $1,241/m^2$ in red fescue, $1,105/m^2$ in crested dog's-tail and $834/m^2$ in naturally regenerated vegetation). Spider (Araneae) density was higher in cock's-foot ($177/m^2$) compared with red fescue ($119/m^2$) and naturally regenerated vegetation ($107/m^2$). Ground beetle (Carabidae) density was 2.5 to 3.5 times higher (significant) in cock's-foot than all other treatments. Boundary-type ground beetles dominated all treatments but were also higher in cock's-foot ($328/m^2$) compared with the other five treatments ($69–126/m^2$). In the first year of the study (third summer after creation) all single grass treatments were dominated by their sown species. In the last year of the study false oat-grass had the highest cover (90%) followed by red fescue (75%), cock's-foot and timothy (70%) and crested dog's-tail (10%). Overall, cock's-foot, false oat-grass and timothy were taller growing and formed denser grass coverage near ground level (0–30 cm) compared with the other treatments. Beetle banks were created in spring 1993, both situated in an 8.6 ha clay soil field. Six treatments (five species of grass and naturally regenerated vegetation) were established with two replicates/bank. Invertebrates were collected from soil samples gathered in January–February 1994–1997. Vegetation was examined visually as well as measured with a graduated board. This study was part of the same experimental set-up as (5, 14, 15, 19).

A randomized, replicated study over seven winters from 1987 to 1988 to 1993–1994 within one beetle bank on a mixed arable estate in Hampshire, UK (21) (an extension of (1)) found that ground beetle (Carabidae) and rove beetle (Staphylinidae) densities were in general highest in cock's-foot *Dactylis glomerata* and Yorkshire fog *Holcus lanatus* respectively, although this was not always significantly higher in comparison with creeping bent *Agrostis stolonifera* or perennial rye grass *Lolium perenne*. Densities of money spiders (Linyphiidae) and wolf spiders (Lycosidae) were also higher, although not always significantly, in these two tussock-forming grasses. The ground beetle species composition changed from dominance by open field species to boundary species over the course of the study. In the last three winters, when sampled, field boundaries had lower densities of predatory invertebrates than the beetle bank, but this was not statistically tested. Percentage cover of the grass species originally sown in plots remained high for all species, except perennial rye grass, plots of which were invaded by cock's-foot by the sixth winter and excluded from sampling in the last two winters because perennial rye grass had become so rare. One beetle bank was created through two-directional ploughing and divided into six blocks in which eight treatments were sown (only the four single grass species treatments were included in this study) in randomized order. Predator communities were sampled through ground-zone searching and destructive sampling in November–February. Vegetation composition was examined in quadrats in October 1992. This study was part of the same experimental set-up as (1–4).

A replicated study in 1999 and 2003 on 256 arable and pastoral fields across 84 farms in East Anglia and the West Midlands, UK (22) found that out of 12 farmland bird species, none were strongly associated (either positively or negatively) with beetle banks. The species analysed were skylark *Alauda arvensis*, corn bunting *Miliaria calandra*, lapwing *Vanellus van-*

ellus, yellow wagtail *Motacilla flava*, chaffinch *Fringilla coelebs*, dunnock *Prunella modularis*, greenfinch *Carduelis chloris*, Eurasian linnet *C. cannabina*, reed bunting *Emberiza schoeniclus*, tree sparrow *Passer montanus*, whitethroat *Sylvia communis* and yellowhammer *E. citrinella*.

A 2007 UK literature review (23) describes a study which found that beetle banks held higher densities of harvest mouse *Micromys minutus* nests than field margins. Other studies found that grey partridge *Perdix perdix* and Eurasian skylark *Alauda arvensis* also nested in beetle banks. Skylarks were found to be more likely than yellowhammer *Emberiza citrinella* to forage in beetle banks. However, a study in Leicestershire, UK, found that lesser marsh grasshoppers *Chorthippus albomarginatus* did not use two species of plant commonly planted in beetle banks (cock's-foot *Dactylis glomerata* and false oat grass *Arrhenatherum elatius*) as food plants.

A replicated site comparison study from 2004 to 2008 in England (24) found that grey partridge *Perdix perdix* overwinter survival was significantly and positively correlated with the presence of beetle banks in 2007–2008. Across all years there was a positive relationship with the ratio of young to old birds. There were no relationships between beetle banks and brood size or year-on-year density changes. Spring and autumn counts of grey partridge were made at 1,031 sites across England as part of the Partridge Count Scheme.

(1) Thomas M.B. (1991) Manipulation of overwintering habitats for invertebrate predators on farmland. Thesis. University of Southampton.
(2) Thomas M.B., Wratten S.D. & Sotherton N.W. (1991) Creation of island habitats in farmland to manipulate populations of beneficial arthropods – predator densities and emigration. *Journal of Applied Ecology*, 28, 906–917.
(3) Thomas M.B., Wratten S.D. & Sotherton N.W. (1992) Creation of island habitats in farmland to manipulate populations of beneficial arthropods – predator densities and species composition. *Journal of Applied Ecology*, 29, 524–531.
(4) MacLeod A. (1994) Provision of plant resources for beneficial arthropods in arable ecosystems. PhD thesis. University of Southampton.
(5) Collins K.L., Wilcox A., Chaney K. & Boatman N.D. (1996) Relationships between polyphagous predator density and overwintering habitat within arable field margins and beetle banks. *Proceedings of the Brighton Crop Protection Conference: Pests & Diseases*, Farnham. pp. 635–640.
(6) Barker A.M. & Reynolds C.J.M. (1999) The value of planted grass field margins as a habitat for sawflies and other chick-food insects. *Aspects of Applied Biology*, 54, 109–116.
(7) Aebischer N.J., Green R.E. & Evans A.D. (2000) From science to recovery: four case studies of how research has been translated into conservation action in the UK. In N.J. Aebischer, A.D. Evans, P.V. Grice & J.A. Vickery (eds.) *Ecology and Conservation of Lowland Farmland Birds*, British Ornithologists' Union, Tring. pp. 43–54.
(8) Holland J.M. & Luff M.L. (2000) The effects of agricultural practices on Carabidae in temperate agroecosystems. *Integrated Pest Management Reviews*, 5, 109–129.
(9) Thomas S.R., Goulson D. & Holland J.M. (2000) The contribution of beetle banks to farmland biodiversity. *Aspects of Applied Biology*, 62, 31–38.
(10) Thomas S.R. (2001) Assessing the value of beetle banks for enhancing farmland biodiversity. PhD thesis. University of Southampton.
(11) Thomas S.R., Goulson D. & Holland J.M. (2001) Resource provision for farmland gamebirds: the value of beetle banks. *Annals of Applied Biology*, 139, 111–118.
(12) Evans A.D., Armstrong-Brown S. & Grice P.V. (2002) The role of research and development in the evolution of a 'smart' agri-environment scheme. *Aspects of Applied Biology*, 67, 253–264.
(13) Moreby S.J. (2002) Permanent and temporary linear habitats as food sources for the young of farmland birds. In D. E. Chamberlain & A. Wilson (eds.) *Avian Landscape Ecology: Pure and Applied Issues in the Large-Scale Ecology of Birds*, International Association for Landscape Ecology (IALE(UK)), Aberdeen. pp. 327–332.
(14) Moreby S.J. & Southway S. (2002) Cropping and year effects on the availability of invertebrate groups important in the diet of nestling farmland birds. *Aspects of Applied Biology*, 67, 107–112.
(15) Murray K.A., Wilcox A. & Stoate C. (2002) A simultaneous assessment of farmland habitat use by breeding skylarks and yellowhammers. *Aspects of Applied Biology*, 67, 121–127.
(16) Stoate C. (2002) Multifunctional use of a natural resource on farmland: wild pheasant (*Phasianus colchicus*) management and the conservation of farmland passerines. *Biodiversity and Conservation*, 11, 561–573.

(17) Thomas S.R. (2002) The refuge role of beetle-banks and field margins for carabid beetles on UK arable farmland: densities, composition and relationships with vegetation. *Conference Proceedings – How to Protect What we Know about Carabid Beetles: From knowledge to application*. Tuczno, Poland. pp. 185–199.

(18) Thomas S.R., Noordhuis R., Holland J.M. & Goulson D. (2002) Botanical diversity of beetle banks: effects of age and comparison with conventional arable field margins in southern UK. *Agriculture, Ecosystems & Environment*, 93, 403–412.

(19) Bence S.L., Stander K. & Griffiths M. (2003) Habitat characteristics of harvest mouse nests on arable farmland. *Agriculture Ecosystems & Environment*, 99, 179–186.

(20) Collins K.L., Boatman N.D., Wilcox A. & Holland J.M. (2003) Effects of different grass treatments used to create overwintering habitat for predatory arthropods on arable farmland. *Agriculture, Ecosystems and Environment*, 96, 59–67.

(21) MacLeod A., Wratten S.D., Sotherton N.W. & Thomas M.B. (2004) 'Beetle banks' as refuges for beneficial arthropods in farmland: long-term changes in predator communities and habitat. *Agricultural and Forest Entomology*, 6, 147–154.

(22) Stevens D.K. & Bradbury R.B. (2006) Effects of the Arable Stewardship Pilot Scheme on breeding birds at field and farm-scales. *Agriculture, Ecosystems & Environment*, 112, 283–290.

(23) Stoate C. & Moorcroft D. (2007) Research-based conservation at the farm scale: development and assessment of agri-environment scheme options. *Aspects of Applied Biology*, 81, 161–168.

(24) Ewald J.A., Aebischer N.J., Richardson S.M., Grice P.V. & Cooke A.I. (2010) The effect of agri-environment schemes on grey partridges at the farm level in England. *Agriculture, Ecosystems & Environment*, 138, 55–63.

Additional references

Aebischer N.J. (1997) Gamebirds: management of the grey partridge in Britain. In M. Bolton (ed.) *Conservation and the Use of Wildlife Resources*. Chapman & Hall, London. pp. 131–151.

Wilson S., Baylis M., Sherrott A. & Howe G. (2000) *Arable Stewardship Project Officer Review*. Farming and Rural Conservation Agency report.

ADAS (2001) *Ecological Evaluation of the Arable Stewardship Pilot Scheme, 1998–2000*. ADAS report.

2.6   Plant nettle strips

* A small study from Belgium found that planting nettle strips in the margins of three arable fields resulted in a higher number of aphid predator species[1]. The number of aphid predators on a natural patch of nettles was higher than on crops, however there were fewer predators on nettle strips than on crops. Three insect families, including green lacewings, were only found on nettles.

Background

In agricultural landscapes, field margins can provide valuable habitats for maintaining and enhancing biodiversity. The common nettle *Urtica dioica* is a perennial species that provides food for a range of insects (Greig-Smith 1948), including 'beneficial' species that help to suppress pests. This intervention involves planting nettle strips, which may enhance invertebrate diversity on arable fields.

Greig-Smith P.W. (1948) Biological flora of the British Isles. *Journal of Ecology*, 36, 343–351.

A small study in 2005 of six nettle *Urtica dioica* strips planted in the margins of three arable crops in Belgium (1) found that planting nettles resulted in a higher number of aphid predator species. Numbers of aphid predator species tended to be higher on nettles (nettle strips: 6–9 species; natural nettle stand: 15) than crop plants (0–4 species). Predator abundance was significantly greater on the natural nettle stand (89 individuals) than wheat and pea crops (predators: 17–20). However predator abundance was lowest on nettle plots (predators: 6–14). Nine ladybird (Coccinellidae) species were observed on nettle, compared to six on crops. Ladybird and hoverfly (Syrphidae) abundance was highest on the natural nettle stand (62 ladybirds and 7 hoverflies), followed by wheat and pea crops (5–19 and 3–5 respectively) and lowest on nettle plots (3–9 and 0–1). Predatory minute pirate bugs (Anthocoridae), plant bugs (Miridae) and green lacewings (Chrysopidae) were only observed on nettle. Nettle strips were planted in two plots (10 x 20 m) in the margins of three arable fields, a nearby large natural nettle stand (1,000 m²) in a natural reserve was also sampled. Ten plants/plot were randomly selected each week to count and identify all aphid predator populations (May–August 2005). Larvae were collected and reared until emerged adults could be identified. Aphid data are not presented here.

(1) Alhmedi A., Haubruge E., Bodson B. & Francis F.R. (2007) Aphidophagous guilds on nettle (*Urtica dioica*) strips close to fields of green pea, rape and wheat. *Insect Science*, 14, 419–424.

2.7   Leave unharvested cereal headlands in arable fields

- We have captured no evidence for the effects on farmland wildlife of leaving unharvested cereal headlands.

Background
Unharvested cereal headlands are strips of cereal crop around the edge of arable fields that are left unharvested throughout the winter. In addition, they are often treated less intensively with few fertilizers and no broadleaved herbicides.

2.8   Leave cultivated, uncropped margins or plots (includes 'lapwing plots')

- Nineteen individual studies looked at the effect of uncropped, cultivated margins or plots on wildlife. Seventeen studies from the UK and northwest Europe (six reviews and seven replicated studies, of which two were site comparisons, one a before-and-after trial and one controlled and randomized) found that leaving uncropped, cultivated margins or plots on farmland provided benefits to some or all target farmland bird species[2, 7, 9, 11, 13, 17, 18, 20], plants[3, 4, 6, 8, 10, 12, 14, 15, 18], invertebrates[1–3, 5] and mammals[2]. These wildlife benefits included increased species richness of plants[4, 6, 8, 15], bumblebees[5], species richness and abundance of spiders[2], abundance of ground-dwelling invertebrates[2] and ground beetles[1,2], increased stone curlew breeding population size[9, 18], northern lapwing hatching success[11], Eurasian skylark nesting success[13] and the establishment, abundance or species richness of rare arable plant species[10, 12, 14, 15, 18]. A replicated study found that northern lapwing, Eurasian skylark, grey partridge and yellow wagtail bred in lapwing plots[17].

- Two studies (a replicated study and a review) from the UK found that leaving uncropped, cultivated margins or plots on farmland had no effect on 11 out of 12 farmland bird species[7] or ground beetles[3]. A replicated site comparison study in the UK found fewer seed-eating birds on fallow plots for ground-nesting birds in two out of three regions[20]. One review from the UK found evidence that pernicious weeds were more commonly found on uncropped, cultivated margins than conservation or conventional headlands[16]. A replicated site comparison study from the UK found that the proportion of young grey partridges in the population was lower in areas with a high proportion of uncropped, cultivated margins and plots[19].

Background

This intervention can be introduced for ground-nesting birds (sometimes called 'lapwing plots'), rare arable plants, or both. It may also provide habitat for insects and foraging sites for mammals and seed- and insect-eating birds (Campaign for the Farmed Environment 2011).

Plots or strips are cultivated, but left undrilled. For ground-nesting birds, the plots are usually at least 2 ha in size. They are different from 'skylark plots', which are much smaller and usually created in groups. If this measure is taken in field margins (6 m strips at the edge of arable fields) no fertilizer is applied and herbicide applications are minimal, with only spot treatment of particular weeds permitted.

See also 'All farming systems: Provide or retain set-aside areas in farmland', for rotational or long-term fallow land at the field scale not cultivated or implemented specifically for ground-nesting birds.

See also 'Sow rare or declining arable weeds' for a study of different management options that could also apply to uncropped, cultivated margins.

Campaign for the Farmed Environment (2011) *Guide to Voluntary Measures 2011 Edition*. Campaign for the Farmed Environment, Warwickshire.

A small study of the margin of an arable field in the Breckland Environmentally Sensitive Area, in the east of England (1) found that the uncropped margin supported more adult ground beetles (Carabidae) than the cropped margin or the crop, and more larvae (uncropped: 38 larvae/trap; cropped: 12 larvae/trap). The ground beetle *Bembidion lampros* was significantly more abundant in the 6 m uncropped margin (reduced pesticides) than the cropped margin (fully sprayed) or crop, and tended to move into the crop. Catches of *Pterostichus melanarius* were consistently higher in the crop than the uncropped and the cropped margin. *Agonum dorsale* abundance was lowest in the uncropped margin, and tended to move from field boundaries into the crop. The uncropped margin had significantly less vegetation than the cropped margin. The field margin was divided into two blocks, each with both treatments (120 m long). Ground beetles were sampled with 5 pitfall traps in each plot (20 m apart), 32 m into the adjacent crop and field boundary block. Directional traps, an 'H' shape (2 m long) barrier with 5 pitfalls on each side, were constructed to investigate movement at the field boundary-margin and margin-crop interface in each replicate strip. Traps were emptied weekly from April–August 1991.

A 1999 review of research into uncropped strips in northwest Europe (2) found that biodiversity was enhanced by establishing uncropped strips. Two studies found that ground-dwelling invertebrates were more abundant in uncropped strips than unsprayed cereal strips (Hawthorne & Hassall 1994, White & Hassall 1994). Another reported that ground beetles (Carabidae) were more abundant in the uncropped strip and adjacent crop than in the crop adjacent to sprayed and unsprayed crop strips (Cardwell *et al.* 1994). Spider (Araneae) species richness and abundance was also reported to be higher in uncropped strips than unsprayed cereal strips by one study (White & Hassall 1994). An additional study found positive results for gamebirds, songbirds and hares *Lepus* spp. (Anon 1990).

A 2002 review (3) of two reports (Wilson *et al.* 2000, ADAS 2001) evaluating the effects of the Pilot Arable Stewardship Scheme in two regions (East Anglia and the West Midlands) in the UK from 1998 to 2001 found that grass margins benefited plants, bumblebees *Bombus* spp., bugs (Hemiptera) and sawflies (Symphyta) but not ground beetles (Coleoptera). The grass margins set of options included uncropped cultivated wildlife strips, sown grass margins, naturally regenerated margins and beetle banks. The review does not distinguish between these options, although the beneficial effects were particularly pronounced on uncropped cultivated wildlife strips for all four groups. The effects of the pilot scheme on plants and invertebrates were monitored over three years, relative to control areas. Grass margins were implemented on total areas of 361 and 294 ha in East Anglia and the West Midlands respectively.

A replicated study in the summers of 1999–2000 comparing ten different conservation measures on arable farms in the UK (4) found that uncropped, cultivated margins appeared to be one of the three best options for conservation of annual herbaceous plant communities. Wildlife seed mix (largely sown for birds) and no fertilizer conservation headlands were the other two options. Uncropped, cultivated margins were dominated by annual plant species. Of the ten measures, they had the highest numbers of annual and herbaceous plant species, unsown crops (crop volunteers), bare ground and litter, and the lowest cover and species richness of grasses. Cultivated spring fallows had fewer plant species than cultivated margins, but relatively high total plant cover, and over 50% cover of monocotyledonous plants (mainly grasses). The average numbers of plant species in the different conservation habitats were uncropped cultivated margins 6.3; wildlife seed mixtures 6.7; undersown cereals 5.9; naturally regenerated grass margins 5.5; no-fertilizer conservation headlands 4.8; spring fallows 4.5; sown grass margins 4.4; overwinter stubbles 4.2; conservation headlands 3.5; grass leys 3.1. Plants were surveyed on a total of 294 conservation measure sites (each a single field, block of field or field margin strip) on 37 farms in East Anglia (dominated by arable farming) and 38 farms in the West Midlands (dominated by more mixed farming). The ten habitats were created according to agri-environment scheme guidelines. Vegetation was surveyed once in each site in June–August in 1999 or 2000, in thirty 0.25 m² quadrats randomly placed in 50–100 m randomly located sampling zones in each habitat site. All vascular plant species rooted in each quadrat, bare ground or litter and plant cover were recorded.

A controlled trial on paired sites in 2003 on Arable Stewardship Pilot Scheme farmland in the UK (5) found that bumblebee *Bombus* spp. foraging activity and species richness were significantly enhanced on uncropped, regularly cultivated field margins where natural regeneration had been allowed to take place for five years, compared to sites of conventionally managed cereal. The uncropped margins had significantly more plant species than either conservation headlands or uncropped margins sown with a wildflower seed mix. However, two species considered to be pernicious weeds, spear thistle *Cirsium vulgare* and creeping thistle *C. arvense*, were key forage plants for the bumblebees, so this option may lead to conflict between agricultural and conservation objectives. Bumblebee numbers were estimated through paired surveys on field margins and conventionally-managed cereal field margins. Foraging bumblebees were recorded along 100 x 6 m transects and the plant species on which bumblebees were observed feeding was noted. Twenty 0.5 x 0.5 m quadrats were used along the bumblebee transects to record the presence of all plant species.

A replicated before-and-after trial in 1997–2000 on cultivated headlands in arable fields at three sites in Suffolk, Hampshire and North Yorkshire, UK (6) found that plant species richness increased when headlands were left uncropped with no inputs. In July 1997, before cropping ceased, the 3 sites had 33, 70 and 19 plant species respectively. When uncropped, the number of species found each year, over the 3 years of the trial, increased to 75–85, 93–94 and 55–59 at the 3 sites respectively. Although the main components of the vegetation were target annual and broadleaved plants, there was also an increase in perennial plants (from 1–3% to 27–40% cover) and monocotyledons (mainly grasses) (1–10% to 18–31%), and the authors note that these may need to be controlled. Treatments were replicated three times (in 6 x 6 m plots) at each site. Plants were surveyed each July in 32 quadrats (0.5 x 0.5 m) in each plot.

A replicated study in 1999 and 2003 on 256 arable and pastoral fields across 84 farms in East Anglia and the West Midlands, UK (7) found that only 1 out of 12 farmland bird species, reed bunting *Emberiza schoeniclus*, was strongly and positively associated with uncropped, cultivated strips. No other species showed a strong association (positive or negative) with the strips.

A replicated site comparison study in the UK (8) found that uncropped, cultivated margins had more plant species than other field margin types, and increased plant species richness over time in one (but not all) areas. In a national survey of field margins under the Countryside Stewardship Scheme, 39 uncropped regularly cultivated field margins had more plant species (31 species/margin) than 72 control margins (8 species/margin) and 78 conservation headlands (11–17). Thirty-nine margins that were uncropped and cultivated for 1 year (called 'spring fallow') had 20 plant species on average. In the pilot Arable Stewardship Scheme in 2 English regions (East Anglia and the West Midlands), 24 uncropped cultivated strips had greater numbers of perennial plants and pernicious weeds after 4 years (measured in 1999 and 2003), but the total number of species did not increase (7–8 plant species/margin). By contrast, there was a substantial increase in number of plant species in 32 uncropped cultivated margins in the Brecklands Environmentally Sensitive Area (ESA) between 1996 (12 species/margin) and 2004 (18 species/margin). Here the number of pernicious weed species did not increase. Plants were surveyed in either thirty 0.025 m^2 quadrats within a 100 m sampling zone or twenty 10 x 10 cm quadrats (only used in Brecklands ESA). Percentage cover and plant species were recorded in each quadrat.

A 2007 review of a Countryside Stewardship Scheme in southern England (9) found that the population of Eurasian thick-knees (stone curlew) *Burhinus oedicnemus* increased from 71 breeding pairs in 2000 to 103 in 2005, following the creation of 156 stone curlew plots over the study period typically located close (<1 km) to pasture, pig farms or other food sources and away from edges of fields. A further 51 plots were created in 2006 under Higher Level Stewardship. The UK stone curlew population increased from 160 pairs in the 1980s to 300 pairs in 2005. Stone curlew plots consisted of 1–2 ha of arable or set aside land cultivated to create a 'rough fallow' in spring.

A 2007 review of published and unpublished literature (10) found experimental evidence of benefits of fallow plots to plants, from one study on Salisbury Plain Training Area, UK (Walker *et al.* 2001). Stone curlew (Eurasian thick-knees *Burhinus oedicnemus*) plots, not sprayed with herbicide, hosted rare arable weed species including dense-flowered fumitory *Fumaria densiflora* and red hemp nettle *Galeopsis angustifolia*.

A replicated, controlled study in the breeding seasons of 1999–2000 on 28 farms in western England (11) found that 85% of 34 northern lapwing *Vanellus vanellus* nests successfully hatched at least 1 chick on fields with cultivated 'lapwing plots', compared to 64% of 154 nests on all other fields types. Nest survival estimates were also significantly higher (99% daily survival vs 95–96% on spring cereals, stubbles and grass habitats), and no nests were lost to agricultural operations, compared to over 50% in other fields.

At Ranscombe Farm, a nature reserve managed for arable plants in the north Kent Downs, UK (12), two to three kilometres of uncropped cultivated margins yielded populations of one or two species of rare arable plants in the first year of establishment. Two kilometres of margins established in autumn 2004 grew populations of hairy mallow *Althaea hirsuta* and broad-leaved cudweed *Filago pyramidata* in 2005. Three kilometres of margins established in spring 2006 supported approximately 10,000 broad-leaved cudweed plants, the second largest population in the UK.

A study in 2003–2005 in Cambridgeshire, UK (13) found that the nesting success of Eurasian skylark *Alauda arvensis* was significantly higher in a field that was fallowed after harvest, compared to in cereal crop fields (84% success in the fallow field vs 35%), whilst the number of nests in the field increased from two to eight following the fallow. Overwinter counts of yellowhammer *Emberiza citrinella*, reed bunting *E. schoeniclus*, linnet *Carduelis cannabina* and skylark on the fallow field were also far higher than in previous years.

A replicated, controlled, randomized site comparison study of field margins at 39 sites in England (14) (same study as (15)) found that uncropped cultivated margins significantly increased rare arable plants. Uncropped cultivated margins had significantly higher numbers of rare arable plants (1.4/sample zone) than conservation headlands (0.1), no-fertilizer conservation headlands (0.7), spring fallow (0.6) and the crop (0.1). A total of 145 records of 34 rare arable plants were found on the 195 field margins, including 4 UK Biodiversity Action Plan species. Three species occurred on 7–10% of sites, a further 19 occurred on 1–5% and the remainder were found on just one margin. In total 25 rare arable plants were recorded on uncropped margins, 18 on no-fertilizer conservation headlands and 16 on spring fallow. There were no significant differences in rare arable plant diversity at 1, 3 or 5 m from the field edge within margin types. There were significant regional differences in diversity. One of each margin type and an adjacent control was randomly selected in thirty-nine 20 x 20 km squares in England. Rare arable plants were sampled in 10 quadrats (0.5 x 0.5 m) at 3 distances (1, 3 and 5 m) from the field edge within a 100 x 6 m sample zone in June–July 2005.

A replicated, controlled, randomized site comparison study in 2005 of Countryside Stewardship Scheme field margin options across England (15) (same study as (14)) found more arable plant species on uncropped cultivated margins (7.5 species on average) than on 'spring fallow' plots (4.3 species), conservation headlands (2.4–4.1 species) or cereal crop control (1.4 species). Thirty-four rare arable plant species were recorded, only 12 of which were found in over 2% of sites. Uncropped margins had significantly more rare species (1.4 species/margin on average) than the other 3 options (0.2–0.8 species/margin). A total of 39 randomly selected 20 x 20 km squares throughout England were visited to sample: uncropped cultivated margins, spring fallow (cultivation of stubble in whole/part field) and conservation headlands with and without fertilizer. A conventionally managed cereal crop (control) was also sampled at each of the farms visited. A total of 195 field margin agreements were surveyed during June and July 2005. All plant species and 86 rare arable plants were investigated.

A 2008 review of control methods for competitive weeds in uncropped cultivated margins managed to maintain uncommon arable plant populations in the UK (16) found that specific management regimes can reduce abundance of pernicious weeds in margins. One study found pernicious weeds were more likely in uncropped cultivated margins than in conservation or conventional headlands (Critchley et al. 2004). Abundance of perennial plants tended to increase if uncropped cultivated margins were not cultivated annually in two studies (Critchley 1996b, Critchley 2000). However five studies found weeds also build up on margins cultivated annually, particularly with the same annual cultivation regime (Critchley 1996a,b, Critchley et al. 2004, (6), Still & Byfield, 2007). One study found cutting twice in spring decreased annual broadleaved plants in uncropped cultivated margins (Marshall 1998).

A replicated study in 2007 (17) found that northern lapwing *Vanellus vanellus* used 39% of 212 lapwing plots on 180 farms across England, with breeding suspected on 25% of plots. In addition, Eurasian skylark *Alauda arvensis*, grey partridge *Perdix perdix* and yellow wagtail *Motacilla flava* were recorded breeding in 73%, 17% and 6% of plots respectively. There were no significant differences in lapwing occurrence or breeding in plots managed under Higher Level Stewardship compared with those under the Countryside Stewardship Scheme.

Lapwing occurrence decreased if there was woodland adjacent, and the probability of breeding increased with the proportion of bare ground present on plots. Skylarks were less likely to be found on plots near hedgerows.

A 2009 literature review of agri-environment schemes in England (18) found that spring and summer fallows provided nesting habitats for northern lapwing *Vanellus vanellus*, with 40% of fallow plots used by lapwings and breeding suspected on 25% plots (17). In addition, the number of breeding pairs of Eurasian thick-knee (stone curlew) *Burhinus oedicnemus* in southern England increased from 63 in 1997 to 103 in 2005 following the implementation of a Countryside Stewardship Scheme 'special project' which included the provision of fallow plots. One study (15) found that 264 plant species typically found in disturbed or arable habitats, including 34 rare and uncommon arable plants, were recorded in 3 agri-environment scheme options: uncropped cultivated margins (highest diversity); spring fallow; conservation headlands (lowest diversity).

A replicated site comparison study from 2004 to 2008 in England (19) found a lower proportion of young grey partridges *Perdix perdix* in the population in 2007 on sites with a high proportion of uncropped cultivated margins and plots. There were no significant relationships with changes in partridge density, brood size or overwinter survival. Spring and autumn counts of grey partridge were made at 1,031 sites across England as part of the Partridge Count Scheme.

A replicated site comparison study in 2008 and 2009 on farms in three regions in England (20) found that in two of the three regions, Higher Level Stewardship fallow plots for ground-nesting birds had significantly fewer seed-eating farmland songbirds than conventional crop fields during summer. On farms in East Anglia and the Cotswolds, there were approximately 2.5 birds/ha on crops compared to 1 bird/ha on fallow plots. However, in a third region, the West Midlands, more seed-eating farmland birds were recorded on fallow plots than in crop fields (1.5 birds/ha on fallow plots compared to <0.5 birds/ha on crops). The group of birds analysed included tree sparrow *Passer montanus* and corn bunting *Emberiza calandra*, but not grey partridge *Perdix perdix*. Surveys were carried out in the summers of 2008 and 2009, on 69 farms with Higher Level Stewardship in East Anglia, the West Midlands or the Cotswolds and on 31 farms across all 3 regions with no environmental stewardship.

(1) Hawthorne A.J., Hassall M. & Sotherton N.W. (1998) Effects of cereal headland treatments on the abundance and movements of three species of carabid beetles. *Applied Soil Ecology*, 9, 417–422.
(2) de Snoo G.R. & Chaney K. (1999) Unsprayed field margins – what are we trying to achieve? *Aspects of Applied Biology*, 54, 1–12.
(3) Evans A.D., Armstrong-Brown S. & Grice P.V. (2002) The role of research and development in the evolution of a 'smart' agri-environment scheme. *Aspects of Applied Biology*, 67, 253–264.
(4) Critchley C., Allen D., Fowbert J., Mole A. & Gundrey A. (2004) Habitat establishment on arable land: assessment of an agri-environment scheme in England, UK. *Biological Conservation*, 119, 429–442.
(5) Pywell R.F., Warman E.A., Carvell C., Sparks T.H., Dicks L.V., Bennett D., Wright A., Critchley C.N.R. & Sherwood A. (2005) Providing foraging resources for bumblebees in intensively farmed landscapes. *Biological Conservation*, 121, 479–494.
(6) Critchley C.N.R., Fowbert J.A. & Sherwood A.J. (2006) The effects of annual cultivation on plant community composition of uncropped arable field boundary strips. *Agriculture, Ecosystems & Environment*, 113, 196–205.
(7) Stevens D.K. & Bradbury R.B. (2006) Effects of the Arable Stewardship Pilot Scheme on breeding birds at field and farm-scales. *Agriculture, Ecosystems & Environment*, 112, 283–290.
(8) Critchley C.N.R., Walker K.J., Pywell R.F. & Stevenson M.J. (2007) The contribution of English agri-environment schemes to botanical diversity in arable field margins. *Aspects of Applied Biology*, 81, 293–300.
(9) Evans A.D. & Green R.E. (2007) An example of a two-tiered agri-environment scheme designed to deliver effectively the ecological requirements of both localised and widespread bird species in England. *Journal of Ornithology*, 148, S279–S286.
(10) Fisher G.P., MacDonald M.A. & Anderson G.Q.A. (2007) Do agri-environment measures for birds on arable land deliver for other taxa? *Aspects of Applied Biology*, 81, 213–219.

(11) Sheldon R., Chaney K. & Tyler G. (2007) Factors affecting nest survival of Northern Lapwings *Vanellus vanellus* in arable farmland: an agri-environment scheme prescription can enhance nest survival. *Bird Study*, 54, 168–175.

(12) Still K.S. (2007) A future for rare arable plants. *Aspects of Applied Biology*, 81, 175–182.

(13) Stoate C. & Moorcroft D. (2007) Research-based conservation at the farm scale: development and assessment of agri-environment scheme options. *Aspects of Applied Biology*, 81, 161–168.

(14) Walker K.J., Critchley C.N.R. & Sherwood A.J. (2007) The effectiveness of new agri-environment scheme options in conserving rare arable plants. *Aspects of Applied Biology*, 81, 301–308.

(15) Walker K.J., Critchley C.N.R., Sherwood A.J., Large R., Nuttall P., Hulmes S., Rose R. & Mountford J.O. (2007) The conservation of arable plants on cereal field margins: an assessment of new agri-environment scheme options in England, UK. *Biological Conservation*, 136, 260–270.

(16) Critchley C.N.R. & Cook S.K. (2008) *Long-term Maintenance of Uncommon Plant Populations in Agri-environment Scheme in England. Phase 1 Scoping Study.* Defra/ADAS BD1630.

(17) Chamberlain D., Gough S., Anderson G., Macdonald M., Grice P. & Vickery J. (2009) Bird use of cultivated fallow 'Lapwing plots' within English agri-environment schemes. *Bird Study*, 56, 289–297.

(18) Natural England (2009) *Agri-environment Schemes in England 2009: A Review of Results and Effectiveness.* Natural England, Peterborough.

(19) Ewald J.A., Aebischer N.J., Richardson S.M., Grice P.V. & Cooke A.I. (2010) The effect of agri-environment schemes on grey partridges at the farm level in England. *Agriculture, Ecosystems & Environment*, 138, 55–63.

(20) Field R.H., Morris A.J., Grice P.V. & Cooke A.I. (2010) Evaluating the English Higher Level Stewardship scheme for farmland birds. *Aspects of Applied Biology*, 100, 59–68.

Additional references

Anon (1990) Öko-Wertstreifen in Ackerbaugebieten [Conservation margins in arable areas]. *Jagd und Hege*, 12, 5–7.

Cardwell C., Hassall M. & White P. (1994) Effects of headland management on carabid beetle communities in Breckland cereal fields. *Pedobiologia*, 38, 50–62.

Hawthorne A. & Hassall M. (1994) Effects of management treatments on carabid communities of cereal field margins. *British Crop Protection Council Monographs*, 58, 313–318.

White P. C.L. & Hassall M. (1994) Effects of management on spider communities of headlands in cereal fields. *Pedobiologia*, 38, 169–184.

Critchley C.N.R. (1996a) Monitoring as a feedback mechanism for the conservation management of arable plant communities. *Aspects of Applied Biology*, 44, 239–244.

Critchley C.N.R. (1996b) Vegetation of arable field margins in Breckland. PhD thesis, University of East Anglia.

Marshall E.J.P. (1998) *Guidelines for the Siting, Establishment and Management of Arable Field Margins, Beetle Banks, Cereal Conservation Headlands and Wildlife Seed Mixtures.* Institute of Arable Crops Research report to MAFF.

Critchley C.N.R. (2000) Ecological assessment of plant communities by reference to species traits and habitat preferences. *Biodiversity and Conservation* 9, 87–105.

Wilson S., Baylis M., Sherrott A. & Howe G. (2000) *Arable Stewardship Project Officer Review.* Farming and Rural Conservation Agency report.

ADAS (2001) *Ecological Evaluation of the Arable Stewardship Pilot Scheme, 1998–2000.* ADAS report.

Walker K., Pywell R.F., Carvell C. & Meek W.R. (2001). *Arable Weed Survey of the Salisbury Plain Training Area.* Centre for Ecology and Hydrology, Abbots Ripton.

Critchley C.N.R., Fowbert J.A. & Sherwood A.J. (2004) *Botanical Assessment of the Arable Stewardship Pilot Scheme, 2003.* ADAS report to the Department for Environment, Food and Rural Affairs April 2004.

Still K. & Byfield A. (2007) *New Priorities for Arable Plant Conservation.* Plantlife, Salisbury.

2.9   Plant crops in spring rather than autumn

- A total of nine studies from Denmark, Sweden and the UK looked at the effects of sowing crops in spring or autumn on farmland wildlife. Five studies (including one replicated controlled trial, and a review) found that planting crops in spring rather than autumn resulted in higher numbers of farmland birds[7, 9], weed diversity[4] or weed density[2] and one arable weed species produced more fruit on spring-sown crops[3].

- A review found one study from the UK showing that four out of five species of arable weed produced more fruits on autumn-sown crops[3]. A second review found one study showing that there were more invertebrates in winter wheat than spring wheat[6].

- A replicated study from the UK[1] found that winter and spring sown crops were used for different broods by Eurasian skylarks. A replicated site comparison found arthropod abundance was higher in autumn barley in early summer and spring barley in late summer[8]. A replicated, controlled study in Sweden[5], found that northern lapwings nested on spring-sown crops more than expected based on their availability, but hatching success on spring crops was lower than on autumn crops.

Background

This intervention involves planting crops in spring rather than autumn. Changes in farming practice in northern Europe have included a shift from sowing crops in spring to sowing them the preceding autumn/winter. This change is considered to have adversely affected farmland biodiversity including invertebrates and farmland birds (see, for example, Donald & Vickery 2000).

Donald P. F. & Vickery J. A. (2000) The importance of cereal fields to breeding and wintering skylarks *Alauda arvensis* in the UK. In N. J. Aebischer, A. D. Evans, P. V. Grice and J. A. Vickery (eds.) *Ecology and Conservation of Lowland Farmland Birds*, British Ornithologists' Union, Tring. pp. 140–150.

A replicated study in 1992 and 1993 within the South Downs Environmentally Sensitive Area, Sussex, UK (1) found that winter and spring-sown crops were used for different broods by Eurasian skylark *Alauda arvensis*. Winter-sown crops tended to be used more for first brood nesting skylarks (first brood period: 8–15 males/km²; second: 4–9) whereas spring-sown crops were used more for the second brood (first brood period: 3–4 males/km2; second: 7–14). Four arable, ten mixed and three pastoral farms were studied. Skylarks were sampled by mapping breeding males during 2 counts along transects on 12–17 farms from April to June.

A replicated, controlled study of arable fields at three sites within the TALISMAN MAFF-funded experiment in England (2) found that weed density tended to be higher in plots with increased spring cropping compared to those with winter dominated cropping. Seed bank density depended on site. At Boxworth, seed bank density was higher on increased spring cropping rotations (8,780–25,824/m²) compared to winter dominated cropping (2,172–2,209/m²). In contrast, at High Mowthorpe, seed densities were higher with winter cropping (11,300–16,231/m² vs 1,764–3,181/m²). Total plant density tended to be higher in plots with increased spring cropping than with winter dominated cropping (4–18 vs 3–9/m²). There were differences between species and at High Mowthorpe, some had significantly higher populations on plots with winter cropping. At Boxworth there were two replicates in two blocks; at the other two sites there was one replicate in three blocks. Seed banks were sampled at Boxworth and High Mowthorpe after harvest from 3 sub-samples (60 combined soil cores) in each plot. Weed density was sampled in 15 quadrats/plot at the 3 sites after harvest (August–September) and in October–November.

A 1998 literature review (3) looked at the effect of agricultural intensification and the role of set-aside on the conservation of farmland wildlife, particularly endangered annual arable wildflowers and gamebirds. It found one UK study comparing arable weeds in spring and autumn-sown cereals showing that rough poppy *Papaver hybridum*, shepherd's-needle *Scandix pecten-veneris*, corn buttercup *Ranunculus arvensis* and common corncockle *Agrostemma githago* produced significantly more fruits/plot in autumn-sown than spring-sown cereals. In contrast broad-fruited cornsalad *Valerianella rimosa* produced significantly more fruits in spring-sown crops (Wilson 1994).

A replicated site comparison study in 1988–1992 in 19 arable fields in Denmark (4) found that weed diversity in unsprayed crop margins was over 25% lower in winter cereals than in spring cereals. Of the 114 weed species found, 97 were present in spring cereals compared to 87 in winter cereals. Of the species with known germination seasons, 94% were able to germinate in spring cereals whilst 64% were able to germinate in winter cereals. In addition, important food plants for arthropod herbivores occurred at greater densities and higher relative abundance in spring cereals. Experimental plots were 6 x 20 m, with crop rotation determined by the farmer. Each plot was sampled in spring each year, using 10 permanent 0.1 m circles/plot. Only data from permanently unsprayed plots in fields that supported at least one winter and one spring cereal sample were used in the analysis (72 plots).

A replicated, site comparison study between 1984 and 1994 in Västmanland, Sweden (5), found that northern lapwing *Vanellus vanellus* nested on spring-sown crops more than expected based on their availability, and on autumn sown crops less than expected. However, hatching success on spring crops was lower than on autumn crops (29–50% for 1,236 nests on spring crops vs approximately 85% for 27 nests on autumn crops).

A 2003 literature review in Europe (6) found one study that reported that winter wheat supported higher numbers of invertebrates than spring wheat (Green 1984).

A before-and-after site-comparison study in 2000–2005 in Bedfordshire, England (7), found that fields sown with wheat in spring held significantly more Eurasian skylark *Alauda arvensis*, seed-eating songbirds and insect-eating birds than winter-sown wheat. In addition, 20 bird species showed significant population increases on a 61 ha site where the area of spring-sown wheat and naturally regenerated set-aside was increased over the study period. Increases were lower or absent on an 80 ha area of farmland adjacent to the experimental area and without the land use change. Five species were recorded breeding for the first time after management started. Ten species showed no significant increase on the study site, whilst none decreased significantly. The biggest increases occurred in the first three years of management and were higher for farmland birds than for woodland birds.

A replicated site comparison study in 2004 of autumn-sown and spring-sown barley on four farms in Scotland (8) found that arthropod abundance was higher in autumn barley in early summer and in spring barley in late summer. Arthropod abundance was significantly higher in autumn barley from April to June (autumn barley: 8–21/sample; spring: 3–14), consistent with earlier crop development. The reverse was true in July and August (autumn barley: 15–23; spring: 20–26/sample). Abundances of individual arthropod orders varied slightly between the two sowing regimes. A total of five spring and five autumn barley fields were selected from four farms (two of each crop type). No insecticides were applied, but fields received one or two herbicide applications. Arthropods were sampled on five occasions in each field (April–August 2004) using a leaf vacuum (15 cm diameter). Sampling was undertaken at intervals (5 or 30 m) along 2–5 parallel transects (100 m apart) across the width of each field.

A replicated, paired site comparison study in 2004 of autumn-sown wheat and spring-sown barley in Sweden (9) found that there were significantly greater numbers of ground-foraging breeding birds in spring-sown cereals. There were 0.8 species/ha in spring-sown compared to 0.5 species/ha in autumn-sown cereal plots. Territory densities of northern lapwing *Vanellus vanellus* and northern wheatear *Oenanthe oenanthe* were also higher in spring-sown (lapwing: 0.08 territories/ha; wheatear: 0.12) compared to autumn-sown cereal plots (lapwing: 0.02; wheatear: 0.05). There was no effect of sowing time on Eurasian skylark *Alauda arvensis* or yellowhammer *Emberiza citrinella* breeding density. In spring-sown plots, numbers of species decreased significantly as the proportion of autumn-sown cereals in the surrounding landscape increased. Forty-one independent pairs of autumn-sown wheat and spring-sown barley plots were selected, each centred on an infield non-crop island. Non-crop islands were surveyed for cover of trees, shrubs and weeds and cereal height was measured on five occasions in each field. All birds were recorded within a radius of 100 m from the centre of each plot during five point counts of seven minutes (mid-May to end of June).

(1) Wakeham-Dawson A. (1995) Hares and skylarks as indicators of environmentally sensitive farming on the South Downs. PhD thesis. Open University.
(2) Jones N.E., Burn A.J. & Clarke J.H. (1997) The effects of herbicide input level and rotation on winter seed availability for birds. *Proceedings of the 1997 Brighton Crop Protection Conference – Weeds*, 3. pp. 1161–1166.
(3) Sotherton N. (1998) Land use changes and the decline of farmland wildlife: an appraisal of the set-aside approach. *Biological Conservation*, 83, 259–268.
(4) Hald A.B. (1999) The impact of changing the season in which cereals are sown on the diversity of the weed flora in rotational fields in Denmark. *Journal of Applied Ecology*, 36, 24–32.
(5) Berg Å., Jonsson M., Lindberg T. & Kallebrink K.-G. (2002) Population dynamics and reproduction of northern lapwings *Vanellus vanellus* in a meadow restoration area in central Sweden. *Ibis*, 144, E131–E140.
(6) Bat Conservation Trust (2003) *Agricultural Practice and Bats: A Review of Current Research Literature and Management Recommendations*. Defra BD2005.
(7) Henderson I.G., Ravenscroft N., Smith G. & Holloway S. (2009) Effects of crop diversification and low pesticide inputs on bird populations on arable land. *Agriculture, Ecosystems & Environment*, 129, 149–156.
(8) Douglas D.J.T., Vickery J.A. & Benton T.G. (2010) Variation in arthropod abundance in barley under varying sowing regimes. *Agriculture, Ecosystems & Environment*, 135, 127–131.
(9) Eggers S., Unell M. & Part T. (2011) Autumn-sowing of cereals reduces breeding bird numbers in a heterogeneous agricultural landscape. *Biological Conservation*, 144, 1137–1144.

Additional references
Green R.E. (1984) The feeding ecology and survival of partridge chicks (*Alectoris rufa* and *Perdix perdix*) on arable farmland in East Anglia. *Journal of Applied Ecology*, 21, 817–830.
Wilson P. J. (1994) Botanical diversity in arable field margins. *British Crop Protection Council Monographs*, 58, 53–58.

2.10   Undersow spring cereals, with clover for example

- A total of fifteen studies from Austria, Denmark, Finland, Switzerland and the UK (including four replicated, controlled and randomized studies and two reviews) looked at the effects of undersowing spring cereals on biodiversity. Eleven studies (including seven replicated trials, of which one was controlled and three were randomized and controlled, and one review) found that undersowing spring cereals benefited some birds[1, 7, 11, 18], plants[12], insects[2, 4, 8, 14–16], spiders[15] and earthworms[3, 9]. These benefits to farmland wildlife included increases in barnacle goose abundance[1], densities of singing Eurasian skylark[7] and nesting dunnock[18], arthropod abundance and species richness[2] and bumble-bee, butterfly, earthworm, ground beetle, spider or springtail abundances[3, 4, 9, 14–16].

- Five studies from Austria, Finland and the UK (including three replicated studies, of which one was also controlled and randomized, and a review) found that undersowing spring cereals did not benefit invertebrates[5, 6, 10, 13], plants[10], grey partridge population indicators[17], or nesting densities of two out of three farmland bird species[18]. One replicated study from the UK found that only one out of five bird species was found more frequently on undersown wheat stubbles than conventionally managed barley[11].

Background

This intervention involves sowing grass or clover beneath a cereal crop. The undersown crop is later ploughed in.

A before-and-after study in Dumfries, southern Scotland (1) found that the number of barnacle geese *Branta leucopsis* on a mixed agricultural site and nature reserve increased from 3,200 in 1970 to 6,000 in 1975 after all cereals sown on the site were undersown from 1970 onwards. The nature reserve consists of 220 ha of salt pasture, whilst the agricultural land is 50 ha arable fields. Most of the increased goose numbers feed on the arable land. In addition to undersowing, the proportion of cereals grown on the arable land decreased and no stock were allowed to graze on the arable land after November.

A replicated, controlled study of arable fields in West Sussex, England (2) found that arthropod abundance, density and species richness was higher in undersown spring barley and undersown grass fields compared to mono-cropped fields. Arthropod abundance and diversity was greater in undersown barley fields (767–874 m^2; 19–23 species), compared to mono-cropped barley (677–714 m^2; 14–18 species) and grass (281–391 m^2; 12–15 species). Only the true bugs (Hemiptera) were found in greater numbers in barley than in undersown barley. On average, 70% more arthropods emerged from undersown grass (555–623 m^2) than cultivated fields (280–391 m^2). Species diversity was also higher in undersown grass (22–28 vs 11–16). Half of the cereal fields in the ley farming area contained over 200 parasitic Hymenoptera/m^2 compared to just 9% in the 'modern arable area' (no grass). Arthropods from the field and ground zone were sampled using a Dietrick vacuum (five sub-samples each 0.09 m^2) at regular intervals across one field from each treatment in 1972 and 1973 and in one grass field in 1974. In addition one sample was taken from 150 fields in June 1972–1974. In 1970 and 1971, two adjacent fields, one undersown with grass and one cultivated, were sampled for emergent arthropods using two rows of five emergence traps (area enclosed 2.8 m^2). Traps were emptied regularly between March and June.

A trial at an experimental farm in 1989 on the Swiss Plateau, Switzerland (3) found that earthworm (Lumbricidae) abundance and biomass were higher in a maize *Zea mays* plot undersown with grass than in conventionally managed maize, although statistical analyses were not presented. Control and undersown plots had averages of 127 and 145 earthworms/m^2 and 45 and 71 g earthworm biomass/m^2, respectively. The proportion of deep-burrowing earthworms was similar with 14 and 12% of individuals in the control and undersown plots respectively. A test strip of maize 14 m long was undersown with grass in summer and compared with a control strip of conventional maize. Earthworms were sampled by hand-sorting 0.1 m^3 of soil from each test strip, to a depth of 40 cm, on 6 dates between April and October 1989. There was no replication.

A replicated, controlled, randomized study of an undersown cereal field from 1989 to 1991 in Helsinki, Finland (4) found that green manuring with undersown rye grass *Lolium* spp. resulted in a 50% increase in new generation adult ground beetles (Carabidae). A total of 33 species of ground beetles were caught. For three of the most common species, plots that had received rye grass green manuring (0.5 t/ha) in autumn 1990 resulted in a 50% increase in new generation adults in 1991 compared to mono-cropped plots (*Trechus discus*: 681 vs 442; *Clivina fossor*: 160 vs 137; *Bembidion guttula*: 108 vs 61). Rye grass provided more green manure than clover *Trifolium* spp. (49–412 vs 216–474 g/m^2). A 1 ha block of the field was divided into 25 x 25 m plots with treatments in a 4 x 4 Latin square design. Clover or grass were sown straight after the cereal and were ploughed into the soil as a green manure in the autumn.

Plant biomass was sampled within 0.25 x 0.5 m quadrats just before ploughing. Emergence rates of ground beetles were sampled using enclosures (0.5 x 0.5 m^2) with four pitfall traps. Traps were emptied every seven days over a sequence of three (1989) or five (1990 and 1991) trapping periods of approximately one month (June–September).

A replicated, controlled, randomized study of ground beetles (Carabidae) in arable fields in Finland (5) (same study as (6)) found no significant difference in beetle abundance between conventional and integrated farming practices (including undersowing with grass/clover *Trifolium* spp.). Abundance was higher with reduced pesticide applications. There were 6 replicate blocks and treatments (in 0.7 ha plots) which were fully randomized within blocks (1 treatment combination/plot). Treatments were conventional pesticide applications, reduced pesticides or no pesticides (control) and customary or integrated (including undersowing) cultivation. Beetles were sampled with pitfall traps at 12, 66 and 120 m into each crop 8–10 times (1 week/sample) between sowing and harvest.

A replicated, controlled, randomized study of spiders (Araneae) in arable fields in Finland (6) (same study as (5)), found no significant difference in spider abundance between conventional and integrated farming practices (including undersowing with grass/clover *Trifolium* spp.). Abundance was higher with reduced pesticide applications. There were 6 replicate blocks and the treatments (in 0.7 ha plots) were fully randomized within blocks (1 treatment combination/plot). Treatments were conventional pesticide applications or reduced pesticides and customary or integrated (including undersowing) cultivation. Spiders were sampled with pitfall traps at 12, 66 and 120 m into each crop 8–10 times (1 week/sample) between sowing and harvest.

A replicated study in summer 1995 in southern England (7) found that the density of singing Eurasian skylark *Alauda arvensis* was higher on undersown spring barley fields than on any other field type (approximately 22 birds/km^2 on 4 spring barley fields vs 2–15 birds/km^2 on 85 other fields). Other field types were arable fields reverted to species-rich or permanent grassland, downland turf (close-cropped, nutrient-poor grassland), permanent grassland, winter wheat, oilseed rape and set-aside. The number and location of singing skylarks were recorded in May–June 1995 on 89 fields.

A 2000 literature review (8) looked at which agricultural practices could be altered to benefit ground beetles (Carabidae). It found just one study (2) showing that some ground beetle species benefited from undersowing spring cereals, and that emergence the following spring was higher than in cereal fields.

A study of paired, intercropped and conventional wheat fields at four sites in the UK (9) found that intercropping resulted in higher earthworm (Lumbricidae) abundance, biomass and species diversity than conventional wheat management. Earthworm populations and biomass were greater in wheat-clover *Trifolium* spp. fields (individuals: 548/m^2; biomass: 137 g/m^2) than conventional wheat fields (194/m^2; 36 g/m^2) from autumn 1995–1997. Abundance varied more between conventional sites (55–408/m^2) than between wheat-clover sites (337–733/m^2). Population size ratios (wheat-clover:conventional wheat) ranged from approximately 2:1 to 9:1 and the overall mean ratio was 4:1. Species diversity was greater in wheat-clover fields (7–10 species) than conventional fields (5–9 species). White clover *Trifolium repens* was established in spring, and winter wheat was direct-drilled into the clover sward. Mono-cropped wheat was drilled at the same time. Intercropped fields received reduced applications. Earthworm communities were sampled in spring and autumn using the formalin method (10–12 quadrats of 0.25 m^2/field) and an electrical sampling method (5–10 samples of 0.125 m^2/field). Community biomass values refer to the live biomass.

A review (10) of two reports (Wilson *et al.* 2000, ADAS 2001) evaluating the effects of the Pilot Arable Stewardship Scheme in two regions (East Anglia and the West Midlands) in the UK from 1998 to 2001 found that undersown spring cereals did not benefit plants or invertebrates. The undersown cereals could be preceded by overwinter stubble or followed by a grass or grass/clover ley. There were 148 ha and 470 ha of this option in total in East Anglia and the West Midlands respectively. The effects of the pilot scheme on plants and invertebrates (bumblebees *Bombus* spp.; true bugs (Hemiptera); ground beetles (Carabidae); sawflies (Symphyta)) were monitored over three years, relative to control areas.

A replicated study in the winters of 1997–1998 and 1998–1999 on 122 stubble fields on 32 farms in central England (11) found that of 5 bird species using stubble fields, only 1 species, woodpigeon *Columba palumbus* was found most frequently on undersown organic wheat *Triticum* spp. stubbles. Eurasian linnet *Carduelis cannabina*, Eurasian skylark *Alauda arvensis*, reed bunting *Emberiza schoeniclus* and corn bunting *Miliaria calandria* were found more frequently on intensively-farmed barley *Hordeum* spp. stubbles than intensive or undersown organic wheat. Weed seed densities in March were highest on undersown organic wheat stubble fields compared to intensive barley or wheat stubbles in March. Weed seed density decreased the least on undersown organic wheat stubbles between October and March compared to intensive barley or wheat stubbles (11% decline on undersown organic wheat stubbles; 23% decline on intensive wheat stubbles; 35% decline on intensive barley). Seventeen stubble fields contained organic wheat with the previous crop undersown with rye grass *Lolium* spp. and white clover *Trifolium repens*. Sixty-seven fields were managed for intensive wheat and 38 fields for intensive barley; both intensively-managed crops received inorganic fertilizer and pesticide applications. Each study field was either overwintering as stubble or entered into the first year of a set-aside scheme. Plants were surveyed in forty 20 x 20 cm quadrats in each field in October. Seed densities were recorded in 27 fields from 10 soil cores/field in October 1997 and March 1998. Birds were surveyed monthly on parallel transects.

A replicated study in the summers of 1999–2000 comparing ten different conservation measures on arable farms in the UK (12) found that undersown spring cereals had more plant species than seven other conservation measures, but were not considered one of the best options for conservation of annual herbaceous plant communities. Average numbers of plant species in the different conservation habitats were undersown cereals 5.9; wildlife seed mixtures 6.7; uncropped cultivated margins 6.3; naturally regenerated grass margins 5.5; no-fertilizer conservation headlands 4.8; spring fallows 4.5; sown grass margins 4.4; overwinter stubbles 4.2; conservation headlands 3.5; grass leys 3.1. Plants were surveyed on a total of 294 conservation measure sites (each a single field, block of field or field margin strip) on 37 farms in East Anglia (dominated by arable farming) and 38 farms in the West Midlands (dominated by more mixed farming). The ten habitats were created according to agri-environment scheme guidelines. Vegetation was surveyed once in each site in June–August in 1999 or 2000, in thirty 0.25 m² quadrats randomly placed in 50–100 m randomly located sampling zones in each habitat site. All vascular plant species rooted in each quadrat, bare ground or litter and plant cover were recorded.

A controlled study in May to September 2000 in the sub-urban area of Vienna, Austria (13) found no difference in the number of ground beetle (Carabidae) species between 1 rye field undersown with a wildflower mix (approximately 20 species) and 4 conventional fields without wildflowers (approximately 12–26 spp.). The number of ground beetle species in the undersown rye field was lower than on two types of set-aside land (unsown or sown with a wildflower mix). No statistical analyses were presented in this paper. Typical crops for the region were sown on five arable fields. One of the fields was under conservation contract growing a wildflower seed mix undersown with rye. Ground beetles were sampled using 4 pitfall traps 10 m apart in each habitat and site. There were five sampling periods each year, each lasting two to three days (2001) or seven days (2000).

A randomized, replicated, controlled trial in 2003 to 2006 on 4 farms in southwest England (14) (same study as (16, 18)) found that 50 x 10 m plots of permanent pasture sown with spring barley *Hordeum vulgare* and a grass and legume mix attracted more bumblebees *Bombus* spp. and adult butterflies (Lepidoptera) than control plots. However undersown barley plots had either similar numbers (for suction trapped beetles (Coleoptera); ground beetles (Carabidae); spiders (Araneae); grasshoppers and crickets (Orthoptera); flies (Diptera); butterfly larvae or sawfly larvae (Hymenoptera: Symphyta); slugs (Gastropoda)) or fewer numbers (true bugs (Hemiptera); planthoppers (Auchenorrhyncha)) of other invertebrate groups than control plots. Control plots were managed as silage, cut twice in May and July, and grazed in autumn/winter. Small insect-eating birds (dunnock *Prunella modularis*, wren *Troglodytes troglodytes* and European robin *Erithacus rubecula*) and seed-eating finches (Fringillidae) and buntings (Emberizidae) preferred undersown cereal plots to control plots for foraging. Dunnock, but not chaffinch *Fringilla coelebs* or blackbirds *Turdus merula*, nested in hedgerows next to the sown plots more than expected, with 2.5 nests/km, compared to less than 0.5 nests/km in hedges next to experimental grass plots. There were 12 replicates of each management type, monitored over 4 years. More information on the use of these plots by bumblebees and butterflies is described in (16).

A replicated, controlled, randomized study of undersown and conventional cereal systems in Denmark (15) found that undersown crops had higher money spider (Linyphiidae) web density, adult *Bathyphantes gracilis* and *Tenuiphantes tenuis* (both money spiders), springtail (Collembola) and vegetation density compared to conventional crops. Web density was higher in undersown crops (unfertilized: peak 250–300/m^2; low fertilizer input: 200–250/m^2) than conventional crops (low fertilizer input: 150–200/m^2; high-input: 100–150/m^2). More adult *Bathyphantes gracilis* were found in undersown crops (5 individuals/m^2) and *Tenuiphantes tenuis* in unfertilized undersown crops (4/m^2) compared with the high-input conventional system (1/m^2). Springtail density was significantly higher in the fertilized (2,350 individuals/m^2) than unfertilized undersown crops (1,600/m^2) and conventional crops (low-input: 1,250/m^2; high-input: 300/m^2). Sixteen experimental plots (12 x 50 m) were established in a randomized block design. Treatments were wheat with clover *Trifolium* spp. undersown, without or with nitrogen fertilization (50 kg/ha), or conventional wheat with low (50 kg/ha) or high nitrogen fertilization (160 kg/ha), only the latter received pesticide applications. Money spider web densities, vegetation density (lower layer only, i.e. clover and weed layer) were sampled between May–October 1995–1997. Money spiders and springtails were sampled in 1996.

A randomized, replicated, controlled trial from 2003 to 2006 in southwest England (16) (same study as (14, 18)) found plots on permanent pasture annually sown with a mix of legumes, or grass and legumes, supported more common bumblebees *Bombus* spp. (individuals and species) than seven grass management options. In the first two years, numbers of common butterflies (Lepidoptera) and common butterfly species were higher in plots sown with legumes than in five intensively managed grassland treatments. No more than 2.2 bumblebees/transect were recorded on average on any grass-only plot in any year, compared to over 15 bumblebees/transect in both sown treatments in 2003. Plots sown with legumes generally had fewer butterfly larvae than all grass-only treatments, including conventional silage and six different management treatments. Experimental plots 50 x 10 m were established on permanent pastures (more than 5-years-old) on 4 farms. There were nine different management types, with three replicates/farm, monitored over four years. The two legume-sown treatments comprised either spring barley *Hordeum vulgare* undersown with a grass and legume mix (white clover *Trifolium repens*, red clover *T. pratense*, common vetch *Vicia sativa*, bird's-foot trefoil *Lotus corniculatus* and black medick *Medicago lupulina*) cut once in July, or a mix of crops including linseed *Linum usitatissimum* and legumes, uncut. Seven management types involved different management options for grass-only plots, including mowing and fertilizer addition. Bumblebees and butterflies were surveyed along a 50 m transect line in the centre of each experimental plot, once a month from June to September annually. Butterfly larvae were sampled on two 10 m transects using a sweep net in April and June–September annually.

A replicated site comparison study from 2004 to 2008 in England (17) found measures of grey partridge *Perdix perdix* populations were negatively related to the proportion of sites covered by undersown spring cereals (following overwinter stubbles). There were significant negative relationships with year-on-year density changes in 2006–2007 and with overwinter survival rates in conjunction with overwinter stubbles (across all years combined and significantly in 2004–2005 and 2005–2006). There were no relationships with brood size or the proportion of young birds to old. Spring and autumn counts of grey partridge were made at 1,031 sites across England as part of the Partridge Count Scheme.

A replicated study from April–July in 2006 on four livestock farms in southwest England (18) (same study as (14, 16)) found that dunnock *Prunella modularis*, but not Eurasian blackbird *Turdus merula* or chaffinch *Fringella coelebs*, nested at higher densities in hedges alongside field margins sown with either barley undersown with grass and clover *Trifolium* spp. or wild bird seed crops, compared to those next to grassy field edges under various management options (dunnock: approximately 2.5 nests/km for seed crops vs 0.3/km for grass margins; blackbird: 1.0 vs 1.3; chaffinch: 1.5 vs 1.4). There were three replicates/farm. Margins were 10 x 50 m and located adjacent to existing hedgerows. Seed crop margins were sown with barley (undersown with grass/legumes) or a kale/quinoa mix. There were 12 replicates of each treatment; 3 replicates on each farm.

(1) Owen M. (1977) The role of wildfowl refuges on agricultural land in lessening the conflict between farmers and geese in Britain. *Biological Conservation*, 11, 209–222.
(2) Vickerman G.P. (1978) The arthropod fauna of undersown grass and cereal fields. *Scientific Proceedings of the Royal Dublin Society, Series A*, 6, 273–283.
(3) Wyss E. & Glasstetter M. (1992) Tillage treatments and earthworm distribution in a Swiss experimental corn field. *Soil Biology & Biochemistry*, 24, 1635–1639.
(4) Helenius J. & Tolonen T. (1994) Enhancement of generalist aphid predators in cereals: effect of green manuring on recruitment of ground beetles. *IOBC/WPRS Bulletin*, 17, 201–210.
(5) Huusela-Veistola E. (1996) Effects of pesticide use and cultivation techniques on ground beetles (Col, Carabidae) in cereal fields. *Annales Zoologici Fennici*, 33, 197–205.
(6) Huusela-Veistola E. (1998) Effects of perennial grass strips on spiders (Araneae) in cereal fields and impact on pesticide side-effects. *Journal of Applied Entomology*, 122, 575–583.
(7) Wakeham-Dawson A., Szoszkiewicz K., Stern K. & Aebischer N.J. (1998) Breeding skylarks *Alauda arvensis* on Environmentally Sensitive Area arable reversion grass in southern England: survey-based and experimental determination of density. *Journal of Applied Ecology*, 35, 635–648.
(8) Holland J.M. & Luff M.L. (2000) The effects of agricultural practices on Carabidae in temperate agroecosystems. *Integrated Pest Management Reviews*, 5, 109–129.
(9) Schmidt O., Curry J.P., Purvis G. & Clements R.O. (2001) Earthworm communities in conventional wheat monocropping and low-input wheat-clover intercropping systems. *Annals of Applied Biology*, 138, 377–388.
(10) Evans A.D., Armstrong-Brown S. & Grice P.V. (2002) The role of research and development in the evolution of a 'smart' agri-environment scheme. *Aspects of Applied Biology*, 67, 253–264.
(11) Moorcroft D., Whittingham M.J., Bradbury R.B. & Wilson J.D. (2002) The selection of stubble fields by wintering granivorous birds reflects vegetation cover and food abundance. *Journal of Applied Ecology*, 39, 535–547.

(12)   Critchley C., Allen D., Fowbert J., Mole A. & Gundrey A. (2004) Habitat establishment on arable land: assessment of an agri-environment scheme in England, UK. *Biological Conservation*, 119, 429–442.
(13)   Kromp B., Hann P., Kraus P. & Meindl P. (2004) Viennese Programme of Contracted Nature Conservation 'Biotope Farmland': monitoring of carabids in sown wildflower strips and adjacent fields. *Deutsche Gesellschaft fur allgemeine und angewandte Entomologie*, 14, 509–512.
(14)   Defra (2007) *Potential for Enhancing Biodiversity on Intensive Livestock Farms (PEBIL)*. Defra BD1444.
(15)   Gravesen E. (2008) Linyphiid spider populations in sustainable wheat-clover bi-cropping compared to conventional wheat-growing practice. *Journal of Applied Entomology*, 132, 545–556.
(16)   Potts S.G., Woodcock B.A., Roberts S.P.M., Tscheulin T., Pilgrim E.S., Brown V.K. & Tallowin J.R. (2009) Enhancing pollinator biodiversity in intensive grasslands. *Journal of Applied Ecology*, 46, 369–379.
(17)   Ewald J.A., Aebischer N.J., Richardson S.M., Grice P.V. & Cooke A.I. (2010) The effect of agri-environment schemes on grey partridges at the farm level in England. *Agriculture, Ecosystems & Environment*, 138, 55–63.
(18)   Holt C.A., Atkinson P.W., Vickery J.A. & Fuller R.J. (2010) Do field margin characteristics influence songbird nest-site selection in adjacent hedgerows? *Bird Study*, 57, 392–395.

Additional references
Wilson S., Baylis M., Sherrott A. & Howe G. (2000) *Arable Stewardship Project Officer Review*. Farming and Rural Conservation Agency report.
ADAS (2001) *Ecological Evaluation of the Arable Stewardship Pilot Scheme, 1998–2000*. ADAS report.

2.11   Create rotational grass or clover leys

• A controlled study in Finland found that creating clover leys resulted in higher spider abundance and fewer pest insects than a barley control plot[3]. A study in the UK[1] found that one-year ley plots had significantly lower earthworm species richness and abundance than three-and-a-half-year leys.

• A replicated study in the UK found that grass leys had fewer plant species than nine other conservation measures[2].

Background

This intervention can be established on arable land by under-sowing spring cereal and leaving the under-sown sward to develop the following year. It can also be established through other means. The results of the intervention are monitored on the grass/clover ley. See also 'Undersow spring cereals, with clover for example'.

A study of grass/clover ley fields on an arable site from 1950 to 1956 in the UK (1) found that one-year ley plots had significantly lower earthworm species richness and abundance than three-and-a-half-year leys. One-year ley plots (within arable rotations) had lower numbers of species and overall abundance of earthworms (8–15/cubic foot) compared to 3-1/2-year ley plots (17–39/cubic foot). Overall earthworm weight showed the same trend (4 vs 8 g/cubic feet). One plot of each ley treatment was established in each of the six years. Leys were ploughed in the autumn and winter wheat sown. Plots were sampled for earthworms when they were ploughed out of leys (1953–1956). Four samples of two cubic feet of soil were sampled in each plot per year.

A replicated study in the summers of 1999–2000 on arable farms in the UK (2) found that grass leys had fewer plant species than nine other conservation measures. Average numbers of plant species in the different conservation habitats were grass leys 3.1; wildlife seed mixtures 6.7; uncropped cultivated margins 6.3; undersown cereals 5.9; naturally regenerated grass margins 5.5; no-fertilizer conservation headlands 4.8; spring fallows 4.5; sown grass margins 4.4; overwinter stubbles 4.2; conservation headlands 3.5. Grass leys had the lowest number of plant species, lower than in undersown cereals, due to the later successional stage of the sown grass and clover Trifolium spp. species that dominated the leys. Plants were surveyed on a total of 294 conservation measure sites (each a single field, block of field or field margin strip) on 37 farms in East Anglia (dominated by arable farming) and 38 farms in the West Midlands (dominated by more mixed farming). The ten habitats were created according to agri-environment scheme guidelines. Vegetation was surveyed once in each site in June–August in 1999 or 2000 in thirty 0.25 m^2 quadrats randomly placed in 50–100 m randomly located sampling zones in each habitat site. All vascular plant species rooted in each quadrat, bare ground or litter and plant cover were recorded.

A controlled trial from 2003 to 2004 in Jokioinen, southern Finland (3) found that fallow plots established by undersowing spring barley with grass or grass and red clover Trifolium pratense had more spiders (Araneae) and fewer pest insects than a control plot of spring barley but similar numbers of ground beetles (Carabidae). For example, there were 28–35 spiders/trap, compared to around 5 spiders/trap in the control plot. The only difference between seed mixtures used was that the plot sown with red clover in the mix had fewer unsown plant species (around 2 species/m^2) but higher plant biomass than control cereal fields, which had around 16 plant species/m^2. The plot undersown with just grasses had around 6 plant species/m^2. There was no difference in the numbers of spiders, beetles, flying insects or unsown plant species between two-year grass or grass-clover fallow plots established by undersowing spring barley and similar plots sown without accompanying cereals. The fallow treatments were established in 2003, each on a 44 x 66 m plot. A control plot was sown with spring barley in 2004. Insects were sampled using a yellow sticky trap and three pitfall traps in the centre of each plot for a week in June, July and August 2004. Unsown plant species were counted in four 50 x 50 cm quadrats in each plot in late August 2004.

(1) Heath G.W. (1962) The influence of ley management on earthworm populations. Grass and Forage Science, 17, 237–244.
(2) Critchley C., Allen D., Fowbert J., Mole A. & Gundrey A. (2004) Habitat establishment on arable land: assessment of an agri-environment scheme in England, UK. Biological Conservation, 119, 429–442.
(3) Huusela-Veistola E. & Hyvanen T. (2006) Rotational fallows in support of functional biodiversity. IOBC/WPRS Bulletin, 29, 61–64.

2.12   Convert or revert arable land to permanent grassland

- All seven individual studies (including four replicated studies, of which two were also controlled and a review) looking at the effects of reverting arable land to grassland found no clear benefit to wildlife. The studies monitored UK birds in winter and summer[3, 4], wading birds in Denmark[6], grey partridges[2, 8], brown hares in the UK[1] and plants in the Czech Republic[7].

- One of the studies, a controlled before-and-after study from the UK, showed that grey partridge numbers fell significantly following the reversion of arable fields to grassland[2].

Background

This intervention involves changing from an arable crop to a sown agricultural grassland, to be used for grazing or silage. It is not the same as the creation of species-rich or other semi-natural grasslands.

See also 'All farming systems: Provide or retain set-aside areas in farmland' for some studies where non-rotational set-aside land was sown with grass, but managed as set-aside rather than as permanent agricultural grassland.

A replicated study from 1992 to 1994 in the South Downs Environmentally Sensitive Area, Sussex, UK (1) found that foraging brown hares *Lepus europaeus* generally avoided farms and areas of farms that had been converted from arable crops to Environmentally Sensitive Area grasslands. Four arable, 10 mixed and 3 pastoral farms were studied. Hares were sampled by spotlight counting over an average of 26% of the area of each farm between November and March (1992–1993, 1993–1994).

A controlled before-and-after study from 1970 to 1994 in Sussex, England (2), found that grey partridge *Perdix perdix* numbers declined rapidly on arable fields following their reversion to grassland, which began in 1987 (average 6.5 coveys (flocks)/km^2 in 1970–1986 vs 1.1 coveys/km^2 in 1987–1994). There was a considerably smaller decline on arable fields that were not reverted to grassland (average 4.9 coveys/km^2 in 1970–1986 vs 2.5 coveys/km^2 in 1987–1994). Fields that were reverted had been favoured by partridges prior to reversion, in comparison to arable fields, but were less favoured after reversion, equating to a 23% per year decrease in relative habitat quality. Fields in a 28 km^2 area were surveyed for grey partridges in late August/early September after the autumn harvest by driving across fields at dawn and dusk and mapping the position of each observation.

A replicated, controlled study in the winters of 1994–1997 in southern England (3) (same study as (4)) found that Eurasian skylarks *Alauda arvensis*, corn buntings *Miliaria calandra* and meadow pipits *Anthus pratensis* were not consistently more abundant on arable land reverted to grassland than on intensively managed permanent grassland or winter wheat fields (4–11 skylarks/km^2 on reverted fields vs 0–10 and 1–8 on permanent grassland and winter wheat; 0.1–0.2 corn buntings/km^2 on reverted fields vs 0 and 0–1; 0–1.1 meadow pipits/km^2 on reverted fields vs 0 and 0–4). Densities of rooks *Corvus frugilegus* did not differ across field types. Reverted arable fields were sown with agricultural grass mixtures and managed under specific guidelines, whilst the permanent grassland fields were mown frequently and fertilized. Fields on 40 farms were surveyed. Birds were surveyed once during December and January on 217 fields in winter 1994–1995, repeated on 205 fields in winter 1995–1996 and on 225 fields in winter 1996–1997.

A replicated, controlled study in spring and summer from 1994 to 1996 in southern England (4) (same study as (3)) found that arable fields reverted to permanent grassland had similar densities of Eurasian skylarks *Alauda arvensis* to winter wheat and intensively managed permanent grassland, except in summer 1994 when they had significantly higher densities, and summer 1995 when they had lower densities than winter wheat. In summer 1994 there were 11.9 birds/km^2 on reverted fields (65 fields) vs 2.6 and 4.4 on permanent grassland (29 fields) and winter wheat (47 fields) respectively. In summer 1995 there were 2.1 birds/km^2 on reverted fields (15 fields) vs 3.0 and 11.0 on permanent grassland (7 fields) and winter wheat (26 fields); in other seasons 5.7–9.1 birds/km^2 on reverted fields vs 3.6–4.0 and 8.5–13.0 on permanent grassland and winter wheat. Densities of carrion crows *Corvus corone* tended to be higher on reverted arable land, significantly so in some seasons (1.8–4.8 birds/km^2 on reverted fields vs 0–3.0 and 0–1.1 on permanent grassland and winter wheat). Rooks *C. frugilegus* were never found on winter wheat. Fields on 40 farms were surveyed. In 1994 and 1996 between 65 and 82 reverted arable fields each sown with agricultural grass mixtures and managed under specific guidelines were studied, as well as 15–29 permanent grassland fields, which were frequently mown and fertilized, and 38–47 winter wheat fields. In 1995, 15 reverted arable fields, 7 permanent grassland fields and 26 winter wheat fields were surveyed. The number and locations of singing skylarks were recorded in April–May and June–July in 1994 and 1996 and in May–June 1995. The locations of foraging carrion crows and rooks were also recorded in 1994 and 1996.

A 2000 literature review (5) looked at grassland management practices in the UK. It reported three studies that found reversion of arable land to permanent grassland resulted in decreased abundance of broad-leaved weed seeds (3) (in downland Environmentally Sensitive Areas), and lower densities of grey partridge *Perdix perdix* and corn bunting *Miliaria calandra* (Potts 1997, Wakeham-Dawson 1997). A further two studies found breeding corn bunting and Eurasian skylark *Alauda arvensis* abandoned arable reversion grassland fields that were mown and grazed during the nesting season (Wakeham-Dawson 1997, (4)).

A replicated, controlled study in 2004 and 2005 in Jutland, Denmark, (6) found that populations of four wading birds (northern lapwing *Vanellus vanellus*, black-tailed godwit *Limosa limosa*, common redshank *Tringa totanus* and Eurasian oystercatcher *Haematopus ostrolagus*) did not increase on newly created grasslands (formerly croplands), whether or not they were under a scheme designed to increase water levels in fields. There were population increases on some other field types. A total of 615 fields were studied, comprising of permanent grassland, reverted grassland and cultivated fields in rotation. The four species were surveyed twice during the breeding season (April–May) and the number of each species and their location recorded.

A site comparison study from 2004 to 2008 in the Czech Republic (7) found that one arable reversion field had the lowest number of plant species out of 47 grassland sites managed under different agri-environment schemes. Only 5 plant species were recorded over 5 years of monitoring on the arable reversion site, compared to a maximum of 26 plant species at other sites. No increase in species richness was observed during the monitoring period. The agri-environment management allowed up to 60 kg Nitrogen/ha fertilizer, 2 cuts and cattle grazing. Forty-seven grassland sites were monitored in May/June and October each year from 2004 to 2008. All plant species on each site were recorded, and plant diversity measured in a permanent 3 x 3 m quadrat.

A replicated site comparison study from 2004 to 2008 in England (8) investigated the impact of restoration of different grasslands on grey partridge *Perdix perdix*. There was a negative relationship between a combined intervention (grassland restoration, scrub restoration and control and rough grazing) and the ratio of young to old partridges in 2008. The study does not distinguish between the individual impacts of grassland restoration, scrub restoration and control and rough grazing. Spring and autumn counts of grey partridge were made at 1,031 sites across England as part of the Partridge Count Scheme.

(1) Wakeham-Dawson A. (1995) Hares and skylarks as indicators of environmentally sensitive farming on the South Downs. PhD thesis. Open University.
(2) Aebischer N.J. & Potts G.R. (1998) Spatial changes in grey partridge (*Perdix perdix*) distribution in relation to 25 years of changing agriculture in Sussex, UK. *Gibier faune sauvage, Game Wildlife*, 15, 293–308.
(3) Wakeham-Dawson A. & Aebischer N.J. (1998) Factors determining winter densities of birds on environmentally sensitive area arable reversion grassland in southern England, with special reference to skylarks (*Alauda arvensis*). *Agriculture, Ecosystems & Environment*, 70, 189–201.
(4) Wakeham-Dawson A., Szoszkiewicz K., Stern K. & Aebischer N.J. (1998) Breeding skylarks *Alauda arvensis* on Environmentally Sensitive Area arable reversion grass in southern England: survey-based and experimental determination of density. *Journal of Applied Ecology*, 35, 635–648.
(5) Wakeham-Dawson A. & Smith K.W. (2000) Birds and lowland grassland management practices in the UK: an overview. *Proceedings of the Spring Conference of the British Ornithologists' Union, March 27–28, 1999.* Southampton, England. pp. 77–88.
(6) Kahlert J., Clausen P., Hounisen J. & Petersen I. (2007) Response of breeding waders to agri-environmental schemes may be obscured by effects of existing hydrology and farming history. *Journal of Ornithology*, 148, 287–293.
(7) Holubec V. & Vymyslický T. (2009) Botanical monitoring of grasslands after the adoption of agro-environmental arrangements. *Grassland Science in Europe*, 14, 128–131.
(8) Ewald J.A., Aebischer N.J., Richardson S.M., Grice P.V. & Cooke A.I. (2010) The effect of agri-environment schemes on grey partridges at the farm level in England. *Agriculture, Ecosystems & Environment*, 138, 55–63.

Additional references

Potts G.R. (1997) Cereal farming, pesticides and grey partridges. In D.J. Pain & M.W. Pienkowski (eds.) *Farming and Birds in Europe: The Common Agricultural Policy and its implications for bird conservation*, Academic Press, London. pp. 150–177.

Wakeham-Dawson A. (1997) Corn buntings *Miliaria calandra* in the South Downs and South Wessex Downs Environmentally Sensitive Areas (ESAs), 1994–1995. In P.F. Donald & N.J. Aebischer (eds.) *The Ecology and Conservation of Corn Buntings Miliaria Calandra*. UK Nature Conservation No. 13, Joint Nature Conservation Committee, Peterborough. pp. 186–190.

2.13   Reduce tillage

- A total of 42 individual studies (including 7 replicated, controlled and randomized studies and 6 reviews) from Belgium, Denmark, France, Germany, Hungary, Ireland, Lithuania, the Netherlands, Slovakia, Switzerland and the UK investigated the effects of reducing tillage on farmland wildlife. Thirty-three studies (of which 21 were replicated and controlled and 7 also randomized and 5 reviews) from 9 European countries found some positive effects on earthworms[1–5, 8, 19, 20, 22, 24, 27, 28, 33, 40, 42, 45, 46], some invertebrates (other than earthworms)[1, 5, 8–10, 15, 16, 18, 21, 23, 24, 27–30, 32, 35], weeds[7, 13, 17, 28] or farmland birds[27, 31, 37, 38, 44], of reducing tillage compared to conventional management. Positive effects included increased biomass[3, 8, 20, 40, 42, 45, 46], species richness[22] or abundance[1–5, 19, 20, 22, 24, 28, 33] of earthworms, greater abundance of some invertebrates other than earthworms [1, 5, 9, 10, 15, 16, 18, 21, 23, 24, 28–30, 32, 35], increased numbers of some weeds and/or weed species[7, 13,17, 28], higher Eurasian skylark nest density, earlier laying date and shorter foraging distances on reduced tillage fields[38], and greater abundance of some birds[44], including Eurasian skylark, seed-eating songbirds and gamebirds in late winter[31], on non-inversion or conservation tillage. A review found tillage had negative effects on invertebrate numbers and no-till systems had more invertebrate bird food resources[34].

- Twenty-seven studies (of which 13 were replicated and controlled and 3 also randomized and 5 reviews) from 9 European countries found reducing tillage had either negative, no effect or no consistent effects on abundance[1, 5, 9, 11, 12, 14–16, 18, 21, 24–28, 30], biomass[8], or species richness[11, 35] of some invertebrates (other than earthworms), earthworm abundance[1, 6, 19, 26, 39, 40, 43, 45, 46], biomass[6], or species richness[46], number of different plant species found as seeds[26], number of some weed species[13, 28], mammal abundance[28], some bird species[28, 31, 44], and one study found bird preferences for conservation tillage fields decreased over time[37]. Two studies found that crop type affected the number of weeds under different tillage regimes[7, 17].

- One small replicated trial in the UK[41] compared bird numbers under two different forms of reduced tillage, and found more birds from species that make up the 'Farmland Bird Index' on broadcast than non-inversion tillage fields.

- Two studies looked at the long-term effects of reduced tillage on earthworms (after ten years). One study found higher earthworm biomass under reduced tillage[42]; the other study found earthworm abundance was the same between conventional and reduced tillage plots[43].

- Three of the studies mentioned above did not distinguish between the effects of reducing tillage and reduced pesticide and/or fertilizer inputs[8, 22, 23, 30].

Background

Conventional ploughing uses a mould-board plough, cultivating to a depth of around 20 cm. This intervention includes various methods to reduce the depth or intensity of ploughing, such as layered cultivation, non-inversion tillage and conservation tillage. It also includes stopping tillage altogether in some areas.

Reduced tillage is often used in the context of 'Integrated Farm Management' (IFM), a whole farm system designed to enhance natural processes within a productive system (for example Hasken & Poehling 1995). These studies have not been included here unless the effects can clearly be related to the tillage practice as opposed to other elements of the system such as crop rotation, field margins, or reduced chemical use.

Conservation tillage, often accompanied by the use of cover crops or retention of crop residues, is widely practiced in the USA. A 2004 review (28) of the broad environmental effects of conservation tillage found the bulk of the evidence came from North American studies, with studies from Europe being sparse. For example, seven American studies found more ground-nesting birds on conservation tillage fields than conventionally tilled fields, and two studies found that rodents were more abundant in conservation tillage fields. However, both bird and rodent faunas in North America are functionally different from those in Europe (there are more ground-nesting birds in arable fields in the US, for example).

A set of studies on the effects of avoiding tillage when re-seeding heathland with grass to improve it for grazing (Grant 1992, Grant et al. 1992a,b) is included under 'Livestock farming: Employ areas of semi-natural habitat for rough grazing'.

Grant M.C. (1992) The effects of re-seeding heathland on breeding whimbrel *Numenius phaeopus* in Shetland. I. Nest distributions. *Journal of Applied Ecology*, 29, 501–508.

Grant M.C., Chambers R.E. & Evans P.R. (1992a) The effects of re-seeding heathland on breeding whimbrel *Numenius phaeopus* in Shetland. II. Habitat use by adults during the pre-laying period. *Journal of Applied Ecology*, 29, 509–515.

Grant M.C., Chambers R.E. & Evans P.R. (1992b) The effects of re-seeding heathland on breeding whimbrel *Numenius phaeopus* in Shetland. III. Habitat use by broods. *Journal of Applied Ecology*, 29, 516–523.

Hasken K.H. & Poehling H.M. (1995) Effects of different intensities of fertilizers and pesticides on aphids and aphid predators in winter-wheat. *Agriculture Ecosystems & Environment*, 52, 45–50.

A replicated, controlled study of arable fields at eight sites in England (1) found that abundance of mites (Acari), springtails (Collembola) and some earthworm (Lumbricidae) species tended to be higher in direct-drilled plots, whereas insects were more numerous in ploughed plots. Direct-drilled plots contained 922–2,665 mites and 106–2,408 springtails, whereas ploughed plots contained 620–2,340 and 77–1,904 respectively. The opposite trend was seen for insects (direct-drilled: 39–123; ploughed: 44–156) as numbers of taxa such as fly (Diptera) larvae, rove beetles (Staphylinidae) and ground beetles (Carabidae) were higher in ploughed plots. Earthworm numbers were higher in direct-drilled plots at all sites (811–1,638 vs 628–1,243). Species such as the earthworm *Lumbricus terrestris* followed this trend (direct: 22–323; ploughed: 4–103), however, other species showed a slight tendency for a higher abundance in ploughed plots. Four replicate plots (6.4 x 18 m) of winter wheat under each treatment were established at Rothamsted Experimental Station (1964–1967) and Woburn (1965–1971). Half of each plot received insecticides. Soil arthropods were sampled every two months by taking soil cores, and earthworms in spring and autumn. In 1974 soil animals were assessed in six additional experiments comparing direct-drilling and ploughing by the Letcombe Laboratory and National Institute of Agricultural Engineering. Results for pest species are not presented here. This study is partly the same study as (4).

A replicated, controlled trial in southern England from 1973 to 1976 (2) found there were always significantly more earthworms (Lumbricidae)/m^2 on the direct-drilled (no-tillage) plots than on the ploughed plots. Numbers on tine-cultivated plots were similar to those on ploughed plots. For example, at one site, there were 145–345 earthworms/m^2 in direct drilled plots (1973–1976) compared to 128–139 earthworms/m^2 in tine cultivated plots (1973 only) and 50–218 earthworms/m^2 in ploughed plots. There were no significant differences in numbers of particular earthworm species between the treatments. Deep-burrowing species were less than 10% of the earthworm communities in this study. Three cultivation treatments were compared in cereal fields (barley or winter wheat): direct drilling (no-tillage); tine cultivation to 8 or 15 cm; conventional ploughing to 20 cm. There were four replicates of each treatment at two separate sites, and for two soil types: clay and sandy loam.

A replicated trial on an experimental farm in eastern Scotland (3) found that the average number and biomass of earthworms (Lumbricidae) was significantly higher in untilled soil (137 earthworms/m^2 and 0.9 tonnes earthworm/ha) than in cultivated treatments (67–93 earthworms/m^2 and 0.3–0.4 tonnes/ha). The experiment was replicated eight times. Spring barley crops were managed from 1967 until 1973 with either deep ploughing (30–35 cm), normal ploughing (15–20 cm), tined cultivation (12–30 cm deep) or no ploughing (untilled, direct drilled). Between 1969 and 1973, the average number of adult and large juvenile earthworms on 2 replicates increased from 37 earthworms/m^2 to 114 worms/m^2 under direct drilling, but did not change significantly under the 3 cultivation treatments (21 to 80 earthworms/m^2).

A replicated trial on three farms in the UK (4) over five years found that one or both species of deep-burrowing earthworm *Lumbricus terrestris* and *Allolobophora longa* were significantly more abundant in untilled than in deep-ploughed plots at all three sites in all five years. After 5 years, untilled plots had 16.8, 8.6 and 1.2 *L. terrestris*/m^2 on average at Woburn, Rothamsted and Boxworth experimental farms respectively, compared to 7.8, 0.3 and 0.1 *L. terrestris*/m^2 on deep ploughed plots. Shallow working earthworm species showed few differences between untilled and ploughed treatments. In two studies with one year of monitoring, earthworms were also more abundant in untilled plots than ploughed plots. There were 250 earthworms/m^2 in plots untilled for 4 years compared to around 50 earthworms/m^2 in annually ploughed plots, and around 100 in plots ploughed for 2 of the 4 years at North Creake, Norfolk. At Lee Farm, Sussex there were between 5 and 70 *L. terrestris*/m^2 in untilled fields, compared to between 1 and 12.5 *L. terrestris*/m^2 in ploughed fields. There were between three and seven replicates of each treatment at each farm. The Woburn experiment on winter wheat ran from 1965 to 1971; plots were 6.4 x 18.0 m. The Rothamsted experiment on winter wheat started in 1972 with sampling from 1975 to 1979; plots were 33 x 13.5 m. The Boxworth experiment also on winter wheat started in 1971 with sampling from 1974 to 1978; plots were 36 x 13.5 m. This study is partly the same study as (1).

A replicated, controlled study at an arable farm over three years in England (5) found that the effect of reduced tillage on soil invertebrate numbers was not consistent, but depended on taxa, site and year. Of the 39 beetle (Coleoptera) species analysed, 10 were more active on conventionally ploughed, 10 on minimal-tillage (tined to 10 cm and disced) and 10 on zero-tillage plots. At one of two sites, numbers of species of ground beetle (Carabidae) were significantly higher on zero-tillage plots (zero tillage: 2.8–7.3; conventional: 2.4–6.4; minimal: 2.6–6.6) and species of rove beetles (Staphylinidae) were higher on conventional plots (conventional: 8–9; minimal: 7; zero: 6–8); other beetles did not differ. Excluding beetles, invertebrate numbers showed some variation between cultivation treatments with year and site; numbers increased in conventional plots following sowing. Crane flies (Tipulidae), spiders (Araneae) and froghoppers (Cercopidae) consistently had significantly higher numbers in zero-tillage plots. Earthworm (Lumbricidae) numbers tended to be higher on zero- or minimal-tillage plots and lower on conventional plots. The replicated (two) block design was established in 1972. Between 1978–1980 the rotation comprised: spring barley/rye grass *Lolium* spp. and clover *Trifolium* spp., rye grass and clover and then winter wheat. Thirty pitfall traps/plot were sampled every 14–28 days. Earthworms were sampled by formalin extraction or hand sorting ten times/plot in April–May and September–October.

A trial at an experimental farm in 1989 on the Swiss Plateau, Switzerland (6) found that earthworm (Lumbricidae) abundance and biomass were not higher in a no-tillage plot than other plots. No-tillage and control plots had averages of 47 and 127 earthworms/m^2, and 57 and 45 g earthworm biomass/m^2, respectively. There was a much higher proportion of deep-burrowing earthworms in the no-tillage plot (67% of individuals, compared to 11–14% of individuals in ploughed plots), which is why there were more individual worms in the control plot. Test strips of maize *Zea mays* 14 m long were either managed with no-tillage (sowing directly into undisturbed stubble) or conventionally ploughed and harrowed. The no-tillage treatment also had rye grass *Lolium* spp. sown after the maize. Earthworms were sampled by hand-sorting 0.1 m^3 of soil from each test strip, to a depth of 40 cm, on 6 dates between April and October 1989. There was no replication.

A controlled study in 1988–1990 in five plots in an arable field (7) found that weed cover was significantly higher in the conservation and minimum tillage regimes than under traditional tillage in most crops (no difference in corn and winter rye after corn). This study was presented at a conference in Germany, no location details were provided. The effect of reduced tillage on weed numbers and cover depended both on the current and previous crop in rotation. Conservation tillage led to higher weed numbers in winter rye after potatoes and in fodder radish (year 5), minimum tillage in winter rye after winter rye, and both reduced tillage systems in winter rye after corn. Weed numbers in traditionally ploughed plots were higher in fodder radish (year 1). Tillage regime also affected weed community composition with some species being more dominant in traditional ploughing, others in reduced tillage systems. The following tillage regimes were used: traditional ploughing (18–30 cm deep), conservation tillage (combination of ploughing and non-ploughing: 10–15 cm) and minimum tillage (combination of ploughing and non-ploughing: 10–15 cm) on a crop rotation with 5 crops (potatoes; winter rye with catch crop; corn; winter rye; winter rye with catch crop). Plants were surveyed on 1 m^2 quadrats with 8–10 replicates/crop. Surveys were conducted two to three times yearly in 4 m^2 unsprayed plots. Number of plants, weed cover, crop cover and species composition (number and frequency of species) were recorded in crops (except potatoes) and catch crops.

A site comparison study at the Lovinkhoeve Experimental Farm, Noordoostpolder, the Netherlands (8) found greater biomass of microbes, protozoa, nematodes (Nematoda) and earthworms (Lumbricidae), but not of mites (Acari) and springtails (Collembola), in the upper 10 cm of an arable soil with reduced tillage and reduced fertilizer and pesticide inputs, than in a conventionally managed soil. At lower depth (10–25 cm), there were no consistent differences in soil fauna. The reduced tillage plot had 8.9 kg C/ha of earthworms in the top 10 cm, and 4.7 kg C/ha at 10–25 cm depth. No earthworms were recorded in conventional plots. Total biomass of nematodes in the upper layer was 0.79 kg C/ha in the reduced tillage plot, and 0.30 kg C/ha in the conventional plot. Reduced tillage plots were cultivated to 12–15 cm depth without inversion of the topsoil, compared to 20–25 cm deep ploughing on conventional plots. They also had reduced nitrogen and pesticide applications. The experiment began in 1985. Soil samples were taken from three areas of each plot under winter wheat in 1986.

A replicated, controlled, randomized study of cultivation treatments from 1989 to 1992 on an arable farm 3 km from Long Ashton Research Station, England (9) found more money spiders (Linyphiidae) and slugs (Gastropoda) on arable soil after direct-drilling than after ploughing. Rove beetle (Staphylinidae) and ground beetle (Carabidae) numbers were not consistently different between treatments. In one field in autumn and winter, money spider numbers tended to be higher following direct-drilling (1–9/trap/week) than non-inversion (1–4) or ploughing (1–4), whereas in summer, numbers were higher on cultivated (16–25/trap/week) compared to direct-drilled plots (9–16). In the second field studied, no difference between treatments was found. Eight beetle (Coleoptera) groups tended to be more prevalent on ploughed plots (smaller beetles), 11 on Dutzi cultivated and/or direct-drilled plots (larger beetles); 9 beetle groups showed no difference between treatments. Slug numbers tended to be higher on direct-drilled (4–9/sample) and non-inversion tillage plots (1–16) than ploughed plots (1–4). Plots of 30 or 50 x 12 m of each treatment were randomized in three or five replicated blocks in two winter cereal fields (3–4 ha). Half of each plot received a selective pesticide for aphids (Aphidoidea) in 1990–1991. Predators were sampled using two pitfall traps/plot for seven days each month from 1989 to 1992. Slugs were monitored by flooding a soil sample from each plot at one to six month intervals. Results for other pest species, crop damage and the effects of incorporating straw are not included here.

A replicated, controlled study of two fields on two farms in Saxony, Germany from 1991 to 1992 (10) found that conservation tillage plots (with catch crops of phacelia *Phacelia tanacetifolia* or white mustard *Sinapis alba*) without seed-bed preparation in the spring resulted in an increase in spiders (Araneae), rove beetles (Staphylinidae) and ground beetles (Carabidae). Spider and ground beetle density was higher in conservation tillage plots without tillage in spring (spiders: $32–85/m^2$; ground beetles: $6–21/m^2$) compared to those with tillage (15–38, $2–14/m^2$ respectively). However rove beetle abundance differed between catch crops: rove beetles, no tillage: $70–95/m^2$ in phacelia, $54–100/m^2$ in white mustard; tillage in spring: $50–83/m^2$ in phacelia, $84–148/m^2$ in white mustard. Plots with conservation tillage had higher numbers of all three taxa than conventional plots (spiders: $10–18/m^2$; rove beetles: $43–62/m^2$; ground beetles: $2–11/m^2$). Numbers tended to be higher when white mustard was used compared to phacelia, particularly for ground beetles ($4–21$ vs $2–17/m^2$). Fields were divided into plots (12–24 x 100 m) with 2 replicates of 5 soil cultivations: conventional (ploughed, tillage, drilling of sugar beet *Beta vulgaris*) or conservation tillage with phacelia or white mustard (ploughed, tillage and drilled) followed by soil tillage and drilling or direct drilling of sugar beet in spring. Insecticides were not applied where predatory arthropods were monitored. Two ground photo-eclectors with a pitfall trap were used in each plot and were emptied and moved one to two times/week from sugar beet drilling until the end of June. Pest data are not included here.

A replicated, controlled study of arable cultivation over one year in Belgium (11) found that reduced tillage did not increase ground beetle (Carabidae) abundance or species richness. Ground beetle abundance was higher in conventionally ploughed plots (30 cm: 4,073–6,166 individuals) than those with reduced tillage (15 cm: 3,361–4,496) or no ploughing (2,604–3,577), largely due to one dominant species *Pterostichus melanarius* in ploughed fields. Abundance varied with crop type. Species richness also varied with crop type (beet: 13–14 species; wheat: 14–15; barley: 14–16; maize: 15–16) but not treatment (ploughed: 13–15; reduced tillage: 13–16; none: 14–16). However, less abundant species in conventionally ploughed plots tended to increase with reduced or no tillage. No-tillage plots received 30 kg/ha nitrogen and herbicide. Ground beetles were sampled using 6 pitfall traps in 2 plots (40 x 20 m) per treatment and crop. Traps were collected weekly from April until harvest in 1982.

A controlled trial at Reinshof experimental farm, Lower Saxony, Germany (12) found that the number of adult rove beetles (Staphylinidae) was similar in ploughed and unploughed wheat field plots, but there were more beetle larvae in unploughed plots. Ten rove beetle species (of a total of 94 species or types) preferred soils with reduced tillage as larvae and adults. The experiment was carried out on four wheat fields, half ploughed and half subject to non-inversion tillage, in 1992 and 1993. Four pitfall and four emergence traps were set in each half of each field and monitored throughout the year, or from April to July respectively. Each field was managed under a different farming system, as part of another experiment, so the four fields were not replicates.

A randomized, replicated, controlled trial from 1990 to 1992 in Suffolk, UK (13) found that abundance of the grass weed sterile brome *Bromus sterilis* increased ten-fold each year in plots with minimum tillage, but did not increase in ploughed plots. This was true on plots where sterile brome was sown alone, with other weed species or control plots with weeds unsown. Numbers of other weeds, common poppy *Papaver rhoeas* and cleavers *Galium aparine*, remained low on most plots and did not show a consistent difference between ploughed and minimum tillage plots. From October 1989 winter wheat plots were either ploughed to a depth of 22 cm or minimum-tilled to a depth of 6 cm. Minimum tilled plots were treated with conventional herbicides used to control grass weeds in cereals. Ploughed plots were selectively weeded and hoed by hand twice a year at most. There were three 9 m^2 replicate plots for each combination of treatments. Weed growth was monitored from 1990 to 1992.

A paired site comparison study on two farms, at Relliehausen and Grossobringen, Germany (14), found significantly more potworms (Enchytraeidae) in conventionally ploughed treatments than in reduced tillage. There were averages of 8,265–8,664 potworms/m^2 under reduced tillage, and 3,620–6,296 potworms/m^2 under conventional tillage. In plots with reduced tillage more than 60% of potworms were in the upper 10 cm of soil at both sites. In deep ploughed plots the potworms were distributed down to 25 cm deep. Conventional treatments were ploughed to 25–30 cm depth at both sites. The reduced tillage treatments were conservation tillage with a rotary harrow to a depth of 12 cm, incorporating mulch at Relliehausen, and shallow ploughing to 12 cm at Grossobringen. The systems had been in place since 1990. In spring 1995, potworms were extracted from fifteen 25 cm deep soil cores, divided into 5 cm layers, in each tillage system.

A replicated, controlled trial at the University of Agriculture, Nitra, Slovakia (15) found that flying insects in an organic wheat crop – both pests and predators – were more abundant after minimum tillage than after ploughing. Pest insects, excluding aphids (Aphidoidea), were generally more abundant under minimal tillage in a given year. This group included 11 different types of thrip (Thysanoptera), bug (Hemiptera), beetle (Coleoptera), sawfly (Hymenoptera), moth (Lepidoptera) and fly (Diptera). Natural enemies, which included flies, wasps (Hymenoptera) and beetles were also generally more abundant after minimal tillage than after ploughing, although the effect was less strong and not true for hoverflies (Syrphidae). Natural enemy insects were more affected by the previous crop, being more abundant in wheat following a maize crop. Two 50 m^2 study plots were ploughed to 24 cm deep, and 2 were ploughed to 15 cm deep (minimal tillage) each year from 1994 to 1996, and planted with winter wheat. Insects were collected with a sweep net in 5 m^2 patches of each plot, weekly from April or May to June or July in 1995, 1996 and 1997.

A 1999 literature review (16) found that reduced tillage (either shallow ploughing, 'conservation' tillage or no tillage) has been shown to enhance ground beetle (Carabidae) numbers in four European studies (including (10)) relative to conventional ploughing. One European study showed no difference in numbers between conventionally ploughed and reduced tillage fields (Paul 1986). One European study (11) showed greater numbers of ground beetles on deep ploughed fields than under reduced tillage. However, different species responded differently. One study (11) listed seven ground beetle species associated with reduced tillage or untilled plots.

A paired sites study in 1993–1999 on arable fields in Gülzow, north Germany (17) found that reduced tillage could lead to higher weed densities and higher weed species numbers compared to ploughing. Single weed species were affected differently by the tillage method in different crops. For example, goosefoot *Chenopodium album* and couch grass *Elymus repens* were observed more frequently under reduced tillage than after ploughing in summer cereals, but less frequently in reduced tillage winter cereals. The opposite was found for others such as knotweed species *Polygonum* spp. and common chickweed *Stellaria media*, which were more frequent in ploughed than in reduced tillage summer cereal fields, whereas in winter cereals they were more frequent under reduced tillage. Fields were divided into one organically and one integrated managed part (0.55–1.1 ha). Within each management system, two types of soil preparation (ploughing and reduced tillage) were compared. The 6-year crop rotation included clover *Trifolium* spp.-grass ley, potatoes/corn, spring barley, fodder peas, winter wheat/rye and oat undersown with red clover *T. pratense*. Mechanical weed control was adopted on the organic fields. Herbicide use in the integrated system was adapted to the actual weed abundance. Weed density (plants/m²), weed cover (%) and species number were recorded yearly before weed control activities on four plots (from 1997) in each field. Note that no statistical analyses have been performed on the data presented in this paper.

A 2000 literature review (18) looked at which agricultural practices can be altered to benefit ground beetles (Carabidae). It found one study from Europe showing more ground beetles after non-inversion tillage (Heimbach & Garbe 1995). One European study found no effect of tillage on ground beetle numbers (Huusela-Veistola 1996). Two studies from Europe showed that different species respond differently (Hance & Gregoire-Wibo 1987, (9)).

A 2001 review of published literature (19) found seven studies showing higher earthworm (Lumbricidae) populations under conservation tillage, with two to nine times more earthworms than under conventional tillage. Three of these studies were European studies considered above (2–4); one was in Australia; two in the USA; and one in the tropics. Two studies in the UK and one in Switzerland (3, 4, 6) found more large-bodied deep-burrowing earthworms under no-tillage, and similar numbers or fewer smaller-bodied, not deep burrowing worms under no-tillage compared with conventional ploughing.

A replicated, controlled trial in Rhine-Hessia, Germany from 1995 to 1998 (20) found that soils managed with layer cultivation (conservation tillage) had more adult and juvenile earthworms (Lumbricidae) and a greater biomass of earthworms than soils that were ploughed or two-layer ploughed. In most cases there were twice as many worms under layer cultivation. For example, there were 22 *Lumbricus terrestris* individuals under layer cultivated winter rye, compared to 9 in ploughed fields and 7 in two-layer ploughed fields. Four earthworm species were found in ploughed fields, five to six species in two-layer ploughed fields and six to seven species in fields under layer cultivation. Ploughing, two-layer ploughing (shallow turning to 15 cm, soil loosening to 30 cm) and layer cultivation (also called conservation tillage, only loosening the soil to 30 cm depth, no turning) were tested on ten 12 x 100 m plots. There were five different crop types in the experiment: green fallow, winter wheat with intercrop, peas, winter rye with intercrop and summer barley. Each crop/tillage combination was replicated twice. Crop type did not have a significant effect on the number or biomass of earthworms.

A replicated, controlled trial at the University of Agriculture, Nitra, Slovakia (21) found that predatory insects were more abundant after minimum tillage than after deep ploughing in a conventionally farmed wheat crop. Most pest insects were less abundant in a given year under minimal tillage than in ploughed plots. This group included thrips (Thysanoptera), bugs (Hemiptera), beetles (Coleoptera), sawflies (Hymenoptera), moths (Lepidoptera) and flies (Diptera). Only sawflies in the family Tenthredinidae and some bugs (Heteroptera) were more abundant on minimal tillage plots. Natural enemies, which included flies (Diptera), wasps (Hymenoptera) and beetles (Coleoptera) were more abundant after minimal tillage than after ploughing, although this was not true for hoverflies (Syrphidae). Pest insects were less abundant after minimum tillage. Two 50 m^2 study plots were ploughed to 24 cm deep, and 2 were ploughed to 15 cm deep (minimal tillage) each year from 1994 to 1996, and planted with winter wheat. Insects were collected with a sweep net in 5 m^2 patches of each plot, weekly from April or May to June or July in 1995, 1996 and 1997.

A replicated, controlled, randomized study of conventional and non-inversion tillage in six fields in Somerset, UK (22) found that earthworm (Lumbricidae) abundance and species diversity were higher in non-inversion regimes using a Dutzi machine than in either non-inversion farming using a Vaderstad drill or conventional ploughing and drilling. From 1990–1994 there was no significant difference between density in Dutzi non-inversion plots (65/m^2) and conventional plots (64/m^2) but biomass was significantly greater in Dutzi plots in 1993 and 1994 (23–40 vs 13–16 g/m^2). From 1995–2000 worm density was significantly greater in Dutzi plots than conventional plots in 1995, 1999 and 2000 (72–155 vs 38–66/m^2); Vaderstad non-inversion plots did not differ from conventional plots (62–72 vs 38–66/m^2). Biomass was significantly greater in Dutzi than conventional plots in all but one year (35–68 vs 16–31 g/m^2); biomass in Vaderstad plots was only greater than conventional plots in two years (33–42 vs 16–19 g/m^2). Thirteen species were recorded from 1995 to 2000, four of which were significantly more abundant in Dutzi than conventional plots; densities in Vaderstad plots were intermediate. There was no significant effect of treatment on the other six common species, although densities of four tended to be higher in Dutzi than conventional plots. Fields were divided into four plots (1 ha) which were assigned randomly to treatments. In autumn 1994–2000 an additional non-inversion tillage regime was included, using a Vaderstad disc coulter drill. Fertilizers and pesticides were also reduced (25–40% and 30–90% respectively) in non-inversion tillage regimes compared to conventional farming. Earthworms were sampled over one hour using diluted formalin on the soil in three quadrats (0.25 m^2) placed at random/plot in March–April and September–October each year.

A small replicated trial in 1997 at an experimental farm in Normandy, France (23) (same study as (30)) found that the biodiversity of small arthropods (mites (Acari), springtails (Collembola) and others) was higher on arable land without deep ploughing than on conventionally ploughed land. This difference was true for five of the six monitoring months, from January to June 1997. The comparison was replicated on two fields. The land not ploughed in 1997 had been managed under integrated farm management for the previous eight years, and had been treated with significantly less insecticide and fungicide on average (but not less herbicide) than the conventional treatment over five cropping years. Another replicate of the integrated and conventional management was not tilled in 1997. Here there was not such a consistent difference in diversity of small arthropods. The authors concluded that tillage had more influence on small soil arthropods than reduced pesticide use.

A replicated, controlled study in the winters of 2000–2003 in 63 experimental and 58 control winter wheat and barley fields in Oxfordshire, Leicestershire and Shropshire, UK (24) found that significantly more beetle (Coleoptera) larvae and earthworms (Lumbricidae) were recorded in non-inversion tillage fields than in conventionally-tilled fields (no data given). The opposite was true for rove beetles (Staphylinidae). Ground beetles (Carabidae) and spiders (Araneae) showed no significant differences between treatments. This study was part of the same experimental set-up as (26, 31).

A replicated, controlled, before-and-after trial on the agricultural research farm at Rugballegaard in East Jutland, Denmark (25) found no difference in the total abundance of springtails (Collembola) between conventionally ploughed and reduced tillage plots. The total number of springtails fell from around 90,000/m^2 to around 30,000/m^2, shortly after both tillage treatments. The distribution of springtails at different depths in the soil differed between treatments. After ploughing, there were significantly fewer springtails in the upper 4 cm of the soil on ploughed plots and an increase in springtail numbers at 16–20 cm depth (statistically significant for some species only). This was thought to be caused by the inversion of soil during ploughing. Two tillage methods were tested on 4 areas of organic wheat fields from 1998 to 1999: conventional mouldboard ploughing to 20 cm depth followed by harrowing, or deep tillage with a non-inverting tine subsoiler to 25–35 cm depth, rotavated at the surface. The first samples were taken in September 1998, before the first tillage treatment. Springtails were extracted from soil samples at 3 locations in each plot, and at 4 depths: 0–4, 8–12, 16–20 and 28–32 cm. Subsequent samples were taken in October 1998 (two samples) and March 1999.

A replicated, controlled study in the winters of 2001–2003 in 20 experimental and 20 control winter wheat fields at 7 farms in Leicestershire and Shropshire, UK (26) found that there was no significant difference in earthworm (Lumbricidae), ground beetle (Carabidae), rove beetle (Staphylinidae) or spider (Araneae) numbers in non-inversion tillage fields compared to conventionally-tilled fields. Beetle (Coleoptera) larvae showed some tendency for higher numbers in conventional (1.2) compared to non-inversion tillage (0.5) in July, but not March or May. The mean number of seed species per field did not differ significantly between treatments in autumn (17–18/m^2) or spring (15–16/m^2). Nine samples for earthworms and seeds were taken in October–November and March and for arthropods in March, May and July. Earthworms were sampled in 10 cm diameter by 10 cm deep cores, seeds in surface soil samples of 25 cm^2 and 1 cm deep and spiders and insects in pitfall traps. This study was part of the same experimental set-up as (24, 31).

A 2004 review of the effects of non-inversion tillage on beetles (Coleoptera), spiders (Araneae), earthworms (Lumbricidae) and farmland birds across the world, but with special reference to the UK and Europe (27), found evidence for some positive responses. It found one three-year study from the UK ((24), Cunningham et al. 2003, (31)) that found Eurasian skylark *Alauda arvensis*, gamebirds and seed-eating songbirds were more abundant on non-inversion tillage fields in late winter compared to conventional tillage. Two studies found more beetles in reduced or no tillage plots (Andersen 1999, Holland & Reynolds 2003); four studies found mixed results. Two out of three studies found positive effects of reduced or non-inversion tillage on spiders ((9), Holland & Reynolds 2003). Ten out of 13 studies found positive effects of reduced or non-inversion tillage on earthworms.

A 2004 review of the effects of conservation tillage relative to conventional ploughing (28) mainly but not exclusively focussing on European studies, found that earthworms (Lumbricidae) almost always benefit from conservation tillage, but effects are more mixed for other organisms, including plants, birds and mammals. Four European experimental studies and two reviews showed that conservation tillage increased earthworm populations, particularly deep-burrowing species such as *Lumbricus terrestris*, with up to six times more earthworms under conservation tillage in the context of integrated farming (including: (4), El Titi & Ipach 1989, Jordan *et al*. 2000, Kladivko 2001). Conservation tillage increased the diversity and abundance of springtails (Collembola) and mites (Acari) in four studies (Bertolani *et al*. 1989, El Titi & Ipach 1989, Vreeken-Buijs *et al*. 1994, Franchini & Rockett 1996). European studies on larger arthropods (beetles (Coleoptera) and spiders (Araneae)) were less consistent, with two studies showing increased numbers under conservation tillage ((9), Purvis & Fadl 1996), one showing no effect (Huusela-Veistola 1996) and two showing both increases and decreases (Andersen 1999, Holland & Reynolds 2003). Different arthropod species were affected differently. Four UK studies showed an increase in grass species classed as weeds under conservation tillage (Theaker *et al*. 1995, Rew *et al*. 1996, Cavan *et al*. 1999, (13)). Other weed species have been shown to decline under conservation tillage in the context of integrated farming (one German study, Albrecht & Mattheis 1998) or remain stable (one UK study, (13)). For birds, one study showed no effect on five bird species in the context of organic farming (Saunders 2000). For mammals, one European study found that wood mice *Apodemus sylvaticus* were more abundant on conventionally ploughed fields than under conservation tillage in the context of organic and integrated farming (Higginbotham *et al*. 2000).

A randomized, replicated, controlled trial in spring 1999 and 2000 at the Rugballegaard Institute of Agricultural Sciences, Denmark (29) found that both soil loosening and non-inversion tillage have adverse effects on ground beetles (Carabidae) and spiders (Araneae), but for non-inversion tillage these are not quite as severe as the effects of ploughing. There were around 20 ground beetles/m^2 immediately after non-inversion tillage, compared to around 12 ground beetles/m^2 after ploughing and around 18 in untreated control plots. There was no difference between ploughing and non-inversion tillage plots in numbers of spiders or rove beetles (Staphylinidae), or in any of the 3 arthropod groups 26 days after the treatment. Overall, neither ploughing nor non-inversion tillage immediately reduced the numbers of predatory arthropods significantly, relative to untreated control plots, but all three groups had lower numbers in ploughed or non-inversion tilled plots 26 days later than in untreated control plots (for example <5 spiders and <20 ground beetles/m^2 in both ploughed and non-inversion tillage plots, compared to around 25 spiders and 130 ground beetles/m^2 in control plots). In a separate experiment, soil-loosening to 8 cm depth with a tined hoe immediately reduced spider numbers by 25% (around 120 spiders/m^2 in control plots and 90 spiders/m^2 in treated plots) and ground beetle numbers by 51% (around 70 ground beetles/m^2 in control plots; 35 ground beetles/m^2 in treated plots) but not rove beetle numbers. These differences were statistically significant and persisted in a second sample 18 days later. The treatments were replicated between 4 and 8 times, on 12 x 40 m plots. Predatory arthropods were sampled using emergence traps.

A small replicated trial at an experimental farm in Normandy, France (30) (same study as (23)) found more spiders (Araneae) and ground beetles (Carabidae), but fewer rove beetles (Staphylinidae) in arable plots managed without deep ploughing than in plots with conventional ploughing. The unploughed plots were also managed with limited used of herbicides and fungicides, and no insecticides, so it is difficult to separate the effects of ploughing from the effects of reducing pesticide use. However, both ground beetles and spiders were also more abundant in sub-plots that restricted pesticide and herbicide use even more, whereas rove beetles were not. There were three replicates of each treatment. Management was over 11 years from 1990 to 2001. Insects and spiders were monitored in May and June from 1999 to 2001.

A replicated, controlled study in the winters of 2000–2003 in 63 experimental and 58 control winter wheat and barley fields in Oxfordshire, Leicestershire and Shropshire, UK (31) found that Eurasian skylark *Alauda arvensis*, seed-eating songbirds and gamebirds occupied a significantly higher proportion of fields managed through non-inversion tillage than conventionally ploughed fields in late winter (January–March). Species richness of seed-eating songbirds was also higher on non-inversion tillage fields (five species vs one on conventionally ploughed fields). No birds showed any preference for field type in early winter (October to December), and crows (Corvidae), pigeons (Columbidae) and insect-eating birds showed no preference across the study period. Field size ranged from 1.6 to 22.3 ha, with similar numbers of non-inversion tillage and conventionally-ploughed farms censused each year. This study was part of the same experimental set-up as (24, 26) and is also described in an additional publication (Cunningham *et al*. 2003).

A replicated, controlled trial at the Oakpark Research Centre, County Carlow, Ireland (32), found that winter wheat plots subjected to a reduced tillage regime for three years had more springtails (Collembola) in the soil than conventionally ploughed plots. Conventional plots had around 100 springtails/m^2 and 'ECOtilled' plots had over 300 springtails/m^2 on average. Sixteen 24 x 30 m plots were established in 2000 and sown with winter wheat every year. 'ECOtillage' plots were cultivated with a shallow cultivator 5–10 cm deep after harvesting. Weeds were sprayed with herbicide, and the crop was sown with a cultivator drill. Control plots were ploughed with a mouldboard plough to a depth of 25 cm and cultivated with a power harrow (10–15 cm) before sowing. At harvest, straw was either baled and removed or chopped and replaced on the soil surface. There were 4 replicates of each treatment combination and 12 m buffer strips around each plot. Springtails were extracted from soil samples in 2003.

A replicated, controlled, randomized study of cultivation techniques at the Lithuanian Institute of Agriculture (33) found that earthworm (Lumbricidae) abundance tended to be higher under reduced tillage than deep ploughing. In 2000, the number of earthworms was higher in plots with reduced tillage (42–97 m^2) than ploughed plots (with and without straw; 38–80/m^2); there was no difference in 2001. Compared to deep ploughing, earthworm population density increased through soil conservation technology using several measures (straw disced-in; catch crop; not ploughed) by 53/m^2 (141%) in wheat stubble and 40/m^2 (103%) in oat stubble; there was no effect in barley stubble. Earthworm numbers in ploughed soil with straw incorporated and a catch crop were significantly larger (by 28/m^2) in one of the three years. Intensive soil tillage (straw; shallow discing; herbicide; deep ploughing) did not affect earthworm density. There were four replicates of eight treatments: conventional and conservation soil tillage in combination with chopped straw mulch (wheat or barley), catch crop (white mustard *Sinapis alba*) and herbicide (Roundup; 3 l/ha) application. Earthworms were counted in 4 replications (0.25 m^2, depth 25 cm) in 3 locations in each plot in April 2000–2002.

A 2006 review (34) of the impact of farm management practices on below-ground biodiversity and ecosystem function found tillage had negative effects on beetles (Coleoptera), springtails (Collembola), mites (Acari), spiders (Araneae) and earthworms (Lumbricidae). The review looked at studies worldwide but here we focus on European studies. One review (Wardle 1995) (location not provided) concluded that tillage tended to reduce large soil organisms (beetles, spiders and earthworms) more than the smallest ones (bacteria, fungi), and that intermediate-sized groups (nematodes (Nematoda), mites and potworms (Enchytraeidae)) showed small population increases. Two studies (one from Sweden; one review) demonstrated the direct negative effects of tillage on mites, springtails, and beetles (Andren & Lagerlöf 1980, Wardle 1995); a further study showed that compaction during tilling can reduce the number of earthworms and microarthropods (Aritajat *et al*. 1977;

location not given). One study from Denmark showed that tillage reduced the springtail population to about 1/3 of the pre-tillage level one week after cultivation (25). Two studies (one study from Switzerland) noted differences in the species composition of arbuscular mycorrhizal fungi, earthworms and nematodes (Nematoda) between tillage and no-till systems (Wardle 1995, Jansa *et al*. 2003). Two studies (one study from Germany) investigated the impact of tillage on the balance between bacteria and fungi, with mixed results (Wardle 1995, Ahl *et al*. 1998). One study from the UK found that invertebrate food resources for birds increased in no-till compared to conventionally tilled systems (Tucker 1992).

A replicated, controlled study in May to July 2003–2004 in two arable regions in central Germany (35) found that the abundance/activity density of both spiders (Araneae) and ground beetles (Carabidae) was higher on fields with reduced tillage (ground beetles: 1,446 individuals (mulched fields), 1,634 (directly sown fields); spiders: 4.75 individuals/day and trap (mulching), 2.9 (direct sown)) than on conventional ploughed fields (ground beetles: 1,241 individuals; spiders: 2.85 individuals/day and trap), but lower than on organic ploughed fields (ground beetles: 2,725 individuals; spiders: 6.05 individuals/day and trap). Species richness of spiders was higher on reduced tillage fields (direct sown: 40 species, mulched: 35 spp.) than on the other field types (organic: 37.5 spp., conventional ploughed: 35 spp.), but the number of ground beetle species was lower on reduced tillage fields (mulched: 35.5 spp., direct sown: 34 spp.) than on the other field types (39 spp. conventionally ploughed, 50 spp. organically ploughed). However, the effect of reduced tillage was species dependent for both spiders and ground beetles, i.e. some species clearly benefited from reduced tillage, whereas others preferred ploughed fields. Four field types were investigated: organic ploughed fields, conventional ploughed fields, conventional mulched fields (no plough), and conventional directly sown fields (no plough). Cereals were grown on all fields during the study years. Spiders and ground beetles were caught using pitfall traps (six replications/field type). Note that no statistical analyses were performed on the data presented in this study.

A replicated site comparison study in 2005 and 2006 on 31 farms in Seine-et-Marne, France (36) reported in the text that the number of plant species was higher on no-till farms than conventional farms, but the data presented on a graph in this paper appeared to show no difference, with 5 plant species on both types of farm. Twenty-six fields from 17 farms were sampled 3 times in 2005 (April, June, September). Sixty-four fields from 31 farms (including all those surveyed in 2005) were sampled twice in 2006 (April and July). Plants were recorded in 10 permanent, regularly spaced, 1 m^2 (0.5 x 2 m) quadrats along the permanent margins of each field. The difference between different ploughing systems was only found in 2006.

A replicated, paired site study from October to March 2003–2006 in 12 pairs of winter wheat fields in Dióskál, Hungary (37) found that the preference of some farmland birds for conservation tillage fields over adjacent ploughed fields decreased over the study period. In the first farm (with eight field pairs) Eurasian skylark *Alauda arvensis* and seed-eating song-birds (mostly goldfinch *Carduelis carduelis*) were more abundant on conservation tillage fields in the first winter (2003–2004), whilst European starling *Sturnus vulgaris* and skylark were more abundant on conservation tillage fields over the second and third winter respectively. In the second farm (four field pairs) skylark and crows (Corvidae) were more abundant on conservation tillage fields in the first winter only. The number of days with snow cover on the ground increased over the three years. The authors suggest such abnormal weather may have confounded the results.

A small replicated, randomized, controlled study from April–July 2005 in two experimental and two control fields of winter wheat in Rutland, England (38) found that Eurasian skylark *Alauda arvensis* nest density was higher in fields managed through conservation tillage than fields that were ploughed (24 out of 32 nests in conservation tillage fields). Average laying date was also significantly earlier on conservation tillage fields by 25 days. The authors suggest the effect was due to conservation tillage fields containing more crop residue than ploughed fields (32% compared to 0% residue respectively). Foraging distance of adult skylarks providing food for nestlings was halved on conservation tillage fields (48 m vs 93 m). However, nest success and nestling size were similar in both field types. Control fields were sown with winter wheat after mould-board ploughing, while conservation tillage fields were direct drilled into oilseed rape residue after light rotary harrow.

A replicated, controlled trial at the University of Kassel experimental farm, Franken-hausen, Germany, (39) found that neither of two methods of reducing tillage suitable for use on organic farms enhanced numbers of earthworms (Lumbricidae) in the soil. Ploughing is important for weed control in organic farming, so both systems involved some soil inversion. A ridge culture system, using a shallow plough that formed ridges and loosened the soil with a spike to 35 cm depth, and a shallow inversion plough to 10 cm depth, were compared with conventional ploughing to 30 cm depth. There was no difference in the abundance or total biomass of earthworms between the conventional ploughing and shallow ploughing. On average between 5 and 30 earthworms/m^2 and between 2 and 40 g earthworm/m^2 were found in the different crops for these treatments. Under the ridge culture system there were significantly fewer earthworms (3–20 earthworms/m^2 on average) and lower biomass of earthworms (0–27 g/m^2 on average). The experiment began in 2003, with 12 replicates of each treatment. Plots were managed organically. Earthworms were monitored by hand sorting and extraction in October 2005.

A replicated trial at Estrées-Mons, France (40), found that reduced tillage plots had a significantly higher average biomass of earthworms (Lumbricidae) but not a greater number of individual worms. Under reduced tillage there were 77 g earthworms/m^2 and 116 earth-worms/m^2. Under conventional tillage there were 37 g earthworms/m^2 and 111 earth-worms/m^2. This difference was because there were more large, deep-burrowing worms such as *Lumbricus terrestris* and *Aporrectodea giardi* and fewer small litter-dwelling worms such as *A. caligosa* in the reduced tillage plots. Soils under reduced tillage had significantly more large pores created by earthworm activity in all size classes. Twelve 0.4 ha arable plots were subject to reduced tillage, prepared only with a rotary or disc harrow to 7 cm depth. Twelve con-trol plots underwent conventional tillage, with a mouldboard plough to a depth of 30 cm, followed by seed bed preparation with a harrow to 7 cm. The management began in 1999. Earthworms were sampled in November after tillage and in April, from November 2003 until April 2006 (six times).

A replicated trial in the winters of 2006–2008 in four (2006–2007) and two (2007–2008) fields (located on one farm) of winter oilseed rape *Brassica napus* crops in Cambridgeshire, UK (41) found that bird densities were similar between oilseed rape established using two different methods of reduced tillage (non-inversion tillage and broadcasting). Neither indi-vidual species nor groups of species (seed-eaters, probers) responded to differences in crop establishment. However, a Farmland Bird Index (which included omnivorous, carnivorous, insect-eating and seed-eating species) was significantly higher on broadcast oilseed rape fields. The authors point out that the overall densities on both treatments were still relat-ively low compared to other interventions (such as wild bird seed and overwinter cereal stubble). Two surveys were made in each field each month between September–March across the whole field area.

A replicated, controlled trial near Welschbillig, southern Eifel, Germany (42) found a higher biomass of large deep-burrowing earthworms (Lumbricidae) in arable soils subject to

4 different types of reduced tillage, compared to ploughed soils, after 10 years. There were 52–79 g deep-burrowing worm/m^2 under reduced tillage, compared to 10 g/m^2 in ploughed treatments. For two non-inversion tillage treatments, there were greater numbers of deep-burrowing worms. One of these treatments had mulched crop residue on the surface. On average there were 5 deep-burrowing earthworms/m^2 in the ploughed treatment, compared to 21–25 deep-burrowing earthworms/m^2 with non-inversion tillage. The total number of earthworms was not significantly different between tillage treatments (113–160 earthworms/m^2 on average) but total mass of all earthworms was significantly higher in the disc harrow treatment than the ploughed treatment (119 g/m^2 compared to 67 g/m^2 under ploughing). Five tillage treatments were carried out on 2 replicate plots each, for 10 consecutive years: conventional ploughing to 25 cm depth, non-inversion loosening of topsoil to 15 cm depth, disc harrowing and slightly loosening soil to 15 cm depth, non-inversion tillage with crop residue mulch on the surface, or no tillage with direct sowing of crop. Earthworms were sampled at the end of the experiment in spring 2008, under a winter barley crop.

A site comparison study in Komturei Lietzen, Brandeburg, Germany (43), found that the average abundance of earthworms (Lumbricidae) in arable soil was almost identical under conventional and reduced tillage over ten years – around 12 earthworms/m^2. From September 1996 until 2006, one half of a 74 ha arable field was conventionally ploughed to a depth of 25 cm. The other half was subject to non-inversion tillage using a precision cultivator to a depth of 15–18 cm. Earthworms were collected by hand-sorting from twenty-one 40 x 50 x 20 cm blocks of soil in each treatment, in September and April–May of each year. Large, deep burrowing earthworms may be underestimated by this method. When paired sample points with similar soil properties were compared, average abundance of earthworms was higher under reduced tillage in soils with fine particles (>7% fine particles) but not in sandy soils.

A replicated, controlled study from April–June in 2006–2007 in 48 conservation tillage, 31 organic and 63 conventional winter barley and wheat fields in Seine-et-Marne, France (44) found that that bird species differed in their responses to management. Two species were more abundant in conservation tillage fields than conventional fields, while seven were more abundant on conservation tillage fields than on organic. One species was more abundant on conventional fields and five on organic, compared to conservation tillage. Specialist species were least abundant on conservation tillage fields, whilst insect-eating birds were more abundant. The authors point out that conservation tillage fields were more intensely managed than conventional fields and experienced much disturbance. Habitat and dietary data were used to construct a species specialization index.

A randomized, replicated, controlled trial on three organic arable farms in different regions of France (45) found that earthworm (Lumbricidae) biomass was higher under no tillage than on the control or other reduced tillage treatments at all three sites, in at least two years. At two of the sites, there was no difference between treatments in earthworm abundance. At the other site (an irrigated farm in the Rhône Alpes region of southeastern France) earthworm abundance was also significantly higher in the no tillage treatment in two of the three sampling years. In general there were more deep-burrowing species in the no tillage treatment than other treatments. This difference was statistically significant at two of the three sites. There was no increase in the number of earthworm burrows (created by deep-burrowing earthworms) under no tillage. Four tillage treatments were compared: conventional mouldboard ploughing to 30 cm; shallow ploughing to 15–20 cm; reduced tillage with tined tools to 12–15 cm; no tillage. On each farm, three replicates of each treatment were randomly located within three blocks. Experiments began between 2003 and 2005 and were monitored annually for two to five years. Earthworms were extracted using formalin in October or April–May.

A small replicated trial near Paris, France (46) found no difference in the total number of earthworms or earthworm (Lumbricidae) species on direct drilled (no-till) plots compared to conventionally farmed plots, but earthworm biomass was always higher in direct drilled plots. These plots had an average of 79 g earthworm/m^2, compared to 32 g/m^2 on conventional plots. There was a much higher proportion of deep-burrowing species (50% of all earthworms were deep-burrowing) in the direct-drilled plots than in conventional plots (13% of all earthworms). There was also a higher proportion of litter-dwelling earthworms in the direct drilling plots (14% of all earthworms, compared to 2% in conventional plots). From 1997 to 2007 treatments were compared on 1 ha arable plots with two replicates of each treatment. The direct drilled treatment involved a continuous plant cover 'living mulch' with herbicides used to control weeds and no tillage. Earthworms were sampled from five sample points in each plot by chemical extraction and hand-sorting, every autumn for three years (2005–2007).

(1) Edwards C.A. (1975) Effects of direct drilling on the soil fauna. *Outlook on agriculture*, 8, 243–244.
(2) Barnes B.T. & Ellis F.B. (1979) Effects of different methods of cultivation and direct drilling, and disposal of straw residues on populations of earthworms. *Journal of Soil Science*, 30, 669–679.
(3) Gerard B.M. & Hay R.K.M. (1979) The effect on earthworms of ploughing, tined cultivation, direct drilling and nitrogen in a barley monoculture system. *Journal of Agricultural Science*, 93, 147–155.
(4) Edwards C.A. & Lofty J.R. (1982) The effect of direct drilling and minimal cultivation on earthworm populations. *Journal of Applied Ecology*, 19, 723–734.
(5) Lee E.E. (1984) The effect of reduced cultivation on selected soil fauna. PhD thesis. University of Salford.
(6) Wyss E. & Glasstetter M. (1992) Tillage treatments and earthworm distribution in a Swiss experimental corn field. *Soil Biology & Biochemistry*, 24, 1635–1639.
(7) Glemnitz M. (1993) Effects of conservation tillage on abundance and dominance in weed communities on arable land. *Proceedings of the EWRS 8th Symposium: Quantitative approaches in weed and herbicide research and their practical application*. Braunschweig, Germany, 14–16 June. pp. 697–704.
(8) Ruiter P.C.D., Moore J.C., Zwart K.B., Bouwman L.A., Hassink J., Bloem J., Vos J.A.D., Marinissen J.C.Y., Didden W.A.M., Lebrink G. & Brussaard L. (1993) Simulation of nitrogen mineralization in the below-ground food webs of two winter wheat fields. *Journal of Applied Ecology*, 30, 95–106.
(9) Kendall D.A., Chinn N.E., Glen D.M., Wiltshire C.W., Winstone L. & Tidboald C. (1995) Effects of soil management on cereal pests and their natural enemies. In D.M. Glen, M.P. Greaves & H.H. Anderson (eds.) *Ecology and Integrated Farming Systems*, John Wiley and Sons, Inc., London. pp. 83–102.
(10) Heimbach U. & Garbe V. (1996) Effects of reduced tillage systems in sugar beet on predatory and pest arthropods. In K. Booij & L. den Nijs (eds.) *Arthropod Natural Enemies in Arable Land ii - Survival, Reproduction and Enhancement*, Aarhus University Press, Aarhus, Denmark. pp. 195–208.
(11) Baguette M. & Hance T. (1997) Carabid beetles and agricultural practices: influence of soil ploughing. *Biological Agriculture & Horticulture*, 15, 185–190.
(12) Krooss S. & Schaefer M. (1998) The effect of different farming systems on epigeic arthropods: a five-year study on the rove beetle fauna (Coleoptera: Staphylinidae) of winter wheat. *Agriculture, Ecosystems & Environment*, 69, 121–133.
(13) McCloskey M.C., Firbank L.G., Watkinson A.R. & Webb D.J. (1998) Interactions between weeds of winter wheat under different fertilizer, cultivation and weed management treatments. *Weed Research*, 38, 11–24.
(14) Rohrig R., Langmaack M., Schrader S. & Larink O. (1998) Tillage systems and soil compaction – their impact on abundance and vertical distribution of Enchytraeidae. *Soil & Tillage Research*, 46, 117–127.
(15) Gallo J. & Pekar S. (1999) Winter wheat pests and their natural enemies under organic farming system in Slovakia: effect of ploughing and previous crop. *Anzeiger Fur Schadlingskunde – Journal of Pest Science*, 72, 31–36.
(16) Kromp B. (1999) Carabid beetles in sustainable agriculture: a review on pest control efficacy, cultivation impacts and enhancement. *Agriculture, Ecosystems & Environment*, 74, 187–228.
(17) Gruber H., Händel K. & Broschewitz B. (2000) Influence of farming system on weeds in thresh crops of a six-year crop rotation. *Zeitschrift für Pflanzenkrankheiten und Pflanzenschutz* Sonderh. 17, 33–40.
(18) Holland J.M. & Luff M.L. (2000) The effects of agricultural practices on Carabidae in temperate agroecosystems. *Integrated Pest Management Reviews*, 5, 109–129.

(19) Chan K.Y. (2001) An overview of some tillage impacts on earthworm population abundance and diversity – implications for functioning in soils. *Soil and Tillage Research*, 57, 179–191.

(20) Emmerling C. (2001) Response of earthworm communities to different types of soil tillage. *Applied Soil Ecology*, 17, 91–96.

(21) Gallo J. & Pekar S. (2001) Effect of ploughing and previous crop on winter wheat pests and their natural enemies under integrated farming system in Slovakia. *Anzeiger fur Schadlingskunde*, 74, 60–65.

(22) Hutcheon J.A., Iles D.R. & Kendall D.A. (2001) Earthworm populations in conventional and integrated farming systems in the LIFE Project (SW England) in 1990–2000. *Annals of Applied Biology*, 139, 361–372.

(23) Cortet J., Ronce D., Poinsot-Balaguer N., Beaufreton C., Chabert A., Viaux P. & de Fonseca J.P.C. (2002) Impacts of different agricultural practices on the biodiversity of microarthropod communities in arable crop systems. *European Journal of Soil Biology*, 38, 239–244.

(24) Cunningham H.M., Chaney K., Wilcox A. & Bradbury R. (2002) The effect of non-inversion tillage on earthworm and arthropod populations as potential food sources for farmland birds. *Aspects of Applied Biology*, 67, 101–106.

(25) Petersen H. (2002) Effects of non-inverting deep tillage vs. conventional ploughing on collembolan populations in an organic wheat field. *European Journal of Soil Biology*, 38, 177–180.

(26) Cunningham H.M. (2004) The effect of non-inversion tillage on farmland birds, soil and surface-active invertebrates and surface seeds. PhD thesis. Open University.

(27) Cunningham H.M., Chaney K., Bradbury R.B. & Wilcox A. (2004) Non-inversion tillage and farmland birds: a review with special reference to the UK and Europe. *Ibis*, 146, 192–202.

(28) Holland J.M. (2004) The environmental consequences of adopting conservation tillage in Europe: reviewing the evidence. *Agriculture, Ecosystems & Environment*, 103, 1–25.

(29) Thorbek P. & Bilde T. (2004) Reduced numbers of generalist arthropod predators after crop management. *Journal of Applied Ecology*, 41, 526–538.

(30) Chabert A. & Beaufreton C. (2005) Impact of some agricultural practices on carabidae beetles. *IOBC/WPRS Bulletin*, 28, 101–109.

(31) Cunningham H.M., Bradbury R.B., Chaney K. & Wilcox A. (2005) Effect of non-inversion tillage on field usage by UK farmland birds in winter. *Bird Study*, 52, 173–179.

(32) Brennan A., Fortune T. & Bolger T. (2006) Collembola abundances and assemblage structures in conventionally tilled and conservation tillage arable systems. *Pedobiologia*, 50, 135–145.

(33) Kinderiene I. (2006) The effect of conservation farming on the abundance of earthworms on eroded soils. *Zemdirbyste/Agriculture*, 93, 96–105.

(34) Stockdale E.A., Watson C.A., Black H.I.J. & Philipps L. (2006) *Do Farm Management Practices Alter Below-ground Biodiversity and Ecosystem Function? Implications for Sustainable Land Management*. Joint Nature Conservation Committee, 364.

(35) Volkmar C. & Kreuter T. (2006) Biodiversity of spiders (Araneae) and carabid beetles (Carabidae) on fields in Saxony. *Mitteilungen Der Deutschen Gesellschaft Fur Allgemeine Und Angewandte Entomologie*, 15, 97–102.

(36) Chateil C., Abadie J.C., Gachet S., Machon N. & Porcher E. (2007) Can agri-environmental measures benefit plant biodiversity? An experimental test of the effects of agri-environmental measures on weed diversity. *Proceedings of the Vingtième conférence du columa journées internationales sur la lutte contre les mauvaises herbes*. Dijon, 11–12 December. pp. 356–366.

(37) Field R.H., Benke S., Badonyi K. & Bradbury R.B. (2007) Influence of conservation tillage on winter bird use of arable fields in Hungary. *Agriculture, Ecosystems & Environment*, 120, 399–404.

(38) Field R.H., Kirby W.B. & Bradbury R.B. (2007) Conservation tillage encourages early breeding by skylarks *Alauda arvensis*. *Bird Study*, 54, 137–141.

(39) Metzke M., Potthoff M., Quintern M., Hess J. & Joergensen R.G. (2007) Effect of reduced tillage systems on earthworm communities in a 6-year organic rotation. *European Journal of Soil Biology*, 43, S209–S215.

(40) Capowiez Y., Cadoux S., Bouchant P., Ruy S., Roger-Estrade J., Richard G. & Boizard H. (2009) The effect of tillage type and cropping system on earthworm communities, macroporosity and water infiltration. *Soil and Tillage Research*, 105, 209–216.

(41) Dillon I.A., Morris A.J. & Bailey C.M. (2009) Comparing the benefits to wintering birds of oil-seed rape establishment by broadcast and non-inversion tillage at Grange Farm, Cambridgeshire, England. *Conservation Evidence*, 6, 18–25.

(42) Ernst G. & Emmerling C. (2009) Impact of five different tillage systems on soil organic carbon content and the density, biomass, and community composition of earthworms after a ten year period. *European Journal of Soil Biology*, 45, 247–251.

(43) Joschko M., Gebbers R., Barkusky D., Rogasik J., Höhn W., Hierold W., Fox C.A. & Timmer J. (2009) Location-dependency of earthworm response to reduced tillage on sandy soil. *Soil and Tillage Research*, 102, 55–66.

(44) Ondine F.C., Jean C. & Romain J. (2009) Effects of organic and soil conservation management on specialist bird species. *Agriculture, Ecosystems & Environment*, 129, 140–143.

(45) Peigne J., Cannavaciuolo M., Gautronneau Y., Aveline A., Giteau J.L. & Cluzeau D. (2009) Earthworm populations under different tillage systems in organic farming. *Soil & Tillage Research*, 104, 207–214.
(46) Pelosi C., Bertrand M. & Roger-Estrade J. (2009) Earthworm community in conventional, organic and direct seeding with living mulch cropping systems. *Agronomy for Sustainable Development*, 29, 287–295.

Additional references

Aritajat U., Madge D.S., & Gooderham P.T. (1977) Effects of compaction of agricultural soils on soil fauna. 1. Field Investigations. *Pedobiologia*, 17, 262–282.
Andren O. & Lagerlöf J. (1980) The abundance of soil animals (microarthropoda, enchytraeids, nematoda) in a crop rotation dominated by ley and in a rotation with varied crops. In D. L. Dindal (ed.) *Soil Biology is Related to Land-use Practices.* Environmental Protection Agency, Washington. pp. 274–279.
Paul W.-D. (1986) Vergleich der epigäischen Bodenfauna bei wendender bzw. nichtwendender Grundbodenbearbeitung [Comparison of epigeic soil fauna under inversion and non-inversion tillage]. *Mitteilungen aus der Biologischen Bundesanstalt für Land und Forstwirtschaft*, Berlin-Dahlem, 232–290.
Hance T. & Gregoire-Wibo C. (1987) Effect of agricultural practices on carabid populations. *Acta Phytopathologica et Entomologica Hungarica*, 22, 147–160.
Bertolani R., Sabatini M.A. & Mola L. (1989) *Effects of Changes in Tillage Practices in Collembola Populations.* Proceedings of the Third International Symposium on Apterygota, Siena, 291–297.
El Titi A. & Ipach A. (1989) Soil fauna in sustainable agriculture: results of an integrated farming system at Lautenbach, FRG. *Agriculture, Ecosystems and Environment*, 27, 561–572.
Tucker G. M. (1992) Effects of agricultural practices on food use by invertebrate feeding birds in winter. *Journal of Applied Ecology*, 29, 779–790.
Vreeken-Buijs M.J., Geurs M., de Ruiter P.C. & Brussaard L. (1994) Microarthropod biomass-c dynamics in the belowground food webs of two arable farming systems. *Agriculture, Ecosystems and Environment*, 51, 161–170.
Heimbach U. & Garbe V. (1995) Effects of reduced tillage systems in sugar beet on predatory and pest arthropods. *Acta Jutlandica*, 71, 195–208.
Theaker A.J., Boatman N.D. & Froud-Williams R.J. (1995) The effect of nitrogen fertiliser on the growth of *Bromus sterilis* in field boundary vegetation. *Agriculture, Ecosystems and Environment*, 53, 185–192.
Wardle D. A. (1995) Impacts of disturbance on detritus food webs in agro-ecosystems of contrasting tillage and weed management practices. *Advances in Ecological Research*, 26, 105–185.
Franchini P. & Rockett C.L. (1996) Oribatid mites as 'indicator' species for estimating the environmental impact of conventional and conservation tillage practices. *Pedobiologia*, 40, 217–225.
Huusela-Veistola E. (1996) Effects of pesticide use and cultivation techniques on ground beetles (Col, Carabidae) in cereal fields. *Annales Zoologici Fennici*, 33, 197–205.
Purvis G. & Fadl A. (1996) Emergence of Carabidae (Coleoptera) from pupation: a technique for studying the 'productivity' of carabid habitats. *Annales Zoologici Fennici*, 33, 215–223.
Rew L.J., Froud-Williams R.J. & Boatman N.D. (1996) Dispersal of *Bromus sterilis* and *Anthriscus sylvestris* seed within arable field margins. *Agriculture, Ecosystems and Environment*, 59, 107–114.
Ahl C., Joergensen R.G., Kandeler E., Meyer B., & Woehler V. (1998) Microbial biomass and activity in silt and sand loams after long-term shallow tillage in central Germany. *Soil and Tillage Research*, 49, 93–104.
Albrecht H. & Mattheis A. (1998) The effects of organic and integrated farming on rare arable weeds on the Forschungsverbund Agrarokosysteme Munchen (FAM) research station in southern Bavaria. *Biological Conservation*, 86, 347–356.
Andersen A. (1999) Plant protection in spring cereal production with reduced tillage. II. Pests and beneficial insects. *Crop Protection*, 18, 651–657.
Cavan G., Cussans G. & Moss S.R. (1999) Modelling strategies to prevent resistance in black-grass (*Alopecurus mysosuroides*). Presented at Brighton Crop Protection Conference on Weeds, 777–782.
Higginbotham S., Leake A.R., Jordan V.W.L. & Ogilvy S.E. (2000) Environmental and ecological aspects of integrated, organic and conventional farming systems. *Aspects of Applied Biology*, 62, 15–20.
Jordan V.W., Leake A.R. & Ogilvy S.E. (2000) Agronomic and environmental implications of soil management practices in integrated farming systems. *Aspects of Applied Biology*, 62, 61–66.
Saunders H. (2000) Bird species as indicators to assess the impact of integrated crop management on the environment: a comparative study. *Aspects of Applied Biology*, 62, 47–54.
Kladivko E.J. (2001) Tillage systems and soil ecology. *Soil and Tillage Research*, 61, 61–76.
Cunningham H.M., Chaney K., Bradbury R.B. & Wilcox A. (2003) *Non-inversion Tillage and Farmland Birds in Winter.* Proceedings of the British Crop Protection Council Congress – Crop Science & Technology. Farnham, UK, 533–536.
Holland J.M. & Reynolds C.J.M. (2003) The impact of soil cultivation on arthropod (Coleoptera and Araneae) emergence on arable land. *Pedobiologia*, 47, 181–191.

Jansa J., Mozafar A., Kuhn G., Anken T., Ruh R., Sanders I.R. & Frossard E. (2003) Soil tillage affects the community structure of mycorrhizal fungi in maize roots. *Ecological Applications*, 13, 1164–1176.

2.14   Add 1% barley into wheat crop for corn buntings

- We have captured no evidence for the effects of adding 1% barley into wheat crop for corn buntings on farmland wildlife.

Background
Adding 1% barley into a wheat crop is a way of providing the preferred food source of corn buntings *Miliaria calandra*.

2.15   Create corn bunting plots

- We have captured no evidence for the effects of creating corn bunting plots on farmland wildlife.

Background
Corn bunting plots are sown patches (normally between 0.15 and 0.6 ha in size) of either grass or a cereal mix designed to provide nesting habitat for corn buntings *Miliaria calandra*.

2.16   Create skylark plots

- All four studies from Switzerland and the UK (two replicated and controlled and one review) investigating the effect of skylark plots on Eurasian skylarks found a positive effect, reporting increases in skylark population size[3], breeding density, duration or success[2, 4, 7] or a lower likelihood of skylarks abandoning their territory[9] relative to fields without plots. A replicated study from Denmark found that skylarks used undrilled patches within cereal fields more than expected by an even distribution across the landscape[1].

- Four studies reported the effect of undrilled patches on wildlife other than skylarks[4–6, 8, 10, 11]. Three studies from the UK (including two replicated studies, of which one was also controlled and a review) found benefits to plants and invertebrates[4–6, 8], while two studies (both replicated, one also controlled) found no significant differences in the number of some invertebrates[10] or seed-eating songbirds[11] between skylark plots and conventional crop fields.

- One replicated study from the UK[8] investigated different skylark plot establishment techniques. Plots that were undrilled had greater vegetation cover and height than plots established by spraying out with herbicide.

Background
Eurasian skylarks *Alauda arvensis* require short vegetation to nest in. Skylark plots are small (usually 4–16 m^2) undrilled patches within cereal fields which provide this short vegetation, with little impact on overall productivity. They are similar to lapwing plots

(see 'Leave cultivated, uncropped margins or plots (includes 'lapwing plots')') but much smaller.

A replicated study from April–May 1990 to 1993 in five spring-sown barley fields in eastern Jutland, Denmark (1) found that Eurasian skylarks *Alauda arvensis* used unsown plots in the fields significantly more than expected by an even distribution across the landscape. Radio-tracked birds were observed more in tramlines and unsown plots and mean dropping density was significantly higher in unsown areas than in crops (1.4 droppings/ha vs 0.1). One 22 ha field with one-hundred 40 m^2 plots had higher densities of skylarks than 4 fields with an average of 7 plots/ha, each of 7 m^2. Tramlines (30 cm wide; 18 m apart) were kept clear of vegetation by driving a truck along them several times a week. Adult male and female skylarks were radio-tracked and observed visually. Dropping counts were made in two 5 x 5 m squares in 8 territories in 1 field (May and June 1991).

A replicated, controlled study from April–August in 2002–2003 in 15 sites in northern, eastern and southern England (2) found that Eurasian skylark *Alauda arvensis* breeding density, duration and success were higher in winter wheat fields with undrilled patches (4 x 4 m) than in fields with widely-spaced (25 cm apart) rows or under conventional management (0.3 nests/ha in fields with undrilled plots vs 0.2 for the other treatments). Fields with undrilled patches also lost fewer territorial and nesting birds over the breeding season and by the end of the breeding season nests in these fields produced an average of one more chick than control nests. Body condition of nestlings decreased in control nests over the breeding season but increased in experimental fields. The proportion of within-treatment foraging flights remained constant in fields with undrilled patches but decreased over time in other treatments. Three treatments were surveyed: winter wheat sown in wide-spaced rows, undrilled patches with a density of 2 patches/ha, and conventional control winter wheat fields. Skylarks were surveyed from April to mid-August, with the number of territorial males, nests, nest productivity, nestling body condition and foraging locations recorded. Ten of the sites were part of the same replicated, controlled study (SAFFIE – Sustainable Arable Farming For an Improved Environment) as (4, 6, 10).

A before-and-after study from 2000 to 2005 in Cambridgeshire, England (3), found that the population of Eurasian skylarks *Alauda arvensis* on an arable farm increased from 10 territorial males in 2000 to 34 in 2005, following the introduction of skylark plots in 2001 (in addition to 6 m margins around fields and set-aside). Nests were also aggregated in fields with skylark plots. The study also reports that fields on 15 experimental farms with skylark plots had 30% more skylarks than control fields. In addition, nests in fields with skylark plots produced 0.5 more chicks/breeding attempt. Skylark plots 4 x 4 m were established at a density of 2 plots/ha. This study was part of the SAFFIE – Sustainable Arable Farming For an Improved Environment research project.

A replicated, controlled study in 2002–2003 on 10 farms in England (4) found that 45% of 159 Eurasian skylark *Alauda arvensis* nests monitored were found in fields with skylark plots. By June, fields with skylark plots had 30% more skylarks and 100% more nests than control winter wheat fields with normal row spacing. At the start of the breeding season there was little difference in success between treatments but by June fields with skylark plots had more nests (1 nest/ha vs 0.4) and more chicks/nest than controls (1.75 chicks/nest vs 0.9). Over the whole season, nests in experimental fields raised 0.5 more chicks than controls and 1.5 more chicks than controls late in the season. Plots had significantly higher undesirable weed cover than surrounding crop (6% vs 4% weed cover) although cover in the field as a whole was no higher (2% vs 1.5%). In 2002, but not 2003, invertebrate species richness and abundance were higher in fields with patches, compared to controls. Invertebrates, plants and skylarks were monitored. This study was part of the same replicated, controlled study (SAFFIE – Sustainable Arable Farming For an Improved Environment) as (2, 6, 10).

A 2007 review of published and unpublished literature (5) found one study (4) with experimental evidence of the benefits of skylark plots to plants (although undesirable plant species were prevalent) and invertebrates (invertebrate abundance was higher in the surrounding crop).

A replicated, controlled study in 2002 and 2003 at ten sites in England (6) found that plant species richness and arthropod species richness in one of two years were higher in wheat fields with undrilled patches than control fields. Weed and crop cover did not differ significantly between treatments (weeds: 1–2%; crops: 33–55%) but plant species richness was higher in fields with undrilled patches (11 species) than control fields (7 species). Weed cover and species richness were also significantly higher on undrilled patches (1–22% cover; 9–10 species) than the surrounding field (1–4%; 5–8 species). In 2002, arthropod species richness and rove beetle (Staphylinidae) abundance were higher in fields with patches than control fields (10 vs 6 arthropod species; 9 vs 6 rove beetles); there was no difference in 2003. In 2002, wolf spiders (Lycosidae) were more abundant in undrilled patches than the surrounding field (1.1 vs 0.4 individuals), whereas the opposite was true for rove beetle abundance (4 vs 9 individuals), species richness (4 vs 6 species) and in 2003 abundance of ground active invertebrates (0.3 vs 104 individuals). In 2002 arthropod abundance was higher in the surrounding fields than undrilled patches in May (23 vs 16 individuals) but higher in undrilled patches than surrounding fields in July (53 vs 36). Undrilled patches (4 x 4 m) were created at a density of 2 patches/ha in an otherwise conventionally managed crop. Vegetation composition was sampled in 24 quadrats (0.25 m^2) in May and July 2002 and 2003. Arthropods were sampled in the same locations using D-Vac suction sampling in May, June, and July and in pitfall traps for 7 days in June. This study was part of the same replicated, controlled study (SAFFIE – Sustainable Arable Farming For an Improved Environment) as (2, 4, 10).

A 2007 study and literature review (7) reports that Eurasian skylarks *Alauda arvensis* were able to raise 49% more young in fields with skylark plots, compared to fields without plots, by prolonging the length of the breeding season.

A replicated study of skylark plot establishment in two wheat fields in 2008–2009 in Cambridgeshire, UK (8) found that plots left undrilled had greater vegetation cover than those established by spraying out with herbicide. Vegetation cover within sprayed plots (sprayed out in December, January or February) tended to remain very low (<30%), particularly in February-sprayed plots (<10%); undrilled plots had 54% cover in July. Increase in cover in undrilled plots was related to a greater abundance of crop (May 5%, July 8%) and blackgrass *Alopecurus myosuroides* (May 10%, July 40%). Cover of both crop and blackgrass remained low in all sprayed treatments (blackgrass: <5%; crop: <2%). Plots following spring beans tended to have greater vegetation cover than those following oilseed rape. Undrilled plots had significantly taller vegetation (17–29 cm) than sprayed plots (July: 2–13 cm). A total of 56–65 plots were established/year by leaving them undrilled during wheat drilling, or by spraying out using a glyphosate herbicide until the density of plots was at least 2 plots/ha. Presence of bare ground, crop, blackgrass and charlock *Sinapis arvensis* were sampled in May, June and July in ten 0.25 m^2 quadrats/plot. Maximum vegetation height was also recorded in June and July.

A replicated, controlled study from March–July 2006 in mixed farmland near Berne, Switzerland (9) found that Eurasian skylarks *Alauda arvensis* with territories that included undrilled patches were significantly less likely to abandon their territory than birds without patches and more likely to use the undrilled patches as nesting and foraging sites. Use of winter wheat fields by skylarks changed through the breeding season; from June to July, the percentage of control fields (without undrilled plots) in skylark territories decreased from 60% to 38%, whilst the percentage of undrilled patches in skylark territories remained approximately 55% from May to July. Nest productivity was identical between control areas and fields with undrilled patches (1.4 chicks/territory) and there was no difference in chick body mass or tarsus length. Undrilled patches were composed of either four 3 x 12 m patches/ha (in 7 fields) or a single strip 2.5 x 80 m (in 14 fields). In autumn 2005 undrilled patches were sown with six annual weed species including common corncockle *Agrostemma githago* in winter wheat fields. Skylark territories were surveyed over 1 breeding season (2006) in 21 experimental sites and 16 control wheat fields.

A replicated, controlled study from April–August in 2002 and 2003 on 10 farms in northern and eastern England (10) found that invertebrate abundance in undrilled patch fields was not significantly different from conventional (control) winter wheat fields. In 2002, mean invertebrate species richness in undrilled patch fields was 9.8 compared to 6.4 in control fields. There were no significant differences in 2003, possibly because weed cover was 50% lower than 2002. Within undrilled patch fields, rove beetle (Staphylinidae) abundance and species richness was higher in crop areas while money spider (Lycosidae) and herbivorous invertebrate species were more abundant in the undrilled patches. There were no significant differences in faecal content between Eurasian skylark nestlings in treatment or control fields. The authors suggest that a critical threshold of weedy cover must be reached before any significant effect on invertebrates is detected. Three treatments were established on each farm: undrilled patches (4 x 4 m) with a density of 2 patches/ha, winter wheat sown in wide-spaced rows (25 cm apart) and conventional control winter wheat fields. Invertebrates were sampled using vacuum sampling (May, June, July) and pitfall traps (June). Plants were surveyed in twenty-four 0.25 m^2 quadrats in May and July. Skylark droppings were collected from nestlings, fledglings and adults for faecal analysis in April–September 2002–2003. This study was part of the same replicated, controlled study (SAFFIE – Sustainable Arable Farming For an Improved Environment) as (2, 4, 6).

A replicated site comparison study in winter 2007–2008 and summers 2008 and 2009 on farms in 3 English regions (11) found that skylark plots were well used (1–3 seed-eating farmland songbirds/ha) but did not have significantly more birds than crop fields or fallow plots. Surveys were carried out on 69 farms with Higher Level Stewardship in East Anglia, the West Midlands or the Cotswolds and 31 farms across all 3 regions with no environmental stewardship. Flush transects were used to record as many birds as possible.

(1) Odderskær P., Prang A., Poulsen J., Andersen P. & Elmegaard N. (1997) Skylark (*Alauda arvensis*) utilisation of micro-habitats in spring barley fields. *Agriculture, Ecosystems & Environment*, 62, 21–29.
(2) Morris A.J., Holland J.M., Smith B. & Jones N.E. (2004) Sustainable Arable Farming For an Improved Environment (SAFFIE): managing winter wheat sward structure for skylarks *Alauda arvensis*. *Ibis*, 146, S155–162.
(3) Donald P. & Morris T. (2005) Saving the sky lark: new solutions for a declining farmland bird. *British Birds*, 98, 570–578.
(4) Ogilvy S.E., Clarke J.H., Wiltshire J.J.J., Harris D., Morris A., Jones N., Smith B., Henderson I., Westbury D.B., Potts S.G., Woodcock B.A. & Pywell R.F. (2006) SAFFIE – research into practice and policy. *Proceedings of the HGCA Conference, Arable Crop Protection in the Balance: Profit and the environment*, 14, 1–12.
(5) Fisher G.P., MacDonald M.A. & Anderson G.Q.A. (2007) Do agri-environment measures for birds on arable land deliver for other taxa? *Aspects of Applied Biology*, 81, 213–219.
(6) Smith B. & Jones N.E. (2007) Effects of manipulating crop architecture on weed and arthropod diversity in winter wheat. *Aspects of Applied Biology*, 81, 31–38.

(7) Stoate C. & Moorcroft D. (2007) Research-based conservation at the farm scale: development and assessment of agri-environment scheme options. *Aspects of Applied Biology*, 81, 161–168.

(8) Dillon I., Morris A.J., Bailey C.M. & Uney G. (2009) Assessing the vegetation response to differing establishment methods of 'Skylark Plots' in winter wheat at Grange Farm, Cambridgeshire, England. *Conservation Evidence*, 6, 89–97.

(9) Fischer J., Jenny M. & Jenni L. (2009) Suitability of patches and in-field strips for sky larks *Alauda arvensis* in a small-parcelled mixed farming area. *Bird Study*, 56, 34–42.

(10) Smith B., Holland J., Jones N., Moreby S., Morris A.J. & Southway S. (2009) Enhancing invertebrate food resources for skylarks in cereal ecosystems: how useful are in-crop agri-environment scheme management options? *Journal of Applied Ecology*, 46, 692–702.

(11) Field R.H., Morris A.J., Grice P.V. & Cooke A.I. (2010) Evaluating the English Higher Level Stewardship scheme for farmland birds. *Aspects of Applied Biology*, 100, 59–68.

2.17   Plant cereals in wide-spaced rows

- One replicated, controlled, randomized study and four reports from the same replicated, controlled study in the UK investigated the effects of planting cereals in wide-spaced rows on birds, invertebrates and plants. Both studies found no or inconsistent differences in plant[2-4] and invertebrate[2-5] abundance and/or species richness between wide-spaced row and control fields. The replicated controlled study found higher undesirable weed cover[2], and no significant difference in weed cover[4] in fields with wide-spaced rows compared to control fields.

- One study[1] found significantly lower invertebrate abundances and fewer Eurasian skylark nests in wide-spaced row fields than control fields or fields with undrilled patches. However it also found an increase in the body condition of nestlings over the breeding season in wide-spaced row fields compared with control fields.

Background
Planting cereals in widely-spaced rows can increase the proportion of farmland habitat for wildlife, including birds, arthropods and plants. Spaces between rows can be left fallow or planted with grass or legumes.

A replicated, controlled study from April–August in 2002 and 2003 in 15 winter wheat fields in northern and eastern England (1) found that Eurasian skylark *Alauda arvensis* nests were significantly less abundant on fields with wide-spaced rows than on control fields or those with undrilled patches (0.16 nests/ha in fields with wide-spaced rows vs 0.18 for controls and 0.31 for fields with undrilled patches). The proportion of within-treatment foraging flights decreased over time in control and wide-spaced row fields but remained constant in fields with undrilled patches. Body condition of nestlings decreased in control nests but increased in the other treatments over the breeding season. Invertebrate abundance, particularly beetles (Coleoptera), was significantly lower on wide-spaced row fields. Bare ground was significantly more extensive in wide-spaced row fields. Three treatments were surveyed: winter wheat sown in wide-spaced rows, undrilled patches (4 x 4 m) with a density of 2 patches/ha, and conventional control winter wheat fields. Skylarks were surveyed from April to mid-August, with the number of territorial males, nests, nest productivity, nestling body condition and foraging locations recorded. Invertebrates were sampled in the crop (30 m from the nearest field boundary) and in undrilled patches using suction sampling (May–July), sweep netting (May–June) and pitfall traps (June). Vegetation and bare ground cover were surveyed in twenty-four 0.25 m^2 permanent quadrats/treatment. Ten of the sites were part of the same replicated, controlled study (SAFFIE – Sustainable Arable Farming For an Improved Environment) as (2, 4, 5).

A replicated, controlled study in 2002 and 2003 on ten farms in England (2) found that wide-spaced rows offered 'significant benefits' to Eurasian skylarks *Alauda arvensis*, but details were not given. The authors note that skylark plots were more consistently beneficial. Across fields as a whole there was a higher proportion of undesirable weed cover in fields with wide-spaced rows than fields with conventional spacing (4% weed cover vs 1.5%). However, a second experiment found no effect of wide-spaced rows on weed diversity, when compared to conventional herbicide treatment. In 2002, but not 2003, there were higher abundances of wolf spiders (Lycosidae) in fields with wide-spaced rows than control fields. Invertebrates, plants and skylarks were monitored. This study was part of the same replicated, controlled study (SAFFIE – Sustainable Arable Farming For an Improved Environment) as (1, 4, 5).

A replicated, controlled, randomized study from 2003 to 2005 of arable fields at three sites in the UK (3) found that sowing in wide-spaced rows had little effect on the abundance or species diversity of weeds or arthropods. Weed species richness was higher under wide-spaced rows (3 species) compared to conventional cultivation (2 species) at one of the 3 sites in one year. There were 3 (2003) or 5 (2004–2005) replicate plots (3 or 4 x 24 m) per treatment at Boxworth, Gleadthorpe and High Mowthorpe. Treatments were: conventional spacing, wide-spaced rows and wide-spaced rows with spring cultivation (hoeing) between rows. Vegetation was sampled in 5 quadrats (0.25 m^2) per plot (June 2003–2005). Arthropods were sampled using a D-Vac suction sampler (5 sub-samples of 10s/plot) in a sub-set of treatments (June).

A replicated, controlled study in 2002 and 2003 of wheat fields at ten sites in England (4) found that sowing crops in wide-spaced rows had little effect on plant cover and species richness or arthropod abundance. There was no significant difference in weed cover (1–2%), crop cover (30–48% vs 33–55%) or plant species richness (7 vs 7 species) between fields with wide–spaced rows and control fields. There was little effect of wide-spaced rows on arthropod abundance. However, in 2002 wolf spiders (Lycosidae) were more abundant in fields with wide-spaced rows (0.9 individuals) than controls (0.4); the opposite was true for rove beetles (Staphylinidae: 4 vs 6 individuals in fields with wide-spaced rows and controls respectively). Wide-spaced rows were sown at double the normal width (25 cm between rows). Vegetation composition was sampled within 24 quadrats (0.25 m^2) in May and July (2002 and 2003). Arthropods were sampled in the same locations using D-Vac suction sampling (May, June, July), sweep netting (May, June) and pitfall traps open for 7 days (June). This study was part of the same replicated, controlled study (SAFFIE – Sustainable Arable Farming For an Improved Environment) as (1, 2, 5).

A replicated, controlled study from April–August in 2002 and 2003 on 10 sites in northern and eastern England (5) found that invertebrate abundance in wide-spaced row fields was not significantly different from conventional (control) winter wheat fields. In 2002, wide-spaced row fields contained an average of 6.0 invertebrate species/pitfall trap compared to 6.4 in control fields. Both grass and broadleaved vegetation cover had a negative relationship with wide-spaced row fields. There were no significant differences in skylark *Alauda arvensis* faecal content between nestlings in treatment or control fields. The authors suggest that the value of wide-spaced row fields for skylarks lies in increased access to other food resources. There were three treatments at each site: wide-spaced rows sown at double the normal width (25 cm between rows) but with the same seed rate as control fields, control fields with 12.5 cm between rows, and undrilled patches (4 x 4 m) with a density of 2 patches/ha. This study was part of the same replicated, controlled study (SAFFIE – Sustainable Arable Farming For an Improved Environment) as (1, 2, 4).

(1) Morris A.J., Holland J.M., Smith B. & Jones N.E. (2004) Sustainable Arable Farming For an Improved Environment (SAFFIE): managing winter wheat sward structure for skylarks *Alauda arvensis*. *Ibis*, 146, S155–162.
(2) Ogilvy S.E., Clarke J.H., Wiltshire J.J.J., Harris D., Morris A., Jones N., Smith B., Henderson I., Westbury D.B., Potts S.G., Woodcock B.A. & Pywell R.F. (2006) SAFFIE – research into practice and policy. *Proceedings of the HGCA Conference, Arable Crop Protection in the Balance: Profit and the environment*, 14, 1–12.
(3) Jones N.E. & Smith B. (2007) Effects of selective herbicide treatment, row width and spring cultivation on weed and arthropod communities in winter wheat. *Aspects of Applied Biology*, 81, 39–46
(4) Smith B. & Jones N.E. (2007) Effects of manipulating crop architecture on weed and arthropod diversity in winter wheat. *Aspects of Applied Biology*, 81, 31–38.
(5) Smith B., Holland J., Jones N., Moreby S., Morris A.J. & Southway S. (2009) Enhancing invertebrate food resources for skylarks in cereal ecosystems: how useful are in-crop agri-environment scheme management options? *Journal of Applied Ecology*, 46, 692–702.

2.18 Sow rare or declining arable weeds

• Two studies from the UK (both replicated, controlled and randomized) found that the establishment of rare or declining arable weeds depended upon cover crop, cultivation, timing of cut and year[1] or a combination of cultivation in autumn and herbicide treatment[2].

Background

This intervention involves sowing rare or declining arable plant species such as common corncockle *Agrostemma githago* to re-establish populations in the farmed landscape.

A replicated, controlled, randomized study in 1993 and 1994 of set-aside at a site in England (1) found that the establishment of common corncockle *Agrostemma githago* and interrupted brome *Bromus interruptus* depended on cover crop, cultivation, management and year. Common corncockle density was significantly higher on the two sown grass covers consisting of 95% or 67% grass (125–135 common corncockle plants/m^2) than on naturally regenerated plots (120/m^2) or wheat crops (105/m^2). The reverse was true in year 2 (natural regeneration: 600/m^2; crop: 1,100/m^2; grass covers: 400–500). Reproductive output was higher on wheat crop cover and natural regeneration plots (275 seed capsules/m^2) than sown grass cover (210–250 seed capsules/m^2). Cover crop did not affect interrupted brome establishment in year 1 (4–5/m^2) but in year 2 density was higher on natural regeneration (24/m^2) and wheat crop (23/m^2) covers than on grass covers (10–15/m^2). Both species increased from year one to two. No cultivation in year 2 resulted in the only decline in density of common corncockle between year 1 and 2 (70/m^2). Interrupted brome density was highest with no cultivation in the second year compared to cultivation or cultivation and re-sowing of rare arable weeds (29 vs 10–15). An early year 1 cut (1 August) resulted in significantly lower densities of common corncockle (440/m^2) compared to a later cut (30 August) or no cut (750/m^2). A late cut increased year 2 interrupted brome density compared to the early cut (28 vs 17/m^2) but no cut significantly decreased the density (9/m^2). Re-sowing common corncockle had no effect on year 2 densities (860–1,000/m^2). The trial comprised a split plot (2 x 2 m) randomized block design with 3 replicate blocks each containing 36 treatment combinations. Wheat (drilled), grass crop and rare weeds (hand sown) were planted in October–November 1993 and 1994. Rare arable weeds were sampled in 1 x 1 m quadrats in the centre of each plot. Seed bearing capsules were counted on 10 individuals/plot.

A randomized, replicated, controlled trial in 2005 and 2007 in Oxfordshire, UK (2) showed that it is possible to establish and maintain new populations of scarce arable plants by combining cultivation and herbicide treatment. Twelve 25 x 14 m plots were cultivated, harrowed and sown with 7 scarce arable plant species in October 2005 and monitored after 2 years in June 2007. Plots annually cultivated in autumn had significantly greater cover and more species of sown scarce arable plants than spring cultivated plots (25–60% cover, >1.5 species/m^2 on average for autumn cultivated plots, compared with 10–30% cover and 0.5–1.7 species/m^2 for spring cultivated plots). The highest cover by scarce arable plants (average 60%) was in plots cultivated and treated with grass-specific herbicide in autumn. Five sown arable species achieved >1% cover on average in year 2. Common corncockle *Agrostemma githago* and cornflower *Centaurea cyanus* both increased significantly under autumn cultivation. There were three replicates of each combination of herbicide and cultivation treatments.

(1) Neve P., Mortimer A.M. & Putwain P.D. (1996) Management options for the establishment of communities of rare arable weeds on set-aside land. *Aspects of Applied Biology*, 44, 257–262.
(2) Pywell R.F., Hulmes L., Meek W.R. & Nowakowski M. (2010) Practical management of scarce arable plant populations. *Aspects of Applied Biology*, 100, 375–380.

2.19   Use new crop types to benefit wildlife (such as perennial cereal crops)

- We have captured no evidence for the effects of using new crop types to benefit wildlife (such as perennial cereal crops).

Background
This intervention may involve sowing new crop types which provide a benefit to farmland wildlife. Perennial crops may provide benefits such as provision of cover and by reducing soil disturbance.

2.20   Plant more than one crop per field (intercropping)

- Three replicated, controlled and randomized studies from the Netherlands, Poland and the UK found that intercropping cabbage with French beans or clover resulted in increased ground beetle abundance[1, 2, 4].

- A trial from Switzerland found increased earthworm abundance in a maize plot immediately followed by a rye grass crop[3].

- A review found ground beetle numbers were enhanced by intercropping relative to single crops[5].

Background
Planting more than one crop in each field increases habitat heterogeneity at a smaller scale than increasing crop diversity at a landscape scale (see 'Increase crop diversity'). Farmland heterogeneity is thought to be key for increasing farmland biodiversity (Benton *et al.* 2003) and so planting multiple crops in a single field may help increase its biological value.

Benton T.G., Vickery J.A. & Wilson J.D. (2003) Farmland biodiversity: is habitat heterogeneity the key? *Trends in Ecology & Evolution*, 18, 182–188.

A replicated, controlled, randomized study of a cabbage crop in 1976 in the UK (1) found that ground beetle (Carabidae) species were almost twice as abundant in cabbage *Brassica oleracea* intercropped with French beans *Phaseolus vulgaris* than in a cabbage monoculture. However, only *Bembidion lampros* (210 vs 90 individuals), *B. quadrimaculatum* (400 vs 260), *Pterostichus madidus* (45 vs 15) and *Harpalus* spp. were significantly more abundant in intercropped plots. Rove beetles (Staphylinidae) did not tend to differ between treatments. Plots of 10 x 10 m were established in 2 randomized blocks, each with 4 treatments: Brussels sprouts, Brussels sprouts inter-planted with two rows of French beans between each row, Brussels sprouts inter-planted with two rows of beans between and across rows and beans alone. Beans were sown in May and Brussels sprouts in mid-June 1976. The abundance of adult ground beetles and rove beetles were sampled using two pitfall traps/plot. Aphid (Aphidoidea) and root fly *Delia brassicae* abundance was also sampled.

A replicated, controlled, randomized study from 1985 to 1990 of a cabbage crop in Poland (2) found that intercropping with white clover *Trifolium repens* resulted in a higher abundance of ground beetles (Carabidae). The total number of ground beetles was greater in inter-cropped cabbage *Brassica oleracea* and clover plots (2,528) than those with cabbage alone (1,753). This was the case for *Amara aulica* (16 individuals vs 52), *Calathus fuscipes* (105 vs 199), *Harpalus rufipes* (586 vs 920), *Pterostichus cupreus* (110 vs 305), *P. vulgaris* (13 vs 53) and 'other species' (360 vs 547) but not *Bembidion properans* (452 vs 563). There was no association between increased cover and predator diversity. White clover was sown in 30 m² plots in May on a 4 replicated block design. Cabbage was sown between the rows within plots in June. Ground beetles were sampled with two pitfall traps in the centre of each plot. Samples were collected weekly through the growing period. Aphid (Aphidoidea) abundance and damage by the cabbage moth *Mamestra brassicae* was also sampled.

A trial at an experimental farm on the Swiss Plateau, Switzerland in 1989 (3) found that earthworm (class: Oligochaeta) abundance but not biomass was higher in a maize *Zea mays* plot immediately followed by a rye grass *Lolium perenne* crop (called a 'catch crop') to provide winter cover. Control and catch crop plots had averages of 127 and 111 earthworms/m² and 45 and 64 g earthworm biomass/m², respectively. The proportion of deep-burrowing earthworms was similar with 14 and 13% of individuals in the control and catch crop plots respectively. A test strip of maize 14 m long was sown with a rye grass catch crop in autumn, and compared with a control strip of conventional maize. Earthworms were sampled by hand-sorting 0.1 m³ of soil from each test strip, to a depth of 40 cm, on 6 dates between April and October 1989. There was no replication.

A replicated, controlled, randomized study in 1991 and 1992 of a cabbage crop near Lienden in the Netherlands (4) found that ground beetle (Carabidae) abundance was higher in cabbage *Brassica oleracea* intercropped with clover *Trifolium* spp. than a cabbage monoculture. The overall activity-density of ground beetles was significantly higher in inter-cropped plots (151–455 caught) than within cabbage alone (85–263). There was no difference in intercropping with white clover *Trifolium repens* (21–22 species/plot) or subterranean clover *Trifolium subterraneum* (20–23 species/plot). A number of species were caught signific-antly more frequently in intercropped plots including the most dominant species *Pterostichus melanarius* (81–153 individuals vs 71) and *Agonum dorsale* (77–152 vs 63) and also *Loricera pilicornis* (6.3–9.3 vs 0.3). In both years, species diversity was higher in intercropped plots (20–23 species/plot) than cabbage alone (17–18) but this was only significant in 1991 (20–22 vs 18). Differences in activity-density between treatments tended to be lost by late summer.

A randomized block experiment with 4 replicates of 3 treatments in plots of 25 x 25 m² was established in 1990. Clover was sown before cabbage and covered approximately half of the soil surface. No pesticides were applied. Six pitfall traps were set 7 m apart in each plot and were emptied weekly between May and September.

A 1999 review of literature (5) reported six studies showing that intercropping enhanced ground beetle (Carabidae) numbers: (1, 2), Armstrong & McKinlay (1994), Carcamo & Spence (1994), Helenius & Tolonen (1994), (4), relative to single crops.

(1) Tukahirwa E.M. & Coaker T.H. (1982) Effect of mixed cropping on some insect pests of Brassicas – reduced *Brevicoryne brassicae* infestations and influences on epigeal predators and the disturbance of oviposition behavior in *Delia brassicae*. *Entomologia experimentalis et applicata*, 32, 129–140
(2) Wiech K. & Wnuk A. (1991) The effect of intercropping cabbage with white clover and French bean on the occurrence of some pests and beneficial insects. *Folia Horticulturae*, 3, 39–45.
(3) Wyss E. & Glasstetter M. (1992) Tillage treatments and earthworm distribution in a Swiss experimental corn field. *Soil Biology & Biochemistry*, 24, 1635–1639.
(4) Booij C.J.H., Noorlander J. & Theunissen J. (1997) Intercropping cabbage with clover: effects on ground beetles. *Biological Agriculture & Horticulture*, 15, 261–268.
(5) Kromp B. (1999) Carabid beetles in sustainable agriculture: a review on pest control efficacy, cultivation impacts and enhancement. *Agriculture, Ecosystems & Environment*, 74, 187–228.

Additional references
Armstrong G. & McKinlay R.G. (1994) *Undersowing Brassicas with Clover to Increase the Activity of Carabid Beetles*. Proceedings – Brighton Crop Protection Conference, Pests and diseases, Bracknell, UK, 3, 1175–1180.
Carcamo H.A. & Spence J.R. (1994) Crop type effects on the activity and distribution of ground beetles (Coleoptera, Carabidae). *Environmental Entomology*, 23, 684–692.
Helenius J. & Tolonen T. (1994) Enhancement of generalist aphid predators in cereals: effect of green manuring on recruitment of ground beetles. *IOBC/WPRS Bulletin*, 17, 201–210.

3 Perennial (non-timber) crops

Key messages

Maintain traditional orchards
A replicated, controlled site comparison from Germany found more plant species in mown orchards than grazed or abandoned ones, but found no effects on wasps or bees. Two replicated site comparisons from Germany and Switzerland found traditional orchards managed under agri-environment schemes either did not have more plant species than controls or offered no clear benefits to birds.

Restore or create traditional orchards
We have captured no evidence for the effects of restoring or creating traditional orchards on farmland wildlife.

Manage short-rotation coppice to benefit wildlife (includes 8 m rides)
We have captured no evidence for the effects of managing short-rotation coppice to benefit wildlife (including 8 m rides).

3.1   Maintain traditional orchards

- Two replicated site comparisons from Germany and Switzerland found that, on average, 12% of traditional orchards in Swiss Ecological Compensation Areas were of 'good ecological quality'[2], and traditional orchards under a German agri-environment scheme did not have more plant species than paired control sites[3]. Traditional orchards in Ecological Compensation Areas appeared to offer little benefit to birds[2].

- A replicated, controlled site comparison study in Germany[1] found that plant species richness was higher on mown orchards than grazed or abandoned ones, but numbers of species and brood cells of bees and wasps did not differ.

Background
Traditionally-managed orchards are low-intensity systems that have the potential to provide unique habitats for wildlife and that tend to hold older and rarer varieties of fruit. However, they are threatened in many countries, with 60% of traditional orchards in Britain having been lost and another 30% converted to intensive production since the 1950s.

A replicated, controlled site comparison study in 1998 and 1999 of 45 orchard meadows in Lower Saxony, Germany (1) found that plant species richness was higher on mown meadows than grazed or abandoned meadows but that numbers of species and brood cells of bees and wasps (Hymenoptera) did not differ. A significantly higher number of plant species was found on mown (24) than abandoned meadows (18); grazed meadows had intermediate numbers (22). A similar trend was found for grasses (8 species vs 5), but not herbs (16 vs 12). Plant height was higher on abandoned meadows (100 cm) than mown meadows (85 cm) and grazed meadows (55 cm). Vegetation cover did not differ significantly between management regimes. The number of species or brood cells for all species, or separately for bees (Apidae), potter wasps (Eumenidae) or sphecid wasps (Sphecidae) did not differ between treatments. However, the abundance of sphecid wasps was significantly higher on abandoned meadows (180 brood cells) than grazed (55) or mown (60) meadows. There was no significant difference in species richness of natural enemies or the rate of parasitism of bees and wasps between management types. Orchards were either mown once or twice a year, grazed (usually by sheep) or had no management for at least five years. Vegetation was sampled on a central plot of 25 m within each site from June–July 1998. Nesting traps (4/location) were set up at regular distances at each site from April–September, 1998 and 1999.

A replicated site comparison between 1998 and 2001 (2) found that, on average, only 12% of traditional orchards in Ecological Compensation Areas on the Swiss plateau were of 'good ecological quality' (based on national guidelines for Ecological Compensation Area target vegetation). Orchard Ecological Compensation Areas appeared to offer little benefit to orchard birds, with territories of only one species (green woodpecker *Picus viridis*) found more frequently in or near Ecological Compensation Area orchards (11 territories) than expected. Plant species and orchard characteristics were recorded for 187 Ecological Compensation Area orchards (total area 108 ha) between 1998 and 2001. Territories of breeding birds were mapped in 23 study areas, based on 3 visits between mid-April and mid-June.

A replicated, paired site comparison study in Bavaria, Germany (3) found that traditional orchards managed under the Bavarian agri-environment scheme 'Agricultural Landscape Programme' (KULAP) did not have more plant species than paired control sites. There were 26 site pairs and around 18–20 plant species/site. Pairs of 25 m^2 grassland plots were selected from 4,400 plots in the Bavarian grassland survey. All plant species within the plot were recorded between April and October (year not given). Plot pairs were in the same natural landscape, 90% within 10 km of each other. In each pair, one plot was under an agri-environment scheme agreement, the other was not – in this case traditional orchard management.

(1) Steffan-Dewenter I. & Leschke K. (2003) Effects of habitat management on vegetation and above-ground nesting bees and wasps of orchard meadows in Central Europe. *Biodiversity and Conservation*, 12, 1953–1968.
(2) Herzog F., Dreier S., Hofer G., Marfurt C., Schüpbach B., Spiess M. & Walter T. (2005) Effect of ecological compensation areas on floristic and breeding bird diversity in Swiss agricultural landscapes. *Agriculture, Ecosystems & Environment*, 108, 189–204.
(3) Mayer F., Heinz S. & Kuhn G. (2008) Effects of agri-environment schemes on plant diversity in Bavarian grasslands. *Community Ecology*, 9, 229–236.

3.2   Restore or create traditional orchards

• We have captured no evidence for the effects of restoring or creating traditional orchards on farmland wildlife.

Background

This intervention may involve restoring or creating traditionally-managed orchards which are low-intensity systems that have the potential to provide unique habitats for wildlife.

3.3   Manage short-rotation coppice to benefit wildlife (includes 8 m rides)

- We have captured no evidence for the effects of managing short-rotation coppice to benefit wildlife (including 8 m rides) on farmland wildlife.

Background

Short-rotation coppice consists of areas of densely planted willow or poplar which are harvested on a 2–5 year cycle (Defra 2004). Eight-metre wide headlands at the end of each area of coppice are required to allow vehicles to turn and rides along the crop edges are required for machinery access. Both rides and headlands may be sown with grass and cut twice each year (Defra 2004).

Defra (2004) *Growing Short Rotation Coppice – Best Practice Guidelines for Applicants to Defra's Energy Crops Scheme*. Department for Environment Food and Rural Affairs.

4 Livestock farming

Maintain species-rich, semi-natural grassland

Nine studies (including two randomized, replicated before-and-after trials) from Switzerland and the UK looked at the effectiveness of agri-environment schemes in maintaining species-rich grassland and all except one found mixed results. All twelve studies (including a systematic review) from six countries looking at grassland management options found techniques that improved or maintained vegetation quality. A site comparison study from Finland and Russia found butterfly communities were more affected by grassland age and origin than present management.

Restore or create species-rich, semi-natural grassland

Twenty studies (including three randomized, replicated, controlled trials) from six countries found restored species-rich, semi-natural grasslands had similar invertebrate, plant or bird diversity or abundance to other grasslands. Seven studies (two randomized, replicated, controlled trials) from five countries found no clear effect on plant or invertebrate numbers; three replicated studies (of which two were site comparisons) from two countries found negative effects. Forty studies (including six randomized, replicated, controlled trials) from nine countries identified effective techniques for restoring species-rich grassland.

Add yellow rattle seed (Rhinanthus minor) to hay meadows

A review from the UK reported that hay meadows had more plant species when yellow rattle was present. A randomized, replicated controlled trial in the UK found yellow rattle could be established by 'slot seeding'.

Reduce management intensity on permanent grasslands (several interventions at once)

Eleven studies (including four replicated site comparisons) from three countries found reducing management intensity benefited plants. Sixteen studies (including four paired site comparisons) from four countries found benefits to some or all invertebrates. Five studies (including one paired, replicated site comparison) from four countries found positive effects on some or all birds. Twenty-one studies (including two randomized, replicated, controlled trials) from six countries found no clear effects of reducing management intensity on some or all plants, invertebrates or birds. Five studies (including two paired site comparisons) from four countries found negative effects on plants, invertebrates or birds.

Raise mowing height on grasslands

Three studies (including one replicated, controlled trial) from the UK or unspecified European countries found raised mowing heights caused less damage to amphibians and invertebrates or increased Eurasian skylark productivity. Two studies (one randomized, replicated, controlled) from the UK found no effect on bird or invertebrate numbers and a replicated study from the UK found young birds had greater foraging success in shorter grass.

Delay mowing or first grazing date on grasslands
Eight studies (including a European systematic review) from the Netherlands, Sweden and the UK found delaying mowing or grazing benefited some or all plants, invertebrates or birds, including increases in numbers or productivity. Three reviews found the UK corncrake population increased following management that included delayed mowing. Six studies (including a European systematic review) from five countries found no clear effect on some plants, invertebrates or birds.

Use mowing techniques to reduce mortality
Seven studies (including two replicated trials, one controlled and one randomized) from Germany, Ireland, Switzerland and the UK found mowing techniques that reduced mortality or injury in amphibians, birds, invertebrates or mammals. A review found the UK corncrake population increased around the same time that Corncrake Friendly Mowing was introduced and a replicated trial found mowing from the field centre outwards reduced corncrake chick mortality.

Reduce grazing intensity on grassland (including seasonal removal of livestock)
Fifteen studies (including three randomized, replicated, controlled trials) from four countries found reducing grazing intensity benefited birds, invertebrates or plants. Three studies (including one randomized, replicated, controlled trial) from the Netherlands and the UK found no benefit to plants or invertebrates. Nine studies (including a systematic review) from France, Germany and the UK found mixed effects for some or all wildlife groups. The systematic review concluded that intermediate grazing levels are usually optimal but different wildlife groups are likely to have different grazing requirements.

Leave uncut strips of rye grass on silage fields
Four studies (including two replicated, controlled trials) from the UK found uncut strips of rye grass benefited some birds, with increased numbers. A randomized, replicated, controlled study from the UK found higher ground beetle diversity on uncut silage plots, but only in the third study year.

Plant cereals for whole crop silage
A replicated study from the UK found cereal-based whole crop silage had higher numbers of some birds than other crops. A review from the UK reported that seed-eating birds avoided cereal-based whole crop silage in winter, but used it as much as spring barley in summer.

Maintain rush pastures
We have captured no evidence for the effects of maintaining rush pastures on farmland wildlife.

Maintain traditional water meadows (includes management for breeding and/or wintering waders/waterfowl)
Four studies (including a replicated site comparison) from Belgium, Germany, the Netherlands and the UK found maintaining traditional water meadows increased numbers of some birds or plant diversity. One bird species declined. Two studies (including a replicated site comparison from the Netherlands) found mixed or inconclusive effects on birds, plants or wildlife generally. A replicated study from the UK found productivity of one wading bird was too low to sustain populations in some areas of wet grassland managed for wildlife.

Restore or create traditional water meadows

Three studies (two before-and-after trials) from Sweden and the UK looked at bird numbers following water meadow restoration; one found increases; one found increases and decreases; one found no increases. Seventeen studies (two randomized, replicated, controlled) from six countries found successful techniques for restoring wet meadow plant communities. Three studies (one replicated, controlled) from four countries found restoration of wet meadow plant communities had reduced or limited success.

Maintain upland heath/moorland

Eight studies (including one randomized, replicated, controlled trial) from the UK found management, including reducing grazing, can help to maintain the conservation value of upland heath or moorland. Benefits included increased numbers of plants or invertebrates. Three studies (including a before-and-after trial) from the UK found management to maintain upland heath or moorland had mixed effects on some wildlife groups. Four studies (including a controlled site comparison) from the UK found reducing grazing had negative impacts on soil organisms, but a randomized, replicated before-and-after study found heather cover declined where grazing intensity had increased.

Restore or create upland heath/moorland

A small trial in northern England found moorland restoration increased the number of breeding northern lapwing. A UK review concluded that vegetation changes were slow during the restoration of heather moorland from upland grassland.

Plant brassica fodder crops (grazed *in situ*)

We have captured no evidence for the effects of planting brassica fodder crops (grazed *in situ*) on farmland wildlife.

Use mixed stocking

A replicated, controlled study in the UK found more spiders, harvestmen and pseudoscorpions in grassland grazed by sheep-only than grassland grazed by sheep and cattle. Differences were only found when suction sampling not pitfall-trapping.

Use traditional breeds of livestock

Three studies (one replicated) from the UK found the breed of livestock affected vegetation structure, invertebrate communities and the amount of plants grazed. A replicated trial from France, Germany and the UK found no difference in the number of plant species or the abundance of birds, invertebrates or mammals between areas grazed by traditional or commercial livestock.

Employ areas of semi-natural habitat for rough grazing (includes salt marsh, lowland heath, bog, fen)

Three studies (two replicated) from the UK and unspecified European countries found grazing had positive effects on birds, butterflies or biodiversity generally. A series of site comparisons from the UK found one bird species used heathland managed for grazing as feeding but not nesting sites. Two studies (one replicated site comparison) from the UK found grazing had negative effects on two bird species.

Maintain wood pasture and parkland
A randomized, replicated, controlled trial in Sweden found annual mowing on wood pasture maintained the highest number of plant species.

Restore or create wood pasture
A replicated, controlled trial in Belgium found survival and growth of tree seedlings planted in pasture was enhanced when they were protected from grazing. A replicated study in Switzerland found cattle browsing had negative effects on tree saplings.

Exclude livestock from semi-natural habitat (including woodland)
Three studies (including one randomized, replicated, controlled trial) from Ireland and the UK found excluding livestock from semi-natural habitats benefited plants and invertebrates. Three studies (one replicated, controlled and one replicated paired site comparison) from Ireland and the UK did not find benefits to plants or birds. Two studies (one replicated, controlled and a review) from Poland and the UK found limited or mixed effects.

Mark fencing to avoid bird mortality
We have captured no evidence for the effects of marking fencing to avoid bird mortality on farmland wildlife.

Create open patches or strips in permanent grassland
A randomized, replicated, controlled study from the UK found more Eurasian skylarks used fields containing open strips, but numbers varied. A randomized, replicated, controlled study from the UK found insect numbers on grassy headlands initially dropped when strips were cleared.

Provide short grass for birds
A replicated UK study found two bird species spent more time foraging on short grass than longer grass.

4.1   Maintain species-rich, semi-natural grassland

- Of 22 studies (including 11 replicated trials, 3 reviews and a systematic review) from the Czech Republic, Finland, Germany, Russia, Slovenia, Switzerland and the UK, 13 identified management regimes that maintained species-rich grassland. Four of these studies were replicated, controlled trials (including two randomized)[3, 14, 20, 21].

- Nine studies (including two randomized, replicated before-and-after trials) from Switzerland and the UK examined the effectiveness of existing or historical agri-environment schemes: seven testing the effectiveness of the Environmentally Sensitive Areas scheme in England[4–9, 16] and two testing the effectiveness of the Ecological Compensation Areas scheme in Switzerland[10, 18]. All except one[7] reported mixed results, with the schemes broadly maintaining plant species richness, but being less effective, for example, in enhancing species richness[4, 9], preserving the highest quality sites[6], or overcoming the effects of past intensive management[10]. One study found six Environmentally Sensitive Areas were of 'outstanding' significance for their lowland grassland, containing >40% of the English resource of a grassland type[4]. A replicated site comparison study[10] found that on average 86% of Swiss Ecological Compensation Area litter meadows were of 'good ecological quality' compared with only 20% of hay meadow Ecological Compensation Areas.

- Twelve studies (including a systematic review, six replicated trials, of which two were also controlled and randomized, and three reviews) from the Czech Republic, Finland, Germany, Slovenia, Switzerland and the UK tested the effects of management treatments on species richness or vegetation quality usually involving combinations of mowing, grazing or no fertilizer[1, 2, 11, 12, 14, 15, 19–21] but some also tested the effectiveness of mulching or burning[3, 17, 22]. All of these studies identified management treatments which benefited or maintained species richness or vegetation quality.

- One site comparison from Finland and northwest Russia found that butterfly species richness, diversity and total abundance did not differ significantly between mown meadows and grazed pastures and that grassland age and origin had a greater impact on butterfly communities than present management[13].

Background

Species-rich, semi-natural grasslands have declined drastically in Europe over the last 100 years (e.g. Poschlod & Bonn 1998, Eriksson *et al.* 2002) and conservation efforts have been directed to maintaining existing areas of these habitats. This section summarizes the results of studies that tested the effectiveness of interventions in maintaining the conservation value of species-rich grasslands, including hay meadows, litter meadows and other semi-natural pastures. The management regimes tested by these studies involve combinations of interventions, typically involving altered mowing or grazing regimes and reduced inputs, while some studies also tested the effectiveness of mulching or burning.

See also 'Delay mowing or first grazing date on grasslands' and 'Agri-chemicals: Reduce chemical inputs in grassland management' for studies which tested the effectiveness of individual (uncombined) interventions in maintaining the conservation value of species-rich grassland.

Poschlod P. & Bonn S. (1998) Changing dispersal processes in the central European landscape since the last ice age: an explanation for the actual decrease of plant species richness in different habitats? *Acta Botanica Neerlandica*, 47, 27–44.

Eriksson O., Cousins S.A.O. & Bruun H.H. (2002) Land-use history and fragmentation of traditionally managed grasslands in Scandinavia. *Journal of Vegetation Science*, 13, 743–748.

A 1994 review (1) describes the results of a controlled trial in 1990–1992 on a species-rich upland meadow with a plant community characterized by sweet vernal grass *Anthoxanthum odoratum* and wood cranesbill *Geranium sylvaticum* (MG3 under the UK National Vegetation Classification) in the Pennines, northern England. The highest species richness (on average 17 species per 25 x 25 cm quadrat) was produced by a combination of a mid-July hay cut, spring and autumn grazing and no fertilizer. The lowest species richness was produced by no grazing, application of fertilizer and either a September (on average 10 species/quadrat) or June (11 species/quadrat) hay cut. Autumn-only grazing produced a vegetation layer with an intermediate species-richness (13–16 species/quadrat). Combinations of the following treatments were compared: cutting in mid-June, mid-July or early September; spring-sheep and autumn-cattle grazing, autumn-cattle grazing or no grazing; nitrogen, phosphorous, potassium (NPK) fertilizer (400 kg/ha) or no fertilizer. The number of replicates is not stated.

The same trial is described in (2), which reports the number of plant species found under single applications of the same treatments. The four treatments that produced the highest species richness were cutting in mid-July, grazing in autumn and spring, grazing in autumn only, and applying no fertilizer. Each of these treatments produced on average 15 plant species/quadrat (25 x 25 cm). Cutting earlier, in mid-June, reduced species richness to 13 species/quadrat. Although grazing reduced the hay yield (from 6 t/ha to 4–5 t/ha), it was necessary to maintain species richness. Under grazed treatments, 15 plant species/quadrat were recorded, compared to 12 species/quadrat with no grazing. Use of NPK fertilizer increased hay yield (from 4 to 6 t/ha), but reduced species richness (from 15 to 13 species/quadrat). Plants were surveyed in June 1992 and dry matter yield was measured at the time of cutting (averaged between 1990–1992). The number of replicates is not stated.

A long-term replicated, controlled trial in the Jura Mountains near Schaufhaussen, Switzerland (3) found plots that were cut annually (in July or October) or every second year (in July) retained a higher number of plant species than those burned, or cut less frequently than every two years. Frequently cut plots had 53 plant species/40 m², and 37 species/m² on average, compared to 45 species/40 m² and 24 species/m² on average for annually burned plots, those cut every fifth year and control unmanaged plots. There were 3 replicate 50 m² plots for each treatment and the experimental management regimes were carried out from 1978 to 1993. The percentage cover of plant species was estimated in 40 m² and 1 m² sample areas in each plot, in the last week of June 1991 and 1993.

A 1997 review (4) concluded that the Environmentally Sensitive Areas scheme made a significant contribution to halting the loss of semi-natural grasslands in England, but was less effective in enhancing and restoring grassland biodiversity. The paper made a broad assessment of the effectiveness of the scheme in protecting England's lowland semi-natural grasslands, a decade after its introduction. Among Environmentally Sensitive Areas of greatest significance for their lowland grassland, 6 were of 'outstanding' significance (containing >40% of the English resource of a grassland type) and 2 were of 'considerable' significance (containing >10–40% of 1–2 grassland types or >5–<10% of 3 or more grassland types). Entry of land supporting semi-natural grassland was generally high (e.g. covering 80% of chalk grassland in the South Downs Environmentally Sensitive Area). However, there was evidence in some Environmentally Sensitive Areas that grassland habitats were declining in quality due to management being insufficiently tailored to biodiversity interest, such as permitting the use of inorganic fertilizers.

A before-and-after trial in England (5) concluded that management prescriptions in the Exmoor Environmentally Sensitive Area are maintaining the condition of unimproved grassland, based on trends in bird populations in parts of the Environmentally Sensitive Area under long-term management agreements. The same study found that five red/amber-listed species of conservation concern (Eurasian linnet *Carduelis cannabina*, Eurasian bullfinch *Pyrrhula pyrrhula*, grey partridge *Perdix perdix*, house sparrow *Passer domesticus* and garden warbler *Sylvia borin*) appeared to be increasing in density within the Cotswolds Environmentally Sensitive Area while declining nationally, suggesting that they benefit from some aspect of Environmentally Sensitive Area management. In each Environmentally Sensitive Area, breeding birds were surveyed in May–August 2002, and results were compared with baseline survey information from 1992 to 1993 (Exmoor) and 1997 (Cotswolds). In the Cotswolds Environmentally Sensitive Area, birds were surveyed in 96 randomly-selected 1 km squares, while the majority (153 km²) of the Exmoor Environmentally Sensitive Area was surveyed.

A long-term replicated, before-and-after trial in 1987–2002 on upland hay meadows in the Pennine Dales Environmentally Sensitive Area, County Durham, England (6) found that plant species richness improved overall between 1987 (when the Environmentally Sensitive Area was introduced) and 1995, then declined to close to its original level by 2002. Condition was broadly maintained and slightly enhanced on semi-improved and improved hay meadows, but there was deterioration among the best quality unimproved hay meadows, with an increase in grass species at the expense of herbs. There were clear relationships between vegetation change and management, with the following practices leading to reduced vegetation quality: early cutting (before 15 July), spring grazing (especially if prolonged after 15 May) and application of inorganic nitrogen. The first two practices led to a decline in herb richness, while the third led to a decline in characteristic hay meadow vegetation. The study built on monitoring carried out between 1987 and 1995, by resurveying a subset of 164 sites between June and August 2002. During the resurvey, vegetation was surveyed in three 1 m^2 quadrats/site, and management information from 1995 to 2002 was collected by interviewing farmers.

A randomized, replicated before-and-after trial in England (7) found that the conservation value of hay meadows in the Dartmoor Environmentally Sensitive Area increased during the nine years following its introduction in 1994. Eighteen randomly chosen Environmentally Sensitive Area hay meadows (with agreements aimed at habitat enhancement) were surveyed in June–July 1995 and 2003. Most hay meadows increased in conservation value, with the biggest improvement seen in poorer quality sites. There was an overall increase in plant species richness and the vegetation became closer to that of meadows characterized by crested dog's-tail *Cynosurus cristatus* and common knapweed *Centaurea nigra* (MG5 under the UK National Vegetation Classification Scheme). This was accompanied by a general trend of declining soil fertility and a narrowing of the difference in nutrient availability between sites (particularly potassium).

A replicated before-and-after trial in England (8) found that the Environmentally Sensitive Area scheme maintained extensively-managed permanent pasture and wet grassland/hay meadows in the Avon Valley and Upper Thames Tributaries Environmentally Sensitive Areas, introduced in 1993 and 1995, respectively. The range of vegetation communities in these Environmentally Sensitive Areas was broadly similar between baseline surveys (following introduction) and resurveys in 2003. However, in the Somerset Levels and Moors Environmentally Sensitive Area (baseline surveys in 1993, 1995 and 1998), management under the 'raised water tier' of the scheme appeared to be encouraging the formation of less species-rich inundation and rush pasture communities, rather than maintaining species-rich grassland. There were 34 replicate sites in Avon Valley, 39 in Upper Thames Tributaries and 25 in Somerset Levels and Moors Environmentally Sensitive Area.

A randomized, replicated before-and-after trial in England (9) found that management under the Environmentally Sensitive Area scheme broadly maintained species richness on enclosed rough grassland in the Blackdown Hills, Shropshire Hills and South West Peak Environmentally Sensitive Areas. In all three, the range of vegetation communities was similar between baseline surveys (in 1994–1995) and resurveys in May–August 2003. The number of species in survey plots was also similar between years, except in the South West Peak Environmentally Sensitive Area, where the average number of species per plot declined by 1.5 (to 23). The number of species at a finer within-plot scale declined in the Shropshire Hills and South West Peak Environmentally Sensitive Area, but not by an extent to indicate loss of ecological quality. While the results suggested that management prescriptions were broadly maintaining species richness, they appeared less effective in encouraging desirable species to colonize improved/semi-improved grasslands. There were 22 replicate sites in the Blackdown Hills, 29 in the Shropshire Hills and 38 in the South West Peak Environmentally Sensitive Area.

A replicated site comparison study in 1998–2001 in Switzerland (10) found that on average 86% of litter meadows in Ecological Compensation Areas on the Swiss plateau were of 'good ecological quality' and attracted wetland birds, which had significantly more territories

(52) than expected (31) in these areas. However, only 20% of hay meadow Ecological Compensation Areas were of good ecological quality, and they did not appear to benefit birds of open cultivated land, containing fewer of these birds' territories (68) than expected (151). Previous intensive management appeared to limit the effectiveness of hay meadow Ecological Compensation Areas, with significantly lower ecological condition in the more intensively farmed 'lowland' zone of the Swiss plateau, compared to 'pre-alpine hills' zone. Under the Ecological Compensation Area scheme farmers must carry out low intensity management on 7% of their land. For litter meadows, this includes traditional litter use, mowing and no fertilizer; and for hay meadows, restrictions on fertilizer use and mowing (late cut). Plant species were recorded in 1,306 hay meadow Ecological Compensation Areas and 104 litter meadow Ecological Compensation Areas, in 11 study areas between 1998 and 2001. Breeding bird territories were mapped in 23 study areas, on 3 visits to each area between mid-April and mid-June.

A 2005 review (11) of seven studies exploring the role of cutting, grazing and fertilizer in maintaining species richness of upland UK hay meadows concluded that the best management involves spring and autumn grazing, a mid-July hay cut and no inorganic fertilizer. The review recommends using only low levels of farmyard manure as two studies found that it can lead to a shift towards improved grassland plant communities (Edwards *et al.* 2002) and is unlikely to assist seed dispersal (Tallowin 2005).

A replicated study at 2 dry limestone grassland sites over 23 years in Switzerland (12) found that compared to traditional management, alternative management regimes resulted in greater changes in plant species composition and abundance. Changes were most evident 12 years after the start and were continuing even after 22 years. By the end of the study at one site, the number of species had declined in the following order compared to plots under traditional management (annual mowing in July; 50 species): mowing annually in October (93% of species in traditionally managed plots), mowing every second year in July (90%), mowing every 5th year in July (79%), annual controlled burning February–March (79%) and abandoned (70%). At the other site, species number did not tend to differ with management (85–120% of traditionally managed plots). Cover of specific species tended to differ with management regime at both sites. Plots with mowing every second year in July, followed by mowing annually in October were the most similar in composition to traditional management. The six management regimes were investigated with 3 replicate plots (5 x 10 m) at each site from 1977 to 2001. Cuttings were removed immediately. Species composition was sampled 4–6 times at each site 1977–1996. The inner 4 x 9 m of each plot was also surveyed at both sites in May, June and September in 1997–1999.

A site comparison study of 12 pastures and meadows (4 ha) in northwest Russia and 4 in Finland (13) found that butterfly (Lepidoptera) species richness, diversity and total abundance did not differ significantly between mown meadows and grazed pastures, although meadows were preferred by more species (46 vs 42). A total of 3,660 individuals were recorded in the meadows and 2,082 in pastures. Butterfly communities were affected more by the origin and age of the grassland than the present management method. Landscape factors such as surrounding habitat, abundance of nectar plants and intensity of tilling were the most important factors differentiating older grasslands from the younger ones. Meadows were mown annually in late July or August and pastures were grazed by cattle, some with sheep or horses temporarily. Tilling and fertilization (manure) tended to occur at intervals of 3–10 years. Butterflies were sampled 11–13 times along transects (640–720 m) in June and July 1997–1999.

A randomized, replicated, controlled trial on an abandoned meadow in northern Finland (14) found that the number of plant species (excluding mosses and lichens) hardly changed over five years in response to regular annual mowing. However, mosses and lichens were maintained by mowing and declined to almost nothing in unmown control plots. There were 14 plant species/plot on average in 1998 and 12 species/plot in 2003, with no differences between mowing treatments. Some species responded to treatments. For example, August mowing plus disturbance favoured harebell *Campanula rotundifolia* but reduced cover of common bent grass *Agrostis capillaris*. The cover of mosses and lichens fluctuated between years but was not different between mowing treatments. The meadow was abandoned in 1985. In 1993, it was divided into forty 50 x 50 cm study plots, each at least 2 m apart, and annually mown in August, without grazing. From 1998, 10 plots were mown in June, 10 in August, and 10 mown in August with bare soil exposed in 25% of the plot area, using a spade. Ten control plots were not mown. The percentage cover of all plant species (including mosses and lichens) in the treatment plots was monitored in June every year from 1998 to 2003.

A 2006 systematic review (15) of the impacts of grazing management on one type of species-rich old meadow in the UK found that intermediate grazing intensities were generally most appropriate for plant conservation. Choice of stock type (sheep, cattle or horses) appeared less critical than intensity but there is some evidence that sheep grazing can result in lower diversity of herbaceous plants than cattle grazing at high stocking rates. Two studies showed that sheep grazing led to lower plant diversity and herbaceous plant cover than cattle grazing on old or restored pastures (one study for each type). Data from Welsh grasslands showed that with sheep grazing, a lower grazing intensity (grass height >10 cm) was needed to maximize the abundance and number of species of herbaceous plants than for horse or cattle grazed sites. There was little information regarding taxa other than higher plants. The review searched for studies from the UK or Ireland comparing the effects of sheep grazing with horse, cattle or no grazing, on pastures characterized by crested dog's-tail *Cynosurus cristatus* and black knapweed *Centaurea nigra*, known as MG5 under the UK National Vegetation Classification. Studies on similar grassland types in northwest Europe were also considered. Forty-two studies were found, of which 22 were reviews.

A before-and-after trial of 16 upland hay meadows in the Pennines, northern England (16) found that 15 years of low intensity management under the Environmentally Sensitive Area agri-environment scheme maintained the total number of plant species, but not the number of broadleaf (forb) species. The meadow plant communities were characterized by sweet vernal grass *Anthoxanthum odoratum* and wood cranesbill *Geranium sylvaticum* (MG3 under the UK National Vegetation Classification). The number of broadleaf species declined from 14 to 10 species/m^2 on average between 1987 and 2002, while the average number of grass species increased from 8 to 12 species/m^2. The authors linked the loss of broadleaf species to inappropriate grazing intensity. Plant species were recorded in three 1 m^2 permanent quadrats at each site, in 1987 (the year the scheme was introduced) and 2002. On 62 sites in the same study with areas of other types of species-rich grassland (not MG3) the average number of plant species increased slightly from 21.6 to 22.8 species/m^2. Again, the increase was predominantly in the number of grass species, not broadleaf herbaceous species.

A long-term unreplicated trial from 1975 to 2000 in Baden-Württemberg, Germany (17) found that all tested treatments: low intensity grazing, mowing-and-mulching (in June or June and August) and winter burning maintained a variety of types of semi-natural grassland. Plants showed similar functional responses to treatments: grazing encouraged species with small seeds and a persistent seed bank (and woody species, though this was probably due to very low grazing pressure), mulching led to an increase in ground-layer species and winter burning benefited species with storage organs. The authors suggest that this functional approach may help to predict vegetation changes following management. Treatments were applied in 20 x 40 m fields from 1975. Plants were surveyed in one 25 m^2 plot/field until 2000 and graded according to 11 functional traits. The study was carried out over 14 sites, characterized by different grassland vegetation types. There were no replicates.

A site comparison study in Switzerland (18) found that 9 alpine meadows managed under the Ecological Compensation Areas agri-environment scheme for at least 5 years had more plant species (around 37 species/meadow on average) than 9 conventionally managed meadows (around 27 species/meadow) but not more grasshopper (Orthoptera) species (3–3.5 grasshopper species/meadow on average). Conventionally managed meadows were cut 2 to 6 times annually, with 20–30 kg N/ha added after each cut as liquid or solid manure. Ecological Compensation Area meadows were first cut in July and had a maximum of 30 kg N/ha applied annually as solid manure. Sites in the study were randomly selected from target regions in the northern and east central Alps. Ecological Compensation Area sites were in higher, more remote locations and had steeper slopes than conventionally managed meadows. Since 79% of sites in the study had been managed in exactly the same way for 10 years, the results show that the Ecological Compensation Area scheme can maintain higher plant species richness in alpine meadows.

A randomized, replicated trial in 1995–2006 at a semi-natural grassland site near Maribor, Slovenia (19) found that occasional late cutting may allow farmers to maintain a stable grass yield, without stimulating the spread of broadleaved plants (forbs). The study applied different cutting frequencies (at 2, 4, 6, 8, 10 and 12 week intervals) on semi-natural grassland between 1995 and 2006, and measured the effect on the dry matter proportions of grasses, legumes and non-legume broadleaved plants in May 1995, 1999, 2002 and 2006. There were four replicates. In the final year (2006) the treatment with the least frequent cuts resulted in the highest proportion of grasses (77%) and lowest proportion of non-legume broadleaved plants (19%) in the dry matter of harvest. Although the digestibility of forage produced by infrequent cutting does not meet the needs of modern livestock production, occasional very late cuts could offer a compromise between maintaining grassland biodiversity, while allowing farmers to maintain a stable grass yield.

A controlled, replicated trial in the Harz mountains of northern Germany (20) found that annual mowing from the end of June onwards was the best method for conserving and restoring mountain meadows. Ideally, this management should occur within a mosaic of areas which are mown less frequently or not at all. Mulching was a less effective form of management due to the development of a thick litter layer. Mowing every two or three years benefited species with more rigorous growth, and abandonment led to the development of stands of tall herb communities. The study was replicated over four meadow sites with treatments (annual mowing; two or three yearly mowing; mulching and abandonment) commencing in 1987–1988 and plant surveys continuing until 2003.

A replicated, controlled, randomized study of previously restored semi-natural grassland in central Finland (21) found that mowing affected woody plants, leguminous broadleaved plants (forbs) and bacterial-feeding nematodes (Nematoda) but not overall plant abundance or species diversity. Mowing significantly decreased cover of woody plants (mown: 2–10%; unmown: 10–45%) and increased cover of leguminous broadleaved plants (16–42% vs 5–18%), selfheal *Prunella vulgaris*, birdeye pearlwort *Sagina procumbens* and white clover *Trifolium repens* in 2004. Grasses and broadleaved plants and horsetails did not differ and there was no significant difference between annual and bi-annual mowing. Bacterial-feeding nematodes were more abundant in annually mown (9/g dry soil) compared to bi-annually mown (7/g) and unmown (4/g) plots in the upper, but not lower, part of the grassland. Fungal, root and omnivorous-feeding nematodes and pot worms (Enchytraeidae) were not affected by mowing. The three mowing treatments were allocated randomly to the 30 plots (1 x 1 m, 60 cm-wide buffer zones). Plants were sampled in June 2002–2004.

An unreplicated trial in 1996–2007 on a species-rich mountain meadow in the Bohemian Forest Mountains, Czech Republic (22) found that mulching once a year in July produced a greater number of plant species and a greater proportion of broadleaved plants (forbs) than traditional mowing once a year in July (without mulching) or abandonment. Mulching promoted many short broadleaved plants, grasses and graminoids, which were suppressed in the fallow treatment by an increasing share of tall grasses. The study concluded that mulching may be a viable alternative for preventing succession in situations where regular mowing is not economically or technically feasible. Treatments were applied from 1996 in 50 x 100 m plots. Plant species composition was measured each year shortly before mowing or mulching (June/July) in five 1 m² quadrats/plot. Aboveground plant biomass and litter (harvested from four 0.33 x 0.33 m subplots/plot) and belowground plant biomass (to 15 cm depth in four 0.15 x 0.15 m subplots/plot) were also measured.

(1)   Smith R.S. (1994) Effects of fertilisers on plant species composition and conservation interest of UK grassland. *British Grassland Society Occasional Symposium*, 28, 64–73.
(2)   Younger A. & Smith R.S. (1994) Hay meadow management in the Pennine Dales, Northern England. *Proceedings of the Joint Meeting between the British Grassland Society and the British Ecological Society: Grassland management and nature conservation. British Grassland Society Occasional Symposium No. 28, 27–29 September 1993.* University of Leeds, England. pp. 137–143.
(3)   Ryser P., Lagenhauer R. & Gigon A. (1995) Species richness and vegetation structure in a limestone grassland after 15 years management with six biomass removal regimes. *Folia Geobotanica and Phytotaxonomica*, 30, 157–167.
(4)   Tilzey M. (1997) Environmentally sensitive areas and grassland conservation in England. *Proceedings of the International Occasional Symposium of the European Grassland Federation (Management for Grassland Biodiversity).* Warszawa-Łomża, Poland, 19–23 May. pp. 379–390.
(5)   Defra (2002) *Breeding Bird Survey of the Cotswold Hills ESA and Exmoor ESA.* Defra MA01006.
(6)   Defra (2004) *Monitoring of Hay Meadows in Pennine Dales ESA.* Defra MA01005.
(7)   Defra (2004) *Survey of Moorland and Hay Meadows in Dartmoor ESA.* Defra MA01016.
(8)   Defra (2005) *Botanical Survey of Wet Grassland in the Avon Valley, Upper Thames Tributaries and Somerset Levels and Moors ESAs.* Defra MA01014.
(9)   Defra (2005) *Botanical Survey of Grassland in the Shropshire Hills, Blackdown Hills and SW Peak ESAs.* Defra MA01013.
(10)   Herzog F., Dreier S., Hofer G., Marfurt C., Schüpbach B., Spiess M. & Walter T. (2005) Effect of ecological compensation areas on floristic and breeding bird diversity in Swiss agricultural landscapes. *Agriculture, Ecosystems & Environment*, 108, 189–204.
(11)   Jefferson R.G. (2005) The conservation management of upland hay meadows in Britain: a review. *Grass and Forage Science*, 60, 322–331.
(12)   Köhler B., Gigona A., Edwards P.J., Krüsi B., Langenauer R., Lüscher A. & Ryser P. (2005) Changes in the species composition and conservation value of limestone grasslands in Northern Switzerland after 22 years of contrasting managements. *Perspectives in Plant Ecology, Evolution and Systematics*, 7, 51–67.
(13)   Saarinen K. & Jantunen J. (2005) Grassland butterfly fauna under traditional animal husbandry: contrasts in diversity in mown meadows and grazed pastures. *Biodiversity and Conservation*, 14, 3201–3213.
(14)   Hellström K., Huhta A.P., Rautio P. & Tuomi J. (2006) Search for optimal mowing regime – slow community change in a restoration trial in northern Finland. *Annales Botanici Fennici*, 43, 338–348.
(15)   Stewart G.B. & Pullin A.S. (2006) Does sheep-grazing degrade unimproved neutral grasslands managed as pasture in lowland Britain? Systematic Review No.15. *Centre for Evidence-Based Conservation.*
(16)   Critchley C.N.R., Fowbert J.A. & Wright B. (2007) Dynamics of species-rich upland hay meadows over 15 years and their relation with agricultural management practices. *Applied Vegetation Science*, 10, 307–314.
(17)   Kahmen S. & Poschlod P. (2008) Effects of grassland management on plant functional trait composition. *Agriculture Ecosystems & Environment*, 128, 137–145.
(18)   Kampmann D., Herzog F., Jeanneret P., Konold W., Peter M., Walter T., Wildi O. & Lüscher A. (2008) Mountain grassland biodiversity: impact of site conditions versus management type. *Journal for Nature Conservation*, 16, 12–25.
(19)   Kramberger B. & Kaligaric M. (2008) Semi-natural grasslands: the effects of cutting frequency on long-term changes of floristic composition. *Polish Journal of Ecology*, 56, 33–43.

(20) Dierschke H. & Peppler-Lisbach C. (2009) Conservation and regeneration of structure and floristic biodiversity of montane meadows – 15 years of scientific contribution to conservation management in the Harz Mountains. *Tuexenia*, 145–179.
(21) Ilmarinen K. & Mikola J. (2009) Soil feedback does not explain mowing effects on vegetation structure in a semi-natural grassland. *Acta Oecologia*, 35, 838–848.
(22) Maskova Z., Dolezal J., Kvet J. & Zemek F. (2009) Long-term functioning of a species-rich mountain meadow under different management regimes. *Agriculture Ecosystems & Environment*, 132, 192–202.

Additional references
Edwards A.R., Younger A. & Chaudry A.S. (2002) The role of farmyard manure in the maintenance of botanical diversity in traditionally managed hay meadows: the effects of rumen digestion on seed viability. In J. Frame (ed.) *Conservation Pays? Reconciling Environmental Benefits with Profitable Grassland Systems. Occasional Symposium No. 36*, British Grassland Society, Reading. pp. 159–162.
Tallowin J.R.B. (2005) *The Impact of Organic Fertilizers on Semi-natural Grasslands*. Defra BD1415.

4.2   Restore or create species-rich, semi-natural grassland

- Twenty-eight studies monitored the effects on wildlife of restoring species-rich grassland. Of these, 20 from Finland, Germany, Lithuania, Sweden, Switzerland and the UK (15 replicated, of which 8 were controlled and 3 also randomized) found restoring species-rich grassland resulted in higher ground beetle abundance[15], increased plant species richness[4, 5, 17, 18, 24, 26, 31, 33, 45, 51, 53, 55, 58, 61], farmland bird abundance[20, 21, 69], pollinating insect density and diversity[27, 29, 64, 71] and earthworm abundance[34] than other types of grassland, or that restored grasslands had similar abundance and species richness of insects to old traditionally managed sites[37, 39, 68].

- Seven studies from Denmark, Finland, Sweden, Switzerland and the UK (five replicated and controlled, two also randomized) found that efforts to restore species-rich grassland had no clear effect on the species richness or abundance of plants[1, 12, 43, 44, 46, 66], beetles[6, 7], or the abundance of butterflies and moths[42]. Three replicated studies from Sweden and the UK (one also controlled and two site comparisons) found that restored grassland had a lower diversity and frequency of certain plant species[36, 41] and attracted fewer foraging queen bumblebees[64] than continuously grazed or unmanaged grasslands.

- We captured 40 studies (including 19 replicated and controlled studies, of which 6 were also randomized and 6 reviews) from 9 European countries that found 10 different techniques used alone or in combinations were effective for restoring species-rich grassland[2, 4, 5, 8–11, 14–24, 26, 28, 30–33, 35, 37–40, 43, 45, 47–51, 53–58, 60–62, 67, 68, 70–72]. Effective techniques included: grazing[2, 4, 5, 9, 11, 18, 33, 37, 43, 48, 61, 68], introducing plant species[8, 14, 16, 20–22, 31, 38, 40, 48, 49, 51, 54, 56, 67, 70, 71], hay spreading[45, 51, 55, 62, 70] and mowing[18, 19, 24, 31, 55, 61].

- We found 22 studies from 7 European countries that included information on the length of time taken to restore grassland communities (including 16 replicated trials, of which 9 were also controlled and 3 reviews). Six studies saw positive signs of restoration in less than 5 years[5, 14, 19, 22, 38, 72], 11 studies within 10 years[11, 17, 19, 24, 31–33, 38, 49, 56, 61] and 2 studies found restoration took more than 10 years[38, 47]. Six studies found limited or slow changes in plant communities following restoration[11, 15, 59, 62, 63, 65]. Two studies from Germany and the UK (one replicated controlled trial) found differences in vegetation between restored and existing species-rich grasslands 9[55] or 60 years[60] after restoration.

Background

The area of species-rich grassland in northwest Europe declined dramatically during the twentieth century in response to agricultural intensification (Walker *et al.* 2004). Restoring this habitat is a high priority for many conservation groups, and in the UK a number of agri-environment schemes support the restoration or creation of species-rich semi-natural grasslands (Walker *et al.* 2004). Measures used to restore or create grasslands of high plant diversity such as chalk grasslands and hay meadows include grazing, mowing, sowing and hay spreading.

See also 'Reduce management intensity on permanent grasslands' for studies where the endpoint is still an economically viable agricultural grassland, rather than a species-rich grassland with nature conservation as its main objective.

See also 'Restore or create traditional water meadows' for studies where the species-rich grassland in question is (or was) either fen meadow, or frequently flooded water meadow.

See also 'Add yellow rattle seed *Rhinanthus minor* to hay meadows' for studies looking at the effects of adding yellow rattle seed to grasslands.

See also 'Agri-chemicals: Reduce chemical inputs in grassland management', for studies where the sole intervention has been to reduce or cease chemical inputs, in an attempt to restore plant communities.

Walker K.J., Stevens P.A., Stevens D.P., Mountford J.O., Manchester S.J. & Pywell R.F. (2004) The restoration and re-creation of species-rich lowland grassland on land formerly managed for intensive agriculture in the UK. *Biological Conservation*, 119, 1–18.

A replicated, controlled trial in Denmark, from 1974 to 1978 (1) found that re-introducing cattle grazing on abandoned agricultural grassland reduced the frequency of heather *Calluna vulgaris* and wavy hair grass *Deschampsia flexuosa* but did not consistently increase the number of plant species. There were increases in the number of plant species under intensive grazing in the area that had the fewest plant species at the start of the experiment, abandoned 14 years before, but colonization was very slow. Wavy hair grass decreased the most under normal grazing, reducing from 92 to 70% cover on average; intensive grazing had a lesser effect. Heather reduced the most under intensive grazing (from 71% to 6% cover on average). A 59 ha area of hill pasture was subjected to 4 levels of grazing intensity: ungrazed, lightly grazed, normally grazed and intensively grazed. Grazing took place two to four times a year, in late spring and late summer. There were 3 fields: one abandoned in 1910 (64 years before), one in 1960 (14 years before) and one grazed every year since 1910. Vegetation was sampled in July each year on 52 x 1 m² quadrats across all treatments.

A small study of species-poor grassland in the Netherlands (2) found that preferential grazing by sheep gave rise to a 'macro-pattern' of various plant communities absent under a hay-making regime. There were short, heavily grazed patches interspersed with taller, lightly grazed patches, a micro-pattern that tended to be stable after initial establishment. Heavily and lightly sheep grazed areas differed little in plant species composition but abundances of species differed considerably. Heavily grazed areas were characterized by equal amounts of monocots (mainly grasses) and broadleaved species, had higher abundances of rosette species and, to a lesser extent, greater persistence of perennial rye grass *Lolium perenne*. Lightly grazed patches were dominated by common bent grass *Agrostis tenuis* and had a large amount of plant litter. Prior to 1972, the grassland was cut for hay and aftermath grazed by cattle. Sheep grazed part of the grassland (3 sheep/ha) from 1972 in July–December and January–July. After 1980, the area was left ungrazed for two months each winter. Vegetation was recorded annually within permanent 2 x 2 m quadrats. Grazing intensity was recorded in February as lightly grazed (>70% litter cover), heavily grazed (<30% litter cover), or intermediate. Species abundance was quantified on a dry weight basis. Vegetation was also mapped from October 1979–1982 in a 10 x 10 m² area and recorded as heavily grazed (<5 cm tall) and lightly grazed (>10 cm tall).

A 1985 review of four techniques aimed at reducing soil fertility for the restoration of semi-natural grassland or heath (3) found that there was some success. Addition of inorganic nitrogen to a crop was found to extract more phosphorus and other elements from the soil than treatments with no fertilizer addition. Fertilizer addition almost completely exhausted the soil phosphorus store in long-term wheat experiments. Three studies investigating nutrient removal by stubble burning were reviewed (Allen 1964, Kenworthy 1964, Chapman 1967). Burning decreases phosphorus availability and increases potassium availability. Topsoil stripping reduces fertility but can result in the loss of a large proportion of the seed bank.

A replicated, controlled study of an abandoned arable field in Oxfordshire, UK (4) found that spring grazing by sheep increased plant species richness. However, this was only evident at intermediate to large sampling scales, suggesting that differences arose from changes in rarer and widely dispersed species. Many annual herbs decreased in ungrazed paddocks in 1985, whilst some annual grasses and other perennial herbs and grasses increased. In grazed (spring or autumn) paddocks, many more species increased their distribution compared to ungrazed plots, particularly annual herbs. The field (12 ha) had been permanent pasture until 1960 and was then cultivated until 1981. In 1985 the central 90 x 90 m blocks of two 1 ha blocks were divided into 9 paddocks, i.e. 3 replicates of 3 treatments. Vegetation was sampled within various quadrats by several methods in April, July, August and October 1984–1985. This study was part of the same experimental set-up as (5, 9, 11).

A replicated, controlled study of an abandoned arable field in Oxfordshire, UK (5) found that plant species establishment was better in grazed areas and species richness, diversity, and abundance of individual species were also higher compared to ungrazed controls, however there were no apparent effects of sheep grazing treatment on the likelihood of new plant species arriving in the grassland. Within 18 months, 57% of plant species restricted to patches of old chalk/limestone (calcareous) grassland within 2 km had colonized the study field. Colonization was not thought to have come via the seed bank. In addition, the area grazed April–November contained many component species of mature chalk/limestone (calcicolous) grassland, unlike ungrazed controls. The 12 ha field had been permanent pasture until 1960 and was then cultivated until 1981. In 1985, 2 blocks of 1 ha were divided into 9 paddocks, 3 replicates of spring grazed, autumn grazed and ungrazed treatments. Additional treatments were grazing from April–November with a short break in summer and continuous grazing August–November. Vegetation was sampled within permanent quadrats (1 m²) from April 1984 to October 1986. This study was part of the same experimental set-up as (4, 9, 11).

A replicated, controlled, randomized study in 1973–1974 of limestone grassland in Cambridgeshire, UK (6) and (7) found that cutting tended to benefit plant-eating beetle (Coleoptera) species, whereas most beetles that feed on decaying matter, detritus-feeding beetles, fungus-eating beetles and some predatory beetles were more abundant in uncut plots. Beetle families that feed on decaying matter, and predatory beetle families were next most abundant in May-cut plots. Overall species diversity was highest on uncut plots and then July-cut plots. No beetle species showed a response to treatments in all five sampling periods. A significant reduction in abundance was recorded for 17 species, whilst 12 increased under one or more treatments. Treatments were an annual cut in May or July, or both, and an uncut control, with 4 replicates (each 16 x 12 m) of each in a randomized block design. The site had previously been unmanaged for several years. Beetles were sampled by vacuum netting 1 m^2 in each plot at 2–4 weekly intervals from October 1972 to December. Berlese-type funnels were also used on one turf sample (each 0.07 m^2) per plot/week over 4 weeks in each of 5 sampling periods in 1973–1974.

A replicated trial in 1986 and 1987 at the Chilworth Research Centre, Hampshire, UK (8) tested survival of twelve native plant species, either grown in pots or sown as seeds on an existing meadow. Three species, black knapweed *Centaurea nigra*, St John's wort *Hypericum perforatum* and musk mallow *Malva moschata* always survived well (83% or more survived). Five species survived better when sown as seeds (oxeye daisy *Leucanthemum vulgare*, self-heal *Prunella vulgaris* and lady's bedstraw *Galium verum*) or when planted as plug plants (betony *Betonia officinalis* and cowslip *Primula veris*). Four species had poor survival in both treatments: yellow rattle, *Rhinanthus minor*, field scabious *Knautia arvensis* (both 0% survival), harebell *Campanula rotundifolia* and bulbous buttercup *Ranunculus bulbosus* (0–8% survival). Plants sown as seeds into cleared plots were almost always larger after two growing seasons than those planted out as plug plants. Sown plots were cleared with herbicide in March 1986. Seeds were covered with 2.5 cm of compost, a perforated polythene cloche, or both. There were three replicates of each treatment. Results showed no difference between the germination or survival of plants under different coverings. Twenty-four plants of each species were planted out into a meadow where the vegetation was 10–15 cm high in July 1986, at 1 m intervals. The growth of sown and planted plants was monitored in September 1987.

A replicated, controlled study in 1986 and 1987 of an abandoned arable field in Oxfordshire, England (9) found that sheep grazing increased seedling establishment compared to ungrazed plots. The most heavily grazed treatment had the highest levels of seedling establishment, whereas few new seedlings were recorded on ungrazed paddocks. Treatments with some autumn grazing had a peak of seedling establishment the following spring. Seedling survival was not affected by grazing treatment or gap size. The two short-grazing treatments (lasting 10 days) had the least bare ground whilst April–November grazed areas had the most. Insecticide use increased seedling establishment in October in ungrazed and spring-grazed paddocks but decreased establishment in autumn-grazed paddocks. In 1985, 3 treatments were applied in 6 replicate (30 x 30 m) paddocks: 10 days grazing in spring or autumn, and ungrazed controls. Two 3 x 3 m permanent quadrats were treated weekly with malathion insecticide in each paddock and 4 permanent 1 x 1 m sampling quadrats were established. Another two paddocks were grazed from April–November with or without a short break in summer, twelve 1 x 1 m quadrats were established in each. Gap type and seedling sampling was undertaken in all quadrats seven times from April 1986 to July 1987. Vegetation height was recorded in September 1986. This study was part of the same experimental set-up as (4, 5, 11).

A replicated, controlled study of grassland restoration within 4 geological regions in southern Germany (10) found that in the first year natural regeneration following topsoil removal had higher species diversity than sown treatments (47 vs 33–38 species) but sowing tended to reduce pioneer species and perennial weeds. In the first year, the most successful establishment was of nutrient-rich meadows and tall herbaceous vegetation in existing arable soils (no soil removal). Topsoil removal enhanced establishment of semi-dry grassland species (37 vs 33 species) but not other communities. Complete soil removal favoured pioneer, semi-dry grassland and neutral (mesotrophic) edge communities. Isolated plots within fields had lower numbers of species than plots adjacent to existing edge habitats for all treatments. Six communities were sown in plots of 10 x 5 m: control (existing seed bank), semi-dry grassland, nutrient-rich grassland, neutral edge communities, pioneer vegetation and tall herbaceous vegetation. Seeds were collected within a 10 km radius. Establishment methods were complete soil removal, topsoil removal or existing arable soil. Results are from year 1 of 15 years of monitoring.

A controlled, replicated trial from 1982 to 1993 on an abandoned ex-arable field at Oxford University Farm at Wytham, Oxfordshire, UK (11) found that 10 years after abandonment, heavily grazed treatments (particularly spring and autumn grazing, at 3–6 sheep/ha) more closely resembled target ancient chalk/limestone (calcicolous) communities than lightly grazed or ungrazed treatments. Over the experimental site, 250 plant species colonized, 77 of which were typical of chalk/limestone grassland. However, the species composition of the site differed markedly from that of nearby ancient chalk/limestone grassland (where later-successional and stress-tolerator species were more common), indicating that restoration may take decades. Arable cultivation was abandoned in 1982 and five grazing treatments began in 1985. Three treatments were replicated 6 times in 30 x 30 m paddocks (ungrazed control, short-period spring and short-period autumn grazing) and two treatments were applied in larger areas (spring-and-autumn grazing and long-period autumn grazing, not replicated). Plants were surveyed 4 times a year in 12 quadrats (1 m^2) in each replicate and in nearby ancient grassland patches. This study was part of the same experimental set-up as (4, 5, 9).

A replicated, controlled trial from 1984 to 1990 at Little Wittenham Nature Reserve, Oxfordshire, UK (12) found that plant composition on a previously improved pasture hardly changed in response to reduced sheep grazing intensity and no fertilizer. Plant species diversity was still low after six years. The vegetation remained dominated by perennial grasses, with 4 species making up 80% of records. Herbaceous plants (non-grasses) made up just 0.4% of records. Seventy percent of seedlings growing in artificial gaps in the grass cover were of two grass species, perennial rye grass *Lolium perenne* and meadow barley *Hordeum secalinum*. Only 4% of seedlings were non-grass species, and none were species not already found in the paddocks. There was no evidence of a seed bank (gaps with original topsoil did not differ from gaps with topsoil replaced by sterile soil). There were eight levels of sheep grazing: summer grazing to a height of either 3 cm (more intensive) or 9 cm (less intensive), with or without winter and/or spring grazing, but grazing intensity had only small effects on the vegetation. Each treatment was replicated in two 50 x 50 m paddocks. Plants were surveyed using a point quadrat at 64 points/paddock in 1990. Vegetation and topsoil or vegetation-only were removed in September 1990 in five 10 cm diameter circles/paddock and seedlings growing in these areas counted and removed regularly until January 1992.

A study of a former improved grassland over 16 years in the Netherlands (13) found that plant species richness was lower in the fertilized compared to unfertilized plot, even after 16 years. One plot was treated with nitrogen fertilizer in an attempt to remove nutrients, other than nitrogen, from the grassland via hay removal. Although there was a peak in above-ground biomass in the fertilized plot in 1977 (900 g/m^2 vs 400 g/m^2), in 1986 biomass in the fertilized plot started to decrease and by 1990 it was the same as in the unfertilized plot (300 g/m^2). There were no significant differences in soil chemical variables after 16 years. The authors conclude that fertilizer application as a conservation measure does not seem appropriate for restoring species-rich grassland. In 1972 the grassland was taken out of production, fertilizer addition ceased and vegetation was mown and removed in late July–early August. Two adjacent 20 x 10 m plots were established in 1973; one received nitrogen-fertilizer (50 kg/ha/year) and the other was unfertilized. Plant species composition and above-ground standing crop was sampled from 1972 onwards.

A randomized, replicated, controlled trial from 1993 to 1994 in Hampshire, UK (14) found that hand-sowing chalk grassland plants on rotavated ex-arable plots created a community partly resembling the target plant community after two years. Computer analysis showed a 45–62% fit to a UK National Vegetation Classification scheme chalk/limestone grassland community (CG2a – sheep's fescue *Festuca ovina*; meadow oat-grass *Avenula pratensis* grassland; dwarf thistle *Cirsium acaule*; squinancy wort *Asperula cynanchica* subcommunity). Higher sowing rates gave a better fit, and a higher percentage cover of chalk grassland plants (from 10% cover at the lowest sowing rate to 100% at the highest rate). The three higher sowing rates had similar numbers of chalk grassland species in the plots (28, 30 and 31 species respectively) by 1994. Control plots and plots sown at 0.1 g/m^2 had around 6 and 20 species of chalk grassland species respectively. Seed was sown at 0.1, 0.4, 1 or 4 g/m^2. The mixture contained 22 grass/sedge species and 25 herb species. Each rate was sown in 4 replicate 6 m^2 plots and 4 control plots were not sown. Plots were rotavated in March 1993, sown by hand, raked and left unmanaged (lightly grazed by rabbits). Plant cover was measured in two 1 m^2 quadrats/plot in August 1993 and 1994.

A trial at the Crichton Royal Farm, Dumfries, Scotland (15) found more ground beetle (Carabidae) species at the restored wildflower sites (18–26 species/line of traps) than in more intensively managed grassland (17–23 species) or unmanaged grassland (15–16 species). However, an index of ground beetle diversity was highest at the unmanaged grassland nature reserve site. Beetles found there were larger, and a different set of species. The authors concluded that the beetle community found in natural grassland habitat in the area had not re-colonized the restored species-rich grasslands, even after five years. Two fields were ploughed and sown with 17 plant species in August 1987 (5 grasses, 2 clovers *Trifolium* spp. and 12 other broadleaved flowering plant species). They were managed without fertilizers, cut once in July and grazed in autumn and winter. Ground beetles were sampled in 18 pitfall traps (laid out in 2 lines) in each treatment area between April and September in 1989 and again in 1993. Ground beetles were also sampled at sites with continuously grazed perennial rye grass *Lolium perenne*, perennial rye grass fertilized with cattle slurry and mineral fertilizer and cut three times a year, and an unmanaged grassland on a nearby National Nature Reserve.

A 1997 review of experimental evidence (16) concluded that a combination of management changes and introducing plant species is necessary to restore species-rich grassland. A four year study from 1991 to 1995 (Steinegger & Koch 1997) in Switzerland found that a late hay cut and no fertilization did not restore a species-poor rye grass *Lolium* spp. sward to a species-rich meadow. The number of plant species gradually increased to 40 on plots where additional plant species were repeatedly sown in. Another study (Lehmann *et al.* 1996) found removing the existing grass sward and re-sowing with a tailored mix of at least 30 species (12 grasses, 5–7 legumes and 13–16 herbs) was the most reliable and quickest way to restore species-rich grassland.

A trial near Clavering in Essex, UK (17) found that plant species richness increased dramatically on an ex-arable field following ten years of traditional hay meadow management. The number of plant species in the restored meadow increased from 19 species in 1988 to 42 species in 1994, after planting with a commercial rye grass *Lolium perenne* and white clover *Trifolium repens* seed mix in 1984. The meadow changed from being dominated by sown rye grass to including grasses such as soft brome *Bromus hordeaceus* and crested dog's-tail *Cynosurus cristatus*. Species usually found on species-rich grasslands, black medick *Medicago lupulina*, not found in 1988, became dominant. Two orchid species colonized (common spotted orchid *Dactylorhiza fuschii* and bee orchid *Ophrys apifera*). The 3 ha site was cut once in July, aftermath grazed with around 5 sheep/ha from August to October and had no fertilizer added. Plants were monitored each June from 1988 to 1994, in 40 randomly placed 0.25 m² quadrats. Importantly, the site was immediately adjacent to 2 established hay meadows (present in 1941) in which the number of plant species also increased from 26 in 1989 to 48–57 species in 1994.

A replicated study in 1975–1995 in pastureland in the Archipelago National Park, south-west Finland (18) found that restoration methods including removing trees and shrubs, grazing, pollarding trees and mowing increased the average number of plant species/plot from 32 to 41. This was mostly due to an increase in common species, but the number of old meadow indicator species also increased slightly. Three out of four locally endangered plant species increased in cover but none colonized new areas. Grasses benefited more than broadleaved flowering plants, increasing their overall average cover from 19% before restoration to 26% after. The study used 41 permanent 10 x 10 m plots, 16 of which were in grazed areas, 11 in thinned areas (some of which were also grazed) and 14 in thinned, mowed and grazed areas. Monitoring took place every three to eight years. Older plots were assessed six times and newer ones twice.

A 1998 review of case studies in France gathered from published and unpublished literature on grasslands (19) found that the restoration process generally takes eight years or more and does not always work as expected. Fifty-one case studies were identified across France (including 16 on wet grassland). Most were not discussed in detail. On chalk grassland, one study showed an increase in a plant species diversity index over three years in response to mowing (Dutoit 1996). On meadows degraded by engineering work or ski tracks, seven studies were found demonstrating that grasslands can be artificially re-created, but that it takes many years. One study, for example, demonstrated that achieving vegetation similar to the original vegetation took nine years (Bédécarrats 1991). Another demonstrated how, over eight years, sown species gradually disappeared to be replaced by a newly created grassland with a different plant community from the original (Maman 1985, Coin 1992).

A replicated, controlled study in the winters of 1994–1997 in southern England (20) (same study as (21)) found that Eurasian skylark *Alauda arvensis*, corn bunting *Miliaria calandra* and meadow pipit *Anthus pratensis* were consistently more abundant on arable fields reverted to species-rich chalk grassland than on land reverted to permanent grassland (sown with agricultural grasses), intensively managed permanent grassland or winter wheat fields (25–230 skylarks/km² on reverted chalk grassland vs 0–11 on other field types; 0.9–4.7 corn buntings/km² on reverted chalk grassland vs 0–1 on other field types; 3.7–6.1 meadow pipits/km² on reverted chalk grassland vs 0–4.3 on other field types). Densities of rooks *Corvus frugilegus* and species richness of plant seeds did not differ across field types. However, there were significantly more plant species on reverted chalk grassland than the other field types (7.8–9.2 species/quadrat vs 1.4–5.1 species/quadrat). Reverted chalk grassland fields were sown with species such as *Festuca* spp. and *Bromus* spp. grasses. Fields on 40 farms were surveyed. Birds were surveyed once during December and January on 217 fields in winter 1994–1995, repeated on 205 fields in winter 1995–1996 and on 225 fields in winter 1996–1997. The numbers of grassland birds and types of grazing livestock were recorded. In November in the winters of 1995–1996 and 1996–1997, seeds lying on the ground in 31 fields were sampled in two 0.25 m² quadrats/field and identified to species. Plant species were surveyed in four 0.25 m² quadrats/field in July–August 1994 and 1996 in 121 and 72 fields.

A replicated, controlled study in spring and summer from 1994 to 1996 in southern England (21) (same study as (20)) found that arable fields reverted to species-rich chalk grassland consistently had higher densities of Eurasian skylark *Alauda arvensis* than land reverted to permanent grassland (sown with agricultural grasses), intensively managed permanent grassland or winter wheat fields (12–23 skylarks/km² on reverted chalk grassland fields

(16–35 fields) vs 3–12 skylarks/km^2 on 16–82 fields of other types). Densities of carrion crows *Corvus corone* and rooks *C. frugilegus* were not consistently higher on any field type (0.5–1.9 crows/km^2 and 0–14 rooks/km^2 on chalk grassland vs 0–1.8 crows/km^2 and 0–90 rooks/km^2 on other fields). Reverted chalk grassland fields were sown with *Festuca* spp. and *Bromus* spp. grasses. The other field types studied were arable fields reverted to permanent grassland, downland turf (close-cropped, nutrient-poor grassland), permanent grassland, winter wheat, barley, oilseed rape and set-aside. Fields on 40 farms were surveyed. The number and location of singing skylarks were recorded in April–May and June–July 1994 and 1996 and in May–June 1995. The locations of foraging carrion crows and rooks were also recorded in 1994–1996.

A replicated, controlled trial on 6 Environmentally Sensitive Area sites in England and Wales (22) found that sowing 35 to 40 plant species increased the number of broadleaved plant species on all sites relative to the control and the number of grass species increased on 5 of the 6 sites by the second year. The sowing treatments with the most soil disturbance, rotavating and de-turfing, established the most species. In year 1, there was an average of 7 broadleaved species and 7 grass species on the control plots, compared to 16 broadleaved species and 10 grass species on the deturfed treatment, which had the strongest increase in species richness. The trial was carried out in 1994 on 6 x 4 m plots monitored for the following two years, with four replicates of each treatment on each farm. Seven species were successfully introduced by some or all treatments on all sites: yarrow *Achillea millefolium*, oxeye daisy *Leucanthemum vulgare*, self-heal *Prunella vulgaris* and ribwort plantain *Plantago lanceolata*, black knapweed *Centaurea nigra*, bird's-foot trefoil *Lotus corniculatus* and cat's ear *Hypochoeris radicata*. Other species, including yellow rattle *Rhinanthus minor*, failed to establish.

A replicated, controlled trial in Trawsgoed Research Farm, Aberystwyth, Wales (23) (partly the same study as (53)) found that seedlings established best and survived best in plots that were cut twice with aftermath grazing by sheep in winter. The lowest rates of seedling establishment and plant survival (lower than the control) were in plots cut twice but without grazing. The authors conclude that winter grazing is very important when re-introducing plants to restore hay meadows. By September 1996, seeds from the local area; yarrow *Achillea millefolium*, purple betony *Stachys officinalis* and self-heal *Prunella vulgaris* had survived better than non-local seeds, with no difference in two other species (black knapweed *Centaurea nigra*, ribwort plantain *Plantago lanceolata*). The five plant species were sown in October 1994, two years after the management experiment began. Fifty seeds of each species and provenance were sown in each of three 1 m^2 quadrats/plot. Seeds were either gathered from within 8 km or purchased from a seed supplier (from elsewhere in the UK). Plants were monitored every month after sowing until April 1995, then in April and September 1996.

Two trials on Monte Generoso, Switzerland (24) investigated the effects of different plant establishment techniques on an abandoned grassland. The controlled trial found that introducing mowing to an abandoned grassland increased the number of plant species after ten years, but the response was slow. The replicated controlled trial found that when 12 native grassland species were sown, fewer seedlings emerged in burned plots (average 5.6 seedlings/species/plot after one year) than in plots mown with litter removal before sowing (average 8 seedlings/species/plot). Plots mown without litter removal were intermediate (6.1 seedlings/species/plot). Mown grassland plots were still dominated by tor grass *Brachypodium pinnatum* at the end of the experiment. Mown plots increased from a total of 46 plant species in 1989 to 68 species in total by 1996–1997. The total frequencies of all species other than tor grass stayed low for six years and increased strongly in the final four years of the experiment. The replicated controlled study also found common rockrose *Helianthemum num-*

mularium had the lowest seedling emergence (0–0.8 seedlings/plot after one year) and rough hawkbit *Leontodon hispidus* had the highest (17–24 seedlings/plot after one year). After 2 years, 10 of the sown species survived, with the grasses having the highest survival rates (27–36% of seeds germinated). No bulbous buttercup *Ranunculus bulbosus* or meadow clary *Salvia pratensis* seedlings survived. In the controlled study, mowing was resumed in the abandoned grassland in two 100 m² plots. From 1988, one plot was mown twice a year (July and September–October); one was mown in early July only. A control plot was not mown. Plant species composition was monitored in September–October at 456 point quadrat samples/plot over 10 years, from 1988 to 1997. In July 1989 and 1995, percentage cover across each whole plot was estimated for plant species. In the replicated controlled study 45 seeds of each species, hand-collected from a nearby meadow, were sown into twenty-four 60 x 60 cm subplots in October 1995. Each plot was subject to one of three treatments before sowing (six replicates of each): mowing with removal of mown vegetation, burning or mowing with complete removal of litter. For each treatment, a control 60 x 60 cm plot was treated the same but not sown. Seedling emergence was recorded every one to five months for the following two years. Plots were mown twice in 1996 and 1997.

A replicated, controlled study in 1988–1997 in temperate grassland and seasonal freshwater marshes in Strathrory Glen, northeast UK (25) found that cattle grazing at a stocking rate of 2.2–2.5/ha prevented further decline in plant species richness, but did not promote recovery. In grazed plots, overall species richness remained static and moss/liverwort richness increased, whilst in control plots species richness declined. However, cattle grazing had no effect on species cover and plants of conservation importance showed no increase. The study used four plots in each of three different vegetation types: acid grassland, rush pasture and vegetation associated with calcareous springs (seepage flush). Each plot had two 10 x 10 m subplots, one of which was fenced to exclude cattle. Sampling was carried out in 1988 before the start of grazing, and again in 1991 and 1997, using 4 permanent 2 x 1 m quadrats within a central 7 x 7 m area in each plot.

A replicated trial from 1990 to 1998 of combined management treatments on an agriculturally-improved meadow in the Pennine Dales Environmentally Sensitive Area, North Yorkshire, England (26) (same study as (31)) found that the highest increase in plant species diversity was achieved with a combination of autumn and spring grazing, 21 July hay cut date and sowing native plant species. The study took place in 6 x 6 m plots on a 2.75 ha meadow within the Ingleborough National Nature Reserve. Plots were either sown with many native grassland species (including yellow rattle *Rhinanthus* minor) or not. The experiment also included three different grazing treatments (sheep and cattle), plots with or without fertilizer and three earliest dates for hay cut. Yellow rattle spread to most plots after its introduction as a constituent of the seed addition treatment. By 1996 it was particularly abundant in treatment combinations that included autumn grazing, no mineral fertilizer and a July hay cut. Populations of over 40 plants/m² were associated with lowest hay yields, presumably as it suppressed grass growth.

A replicated study in 1995–2000 in 11 sites with newly created flower-rich meadows on set-aside land in central Switzerland (27) found generally higher butterfly (Lepidoptera) densities on the created meadows (e.g. 170 individuals/ha in Riedikon) than on intensively managed arable land. However, the highest densities were found in flower-rich field margins (e.g. 440 individuals/ha in Seewadel). Adult butterfly abundance was positively correlated to the number of flower units and up to 98% of flower visits were recorded on only 5 plant species. Note that no statistical analyses are presented in this study. On three main sites (Berg, Riedikon and Seewadel) flower-rich meadows were established on 0.5–0.6 ha experimental plots. Butterflies were recorded during seven visits from May–September 1999 along fixed transects. In an additional eight sites, butterflies and grasshoppers (Orthoptera) were monitored in 1995–2000 approximately twice a year.

A randomized, replicated, controlled trial at Hill Farm, Little Wittenham, Oxfordshire, England (28) found that both yellow rattle *Rhinanthus minor* and oxeye daisy *Leucanthemum vulgare* could be effectively established on a pasture field by 'slot seeding'. Different management treatments, cutting, grazing or both, did not affect the survival or establishment of either species, but yellow rattle seeds were spread a greater distance when hay was cut in July than without a hay cut. Seeds were sown in strips previously sown with herbicide by a tractor-mounted slot seeder in October 1995. Four management treatments were replicated 5 times in 20 x 10 m plots. The treatments were cut once (July), cut twice (July and September), cut July and autumn grazed. Monitoring of plant dispersal was carried out using seed traps at the soil surface from June to October 1997.

A study in eastern England of the pollinator community on a species-rich grassland restoration experiment compared to native grassland of the same plant community (29) found a greater diversity of pollinating insects on the restored hay meadow site than on the ancient meadow. Six common species of bumblebee *Bombus* spp. were recorded at both sites, and the most abundant insect visitor was a bumblebee on both meadows: white-tailed bumblebees *Bombus terrestris/lucorum* at the restored site; red-tailed bumblebees *B. lapidarius* at the ancient meadow site. Seven and five species of solitary bee were recorded at restored and ancient sites respectively.

A randomized, replicated study in 1994–1998 in arable fields in five lowland areas in the UK (30) found that ploughing to 30–40 cm depth and sowing with a species-rich seed mixture created a community similar to the target community on neutral soils. This was significantly more successful than natural regeneration or sowing with a species-poor mix. Sites on acidic or calcareous soils were less similar to their specific target communities. Sowing a nurse crop had no beneficial effects. All treatments reduced nutrient levels. The five sites had four replicate blocks each containing seven experimental plots with different treatments. Vegetation was cut and removed each year in June or July, and sheep were grazed between October and December at 25–40 sheep/ha for 6 to 8 weeks. Vegetation sampling used three 40 x 40 cm quadrats randomly placed within each plot in June each year. Nutrient sampling used ten soil samples per plot in September 1994 and 1998.

In the same replicated study described in (26) in North Yorkshire, UK (31) plots sown with grassland species had more plant species in the vegetation, and plots cut in July had more plant species in the seed bank, ten years after the restoration began. There were 22 species/m^2 in sown plots compared to 18 species/m^2 on unsown plots. Of three species sown in 1999, only bird's-foot trefoil *Lotus corniculatus* was found frequently in 2000. A single plant each of quaking grass *Briza media* and bulbous buttercup *Ranunculus bulbosus* (both also sown in 1999) were found. The upper 5 cm of soil held viable seeds of on average 13 species in plots always cut in July, <12 species in plots previously cut in September and 10 species in plots previously cut in June. From September 1990–1998, the experiment combined grazing, cutting date and fertilizer level treatments in 6 x 3 m plots (3 replicates of each). From 1998 onwards, all plots were cut on 21 July, with both autumn and spring grazing. From 1999 plots were either treated with farmyard manure at 12 tonnes/ha/yr in April, or not. The soil seed bank was monitored from soil cores in July 1998 and the vegetation surveyed in June/July 2000.

A replicated trial from 1994 to 1990 in Aberdeenshire, Scotland (32) found that sowing meadow plants followed by summer cattle grazing achieved the best results in terms of recreating a species-rich grassland on a former arable field last ploughed in 1993. After six years sown plots on average contained more sown species (4.9 species/m²) and had greater cover of sown species (97%) than non-sown plots (1.8/m², 43%). Cattle grazed plots contained more sown species (4.8/m²) than sheep grazed plots (2.2/m²) but sheep grazed plots had greater sown species cover (91% vs 46%). August cut plots with cuttings removed had more sown species but lower cover than plots with cuttings left (3.4 vs 2.3/m², 63% vs 76%, respectively). Twenty-four 20 x 40 m fenced plots were either sown with 10 herb and 4 grass species from a crested dog's-tail *Cynosurus cristatus*, lesser knapweed *Centaurea nigra* (UK National Vegetation Classification MG5) grazed hay meadow community or not sown after ploughing in April 1993. Within these treatments, six management treatments were tested: sheep or cattle summer grazing, cut in June with aftermath grazing by sheep or cattle, or cut in August with cuttings either left or removed. No plots were cut in August with aftermath grazing.

A replicated site comparison study in 1999 and 2000 in southwest Finland (33) (the same study as (37, 42, 43)) found that resuming grazing on abandoned species-rich grasslands began to enhance the number of plant species after around five years. The number of plant species was higher on restored pastures with resumed grazing than on old abandoned pastures (for example, 16.4 species/m² on average, compared to 11.2 species/m²) but the difference was not statistically significant at any scale. Old grazed pastures had significantly more plant species than restored or abandoned pastures at the 1 m² scale, and significantly more species than abandoned pastures but not significantly more than restored pastures at the whole site scale (0.25–0.8 ha). However, the number of rare native plant species had not increased in response to resumed grazing. Plants were monitored in 1999 or 2000 on 11 old grazed pastures, 12 abandoned pastures (no grazing for more than 10 years) and 10 restored pastures abandoned for more than 10 years with grazing re-started 3 to 8 years (average 5 years) before the study.

A replicated, controlled study of former agricultural land in four European countries (34) (same study as (56)) found that the abundance of earthworms (Lumbricidae) was higher in sown and naturally colonized grassland than in an agricultural rotation in two of four European countries. Numbers and biomass of earthworms were significantly higher in the restoration plots (Netherlands: high plant diversity plots 43 individuals/m², low diversity plots 52, naturally colonized plots 95; Sweden: high 254, low 289, natural 169) than in the agricultural rotation treatment (Netherlands: 5; Sweden: 15 individuals/m²). In the Netherlands, numbers were significantly higher in the naturally colonized plots than in the sown treatments. Differences between treatments did not differ at the UK or Czech Republic sites. In Sweden, species diversity was lower in the agricultural plots (2 vs 3–5 species) and worm biomass increased with legume biomass. In the UK worm biomass increased with grass biomass. In each country there were 5 blocks each with 4 treatment plots (10 x 10 m): seed sown to give high (15 species) and low diversity (4 species), natural colonization and a continued agricultural rotation treatment. A target plant community was also sampled. Earthworms were collected within 30 minutes from 4 to 5 squares in each plot (July or August 1998), identified to species or genus and wet weight recorded. Vegetation was sampled within 12 permanent subplots in each plot.

A replicated, controlled trial near Göttingen Germany (35) found that harrowing before sowing enhanced the survival of most grassland plants sown into existing grassland. For six of eight species tested seedling emergence was highest if seeds were sown after harrowing. All species except red clover *Trifolium pratense* were more likely to have survived after one year if sown after harrowing than on control plots without harrowing. Complete removal of vegetation with herbicide before sowing, or frequent cutting before or after, did not enhance seedling emergence any further. For red clover, cutting and pre-sowing disturbance made no difference to seedling emergence. Autumn hawkbit *Leontodon autumnalis* had the highest emergence with the highest disturbance (cutting every week, or complete removal of vegetation before planting). All tested species survived better in plots cut every one or three weeks after sowing than in those cut every nine weeks. Eight plant species were sown in May 1998 in 0.5 x 0.5 m plots on a previously intensively managed, species-poor grassland. Plots were cut every one, three or nine weeks in the nine weeks before sowing and every one, three or nine weeks after sowing. They were either left undisturbed, harrowed or treated with herbicide before sowing. There were three replicates of each treatment combination. Seedling survival was recorded until July 1999.

A replicated, controlled, site comparison study of 26 restored semi-natural grasslands in southeastern Sweden (36) found that continuously grazed control sites had higher plant species diversity and a higher proportion of typical grassland species in the community than restored grasslands. Plant species diversity at restored sites was 16–20 species/m^2 compared to 24–30 species/m^2 at continuously grazed control sites. Total species richness was positively associated with time since restoration (1–7 years) and the abundance of trees and shrubs. Overall species composition differed between restored and control sites, with control sites having a higher proportion of typical grassland species than restored sites. However within grassland types (dry, dry to damp (mesic) or damp to wet) species composition was similar between each pair of restored and control sites. Restored damp to wet grassland was dissimilar in species composition to all other plots. Abundance of 10 grazing-indicator species tended to be lower at restored sites. Restored site area (3–35 ha), time between abandonment and restoration, time since restoration and abundance of trees and shrubs were not related to species composition among restored sites or the 10 grazing-indicator species. Restored sites were grazed before abandonment and after restoration, control sites had been grazed continuously. The six control sites were compared to restored sites in the same region. Plants were sampled within 10 randomly distributed plots (1 m^2) in July–August 2001. Trees and shrubs were counted within a 40 m diameter circle at each site.

A replicated site comparison study in 1999 and 2000 in southwest Finland (37) (same study as (33, 42, 43)) found that abandoned species-rich grasslands restored with cattle grazing had similar butterfly and day-flying moth (Lepidoptera) numbers to old grazed pastures. The abundance of butterflies and moths, and the number of species, did not differ between grazed and restored pastures. There were 22–26 species and 126 individuals/site in restored and old grazed pastures. Restored pastures varied more in the identities of species found than old grazed pastures. Some restored pastures had less 'diverse' butterfly and moth communities than old pastures because they were more likely to be dominated by abundant, common species. Some species only occurred in old pastures. Butterflies and moths were monitored in 1999 or 2000 on 11 old grazed pastures and 10 restored pastures abandoned for more than 10 years, with grazing re-started 3 to 8 years before the study. All restored pastures received support under the Finish agri-environment support scheme for managing semi-natural grassland. Insects were counted along walked transects between four and seven times between May and August. Either transect length (2000) or searching time (1999) were standardized across sites.

A 2004 review of published and unpublished literature from the UK (38) found that introducing plant species and removing nutrients are important to effective grassland restoration. The review identified eight studies that tested effects of reinstating cutting and grazing management on grassland and ceasing fertilizer use and concluded that this could enhance the number of plant species, but it was slow and did not always work. Just one study, in west Wales, found a marked increase in the number of plant species over eight years (Hayes

& Sackville Hamilton 2001). Five found a slight or gradual increase, over 4–14 years ((12), Oomes 1990, Olff & Bakker 1991, Hayes *et al.* 2000, (26)). Two studies found no change or a decrease (Oomes 1990, Mountford *et al.* 1994). Increases in the number of species were modest and slow. Cutting and grazing together were generally more effective than cutting or grazing alone. On existing grassland, the review found nine studies that tested various methods of adding plant species: adding seed (over-sowing), drilling seed (slot-seeding) and planting small plants (plug-planting), with effects monitored over two to eight years. Three studies found that over-sowing was most effective when combined with either cutting and grazing or de-turfing (22, 23, 26). Five studies found that slot-seeding was not very effective (Wells *et al.* 1989, (22, 28)). Of five studies that tested plug-planting, three found it was initially effective, but survival of the introduced plants fell after two to five years (Wells *et al.* 1989, (22), Barratt *et al.* 2000). Two found that 60–70% of plants established (over 2 or 5 years) (Boyce 1995, (22)). On ex-arable land, 10 studies tested sowing grassland species, monitoring effects for between 1 and 20 years. All found increased plant species diversity and enhanced similarity to the target plant community, which was either upland, chalk or neutral (mesotrophic) grassland. Similarity to the target community was quantified for 6 of these studies and fell between 50% and 81%, usually after 2 to 5 years (20 years in one case) (McDonald 1992, Wells *et al.* 1994, (14), Pywell *et al.* 2000, (30)). Cultivation, followed by a relatively high seeding rate, seemed the most effective approach. On upland grassland, adding sulphur to acidify the soil prior to seeding led to effective establishment of sown species in two studies.

A small site comparison study of two restored hay meadows with two ancient hay meadows in the Bristol area, UK (39) found no consistent differences in the abundance or diversity of pollinating insects (dominated by bees (Hymenoptera) and flies (Diptera)) between ancient and restored sites, and considered the pollinator community to be effectively restored. The 4 hay meadows were 1–9 ha and managed with an annual hay cut, no grazing and no artificial fertilizer. One restored meadow was previously a golf course, restored in the early 1990s using a traditional annual hay cut. The other restored meadow was previously part of an urban park, restored in 1981 by translocating turf from an ancient meadow, planting and sowing wildflowers and using traditional meadow management. Flower-visiting insects were sampled on two 50 x 2 m transects/meadow every 2 weeks from early May to the end of July 2000.

A 2005 review of six studies exploring the best management for restoring upland hay meadow vegetation on semi-improved grassland in the UK (40) suggested that the highest plant species richness is produced by spring and autumn grazing, a mid-July hay cut and no inorganic fertilizer. Addition of seed from outside the site (either from natural dispersal or sowing) is also likely to be necessary. Three studies found that adding *Rhinanthus minor* hay rattle seed can help the colonization of other sown species (Smith *et al.* 2003, Pywell *et al.* 2004, Smith 2005). One study in North Yorkshire (Smith 2005) found that adding farmyard manure had a generally harmful effect on restoration of upland hay meadow communities, and recommended that this should be avoided, at least in the early stages of restoration. However, results were based on using larger quantities of manure than under traditional management.

A replicated, controlled study of semi-natural grasslands at two sites in southeast Sweden (41) found that the frequency of harebell *Campanula rotundifolia*, cowslip *Primula veris* and yellow rattle *Rhinanthus minor* was lower in restored than traditionally managed, continuously grazed (from 1945 to 1998) grasslands and in one case lower than abandoned grasslands. Cowslip had significantly lower frequency in restored (6% plots contained cowslip) than continuously grazed (22%) and abandoned grasslands (20%) at one site. At the other site cowslip frequency did not differ significantly (restored 24%, grazed 38%, abandoned 16%).

Yellow rattle was absent or less frequent in restored (0–6%) than continuously grazed grasslands (6–27%) and was absent from abandoned grasslands. Harebell had a lower frequency in restored (28–54%) and abandoned grasslands (28–62%) than continuously grazed sites (42–80%). There was no significant difference in seedling emergence for cowslip (0.2–2.0/plot) or harebell (5–8/plot) within continuously grazed grasslands or those restored 3 or 40 years ago or abandoned, however harebell did not emerge in abandoned grasslands. At each site, 100 randomly distributed 1 m² plots were sampled for presence/absence of the 3 species within each grassland type. At one site, 50 seeds of each species (collected in the area in autumn 2002) were sown (within 2 weeks) in each of 8 randomly distributed 1 dm² plots in each grassland. Emergence was recorded in 2003.

A further report (42) from the same replicated site comparison study in southwest Finland as (33, 37, 43) looking at the responses of individual butterfly and moth (Lepidoptera) species showed that three species were most abundant in old pastures and did not recover in pastures where grazing had been reintroduced following abandonment. These were the purple-edged copper *Lycaena hippothoe*, the common blue *Polyommatus icarus* and the yellow shell moth *Camptogramma bilineatum*. Three moth species were more abundant in restored pastures than old pastures: *Epirrhoe hastulata* (no common name), the silver-ground carpet *Xanthorhoe montanata* and the latticed heath *Chiasmia clathrata*. Two species, the scarce copper butterfly *Lycaena virgaureae* and the black-veined moth *Siona lineata* were less abundant in restored pastures than in old grazed pastures or abandoned pastures (12 pastures abandoned for more than 10 years were monitored for comparison), implying that they were negatively affected by the reintroduction of grazing.

A further report (43) from the same replicated site comparison study in southwest Finland as (33, 37, 42) found that the frequency of 31 of 76 plant species increased or recovered following the reintroduction of grazing on abandoned species-rich grasslands. Twenty-nine species seemed to have increased in response to grazing, having higher frequencies on restored pastures than on abandoned pastures but lower than old pastures. Species included common bent grass *Agrostis capillaris* and harebell *Campanula rotundifolia*. Just two species, red fescue *Festuca rubra* and Goldilocks buttercup *Ranunculus auricomus* had similar frequencies in old and restored pastures. Eight species, such as meridian fennel *Carum carvi*, cowslip *Primula veris* and self-heal *Prunella vulgaris*, showed no recovery in response to resumed grazing, having similar frequencies in restored and abandoned pastures. Thirty one species did not differ between three types of grassland.

A randomized, replicated, controlled trial from 1998 to 2003 on a meadow in northern Finland (44) (same study site as (63)) found that the number of plant species did not increase in response to conservation-oriented mowing regimes. There were 14 plant species/plot on average in 1998 (not including mosses and lichens), 12 species/plot in 2003 and no difference between mowing treatments. Some species responded to treatments. For example, late mowing plus disturbance favoured harebell *Campanula rotundifolia* but reduced the cover of common bent grass *Agrostis capillaris*. The Kiiminki Haaraoja Meadow in northern Finland was traditionally grazed, but abandoned in 1985. In 1993, the meadow was divided into forty 50 x 50 cm study plots, each at least 2 m apart and annually mown in August, without grazing. From 1998, 10 plots were mown in June, 10 in August, and 10 mown in August with bare soil exposed in 25% of the plot area, using a spade. Ten control plots were not mown. The percentage cover of all plant species (including mosses and lichens) in the treatment plots was monitored in June every year from 1998 to 2003.

A replicated, controlled trial from 1992 to 2003 in southeast Germany (45) (same study as (55)) found that hay spreading enhanced plant species richness but not grasshopper

(Orthoptera) species richness on hay meadows, and that topsoil removal enhanced some grasshopper and plant species but not the total number of plant species. For example, in 2001 there were 20–25 plant species/plot with hay transfer and 10–20 species/plot without hay transfer. The number of plant species from the target plant community and the number of Red-Listed plant species were also higher on sites with hay transfer. Hay transfer had no effect on the number of grasshopper species (5–6 species/100 m on all plots), or grasshopper species associated with dry grassland, or Red-listed grasshopper species. Topsoil removal enhanced the number of dry grassland grasshopper species, plant species from the target plant community, Red-listed species of plant and grasshopper, and the total number of grasshopper species, but not the total number of plant species. Four ex-arable fields were half spread with hay from a nearby nature reserve between July and September 1993. The other half had no hay added. One field had the topsoil removed (to 40 cm depth) in 1993. Plant species were monitored every year from 1993 to 2002 on thirty 4 m^2 plots/field. Grasshoppers were counted 4 times between July and September 2001 on 18 transects at the restoration experiment.

A site comparison study of semi-natural grasslands near Lund, Sweden (46) found no significant difference in plant species richness or abundance in recently restored, abandoned or continuously grazed grasslands. There was a decrease of management-dependent plant species with increasing tree and shrub cover at abandoned sites. Present management significantly affected butterflies (Lepidoptera) and plants; their species richness increased with increasing vegetation height, but this differed between sites depending on whether they were grazed by cattle, horses or sheep. Sheep grazing negatively impacted species richness compared to cattle or horses. There were 12 grasslands of each type and current management comprised 12 sites cattle grazed, 6 horse grazed, 8 sheep grazed and 10 with no grazing. Butterflies and burnet moths (Zygaenidae) were sampled using a transect count method (150 m/ha) 6 to 7 times in May–August 2003 or June–August 2004. Plant presence was sampled in ten 0.25 m^2 quadrats (divided into twenty-five 10 x 10 cm squares) at each site in June–August 2004. Vegetation height was also measured.

A replicated trial from 1987 to 2004 at Somerford Mead, Oxfordshire, UK (47) found that both plant and beetle (Coleoptera) communities on an experimentally restored meadow were closest to the flood meadow restoration target under a regime of hay cutting and aftermath grazing. For plants, sheep grazing was slightly better, but for beetles, cattle grazing was better. There were fewer beetles and beetle species on plots cut for hay but without aftermath grazing. After 18 years, neither the plant nor the beetle communities were fully restored to floodplain meadow species assemblages. The site was characterized by a high percentage cover of red fescue *Festuca rubra*. A former arable field was sown with seed harvested from a local floodplain meadow in 1985. From 1987 it was cut in July and aftermath grazed. From 1989, three aftermath grazing treatments were tested: sheep, cattle or no grazing, on three 0.4 ha plots each. Plants and invertebrates were monitored in 2004 and compared with communities on two nearby floodplain meadows.

A replicated, controlled seed addition trial in 2000–2001 in four grasslands subject to different management practices in southeastern Sweden (48) found that seedling emergence was higher in grazed grasslands than in an abandoned grassland. Seedling emergence was similar for six plant species favoured by grazing (target species) and six plant species favoured by no grazing (generalist species) in all four grasslands. The proportion of sown seeds emerging differed among species (range: 1.2–12.6%). The highest proportion of seeds germinated at an intermediate sowing density (20–50 seeds/dm^2). Target species recruited well in the former arable fields (grazed for 10 or 30 years) and generalists also recruited well at grazed sites. All sown species performed poorly in the abandoned (40 years ago) grassland. The two grasslands with the longest grazing history (continuously grazed grassland and former arable grazed for 30 years) were positively associated with the emergence of target species. There were 4 grassland types: former arable field grazed for 30 years; former arable field grazed for 10 years; continuously grazed (since 17th century) semi-natural grassland; abandoned semi-natural grassland (previously grazed, abandoned 40 years ago). Seed was collected locally in autumn 2000 and sown within two weeks. Each species was sown in ten 10 x 10 cm plots/ grassland at 4 densities: 10, 21, 46 and 100 seeds. Ten plots were left unsown as controls. Seedling emergence was recorded in June 2001 in randomly placed 1 m² quadrats.

A replicated study in 1995–1998 and 2002 of former arable fields at two sites in England (49) found that after eight years plots sown with species-rich mixtures resembled target grassland community types. Plots sown with a species-poor mix, although colonized by some additional species, had fewer grass, legume and other broadleaved species. Hay yield increased in the species-rich plots in the first years of the experiment and the increased yield was still apparent after eight years (43% higher yield than species-poor plots). This was largely due to differences in numbers of non-leguminous broadleaved plants. There were four replicate blocks of plots (6 x 4 m). The species-rich mixture comprised 11 grasses and 28 broadleaved species, to resemble species-rich hay meadows. The species-poor mixture comprised seven grasses to establish moderately diverse grassland. Vegetation was sampled in early June in three quadrats (0.4 x 0.4 m) per plot in 1995–1998 and 2002. During the hay cut in July, a 6 x 1.2 m sample of hay was removed from each plot and weighed and a 500 g subsample was dried to calculate hay yield.

A 2007 review of experimental evidence on how to restore species-rich grassland on old arable fields (50) found that removing excess nutrients is very slow if done simply by grazing and cutting hay (2 studies), with only 3–5% of the soil nutrient pool removed each year. Removing topsoil can effectively remove nitrogen but not phosphorus (one study; Verhagen et al. 2001). The authors argued that it is necessary to introduce plants by sowing because rare grassland species are under-represented in the seed bank. They found one review (Pywell et al. 2003) showing that plants that were good colonizers and competitors, associated with fertile soils, were most likely to establish in restoration experiments. Two sets of experiments demonstrated that seedlings of grassland or wet grassland plants survive less well in low light conditions (as in a dense productive grassland).

A randomized, replicated, controlled trial from 2000 to 2004 at a farm in East Sussex, UK (51) found that hay spreading was the most effective technique for restoring a hay meadow plant community similar to the seed donor site. Both hay spreading and the addition of brush-harvested seed increased plant species richness, and harrowing increased the effectiveness of the seed addition treatments. Hay spreading was thought more effective because it captured seeds from a greater range of heights in the sward, and allowed for seeds to mature on the restored site after the restoration activity. Eight different combinations of harrowing and the two methods of applying seed were tested, on land that had been improved agricultural grassland, with two different rates of hay application. There were four replicates of each combination of treatments. Plants were monitored before treatment (July 2000) in two random plots from each block and every June from 2001 to 2004, in ten 50 x 50 cm quadrats in each plot.

A replicated, randomized study of sown grassland species on an intensive cattle farm in Ireland (52) found that for individual species, the maximum growth rate was either higher at a fertilization rate of 225 kg nitrogen/ha than at 90 kg N/ha, or did not differ. On dry soils, species either experienced slightly more competition with members of the same species at 90 kg N/ha than at 225 kg N/ha, or there was no difference between treatments. Twelve plant species were sown in mixtures of one, two or three species in equal amounts, comprising of one species plus perennial rye grass *Lolium perenne* plus a production/weed grass or legume. Sowing was in April 2004 at a rate of 1.5 or 3 g/m^2 in plots of 1.5 x 1.5 m at a dry and wet site. High (225 kg N/ha) or low (90 kg N/ha) fertilization rates were applied with three replicates for each seed mixture and soil type in a randomized block design. Biomass was recorded weekly for four to five weeks following cuts in alternate months. Botanical composition and weekly changes in green plant area and relative growth rates were also recorded.

A randomized, replicated, controlled trial at two experimental farms in Wales (53) (results from the two farms are presented in (23) and (61)) found that plant species richness on grasslands increased over 10–13 years in response to imposing traditional management practices, providing the management involved both hay cutting and aftermath (autumn–winter) grazing. In the final year these sites had over 13 plant species/quadrat (over 15 at the upland site) and over 40 plant species on each plot (over 50 at the upland site) compared to 8–9 species/quadrat and 24 species/plot in control plots. They were colonized by desirable plant species, such as yellow rattle *Rhinanthus minor* and heath spotted orchid *Dactylorhiza maculata*. Plots managed with just hay cutting, or just grazing, did not show strong increases in plant species richness. The experiments took place in Ceredigion (lowland) from 1992 to 2005 and the Cambrian Mountains (upland fringe) from 1995 to 2005. Six or seven treatments were each replicated 3 times on 0.15 ha plots. The control plot was species poor pasture sheep-grazed from April to November, fertilized (with nitrogen, phosphorous, potassium NPK fertilizer) at a rate of 150 kg nitrogen/ha, and limed once, in the second or third experimental year. Other plots were not fertilized at all, but had various combinations of cutting, grazing and liming. Adding lime slightly enhanced plant species richness in summer grazed treatments at the upland site. Plants were monitored in summers of 1992–1997 and 2000, 2003 and 2005 in ten quadrats in each plot.

A replicated trial in the White Carpathians Protected Landscape Area, in the eastern Czech Republic (54), found that sowing a regional seed mixture over the entire plot was the most effective treatment for establishing hay meadow vegetation. Four restoration treatments were tested, each in four 55 x 20 m plots, replicated within a single 3 ha arable field. The experimental treatments were sowing 7 grasses and 20 herb species throughout the plot, or sowing 2.5 m-wide strips of just the herb species with or without a commercial grass mix. Control plots were left to naturally regenerate. In the fully sown plots, 19 of the 20 herb species, and all 7 grass species had established by 2004, providing 30% and 55% cover on average. The cover of sown herb and grass species in the strip-sown or unsown treatments were less than 5% and 2–9% respectively. Plots were sown in spring 1999, and vegetation monitored in June 2002–2004. All plots were cut once in July and the hay removed, following restoration.

A replicated, controlled trial near Munich, southeast Germany from 1993 to 2002 (55) (same study as (45)), found that spreading hay from a nearby nature reserve rapidly increased the number of plant species and the number of target hay meadow species in ex-arable fields managed to restore hay meadow vegetation. The removal of topsoil combined with hay spreading increased the proportion of target species and the persistence of species, but led to a very low hay crop even after nine years. Mowing (once or twice) also increased plant species richness and the number of target plant species. Nine years after restoration, the best plots in this experiment (mown, with hay spreading) still had a different plant community from species-rich grassland on a nearby nature reserve (Garchinger Heide). Restoration was tested on 4 ex-arable fields, 1.3–3.2 ha in size, beginning in 1993. Half of each field had hay added between July and September 1993 (once only) and the other half did not. Experimental plots within these treatments were either mown once, mown twice, mown with cuttings left as mulch, or grazed through spring and summer. One field had the upper 40 cm of topsoil removed. This field was either mown once in July or left unmanaged. Plant species were monitored every year on thirty 4 m² plots per field.

A replicated, controlled experiment in five European countries (56) from 1996 to 2003 (the same study as (34)) found that most hay meadow species sown on plots of abandoned arable land established well at four locations (all except Sweden, where less than half of the sown species established). In the UK, the Czech Republic and the Netherlands, more than 70% of the sown species were established after eight years. Grasses, bird's-foot trefoil *Lotus corniculatus* and red clover *Trifolium pratense* established well at almost all sites, but small legumes, black medick *Medicago lupulina* and lesser hop trefoil *Trifolium dubium* disappeared quickly at all except the UK site. The success of other plant species varied between sites. Plots sown with 15 plant species always established some of the sown species, while some plots sown with just 4 plant species failed to establish any. Plots (10 x 10 m) were either left to naturally regenerate, or sown with 4 or 15 plant species in autumn 1995. The experiment was repeated in 5 countries: the Czech Republic, the Netherlands, Spain, Sweden and the UK, with 5 replicates of each treatment. Plant species naturally occurring in local grassland systems were sown. Red fescue *Festuca rubra*, Timothy grass *Phleum pratense*, bird's-foot trefoil *Lotus corniculatus*, red clover, and ribwort plantain *Plantago lanceolata* were sown at all sites. All plots were mown once or twice a year and separated by 2 m borders. Plant cover was measured from 1996 to 1998 and 2002 to 2003, in ten 1 m² quadrats/plot.

A randomized, replicated, controlled trial in 1999–2003 of restoration methods at two sites in the UK (57) found that turf removal followed by seed addition was the most effective means of increasing plant diversity. Multiple harrowing was moderately effective and was enhanced by applying snail/slug pesticide and sowing yellow rattle *Rhinanthus minor* (which reduced competition from grasses). Grazing, slot-seeding and inoculation with soil microbial communities from species-rich grasslands did not increase botanical diversity, and different grazing management regimes had little impact. Thirteen treatments were applied to 15 x 15 m plots at sites in Devon and Buckinghamshire, with 8 replicates of each treatment. All treatments were managed with a single July hay cut.

A controlled study in 2001–2005 of two sown meadows in the Kedainiai District, Lithuania (58) found that an extensively managed meadow was restored faster than an intensively managed meadow. The total number of plant species was higher in the extensively managed meadow compared to the intensively managed meadow (79 species, annual range 27–40 species in the extensively managed meadow vs 39 species, annual range 22–30 species in the intensive meadow). The same trend was seen for biomass (850–1,480 vs 720–1,340 g/m²), moss/liverwort/hornwort (bryophyte) content (6–26% vs <1.4%) and dead plant matter (2–22% vs 1–12%). A grassland/clover *Trifolium* spp. mixture was sown (27 kg/ha) in 2 arable fields in 1991. One received intensive management: fertilization, hay making and grazing (June, August, September). The other received extensive management: no fertilization, annual hay cutting and occasional grazing (once in July 2001–2005). Botanical composition was sampled in three permanent plots (100 m²) in each field in June, July and August from 2001 to 2005. Above ground biomass was sampled in three 1 m² quadrats in representative areas (by composition and cover) in each plot.

A site comparison study in 2006 on a 35 ha area of an ancient hay field in Lääne, western Estonia (59) found that recovery of the plant community was slow after restoring regular annual mowing. Even though sites were located alongside one another, there were differences in the plant community between plots where annual mowing had been reinstated for five or thirteen years, and sites with continuous annual mowing since the 1960s. Areas without continuous management had fewer plant species. Management history, not soil conditions, was the most important factor determining the number and identity of plant species. Plants were counted in five 1 m^2 quadrats at 30 sites of known management history in 2006. Three management histories were identified: continuously managed for 200 years and annually mown since the 1960s, irregularly mown every 2 or 3 years from the early 1980s to 1993 then annually since 1993 (regular mowing restored 13 years before the study), or unmanaged from the early 1980s until 2000 or 2001 when annual mowing was restored (5 or 6 years before the study). The latter group of sites had become overgrown with trees. Mowing was in late June or early July in each case.

A randomized, paired site comparison in five areas of southern England (60) found distinct differences in vegetation between restored and ancient chalk/limestone (calcareous) grasslands, even after 60 years. Sites seeded with just grasses remained dominated by a few grass species. Sites allowed to regenerate naturally moved towards the target plant community over time, although success was limited by proximity to ancient grasslands. Some features of restored grassland (such as the proportion of perennial plants) became more like ancient grasslands with increasing age. High soil phosphorus concentration (due to former fertilizer application) was detrimental to restoration. Forty restored grassland sites were randomly selected from all those available, to give equal representation in four age classes and the five areas (North Downs; South Downs; South Wessex Downs; Chilterns; Cotswolds). Sites were one to 103 ha in size. They were restored either by natural regeneration, seeding with grasses, or seeding with a flower-rich seed mix. All sites were grazed and some occasionally mown. Each was paired with an ancient grassland no more than 9.3 km away. Plants were surveyed in ten 0.25 m^2 quadrats at each site, in summer 2004, and soil analysed in September 2004.

A replicated, controlled trial at the Pwllpeiran Research Centre, in the Cambrian Mountains, west Wales (61) (partly the same study as (53)) found that plant species richness increased and rye grass *Lolium perenne* cover declined on improved upland grassland after ten years of management with hay cutting and/or grazing but no fertilizer addition. In the restoration plots, rye grass cover declined from 58% to just under 10% on average. All treatments enhanced plant species richness but the hay cut and grazing combined treatments were the most effective. These plots had an average of 51 species/plot by 2005 compared to 24 species in control plots. They also had almost 50% cover by non-grass, desirable herbaceous species (forbs). Treatments with hay cut but no grazing had 29–30 species on average in 2005 and those with grazing only had 31–35 plant species in 2005. Both had an increase in weedy, undesirable species. Seven management treatments were set up in 1994 on 0.15 ha plots with 3 replicates of each treatment. Control plots had standard intensive management, fertilized with nitrogen, phosphorous, potassium (NPK) fertilizer, limed and grazed by sheep. Six restoration treatments were either grazed from April to November, cut for hay in July/August without grazing, or hay cut and grazed from September to November, each with or without lime added in 1998. Plots with an application of lime had more desirable species by 2005 than those without lime.

A replicated, controlled study of chalk grassland restoration on land taken out of arable production in Oxfordshire, UK (62) found that plant and beetle (Coleoptera) species richness tended to be higher when seeds from the local area were applied by brush harvesting or hay spreading. Sowing of a grass-only seed mix, reduced the overall number of plant (36 vs 40 species) and beetle species (5–7 vs 6–8 species), independent of whether or not it was used in combination with the addition of local seeds. Plant species richness tended to be higher in plots receiving brush-harvested seeds or hay than controls, although there was limited change in those plots from 2002 to 2004 (at low or high seed application rates). The highest beetle species richness was found on plots with high rates of hay application without grass-only seed mix. Changes in beetle community structure were significant for control, high rate hay spreading, grass-only and low rate brush harvesting with grass-only mix. Plots receiving local seeds tended to become more similar to the donor plant community over time. The similarity was greatest in plots without the grass-only mix and with brush harvesting at a high rate for plants and high rate hay spreading for beetles. Changes in beetle assemblage were much greater than for plants. Forty plots of 10 x 10 m were established in 4 blocks. Grass-only seed mixture was sown in half the plots in August. Seeds from the local area were obtained from an adjacent unimproved chalk grassland and were applied by brush harvesting or hay at high or low rates. Plots were grazed by sheep but were not cut. Plants were sampled in ten 0.5 x 0.5 m randomly located quadrats/plot in August 2002–2004; the donor site was sampled in 2004. Plant-eating beetles were sampled using a Vortis suction sampler (15 positions/plot) in May, July and September in 2002–2004; the donor site was sampled in 2001.

A randomized, replicated, controlled trial from 2003 to 2006 on a meadow in northern Finland (63) (same study as (44)) found that of eight plant species sown, only two had established themselves after three years. The two species, maiden pink *Dianthus deltoides* and self-heal *Prunella vulgaris*, were only growing well on plots mown in August, with soil disturbance. Four plant species, small mousetail *Myosurus minimus*, water avens *Geum rivale*, northern dock *Rumex longifolius* and tansy *Tanacetum vulgare* did not establish, with 0–3 seedlings observed in the entire experiment and none on any plots by 2006. Longleaf speedwell *Veronica longifolia* and sticky catchfly *Lychnis viscaria* grew well on the August-mown disturbed plots (10–18 seedlings in total) in the first year but not in subsequent years. No species grew well on the other treatments (0–6 seedlings in total of each species/year). The meadow was divided into forty 50 x 50 cm study plots. In September 2003 half of each plot was sown with 30 locally-collected seeds of 8 plant species. Seedlings were counted in June–July 2004–2006. Ten plots were mown in June, ten in August, and ten mown in August with bare soil exposed in 25% of the plot area, using a spade. Ten control plots were not mown.

A replicated site comparison study in Scotland (64) found that Rural Stewardship scheme species-rich grassland attracted more nest-searching queen bumblebees *Bombus* spp. but fewer foraging queens than areas of naturally regenerated, largely unmanaged grasslands. Five Rural Stewardship Scheme farms participating in the species-rich grassland management or restoration option were paired with five conventional farms. Across all farms, unmanaged grassland on conventional farms attracted the highest abundance of foraging queen bumblebees (over 4 queens/100 m transect on unmanaged grassland vs less than 3 on species-rich grassland), also in comparison with hedgerow and field margin transects. Unmanaged grassland transects had more nectar and pollen-providing flowers than species-rich grassland in April and May, when queen bumblebees are on the wing. Bees were surveyed once a week for five weeks in April–May 2008 using a transect walk method.

A 2009 literature review of agri-environment schemes in England (65) found evidence that plant diversity was higher on Countryside Stewardship scheme plots sown with a chalk grassland mix than on Environmentally Sensitive Area sites sown with a basic grass mix (CABI 2003). However the same study also reported that few of the sown sites were classed as Biodiversity Action Plan habitats. One study found that few sites that had undergone arable reversion for at least five years could be classed as lowland chalk/limestone (calcareous) grassland or lowland meadow under Biodiversity Action Plan definitions (Kirkham *et al.* 2006). Instead many of the sites were comparable to semi-improved grassland.

A replicated, controlled study of ten nitrogen-rich grasslands heavily populated with white hellebore *Veratrum album* in Switzerland (66) found that sawdust addition had limited effects on mountain grassland communities. Sawdust addition reduced grass cover slightly (grazed control 51%; grazed plus sawdust 48%; ungrazed control 48%; ungrazed plus sawdust 42%) but plant diversity and species richness were unaffected, with species richness generally increasing with decreasing productivity in grazed and ungrazed areas. Above-ground grass and broadleaved plant biomass (excluding white hellebore) was 20–25% lower in sawdust plots and biomass of white hellebore was slightly higher compared to controls. Number of shoots was not influenced by sawdust (control: 21 shoots; sawdust: 27 shoots). Paired plots (6 x 3 m) were established, one cattle-grazed, the other not. From 2002 to 2004, sawdust (from local beech trees *Fagus sylvatica*) was hand spread over half of each plot (0.5 kg/m^2/month) over 3 months. Above ground (1 cm) biomass samples were taken in autumn 2004 (white hellebore shoots also counted) and spring 2005. Cover of each plant species in central 2 x 2 m quadrats was also recorded in summer 2002 and spring 2005.

A replicated, controlled study of five wooded hay meadows on the island of Gotland in Sweden (67) found that plug plants were over twice as effective as sowing for plant establishment. Devil's-bit scabious *Succisa pratensis* plugs established in all plots after two growing seasons and seeds in 45% plots. Spotted cat's-ear *Hypochoeris maculata* plugs established in 81% of plots and seeds in 33%. Germination rate of seeds varied between donor sites, particularly for devil's-bit scabious (0–11% germination rate); spotted cat's-ear varied only slightly (7–10%). Litter removal did not affect devil's-bit scabious germination or survival; for spotted cat's-ear the effect of raking on survival depended on donor site. The four donor sites were species-rich, traditionally managed meadows. At the three recipient sites eight 72 x 72 cm plots were established, each divided into sixteen 18 x 18 cm sub-plots. For each species, 6 sub-plots had seeds (50 seeds/sub-plot) and 4-month-old plugs (2 plugs/sub-plot) introduced. Four control sub-plots had no seed or plugs. Seeds were sown in October 2003 and plugs planted in May 2004. Litter was removed from half of each plot (randomly selected). Emerging seedlings were recorded in May 2004, survival in October 2005 and plug survival in October 2005.

A replicated site comparison study of 16 restored and 6 traditionally managed semi-natural grasslands in southern Sweden (68) found no significant difference in ant (Formicidae) species richness between restored sites and continuously grazed traditional sites. Total species richness, richness of forest species and of open-habitat species did not differ between restored and traditional sites. There were 1–12 ant species per site (average 8 species, 2 forest species, 3 open habitat species). Total species richness increased with time since restoration (up to 12 years), largely due to increasing open habitat species richness. However, the proportion of rare species was higher at younger restored sites. Vegetation height, size of study site and numbers of trees and shrubs did not affect species richness. Sites were restored from 1994 to 2001, trees and shrubs were removed and regular grazing resumed. Ants were sampled along a transect of 15 pitfall traps (10 m apart) at each site over 7 days in June 2006. Vegetation height was measured around randomly selected traps.

A replicated site comparison study in 2008 and 2009 on farms in three regions in England (69) found that land managed under Higher Level Stewardship grassland creation/restoration options was used significantly more by seed-eating farmland songbirds than improved grassland in two of the three regions. The strongest difference was in the Cotswolds, where almost 4 birds/ha were recorded on restored grassland, compared to around 1 bird/ha on improved grassland. In East Anglia Higher Level Stewardship creation/restoration grassland did not have more birds than on improved grassland. Surveys were done in the summers of

2008 and 2009 on 69 farms with Higher Level Stewardship in East Anglia, the West Midlands or the Cotswolds and on 31 farms across all 3 regions with no environmental stewardship.

A 2010 review of studies of scientific knowledge about how to re-establish plant communities in grasslands by reintroduction (70) found that direct seeding and hay transfer have been shown to be effective methods. The review found 38 studies; 28 of which provided enough information to evaluate the outcome; 21 of these from European countries (of which some also looked at the effects on wet meadows). Studies were graded as: successful, of limited success, failed introductions, or without the necessary information to evaluate the outcome. Direct seeding had success or limited success in 10 European studies. Hay spreading had success or limited success in seven European studies (four of which were on wet meadows) and was not shown to fail ((26), Patzelt *et al.* 2001, Hölzel & Otte 2003, (55), Rasran *et al.* 2007, Schmiede *et al.* 2009, Klimkowska *et al.* 2010). Plug planting had success/limited success in two European studies (one on wet grassland) ((8), Tallowin & Smith 2001). Strip seeding did not reintroduce species in two studies (both recorded in (57)).

A randomized, replicated, controlled trial in Berkshire, UK started in 2008 (71) found that grasslands sown with a seed mix containing legumes and other herbaceous plant species attracted significantly more pollinators and pollinator species than those sown with a mix of grasses only in the first year. Between six and eight bee (Apidae), butterfly (Lepidoptera) or hoverfly (Syrphidae) species were recorded/plot, compared to around two species on plots sown with grasses. The abundance of pollinators was strongly related to the cover of legumes and other non-grass plants. The cover of sown grasses and non-leguminous broadleaved plants was higher when sown following deep cultivation and herbicide treatment than after shallow cultivation alone. The cover with sown non-grass species (legumes and other broadleaved plants) was significantly higher in plots that were cut twice or three times for silage than in grazed treatments. There were four replicates of each treatment combination, on plots either 16 x 32 m (those cut for silage) or 25 x 50 m (grazed plots).

A replicated trial from 2005 to 2008 in the Hortobágy National Park, eastern Hungary (72) found that perennial meadow grass species sown on ex-arable fields established well after two years of management as hay meadows, but created a dense cover that may prevent more specialist meadow species from establishing. In the first year, weedy annual herbs and grasses dominated (63–82% of vegetation). By the second year, sown grasses had increased, accounting for 16 to 86% of the plant cover. Ten fields were sown with two or three grass species in October 2005, at 25 kg/ha (*Festuca rupicola*, narrow-leaved meadow-grass *Poa angustifolia* and smooth brome *Bromus inermis* on 6 fields, and *Festuca pseudovina* and narrow-leaved meadow grass on 4 fields). The fields covered 93 ha in total. They were mown in late June 2007 and 2008 and hay removed. Plants were surveyed in four permanent plots in each field, from 2006 to 2008.

(1) Bülow-Olsen A. (1980) Changes in the species composition in an area dominated by *Deschampsia flexuosa* (L.) trin. as a result of cattle grazing. *Biological Conservation*, 18, 257–270.

(2) Bakker J.P., de Leeuw J. & van Wieren S.E. (1983) Micro-patterns in grassland vegetation created and sustained by sheep-grazing. *Vegetatio*, 55, 153–161.

(3) Marrs R.H. (1985) Techniques for reducing soil fertility for nature conservation purposes: a review in relation to research at Roper's Heath, Suffolk, England. *Biological Conservation*, 34, 307–332.

(4) Gibson C.W.D., Dawkins H.C., Brown V.K. & Jepsen M. (1987) Spring grazing by sheep: effects on seasonal changes during early old field succession. *Vegetatio*, 70, 33–43.

(5) Gibson C.W.D., Watt T.A. & Brown V.K. (1987) The use of sheep grazing to recreate species-rich grassland from abandoned arable land. *Biological Conservation*, 42, 165–183.

(6) Morris M.G. & Rispin W.E. (1987) A beetle fauna of oolitic limestone grassland, and the responses of species to conservation management by different cutting regimes. *Biological Conservation*, 43, 87–105.

(7) Morris M.G. & Rispin W.E. (1987) Abundance and diversity of the coleopterous fauna of a calcareous grassland under different cutting regimes. *Journal of Applied Ecology*, 24, 451–465.

(8)   Fenner M. & Spellerberg I.F. (1988) Plant species enrichment of ecologically impoverished grassland a small scale trial. *Field Studies*, 7, 153–158.
(9)   Watt T.A. & Gibson C.W.D. (1988) The effects of sheep grazing on seedling establishment and survival in grassland. *Vegetatio*, 78, 91–98.
(10)   Kaule G. & Krebs S. (1989) Creating new habitats in intensively used farmland. In G. P. Buckley (ed.) *Biological Habitat Reconstruction*, Belhaven Press, London. pp. 161–170.
(11)   Brown V.K. & Gibson C.W.D. (1994) Recreation of species-rich calcicolorous grassland communities. *British Grassland Society Occasional Symposium*, 28, 125–136.
(12)   Bullock J.M., Hill B.C., Dale M.P. & Silvertown J. (1994) An experimental study of the effects of sheep grazing on vegetation change in a species-poor grassland and the role of seedling recruitment into gaps. *Journal of Applied Ecology*, 31, 493–507.
(13)   Van Der Woude B.J., Pegtel D.M. & Bakker J.P. (1994) Nutrient limitation after long-term nitrogen fertilizer application in cut grasslands. *Journal of Applied Ecology*, 31, 405–412.
(14)   Stevenson M.J., Bullock J.M. & Ward L.K. (1995) Re-creating semi-natural communities: effect of sowing rate on establishment of calcareous grassland. *Restoration Ecology*, 3, 279–289.
(15)   Blake S., Foster G.N., Fisher G.E.J. & Ligertwood G.L. (1996) Effects of management practices on the carabid faunas of newly established wildflower meadows in southern Scotland. *Annales Zoologici Fennici*, 33, 139–147.
(16)   Nösberger J. & Kessler W. (1997) Utilisation of grassland for biodiversity. *Proceedings of the International Occasional Symposium of the European Grassland Federation: Management for grassland biodiversity*. Warszawa-Lomza, Poland, 19–23 May. pp. 33–42.
(17)   Snow C.S.R., Marrs R.H. & Merrick L. (1997) Trends in soil chemistry and floristics associated with the establishment of a low-input meadow system on an arable clay soil in Essex. *Biological Conservation*, 79, 35–41.
(18)   Kotiluoto R. (1998) Vegetation changes in restored semi-natural meadows in the Turku Archipelago of SW Finland. *Plant Ecology*, 136, 53–67.
(19)   Muller S., Dutoit T., Allard D. & Grevilliot F. (1998) Restoration and rehabilitation of species-rich grassland ecosystems in France: a review. *Restoration Ecology*, 6, 94–101.
(20)   Wakeham-Dawson A. & Aebischer N.J. (1998) Factors determining winter densities of birds on environmentally sensitive area arable reversion grassland in southern England, with special reference to skylarks (*Alauda arvensis*). *Agriculture, Ecosystems & Environment*, 70, 189–201.
(21)   Wakeham-Dawson A., Szoszkiewicz K., Stern K. & Aebischer N.J. (1998) Breeding skylarks *Alauda arvensis* on Environmentally Sensitive Area arable reversion grass in southern England: survey-based and experimental determination of density. *Journal of Applied Ecology*, 35, 635–648.
(22)   Hopkins A., Pywell R.F., Peel S., Johnson R.H. & Bowling P.J. (1999) Enhancement of botanical diversity of permanent grassland and impact on hay production in Environmentally Sensitive Areas in the UK. *Grass and Forage Science*, 54, 163–173.
(23)   Jones A.T. & Hayes M.J. (1999) Increasing floristic diversity in grassland: the effects of management regime and provenance on species introduction. *Biological Conservation*, 87, 381–390.
(24)   Stampfli A. & Zeiter M. (1999) Plant species decline due to abandonment of meadows cannot easily be reversed by mowing. A case study from the southern Alps. *Journal of Vegetation Science*, 10, 151–164.
(25)   Humphrey J.W. & Patterson G.S. (2000) Effects of late summer cattle-grazing on the diversity of upland pasture vegetation in an upland conifer forest in Strathrory Glen, Easter Ross, Scotland. *Journal of Applied Ecology*, 37, 986–996.
(26)   Smith R.S., Shiel R.S., Millward D. & Corkhill P. (2000) The interactive effects of management on the productivity and plant community structure of an upland meadow: an 8-year field trial. *Journal of Applied Ecology*, 37, 1029–1043.
(27)   Bosshard A. & Kuster D. (2001) The significance of restored flower-rich hay meadows on set-aside land for butterflies and grasshoppers. *Agrarforschung*, 8, 252–257.
(28)   Coulson S.J., Bullock J.M., Stevenson M.J. & Pywell R.F. (2001) Colonization of grassland by sown species: dispersal versus microsite limitation in responses to management. *Journal of Applied Ecology*, 38, 204–216.
(29)   Dicks L.V. (2002) The structure and functioning of flower-visiting insect communities on hay meadows. PhD thesis. University of Cambridge.
(30)   Pywell R.F., Bullock J.M., Hopkins A., Walker K.J., Sparks T.H., Burke M.J.W. & Peel S. (2002) Restoration of species-rich grassland on arable land: assessing the limiting processes using a multi-site experiment. *Journal of Applied Ecology*, 39, 294–309.
(31)   Smith R.S., Shiel R.S., Millward D., Corkhill P. & Sanderson R.A. (2002) Soil seed banks and the effect of meadow management on vegetation change in a 10-year meadow field trial. *Journal of Applied Ecology*, 39, 279–293.
(32)   Warren J., Christal A. & Wilson F. (2002) Effects of sowing and management on vegetation succession during grassland habitat restoration. *Agriculture Ecosystems & Environment*, 93, 393–402.
(33)   Pykala J. (2003) Effects of restoration with cattle grazing on plant species composition and richness of semi-natural grasslands. *Biodiversity and Conservation*, 12, 2211–2226.

(34)   Gormsen D., Hedlund K., Korthals G.W., Mortimer S.R., Pizl V., Smilauerova M. & Sugg E. (2004) Management of plant communities on set-aside land and its effects on earthworm communities. *European Journal of Soil Biology*, 40, 123–128.

(35)   Hofmann M. & Isselstein J. (2004) Seedling recruitment on agriculturally improved mesic grassland: the influence of disturbance and management schemes. *Applied Vegetation Science*, 7, 193–200.

(36)   Lindborg R. & Eriksson O. (2004) Effects of restoration on plant species richness and composition in Scandinavian semi-natural grasslands. *Restoration Ecology*, 12, 318–326.

(37)   Poyry J., Lindgren S., Salminen J. & Kuussaari M. (2004) Restoration of butterfly and moth communities in semi-natural grasslands by cattle grazing. *Ecological Applications*, 14, 1656–1670.

(38)   Walker K.J., Stevens P.A., Stevens D.P., Mountford J.O., Manchester S.J. & Pywell R.F. (2004) The restoration and re-creation of species-rich lowland grassland on land formerly managed for intensive agriculture in the UK. *Biological Conservation*, 119, 1–18.

(39)   Forup M.L. & Memmott J. (2005) The restoration of plant-pollinator interactions in hay meadows. *Restoration Ecology*, 13, 265–274.

(40)   Jefferson R.G. (2005) The conservation management of upland hay meadows in Britain: a review. *Grass and Forage Science*, 60, 322–331.

(41)   Lindborg R., Cousins S.A.O. & Eriksson O. (2005) Plant species response to land use change – *Campanula rotundifolia*, *Primula veris* and *Rhinanthus minor*. *Ecography*, 28, 29–36.

(42)   Poyry J., Lindgren S., Salminen J. & Kuussaari M. (2005) Responses of butterfly and moth species to restored cattle grazing in semi-natural grasslands. *Biological Conservation*, 122, 465–478.

(43)   Pykala J. (2005) Plant species responses to cattle grazing in mesic semi-natural grassland. *Agriculture, Ecosystems & Environment*, 108, 109–117.

(44)   Hellström K., Huhta A.P., Rautio P. & Tuomi J. (2006) Search for optimal mowing regime – slow community change in a restoration trial in northern Finland. *Annales Botanici Fennici*, 43, 338–348.

(45)   Kiehl K. & Wagner C. (2006) Effect of hay transfer on long-term establishment of vegetation and grasshoppers on former arable fields. *Restoration Ecology*, 14, 157–166.

(46)   Öckinger E., Eriksson A.K. & Smith H.G. (2006) Effects of grassland abandonment, restoration and management on butterflies and vascular plants. . *Biological Conservation*, 133, 291–300.

(47)   Woodcock B.A., Lawson C.S., Mann D.J. & McDonald A.W. (2006) Effects of grazing management on beetle and plant assemblages during the re-creation of a flood-plain meadow. *Agriculture, Ecosystems & Environment*, 116, 225–234.

(48)   Lindborg R. (2006) Recreating grasslands in Swedish rural landscapes – effects of seed sowing and management history. *Biodiversity and Conservation*, 15, 957–969.

(49)   Bullock J.M., Pywell R.F. & Walker K.J. (2007) Long-term enhancement of agricultural production by restoration of biodiversity. *Journal of Applied Ecology*, 44, 6–12.

(50)   van Diggelen R. (2007) Habitat creation: nature conservation of the future? *Aspects of Applied Biology*, 82, 1–11.

(51)   Edwards A.R., Mortimer S.R., Lawson C.S., Westbury D.B., Harris S.J., Woodcock B.A. & Brown V.K. (2007) Hay strewing, brush harvesting of seed and soil disturbance as tools for the enhancement of botanical diversity in grasslands. *Biological Conservation*, 134, 372–382.

(52)   Geijzendorffer I.R. (2007) Integrating botanical diversity and management of agricultural grassland. PhD thesis. University College, Dublin.

(53)   Hayes M.J. & Tallowin J.R.B. (2007) Recreating biodiverse grasslands: long-term evaluation of practical management options for farmers. In J.J. Hopkins, A.J. Duncan, D.I. McCracken, S. Peel & J.R.B. Tallowin (eds.) *High Value Grassland: Providing biodiversity, a clean environment and premium products. British Grassland Society Occasional Symposium No.38*, British Grassland Society (BGS), Reading. pp. 135–140.

(54)   Jongepierova I., Mitchley J. & Tzanopoulos J. (2007) A field experiment to recreate species rich hay meadows using regional seed mixtures. *Biological Conservation*, 139, 297–305.

(55)   Kiehl K. & Pfadenhauer J. (2007) Establishment and persistence of target species in newly created calcareous grasslands on former arable fields. *Plant Ecology*, 189, 31–48.

(56)   Leps J., Dolezal J., Bezemer T.M., Brown V.K., Hedlund K., Igual A.M., Jorgensen H.B., Lawson C.S., Mortimer S.R., Peix Geldart A., Rodriguez Barrueco C., Santa Regina I., Smilauer P. & van der Putten W.H. (2007) Long-term effectiveness of sowing high and low diversity seed mixtures to enhance plant community development on ex-arable fields. *Applied Vegetation Science*, 10, 97–110.

(57)   Pywell R.F., Bullock J.M., Tallowin J.B., Walker K.J., Warman E.A. & Masters G. (2007) Enhancing diversity of species-poor grasslands: an experimental assessment of multiple constraints. *Journal of Applied Ecology*, 44, 81–94.

(58)   Sendžikaite J., Pakalnis R. & Avižiene D. (2007) Restoration of botanical diversity by extensive management of sown meadows. In J.J. Hopkins, A.J. Duncan, D.I. McCraken, S. Peel & J.R.B. Tallowin (eds.) *High Value Grassland*, British Grassland Society Occasional Symposium No.38, British Grassland Society (BGS), Reading. pp. 313–316.

(59)   Aavik T., Jogar U., Liira J., Tulva I. & Zobel M. (2008) Plant diversity in a calcareous wooded meadow – the significance of management continuity. *Journal of Vegetation Science*, 19, 475–484.

(60)   Fagan K.C., Pywell R.F., Bullock J.M. & Marrs R.H. (2008) Do restored calcareous grasslands on former arable fields resemble ancient targets? The effects of time, methods and environment on outcomes. *Journal of Applied Ecology*, 45, 1293–1303.

(61)   Morgan M., McLean B.M. & Davies O.D. (2008) Long term studies to determine management practices to enhance biodiversity within semi-natural grassland communities. In *Grassland Science in Europe, Volume 13*, Swedish University of Agricultural Sciences, Uppsala. pp. 992–994.

(62)   Woodcock B.A., Mortimer S.R., Edwards A.R., Lawson C.S., Westbury D.B., Brook A.J., Harris S.J. & Brown V.K. (2008) Re-creating plant and beetle assemblages of species-rich chalk grasslands on ex-arable land. In H. G. Schroder (eds.) *Grasslands: Ecology, management & restoration*, Nova Science Publishers, Inc. pp. 133–152.

(63)   Hellström K., Huhta A.-P., Rautio P. & Tuomi J. (2009) Seed introduction and gap creation facilitate restoration of meadow species richness. *Journal for Nature Conservation*, 17, 236–244.

(64)   Lye G.C., Park K., Osborne J., Holland J. & Goulson D. (2009) Assessing the value of Rural Stewardship schemes for providing foraging resources and nesting habitat for bumblebee queens (Hymenoptera: Apidae). *Biological Conservation*, 142, 2023–2032.

(65)   Natural England (2009) *Agri-environment Schemes in England 2009: A review of results and effectiveness.* Natural England, Peterborough.

(66)   Spiegelberger T., Muller-Scharer H., Matthies D. & Schaffner U. (2009) Sawdust addition reduces the productivity of nitrogen-enriched mountain grasslands. *Restoration Ecology*, 17, 865–872.

(67)   Wallin L., Svensson B.M. & Lönn M. (2009) Artificial dispersal as a restoration tool in meadows: sowing or planting? *Restoration Ecology*, 17, 270–279.

(68)   Dahms H., Lenoir L., Lindborg R., Wolters V. & Dauber J. (2010) Restoration of seminatural grasslands: what is the impact on ants? *Restoration Ecology*, 18, 330–337.

(69)   Field R.H., Morris A.J., Grice P.V. & Cooke A.I. (2010) Evaluating the English Higher Level Stewardship scheme for farmland birds. *Aspects of Applied Biology*, 100, 59–68.

(70)   Hedberg P. & Kotowski W. (2010) New nature by sowing? The current state of species introduction in grassland restoration, and the road ahead. *Journal for Nature Conservation*, 18, 304–308.

(71)   Pywell R.F., Woodcock B.A., Orr R., Tallowin J.R.B., McEwen I., Nowakowski M. & Bullock J.M. (2010) Options for wide scale enhancement of grassland biodiversity under the Entry Level Scheme. *Aspects of Applied Biology*, 100, 125–132.

(72)   Török P., Deák B., Vida E., Valkó O., Lengyel S. & Tóthmérész B. (2010) Restoring grassland biodiversity: Sowing low-diversity seed mixtures can lead to rapid favourable changes. *Biological Conservation*, 143, 806–812.

Additional references

Allen S.E. (1964) Chemical aspects of heather burning. *Journal of Applied Ecology*, 1, 347–367.

Kenworthy J.B. (1964) A study of the changes in plant and soil nutrients associated with moorburning and grazing. PhD thesis. University of St Andrews.

Chapman S.B. (1967) Nutrient budgets for a dry heath ecosystem in the south of England. *Journal of Ecology*, 55, 677–689.

Maman L. (1985) La dynamique de la vegetation sur les ouvrages des aménagements hydroélectriques du Rhône [Vegetation dynamics on embankments of hydroelectric facilities in the Rhône valley]. PhD thesis. University of Grenoble.

Wells T.C.E., Cox R. & Frost A. (1989) Diversifying grasslands by introducing seed and transplants into existing vegetation. In G.P. Buckley (ed.) *Biological Habitat Reconstruction*. Belhaven, London. pp. 283–298.

Oomes M.J.M. (1990) Changes in dry matter and nutrient yields during the restoration of species-rich grasslands. *Journal of Vegetation Science*, 1, 333–338.

Bédécarrats A. (1991) Dynamique des enherbements des pistes de ski en Savoie et leur gestion pastorale [Dynamics of grass growing on ski pistes in Savoy and their pastoral management]. In IVth International Rangeland Congress, Association Française de Pastoralisme, Montpellier, France. pp. 77–80.

Olff H. & Bakker J.P. (1991) Long-term dynamics of standing crop and species composition after the cessation of fertiliser application to mown grassland. *Journal of Applied Ecology*, 28, 1040–1052.

Coin R. (1992) Variabilité spatio-temporelle des communautés végétales artificielles sur les ouvrages des aménagements hydroélectriques: enseignements en vue d'améliorer les techniques de végétalisation [Spatio-temporal variability of artificial vegetation communities on revegetated areas near hydroelectric dams: lessons to improve revegetation techniques]. PhD thesis. University of Grenoble, Grenoble, France.

McDonald A.W. (1992) Succession in a 3-year-old flood-meadow near Oxford. *Aspects of Applied Biology*, 29, 345–352.

Mountford J.O., Tallowin J.R.B., Kirkham F.W. & Lakhani K.H. (1994) Effects of inorganic fertilisers in flower-rich hay-meadows in the Somerset Levels. In R.J. Haggar & S. Peel (eds.) *Grassland Management and Nature Conservation*. British Grassland Society Occasional Symposium, 28. British Grassland Society, Reading. pp. 74–85.

Wells T.C.E., Pywell R.F. & Welch R.C. (1994) *Management and Restoration of Species-rich Grassland*. Report to the Ministry of Agriculture, Fisheries and Food (BD0306), Institute of Terrestrial Ecology, Monks Wood.

Boyce D.V.M. (1995) Survival and spread of wildflowers planted into ex-agricultural grassland. *Journal of Practical Ecology and Conservation*, 1, 38–44.

Dutoit T. (1996) *Dynamique et gestion des pelouses calcaires de Haute-Normandie* [Dynamics and management of calcareous grasslands in Haute-Normandie]. Presses Universitaires de Rouen, Rouen.

Lehmann J., Rosenberg E., Bassetti P. & Mosimann E. (1996) Standardmischungen fur den Futterbau [Standard mixtures for forage production]. *Agrarforschung*, 3, 489–500.

Steinegger R. & Koch B. (1997) Naturschutzerische optimierung okologischer ausgleichsflachen [Optimizing the ecological functions of extensively managed meadows on former arable land]. *Agrarforschung*, 4, 35–38.

Barratt D.R., Mountford J.O., Sparks T.H., Walker K.J., Warman E.A. & Garbutt A. (2000) The effects of field elevation and water levels on the establishment of plug-plants in an ex-arable grassland. *Aspects of Applied Biology*, 58, 425–430.

Hayes M.J., Sackville Hamilton N.R., Tallowin J.R.B., Buse A. & Davies O. (2000) *Methods of Enhancing Diversity in Upland Environmentally Sensitive Area Swards*. Report to the Ministry of Agriculture, Fisheries and Food (BD1424), Institute of Grassland and Environmental Research, Aberystwyth.

Pywell R.F., Warman L., Walker K.J. & Sparks T.H. (2000) *Reversion of Intensive Arable Land to Grass Heath and Calluna Heath: Vegetation aspects*. Report to the Ministry of Agriculture, Fisheries and Food (BD1502), Institute of Terrestrial Ecology, Monks Wood.

Hayes M.J. & Sackville Hamilton N.R. (2001) *The Effect of Sward Management on the Restoration of Species-rich Grassland: A reassessment of IGER's grassland restoration experiment, Trawsgoed*. Countryside Council for Wales Contract Science Report No. 438, Bangor.

Patzelt A., Wild U. & Pfadenhauer J. (2001) Restoration of wet meadows by topsoil removal: vegetation development and germination biology of fen species. *Restoration Ecology*, 9, 127–136.

Tallowin J.R.B. & Smith R.E.N. (2001) Restoration of a *Cirsio-Molinietum* fen meadow on an agriculturally improved pasture. *Restoration Ecology*, 9, 167–178.

Verhagen R., Klooker J., Bakker J.P. & van Diggelen R. (2001) Restoration success of low-production plant communities on former agricultural soils after topsoil removal. *Applied Vegetation Science*, 4, 75–82.

CABI (2003) *Chalk Grassland: Enhancement of plant and invertebrate diversity through the use of environmental land management schemes*. Defra project report BD1414, London.

Hölzel N. & Otte A. (2003) Restoration of a species-rich flood meadow by topsoil removal and diaspora transfer with plant material. *Applied Vegetation Science*, 6, 131–140.

Pywell R. F., Bullock J. M., Roy D. B., Warman E. A., Walker K. J. & Rothery P. (2003) Plant traits as predictors of performance in ecological restoration. *Journal of Applied Ecology*, 40, 65–77.

Smith R.S., Shiel R.S., Bardgett R.D., Millward D., Corkhill P., Rolph G., Hobbs P.J. & Peacock S. (2003) Soil microbial community, fertility, vegetation and diversity as targets in the restoration management of meadow grassland. *Journal of Applied Ecology*, 40, 51–64.

Pywell R.F., Bullock J.M., Walker K.J., Coulson S.J., Gregory S.J. & Stevenson M.J. (2004) Facilitating grassland diversification using the hemiparasitic plant *Rhinanthus minor*. *Journal of Applied Ecology*, 41, 880–887.

Smith R.S. (2005) *Ecological Mechanisms Affecting the Restoration of Diversity in Agriculturally Improved Meadow Grassland*. Defra BD1439.

Kirkham F.W., Davis D., Fowbert J.A., Hooke D., Parkin A.B. & Sherwood A.J. (2006) *Evaluation of Arable Reversion Agreements in the Countryside Stewardship and Environmentally Sensitive Areas Scheme*. Defra project report MA0105/RMP 1982, London.

Rasran L., Vogt K. & Jensen K. (2007) Effects of topsoil removal, seed transfer with plant material and moderate grazing on restoration of riparian fen grasslands. *Applied Vegetation Science*, 10, 451–460.

Schmiede R., Donath T.W. & Otte A. (2009) Seed bank development after the restoration of alluvial grassland via transfer of seed-containing plant material. *Biological Conservation*, 142, 404–413.

Klimkowska A., Kotowski W., van Diggelen R., Grootjans A. P., Dzierża P., & Brzezińska K. (2010) Vegetation re-development after fen meadow restoration by topsoil removal and hay transfer. *Restoration Ecology*, 18, 924–933.

4.3   Add yellow rattle seed (*Rhinanthus minor*) to hay meadows

- A review of studies from the UK[2] found that adding hay rattle seed helped other sown target meadow species to colonize and that more plant species were found when yellow rattle was present. A randomized, replicated controlled trial in the UK[1] found that yellow rattle could be established on a pasture field by 'slot seeding'.

Background

This intervention involves adding yellow rattle *Rhinanthus minor* seed to grassland to help establish meadow plant communities. Yellow rattle is an annual wildflower native to the UK which parasitizes other wildflowers and grasses (Natural England 2009). By parasitizing other plants, particularly grasses, yellow rattle can reduce the dominant plant biomass in grasslands allowing other wildflower species to establish (Natural England 2009). See also 'Restore or create species-rich, semi-natural grassland' for studies that used yellow rattle in the restoration or creation of semi-natural grassland.

Natural England (2009) *The Use of Yellow Rattle to Facilitate Grassland Diversification*. Natural England Technical Information Note TIN060.

A randomized, replicated, controlled trial in 1995–1997 in Oxfordshire, UK (1) found that yellow rattle *Rhinanthus minor* could be effectively established on a pasture field by 'slot seeding'. Different management treatments, cutting, grazing or both, did not affect survival or establishment. However yellow rattle seeds were spread a greater distance when hay was cut in July than without a hay cut. Seeds were sown in strips previously sprayed with herbicide by a tractor-mounted slot seeder in October 1995. Four management treatments were replicated 5 times in 20 x 10 m plots. The treatments were cut once (July); cut twice (July and September); cut in July and autumn grazed. Monitoring of plant dispersal was carried out using seed traps at the soil surface, from June to October 1997.

A 2005 review (2) found three studies looking at the role of yellow rattle *Rhinanthus minor* as a tool when restoring upland hay meadow vegetation on semi-improved grassland. One study in North Yorkshire, UK (Smith *et al.* 2003) found that sowing key functional species (legumes and yellow rattle) helped other sown target meadow species to colonize. At the same site, Smith (2005) found that when more yellow rattle was present, herbaceous species increased at the expense of perennial rye grass *Lolium perenne*. The rate of nitrogen mineralization was also faster in the presence of yellow rattle. One study (Pywell *et al.* 2004) found that when restoring species-rich grassland on a semi-improved grassland site, more plant species were found when yellow rattle was present.

(1) Coulson S.J., Bullock J.M., Stevenson M.J. & Pywell R.F. (2001) Colonization of grassland by sown species: dispersal versus microsite limitation in responses to management. *Journal of Applied Ecology*, 38, 204–216.
(2) Jefferson R.G. (2005) The conservation management of upland hay meadows in Britain: a review. *Grass and Forage Science*, 60, 322–331.

Additional references
Smith R.S., Shiel R.S., Bardgett R.D., Millward D., Corkhill P., Rolph G., Hobbs P.J. & Peacock S. (2003) Soil microbial community, fertility, vegetation and diversity as targets in the restoration management of meadow grassland. *Journal of Applied Ecology*, 40, 51–64.
Pywell R.F., Bullock J.M., Walker K.J., Coulson S.J., Gregory S.J. & Stevenson M.J. (2004) Facilitating grassland diversification using the hemiparasitic plant *Rhinanthus minor*. *Journal of Applied Ecology*, 41, 880–887.
Smith R.S. (2005) *Ecological Mechanisms Affecting the Restoration of Diversity in Agriculturally Improved Meadow Grassland*. Defra BD1439.

4.4   Reduce management intensity on permanent grasslands (several interventions at once)

• A total of 32 individual studies from the Czech Republic, Germany, Ireland, the Netherlands, Switzerland and the UK looked at the effects on farmland wildlife of reducing management intensity on permanent grasslands. Twenty-two studies found benefits to some or all wildlife groups studied. Eleven studies (including four replicated site comparisons and three reviews) found reduced management intensity on permanent grassland benefited plants[1, 3, 5, 18, 19, 22, 25, 31, 32, 35–37]. Sixteen studies (including eight site comparisons of which four were paired and three reviews) found benefits to some or all invertebrates[3, 8–15, 18–21, 24, 28, 29, 32, 34–37, 39]. Five studies (including two replicated site comparisons, of which one was paired and a review) found positive effects on some or all birds[13, 18, 24, 26, 36].

- Twenty-one studies from six European countries found no clear effects of reducing management intensity on some or all plants, invertebrates or birds. Seven studies (including two replicated paired site comparisons and a review) found no clear effect on plants[2, 3, 8, 13, 27, 31, 33, 39]. Ten studies (including four site comparisons and one paired site comparison) found mixed or no effects on some or all invertebrates[6, 9, 10, 12, 14, 15, 21, 27, 30, 32, 34]. Two studies (one review, one site comparison) found that invertebrate communities on less intensively managed grasslands were distinct from those on intensively managed grasslands[16, 17]. Four studies (including three site comparisons, of which one was paired and two replicated) found no clear effects on bird numbers or species richness[8, 13, 18, 26, 38].

- Five studies from four European countries found negative effects of reducing management intensity on plants, invertebrates or birds. Two studies (one review, one replicated trial) found some plant species were lost under extensive management[4, 23]. Two studies (one paired site comparison) found more invertebrates in grasslands with intensive management[30, 35]. One paired site comparison found fewer wading birds on grasslands with reduced management intensity than on conventionally managed grassland[8, 13].

Background

Reducing the intensity of grassland management involves reducing or stopping the use of fertilizers, herbicides and pesticides and delaying the mowing date until later in the summer. Studies included here have monitored the effects of carrying out two or more of these management interventions at the same time. All of the studies included here monitor the effects of a change of grassland management, or of implementing a regime of reduced management intensity for the purpose of nature conservation, such as an agri-environment scheme.

Studies that only carry out one of these management interventions, rather than several combined, would be included under the relevant separate intervention. See: 'Reduce chemical inputs in grassland management', 'Raise mowing height on grasslands' and 'Delay mowing or first grazing date on grasslands'.

See also 'All farming systems: Pay farmers to cover the cost of conservation measures (as in agri-environment schemes)' for a study that records the decline in species richness on upland hay meadows over time, following introduction of two main agri-environment schemes in the Yorkshire Dales, UK (Pacha and Petit 2008).

There is also a literature comparing sites without direct intervention, such as Di Giulio et al. (2001), in which wildlife on species-rich grasslands is compared with wildlife on intensive grasslands. Whilst such studies are relevant, they do not demonstrate the effects of direct intervention (reducing management intensity or imposing a reduced management intensity regime) and as such we do not include them.

Di Giulio M., Edwards P.J. & Meister E. (2001) Enhancing insect diversity in agricultural grasslands: the roles of management and landscape structure. *Journal of Applied Ecology*, 38, 310–319.

A replicated site comparison study in Norfolk, UK, in 1987 (1) found higher diversity of plants on 20 grazing fields under the Broads Grazing Marshes Scheme than on 5 more intensively managed fields outside the scheme. Fields within the Scheme had 14–31 plant species; those outside it 10–16 species. Plant diversity (the Shannon-Weiner index) was also higher in fields with less than 125 kg nitrogen/ha applied, fields with less than 1.5 livestock units/ha and unlimed fields. The Broads Grazing Marshes Scheme was set up in 1985. Farmers were paid to retain permanent grazed grassland at a stocking rate from 0.5 to 1.5 livestock units/ acre. Only one silage cut per year was permitted, with aftermath grazing. Farmers needed permission from the Broads Grazing Scheme Unit if they wished to apply more than 125 kg N/ha, apply herbicide other than spot treatment for thistles, or carry out drainage, levelling or re-seeding. Plants were monitored in 1987 in two-hundred 25 x 25 cm quadrats placed along two diagonal axes across each field. Management practices were recorded for each field, using farmer questionnaires.

A small study from 1967 to 1993 of a grassland at the University experimental farm, Meenthoeve in the Netherlands (2) found that plant species increased for six years following extensification, but then decreased in unfertilized plots. Species increased following sowing (1966) and extensification (1971), from 19 species in 1969 to 37 in 1977. Numbers then declined to below 25 species in unfertilized plots as weeds typical of intensive grassland decreased. In 1985, a botanical change characteristic of a reduction in soil fertility occurred. Dominant species were common bent *Agrostis capillaris* and Yorkshire-fog *Holcus lanatus* in all three treatments, sweet vernal grass *Anthoxantum odoratum* in unfertilized treatments and red fescue *Festuca rubra* in the mown treatment. The 0.8 ha area of grassland had received 300 kg N/ha, phosphorous P and potassium K and grazing for 5 years. In 1971, it was divided into a 0.26 ha plot with no fertilizer, sub-divided into intermittent grazing or mown for hay in June and October, and a 0.56 ha plot with 20 kg/ha P_2O_2 and 40 kg/ha K_2O per year and 5 steers/ha until 1975 followed by 3 steers/ha. Botanical composition of plots was sampled in May in ten years (1967–1993). Species were sampled by hand sorting 100 handfuls of grass/ plot and taking 100 soil cores.

A review in 1997 (3) found that reduced management intensity on grassland can benefit plant and insect diversity, but it does not always. Five European studies monitored the biodiversity effects of reducing management intensity on lowland grasslands. Two of these studies (Bakker 1994, (2)) showed an increase in plant diversity, and one an increase in invertebrate diversity (Blake *et al.* 1996) following restoration of the plant community by sowing. In all three cases, the changes took many years. They covered periods of between 7 and 20 years and none of the studies measured diversity equivalent to species-rich semi-natural grasslands after this time period. Two other studies (Marriott *et al.* 1994, Neuteboom *et al.* 1994) showed that reduced management intensity did not improve plant diversity. The review found no direct evidence for effects of extensification in upland grasslands.

A 1997 review of experimental evidence (largely published in German language) (4) described an experiment on reduced management intensity in upland grassland in Wales. This showed that without cutting or grazing, important pasture species such as perennial rye grass *Lolium perenne* and white clover *Trifolium repens* were lost within two years (Fothergill *et al.* 1994). Both species were maintained by an annual summer cut.

A 1998 review of two studies (5) concluded that reducing management intensity on permanent grassland is likely to benefit two important pasture species, white clover *Trifolium repens* and meadow fescue *Festuca pratensis*. The first study (Schwank *et al.* 1986) found that frequent cutting and fertilization reduced the yield of white clover from 5% (under traditional management) to 2%. This was probably due to clover leaves growing closer to the ground and being shaded by taller plants after fertilization. A second study (Carlen 1994) concluded that the decline in meadow fescue found under intensive management results from its low competitive ability (due to small leaves and low root activity), rather than direct effects of cutting or fertilization. The study locations are not given.

A replicated, controlled study in summer 1997 and 1998 in 109 sites of 2 arable regions in Switzerland (6) found mixed effects of reduced management intensity on grasslands on butterflies (Lepidoptera) and spiders (Araneae). For butterflies, species richness did not differ significantly between low-intensity or extensively managed meadows and intensively managed meadows, however butterfly species richness was higher in extensively managed meadows (but not in low-intensity meadows) than in cereal fields. The number of spider species was found to be higher in low-intensity meadows (but not in extensively managed meadows) than in cereal fields, but no difference was detected between extensive, low- and high-intensively managed meadows. Spider community composition differed between the intensively managed meadows and the two meadow types with reduced management intensity. The investigated habitat types were forest edges, arable fields (winter wheat and intensively managed meadow) and ecological compensation areas including hedgerows, extensively managed and low-intensity meadows, wildflower strips on set-aside land and orchard meadows. Spiders were collected in pitfall traps in May and June 1997. Butterflies were observed during 6 visits (10 minutes, covering 0.25 ha) in each site in 1998. This was part of the same study as (10, 11).

The initial findings of a controlled, replicated site comparison study of the Swiss Ecological Compensation Area scheme in 1999–2005 (7) found endangered plant species present on 42% of 582 grassland Ecological Compensation Areas (extensively managed meadows, litter meadows and pastures) examined in 1999. Although the number of spider (Araneae) species was similar (around 20 species) there were significant differences in spider species composition in 23 Ecological Compensation Area and 15 non-Ecological Compensation Area meadows. Butterfly (Lepidoptera) species composition also differed between Ecological Compensation Area and non-Ecological Compensation Area sites. Vascular plants, ground beetles (Carabidae), spiders, butterflies, grasshoppers (Orthoptera) and breeding birds were monitored on grasslands in 3 case study areas of around 5 km^2.

A paired site comparison in the Netherlands (8) found more species of bee (Apidae) and hoverfly (Syrphidae) on grassland fields with management agreements to benefit birds or plants than on conventionally managed fields, but not more species of bird or plant. For hoverflies (but not bees) this difference was mostly in May, and could be related to vegetation height, because conventional fields were cut earlier. There were around 50 plant species/field on both field types. The number of bee species was low (average 1.7 bee species/field overall; 85% from just 3 species). The density of breeding bird territories was not significantly different between 20 fields with meadow bird agreements and 20 control fields, both for all bird species and just for waders. Eurasian oystercatcher *Haematopus ostralegus*, black-tailed godwit *Limosa limosa*, common redshank *Tringa totanus* and northern lapwing *Vanellus vanellus* were all significantly less abundant on management agreement fields than on control fields. There was no significant difference in the number of territories between field types for three of these species, but oystercatchers had significantly fewer territories on management agreement fields than on control fields (0.13 vs 0.52). The study involved 39 field pairs, one with either a 'botanical agreement' (22 field pairs) and/or a 'meadow bird agreement' (20 field pairs), the other managed conventionally. Fields were 2 ha on average. Paired fields were within 1 km of each other, similar in size and soil type. Fertilizer inputs were significantly lower and mowing dates later on fields with management agreements than on conventionally managed fields. Fields were surveyed between March and September 2000 (birds five times between March and June; plants and insects four times May to August). More detailed results are presented in a later paper (13).

A 2003 literature review in Europe (9) found one review and two studies that reported that invertebrates are affected by the frequency and timing of mowing in grassland. The one review suggested that cutting grassland twice a year was detrimental to hoverflies (Syrphidae), although responses for other families were mixed (Gerstmeir & Lang 1996). One study (Fuller *et al.* 2003) found that several cuts within grassland each year had a greater effect on beetles (Coleoptera) than one late cut. Another reported that responses of true bugs (Heteroptera) and plant/leafhoppers (Auchenorrhyncha) to mowing vary between species and timing of the cut (Morris 1981a,b).

A site comparison study in the regions of Ruswil, northwest of Lucerne, Switzerland (10) found that the number of butterfly (Lepidoptera) species, but not spider (Araneae) or ground beetle (Carabidae) species, was significantly higher in low input meadows than in intensively managed meadows. Twenty-three low input grasslands, around 400 m^2 in size and managed as Ecological Compensation Areas were surveyed. There were 2 types: 16 'extensively used meadows' with late mowing and no fertilizer and 7 'low-input meadows', with late mowing and restricted fertilization (up to 60 kg N/ha/year). For comparison, 15 intensively managed meadows were surveyed: 7 conventional grasslands and 8 Ecological Compensation Area meadows in traditional orchards with no restrictions on cutting or fertilizer use. Spiders and ground beetles were monitored using pitfall traps set over five weeks of May and June 1997. Butterflies were observed for 10 minute periods on 0.25 ha of each site on 5 occasions from May to August 1998, between 10:00 and 17:30 on sunny days with temperatures of at least 18°C. More detailed results (in German) are presented in (6).

The same team as in (10) combined the results from Ruswil with another region – Rafz, northwest of Zurich, Switzerland (11). It also found that butterfly (Lepidoptera) species richness was significantly higher in extensively managed and low input meadows than in intensively managed meadows. The same comparison was not possible for spider (Araneae) and beetle (Coleoptera) data because pitfall traps were set for one week longer in one of the regions. Thirty-five low input grasslands managed as Ecological Compensation Areas were included, 23 in Ruswil and 12 in Rafz. There were 2 types: 19 'extensively used meadows' with late mowing and no fertilizer and 16 'low-input meadows', with late mowing and restricted fertilization (up to 60 kg N/ha/year). In Ruswil, seven permanent intensively managed meadows and eight Ecological Compensation Area meadows in traditional orchards (intensively managed because they had no restrictions on cutting or fertilizer use) were sampled for comparison. Butterflies were observed for 10 minute periods on 0.25 ha of each site, on 5 occasions from May to August 1998. More detailed results (in German) are presented in (6).

A replicated, controlled, paired sites study in August in 21 field pairs in 3 farmland regions (Bauma, Ruswil and Flühli) in Switzerland (12) found higher species richness and abundance of grasshoppers (Orthoptera: Caelifera) in low-intensity meadows than in intensively managed meadows. Species richness and the number of rare species were higher in low intensity meadows in 2 regions (Bauma: average 4.4 vs 2.9 species and Flühli: 5.9 vs 4.6 spp.) but no difference was found between the meadow types in Ruswil (2.6 spp. both). Grasshopper abundance was higher in low intensity meadows than in intensively managed meadows in all regions (Bauma: average 24.07 vs 24.14 individuals; Ruswil: 22.29

vs 19.71 ind.; Flühli: 26.21 vs 21.93 ind.). Transect location (field edge or centre) did not affect grasshopper species richness or abundance. Seven field pairs (each consisting of one low intensity and one intensive meadow with similar environmental conditions) were investigated in each region. Grasshoppers were monitored in August along two 95 m long transects in each meadow (one along the field edge and one in the field centre). Transects were walked slowly for 15 minutes and all observed grasshoppers were caught to determine species richness and grasshopper density.

Further analysis of data (13) from the same replicated paired site comparison in the Netherlands as (8) found that meadow songbirds, bees (Apidae) and hoverflies (Syrphidae) were more abundant on fields with management agreements to benefit plants or birds (only fields with bird agreements were analysed for birds). Wading birds were less abundant on 20 fields with meadow bird agreements (average 7 birds and 1.3 territories/field compared to 12 birds and 2.1 territories on 20 conventional fields). Meadow songbirds were more abundant on agreement fields at a 12.5 ha scale (9.9 birds/12.5 ha plot surrounding the field, compared to 7.7 on conventional fields). Duck and non-meadow bird breeding densities did not differ between management types at either the field or 12.5 ha scale. There were 10–15 hoverfly species and 1.5–4.0 bee species/field on fields with agreements, compared to 8–13 hoverfly species and 0.5–2.0 bee species on conventional fields. Hoverfly abundance was also higher (50–125 hoverflies/agreement field vs 50–60 hoverflies/conventional field). Bee abundance was higher on agreement fields for 2 out of 3 soil types (3–7 vs 1–8 bees/field). Numbers of plant species were not higher on agreement fields (approximately 14–16 plant species/20 m^2 on fields under botanical agreements vs 14–15/20 m^2 on conventional fields).

A site comparison study in three regions of the Swiss Plateau (14) found more butterfly (Lepidoptera) species on low input than on intensively managed grasslands in one of two study years. In 2002, but not in 2000, low input grasslands had more butterfly species than intensively managed grasslands (actual numbers not given). The identity of the butterfly species found was not significantly influenced by management intensity, but was different in different regions. Butterflies were recorded in 56 low input grasslands and 48 intensively managed grasslands during the summers of 2000 and 2002. The low input grasslands were managed as Ecological Compensation Areas. They had restricted fertilizer and pesticide use, and late mowing.

A replicated, controlled study of pasture in County Meath, Ireland (15) found that reduced management intensity resulted in increased species richness of spiders (Araneae) and true bugs (Hemiptera) but not beetles (Coleoptera), flies (Diptera) or bees/wasps/ants (Hymenoptera). Plots compatible with the Rural Environment Protection Scheme (0.82 ha/cow units, 88 kg N/ha) had a significantly higher species richness of spiders (3 species/paddock) and true bugs (7 species) compared to conventional plots (0.65 ha/cow units, 220 kg N/ha; 2 and 5 species respectively). There was no significant difference between treatments for beetles, flies or bees/wasps/ants. Four blocks were established, each with one replicate of the two management systems. Each treatment was sub-divided into three grazing paddocks grazed in a fixed sequence. Arthropods were sampled using a Vortis suction sampler once per paddock in August 2003.

A site comparison study in the regions of Nuvilly and Ruswil, Switzerland (16) found that spider (Araneae), ground beetle (Carabidae) and butterfly (Lepidoptera) communities on low input grasslands were distinct and different from those on intensively managed grasslands. The study used some of the same sites as (11). The difference was strongest for spider communities. The study was carried out on 33 low input grasslands managed as Ecological Compensation Areas – 23 'extensively used meadows' with late mowing and no fertilizer and

10 'low-input meadows' with late mowing and restricted fertilization (up to 60 kg N/ha/year). For comparison, there were 24 intensive meadows: 8 permanent intensively managed meadows, 14 Ecological Compensation Area meadows in traditional orchards and 2 seeded Ecological Compensation Areas. These latter Ecological Compensation Area grassland types were considered intensively managed because they had no restrictions on cutting or fertilizer use. Spiders and ground beetles were monitored for three or four years between 1997 and 2003. Butterflies were monitored in three years between 2000 and 2004.

A 2006 review on the effects of the Swiss Ecological Compensation Area scheme in Switzerland (17) found that out of 1,401 Ecological Compensation Area meadows investigated only around 25% reached the required minimum quality, containing indicator plant species of species-rich semi-natural grasslands. The remaining 75% of Ecological Compensation Area meadows were species-poor with a simple vegetation structure. Several case studies showed that the community composition of spiders (Araneae) differs between extensively and conventionally managed meadows. No details on study design, monitoring techniques or other methods were given.

A replicated site comparison study of 42 fields in Switzerland (18) (same study as (19)) found that a number of wildlife groups benefited from fields participating in the Ecological Compensation Area scheme. There were more birds, but not more bird breeding territories in fields participating in the Ecological Compensation Area scheme than in conventionally farmed fields. There was no difference in the number of bird species on each type of farmland. There were also more uncommon species of arthropod (not endangered), significantly more bee (Apidae) and plant species and a greater density of uncommon plant species on Ecological Compensation Area grasslands than conventionally managed grassland. Ecological Compensation Areas are typically hay meadows farmed at low intensity: no fertilizers or pesticides (except for patch-wise control of problem weeds) are permitted, and vegetation must be cut and removed at least once a year – but not before 15 June (lowlands) or early July (mountains). The study surveyed seven pairs of fields (one within an Ecological Compensation Area, one conventionally farmed) from each of three different parts of Switzerland. Diversity and abundance of vascular plants, arthropods and birds were measured using standard sampling methods in late spring and summer 2003. Surveys of observed and territory-holding birds were made at the field scale and at the 1 ha plot scale.

A replicated trial with paired sites in Switzerland (19) (same study as (18)) found 21 hay meadows managed under the Ecological Compensation Areas agri-environment scheme for 3 to 10 years had more species of plant, wild bee (Apidae) and grasshopper (Orthoptera) than 21 paired conventionally managed hay meadows. There were 13 wild bee species/field, 11 individual bees/survey under the agri-environment schemes, compared to 11 bee species/field and 8 individuals/survey on conventional meadows. This agri-environment scheme requires a postponed first cut, in June or later, and no additions of fertilizer or pesticide to the meadow, although in the study three of the trial meadows were fertilized a little, despite the regulations.

A replicated trial in 2004 of 13 meadows managed under the Swiss Ecological Compensation Area agri-environment scheme for at least 5 years (20) found that the species richness and abundance of hoverflies (Syrphidae), solitary bees (Apidae) and large-sized pollinators (mainly social bees and butterflies (Lepidoptera)) visiting potted flowering plants were higher in meadows under the scheme than in adjacent, intensively managed meadows. The total area of each Ecological Compensation Area meadow (0.48–2.15 ha) had no significant influence on the wild pollinator communities in this study.

A site comparison study from 1998 to 2004 in two areas of the Swiss Plateau region (21) found significantly more species of butterfly (Lepidoptera) on Ecological Compensation Area grassland than conventional grassland in one of the two areas. In Nuvilly, there was an average of 12 species on Ecological Compensation Area grasslands and 11 species on conventional grasslands. In Ruswil, there was an average of 3.4 species on Ecological Compensation Area grasslands and 2.6 species on conventional grasslands. When other factors such as number of plant species, coverage of woody plants or distance to forest were taken into account, this difference was only statistically significant in Ruswil, and not in Nuvilly. Ecological Compensation Area sites tended to have more 'specialist' species – those with only one generation per year, poor dispersal ability or larvae that eat only one type of plant. There were 20–22 Ecological Compensation Area meadows and 6–16 conventional grasslands. The conventional grasslands were fertilized with an average of 206 kg N/ha and cut on average 3 times each year. The Ecological Compensation Area grasslands were fertilized with an average of 7 kg N/ha and cut on average twice a year.

A replicated before-and-after trial of 116 upland hay meadows in the Pennines, northern England (22) found that reduced management intensity prescribed under the Environmentally Sensitive Areas agri-environment scheme led to increased plant species richness in areas where the plant community was degraded. At sites with areas of degraded or modified plant communities, the average number of plant species increased slightly (from 21.6 to 22.8 species/m^2 for areas with species-rich but not MG3 (sweet vernal grass *Anthoxanthum odoratum*-wood crane's bill *Geranium sylvaticum* grassland) plant communities (data from 62 sites) and from 17.6 to 19.1 species/m^2 for degraded areas (data from 90 sites)). This increase was predominantly in the number of grass species, not broadleaf herbaceous species. Plant species were recorded in three 1 m^2 permanent quadrats at each site in 1987 (the year the scheme was introduced) and 2002.

A replicated trial in the Pennine Dales, UK (23) found that plant species richness declined on 147 upland hay meadows managed under the Environmentally Sensitive Areas scheme between 1995 and 2002. Eighty-seven sites under Tier 1 of the scheme allowed 125 kg/ha of nitrogen, phosphorous, potassium (NPK) fertilizer/year, 12.5 tonnes/ha of manure, cutting after 8 July, with grazing allowed until 7 weeks before cutting. Sixty sites under Tier 2 of the scheme allowed no mineral fertilizer, 12.5 tonnes/ha of manure, cutting after 15 July, with no grazing after 15 May. Lime addition and herbicide were allowed with written approval under both Tiers. Tier 1 sites had an average of 3 fewer plant species in 2002 than in 1995. Tier 2 sites had on average 1.5 fewer species. The fall in species richness was significant for herbaceous (forb) species, but not for grass species. Sweet vernal grass *Anthoxanthum odoratum*, wood cranesbill *Geranium sylvaticum*, meadow buttercup *Ranunculus acris* and yellow rattle *Rhinanthus minor* were all less frequently found in 2002 than in 1995.

A randomized, replicated, controlled trial from 2003 to 2006 in southwest England (24) found that plots of permanent pasture cut just once in May or July or not at all during the summer and left unfertilized supported greater numbers and more species of beetles (Coleoptera) in suction traps, true bugs (Hemiptera) and planthoppers (Auchenorrhyncha), greater abundances of spiders (Araneae), crane flies (Tipulidae) and St Mark's flies *Bibio marci* and more species of woodlice (Isopoda) than control fertilized plots cut in May and July, as in conventional silage management. Small insectivorous birds (dunnock *Prunella modularis*, wren *Troglodytes troglodytes* and robin *Erithacus rubecula*) and seed-eating finches (Fringillidae) and buntings (Emberizidae) preferred extensively managed treatments (particularly the plots uncut in summer) to control plots for foraging. Experimental plots (50 x 10 m) were sown on

4 farms. There were 12 replicates of each management type, monitored over 4 years. Results from the same study are also presented in (28, 29, 34).

A site comparison study of alpine meadows in the Albula and Surses Valleys in the Canton of Grisons, Switzerland (25) found that low intensity meadows and extensively managed meadows had significantly more plant species than intensively managed meadows. Low intensity meadows had on average 50 and 55 plant species for moist and dry meadows respectively. Extensive meadows had averages of 53 and 58 plant species for moist and dry meadows respectively. Intensively managed meadows had 37 plant species on average (none were dry meadows). The difference in species number between low intensity and extensive meadows was not statistically significant. Sixty-nine sites were surveyed. Thirty extensively managed meadows had no fertilizer input. Twenty-five meadows managed with low intensity had manure inputs equivalent to 30 kg N/ha/year. Both these types of meadow were managed under agri-environment management agreements, and were cut once after 15 July, with autumn grazing allowed. Fourteen intensively managed meadows had fertilizer inputs of around 90 kg N/ha, and were cut three or four times a year without restrictions. The authors suggest that low intensity management retains species richness in alpine meadows (unlike lowland grasslands in Switzerland) because their degradation due to intensive management has been relatively recent.

A replicated, controlled study in 1999–2001 and 2004–2005 in Jutland, Denmark (26), found that permanent grassland fields under an agri-environment scheme designed to increase water levels had significantly higher numbers of three species of wading bird (northern lapwing *Vanellus vanellus*; black-tailed godwit *Limosa limosa*; common redshank *Tringa totanus*) after the scheme was implemented (2004–2005) compared to numbers before the scheme (1999–2001). Eurasian oystercatchers *Haematopus ostrolagus* did not increase. Effects of the scheme varied between restored and permanent grasslands and between wet and dry fields. The scheme involved promoting wet grasslands as well as reducing fertilizer inputs, grazing pressure and the period of mowing. A total of 615 fields were studied. The four species were surveyed twice during the breeding season (April–May) and the number of each species and their location recorded.

A replicated trial in the Netherlands (27) found that an agri-environment scheme aimed at enhancing habitat for birds by reducing fertilizer and pesticide input and delaying cutting or grazing had no impact on diversity of non-*Apis* spp. bee or plant species in 21 Dutch wet meadow fields when compared with paired conventionally managed fields. Bee diversity and abundance was low in both field types (average <3 species/field; <6 individuals per field). This agri-environment scheme allowed application of nitrogen fertilizer at 206 kg/ha, which is 75% of the standard fertilizer application rate (269 kg/ha). The meadows had been under the scheme for between 3 and 10 years.

In the same randomized, replicated, controlled trial as (24) in southwest England, (28) found that 50 x 10 m plots of permanent pasture cut just once in May or July or not at all during the summer and left unfertilized had a greater total diversity of invertebrates than control fertilized plots cut in May and July. There were 12 replicates of each management type, monitored over 5 years (2002–2006). This study was also part of the same study as (29, 34).

In the same randomized, replicated, controlled study as (24) in southwest England, (29) found that 50 x 10 m plots of permanent pasture cut just once in July or not at all during the summer and left unfertilized attracted a greater abundance and more species of beetle (Coleoptera) than control fertilized plots cut in May and July, in the third year of monitoring. Plots without fertilizer added also had higher proportions of seed- and flower-feeding beetle species in the community. There were 12 replicates of each management type, monitored over 3 years (2003–2005). This study was also part of the same study as (28, 34).

A replicated, randomized study of pasture at three sites in Ireland (30) found that overall, earthworm (Lumbricidae) density and biomass tended to increase with management intensity (nitrogen fertilizer input and stocking rate) but that results varied with site. At Solohead there was no significant relationship between management intensity (N fertilizer input: 80, 175, 225 and 350 kg N/ha; stock rate: 1.75–2.5 cows/ha) and density or biomass. At Johnstown Castle, density and biomass increased with fertilizer input (0, 225 and 390 kg N/ha/year) but the effect was only significant in spring (density: 140, 235, 280/m^2; biomass: 60, 115, 160 g/m^2 respectively). At Grange, biomass, but not density, was significantly higher in the 225 kg N and 2.4 cow units/ha treatment than the 100 kg N and 1.7 cow units/ha plots in autumn (92 vs 60 g/m^2), but not spring (69 vs 58 g/m^2). However, out of ten species at Grange, only two were significantly more abundant with higher inputs; species abundance did not vary with input at the other sites. Treatments were laid out as randomized blocks with 5, 3 and 4 replicates (of 1–2 ha) for each treatment at Solohead, Johnstown Castle and Grange respectively. Earthworms were sampled using the formalin method within six 0.5 x 0.5 m quadrats/plot during one spring and autumn at each site from 2003 to 2005.

A replicated, paired site comparison in Bavaria, Germany (31) found that most grasslands managed extensively under the Bavarian 'Agricultural Landscape Programme' (KULAP) did not have more plant species than paired control sites, but sites with a strict regime of no chemical fertilizers or pesticides as part of the agreement did have more plant species. There were 17–20 plant species/site on both agreement and control sites (314 site pairs). These agreements restricted chemical pesticide use and livestock stocking rates (0.5–2.5 units/ha). Fertilizer use was allowed on 189 of the sites; mineral fertilizer was not allowed on 125 sites, but these 2 types were analysed separately and neither showed a difference in numbers of plant species. Another set of 91 site pairs where agreement sites had delayed cutting date (first cutting dates of 1 July or 15 June, combined with a maximum of 2 livestock units/ha, no mineral fertilizer) were also not different from paired control sites (21–24 plant species/site). There were more plant species on grasslands with no chemical pesticide or fertilizer inputs allowed (also limited to 2 livestock units/ha; 57 site pairs). These sites had around 22 plant species/site, compared to around 18 species/site on paired control sites. Pairs of 25 m^2 grassland plots were selected from 4,400 plots in the Bavarian grassland survey. All plant species within the plot were recorded between April and October (year not given). Plot pairs were in the same natural landscape, 90% within 10 km of each other. In each pair, there was one with and one without an agri-environment scheme agreement.

A site comparison study between 1997 and 2004 in two regions of central Switzerland (32) found that Ecological Compensation Area meadows contained significantly more species of plant, butterfly (Lepidoptera) and ground beetle (Carabidae) than conventionally managed meadows, but not more species of spider (Araneae). Estimated total numbers of species were 118 plant, 36 butterfly (Lepidoptera), 98 ground beetle (Carabidae) and 156 spider on Ecological Compensation Area meadows and 83, 34, 88 and 124 on conventional meadows respectively. The study sampled 315 Ecological Compensation Area meadows and 216 conventionally managed grasslands between 1997 and 2004. Rare or threatened species were not found more frequently on Ecological Compensation Area sites. The increased number of species was a response of common species.

A replicated trial from 2004 to 2008 in the Czech Republic (33) found that the number of plant species did not increase or decrease over five years of monitoring on grasslands under various agri-environment schemes, but the proportion of weedy nitrogen-loving species, such as nettle *Urtica dioica*, fell relative to typical meadow species. This change was considered a botanical improvement. Forty-seven grassland sites were monitored in May/June and October each year from 2004 to 2008. Sixteen sites were managed under 'ecological agriculture' and nine under grassland management agri-environment schemes. These allowed up to 60 kg N/ha fertilizer, two cuts and cattle grazing. Nine sites were wet and peaty meadows, on which no fertilizer was allowed. Twelve sites were known locations of nesting corncrake *Crex crex*. One site was an arable field reverted to grassland.

In the same randomized, replicated, controlled trial as (24) from 2003 to 2006 on four farms in southwest England, (34) found that plots of unfertilized permanent pasture cut just once in July or not cut at all during the summer attracted more adult butterflies (Lepidoptera) (but not more butterfly species or common bumblebees *Bombus* spp.) than control fertilized plots cut in May and July, managed as in conventional silage management. Plots cut just once in May, plots cut twice either unfertilized or ungrazed, and plots with a higher cutting height did not support more adult butterflies than control plots. Butterfly larvae were more abundant in unfertilized plots cut just once in May or July, or not at all in summer, than in other treatments. None of the grass treatments supported more common bumblebee species or individuals than control plots. No more than 2 bumblebees/transect were recorded on average on any grassy plot in any year. Experimental plots 50 x 10 m were established on permanent pastures (more than 5-years-old) on 4 farms. There were nine different management types, with three replicates/farm, monitored over four years. Seven management types involved different management options for grass-only plots, including conventional silage practices; no cutting in summer; early summer cut (May); late summer cut (July); raised mowing height. Bumblebees and butterflies were surveyed along a 50 m transect line in the centre of each experimental plot, once a month from June to September annually. Butterfly larvae were sampled on two 10 m transects using a sweep net in April and June–September annually. This study was also part of the same study as (28, 29).

A paired site comparison in Switzerland in 2003 and 2004 (35) found that 24 Ecological Compensation Area meadows supported more species of plant and arthropod on average than adjacent intensively managed meadows. Ecological Compensation Area meadows were cut an average of two times during the year after 15 June and not fertilized. They had been managed in this way for at least five years. Intensive meadows were cut on average over four times, usually beginning in early May, and treated with liquid manure. Around 16 plant species and 75 arthropod species per site were recorded in Ecological Compensation Area meadows, compared to around 10 plant species and 60–65 arthropod species in intensively managed meadows. Of the 234 arthropod species for which more than 5 individuals were recorded 147 (63%) were more abundant in Ecological Compensation Area meadows than in adjacent intensively managed meadows. Spiders (Araneae) and beetles (Coleoptera) (the two most abundant taxa), and consequently total arthropod counts, were more abundant in intensively managed meadows.

A 2010 review of four experiments on the effects of agri-environment measures on livestock farms in the UK (36) found two replicated trials in southwest England showing that reduced management intensity on permanent grasslands benefits foraging birds. Both found higher numbers of invertebrates, seed heads and foraging birds at lower management intensity (less fertilizer, less cutting, less grazing or a combination of these). One study was the Potential for Enhancing Biodiversity on Intensive Livestock Farms (PEBIL) project, also reported in (24) (Defra report BD1444). The other was part of a Defra-funded project focussed largely on the effects of reduced grazing pressure (BD1454) for which no reference is given in the review.

A replicated, controlled study in 2006–2009 on four intensively managed fields in Herefordshire and Yorkshire, UK (37) found that plant seed heads and invertebrates, including important bird food invertebrates, were more abundant on plots with reduced management intensity (lighter grazing, seasonal removal of livestock, reduced fertilizer inputs) than on intensively managed plots. The cover of injurious weeds remained low on all plots throughout the study. Grasses were more abundant and broadleaved plants (forbs) less abundant on reduced management intensity plots with light grazing compared to controls. There were four 0.6 ha treatment plots/field: grass height maintained at either 12–16 cm (lenient) or 6–9 cm (moderate) by cattle April to mid-July with livestock removal from mid-July to the following spring and reduced fertilizer inputs (50 kg N/ha/year); moderate grazing with no livestock removal and either reduced (control) or normal (fertilized control: 150 kg/ha/year) fertilizer inputs. Vegetation cover was sampled in four 1 m^2 quadrats/plot in July 2006 and 2009; weed cover was sampled within a 10 m radius of vegetation quadrats. Seed heads were counted in September–October 2007 and 2009 at 20 points/plot in 0.25 m^2 quadrats. Invertebrates were sampled on transects 10, 30, 50, 70 and 90 m from the field boundary using Vortis sampling, pitfall traps and sweep nets.

A replicated site comparison study in 2008 and 2009 on farms in three regions in England (38) found that grassland managed under Higher or Entry Level Stewardship Schemes with low or very low inputs was not used significantly more by seed-eating farmland songbirds than improved grassland or open rough grassland. Between 0.5 and 2 birds/ha were recorded on average on the different types of grassland. The stewardship grassland category also included land being maintained as semi-natural grassland under the schemes. It is not clear how many sites of the different management types were used in the analysis. Surveys were done in the summers of 2008 and 2009 on 69 farms with Higher Level Stewardship in East Anglia, the West Midlands or the Cotswolds and on 31 farms across all 3 regions with no environmental stewardship.

A randomized, replicated, controlled trial in Berkshire, UK, started in 2008 (39), found that grassland plots sown with a seed mix containing legumes and other broadleaved plant species had more pollinators (bees (Apidae), butterflies (Lepidoptera) and hoverflies (Syrphidae)) and pollinator species in the first summer if left uncut in July and August. Plots were either cut twice in May–June and September, with a summer rest period, or cut three times, also in July–August. The number of silage cuts (two or three) did not affect the cover of sown plant species. In this study, a summer rest from grazing, between May and September, had a similar effect. There were 4 replicates of each treatment on plots either 16 x 32 m (those cut for silage) or 25 x 50 m (grazed plots).

(1) Dodd A.P. (1987) The future for the management of environmentally sensitive land in Britain: the effectiveness of the Broads Grazing Marshes Scheme. MSc thesis. University of Manchester.

(2) Wind K., Neuteboom J.H. & t'Mannetje L. (1994) Effect of extensification on yield and botanical composition of grassland on dry sandy soil. *British Grassland Society Occasional Symposium*, 28, 217–224.

(3) Fisher G.E.J. & Rahmann G. (1997) Extensification - benefits and disadvantages to grassland biodiversity. *Grassland Science in Europe*, 2, 115–123.

(4) Nösberger J. & Kessler W. (1997) Utilisation of grassland for biodiversity. *Proceedings of the International Occasional Symposium of the European Grassland Federation: Management for grassland biodiversity*. Warszawa-Lomza, Poland, 19–23 May. pp. 33–42.

(5) Nösberger J., Messerli M. & Carlen C. (1998) Biodiversity in grassland. *Annales De Zootechnie*, 47, 383–393.

(6) Jeanneret P., Schüpbach B., Steiger J., Waldburger M. & Bigler F. (2000) Evaluation of ecological measures: biodiversity. Spiders and butterflies. *Agrarforschung*, 7, 112–116.

(7) Herzog F., Gunter M., Hofer G., Jeanneret P., Pfiffner L., Schlapfer F., Schupbach B. & Walter T. (2001) Restoration of agro-biodiversity in Switzerland. Pages 397–406 in: Y. Villacampa, C. A. Brebbia & J. L. Uso (eds.) *Ecosystems and Sustainable Development III*, 10, WIT Press, Southampton.

(8) Kleijn D., Berendse F., Smit R. & Gilissen N. (2001) Agri-environment schemes do not effectively protect biodiversity in Dutch agricultural landscapes. *Nature*, 413, 723–725.

(9) Bat Conservation Trust (2003) *Agricultural Practice and Bats: A review of current research literature and management recommendations*. Defra BD2005.

(10) Jeanneret P., Schupbach B. & Luka H. (2003) Quantifying the impact of landscape and habitat features on biodiversity in cultivated landscapes. *Agriculture Ecosystems & Environment*, 98, 311–320.

(11) Jeanneret P., Schüpbach B., Pfiffner L. & Walter T. (2003) Arthropod reaction to landscape and habitat features in agricultural landscapes. *Landscape Ecology*, 18, 253–263.

(12) Jöhl R., Knop E., Herzog F., Jeanneret P., Walter T., Duelli P. & Ewald K.C. (2004) Low intensity meadows harbour endangered grasshopper. *Gefährdete Heuschrecken in extensiv genutzten Wiesen.*, 11, 156–161.

(13) Kleijn D., Berendse F., Smit R., Gilissen N., Smit J., Brak B. & Groeneveld R. (2004) Ecological effectiveness of agri-environment schemes in different agricultural landscapes in The Netherlands. *Conservation Biology*, 18, 775–786.

(14) Aviron S., Berner D., Bosshart S., Buholzer S., Herzog F., Jeanneret P., Klaus I., Pozzi S., Schneider K., Schüpbach B. & Walter T. (2005) Butterfly diversity in Swiss grasslands: respective impacts of low-input management, landscape features and region. In R. Lillak, R. Viiralt, A. Linke & V. Geherman (eds.) *Grassland Science in Europe Volume 10*, Estonian Grassland Society, Tartu. pp. 340–343.

(15) Helden A.J., Anderson A. & Purvis G. (2005) Grassland arthropod species richness in a conventional suckler beef production system and one compatible with the Irish agri-environment scheme (REPS). *Proceedings of the British Society of Animal Science Annual Conference 2005*. York, UK. 4–6 April. pp. 236.

(16) Jeanneret P., Aviron S., Herzog F., Luka H., Pozzi S. & Walter T. (2005) Temporal trends of arthropod diversity in conventional and low-input meadows. In *Grassland Science in Europe Volume 10*, Estonian Grassland Society, Tartu. pp. 344–347.

(17) Herzog F., Buholzer S., Dreier S., Hofer G., Jeanneret P., Pfiffner L., Poiger T., Prasuhn V., Richner W., Schüpbach B., Spiess E., Spiess M., Walter T. & Winzeler M. (2006) Effects of the Swiss agri-environmental scheme on biodiversity and water quality. *Mitteilungen der Biologischen Bundesan-stalt für Land-u. Forstwirtschaft*, 403, 34–39.

(18) Kleijn D., Baquero R.A., Clough Y., Díaz M., Esteban J.d., Fernández F., Gabriel D., Herzog F., Holzschuh A., Jöhl R., Knop E., Kruess A., Marshall E.J.P., Steffan-Dewenter I., Tscharntke T., Verhulst J., West T.M. & Yela J.L. (2006) Mixed biodiversity benefits of agri-environment schemes in five European countries. *Ecology Letters*, 9, 243–254.

(19) Knop E., Kleijn D., Herzog F. & Schmid B. (2006) Effectiveness of the Swiss agri-environment scheme in promoting biodiversity. *Journal of Applied Ecology*, 43, 120–127.

(20) Albrecht M., Duelli P., Müller C., Kleijn D. & Schmid B. (2007) The Swiss agri-environment scheme enhances pollinator diversity and plant reproductive success in nearby intensively managed farmland. *Journal of Applied Ecology*, 44, 813–822.

(21) Aviron S., Jeanneret P., Schüpbach B. & Herzog F. (2007) Effects of agri-environmental measures, site and landscape conditions on butterfly diversity of Swiss grassland. *Agriculture, Ecosystems & Environment*, 122, 295–304.

(22) Critchley C.N.R., Fowbert J.A. & Wright B. (2007) Dynamics of species-rich upland hay meadows over 15 years and their relation with agricultural management practices. *Applied Vegetation Science*, 10, 307–314.

(23) Critchley C.N.R., Martin D., Fowbert J.A. & Wright B. (2007) Providing the evidence base to improve the efficacy of management guidelines for upland hay meadows. In J. J. Hopkins, A. J. Duncan, D. I. McCracken, S. Peel & J. R. B. Tallowin (eds.) *High Value Grassland: Providing biodiversity, a clean environment and premium products. British Grassland Society Occasional Symposium No. 38*, British Grassland Society (BGS), Reading. pp. 129–134.

(24) Defra (2007) *Potential for Enhancing Biodiversity on Intensive Livestock Farms (PEBIL)*. Defra BD1444.

(25) Dietschi S., Holderegger R., Schmidt S.G. & Linder P. (2007) Agri-environment incentive payments and plant species richness under different management intensities in mountain meadows of Switzer-land. *Acta Oecologica*, 31, 216–222.

(26) Kahlert J., Clausen P., Hounisen J. & Petersen I. (2007) Response of breeding waders to agri-environmental schemes may be obscured by effects of existing hydrology and farming history. *Journal of Ornithology*, 148, 287–293.

(27) Kohler F., Verhulst J., Knop E., Herzog F. & Kleijn D. (2007) Indirect effects of grassland extens-ification schemes on pollinators in two contrasting European countries. *Biological Conservation*, 135, 302–307.

(28) Pilgrim E.S., Potts S.G., Vickery J., Parkinson A.E., Woodcock B.A., Holt C., Gundrey A.L., Ramsay A.J., Atkinson P., Fuller R. & Tallowin J.R.B. (2007) Enhancing wildlife in the margins of intensively managed grass fields. In J.J. Hopkins, A.J. Duncan, D.I. McCracken, S. Peel & J.R.B. Tallowin (eds.) *High Value Grassland: Providing biodiversity, a clean environment and premium products. British Grassland Society Occasional Symposium No. 38*, British Grassland Society (BGS), Reading. pp. 293–296.

(29) Woodcock B.A., Potts S.G., Pilgrim E., Ramsay A.J., Tscheulin T., Parkinson A., Smith R.E.N., Gundrey A.L., Brown V.K. & Tallowin J.R. (2007) The potential of grass field margin management for enhancing beetle diversity in intensive livestock farms. *Journal of Applied Ecology*, 44, 60–69.
(30) Curry J.P., Doherty P., Purvis G. & Schmidt O. (2008) Relationships between earthworm populations and management intensity in cattle-grazed pastures in Ireland. *Applied Soil Ecology*, 39, 58–64.
(31) Mayer F., Heinz S. & Kuhn G. (2008) Effects of agri-environment schemes on plant diversity in Bavarian grasslands. *Community Ecology*, 9, 229–236.
(32) Aviron S., Nitsch H., Jeanneret P., Buholzer S., Luka H., Pfiffner L., Pozzi S., Schüpbach B., Walter T. & Herzog F. (2009) Ecological cross compliance promotes farmland biodiversity in Switzerland. *Frontiers in Ecology and the Environment*, 7, 247–252.
(33) Holubec V. & Vymyslický T. (2009) Botanical monitoring of grasslands after the adoption of agro-environmental arrangements. *Grassland Science in Europe*, 14, 128–131.
(34) Potts S.G., Woodcock B.A., Roberts S.P.M., Tscheulin T., Pilgrim E.S., Brown V.K. & Tallowin J.R. (2009) Enhancing pollinator biodiversity in intensive grasslands. *Journal of Applied Ecology*, 46, 369–379.
(35) Albrecht M., Schmid B., Obrist M.K., Schüpbach B., Kleijn D. & Duelli P. (2010) Effects of ecological compensation meadows on arthropod diversity in adjacent intensively managed grassland. *Biological Conservation*, 143, 642–649.
(36) Buckingham D.L., Atkinson P.W., Peel S. & Peach W. (2010) New conservation measures for birds on grassland and livestock farms. *Proceedings of the BOU – Lowland Farmland Birds III: Delivering solutions in an uncertain world*. British Ornithologists' Union. pp. 1–13.
(37) Defra (2010) *Modified Management of Agricultural Grassland to Promote In-field Structural Heterogeneity, Invertebrates and Bird Populations in Pastoral Landscapes*. Defra BD1454.
(38) Field R.H., Morris A.J., Grice P.V. & Cooke A.I. (2010) Evaluating the English Higher Level Steward-ship scheme for farmland birds. *Aspects of Applied Biology*, 100, 59–68.
(39) Pywell R.F., Woodcock B.A., Orr R., Tallowin J.R.B., McEwen I., Nowakowski M. & Bullock J.M. (2010) Options for wide scale enhancement of grassland biodiversity under the Entry Level Scheme. *Aspects of Applied Biology*, 100, 125–132.

Additional references

Morris M. G. (1981a) Responses of grassland invertebrates to management by cutting. III. Adverse effects on Auchenorrhyncha. *Journal of Applied Ecology*, 18, 107–123.
Morris M. G. (1981b) Responses of grassland invertebrates to management by cutting. IV. Positive responses of Auchenorrhyncha. *Journal of Applied Ecology*, 18, 763–771.
Schwank O., Blum H. & Nösberger J. (1986) The influence of irradiance distribution on the growth of white clover (*Trifolium repens* L.) in differently managed canopies of permanent grassland. *Annals of Botany*, 57, 273–281.
Bakker J.P. (1994) Nature management in Dutch grasslands. *Proceedings of the British Grassland Society Occasional Symposium*, 28, 115–124.
Carlen C. (1994) Root competition and shoot competition between *Festuca pratensis* Huds. and *Dactylis glomerata* L. PhD thesis No. 10512, Eidgenössische Technische Hochschule Zürich.
Fothergill M., Davies D.A. & Morgan C.T. (1994) Extensification of upland pasture in Britain. *Proceedings of the New Zealand Grassland Association*, 56, 219–222.
Marriott C.A., Bolton G.R., Common T.G., Small J.L. & Barthram G.T. (1994) Effects of extensification of sheep grazing systems on animal production and species composition of the sward. *Proceedings of the 16th Meeting of the European Grassland Federation*, 505–509.
Neuteboom J.H., t'Mannetje L., Lantinga E.A. & Wind K. (1994) Botanical composition, yield and herbage quality of swards of different age on organic meadowlands. *Proceedings of the 15th Meeting of the European Grassland Federation*, 320–323.
Blake S., Foster G.N., Fisher G.E.J. & Ligertwood G.L. (1996) Effects of management practices on the carabid faunas of newly established wildflower meadows in southern Scotland. *Annales Zoologici Fennici*, 33, 139–147.
Gerstmeir R. & Lang C. (1996). Beitrag zu Auswirkungen der Mahd auf Arthropoden [Effects of mowing on arthropods]. *Zeitschrift fur Okologie un Naturschutz*, 5, 1–14.
Fuller R., Atkinson P.W., Asteraki E. J., Conway G. J., Goodyear J., Haysom K., Ings T., Smith R.E.N., Tallowin J. R. & Vickery J. A. (2003) *Changes in Lowland Grassland Management: Effects on invertebrates and birds*. Defra BD1435.

4.5   Raise mowing height on grasslands

- A replicated controlled study and a review from the UK found that raised mowing heights provided benefits to Eurasian skylark including increased productivity[6, 7]. A review found raised cutting heights were less damaging to amphibians and invertebrates[4].

- A randomized, replicated, controlled trial from the UK found that raising mowing height on grasslands had no effect on numbers of foraging birds or invertebrates[2, 3, 5]. One replicated controlled study found no difference in invertebrate abundance[7].

- One replicated study from the UK found that northern lapwing and common starling chicks had greater foraging success in shorter grass[1].

Background
Vegetation height is important in determining the value of a grassland to wildlife. High vegetation can provide more complex environments and more habitats, but short vegetation can allow birds access to the ground, which can help foraging and reduce the risk of predation.

A replicated study from January to May 2002 of 15 northern lapwing *Vanellus vanellus* chicks on one grassland site in the Isle of Islay and 20 common starlings *Sturnus vulgaris* on one grassland site in Oxfordshire, UK (1) found that both species experienced significantly greater foraging success in shorter grass. For lapwing chicks, foraging rate declined as grass height increased. Starlings spent 30% more time actively foraging and captured 33% more prey in short grass, although intake rate (captures per second of active foraging) did not differ between grass heights. Invertebrate abundance did not differ between long and short grass. Fertilizer application and water level was manipulated to provide a range of grass heights on the lapwing site. Starlings were observed in enclosures placed within intensively managed permanent pasture that was mown to either 3 cm (short grass) or 13 cm (tall grass).

A randomized, replicated, controlled trial from 2003 to 2006 on four farms in southwest England (2) (same study as (3, 5)) found that 50 x 10 m plots of permanent pasture cut to 10 cm in May and July did not attract more invertebrates or foraging birds than control plots cut to 5 cm. Plots were cut twice in May and July and grazed in autumn/winter. There were 12 replicates of each management type, monitored over 4 years.

A randomized, replicated, controlled trial from 2003 to 2005 on 4 farms in southwest England (3) (same study as (2, 5)) found that 50 x 10 m plots of permanent pasture cut to 10 cm in May and July had similar numbers of beetles (Coleoptera) and beetle species to control plots cut to 5 cm. There were 12 replicates of each management type, monitored over 3 years.

A 2009 literature review (4) found that cutting height has a large influence on mowing impact, with a raised cutting height being less damaging to field-dwelling animals. Three studies reported that higher cutting heights were less damaging to invertebrates and amphibians (Löbbert *et al.* 1994, Classen *et al.* 1996, Oppermann *et al.* 2000).

A randomized, replicated, controlled trial from 2003 to 2006 in southwest England (5) (same study as (2, 3)) found plots of permanent pasture cut to 10 cm in May and July did not attract more butterflies (Lepidoptera), butterfly larvae or common bumblebees *Bombus* spp. than control plots cut to 5 cm. Experimental plots 50 x 10 m were established on permanent pastures (more than 5-years-old) on 4 farms. There were nine different management types, with three replicates/farm, monitored over four years. Bumblebees and butterflies were surveyed along a 50 m transect line in the centre of each experimental plot, once a month from June to September annually. Butterfly larvae were sampled on two 10 m transects using a sweep net in April and June–September annually.

A 2010 review of four experiments on the effects of agri-environment measures on livestock farms in the UK (6) found one trial from 2006 to 2008 that tested the effect of mowing height on Eurasian skylark *Alauda arvensis* nesting in silage fields. Preliminary results showed that chick survival was not affected by raised cutting height. However, the number of new birds produced each year (productivity) was more sensitive to re-nesting rates than chick survival. Raised cutting height slightly increased productivity because skylarks re-nested sooner after cutting but this was not enough to maintain a local population given survival rates. This study formed part of a Defra-funded project (BD1454) for which no reference is given in the review.

A replicated, controlled study from 2006 to 2008 on silage fields in Dorset, UK (7) found that raising the cutting height on grasslands benefited Eurasian skylark *Alauda arvensis*. Daily failure rates of skylark nests were lower and the likelihood of chicks fledging higher on fields with raised cutting heights (probability of chicks fledging in May: 0.6 vs 0.4 on high and low cut plots respectively). Annual skylark productivity on two-cut silage plots was higher when the first or both the first and second cuts were raised compared to normal low cuts (21 independent fledglings/100 pairs on plots where the first cut was raised; 24 where both cuts were raised vs 4 on plots where both cuts were low). Skylarks preferentially nested on raised cutting height plots in the 2 week period after mowing in both 2006 and 2007, and only nested in vegetation which was at least 10 cm tall. Invertebrate abundance was not significantly different between cutting heights. A split-plot set-up was used to test the effects of different mowing heights: average raised cutting height 12 cm; low cutting height approximately 6–7 cm (2006: 12 trial fields, 11 controls; 2007: 8 trials, 4 controls). Entire fields were subject to raised or control mowing heights in 2008 (10 trial fields, 15 controls). All plots were cut using disc mowers. Skylark nests were monitored to assess daily productivity, and survival rates and chicks were radio-tagged to assess their survival after fledging. Stochastic simulation modelling was used to investigate the effects on skylark productivity.

(1) Devereux C.L., McKeever C.U., Benton T.G. & Whittingham M.J. (2004) The effect of sward height and drainage on common starlings *Sturnus vulgaris* and northern lapwings *Vanellus vanellus* foraging in grassland habitats. *Ibis*, 146, 115–122.
(2) Defra (2007) *Potential for Enhancing Biodiversity on Intensive Livestock Farms (PEBIL)*. Defra BD1444.
(3) Woodcock B.A., Potts S.G., Pilgrim E., Ramsay A.J., Tscheulin T., Parkinson A., Smith R.E.N., Gundrey A.L., Brown V.K. & Tallowin J.R. (2007) The potential of grass field margin management for enhancing beetle diversity in intensive livestock farms. *Journal of Applied Ecology*, 44, 60–69.
(4) Humbert J.Y., Ghazoul J. & Walter T. (2009) Meadow harvesting techniques and their impacts on field fauna. *Agriculture Ecosystems & Environment*, 130, 1–8.
(5) Potts S.G., Woodcock B.A., Roberts S.P.M., Tscheulin T., Pilgrim E.S., Brown V.K. & Tallowin J.R. (2009) Enhancing pollinator biodiversity in intensive grasslands. *Journal of Applied Ecology*, 46, 369–379.
(6) Buckingham D.L., Atkinson P.W., Peel S. & Peach W. (2010) New conservation measures for birds on grassland and livestock farms. *Proceedings of the BOU – Lowland Farmland Birds III: Delivering solutions in an uncertain world*. British Ornithologists' Union. pp. 1–13.
(7) Defra (2010) *Modified Management of Agricultural Grassland to Promote In-field Structural Heterogeneity, Invertebrates and Bird Populations in Pastoral Landscapes*. Defra BD1454.

Additional references
Löbbert M., Kromer K.-H., Wieland C.C. (1994) Einfluss von Mäh- und Mulchgeräten auf die bodennahe Fauna [Influence of mowing and mulching equipment on ground fauna]. In Forschungsberichte 'Integrative Extensivierungs-und Naturschutzstrategien' H. 15. Universität Bonn, Institut für Landtechnik, Bonn. pp. 7–26.

Classen A., Hirler A., Oppermann R. (1996) Auswirkungen unterschiedlicher Mähgeräte auf die Wiesenfauna in Nordost-Polen [Effects of different mowing equipment on meadow fauna in northeast Poland]. *Naturschutz und Landschaftsplanung*, 28, 139–144.

Oppermann R., Handwerk J., Holsten M., & Krismann A. (2000) Naturverträglich Mähtechnik für das Feuchtgrünland, Voruntersuchung für das E & E – Vorhaben [Nature-friendly mowing for wet grassland, preliminary study for E & E projects]. ILN Singen, Bonn.

4.6   Delay mowing or first grazing date on grasslands

• Eight studies from the Netherlands, Sweden and the UK (three replicated and controlled, of which one was also randomized and one European systematic review) found that delaying mowing or grazing dates resulted in benefits to some or all plants[1, 4, 5, 15], invertebrates[10, 14, 15] or birds[11, 13] studied. These benefits included: higher plant species richness[1], higher densities of two rare arable weeds[4], more insect species and individuals visiting flowers[10], greater abundance of some spiders and ground beetles[14], increased breeding wading bird densities[11] and increased Eurasian skylark productivity[13].

• Three reviews found the UK population of corncrake increased after measures including delaying mowing dates were introduced[6–8].

• Six studies from Finland, the Netherlands, Sweden, Switzerland and the UK (including three replicated controlled trials, of which one was also randomized and a European systematic review) found that delaying mowing or grazing dates on grassland had no clear effect on plant species richness[2, 9, 15], ground beetle communities[3], abundance of some insects and spiders[14], or population trends of wading bird species[12].

Background

This intervention involves delaying the first mowing or grazing date on grasslands. Early-season, mechanized mowing is thought to be responsible for declines in the UK and elsewhere of species such as the corncrake *Crex crex*, with chicks killed and nests destroyed by mowing machinery (Green & Gibbons 2000). Delaying mowing until after chicks can escape is therefore a part of many agri-environment schemes. Delaying mowing or grazing may also provide benefits to other farmland wildlife such as plants and invertebrates.

Green R.E. & Gibbons D.W. (2000) The status of the Corncrake *Crex crex* in Britain in 1998. *Bird Study*, 47, 129–137.

A controlled trial in 1990–1992 on a species rich upland meadow in the Pennines, northern England (1) found that plant species richness was higher in plots cut in mid-July (on average 15 species/quadrat) than in plots cut in mid-June (13 species/quadrat). The meadow plant community was characterized by sweet vernal grass *Anthoxanthum odoratum* and wood cranesbill *Geranium sylvaticum* (MG3 under the UK National Vegetation Classification). Plants were surveyed in June 1992 in 25 x 25 cm quadrats, following 2 years of treatment. The number of replicates is not stated.

A long-term replicated, controlled trial from 1978 to 1993 in the Jura Mountains, Switzerland (2) found that timing of mowing only slightly affected the number of plant species. Plots that were cut annually in October did not have fewer plant species than those cut annually in July or every second year (in July). October-cut plots had 59 plant species/40 m^2 and 45 species/m^2 on average, compared to 56 species/40 m^2 and 38 species/m^2 on average for annually July-cut plots. Mowing in October changed the species composition, for example, by reducing the cover of one of the three most abundant species, meadow brome *Bromus erectus*, from around 40% in July-cut plots to around 20%. There were three replicate 50 m² plots for each treatment, and the experimental management regimes were carried out from 1978 to 1993. The percentage cover of plant species was estimated in 40 m² and 1 m² sample areas in each plot, in the last week of June 1991 and 1993.

A trial at the Crichton Royal Farm, Dumfries, Scotland (3) found no detectable difference in the ground beetle (Carabidae) community between different cutting treatments on experimentally restored flower-rich grassland plots. A field was ploughed and sown with 17 plant species in August 1987 (5 grasses, 2 clovers *Trifolium* spp. and 12 other flowering broadleaved species). It was managed without fertilizers. Half the field was cut once each July. The other half was cut twice, in May and July. Both were grazed in autumn and winter. Ground beetles were sampled in 18 pitfall traps (laid out in 2 lines) in each treatment area between April and September in 1989 and again in 1993.

A replicated, controlled, randomized study of set-aside at the University of Liverpool Horticulture and Environmental Research Station, England (4) found that delaying cutting resulted in higher densities of common corncockle *Agrostemma githago* and interrupted brome *Bromus interuptus*. An early year 1 cut (1 August) resulted in significantly lower densities of common corncockle (440/m²) compared to a later cut (30 August) or no cut (750/m²). A late cut increased year 2 interrupted brome density compared to the early cut (28 vs 17/m²); no cut significantly decreased the density (9/m²). The trial comprised a split plot (2 x 2 m) randomized block design with 3 replicate blocks each containing 36 treatment combinations. Wheat (drilled), grass crop and rare weeds (hand sown) were planted in October–November 1993 and 1994. Plots were cut (to 10 cm, vegetation left *in situ*) on 1 or 30 August, or not at all. Rare arable weeds were sampled in 1 x 1 m quadrats in the centre of each plot.

A replicated study over one year of a neutral meadow grassland at a farm in England (5) found that delaying mowing resulted in an increase of grass and broadleaved flowering plant seeds. Numbers of grass seeds increased from the June to September cut; the most flowering plant seeds were present in the July (traditional) cut and the least in the September cut. The overall broadleaved flowering plant/grass seed quotient decreased from 2.9 in June, to 0.8 in July, to 0.2 in September. Seventeen species showed significant differences in the amount of seed extracted according to cut date. The meadow was divided into 9 contiguous 20 x 30 m plots in 1990. Treatments were a June cut (mineral fertilizer); July cut (no fertilizer, autumn cattle grazing, spring sheep grazing); September cut. Vegetation was cut at 3 cm in 3 randomly placed 0.06 m² quadrats and seed was collected.

A 2000 literature review (6) found that the UK population of corncrakes *Crex crex* increased from 480 to 589 males between 1993 and 1998 (an average rise of 3.5%/year) (Green & Gibbons 2000) following the introduction of 'Corncrake Friendly Mowing' schemes to increase the number of chicks that survive mowing. Management includes delaying mowing dates and leaving unmown 'corridors' to allow chicks to escape to field edges. The reviewers acknowledge that the corncrake population increase and the introduction of these schemes may be coincidental and a longer monitoring period is required to assess the effects of these schemes on corncrake numbers.

A 2000 literature review of grassland management practices in the UK (7) discussed a collaborative study in the Scottish Islands in which management for corncrake *Crex crex*, including delayed mowing and grazing resulted in increased corncrake numbers (Scottish Biodiversity Group 1998). Management included 'Corncrake Friendly Mowing', i.e. delayed mowing and grazing until August, mowing in strips or mowing from the middle of fields outwards. These practices were encouraged using financial incentives. Although success varied between islands, overall corncrake numbers increased since the widespread implementation of the programme.

A 2002 review (8) states that the UK population of corncrakes *Crex crex* increased by 34% between 1993 and 2001, following the implementation of the 'Corncrake Initiative' which financially compensates farmers who agree to delay mowing until after chicks can escape machinery. A second programme, which began in 1999, also included the provision of suitable cover. Both were based in western Scotland where the remaining British population was found.

A randomized, replicated, controlled trial in northern Finland (9) found that the number of plant species (not including mosses and lichens) and the cover of mosses and lichens were not different in meadow plots mown in August and those mown in June over five years. There were 14 plant species/plot on average in 1998 (not including mosses and lichens) and 12 species/plot in 2003, with no differences between mowing treatments. Some species responded to treatments. For example, August mowing plus disturbance favoured harebell *Campanula rotundifolia* but reduced cover of common bent grass *Agrostris capillaris*. The cover of mosses and lichens fluctuated between years, but was not different between mowing treatments. The meadow was abandoned in 1985. In 1993, it was divided into forty 50 x 50 cm study plots, each at least 2 m apart and annually mown in August without grazing. From 1998, 10 plots were mown in June, 10 in August, and 10 mown in August with bare soil exposed in 25% of the plot area, using a spade. Ten control plots were not mown. The percentage cover of all plant species (including mosses and lichens) in the treatment plots was monitored in June every year from 1998 to 2003.

A small study of two semi-natural grasslands in central Sweden (10) found that more insect species and individuals visited flowers under a delayed grazing regime, from mid-July to September, compared to grazing from mid-May. This was likely to be due to the higher abundance of flowers in the late grazing treatment. The only species with higher abundances under the extended grazing regime was the red-tailed bumblebee *Bombus lapidarius*. Visitation rate and flower constancy did not differ between treatments. Flower-visiting insects exhibited a broader range of activities in the late grazing treatment. Pastures were divided into 2 treatment areas from 1997 in Pustnäs (10 ha) and 2001 in Harpsund (12 ha). Insect flower visitors were sampled in 7 pairs of 5 x 5 m plots/treatment for 9 weeks in summer 2003.

A replicated, controlled, paired-sites study in the western Netherlands in 2003 (11) found that 19 grassland plots with delayed mowing had significantly higher breeding densities of wading birds, compared to 19 paired, control plots (approximately 8 territories/plot for delayed-mowing plots vs approximately 3 territories/plot for controls). This difference was not apparent when delayed mowing was combined with per-clutch payment, and there were no differences in abundances of waders or all bird species. However, when delayed mowing was combined with per-clutch payment, breeding densities of all bird species were significantly higher (13 territories/plot for combined schemes; 11 territories/plot for controls). There were higher numbers of redshank *Tringa totanus* on combined plots (approximately 5 birds/plot for combined schemes, 5 birds/plot for per-clutch payment and 3 birds/plot for controls) but not on delayed-mowing plots. There were higher abundances of northern lapwing *Vanellus vanellus* on control plots, compared to delayed-mowing plots, but this difference was not significant (approximately 18 birds/plot for controls vs 13 birds/plot for delayed-mowing plots). There were no significant differences in breeding densities for redshank, northern lapwing, Eurasian oystercatcher *Haematopus ostralegus* or black-tailed godwit *Limosa limosa*. The authors suggest that groundwater depth, soil hardness and prey density drove these patterns. All farms had been operating the schemes for an average of four years before the study.

A replicated, controlled, before-and-after study in 1,040 grassland areas in the Netherlands between 1990 and 2002 (12) found that nesting densities of black-tailed godwit *Limosa limosa* and redshank *Tringa totanus* were higher in areas with management agreements with postponed mowing, but these differences were present before the agreements came into effect. Population trends were similar between management and control areas for black-tailed godwit and Eurasian oystercatcher *Haematopus ostralegus*, but northern lapwing *Vanellus vanellus* and redshank declined on management areas, relative to controls. Mowing was postponed on management areas to the end of May or beginning of June.

A replicated, controlled study from 2006 to 2008 on silage fields in Dorset, UK (13) found delayed mowing increased annual productivity of Eurasian skylark *Alauda arvensis* on plots cut once and that skylark nest abandonment was sensitive to mowing date. Plots cut once with either a low or a raised cutting height in late July had higher annual skylark productivity (94–95 independent fledglings/100 adult pairs) than plots cut at the beginning of June (low cutting height: 22 fledglings/100 pairs; raised cutting height: 50 fledglings/100 pairs). Delaying first mowing date by one week on plots cut at a low cutting height resulted in significant increases in skylark productivity; plots cut at a low cutting height once in early June had significantly lower annual skylark productivity than plots cut in mid-June (22 vs 31 fledglings/100 pairs). Between 2006 and 2008 silage fields were subject to either different mowing regimes or normal (control) management. In 2006 and 2007 a split-plot set up was used to test the effects of different mowing heights; average raised cutting height was 12 cm and the low height approximately 6–7 cm (12 trial fields, 11 controls in 2006; 8 trial fields, 4 controls in 2007). Entire fields were subject to raised or control mowing heights in 2008 (10 trial fields, 15 controls). All plots were cut using disc mowers. Skylark nests were monitored to assess daily productivity, and survival rates and chicks were radio-tagged to assess their survival after fledging. Stochastic simulation modelling was used to investigate the effects on skylark productivity.

A replicated study of spider (Araneae) and insect communities in two semi-natural grasslands in Sweden (14) found that the effect of delayed grazing depended on taxa. Small spiders, some ground beetles (Carabidae) and ants (Formicidae) were more abundant in conventional, continuous grazing (May–September) than in traditional late grazing (mid-July–September) while larger spiders and some ground beetles were more abundant in late grazing. Overall, abundance of ground beetles was higher in continuous grazing in the early summer but higher in late grazing in the late summer. Pitfall traps were used within and outside one grazing exclosure (1–4 ha) at each site, 7–10 times from May–August 2002–2005. Ant abundance was also measured by annually mapping nest density.

A 2012 systematic review looked at the effects of delaying mowing on plants and invertebrates in European meadows (15) and it found that delaying the first mowing date had a positive effect on the number of invertebrate species but a variable effect on the number of plant species. Delaying early season cutting (from spring to summer) had a positive effect on the number of plant species. However delaying first cuts from spring and summer to autumn, or from early to late summer had a negative effect on the number of plant species. The number of invertebrates was also positively influenced when the first cut was delayed, but this finding was only true when two studies were excluded (out of nine studies looking at invertebrate species richness). The review looked at 24 studies comprising 54 experiments where the effects of delayed cutting had been tested for plants or invertebrates.

(1) Younger A. & Smith R.S. (1994) Hay meadow management in the Pennine Dales, Northern England. *Proceedings of the Joint Meeting between the British Grassland Society and the British Ecological Society: Grassland management and nature conservation. British Grassland Society Occasional Symposium No. 28, 27–29 September 1993.* University of Leeds, England. pp. 137–143.

(2) Ryser P., Lagenhauer R. & Gigon A. (1995) Species richness and vegetation structure in a limestone grassland after 15 years management with six biomass removal regimes. *Folia Geobotanica and Phytotaxonomica*, 30 157–167.

(3) Blake S., Foster G.N., Fisher G.E.J. & Ligertwood G.L. (1996) Effects of management practices on the carabid faunas of newly established wildflower meadows in southern Scotland. *Annales Zoologici Fennici*, 33, 139–147.

(4) Neve P., Mortimer A.M. & Putwain P.D. (1996) Management options for the establishment of communities of rare arable weeds on set-aside land. *Aspects of Applied Biology*, 44, 257–262.

(5) Smith R.S., Pullan S. & Shiel R.S. (1996) Seed shed in the making of hay from mesotrophic grassland in a field in northern England: effects of hay cut date, grazing and fertilizer in a split-split-plot experiment. *Journal of Applied Ecology*, 33, 833–841.

(6) Aebischer N.J., Green R.E. & Evans A.D. (2000) From science to recovery: four case studies of how research has been translated into conservation action in the UK. In N.J. Aebischer, A.D. Evans, P.V. Grice & J.A. Vickery (eds.) *Ecology and Conservation of Lowland Farmland Birds*, British Ornithologists' Union, Tring. pp. 43–54.

(7) Wakeham-Dawson A. & Smith K.W. (2000) Birds and lowland grassland management practices in the UK: an overview. *Proceedings of the Spring Conference of the British Ornithologists' Union*, March 27–28, 1999. Southampton, England. pp. 77–88.

(8) Green R.E. (2002) Corncrakes, conservation management and agri-environment schemes. *Aspects of Applied Biology*, 67, 189.

(9) Hellström K., Huhta A.P., Rautio P. & Tuomi J. (2006) Search for optimal mowing regime – slow community change in a restoration trial in northern Finland. *Annales Botanici Fennici*, 43, 338–348.

(10) Sjödin N.E. (2007) Pollinator behavioural responses to grazing intensity. *Biological Conservation*, 16, 2103–2121.

(11) Verhulst J., Kleijn D. & Berendse F. (2007) Direct and indirect effects of the most widely implemented Dutch agri-environment schemes on breeding waders. *Journal of Applied Ecology*, 44, 70–80.

(12) Breeuwer A., Berendse F., Willems F., Foppen R., Teunissen W., Schekkerman H. & Goedhart P. (2009) Do meadow birds profit from agri-environment schemes in Dutch agricultural landscapes? *Biological Conservation*, 142, 2949–2953.

(13) Defra (2010) *Modified Management of Agricultural Grassland to Promote In-field Structural Heterogeneity, Invertebrates and Bird Populations in Pastoral Landscapes.* Defra BD1454.

(14) Lenoir L. & Lennartsson T. (2010) Effects of timing of grazing on arthropod communities in semi-natural grasslands. *Journal of Insect Science*, 10, 1–24.

(15) Humbert J.Y., Pellet J., Buri P. & Arlettaz R. (2012) Does delaying the first mowing date benefit biodiversity in meadowland? *Environmental Evidence*, 1:9.

Additional references

Scottish Biodiversity Group (1998) *Corncrake Newsletter and Initiative Report 1998*. Royal Society for the Protection of Birds, Inverness.

Green R.E. & Gibbons D.W. (2000) The status of the corncrake *Crex crex* in Britain in 1998. *Bird Study*, 47, 129–137.

4.7   Use mowing techniques to reduce mortality

- Eight studies investigated the effects of different mowing techniques on wildlife. Seven studies (including four replicated trials, of which one was randomized, one controlled and three reviews) from Germany, Ireland, Switzerland and the UK found that using specific mowing techniques can reduce mortality or injury in birds, mammals, amphibians or invertebrates[1, 3–8]. A review found that the UK corncrake population increased around the same period that Corncrake Friendly Mowing schemes were introduced[2].

- One replicated trial[1] found that changing the mowing pattern reduced the number of corncrake chicks killed. Sixty-eight percent of chicks escaped mowing when fields were mown from the centre outwards, compared to 45% during conventional mowing from the field edge inwards.

- Six studies looked at the effects of using different mowing machinery. Two studies (one review; one randomized, replicated trial) found bar mowers[5, 8] and one report found double chop mowers[4] caused less damage or lower mortality among amphibians and/or invertebrates than other types of mowing machinery. A review found evidence that twice as many small mammals were killed by rotary disc mowers with conditioners compared to double blade mowers[5]. Two studies found that using a mechanical processor or conditioner killed or injured more invertebrates[3, 8] than without a conditioner, however one replicated controlled study found mower-conditioners resulted in higher Eurasian skylark nest survival than using a tedder. A review of studies found that skylark chick survival was four times higher when wider mowing machinery was used[6]. One replicated controlled trial found skylark nest survival was highest when swather mowers and forage harvesters were used[7].

Background

Mowing and harvesting operations may have a negative impact on farmland wildlife. This intervention involves using different mowing machinery or mowing patterns to reduce the impact of mowing on field-dwelling animals. Adjusting mowing techniques may benefit ground-nesting birds that frequently remain in long grass or crops for as long as possible; mowing from the centre of the field outwards, rather than from the field edge inwards, may allow birds to escape the mowing machinery before the last patch of long grass is harvested.

See also 'Raise mowing height on grasslands' for studies that examine the effects of raising the mowing height on grasslands, and 'All farming systems: Provide refuges during harvest or mowing' for studies investigating the effects of uncut refuges in mown grasslands.

A replicated trial from 1992 to 1995 in Scotland and the Republic of Ireland (1) found that a larger proportion of corncrake *Crex crex* chicks escaped when meadows were mown using a corncrake-friendly mowing pattern, where fields were mown from the centre outwards (inside-outwards) (68% of 76 chicks escaped) compared to the standard outside-inwards mowing pattern (45% of 31 chicks escaped). Fewer chicks were killed by inside-outwards mowing, even when the grass surrounding the plot being mown had already been cut. In plots mown from the inside-outwards, the proportion of chicks that escaped mowing declined as the distance to cover (vegetation over 20 cm tall) increased, and was higher for older chicks. Corncrakes were able to move away from the mower fast enough to escape mowing if an escape route to a refuge area was available, except for the youngest chicks (i.e. less than two-days-old). Hay and silage meadows in one area in Scotland were studied in 1994 and in two areas in Ireland from 1992 to 1995. Corncrakes were seen on a total of 59 meadows. Female corncrakes were radio-tagged to assess movement patterns. Fields were observed during both mowing patterns, and the number of chicks that escaped or were killed recorded. Meadows were searched following mowing to record the number of dead chicks and nest remains.

A 2000 literature review (2) found that the UK population of corncrakes *Crex crex* increased from 480 to 589 males between 1993 and 1998 (an average rise of 3.5%/year) (Green & Gibbons 2000) following the introduction of Corncrake Friendly Mowing schemes to increase the number of chicks that survive mowing. Management includes leaving unmown 'corridors' to allow chicks to escape to field edges. The reviewers acknowledge that the corncrake population increase and the introduction of these schemes may be coincidental and a longer monitoring period is required to assess the effects of these schemes on corncrake numbers.

A replicated trial in Switzerland from 1996 to 1999 (3) found seven times more honey bees *Apis mellifera* were killed or unable to fly when white clover *Trifolium repens* plots were mown with a rotary mower and mechanical processor (which crushes mowings to accelerate drying) than without a processor (14,000 vs 2,000 honey bees/ha dead or unable to fly, respectively). The height of flowers in relation to mower height affected bee survival. Honey bees foraging on phacelia *Phacelia tanacetifolia* flowers taller than the upper edge of the mower (70 cm) were shaken off and able to escape. Bee losses were higher on two white clover plots where the flower height was 25–30 cm (53% and 62% bees lost after mowing) than on one phacelia plot (flower height not given, but taller than the mower) (35% bees lost). Mowing speed did not have a significant impact on bee losses. Three plots were located on one trial farm; plots measured approximately 0.3 ha. One plot was sown with phacelia in 1996. Two plots were sown with 50% white clover in 1998 and 1999. Five to six honey bee colonies were established adjacent to the plots several days before surveying. Plots were mown with a 1.8 m-wide drum mower with integrated conditioner fixed to the side of a tractor. In 1999 the clover plot was mown with and without the processor. The number of bees before and after mowing was recorded in 1–4 m² quadrats.

A 2003 report documenting the results of a large-scale agro-conservation project in northeast Germany (4) included a review of five studies investigating the impacts of meadow mowing techniques on animal mortality. One study found that a double chop mower killed or injured fewer amphibians than the higher performance rotary drum or disc mowers. Cutting using scythes, considered a more cautious method, still injured more amphibians than a double-chop mower, and killed a similar number of amphibians. Another study found that double-chop mowers preserved more ground beetles (Carabidae), perhaps due to their higher (and variable) cutting height. A further three studies demonstrated the importance of mowing technology for amphibians, invertebrates and nesting birds.

A 2009 literature review (5) of thirteen studies found marked differences in the number of small mammals, reptiles, amphibians and invertebrates killed or damaged by different meadow harvesting techniques. Highest mortality was caused by flail and suction flail mowers, which killed or damaged on average 60% and 49% of the invertebrates studied, respectively. Rotary mowers with a conditioner were the next most damaging, killing/damaging on average 21% of amphibians and 35% of invertebrates studied. Bar mowers were the least damaging, with on average 11% mortality among amphibians and 18% among invertebrates. A single study showed that rotary disc mowers with conditioners caused double the number of small mammal deaths compared to double blade mowers (Oppermann *et al.* 2000). Three studies reported that higher cutting heights were less damaging to field-dwelling animals (Löbbert *et al.* 1994, Classen *et al.* 1996, Oppermann *et al.* 2000). The review notes that later harvesting stages also have a considerable impact, especially the removal of baled grass. The collection of lines of mown grass for baling had a greater impact on grasshopper (Orthoptera) populations than mowing (Oppermann *et al.* 2000). The review recommends leaving uncut grass strips when mowing to benefit field-dwelling organisms. It also notes that several of the studies reviewed were poorly replicated or designed.

A 2010 review of four experiments on the effects of agri-environment measures on livestock farms in the UK (6) found one trial from 2006 to 2008 that tested the effect of mowing techniques in reducing the mortality of Eurasian skylarks *Alauda arvensis* nesting in silage fields. Preliminary results showed that chick survival was strongly affected by the type of machinery used. Survival was four times higher using wider machinery and reducing the number of machinery passes than without these changes. However, the number of new birds produced each year (productivity) was more sensitive to re-nesting rates than chick survival. This study formed part of a Defra-funded project BD1454 which is also summarized in (7).

A replicated, controlled study from 2006 to 2008 in Dorset, UK (7) found that the type of silaging machinery used affected Eurasian skylark *Alauda arvensis* nest survival. Survival rates were highest when swather mowers and forage harvesters were used. Fewer nests were covered by grass when using a swather mower (23% of nests covered) compared with bar mowers (60% covered) and fewer nests were abandoned at the egg stage. Survival rates (clutches: 43%, older nestlings: 62%) were highest with swather mowers across the silage harvest. Survival was higher using a mower-conditioner to spread cut grass rather than using a tedder separately. More nests survived forage harvesting than baling (probability of survival 0.83 vs 0.67). During silage collection, most nest failures were caused by being run over by machinery. Between 2006 and 2008, silage fields were subject to either different mowing regimes or normal (control) management. Disc mowers with one, two or three cutting bars were used on all farms, whilst two farms used swather mowers (single bar). Cut grass was normally spread to allow the grass to wilt or placed in rows which were later collected. Machinery traffic was high during the collection process. Skylark nests were monitored to assess daily productivity and survival rates and chicks were radio-tagged to assess their survival after fledging. Stochastic simulation modelling was used to investigate the effects on skylark productivity.

A randomized, replicated trial in 2007 and 2008 in Switzerland (8) found that harvesting meadow plots using a hand-pushed bar mower killed or injured on average 20% of caterpillars (Lepidoptera) added to plots before mowing compared to 37% when using a tractor-pulled rotary drum mower and 69% if a conditioner was attached to the rotary mower. Using a conditioner also increased the proportion of damaged wax invertebrate models from on average 11% (bar mower, no conditioner) to 30% (rotary mower with conditioner). Large (4 cm) invertebrate models were damaged more often than small (2 cm) models. Caterpillars and models placed on the ground or in vegetation (30 cm high) before mowing were affected differently by mowing treatments. Organisms on the ground were strongly impacted by tractor wheels, whilst those in vegetation were damaged by the mower/conditioner. Cutting height did not affect mortality in this study but the authors note it is likely to be important for larger animals. Nine meadows were studied, with four 2.5 m-long plots each. There were four mowing treatments: hand-pushed bar mower (cutting height 6 cm, 1.7 m wide), tractor-pulled rotary drum mower (2.5 m wide) with two cutting heights 9 cm and 6 cm (one with, one without conditioner). Before mowing, 50 small and 50 larger wax invertebrate models were placed either on the ground or tied to vegetation 30 cm high. In 2008 on five meadows, large white butterfly caterpillars *Pieris brassicae* were placed on the ground (50 caterpillars) and in the vegetation (50 caterpillars). After mowing, both wax models and caterpillars that survived mowing were checked for injuries.

(1) Tyler G.A., Green R.E. & Casey C. (1998) Survival and behaviour of Corncrake *Crex crex* chicks during the mowing of agricultural grassland. *Bird Study*, 45, 35–50.
(2) Aebischer N.J., Green R.E. & Evans A.D. (2000) From science to recovery: four case studies of how research has been translated into conservation action in the UK. In N.J. Aebischer, A.D. Evans, P.V. Grice & J.A. Vickery (eds.) *Ecology and Conservation of Lowland Farmland Birds*, British Ornithologists' Union, Tring. pp. 43–54.
(3) Fluri P. & Frick R. (2002) Honey bee losses during mowing of flowering fields. *Bee World*, 83, 109–118.
(4) Flade M., Platcher H., Schmidt R. & Werner A. (2003) *Nature Conservation in Agricultural Ecosystems: Results of the Schorfheide-Chorin Research Project.* Quelle & Meyer, Wiebelsheim, Germany.
(5) Humbert J.Y., Ghazoul J. & Walter T. (2009) Meadow harvesting techniques and their impacts on field fauna. *Agriculture Ecosystems & Environment*, 130, 1–8.

(6) Buckingham D.L., Atkinson P.W., Peel S. & Peach W. (2010) New conservation measures for birds on grassland and livestock farms. *Proceedings of the BOU – Lowland Farmland Birds III: Delivering solutions in an uncertain world.* British Ornithologists' Union. pp. 1–13.

(7) Defra (2010) *Modified Management of Agricultural Grassland to Promote In-field Structural Heterogeneity, Invertebrates and Bird Populations in Pastoral Landscapes.* Defra BD1454.

(8) Humbert J.Y., Ghazoul J., Sauter G.J. & Walter T. (2010) Impact of different meadow mowing techniques on field invertebrates. *Journal of Applied Entomology,* 134, 592–599.

Additional references

Löbbert M., Kromer K.-H. & Wieland C.C. (1994) Einfluss von Mäh- und Mulchgeräten auf die boden-nahe Fauna [Influence of mowing and mulching equipment on ground fauna]. In *Forschungsberichte 'Integrative Extensivierungs-und Naturschutzstrategien' H. 15.* Universität Bonn, Institut für Landtechnik, Bonn. pp. 7–26.

Classen A., Hirler A. & Oppermann R. (1996) Auswirkungen unterschiedlicher Mähgeräte auf die Wiesenfauna in Nordost-Polen [Effects of different mowing equipment on meadow fauan in northeast Poland]. *Naturschutz und Landschaftsplanung,* 28, 139–144.

Green R.E. & Gibbons D.W. (2000) The status of the corncrake *Crex crex* in Britain in 1998. *Bird Study* 47, 129–137.

Oppermann R., Handwerk J., Holsten M., & Krismann A. (2000) Naturverträgliche Mähtechnik für das Feuchtgrünland, Voruntersuchung für das E & E – Vorhaben [Nature-friendly mowing for wet grassland, preliminary study for E & E projects]. ILN Singen, Bonn.

4.8   Reduce grazing intensity on grassland (including seasonal removal of livestock)

- Of 27 individual studies (including 10 replicated, controlled trials, 4 reviews and 1 systematic review) from France, Germany, Ireland, the Netherlands and the UK, 15 (including 3 randomized, replicated, controlled trials) from 4 countries found benefits to birds, plants or invertebrates in response to reducing grazing intensity on permanent grassland (including seasonal removal of livestock)[1–3 ,5, 8, 11–13, 15, 17, 18, 26, 28–30].

- Of these 15 studies, 6 (including 1 randomized, replicated, controlled trial) found that reducing grazing intensity throughout the year increased the abundance and diversity of plants[13, 26], frequency of certain plant species[18], invertebrate diversity[3, 13], usage by geese[1] and the number of northern lapwing and common redshank[12]. Six studies (including three replicated controlled trials, of which two were randomized) found that excluding or delaying summer grazing increased plant species diversity[17], invertebrate abundance[2, 8, 15, 30] and benefited breeding Eurasian skylark[5]. A review[11] found a study that showed that removing autumn grazing after a silage cut increased the winter abundance of seed-eating birds. A review and a replicated controlled study from the UK[28, 29] found that reduced grazing intensity or seasonal removal of livestock increased the number of invertebrates, plant seed heads and foraging skylark, and that some bird species preferred plots with seasonal removal of livestock.

- Three studies (including one randomized, replicated, controlled trial) from the Netherlands and the UK found no benefit to plants or invertebrates from reduced grazing intensity[4, 16, 22, 23, 27]. One randomized, replicated controlled trial excluded grazing in autumn/winter[16, 22, 27] and another study excluded grazing in the summer[23]. A further study found that reducing grazing intensity throughout the year did not increase plant diversity[4].

- Nine studies from France, Germany and the UK reported mixed results for some or all species of wildlife groups considered (including one randomized, replicated, controlled trial, two reviews and a systematic review). Of these, eight studies found that reduced grazing intensity throughout the year benefited some species but not others[6, 7, 9, 14, 20, 21, 24,]

[25], one found that the impact depended on the type of vegetation grazed[19] and one found benefits to bee and wasp abundance but not species richness[10]. One study[20] found that the response of birds to removal of summer grazing varied between functional groups and depended on time of year. A UK review found that reduced grazing benefited invertebrates, plants, rodents and some but not all birds[9].

- A systematic review of the effects of grazing intensity on meadow pasture[24] concluded that intermediate levels of grazing are usually optimal for plants, invertebrates and birds but that trade-offs are likely to exist between the requirements of different taxa.

Background

Studies summarized within this section tested whether reducing livestock grazing intensity on permanent grassland (including seasonal removal of livestock) increased the abundance or species richness of plants, invertebrates or other wildlife. Permanent grassland habitats studied include species-poor improved grass leys, species-rich grasslands and grazed marshes or fens. Studies tested the effectiveness of reducing grazing intensity throughout the year or seasonally removing grazing, for example during or after silage-making or by seasonally excluding grazing in fenced headlands. See also 'Maintain upland heath/moorland' for the effects of reduced grazing intensity on unenclosed upland grassland areas (including upland acid grasslands).

A before-and-after study in Gloucestershire, England (1) found that the proportion of geese on a grassland site using a specifically managed 130 ha area increased from 33% in the winter of 1970–1971 to 87% by 1975–1976 following a reduction in grazing intensity over this period. Starting in 1970, stock were sequentially removed from three sections of the area: the first was ungrazed from 30 September, the second from 31 October and the third from 30 November. A fourth area was not grazed at all.

A study of a perennial rye grass *Lolium perenne* and white clover *Trifolium repens* ley grassland in Ireland (2) found that management-induced changes in grass height had immediate effects on arthropod abundance. Arthropod abundance was greatest (up to $10,351/m^2$) in taller/denser silage grass and lowest (as few as $394/m^2$) in short grass subject to periodic heavy sheep-grazing. The abundance of most groups, particularly larger insects, increased in areas left for silage, whilst numbers of mites (Acari), springtails (Collembola) and many other taxa decreased most under heavy grazing and after cutting. The cropping systems had no overall influence on the range of dominant taxa. The five treatments were intermittent grazing, silage cut only, silage cut followed by grazing, grazing and silage, and continuous grazing. Inorganic fertilizers were applied as appropriate. Plant-dwelling arthropods were sampled at random (10 samples of 0.1 m²) in each plot using a D-Vac suction net monthly from May to September 1976 and in March 1977.

A controlled, replicated trial in 1985–1989 on grassland in Oxfordshire, England (3) found that as sheep grazing intensity increased, the number of species of bugs (Heteroptera), herbivorous beetles (Coleoptera), leafhoppers (Auchenorrhyncha), leaf miners and spiders (Araneae) decreased. In the most intensively grazed treatments, the total number of species of bugs, herbivorous beetles, leafhoppers, leaf miners and spiders was 11, 16, 17, 16 and 17 respectively, compared to 19, 24, 20, 34 and 25 in the least grazed treatment. Grazing treatments began in 1985. Three treatments were replicated 6 times in 30 x 30 m paddocks (ungrazed control, short-period spring and short-period autumn grazing) and 2 treatments were applied in larger areas (spring-and-autumn grazing and long-period autumn grazing, not replicated). Plants were surveyed 4 times a year in 12 quadrats (1 m^2) in each replicate.

A trial from 1986 to 1989 on a grassland in the Netherlands (4) found no difference in the number of plant species between three different cattle stocking rates. No fertilizer was applied during the study. By 1990, only 2 herb species (cuckoo flower *Cardamine pratensis* and meadow buttercup *Ranunculus acer* now *R. acris*) were present in more than 5% of 100 vegetation samples. The authors suggest that plant diversity did not increase because of the dense growth of a small number of competitive grass species. Species that could have colonized were present on the field borders and ditches. The 6.6 ha grassland had been grazed at a rate of 5 steers/ha from April to July, then 3.5 steers/ha until October, from 1973 until 1985. From 1986 onwards it was divided into three 2.2 ha paddocks, each grazed at a single fixed stocking rate – either 2.3, 3.6 or 4.9 steers/ha from April to October.

A replicated study in 1992 and 1993 within the South Downs Environmentally Sensitive Area, Sussex, UK (5) found that breeding Eurasian skylarks *Alauda arvensis* avoided heavily grazed pasture. Grassland that was heavily grazed by sheep was generally not used by skylarks during the second brood period, whereas areas that were not grazed or mown until mid-July onwards were used for both first and second broods. Four arable, 10 mixed and 3 pastoral farms were studied. Skylarks were sampled by mapping breeding males during 2 counts along transects on 12–17 farms from April to June.

A small randomized study over one year of a neutral meadow grassland at a farm in England (6) found that only four plant species exhibited significant differences in seed number as a result of grazing intensity. More Yorkshire fog *Holcus lanatus* was recorded when there was no grazing, more yellow oat grass *Trisetum flavescens* with autumn grazing and more downy oat-grass *Avenula pubescens* and creeping buttercup *Ranunculus repens* with autumn and spring grazing. The three grazing treatments were randomly applied to three plots. Vegetation was cut at 3 cm in 3 randomly placed 0.06 m^2 quadrats and seed collected.

A randomized, replicated, controlled study in spring and summer 1995 and 1996 in Sussex, England (7), found that the densities of Eurasian skylark *Alauda arvensis*, grass seed heads and invertebrates were significantly higher on fields grazed at lower intensities (4.4–14.3 skylarks/km^2, 155 seed heads/m^2 and 17–112 invertebrates/sample on 6 lightly-grazed fields vs 1.3–2.4 skylarks/km^2, 9.6 seed heads/m^2, 10–68 invertebrates/sample and 4–16 invertebrate taxa/sample on 6 intensively-grazed fields). The density of carrion crows *Corvus corone* and rooks *C. frugilegus* did not vary between grazing intensities, nor did the number of invertebrate taxa in 1996 (1995: 12–17 invertebrate taxa/sample on lightly grazed vs 6–11 on intensively-grazed fields; 1996: 6–16 on lightly-grazed vs 4–16 on intensively-grazed fields). In 1996 there were significantly more individuals and taxa of spiders (Araneae) in lightly-grazed fields but no differences in beetles (Coleoptera), flies (Diptera), bees, wasps and ants (Hymenoptera) or larvae. Twelve reverted permanent grassland fields (each field 5 ha) sown with perennial rye grass *Lolium perenne*, cock's-foot *Dactylis glomerata* and white clover *Trifolium repens* were studied. Sheep were used to control the grass height in fields between April and July each year: intensively-grazed fields were managed to keep the grass under 10 cm long; less intensively managed fields had a grass height of 15–25 cm. Skylarks were counted and their locations recorded every 10 days from April–June 1995 and April–July 1996. Foraging rooks were surveyed in 1995 and 1996; carrion crows were surveyed in 1996. Invertebrates were sampled at five locations/field in 1995 and 1996 using a D-Vac suction trap, and in pitfall traps in 1996.

A randomized, replicated, controlled trial in 1997–1998 on permanent pasture at three sites in Dumfries and Galloway, Scotland, UK (8) found that fencing of field headlands to prevent grazing during the summer substantially increased the abundance of three key chick-food insect groups. After one year headlands protected from summer grazing had 19–32 times more chick-food insects (609 true bugs (Hemiptera), 75 sawfly larvae (Symphyta) and 18 caterpillars (Lepidoptera); on average per 10 sample) than grazed headlands (19 true bugs, 2 sawfly larvae and 1 caterpillar). Treatments were carried out from spring 1997 in adjacent plots (10 x 50 m long) on the boundaries of 7 pasture fields: unfenced unsprayed, unfenced sprayed, fenced (May–September) unsprayed, and fenced (May–September) sprayed. In sprayed plots, herbicide was applied in April 1997 to clear strips to trial a method for increasing foraging access for birds. Unfenced plots were grazed by cattle and sheep during summer, and all plots were intermittently grazed by sheep during winter. Insects were sweep net sampled in June and July in 1997 and 1998.

A 2000 literature review of grassland management practices in the UK (9) found one study that reported that grass maintained at a height of 15–25 cm supported twice the number of invertebrates compared with grass grazed by sheep to less than 10 cm. In particular, reduced grazing resulted in significantly more web-spinning spiders (Araneae) and the number of grass seed heads in July was 15 times that in heavily grazed fields (7). Two studies reported that taller or ungrazed grassland supported greater densities of rodents and predatory birds (Dodds *et al.* 1995, Shaw 1995). In contrast, increased vegetation height following a reduction in grazing by livestock and or rabbits *Oryctolagus cuniculus* has been found to result in a decrease in the number of breeding Eurasian thick-knee (stone curlew) *Burhinus oedicnemus*, woodlark *Lullula arborea* and northern wheatear *Oenanthe oenanthe*, however the number of Eurasian curlew *Numenius arquata* increased (Dolman & Sutherland 1992, Green & Taylor 1995, Bealey *et al.* 1999).

A replicated comparison of 6 intensively (5.5 cattle/ha) and 6 lightly (1.5 cattle/ha) cattle-grazed meadows with 6 ungrazed meadows in Germany (10) found that meadows with light grazing had a greater number of individual cavity-nesting bees, wasps (Hymenoptera) and their brood parasites than meadows with intensive grazing. There was an average of 47 emerging individuals/lightly grazed site, compared to 27 emerging individuals/intensively grazed site. Reduced intensity of grazing did not significantly increase the number of bee and wasp species. Both abundance and total species richness of these insects were significantly higher on ungrazed grassland (11.5 species) than on intensively (4.7 species) or lightly (6.2 species) grazed pastures. These results were linked to an increase in vegetation height as grazing intensity is reduced.

A 2004 review of experiments on the effects of agri-environment measures on livestock farms in the UK (11) found one randomized, replicated trial in Scotland showing that grass field headlands left ungrazed in summer had more insects in groups known to be food for game birds (true bugs (Heteroptera), butterflies and moths (Lepidoptera) and sawflies (Symphyta)) than grazed plots (this study was also reported in (8)). Another experiment (no reference given in the review) showed seed-eating birds were more abundant in winter on uncut silage plots left ungrazed than on plots grazed from September. Two other experiments examined effects of reduced grazing intensity, but bird numbers using experimental plots were probably too low for analysis. The review assessed results from seven experiments (some incomplete at the time of the review) in the UK and Europe.

A before-and-after study of grazing marshes in eastern England (12) investigated the effect of reducing grazing intensity and improving footdrain management on breeding wading bird numbers from 1993 to 2003. Northern lapwing *Vanellus vanellus* numbers increased from 19 pairs in 1993 to 85 pairs in 2003 and common redshank *Tringa totanus* rose from 4 to 63 pairs. Numbers of winter wildfowl also increased over the period and changes in vegetation communities to those more tolerant of flooding occurred. Grazing intensity was reduced from 1.5–2 head of cattle to 0.7 head/ha and fertilizer inputs were stopped. In 1993 water levels were raised by 45 cm. From 1995, approximately 600 m of footdrains were opened/year; from 2000 onwards, approximately 2,000 m of footdrains were opened or added.

A controlled study from 2000 to 2004 in the UK (13) found that reduced grazing pressure maintained botanical diversity and abundance and enhanced invertebrate diversity and abundance, but increased pernicious weeds on species-rich grasslands. The cover of positive indicator species of grasslands of high nature conservation value remained stable under lenient grazing pressure (8–10% cover) but decreased under severe (from 9% to 5%) and decreased (from 9% to 5%) but then recovered (8%) under moderate grazing pressures. Competitive grass and legume species increased across all treatments. Abundance and diversity of bumblebee *Bombus* spp. species was higher under moderate (0.34 individuals counted/minute) and lenient grazing (0.38) than severe grazing pressure (0.15). Spider numbers were also significantly higher under lenient (118 individuals/m^2) than severe or moderate grazing (68 individuals/m^2). Percentage cover of spear thistle *Cirsium vulgare* and creeping thistle *Cirsium arvense* increased under all three grazing pressures but did not differ between treatments (2000: 0.4–1.0% cover; 2004: 2.9–3.6). Species-rich grasslands were grazed over 5 years with severe (grass height 6 to 8 cm), moderate (8–10 cm) or lenient (10–12 cm) cattle grazing pressures. A fertilized improved pasture was maintained at 6–8 cm as a control.

A 2006 review of UK studies (14) on the impact of farm management practices on below-ground biodiversity and ecosystem function indicates that the impact of reduced grazing intensity on the diversity of soil organisms is likely to vary among taxa. Three studies found that increased grazing pressure had a negative impact, in terms of species richness or abundance, on ground beetles (Carabidae) (Ni Bhriain *et al.* 2002), spiders (Araneae) (Macaulay Institute 2006) and nematodes (Nematoda) in lowland habitats (Mulder *et al.* 2003). One study found that grazing pressure had no impact on the profile of soil fungal communities (Clegg 2006).

A replicated, controlled study of five grassland headlands on four intensively managed pastoral farms across Scotland (15) investigated the effect of conservation headlands, with no grazing from April–August and no fertilizer or pesticide applications on the abundance of ground active invertebrates and found that aphids/leafhoppers/planthoppers (Homoptera) and true bugs (Heteroptera) were more abundant in conservation headlands (no fertilizers, pesticides or grazing April–August) than conventional headlands and open fields. Aphids/leafhoppers/planthoppers had higher activity densities in conservation headlands (2.1) and field edges (conventional: 2.0; conservation: 1.9) than in conventional headlands (0.8) and open fields (0.6). Roundback slugs (Arionidae) showed the same pattern (2.3 conservation headlands; 2.1 conventional field edges; 2.4 conservation field edges; 0.7 conventional headlands; 0.3 open fields). True bugs were more abundant in conservation headlands (0.7) and field edges (1.1–1.2) than in open fields (0.2). Keelback slug (Limacidae) activity density was greater in both headlands (conventional: 1.9; conservation: 2.8) and field edges (2.3–2.7) than in open fields (1.1). Butterfly/moth (Lepidoptera) and sawfly (Symphyta) larvae showed a similar trend, whereas ground beetle (Carabidae) abundance did not differ with treatment (3.5–3.6). Ground beetle activity density was highest in open fields (4.0). One headland in each field was divided into two areas of 6 x 100 m, a conventional and conservation headland. In each field, invertebrates were sampled with five pitfall transects of nine traps in the conservation headland, conservation field edge, conventional headland, conventional field edge and open field. Traps were set for 3–4 weeks in May–June and July–August 2000–2003.

A randomized, replicated, controlled trial from 2003 to 2006 on 4 farms in southwest England (16) (same study as (22, 27)) found that 50 x 10 m plots of permanent pasture managed just like conventional silage but without autumn/winter grazing did not attract more invertebrates or foraging birds than control plots. Plots were fertilized and cut twice in May and July. Control plots were grazed in autumn/winter. There were 12 replicates of each management type, monitored over 4 years.

A five year trial on a species-rich mountain meadow in France (17) found that biodiversity and structure of the grass layer (sward) were affected not only by extensification of grazing, but also by the way in which the reduced stocking rate was applied. Extending the grazing season after the period of active grass growth, or using a rotational grazing system that excluded part of the area during the main flowering period benefited plant species diversity. At the same time, negative effects on livestock performance were limited.

A replicated, controlled study in a grazed fen area in northern Germany (18) investigated the effects of reduced grazing, topsoil removal and hay transfer on plant species diversity and abundance and found that grazing had minimal effects, but did result in a significant increase of cumulative frequency of wet meadow plant species. Four blocks (12 x 24 m) were established that each combined 3 treatments: moderate grazing (yes/no), topsoil removal (yes/no, to a depth of 30 cm) and hay transfer from a species-rich fen meadow (yes/no, layer of 1–3 cm). Plant cover and species dominance was sampled in 16 permanent squares (1 m²) within each sub-plot in each combination of treatments in 2002–2005. Ten soil seed bank samples were taken from each plot in 2002.

A randomized, replicated trial from 2002 to 2004 in the UK, Germany and France (19) (same study as (21)) found that reduced grazing intensity led to a reduction in the number of plant species at a productive species-poor grassland site (UK) but not at the other two sites: species-rich, semi-natural grassland (France) and moderately species-rich 'mesotrophic' grasslands (Germany). At the UK site, moderately grazed plots had on average 11.4 plant species while lightly grazed plots had 10.2 species and were dominated by grasses. Under reduced grazing, the structural diversity (patchiness) of vegetation decreased at the UK site but increased at the German site. Paddocks 0.4 to 3.6 ha in size were either moderately or leniently grazed with a commercial livestock breed, with treatments replicated 3 times. Actual grazing rates differed according to local conditions, but lenient grazing treatments had 0.3–0.4 fewer animals/ha than the moderate grazing rates. Sites were grazed continuously with cattle. Plants were monitored in 10 fixed 1 m² quadrats in each paddock in April–May, June–July and August–September from 2002 to 2004. Other plant species seen within 5 m of the quadrats were also recorded. An additional study site grazed by sheep in Italy was also included in the analysis but is not reported here because it falls outside the geographical range of this synopsis.

A replicated, controlled trial in 2005–2006 on upland improved grassland in the UK (20) found that the response of bird populations to seasonal removal of grazing (late May–July) for silage making varied between functional groups and depended on the time of year. Plots with seasonal removal of grazing had the greatest number of birds of songbird species between May and July (126 birds compared to 60 in continuously grazed control plots) and between July and September (312 birds compared to 169 in control plots), but numbers were similar to those in continuously grazed plots between October and January (13 and 11, respectively). There were more invertebrate-feeding birds between July and September (105 birds, compared to 41 on control plots) but between October and January there were more invertebrate-feeding birds on continuously-grazed control plots (5,833 birds, compared to 1,458 on plots with seasonal removal of grazing). At all times of year crows (Corvidae) were more abundant on control plots. There were ten replicates. Bird numbers and species were recorded in plots with and without seasonal removal of grazing for silage making in May–July 2005, July–September 2005 and October 2005–January 2006.

A randomized, replicated study from 2002 to 2004 in the UK, Germany and France (21) (same study as (19)), found more butterflies (Lepidoptera) and grasshoppers (Orthoptera) and more butterfly and grasshopper species under lenient than moderate grazing. There were on average 42–47 vs 31 butterflies and 10 vs 8 butterfly species/paddock/year under lenient compared to moderate grazing, and 56–60 vs 34 grasshoppers from 5 vs 4 grasshopper species/paddock/year under lenient compared to moderate grazing. Some groups of insects caught in pitfall traps were more abundant under lenient grazing at some, but not all, sites (for example, ground beetles (Carabidae) and rove beetles (Staphylinidae) at the UK site). Numbers of birds, bird species and European hares *Lepus europaeus* were not different between grazing treatments. Paddocks 0.4 to 3.6 ha in size were either moderately or leniently grazed with a commercial livestock breed, with treatments replicated 3 times. Actual grazing rates differed according to local conditions, but lenient grazing treatments were 0.3–0.4 fewer animals/ha than the moderate grazing rates. Sites were grazed continuously with cattle. Animals were monitored in 2002, 2003 and 2004. Butterflies were counted once a fortnight from May to September and grasshoppers sampled with a sweep net each month from June to October, on three 50 m-long transects. Birds and European hares were counted fortnightly in the early morning, from May to October, with a 7 minute observation period and a walking transect. Ground arthropods were sampled in 12 pitfall traps at each paddock, in spring, summer and early autumn. An additional study site grazed by sheep in Italy was also included in the analysis but is not reported here because it falls outside the geographical range of this synopsis.

A randomized, replicated, controlled trial on 4 farms in southwest England (22) (same study as (16, 27)) found that 50 x 10 m plots of permanent pasture managed without autumn/winter grazing had similar numbers of beetles (Coleoptera) and beetle species to control grazed plots. All plots were cut for silage in May and July. There were 12 replicates of each management type, monitored over 3 years (2003–2005).

A replicated trial of two intensities of cattle grazing on permanent improved grassland plots in the UK (23) found no strong difference in numbers of butterflies (Lepidoptera) between partially grazed and permanently grazed plots. On average there were 7–11 butterfly species (42–156 individuals) on permanently grazed plots and 5–10 butterfly species (21–67 individuals) on partially grazed plots of improved pasture. Ten experimental plots of improved perennial rye grass *Lolium perenne*/white clover *Trifolium repens* were grazed throughout the year by livestock. Ten similar plots were grazed in spring and autumn, but had livestock excluded from May to September and one silage cut taken. No location is given for the experiment. Butterfly transect counts were conducted weekly between mid-April and mid-September in 2005, 2006 and 2007.

A 2008 systematic review (24) identified several studies that suggest that intermediate grazing levels on neutral grassland (MG5 crested dog's-tail *Cynosurus cristata*; lesser knapweed *Centaurea nigra* grassland under the UK National Vegetation Classification) benefit plants, invertebrates or birds, but noted that trade-offs probably exist between the grazing requirements of different taxa. Four studies suggested that intermediate grazing levels are most appropriate for plant conservation objectives. Where grazing levels are very low, or grazing is abandoned, ecological succession leads to a reduction in the number of plant species. High grazing intensity also reduces plant species richness, because a limited number of competitive species become dominant. Six studies found invertebrate abundance or species richness to be greatest at intermediate grazing (or cutting) intensity. Similarly, it is assumed that bird diversity peaks at intermediate grazing levels because of reliance on invertebrates as a food source (three studies), or because intensive grazing increases the risk of nest predation/trampling for ground-nesting birds (three studies). However, one study found that insect diversity was greatest on ungrazed sites and another found no relationship between plant species richness at the local scale and bird diversity. The authors conclude that trade-offs may exist between the grazing requirements of different taxa, and that the limited evidence base necessitates a flexible, site-based approach. Forty-two studies were included in the review.

A randomized, replicated trial of different winter cutting regimes, designed to simulate grazing intensity on grasslands in Oxfordshire, England (25) found that different groups of birds prefer different treatments. Foraging song thrushes *Turdus philomelos*, common starlings *Sturnus vulgaris*, crows (Corvidae) and common kestrels *Falco tinnunculus* preferred mown (grazed) plots to unmown (ungrazed) plots. Grey heron *Ardea cinerea* and meadow pipit *Anthus pratensis* preferred unmown (ungrazed) plots to plots that were mown (grazed) once or twice. For gamebirds, wood pigeon *Columba palumbus* and hedgerow species, there was no significant difference in numbers between the different mowing regimes. Seventeen grass fields (average size 5 ha) were used in the experiment with 2 treatments (mown once vs unmown) or all 4 treatments in each. Winter mowing was used to simulate the effects of grazing or cutting for silage. Grass height did not differ between the 14 replicate plots mown once in November/December, once in January or twice during winter, so one winter cut or grazing period was sufficient to create the habitat advantage for bird groups that prefer short grass.

A long-term randomized, replicated, controlled trial from 1990 to 2005 at two upland grassland sites in North Lanarkshire and Scottish Borders, Scotland, UK (26) found that a reduction in the level of grazing on intensively grazed and fertilized grassland led to an increase in the number and diversity of plant species. However, these changes were very slow and, in some cases, still ongoing after 16 years, suggesting that recovery from intensive management can take many years. Four grazing regimes were applied from spring 1990: conventional management (grazed to maintain grass at 4 cm, with fertilizer), two reduced grazing treatments (grazed to maintain grass at 4 cm or 8 cm, with no fertilizer) and an ungrazed control (with no fertilizer). Each treatment was replicated twice at each site in roughly 0.5 ha plots. Plants were sampled annually in June–July from 1990 to 2005 using an inclined point quadrat (18 quadrats/plot), except in 1991, 1993, 1995, 2002 and 2004.

A randomized, replicated, controlled trial from 2003 to 2006 in southwest England (27) (same study as (16, 22)) found plots of permanent pasture managed without autumn/winter grazing did not attract more butterflies (Lepidoptera), butterfly larvae or common bumblebees *Bombus* spp. than control grazed plots. All plots were cut for silage in May and July. Grazed plots were grazed with cattle from September until the grass height was 5–7 cm. Experimental plots 50 x 10 m were established on permanent pastures (more than 5-years-old) on 4 farms. There were nine different management types, with three replicates/farm, monitored over four years. Bumblebees and butterflies were surveyed along a 50 m transect line in the centre of each experimental plot, once a month from June to September annually. Butterfly larvae were sampled on two 10 m transects using a sweep net in April and June–September annually.

A 2010 review of UK experiments on the effects of agri-environment measures on livestock farms in the UK (28) found two replicated controlled trials that reduced grazing pressure (fewer cattle, cattle removed from July onwards, or both) in pastures over two to four years. One of the studies also reduced fertilizer input from 150 to 50 kg N/ha. Lenient grazing (grass height 12–16 cm) significantly increased the numbers of insects (bugs (Hemiptera)), the production of seed heads in the grasslands and the number of foraging Eurasian skylark

Alauda arvensis on trial fields in both studies. Birds that eat only seeds – European goldfinch *Carduelis carduelis* and Eurasian linnet *C. cannabina* – preferred plots with cattle removed in July. There were eight replicates in the study with additional low fertilizer input and 14 replicates in the other study. These studies formed part of a Defra-funded project (BD1454) for which no reference is given in the review.

A replicated, controlled study in 2006–2009 on 13 semi-improved and 12 intensively managed fields in Devon, Herefordshire and Yorkshire, UK (29) found that lenient grazing or seasonal removal of livestock had positive effects on numbers of invertebrates and birds. Invertebrate abundance (including important bird food invertebrates) on semi-improved grassland was 34–78% higher in July on leniently grazed compared to moderately grazed plots. Seasonal removal of livestock on semi-improved grassland also resulted in positive but smaller increases in invertebrate abundance (5–35%) compared to controls. Leniently grazed plots with grazing removal were used by Eurasian skylark *Alauda arvensis* significantly more than controls (3.6–6.6 times higher than controls). Yellowhammer *Emberiza citrinella*, cirl bunting *E. cirlus* and seed-eating birds with mixed diets showed a preference for plots with seasonal livestock removal. Bird species that feed only on seeds used plots on intensive grassland with grazing removal more than controls but showed no clear preference on semi-improved grassland. During the breeding season, other bird species showed no clear preference or did not prefer plots with reduced grazing. The number of plant species decreased on plots with reduced grazing intensity on semi-improved grassland compared with controls. Plant seed heads were more abundant on plots with reduced grazing intensity or early closure than controls. Weed cover remained low across the treatments. The treatments in approximately 0.6–1 ha plots were grass height maintained at 10–16 cm (lenient) or 6–9 cm (moderate) by cattle; plots were grazed April to mid-July with livestock removal from mid-July to the following spring. Controls were moderately grazed in April–October or until grass growth stopped. Semi-improved grassland had historically received inputs of 50 kg N/ha/year and intensively managed grassland 150 kg N/ha/year plots (reduced fertilizer input was also tested on intensively managed grassland).

A randomized, replicated, controlled trial in Berkshire, UK started in 2008 (30) found that grassland plots sown with a seed mix containing legumes and other herbaceous species had more pollinators (bees (Apidae), butterflies (Lepidoptera) and hoverflies (Syrphidae)) and pollinator species in the first summer with no summer grazing (rested from May to September) than on plots continuously grazed from May to October. The type of grazing (continuous or with a summer rest period) did not affect the cover of sown plant species. There were 4 replicates of each treatment combination, and grazed plots were 25 x 50 m.

(1) Owen M. (1977) The role of wildfowl refuges on agricultural land in lessening the conflict between farmers and geese in Britain. *Biological Conservation*, 11, 209–222.
(2) Purvis G. & Curry J.P. (1981) The influence of sward management on foliage arthropod communities in a ley grassland. *Journal of Applied Ecology*, 18, 711–725.
(3) Gibson C.W.D., Brown V.K., Losito L. & McGavin G.C. (1992) The response of invertebrate assemblies to grazing. *Ecography*, 15, 166–176.
(4) Neuteboom J.H., t'Mannetje L., Lantinga E.A. & Wind K. (1994) Botanical composition, yield and herbage quality of swards of different age on organic meadowlands. *Proceedings of the 15th Meeting of the European Grassland Federation*. pp 320–323.
(5) Wakeham-Dawson A. (1995) Hares and skylarks as indicators of environmentally sensitive farming on the South Downs. PhD thesis. Open University.
(6) Smith R.S., Pullan S. & Shiel R.S. (1996) Seed shed in the making of hay from mesotrophic grassland in a field in northern England: effects of hay cut date, grazing and fertilizer in a split-split-plot experiment. *Journal of Applied Ecology*, 33, 833–841.
(7) Wakeham-Dawson A., Szoszkiewicz K., Stern K. & Aebischer N.J. (1998) Breeding skylarks *Alauda arvensis* on Environmentally Sensitive Area arable reversion grass in southern England: survey-based and experimental determination of density. *Journal of Applied Ecology*, 35, 635–648.

(8)   Haysom K.A., McCracken D.I., Roberts D.J. & Sotherton N.W. (2000) Grassland conservation head-lands: a new approach to enhancing biodiversity on grazing land. *Proceedings of the British Grassland Society Conference. Grazing Management: The principles and practice of grazing, for profit and environmental gain, within temperate grassland systems.* Harrogate, UK, 29 February–2 March. pp. 159–160.

(9)   Wakeham-Dawson A. & Smith K.W. (2000) Birds and lowland grassland management practices in the UK: an overview. *Proceedings of the Spring Conference of the British Ornithologists' Union, March 27–28, 1999.* Southampton, England. pp. 77–88.

(10)   Kruess A. & Tscharntke T. (2002) Grazing intensity and the diversity of grasshoppers, butterflies, and trap-nesting bees and wasps. *Conservation Biology*, 16, 1570–1580.

(11)   Buckingham D.L., Atkinson P.W. & Rook A.J. (2004) Testing solutions in grass-dominated landscapes: a review of current research. *Ibis*, 146, S2 163–170.

(12)   Smart M. & Coutts K. (2004) Footdrain management to enhance habitat for breeding waders on lowland wet grassland at Buckenham and Cantley Marshes, Mid-Yare RSPB Reserve, Norfolk, England. *Conservation Evidence*, 1, 16–19.

(13)   Tallowin J.R.B., Rook. A.J. & Rutter. S.M. (2005) Impact of grazing management on biodiversity of grasslands. *Animal Science*, 81, 193–198.

(14)   Stockdale E.A., Watson C.A., Black H.I.J. & Philipps L. (2006) *Do Farm Management Practices Alter Below-ground Biodiversity and Ecosystem Function? Implications for sustainable land management.* Joint Nature Conservation Committee 364.

(15)   Cole L.J., McCracken D.I., Baker L. & Parish D. (2007) Grassland conservation headlands: their impact on invertebrate assemblages in intensively managed grassland. *Agriculture, Ecosystems & Environment*, 122, 252–258.

(16)   Defra (2007) *Potential for Enhancing Biodiversity on Intensive Livestock Farms (PEBIL).* Defra BD1444.

(17)   Dumont B., Farruggia A. & Garel J.P. (2007) Biodiversity of permanent pastures within livestock farming systems. *Rencontre Recherche Ruminants*, 14, 17–24.

(18)   Rasran L., Vogt K. & Jensen K. (2007) Effects of topsoil removal, seed transfer with plant material and moderate grazing on restoration of riparian fen grasslands. *Applied Vegetation Science*, 10, 451–460.

(19)   Scimone M., Rook A.J., Garel J.P. & Sahin N. (2007) Effects of livestock breed and grazing intensity on grazing systems: 3. Effects on diversity of vegetation. *Grass and Forage Science*, 62, 172–184.

(20)   Vale J.E. & Fraser M.D. (2007) Effect of sward type and management on diversity of upland birds. In J.J. Hopkins, A.J. Duncan, D.I. McCracken, S. Peel & J.R.B. Tallowin (eds.) *High Value Grassland: Providing biodiversity, a clean environment and premium products. British Grassland Society Occasional Symposium No. 38,* British Grassland Society (BGS), Reading. pp. 333–336.

(21)   Wallis De Vries M.F., Parkinson A.E., Dulphy J.P., Sayer M. & Diana E. (2007) Effects of livestock breed and grazing intensity on biodiversity and production in grazing systems. 4. Effects on animal diversity. *Grass and Forage Science*, 62, 185–197.

(22)   Woodcock B.A., Potts S.G., Pilgrim E., Ramsay A.J., Tscheulin T., Parkinson A., Smith R.E.N., Gundrey A.L., Brown V.K. & Tallowin J.R. (2007) The potential of grass field margin management for enhancing beetle diversity in intensive livestock farms. *Journal of Applied Ecology*, 44, 60–69.

(23)   Fraser M.D., Evans J.G., Davies D.W.R. & Vale J.E. (2008) Effect of sward type and management on butterfly numbers in the uplands. *Aspects of Applied Biology*, 85, 15–18.

(24)   Stewart G.B. & Pullin A.S. (2008) The relative importance of grazing stock type and grazing intensity for conservation of mesotrophic 'old meadow' pasture. *Journal for Nature Conservation*, 16, 175–185.

(25)   Whittingham M.J. & Devereux C.L. (2008) Changing grass height alters foraging site selection by wintering farmland birds. *Basic and Applied Ecology*, 9, 779–788.

(26)   Marriott C.A., Hood K., Fisher J.M. & Pakeman R.J. (2009) Long-term impacts of extensive grazing and abandonment on the species composition, richness, diversity and productivity of agricultural grassland. *Agriculture, Ecosystems & Environment*, 134, 190–200.

(27)   Potts S.G., Woodcock B.A., Roberts S.P.M., Tscheulin T., Pilgrim E.S., Brown V.K. & Tallowin J.R. (2009) Enhancing pollinator biodiversity in intensive grasslands. *Journal of Applied Ecology*, 46, 369–379.

(28)   Buckingham D.L., Atkinson P.W., Peel S. & Peach W. (2010) New conservation measures for birds on grassland and livestock farms. *Proceedings of the BOU – Lowland Farmland Birds III: Delivering solutions in an uncertain world.* British Ornithologists' Union. pp. 1–13.

(29)   Defra (2010) *Modified Management of Agricultural Grassland to Promote In-field Structural Heterogeneity, Invertebrates and Bird Populations in Pastoral Landscapes.* Defra BD1454.

(30)   Pywell R.F., Woodcock B.A., Orr R., Tallowin J.R.B., McEwen I., Nowakowski M. & Bullock J.M. (2010) Options for wide scale enhancement of grassland biodiversity under the Entry Level Scheme. *Aspects of Applied Biology*, 100, 125–132.

Additional references

Dolman P.M. & Sutherland W.J. (1992) The ecological changes of Breckland grass heaths and the consequences of management. *Journal of Applied Ecology*, 29, 402–413.

Dodds G.W., Appleby M.J. & Evans A.D. (1995) *A Management Guide to Birds of Lowland Farmland.* Royal Society for the Protection of Birds, Sandy.

Green R. E & Taylor C.R. (1995) Changes in stone-curlew *Burhinus oedicnemus* distribution and abundance and vegetation height on chalk grassland at Porton Down, Wiltshire. *Bird Study*, 42, 177–181.

Shaw G. (1995) Habitat selection by short-eared owls *Asio flammeus* in young coniferous forests. *Bird Study*, 42, 158–164.

Bealey C E., Green R.E., Robson R., Taylor C.R. & Winspear R. (1999) Factors affecting the numbers and breeding success of stone-curlews *Burhinus oedicnemus* at Porton Down, Wiltshire. *Bird Study*, 46, 145–156.

Ni Bhriain B., Skeffington M.S. & Gormally M. (2002). Conservation implications of land use practices on the plant and carabid beetle communities of two turloughs in Co. Galway, Ireland. *Biological Conservation*, 105, 81–92.

Mulder C., De Zwart D., Van Wijnen H. J., Schouten A.J. & Breure A.M. (2003) Observational and simulated evidence of ecological shifts within the soil nematode community of agroecosystems under conventional and organic farming. *Functional Ecology*, 17, 516–525.

Clegg C.D. (2006) Impact of cattle grazing and inorganic fertiliser additions to managed grasslands on the microbial community composition of soils. *Applied Soil Ecology*, 31, 73–82.

Macaulay Institute (2006) *Grazing and Upland Birds*. Available at www.macaulay.ac.uk/projects/project-details.php?302797. Accessed 13 March 2006.

4.9   Leave uncut strips of rye grass on silage fields

- Two reviews and two replicated controlled trials from the UK found that leaving uncut strips of rye grass on silage fields resulted in benefits to birds including increased numbers[2, 4–6]. One of these studies[6] found that whilst seed-eating birds preferred rye grass cut once only, birds that fed on different food resources such as insect-eaters showed more variable results with some preferring plots cut two or more times.

- One replicated, controlled, randomized study from the UK found no difference in ground beetle abundance and diversity between cut and uncut silage field headlands in the first two years of the experiment[1] but higher species diversity in uncut plots in the third year[3].

Background

This intervention involves leaving areas of uncut rye grass in silage fields. In the UK, seed-eating songbirds have declined across farmland, probably in part because of a lack of winter food. Rye grass *Lolium perenne* seeds are a potential food source but cutting rye grass fields multiple times a year for silage removes seed heads before they can ripen and so reduces the food available to birds the following winter. Leaving fields or plots uncut may provide valuable overwinter food for birds and may also provide suitable habitat away from damaging harvesting machinery for other farmland wildlife such as invertebrates.

A replicated, controlled, randomized study in 1996–1997 of a silage field headland in Scotland (1) found that ground beetle (Carabidae) abundance and diversity did not differ significantly in cut and uncut headlands. Total abundance of ground beetles was highest in headlands cut 3 times each year (105–201), followed by those cut annually (43–157), and those uncut (44–132). Specific species differed in their response to cutting regimes. In 1996 species diversity was highest in headlands cut 3 times each year (12), followed by those cut once (10) and uncut (9). In 1997 diversity in headlands cut once (14) was higher than uncut (11) and headlands cut 3 times (10) (results from 1997 are also presented in (3)). Cutting regimes were assigned randomly within blocks to three 10 x 10 m plots. Cuts were in August and May, June and August. Cattle were excluded April–October and plots were intermittently grazed by sheep October–February. No pesticides were used and fertilizers were not used in the headland. Ground beetles were sampled using pitfall traps in late May–mid-July and late

August–early October. Three pitfall traps parallel to the field edge were placed in the centre of each plot. Results from the third year of the experiment are presented in (3).

A 2004 review of experiments on the effects of agri-environment measures on livestock farms in the UK (2) found that leaving perennial rye grass *Lolium perenne* silage uncut was shown to benefit seed-eating birds in winter in one experiment. No reference was given in the review for these results. The birds were only found in any numbers on plots left unmown and were more abundant on plots left ungrazed than plots grazed from September. Yellowhammer *Emberiza citrinella* and reed bunting *E. schoeniclus* reached densities of 132 and 52 birds/ha respectively on unmown, ungrazed plots.

A continuation of the replicated, controlled, randomized study in Scotland in (1), (3) found that ground beetle (Carabidae) species diversity was significantly greater in the third year (1998) in uncut plots than those with one or three annual cuts. Species diversity was reported to be significantly higher in uncut plots (27 species) than cut plots (cut annually: 26 individuals; cut 3 times each year: 23) in 1998. It did not differ in 1997 (uncut: 20; cut: 26–27). The total abundance of ground beetles did not differ between treatments in 1997–1998 (uncut: 559–791 individuals; annual cut: 611–890; cut three times: 927–1,053). Specific species differed in their response to cutting regimes. Plots cut three times tended to have a similar species composition to the main field, whereas the uncut and annually cut plots tended to be more similar to the field boundary.

A replicated, controlled study of four silage fields on separate dairy farms in England (4) found that numbers of yellowhammer *Emberiza citrinella*, reed bunting *Emberiza schoeniclus*, winter wren *Troglodytes troglodytes*, song thrush *Turdus philomelos* and Eurasian skylark *Alauda arvensis* were higher in plots left to set seed compared to mown plots, and in ungrazed seeded plots compared to grazed seeded plots. Significantly higher numbers of yellowhammer were observed in seeded plots (458 birds seen) compared to mown (1 bird) and in ungrazed seeded plots (423) than grazed seeded plots (35). Reed bunting showed a similar response (seeded ungrazed: 160, grazed: 29; mown ungrazed: 3, grazed: 0). As did wren (seeded ungrazed: 22, grazed: 1; mown ungrazed: 2, grazed: 0) and song thrush (seeded ungrazed: 7, grazed: 3; mown ungrazed: 4, grazed: 0). There were more skylark in seeded than mown plots (18 vs 0), and more in grazed (17) than ungrazed seeded plots (1). Two of four plots (0.5 ha) in each field were left uncut when the third silage cut was taken in July–August 2002 so that the grass set seed. One mown and one seeded plot was grazed by cattle until October; cattle were excluded from the other two plots. Numbers and species of birds using each plot were recorded over eight 1 hour periods between November 2002 and February 2003.

A 2010 review of four experiments on the effects of agri-environment measures on livestock farms in the UK (5) found that leaving perennial rye grass *Lolium perenne* silage uncut was shown to benefit seed-eating birds in winter in one experiment. These are further results from a study discussed in (2), with no reference given (Defra project BD1455). Only plots cut once during the previous season produced large seed crops and attracted yellowhammer *Emberiza citrinella* (5.5 birds/visit on average) and reed bunting *E. schoeniclus*, (approximately 2 birds/visit on average) but not finches. Plots cut twice or three times (control) did not attract these birds. Birds were observed over two winters.

A replicated, controlled study on 12 farms in the West Midlands, UK (6) in the winters of 2007–2009 found that seed-eating birds (yellowhammer *Emberiza citrinella* and reed bunting *E. schoeniclus*) preferentially foraged in perennial rye grass *Lolium perenne* fields that were only cut once for silage and ungrazed, compared to twice cut (ungrazed) or control (two or more cuts and grazed) plots. Meadow pipit *Anthus pratensis* (which eat seeds and insects) did not show a preference for perennial rye grass fields under different treatments and showed a weak preference for other rye grasses that were only cut once. Insect-eating winter wren *Troglodytes troglodytes* preferentially foraged in all treatments except controls. Insect-eating European robin *Erithacus rubecula* preferentially foraged on control plots.

(1)   Haysom K.A., McCracken D.I., Foster G.N. & Sotherton N.W. (1999) Grass conservation headlands – adapting an arable technique for the grassland farmer. *Aspects of Applied Biology*, 54, 171–178.
(2)   Buckingham D.L., Atkinson P.W. & Rook A.J. (2004) Testing solutions in grass-dominated landscapes: a review of current research. *Ibis*, 146, S2 163–170.
(3)   Haysom K.A., McCracken D.I., Foster G.N. & Sotherton N.W. (2004) Developing grassland conservation headlands: response of carabid assemblage to different cutting regimes in a silage field edge. *Agriculture, Ecosystems & Environment*, 102, 263–277.
(4)   Buckingham D.L. & Peach W.J. (2006) Leaving final-cut grass silage *in situ* overwinter as a seed resource for declining farmland birds. *Biodiversity and Conservation*, 15, 3827–3845.
(5)   Buckingham D.L., Atkinson P.W., Peel S. & Peach W. (2010) New conservation measures for birds on grassland and livestock farms. *Proceedings of the BOU – Lowland Farmland Birds III: Delivering solutions in an uncertain world*. British Ornithologists' Union. pp. 1–13.
(6)   Defra (2011) *Grass Silage as a New Source of Winter Food for Declining Farmland Birds*. Defra BD1455.

4.10   Plant cereals for whole crop silage

- Two studies (one review, one replicated trial) from the UK investigated the effects of cereal-based whole crop silage. One replicated study found that cereal-based whole crop silage fields were used more by farmland birds[3] and supported a higher abundance of seed-eating songbirds, swallows and martins[1] than other crop types. The same study also found that important bird food plants were more abundant in cereals than other crop types and more invertebrates were found in wheat, barley and grass silage fields compared to maize[1].

- A review found one study in which cereal-based whole crop silage fields were avoided by seed-eating birds during winter but used as much as a control during summer[2].

Background
Cereal-based whole crop silage is an intervention that involves growing crops, not grass, to turn into silage. This may provide seed resources for grain-eating farmland birds throughout the year and provide habitat for invertebrates.

A replicated trial in 2004–2006 in northwest England (1) found that seed-eating songbirds and swallows and martins (Hirundinidae) were more abundant on cereal (wheat and barley) fields planted in livestock areas compared to grass silage and maize fields. For example, in winter 2005–2006, 1,390–1,564 seed-eating birds were recorded on barley stubbles compared to 48 on grass fields and 406 on maize. Large insect-eating birds (thrushes: Turdidae) were far more abundant on grass fields in winter (2,272 birds in total, compared to 28–789 on other field types). Important bird food plants – annual meadow grass *Poa annua*, field pansy *Viola arvensis* and chickweed *Stellaria media* – were more abundant in cereal crops than in maize, and in November and February were more abundant in barley stubbles than replanted wheat, maize or grass fields (only in November for chickweed). Beetles (Coleoptera), flies (Diptera) and bees, wasps, sawflies and ants (Hymenoptera) were more abundant in wheat, barley and grass fields than in maize. Winter wheat and spring barley were sown in 16 trial fields, each on a separate farm in Cheshire, Staffordshire and north Shropshire. Neighbouring maize or short-term grass silage fields were monitored for comparison. Plants, invertebrates and birds were monitored on each field, in summer 2005 and winter 2005–2006.

A review of four experiments on the effects of agri-environment measures on livestock farms in the UK (2) found one study on cereal-based whole crop silage in which winter wheat planted for silage was avoided by seed-eating birds during winter, but used as much as a control spring barley crop during summer. Maize planted for silage was little used by birds in summer or winter. Fields on 16 livestock farms were studied for two growing seasons from 2004 and 2006. Four crop-types were studied on each farm: grass, winter wheat, maize, spring barley. These results are reported in more detail by (3). This study formed part of a Defra-funded project BD1448 (Defra 2007).

An update of (1) included data from winter 2004–2005 (3) and found that cereal-based whole crop silage fields were used significantly more by farmland birds than other crop types. Each farm contained two cereal-based whole crop silage fields (autumn-sown wheat, 5.3 ha and spring-sown barley, 4.4 ha), one maize field (6.1 ha) and one grass field (2.1 ha). During summer, a total of 1,535 seed-eating birds and 1,901 swallows and martins (Hirundinidae) were found on barley cereal-based whole crop silage fields, compared with 847 and 197 for wheat cereal-based whole crop silage fields, 441 and 95 for maize fields, and 41 and 480 for grass fields. Northern lapwing *Vanellus vanellus*, insect-eating species, and crows (Corvidae) did not use cereal-based whole crop silage fields more than other types in summer. In winter, seed-eating species (seed-eating songbirds, Eurasian skylark *Alauda arvensis*, meadow pipit *Anthus pratensis*) used barley stubbles extensively, whilst insect-eating species used other crop stubbles more. The authors argue that cereal-based whole crop silage (with selectively applied herbicide, retention of over-winter stubbles and delayed harvesting) offers a practical conservation measure for seed-eating farmland birds. This study uses data from Defra report number BD1448 (Defra 2007).

(1) Mortimer S.R., Westbury D.B., Dodd S., Brook A.J., Harris S.J., Kessock-Philip R., Chaney K., Lewis P., Buckingham D.L. & Peach W.J. (2007) Cereal-based whole crop silages: potential biodiversity benefits of cereal production in pastoral landscapes. *Aspects of Applied Biology*, 81, 77–86.
(2) Buckingham D.L., Atkinson P.W., Peel S. & Peach W. (2010) New conservation measures for birds on grassland and livestock farms. *Proceedings of the BOU – Lowland Farmland Birds III: Delivering solutions in an uncertain world*. British Ornithologists' Union. pp. 1–13.
(3) Peach W.J., Dodd S., Westbury D.B., Mortimer S.R., Lewis P., Brook A.J., Harris S.J., Kessock-Philip R., Buckingham D.L. & Chaney K. (2011) Cereal-based wholecrop silages: A potential convservation measure for farmland birds in pastoral landscapes. *Biological Conservation*, 144, 836–850.

Additional reference
Defra (2007) *Cereal-based Whole Crop Silages: A potential conservation mechanism for farmland birds in pastoral landscapes*. Defra BD1448.

4.11   Maintain rush pastures

- We have captured no evidence for the effects of maintaining rush pastures on farmland wildlife.

Background
Rush pastures are found in western Europe in lowland areas with poorly draining, typically acidic soils which receive high rainfall. Characteristic vegetation includes rushes such as sharp-flowered rush *Juncus acutiflorus*, as well as purple moor grass *Molinia caerulea*. This habitat is susceptible to agricultural modification and reclamation (UK BAP 2008).

UK BAP (2008) *Purple Moor Grass and Rush Pastures*. UK Biodiversity Action Plan Priority Habitat Descriptions. Biodiversity Reporting and Information Group.

4.12   Maintain traditional water meadows (includes management for breeding and/or wintering waders/waterfowl)

- Four studies from Belgium, Germany, the Netherlands and the UK (including two site comparisons, of which one was also replicated) found that maintaining traditional water meadows resulted in an increased population size or number of territories of northern lapwing, common redshank and black-tailed godwit[2, 4] and increased plant species richness[3, 7]. However one of these studies also found common snipe declined on all sites under management to maintain traditional water meadows[2] and another of the studies found that differences in numbers of birds were present before meadow bird management[4].

- Two studies (a replicated study and a review of European studies) found that managing traditional water meadows by grazing had mixed impacts on wildlife[6] and that the productivity of northern lapwings was too low to sustain populations on three of the four water meadows managed for waders[8]. A randomized, replicated, controlled trial in the Netherlands found that cutting in June maintained relatively stable vegetation[1] and a review found that mowing could be used to maintain water meadows but had variable effects on plant species richness[6].

- One replicated site comparison from the Netherlands found more birds bred on 12.5 ha plots with management for wading birds (in combination with per-clutch payments), however at the field scale there was no difference in bird abundance or species richness between conventionally managed fields and those managed for birds[5].

Background

This intervention may involve using management such as mowing or grazing to maintain plant communities and wildlife typically associated with traditional water meadows. Water meadows are areas of grazing land or hay meadow that have carefully controlled water levels to keep the soil damp. In Europe they provide valuable breeding habitats for wading birds and other biodiversity.

In the Netherlands, the 'meadow bird agreements' agri-environment scheme is designed to protect birds associated with wet meadows (4). Fields managed under 'meadow bird agreements' must not have any agricultural activities from 1 April to June–July, pesticide use is prohibited as is cultivation and re-seeding, the areas must be mown or grazed at least once a year and water levels must be maintained (4).

See also 'Restore or create traditional water meadows' for studies that look at the effects of creating or restoring water meadows.

A randomized, replicated, controlled trial in 1972–1979 on a wet Arrhenatherion elatioris grassland site in the Netherlands (1) found that cutting in June, with or without a second cut in September, maintained a relatively stable vegetation. Cutting once in August had a similar effect, but is expected in the long run to produce denser vegetation with a loss of some species. Cutting in May, May and September, or June increased the number of plant species from 52/plot (at the beginning of the experiment) to 55/plot after 8 years. Never cutting, or cutting every other June, were the only treatments to reduce species richness, to

38 and 49 species/plot, respectively. Fertilizer treatments (NPK: nitrogen, phosphorous, potassium) (50:20:20 kg NPK/ha and 50 kg N/ha, with a June cut) led to dominance by a few species but did not reduce species richness. The Arrhenatherion elatioris grassland included false oat grass *Arrhenatherum elatius*, cock's-foot *Dactylis glomerata* and Queen Anne's lace *Daucus carota*. Treatments were replicated 4 times in 100 m² plots. Plants were surveyed each May in 50 quadrats/plot of 25 cm² for monocotyledons (mainly grasses) and 400 cm² for dicotyledons (broadleaved plants).

A replicated study in 19 lowland wet grassland nature reserves established across England between 1983 and 1999 (2) found that the number of northern lapwing *Vanellus vanellus* and common redshank *Tringa totanus* on 13 nature reserves increased by 300% and 500% respectively in the first 7 years following the initiation of management aimed at wading birds. Numbers then declined but were still higher than before the initiation of management. However, across all reserves, common snipe *Gallinago gallinago* declined, largely due to population collapses on reserves with mineral soils. Management included immediate changes to grazing (reduced during breeding seasons and adjusted to produce a favourable sward) and mowing (delayed until after nesting) and hydrological changes (raising water levels, surface flooding) introduced over two or more years.

A replicated, site comparison study of 16 paired (adjacent) grazed and mown fens in southern Germany (3) found that plant species richness was significantly higher in mown compared to grazed sites, but the percentage of typical fen species or Red Data Book species did not differ. Mown sites had 206 species (50/25 m² plot) compared to 195 (43/25 m² plot) in grazed sites. Numbers of typical fen species (18–19/25 m² plot) and Red Data Book species (6–7/25 m² plot) were similar. Species tended to differ in frequency and abundance in the two treatments. Grazing favoured grasses (35 vs 29%), small plants and meadow plants (55 vs 49%) which were taller. Tree and moss species did not differ (2% and 14–15% respectively). Vegetation height and above-ground biomass did not differ between treatments, whereas species traits tended to differ in their response to management. More indicator species of wet soil conditions and species adapted to flooding were found on pasture. Sites were mown from September or moderately grazed (<0.5 young cattle/ha) from May–October; grazing had been continuous for at least 10 years. Cover of plant species was sampled in 109 vegetation plots (5 x 5 m) with the number of plots/site depending on fen size. Three biomass samples (25 x 25 cm) were also taken from each site.

A site comparison study in 1989, 1992, and 1995 of 34 fields in Zeeland, the Netherlands (4) found no conclusive evidence that meadow bird conservation efforts resulted in higher territory numbers. Although there were significantly more meadow birds and territories of northern lapwing *Vanellus vanellus* and black-tailed godwit *Limosa limosa* on fields managed for meadow bird conservation than on conventionally farmed fields in 1995, these differences were at least partly because those meadows in the bird agreements scheme also had higher groundwater levels. Moreover, population trends between 1989 and 1995 were similar for fields with and without meadow bird agreements, and the observed difference in settlement density in 1995 was also already present in 1989. Seventeen pairs of fields were matched for landscape structure and were surveyed in 1989, 1992 and 1995.

A replicated site comparison study of 42 fields in the Netherlands (5) found that more birds bred on 12.5 ha scheme plots consisting of a mixture of fields with postponed agricultural activities and fields with a per-clutch payment scheme than on conventionally farmed plots. A survey of individual fields found there was no difference in bird abundance and breeding on those fields with postponed agricultural activities only and on conventionally farmed fields. The number of bird species on each type of farmland also did not differ between agri-environment scheme and non-agri-environment scheme plots. The agri-environment scheme, which intended to promote the conservation of Dutch meadow birds, prohibited changes in field drainage, pesticide application (except for patch-wise control of problem weeds) and any agricultural activity between 1 April and early June. Additionally, farmers of surrounding fields were paid for each meadow bird clutch laid on their land (though no agricultural restrictions were in place on these fields). The study surveyed 7 pairs of fields (1 within the agri-environment scheme; 1 conventionally farmed) and the 12.5 ha area surrounding each field, from each of 3 different parts of the Netherlands, 4 times during the breeding season.

A 2006 review (6) describes eight studies on the impact of cattle-grazing or mowing on wet meadows (or sedge or fen meadows) in Europe. Impacts of grazing were mixed. One study found that the abundance of small mammals in wet meadows was increased by cattle grazing at intermediate intensities (around 0.5 cattle/ha) and reduced by much higher or lower stocking densities (Schmidt *et al.* 2005). Another study found that reintroducing cattle in a mosaic landscape with wet sedge meadows and drier grasslands reduced species richness, as the dry sites were overgrazed and the wet sites were avoided and remained ungrazed (Bakker & Grootjans 1991). The review describes six studies that looked at the impact of mowing. Four studies found that hand mowing could be used to maintain sedge/water meadows (or fens) (Hansson & Fogelfors 2000, Øien & Moen 2001, Mitlacher *et al.* 2002, Billeter *et al.* 2003) and another found that re-instating mowing increased the population of the fen-orchid *Nigritella nigra* (Moen & Øien 2002). One study found that mowing resulted in higher species richness than grazing on an abandoned fen meadow (Hald & Vinther 2000) while another found that mowing only increased species richness if the species were part of the standing crop, in the seed bank or could disperse to the site (Billeter *et al.* 2003).

A study of two floodplain meadows in adjacent nature reserves on former natural floodplains of the River Demer, Belgium (7) found that mowing and flooding meadows resulted in increased plant diversity. Mown, frequently flooded plots had higher plant species richness (average 16 species/plot) than non-mown, frequently flooded (10 species) or mown, infrequently flooded (12 species) plots. Mown sites had higher numbers of smaller species; non-mown sites had more tall grasses. Average standing crop (range: 492–654g/m^2/year) was significantly lower in mown sites compared to non-mown sites. Overall there was a significant negative correlation between species richness and standing crop. Data were obtained from two reserves: one frequently flooded (150 ha, flooded at least once a year for 5–50 days) and one (600 ha) in which part is infrequently inundated (about once every 5 years). In each reserve 10 plots (2 x 2 m) were randomly selected in annually June-mown fields, and 5 in non-mown fields. Plant species composition was recorded in each in early July 2005 and standing biomass in mid–end July.

A replicated study in 2010 on 4 areas of wet grassland managed for wildlife in Kent, England (8) found that productivity of northern lapwing *Vanellus vanellus* was too low to sustain populations on 3 of the 4 areas (i.e. below 0.7 chicks/pair/year, which is thought to be the level necessary to maintain populations). The author identifies five management practices thought to be important for lapwing success: grazing regime, water availability, 'micro-topography' (changes in ground level to provide a range of habitats), reduced fertilizer inputs and predator control. At least one of these was rated as 'fair' or 'poor' in all three sites with low productivity.

(1) Oomes M.J.M. & Mooi H. (1981) The effect of cutting and fertilizing on the floristic composition and production of an Arrhenatherion elatioris grassland. *Vegetatio*, 47, 233–239.
(2) Ausden M. & Hirons G.J.M. (2002) Grassland nature reserves for breeding wading birds in England and the implications for the ESA agri-environment scheme. *Biological Conservation*, 106, 279–291.

(3) Stammel B., Kiehl K. & Pfadenhauer J. (2003) Alternative management on fens, response of vegetation to grazing and mowing. *Applied Vegetation Science*, 6, 245–254.
(4) Kleijn D. & van Zuijlen G.J.C. (2004) The conservation effects of meadow bird agreements on farmland in Zeeland, The Netherlands, in the period 1989–1995. *Biological Conservation*, 117, 443–451.
(5) Kleijn D., Baquero R.A., Clough Y., Díaz M., Esteban J.d., Fernández F., Gabriel D., Herzog F., Holzschuh A., Jöhl R., Knop E., Kruess A., Marshall E.J.P., Steffan-Dewenter I., Tscharntke T., Verhulst J., West T.M. & Yela J.L. (2006) Mixed biodiversity benefits of agri-environment schemes in five European countries. *Ecology Letters*, 9, 243–254.
(6) Middleton B.A., Holsten B. & van Diggelen R. (2006) Biodiversity management of fens and fen meadows by grazing, cutting and burning *Applied Vegetation Science*, 9, 307–316.
(7) Gerard M., El Kahloun M., Rymen J., Beauchard O. & Meire P. (2008) Importance of mowing and flood frequency in promoting species richness in restored floodplains. *Journal of Applied Ecology*, 45, 1780–1789.
(8) Merricks P. (2010) Lapwings, farming and environmental stewardship. *British Wildlife*, 22, 10–13.

Additional references
Bakker J.P. & Grootjans A.P. (1991) Potential for vegetation regeneration in the middle course of the Drentsche Aa brook valley (The Netherlands). *Verhandlungen der Gesellschaft für Ökologie*, 20, 249–263.
Hald A.B. & Vinther E. (2000) Restoration of a species-rich fen-meadow after abandonment: response of 64 species to management. *Applied Vegetation Science*, 3, 15–25.
Hansson M. & Fogelfors H. (2000) Management of a seminatural grassland: results from a 15-year-old experiment in southern Sweden. *Journal of Vegetation Science*, 11, 31–38.
Øien D.-I. & Moen A. (2001) Nutrient limitation in boreal plant communities and species influenced by scything. *Applied Vegetation Science*, 4, 197–207.
Mitlacher K., Poschlod P., Rosén E. & Bakker J.P. (2002) Restoration of wooded meadows: a comparative analysis along a chronosequence on Öland (Sweden). *Applied Vegetation Science*, 5, 63–73.
Moen A. & Øien D.-I. (2002) Ecology and survival of *Nigritella nigra*, a threatened orchid species in Scandinavia. *Nordic Journal of Botany*, 22, 435–461.
Billeter R.D., Hooftman A.P. & Diemer M. (2003) Differential and reversible responses of common fen meadow species to abandonment. *Applied Vegetation Science*, 6, 3–12.
Schmidt N.M., Olsen H., Bildsøe M., Sluydts V. & Leirs H. (2005) Effects of grazing intensity on small mammal population ecology in wet meadows. *Basic Applied Ecology*, 6, 57–66.

4.13   Restore or create traditional water meadows

• Of three studies from Sweden and the UK (two before-and-after trials) looking at bird numbers or densities following water meadow restoration, one study found increases[15], one study found increases and decreases[8] and one found northern lapwing populations did not increase despite an increase in the area of managed water meadows[12].

• Seventeen studies from France, Germany, the Netherlands, Poland, Switzerland and the UK (seven replicated controlled studies, of which two were also randomized and two reviews) found one or more management techniques that were successful in restoring wet meadow plant communities[1-3, 5-7, 9-11, 13, 15, 16, 18-22]. The techniques were topsoil removal[1, 5, 11], introduction of target plant species[3, 20, 21], raising water levels[6, 15], grazing[7, 15], mowing[2, 16, 19] or a combination of removing topsoil and introducing target plant species[9, 10, 13, 18], plus livestock exclusion[22]. Three studies (one replicated controlled study and two reviews) from the Netherlands, Sweden, Germany and the UK found restoration of wet meadow plant communities had reduced or limited success[4, 14, 17].

• Thirteen studies (five replicated and controlled, of which two were randomized) from France, Germany, the Netherlands, Switzerland and the UK monitored the effects of methods to restore or create wet meadow plant communities over a relatively short time period after restoration and found some positive effects within five years[1-3, 5-7, 9-11, 13, 15, 16, 18, 20]. Three replicated studies (one controlled, one a site comparison) from the Netherlands and Germany found restoration was not complete 5, 9 or 20 years later[11, 19, 20].

A replicated controlled site comparison study from Sweden found plant species richness increased with time since restoration[14].

Background

This intervention involves restoring or creating traditional water meadows, also known as wet meadows. Water meadows are areas of grazing land or hay meadow that have carefully controlled water levels to keep the soil damp. In Europe they provide valuable breeding habitats for wading birds and other biodiversity.

Techniques used to restore or create plant communities associated with wet meadows may include topsoil removal, which rapidly reduces nutrient levels in nutrient-enriched soils and increases flooding frequency by lowering the ground surface (Hölzel & Otte 2003). In areas where target plant species are not present in the seed bank or plant species have limited dispersal, it may also be important to introduce plant species through sowing, planting or spreading hay from existing wet meadows (Hölzel & Otte 2003).

See also 'Natural system modification: Raise water levels in ditches or grassland' for studies looking at the effects of raising water levels on grassland.

Hölzel N. & Otte A. (2003) Restoration of a species-rich flood meadow by topsoil removal and diaspore transfer with plant material. *Applied Vegetation Science*, 6, 131–140.

A trial from 1985 to 1990 on what was once a species-rich wet meadow at the Veenkampen, near Wageningen, the Netherlands (1) found that topsoil removal increased the number of plant species on plots managed as hay meadows and allowed rare sedge (Cyperaceae) species to establish. Three 15 x 25 m plots had topsoil removed to 5 cm depth. Each had a different water level, so they were not replicated. The number of plant species in these plots remained relatively stable or increased between 1987 and 1990 regardless of water level (dry plots increased from 21 to 28 species). The number of plant species was greater than in plots without soil removal. From 1988, carnation sedge *Carex panicea* and three other sedge species usually restricted to nature reserves in the region were found. Plots that were mown once or twice each year without soil removal (five replicates of four different mowing and hay removal treatments on 10 x 15 m plots) lost plant species. They had 18–20 species in 1987 and 14–17 in 1990. No fertilizer was applied during the experiment.

A replicated, controlled, randomized study of two wet meadows over four years in Switzerland (2) found that overall winter and summer (August) cuts had positive effects on plant species densities in terms of individuals, leaves, shoots and flowers. However individual species were affected differently by cutting regime. For example, an annual winter cut caused an increase in the number of flowers for common reed *Phragmites communis* (now *P. australis*), whilst a summer cut reduced them. Some drier vegetation communities were damaged when cut in summer, whereas wetter communities were more resilient to summer mowing. In general, annual winter cuts tended to improve the vitality of vascular plants (in terms of increased number of individuals, flowering and biomass). Plant vitality was lowest in uncut plots and intermediate with an annual summer cut and winter cut every three years. Vegetation structure differed with treatments (hay removed) and uncut controls. The meadows had been abandoned for many years and treatments were applied in three blocks with three replicates. Vegetation was sampled in July–August within 11 x 11 m and 13 x 13 m permanent plots from 1983 to 1986.

A before-and-after study over one year on agricultural land that had been intensively managed in Oxfordshire, UK (3) found that sowing seeds from a nearby species-rich flood meadow aided restoration of the flood meadow. The existing seed bank contained 38 species (66% perennial and 34% annual species), of which only 55% were grassland species, including 9 species of wet grassland. Following seed sowing in 1986, 43 species were recorded at the site, of which 61% were perennial and 39% annual species. Of the 53 species that did not germinate, 77% were grassland species, of which 6 were wet grassland species. To determine the existing seed bank, 12 stratified random soil samples (5,000 cm^3) were taken from the top 10 cm of recently ploughed soil in April 1986. Twenty-five subsamples (200 cm^3) were taken from each and seeds were germinated and identified. Following harvest, seeds that had been harvested from a nearby species-rich flood meadow (Oxey Mead SSSI) in July were sown in October 1986. Plant species presence/absence was recorded in twelve 5 x 5 m quadrats in June 1987.

A 1994 review of methods to restore grasslands in the Netherlands (4) reports one experiment in which the plant community on a wet peaty grassland changed away from the desired plant community over 14 years, after cessation of fertilizer input and the introduction of grazing. Characteristic species, such as marsh marigold *Caltha palustris*, were replaced by tall plants, such as lesser pond sedge *Carex acutiformis* and reed sweetgrass *Glyceria maxima*. The restoration was considered unsuccessful. The authors argued that continued agricultural drainage on surrounding areas was responsible for the failure of the restoration, because the water table was not high enough to restore the plant community. Fertilizer applications were stopped in 1971. Plants were monitored from 1978 to 1992. Site location and details were not given.

A study of a degraded wet meadow in eastern Netherlands (5) found that a wet meadow plant community (*Cirsio-Molinietum* community) established within 5 years after topsoil removal (5–15 cm topsoil layer removed) on former agricultural land and alder carr. Almost all of the highly productive plant species had disappeared in that time. In contrast, species of the same wet meadow plant community remained rare or absent in the adjacent nutrient-rich (eutrophicated) wet meadow, in areas with and without topsoil removal. This may have been due to prolonged inundation resulting from the topsoil removal. Topsoil removal was undertaken in three areas of Lemselermaten nature reserve: a section of the nutrient-rich wet meadow (1991; 'old reserve'), the adjacent former agricultural grassland and alder carr (1989; 'new reserve'). Vegetation was then mown annually in these areas and in the remainder of the nutrient-rich wet meadow. Three transects were established, one in the new reserve and two in the old reserve (with and without sod cutting). Vegetation cover and abundance was surveyed in several plots (4 m^2) along each transect annually (July–August 1992–1994).

A controlled, randomized study of a former improved pasture in the Netherlands (6) found that raising the water level resulted in a more rapid establishment of species typical of wet conditions, than vegetation management (cutting and removing hay; cutting, mulching and leaving hay; topsoil removal to 5 cm followed by cutting and removing hay). Hay removal plots had more new species than the mulched plots in the wet field (7 vs 2 species; 3 in the dry field). Two years after topsoil removal, there were 37 species established in the wet field and 49 in the dry field; 5 years later there were 13 new species on wet fields and 22 on dry fields. In 1985, the water level was raised to its former level in one area (1.5–2 ha); the other area was left dry. The 3 management practices were implemented in each area: sod cutting (to remove topsoil) in one plot (375 m^2) and hay removal and mulching each in 5 replicate plots (100 m^2). Plant species composition was recorded annually (20–50 samples/plot).

A 1998 review of case studies in France, gathered from published and unpublished literature (7) reported one French study showing an increase in plant species richness (from 5 to

10–25 species over 4 years) on a wet grassland in Brittany following introduction of grazing by Camargue horses (Rozé 1993).

A replicated, before-and-after study of 15 restored wet meadows in agricultural land-scapes in southern Sweden (8) found that the density of 7 bird species increased and 2 decreased significantly following restoration. Three species showed a non-significant tend-ency to decrease and 11 showed no significant difference following restoration. A population was more likely to increase if it was present at the site pre-restoration. No single management regime (mowing; grazing; mowing/grazing; unmanaged) was favoured by a large number of species (but by just 2–4 species). Seven species were associated with site/meadow size (five negatively) and seven with surrounding habitat. Breeding bird survey data was obtained (fol-lowing requests to relevant groups) using three survey methods: territory mapping, counts of duck broods and at two sites thorough field counts. The majority of meadows were located along lakeshores or rivers.

A controlled study in 1992–1997 of wet fen meadows in southern Germany (9) found that topsoil removal and the introduction of target species aided meadow restoration. The removal of the nutrient-rich topsoil (to depths of 20 cm, 40 cm or 60 cm) and introduction of target species in hay cut from 4 fen meadows (layer 5–10 cm thick) resulted in success-ful establishment of 57 fen meadow plant species over 6 years, including 13 regional Red List species. The total cover of hay species from the donor areas reached up to 70% on plots where 20 cm of topsoil was removed, 30% when 40 cm was removed and 5% on the 60 cm removal plots. Plots without hay were established for each level of topsoil removal as con-trols for comparison. Monitoring of vegetation was carried out several times each year on permanent 4 m^2 plots.

A replicated, controlled, randomized study of a species-poor agriculturally improved pasture in the UK (10) found that topsoil removal and planting of seedlings, rather than seeds, resulted in establishment of species typical of a fen meadow plant community (*Cirsio-Molini-etum*: purple moor grass *Molinia caerulea*, meadow thistle *Cirsium dissectum* community) over four years. When seedlings were planted, combined cover by *Cirsio-Molinietum* species was highest in treatments with topsoil removal (up to 75% in year 4). Where topsoil was not re-moved, vegetation was dominated by a few competitive species such as common knapweed *Centaurea nigra* (up to 60% cover). Two years after sowing seeds from a *Cirsio-Molinietum* meadow, only 3 of the 17 species had established at more than trace amounts (combined cover of 8%). Treatments to reduce site fertility included cutting and removal of vegetation, cultiv-ation, fallowing and topsoil removal (10–20 cm) and addition of straw and/or lignitic clay. Randomized block experiments were established with treatments applied to plots of 9 x 2 m where seeds were sown (1989–1992) and 2 x 2 m where seedlings of 14 species were planted (1994–1999). Plant composition of plots was sampled in June 1992 and 1997–1999.

A replicated site comparison study of former arable and pasture fields in the Netherlands (11) found that topsoil removal aided wet meadow restoration. Topsoil removal resulted in increasing similarity (up to 29%) to five target communities: small sedge *Caricion nigrae* community, *Ericion tetralicis* heathland and three nutrient-poor grassland communities (*Junco-Molinion, Nardo-Galion saxatilis* and *Thero-Airio*). Nutrient poor fen communities and a heathland community (*Calluno-Genistion pilosae*) did not establish. Target species increased steadily over time, but 50–100% were still missing from target communities after 9 years. Environmental conditions were suitable or within the range for establishment for all com-munities, apart from two grassland communities: *Thero-Airion* and *Junco-Molinion* grasslands (only one site was suitable for each). Local species pools were good for all but nutrient-poor fen communities. Topsoil was removed (to depths up to 50 cm in 1989–1995) from 8 sites. Vegetation was monitored (July–August) annually at 3–12 plots (2 x 2 m) at each site from 1993 to 1995 until 1998. Vegetation and environmental conditions were compared to five reference plots for each community.

A before-and-after study in 1984–1994 in Västmanland, Sweden (12) found no increase in northern lapwing *Vanellus vanellus* population in the study area despite an increase in the area of managed wet meadows from 163 ha to 530 ha over the study period (approximately 220 pairs in 1985 vs 200 in 1994; range of 152–297 pairs). Both managed and unmanaged meadows were used less for nesting than expected based on their availability. However, average hatching success was significantly higher in meadows (78–90% for 54 nests in meadows) compared to spring-sown crops (29–50% of 1,236 nests). There were no differences between meadows and autumn-sown crops or cultivated grassland (approximately 85% and 75% success respectively). Before 1984, the majority of meadows in the area were overgrown and abandoned. The meadows and arable fields became flooded from mid-April to mid-May in years with high spring flooding.

A replicated, controlled study of a flood meadow (a former arable field) in Germany between 1998–2001 (13) found that the removal of nutrient-rich topsoil and introduction of meadow seeds aided meadow restoration. Topsoil removal (to depths of 30 and 50 cm) and introduction of plant material from nearby species-rich flood meadows (alluvial *Molinion* and *Cnidion* meadows) resulted in a decline of arable weeds and ruderal species and an increase in resident grassland species and transferred species. After 4 years, 64% of all species found in established vegetation were from transferred plant material and 82% of the entire species pool at donor sites was transferred (including 31 species in the national and regional Red Data Book). Transfer rates ranged from 64 to 72%/strip for the flooded strips and 53 to 56% for the dry strips. Following soil removal in 1997, 6 strips (20 x 50 m) were covered with freshly mown plant material (5–10 cm thick) from nearby flood-meadows and 2 were left as controls. Plants were recorded annually in ten 10 x 10 m quadrats/strip and in 6 quadrats/ donor meadow. Twenty soil cores (10 x 3 cm diameter) were taken from 6 plots with and 2 without soil removal and germinated seedlings were identified. Two samples (each 6 quadrats: 32 x 32 cm) of plant material (at the surface and 2 cm of the topsoil) were taken from 4 strips (February 1998 and 1999) to analyse transferred seeds.

A replicated, controlled, site comparison study of 26 restored semi-natural grasslands in southeastern Sweden (14) found that continuously grazed control sites had higher plant species diversity and a higher proportion of typical grassland species in the community than restored grasslands. Plant species diversity at restored sites was 16–20 species/m^2 compared to 24–30 species/m^2 at continuously grazed control sites. Total species richness was positively associated with time since restoration (1–7 years) and the abundance of trees and shrubs. Overall species composition differed between restored and control sites, with control sites having a higher proportion of typical grassland species than restored sites. However within grassland types (dry, dry to damp (mesic) or damp to wet) species composition was similar between each pair of restored and control sites. Restored damp to wet grassland was dissimilar in species composition to all other plots. Abundance of 10 grazing-indicator species tended to be lower at restored sites. Restored site area (3–35 ha), time between abandonment and restoration, time since restoration and abundance of trees and shrubs were not related to species composition among restored sites or the 10 grazing-indicator species. Restored sites were grazed before abandonment and after restoration, control sites had been grazed continuously. The six control sites were compared to restored sites in the same region. Plants were sampled within 10 randomly distributed plots (1 m^2) in July–August 2001. Trees and shrubs were counted within a 40 m diameter circle at each site.

A study in 84 ha of arable land adjoining Berney Marshes RSPB Reserve, Norfolk, England describes their restoration to grazing marsh (15). The fields were acquired in 1998, water levels were raised, foot drains were added, and grazing by sheep (and then cattle) was introduced. By 2003 plant communities had shifted towards those characteristic of lowland wet grassland. Breeding wading bird numbers increased, with 15–20 pairs of northern lapwing *Vanellus vanellus* and 5–10 pairs of common redshank *Tringa totanus* (depending on year). The fields are regularly used for foraging by a large proportion of the estimated 100,000 wintering waterfowl (e.g. Eurasian wigeon *Anas penelope*) now using the reserve.

A replicated, controlled study of 15 abandoned fen meadows in Switzerland (16) found that mowing resulted in an increase in plant biodiversity. Mowing resulted in an increase in plant species richness (control: 28 species/2 m^2; mown: 33), number of indicator species (control: 16; mown: 18), broadleaved plants (control: 18/plot; mown: 21/plot), woody species (control: 1/plot; mown: 2/plot), mosses/liverworts (bryophyte) biomass (control: 55 g/m^2; mown: 85 g/m^2), seedling density of Davall's sedge *Carex davalliana* (control: 0.40/m^2; mown 1.49/m^2) and Devil's bit scabious *Succisa pratensis* (control: 1.04/m^2; mown: 0.53/m^2) and a decrease in total biomass (control: 225 g/m^2; mown: 193 g/m^2). Two indicator species increased substantially with mowing: bog-star *Parnassia palustris* (+10 plots) and heath spotted-orchid *Dactylorhiza maculata/majalis* (+13 plots). Two control and two mown (mid-September) plots (2 m^2) were randomly established in each meadow (4–35 years since abandonment). Plant species richness (1998) and cover (2000) were recorded in late July–early August. Plant biomass (dry weight) was sampled in sub-plots (20 x 20 cm) in early August 2000 and separated into vascular plant, litter and moss/liverwort biomass. Life history traits were investigated for two abundant plant species: Davall's sedge and Devil's bit scabious (3/plot) and seedlings of the species were counted in 5 sub-plots/plot (10 x 10 cm) in May–June 2000.

A 2007 review of data from 36 wet meadows in the Netherlands (majority), Germany and the UK (17) found that restoration attempts have largely had limited success. On average, projects have resulted in an average increase in species richness below 10% of the target community. The more species-rich the meadow was at the start, the closer it was to the target community after restoration, however, there was a corresponding smaller increase in the number of target species. A combination of topsoil removal (deeper than 20 cm) and introducing seedlings (e.g. see (9, 11, 13)) and a combination of these with rewetting, appeared most effective (an increase in 'saturation index' of up to 16%; this index reflects the completeness of restored communities in comparison to target communities). Rewetting alone appeared an ineffective restoration method. Data were obtained from professional networks, experts, peer-reviewed published sources and project records.

A replicated, controlled study in a grazed fen area in northern Germany (18) found that wet meadow species increased and agricultural grassland species decreased following topsoil removal and hay transfer. Target species reached their maximum in the second year of the experiment where topsoil had been removed and hay transferred. Where topsoil had been removed but no hay introduced the species increased slowly over four years. Most species transferred with the hay were only present in areas with removed topsoil, not on intact soil. Grazing had minimal effects but did result in a significant increase of cumulative frequency of wet meadow species. Four blocks (12 x 24 m) were established that each combined three treatments: moderate grazing (yes/no), topsoil removal (yes/no, to a depth of 30 cm) and hay transfer from a species-rich fen meadow (yes/no, layer of 1–3 cm). Plant cover and species dominance was sampled in 16 permanent squares (1 m^2) within each sub-plot in each combination of treatments in 2002–2005. Ten soil seed bank samples were taken from each plot in 2002.

A long-term replicated trial in 1987–2007 on seven semi-natural wet meadow sites in Münsterland, Germany (19) found that mowing twice a year (in June/July and September) without fertilizer was the most effective regime for restoring target wet meadow plant communities and resulted in highest species richness. However, successional changes were still happening 20 years after the start of the trial, probably due to slow immigration of new species. Management regime had a stronger effect on the pattern of succession than other environmental or historic factors. Treatments were carried out from 1987 to 2007 in 200–250 m² fields and plants were surveyed in four 2 x 2 m plots/field at least every second year. Above ground biomass was measured in 1989, 1993, 1998 and 2007 by harvesting eight 0.5 x 0.5 m plots/field.

A replicated, controlled trial near Riedstadt, southwest Germany (20), found that five or six years after restoration of flood meadows by hay spreading the seed bank was still dominated by weedy species of the former arable land use. However the seed bank in control areas without hay spreading had significantly lower numbers of transferred flood meadow species. Five meadows were restored in 2000 and 2001. Plants present above ground and in the seed bank were sampled in 2006, in both restored areas and control areas that had been left to naturally regenerate.

A 2010 review of studies of scientific knowledge about how to re-establish plant communities in grasslands by reintroduction (21) found that hay transfer has been shown to be an effective method in wet meadows. The review found 38 studies, 28 of which provided enough information to evaluate the outcome, 21 of these from European countries, 6 on wet meadows or fens. Studies were graded as: successful, of limited success, failed introductions, or without the necessary information to evaluate the outcome. All four studies on hay spreading in wet meadows or fens were successful ((9, 13), Rasran et al. 2007, (22)). Plug planting had limited success in one UK study (10). Only one study looked at the effects of direct seeding on wet meadows and found the technique was not successful (10).

A replicated, controlled study in 2004–2007 of a degraded species-poor meadow in central Poland (22) found that deep topsoil removal (40 cm), hay transfer from a species-rich meadow and exclusion of livestock resulted in a community most similar to the target vegetation. Shallow soil removal (20 cm) with hay transfer resulted in a community more similar to the degraded meadows. Hay transfer appeared to speed up the establishment of the target vegetation. Two plots (35 x 35 m) were subdivided to test combinations of the following treatments: topsoil removal (to 20 or 40 cm), hay transfer from a nearby meadow (collected mid-July 2004–2005; partly dried; stored for 1.5 months; spread in 5–7 cm layer) and livestock exclusion. Data were obtained from plots on plant species distribution and abundance (2004–2007) and biomass (2006–2007); species composition of degraded meadows and donor meadow were also collected (2004, 2006, 2007). The soil seed bank (top 5 cm) at the 2 topsoil removal depths and seed content of hay were also sampled in 2004.

(1) Berendse F., Oomes M.J.M., Altena H.J. & Elberse W.T. (1992) Experiments on the restoration of species-rich meadows in the Netherlands. *Biological Conservation*, 62, 59–65.
(2) Buttler A. (1992) Permanent plot research in wet meadows and cutting experiment. *Vegetatio*, 103, 113–124.
(3) McDonald A.W. (1993) The role of seed-bank and sown seeds in the restoration of an English flood-meadow. *Journal of Vegetation Science*, 4, 395–400.
(4) Bakker J.P. (1994) Nature management in Dutch grasslands. *Proceedings of the British Grassland Society Occassional Symposium*. 28, pp 115–124.
(5) Jansen A.J.M. & Roelofs J.G.M. (1996) The restoration of *Cirsio-Molinietum* wet meadows by sod cutting. *Ecological Engineering*, 7, 279–298.
(6) Oomes M.J.M., Olff H. & Altena H.J. (1996) Effects of vegetation management and raising the water table on nutrient dynamics and vegetation change in a wet grassland. *Journal of Applied Ecology*, 33, 576–588.

(7)   Muller S., Dutoit T., Allard D. & Grevilliot F. (1998) Restoration and rehabilitation of species-rich grassland ecosystems in France: a review. *Restoration Ecology*, 6, 94–101.
(8)   Hellström M. & Berg Å. (2001) Effects of restoration and management regime on the avifaunal composition on Swedish wet meadows. *Ornis Svecica*, 11, 235–252.
(9)   Patzelt A., Wild U. & Pfadenhauer J. (2001) Restoration of wet fen meadows by topsoil removal: vegetation development and germination biology of fen species. *Restoration Ecology*, 9, 127–136.
(10)   Tallowin J.R.B. & Smith R.E.N. (2001) Restoration of a *Cirsio-Molinietum* fen meadow on an agriculturally improved pasture. *Restoration Ecology*, 9, 167–178.
(11)   Verhagen R., Klooker J., Bakker J.P. & van Diggelen R. (2001) Restoration success of low-production plant communities on former agricultural soils after topsoil removal. *Applied Vegetation Science*, 4, 75–82.
(12)   Berg Å., Jonsson M., Lindberg T. & Kallebrink K.-G. (2002) Population dynamics and reproduction of northern lapwings *Vanellus vanellus* in a meadow restoration area in central Sweden. *Ibis*, 144, E131–E140.
(13)   Hölzel N. & Otte A. (2003) Restoration of a species-rich flood meadow by topsoil removal and diaspore transfer with plant material. *Applied Vegetation Science*, 6, 131–140.
(14)   Lindborg R. & Eriksson O. (2004) Effects of restoration on plant species richness and composition in Scandinavian semi-natural grasslands. *Restoration Ecology*, 12, 318–326.
(15)   Lyons G. & Ausden M. (2005) Raising water levels to revert arable land to grazing marsh at Berney Marshes RSPB Reserve, Norfolk, England. *Conservation Evidence*, 2, 47–49.
(16)   Billeter R., Peintinger M. & Diemer M. (2007) Restoration of montane fen meadows by mowing remains possible after 4–35 years of abandonment. *Botanica Helvetica*, 117, 1–13.
(17)   Klimkowska A., Van Diggelen R., Bakker J.P. & Grootjans A.P. (2007) A review of the success of commonly used wet meadow restoration methods (rewetting, topsoil removal, and diaspore addition) in Western Europe. *Biological Conservation*, 140, 318–328.
(18)   Rasran L., Vogt K. & Jensen K. (2007) Effects of topsoil removal, seed transfer with plant material and moderate grazing on restoration of riparian fen grasslands. *Applied Vegetation Science*, 10, 451–460.
(19)   Poptcheva K., Schwartze P., Vogel A., Kleinebecker T. & Hölzel N. (2009) Changes in wet meadow vegetation after 20 years of different management in a field experiment (North-West Germany). *Agriculture, Ecosystems & Environment*, 134, 108–114.
(20)   Schmiede R., Donath T.W. & Otte A. (2009) Seed bank development after the restoration of alluvial grassland via transfer of seed-containing plant material. *Biological Conservation*, 142, 404–413.
(21)   Hedberg P. & Kotowski W. (2010) New nature by sowing? The current state of species introduction in grassland restoration, and the road ahead. *Journal for Nature Conservation*, 18, 304–308.
(22)   Klimkowska A., Kotowski W., Van Diggelen R., Grootjans A.P., Dzierża P. & Brzezińska K. (2010) Vegetation re-development after fen meadow restoration by topsoil removal and hay transfer. *Restoration Ecology*, 18, 924–933.

Additional references
Rozé F. (1993) Successions végétales après pâturage extensif par des chevaux dans une roselière [Successional patterns after extensive grazing by horses in a *Phragmites australis* plant community]. *Bulletin d'Ecologie*, 24, 203–209.
Rasran L., Vogt K. & Jensen K. (2007) Effects of topsoil removal, seed transfer with plant material and moderate grazing on restoration of riparian fen grasslands. *Applied Vegetation Science*, 10, 451–460.

4.14   Maintain upland heath/moorland

- Of 15 individual studies from the UK, 8 (including 3 replicated, controlled trials, of which 1 was also randomized) found that appropriate management can help to maintain the conservation value of upland heath or moorland.

- Of these eight studies, four tested the effectiveness of excluding or reducing grazing. Impacts included increases in the abundance of Scottish primrose and other broadleaved plant species[6], heather cover and numbers of true bugs[8], biomass of arthropods associated with the bird diet[14], number and diversity of moths[15] and benefits to black grouse[7]. Among other treatments, repeated cutting[3] and grazing by goats[4] were found to be effective in controlling the dominance of certain grass species. A review found management under the Environmentally Sensitive Areas scheme had broadly positive effects on moorland birds and that a reduction in grazing benefited most bird species and increased heath vegetation and heather cover[16]. A replicated before-and-after study found that moorland management under the Environmentally Sensitive Area scheme maintained the number of plant species in two out of three areas[11].

- Three studies (including one before-and-after trial) reported mixed results for invertebrates or birds, where management to maintain upland heath or moorland benefited some but not all species[10, 13, 17] or where the effect depended on the vegetation type[2, 17]. Treatments tested included reducing grazing intensity and grouse moor management (burning and predator control).

- Four studies (including one controlled site comparison and two reviews) found that reducing the intensity of livestock grazing reduced the abundance of soil organisms including invertebrates, bacteria or fungi[1, 5, 10, 12]. A randomized, replicated before-and-after study found that heather cover declined over nine years on a moorland site managed under the Environmentally Sensitive Areas scheme in which grazing intensity had increased[9].

Background

This intervention involves using management techniques to maintain the conservation value of upland heath or moorland, including unenclosed rough grazing. These semi-natural habitats are predominantly found within unenclosed, extensive landscapes; their 'open' nature typically maintained by practices such as grazing, cutting and burning. Treatments tested include removing or reducing grazing, controlling the dominance of grass species by cutting or grazing by goats, and carrying out grouse moor management (rotational burning combined with predator control).

See also 'Natural system modification: Manage heather by swiping to simulate burning' and 'Natural system modification: Manage heather, gorse or grass by burning' for studies that investigated specific techniques for managing heather.

An unreplicated, controlled site comparison study in 1987–1990 at two upland grassland sites in Cumbria, England (1) found that the number of soil-dwelling invertebrates and quantity of fungal mycelium was greater in plots with more intensive sheep grazing. Three adjacent upland grassland plots and an ungrazed control were compared between November 1987 and April 1990: heavily grazed (5–8 ewes/ha, limed and fertilized), moderately grazed (3–5 ewes/ha, limed) and lightly grazed (1 ewe/ha). The average number of springtails (Collembola) in the surface soil was significantly higher in the heavily grazed plot ($44 \times 10^3/m^2$) than in the lightly grazed plot ($20 \times 10^3/m^2$), as were number of earthworms (Lumbricidae, $18 \times 10^3/m^2$ compared to $8 \times 10^3/m^2$) and crane fly larvae (Tipulidae, $38 \times 10^3/m^2$ compared to $1 \times 10^3/m^2$). The quantity of fungal mycelium in the surface soil was also greater in the heavily grazed plot. The trial did not separate grazing impacts from liming and fertilizer use, which may also have influenced the soil biota.

An unreplicated, controlled trial from 1986 to 1990 at a site in County Antrim, Northern Ireland (2) found that the impact of grazing on upland ground-dwelling spider (Araneae) communities varied between vegetation types. The study compared the impact of no grazing, grazing by rabbits/hares (Lagomorpha) and grazing by all herbivores (up to three sheep/ha) on different vegetation types: grass heath, upland grass, wet heath, heather moorland and re-seeded pasture. On grass heath, most spiders were found on ungrazed plots (91) while on wet heath most spiders were found on heavily grazed plots (137). Highly mobile and 'pioneer' spider species were most abundant in heavily grazed plots. Litter-dwelling spider species were relatively rare on heavily grazed grassland plots and one litter dwelling species was significantly more abundant on ungrazed or lightly grazed plots. Plants were surveyed in July 1989 and July 1990, and spiders were sampled using pitfall traps emptied at 2–4 week intervals between October 1988 and September 1990.

A replicated, controlled study of two hill pastures in Scotland (3) found that repeated cutting within a growing season over three successive years resulted in decreased purple moor grass *Molinia caerulea* leaf production, tussock size and productivity. Frequency and severity of defoliation were more important than timing in their effects. Weights of purple moor grass clippings from cut tussocks declined each year in treatments that involved repeated within-season cutting. Three years of repeated light cutting (33% leaf blade length removed each June, July and August), compared with uncut controls, reduced leaf production (numbers and size) in the fourth uncut growing season by 40%. Repeated heavy cutting (66% leaf blade removal) reduced production by 78%. Single annual cuts only reduced leaf production at 66% leaf blade removal when the cut took place late in the season (rather than in June or July). There were 4–6 replicates of each treatment at each site. Clipped vegetation at each cut was dried and weighed. Harvested tussocks were dissected to determine numbers of live vegetative, flowering and dead tillers.

A small study of acid grassland with 10–15% matgrass *Nardus stricta* cover over 4.5 years in Scotland (4) found an increase in use of matgrass by goats as the height of preferred between-tussock grasses decreased. Use declined over successive seasons under sheep grazing but was sustained by goat grazing. Matgrass growth rates were reduced as grazing severity increased, being lowest on the plot grazed to 4–5 cm by goats and highest on the 6–7 cm goat plot. Goats grazed more matgrass at 4.5 and 5.5 cm than the sheep did at 4.5 cm. The 4 treatments (each 0.15 ha) were between-tussock grasses maintained at 4–5 cm, 5–6 cm and 6–7 cm by goats or a sheep control at 4–5 cm. Matgrass utilization was estimated by the proportion of grazed stems (tillers) and leaves and grazing severity sampled by measuring 20–40 random leaf lengths of grazed leaves in July and October 1984–1987. Leaf growth was sampled on 30 stems/plot at weekly intervals from April–September 1988. Vegetation composition was recorded in May 1989 using an inclined point quadrat.

A 1998 review of how soil animals change according to management of agricultural grasslands (5) found one UK study that assessed the effects of lower grazing pressure on soil animals in upland grasslands in Cumbria (1). It found numbers of springtails (Collembola) and mites (Acari) were lower with less sheep grazing. For example, there were over 60,000 springtails/m^2 on heavily grazed land, compared to around 40,000/m^2 on moderately or lightly grazed grassland and 20–30,000/m^2 on ungrazed land. Removing sheep for two years rapidly reduced numbers of springtails.

A trial from 1987 to 1998 at a grassland/heath study site on Hoy, Orkney Islands, Scotland (6) found that under a low intensity seasonal grazing regime, the abundance of Scottish primrose *Primula scotica* increased from 659 plants in 1987 to 3,980 plants in 1998. The abundance and flowering of 40 other broadleaved species also increased. Under the regime, grazing was removed during spring and summer to permit flowering and seed setting, followed by heavy grazing in the autumn, extending into winter, to produce niches for seedling establishment (no artificial fertilizers or pesticides were used). The resulting species-rich vegetation layer included species which had previously been unable to colonize due to the summer grazing pressures. Sheep productivity was comparable with a commercial system but stocking density was lower (around 2.5 ewes/ha). The impact on vegetation was recorded during a tiered monitoring programme, not described in the paper.

A paired sites study on moorland in 1996–2000 in northern England (7) found that the number of displaying black grouse *Tetrao tetrix* males increased by an average of 5% each year at 10 sites where levels of sheep grazing were reduced, compared with average declines of 2% each year at 10 control sites. Changes were most positive in the first years after grazing reduction. The proportion of females with chicks was also significantly higher at treatment sites (average of 54%) than at control sites (32%). However, there were declines in female densities at sites where restricted grazing areas exceeded approximately 1 km^2. Grazing was reduced to below 1.1 sheep/ha in summer and 0.5 sheep/ha in winter for between 1 and 5 years on treatment sites. Densities were two or three times higher on control sites.

A controlled, replicated, before-and-after trial in 1993–1995 on four heather moorland sites in northeast Scotland (8) found that removing grazing increased heather *Calluna vulgaris* cover/height/canopy occupancy and reduced grass cover, increasing the number of true bugs (Hemiptera) and altering their species composition. However, site-specific factors like soil type and number of plant species had a greater impact on true bug communities than management treatments. Sites with more mineral soils had the most plant species and most true bugs and true bug species. Grazing treatments were applied in spring/summer 1993–1995 in 4 plots (5 x 3 m) in each of 4 fenced/unfenced treatment blocks at each site. Every spring/summer two blocks at each site received a variety of nitrogen/phosphorous/potassium NPK fertilizer treatments. Heather growth structure was assessed in May and August 1993–1995. True bugs were sampled in June and July in 1993 and 1995 and plant species were counted.

A randomized, replicated before-and-after trial in England (9) found that the average heather *Calluna vulgaris* cover on 50 moorland sites in the Dartmoor Environmentally Sensitive Area decreased from 10.3% in 1994, when the scheme was introduced, to 7.7% in 2003. This was accompanied by an increase in grazing pressure, as measured by a heather grazing index. These trends were most pronounced on acid grassland habitats, where heather cover was lowest. The tier of Environmentally Sensitive Area managements (indicating whether management is aimed at maintenance or enhancement of habitats) had little effect on these changes, and there was little evidence of any heather recovery.

A replicated trial from 2001 to 2004 at two upland grassland sites in Perthshire and the Scottish Borders, Scotland, UK (10) (partly the same study as (17)) found that sheep grazing intensity influenced the structure of ground beetle (Carabidae) assemblages. There were 2–5 times as many large beetles of the genus *Carabus* on extensive summer-only grazed plots (1–3 sheep/ha, June–September/October) than on intensive year-round grazed plots (1–4 sheep/ha). At one of the two sites, the intensively grazed plot had four times as many springtail-specialist beetles. Grazing intensity did not influence the number of ground beetle species at either site, with 23–33 species/plot. One intensive and one extensive plot (each >40 ha) was established at each study site (2 replicates). Grazing treatments began in 2001–2002. In each plot, ten locations were sampled in 2003 and five in 2004. Beetles were sampled using pitfall traps in May–June and plants were surveyed in June–August.

A replicated before-and-after trial in Northern Ireland (11) found that the number of plant species on heather *Calluna vulgaris* moorland managed under the Environmentally Sensitive Area scheme was maintained in two of three areas for which results were reported. Average cover of heather increased in 1 of the 5 Environmentally Sensitive Areas (13 sites in West Fermanagh) but did not change at 2 others (43 sites in the Sperrins Environmentally Sensitive Area; 6 sites in the Antrim Coast Environmentally Sensitive Area). The number of plant species on heather moorland was maintained at these two Environmentally Sensitive Areas but declined between 1994 and 2004 in the Slieve Gullion Environmentally Sensitive

Area (13 sites). Values are not given for heather cover or numbers of plant species on heather moorland. The study monitored plant diversity at 93 heather moorland sites in Northern Ireland, first in 1993–1994 before the Environmentally Sensitive Area management began and again 10 years later. The sites were randomly selected from a database of farmers joining the Environmentally Sensitive Area scheme in 1993.

A 2006 review of UK studies (12) on the impact of farm management practices on below-ground biodiversity and ecosystem function reported four studies which found positive impacts of increased grazing on soil bacteria (Yeates *et al.* 1997), earthworms (Lumbricidae) (Muldowney *et al.* 2003) and on mites (Acari), springtails (Collembola) and nematodes (Nematoda) (Bardgett *et al.* 1993, 1997) in upland habitats.

A before-and-after study in 2000–2006 on a grouse moor in Dumfries and Galloway, south Scotland (13) found that five bird species decreased following the discontinuation of moor management in 2000, whilst four more increased. Before 2000, the moor underwent rotational burning and red foxes *Vulpes vulpes*, carrion crows *Corvus corone*, stoats *Mustela erminea* and weasels *M. nivalis* were controlled.

A randomized, replicated, controlled trial in 2002–2005 on an upland grassland site in Perthshire, Scotland, UK (14) found that after 18 months of grazing the biomass of arthropods associated with the bird diet was nearly twice as high on ungrazed/lightly grazed plots (sheep and cattle) than on plots grazed at a commercial stocking rate (sheep). The study also found more spiders (Araneae), true bugs (Hemiptera), beetles (Coleoptera) and caterpillars (Lepidoptera) in ungrazed or lightly grazed plots than in intensively grazed plots, but there was no straightforward relationship between grazing intensity and the number of crane fly (Tipulidae) adults and brachyceran flies (Brachycera). From January 2003 three grazing regimes (sheep at 2.7 ewes/ha; sheep at 0.9 ewes/ha; sheep and cattle equivalent to 0.9 ewes/ha) and an ungrazed control were replicated 6 times in 3.3 ha plots (in 3 pairs of adjacent blocks). Arthropods were sampled by suction sampler in spring/summer 2002–2005 (spiders, true bugs, beetles and brachyceran flies) and by sweep net in 2003–2005 (moth caterpillars and crane fly larvae and crane fly adults in 2005).

In the same randomized, replicated, controlled trial (described in (14)) more moths (Lepidoptera) and moth species were found on ungrazed and lightly-grazed plots than on plots grazed at a commercial stocking rate (15). Low-intensity sheep grazing and ungrazed treatments produced the highest number of moths (on average 52 moths/night and 48 moths/night, respectively) and moth species (on average 12.3 species/night; 13.2 species/night). Fewest moths (on average 34 moths/night) and moth species (on average 10.6 species/night) were found under the commercial grazing treatment. Grazing treatments began in January 2003 and moths were sampled between June and October 2007 using a randomly-placed 15W light trap for 6 or 7 sample nights per plot.

A 2009 literature review of agri-environment schemes in England (16) reported a study that concluded that Environmentally Sensitive Area management prescriptions were having positive effects on moorland bird populations in the Dartmoor Environmentally Sensitive Area, Devon (Geary 2002). However, the same study warned that localized problems such as overgrazing, burning or scrub encroachment were negatively affecting birds such as tree pipit *Anthus trivialis*, whinchat *Saxicola rubetra* and ring ouzel *Turdus torquatus*. One study from the north of England (reported in (7)) found that reduced grazing intensity benefited black grouse *Tetrao tetrix*. The review also describes a case study that found that three bird species increased on a farm in Exmoor, Devon from 1993 to 2003, following a reduction in grazing intensity on moorland areas (Eurasian skylark *Alauda arvensis* increased from 0 to 13 birds; Eurasian linnet *Carduelis cannabina* from 0 to 9 birds; common stonechat *Saxicola torquata* from

0 to 1 territory). One species (meadow pipit *A. pratensis*) showed little change (nine birds vs eight) and another (northern wheatear *Oenanthe oenanthe*) declined slightly, from one territory to none. In the same case study heath vegetation increased from 9% to 52%, bent-fescue/ rough acid grassland decreased from 89% to 39%, mean heather cover increased significantly from 5% to 29% and average dwarf shrub height increased from 5 cm to 23 cm.

An unreplicated trial in 2002–2004 at an upland semi-natural grassland site in the Scottish Borders, Scotland, UK (17) (partly the same study as (10)) found that grazing intensity and area of fine- and broad-leaved grasses influenced the assemblage structure of mobile arthropods, immobile invertebrates and ground beetles (Carabidae) at a range of spatial scales. Intensively grazed sites were associated with smaller mobile arthropods (e.g. money spiders (Linyphiidae)), crane fly (Tipulidae) larvae and earthworms (Lumbricidae), while less intensively grazed sites were associated with larger mobile arthropods (e.g. wolf spiders (Lycosidae) and ground beetles of the genus *Carabus*), sawfly larvae (Symphyta) and caterpillars (Lepidoptera). The effect of grazing became less apparent at smaller spatial scales (≤1 m radius) where fine-scale habitat characteristics like vegetation structure and composition were more important. Two large (>40 ha) plots were grazed by 3–4 sheep/ha from autumn 2002: one during June–September (low intensity grazing), the other year round (high intensity grazing). Invertebrates were sampled using pitfall transects (9 traps) at 15 locations/plot during May–June 2004. Vegetation patches were mapped for GIS analysis in a 30 m diameter circle around each transect and 25–50 vegetation height/species measurements were made in each patch in June and August 2004.

(1) Bardgett R.D., Frankland J.C. & Whittaker J.B. (1993) The effects of agricultural management on the soil biota of some upland grasslands. *Agriculture, Ecosystems & Environment*, 45, 25–45.
(2) McFerran D.M., Montgomery W.I. & McAdam J.H. (1994) The impact of grazing on communities of ground-dwelling spiders (Araneae) in upland vegetation types. *Biology and Environment-Proceedings of the Royal Irish Academy*, 94B, 119–126.
(3) Grant S.A., Torvell L., Common T.G., Sim E.M. & Small J.L. (1996) Controlled grazing studies on *Molinia* grassland: effects of different seasonal patterns and levels of defoliation on *Molinia* growth and responses of swards to controlled grazing by cattle. *Journal of Applied Ecology*, 33, 1267–1280.
(4) Grant S.A., Torvell L., Sim E.M., Small J.L. & Armstrong R.H. (1996) Controlled grazing studies on *Nardus* grassland: effects of between-tussock sward height and species of grazer on *Nardus* utilization and floristic composition in two fields in Scotland. *Journal of Applied Ecology*, 33, 1053–1064.
(5) Bardgett R.D. & Cook R. (1998) Functional aspects of soil animal diversity in agricultural grasslands. *Applied Soil Ecology*, 10, 263–276.
(6) Harris R.A. & Jones R.M. (2000) The Loft and Hill of White Hamars Grazing Project. *Proceedings of the British Grassland Society Conference: Grazing management: The principles and practice of grazing, for profit and environmental gain, within temperate grassland systems.* Harrogate, 29 February–2 March. pp. 157–158.
(7) Calladine J., Baines D. & Warren P. (2002) Effects of reduced grazing on population density and breeding success of black grouse in northern England. *Journal of Applied Ecology*, 39, 772–780.
(8) Hartley S.E., Gardner S.M. & Mitchell R.J. (2003) Indirect effects of grazing and nutrient addition on the hemipteran community of heather moorlands. *Journal of Applied Ecology*, 40, 793–803.
(9) Defra (2004) *Survey of Moorland and Hay Meadows in Dartmoor ESA*. Defra MA01016.
(10) Cole L.J., Pollock M.L., Robertson D., Holland J.P. & McCracken D.I. (2006) Carabid (Coleoptera) assemblages in the Scottish uplands: the influence of sheep grazing on ecological structure. *Entomologica Fennica*, 17, 229–240.
(11) McEvoy P.M., Flexen M. & McAdam J.H. (2006) The Environmentally Sensitive Area (ESA) scheme in Northern Ireland: ten years of agri-environment monitoring. *Biology and Environment: Proceedings of the Royal Irish Academy, Section B*, 106B, 413–423.
(12) Stockdale E.A., Watson C.A., Black H.I.J. & Philipps L. (2006) *Do Farm Management Practices Alter Below-ground Biodiversity and Ecosystem Function? Implications for sustainable land management.* Joint Nature Conservation Committee 364.
(13) Baines D., Redpath S., Richardson M. & Thirgood S. (2008) The direct and indirect effects of predation by Hen Harriers *Circus cyaneus* on trends in breeding birds on a Scottish grouse moor. *Ibis*, 150, 27–36.

(14)   Dennis P., Skartveit J., McCracken D.I., Pakeman R.J., Beaton K., Kunaver A. & Evans D.M. (2008) The effects of livestock grazing on foliar arthropods associated with bird diet in upland grasslands of Scotland. *Journal of Applied Ecology*, 45, 279–287.

(15)   Littlewood N.A. (2008) Grazing impacts on moth diversity and abundance on a Scottish upland estate. *Insect Conservation and Diversity*, 1, 151–160.

(16)   Natural England (2009) *Agri-environment Schemes in England 2009: A review of results and effectiveness.* Natural England, Peterborough.

(17)   Cole L.J., Pollock M.L., Robertson D., Holland J.P., McCracken D.I. & Harrison W. (2010) The influence of fine-scale habitat heterogeneity on invertebrate assemblage structure in upland semi-natural grassland. *Agriculture, Ecosystems & Environment*, 136, 69–80.

Additional references

Bardgett R.D., Whittaker J.B. & Frankland J.C. (1993) The diet and food preferences of *Onychiurus procampatus* (Collembola) from upland grassland soils. *Biology and Fertility of Soils*, 16, 296–298.

Bardgett R.D., Leemans D.K., Cook R., & Hobbs P.J. (1997) Seasonality of the soil biota of grazed and ungrazed hill grasslands. *Soil Biology & Biochemistry*, 29, 1285–1294.

Yeates G. W., Bardgett R. D., Cook R., Hobbs P. J., Bowling P. J. & Potter J. F. (1997) Faunal and microbial diversity in three Welsh grassland soils under conventional and organic management regimes. *Journal of Applied Ecology*, 34, 453–470.

Geary S. (2002) *Exmoor Moorland Breeding Bird Survey 2002.* RSPB, Exeter.

Muldowney J., Curry J.P., O'Keeffe J. & Schmidt O. (2003) Relationships between earthworm populations, grassland management and badger densities in County Kilkenny, Ireland. *Pedobiologia*, 47, 913–919.

4.15   Restore or create upland heath/moorland

- A small unreplicated trial of heather moorland restoration in northern England[2] found that mowing and flail cutting along with grazing could be used to control the dominance of purple moor grass. The same study found moorland restoration benefited one bird species, with one or two pairs of northern lapwing found to breed in the area of restored moorland where none had bred prior to restoration.

- A review from the UK concluded that vegetation changes took place very slowly following the removal of grazing to restore upland grassland to heather moorland[1].

Background

Upland heath/moorland typically consists of dwarf shrubs such as heather *Calluna vulgaris* (Natural England 2011). These areas are important for wildlife including rare birds such as black grouse *Tetrao tetrix*, as well as invertebrates (Natural England 2011). Moorland can become dominated by scrub or grasses such as purple moor grass *Molinia caerulea* (Natural England 2011). This intervention involves restoring or creating areas of upland heath/moorland which may benefit birds and other wildlife.

See also 'Natural system modification: Manage heather by swiping to simulate burning' and 'Natural system modification: Manage heather, gorse or grass by burning' for studies that investigated specific techniques for managing heather.

Natural England (2011) *Upland Heath.* Available at www.naturalengland.org.uk/ourwork/conservation/biodiversity/englands/habitatofthemonth/uplandheath.aspx. Accessed 23 October 2012.

A 1984 review of studies in the UK (1) concluded that, after removal of grazing, reversion from upland grassland to heather *Calluna vulgaris* moorland happens very slowly. The review describes two long-term studies (1950s–1970s) that monitored botanical changes following exclusion of sheep from upland grassland plots in England (Welch & Rawes 1964, Rawes 1981, 1983) and Wales (Hughes *et al.* 1975, Hill 1983). By 1983, early vegetation changes were slow, mainly involving an altered balance of plant species already present on plots, and entry of heath species was limited. This may have been due to a lack of local seed sources or because seeds were unable to germinate in the close grass layer.

A small before-and-after study in 2004–2005 on an area of purple moor grass *Molina caerulea*-dominated moorland in northern England (2) found that mowing and flail cutting along with livestock and wild red deer *Cervus elaphus* grazing could be used to control the dominance of purple moor grass and help restore heather *Calluna vulgaris* moorland. Mowing and flail cutting resulted in strong purple moor grass re-growth in spring, which was then heavily grazed, suppressing grass growth. The study also found one or two pairs of northern lapwing *Vanellus vanellus* bred on the area of restored moorland, whereas none had previously bred in the area. An area re-seeded with heather had heather seedlings from mid-June onwards. One hundred hectares of moorland were fenced to exclude livestock. Half of the grassland within the exclosure was burned on 9 March 2004 to reduce the dominance of purple moor grass and 14 ha of the burned area was flail cut on 17 March 2004 to remove the burnt grass tussocks. A 3 ha area was sown with heather seed (40 kg/ha) in May 2004. In spring 2005 a further 4.5 ha area was flail cut and re-sown with heather at 40–60 kg/ha. In 2005 sheep grazed the exclosure until the beginning of June, following which the exclosure gates were opened until mid-June to allow free grazing. Cattle (63 livestock units) then grazed the area from late June. Sheep are now permanently excluded from the area.

(1) Ball D.F. (1984) Studies by ITE on the impact of agriculture on wildlife and semi-natural habitats in the uplands. In D. Jenkins (ed.) *Agriculture and the Environment*, NERC/ITE, Cambridge. pp. 155–162.
(2) Smith D. & Bird J. (2005) Restoration of degraded *Molinia caerulea* dominated moorland in the Peak District National Park Eastern moorlands, Derbyshire, England. *Conservation Evidence*, 2, 101–102.

Additional references
Welch D. & Rawes M. (1964) The early effects of excluding sheep from high-level grasslands in the northern Pennines. *Journal of Applied Ecology*, 1, 281–300
Hughes R.E., Dale J., Lutman J. & Thomson A.G. (1975) Effects of grazing on upland vegetation in Snowdonia. *Annual Report of Institute of Terrestrial Ecology*, 1974, 46–50.
Rawes M. (1981) Further results of excluding sheep from high level grasslands in the northern Pennines. *Journal of Ecology*, 69, 651–669.
Hill M.O. (1983) Effects of grazing in Snowdonia. *Annual Report of Institute of Terrestrial Ecology*, 1982, 31–32.
Rawes M. (1983) Changes in two high altitude blanket bogs after the cessation of sheep grazing. *Journal of Ecology*, 71, 219–235.

4.16   Plant brassica fodder crops (grazed *in situ*)

- We have captured no evidence for the effects of planting brassica fodder crops (grazed *in situ*) on farmland wildlife.

Background
Brassica fodder crops such as swedes, turnips, fodder rape and kale are grazed *in situ* by livestock.

4.17   Use mixed stocking

- A replicated, controlled study in the UK found more spiders, harvestmen and pseudoscorpions on sheep-grazed grassland than on mixed livestock-grazed grassland when suction sampling, but not when pitfall-trapping[1].

Background
Different species of livestock forage differently, therefore stocking multiple species in one area may help to create a more diverse habitat.

A replicated, controlled study from 1991 to 1994 in the UK (1) in a matgrass *Nardus stricta*-dominated grassland found that arachnid (spiders (Araneae), harvestmen (Opiliones) and pseudoscorpions (Pseudoscorpionida)) abundance and species richness (using suction but not pitfall sampling) were higher on single rather than mixed livestock grazed grasslands. Arachnid abundance was significantly higher in ungrazed and taller (6.5 cm) sheep-grazed swards than in both tall and short (4.5 cm) mixed stocking treatments (suction traps: 82 individuals in ungrazed, 26 in sheep-grazed to 6.5 cm, 13–16 in mixed stocking; pitfall traps: 3.5 individuals in ungrazed, 3.5 sheep-grazed to 6.5 cm, 3.2–3.3 mixed stocking treatments). Arachnid numbers in short sheep-grazed plots were intermediate (suction trap: 21 individuals; pitfall traps: 3.3). Individual species tended to show similar patterns. Pitfall traps indicated no significant difference in total arachnid species richness in ungrazed (49 species), sheep-grazed (45–50) or mixed-grazed treatments (49–50) or numbers of money spider (Linyphiidae) species (37 species in ungrazed treatment; 32–37 in sheep-grazed treatment; 35–38 in mixed stocking). Suction sampling showed species richness was significantly higher in ungrazed and taller sheep-grazed swards than in mixed livestock treatments for total arachnid species (31 species in ungrazed, 21 in taller sheep-grazed swards, 15–16 in mixed stocking) and money spider species (18, 15 and 9–11 respectively). Significantly more spider webs were counted in ungrazed (1–14 webs/m²) than in sheep-only (1–4 webs/m²) and mixed-stocking treatments (1–4 webs/m²). More money spider species were sampled from webs in tall mixed-stocking (6 species) and shorter sheep-only (4.5) plots than in other treatments (2.0–2.5). There were 2 replicates of each treatment in 10 plots across 22 ha. The 5 treatments were: ungrazed; sheep-grazed (sward 4.5 cm tall); sheep-grazed (6.5 cm); sheep and cattle-grazed (4.5 cm); sheep and cattle-grazed (6.5 cm). Sheep were grazed continuously at varied rates to achieve the different sward heights from May–October (1991–1994). Six cattle were grazed from June–August in the two mixed livestock treatments. Twelve pitfall traps/plot were used to sample continuously from April to October, 1993–1994. At monthly intervals, 6 suction samples were taken and webs were counted within a 1 m² rectangle on the grass near 6–12 pitfall traps/plot.

(1)   Dennis P., Young M.R. & Bentley C. (2001) The effects of varied grazing management on epigeal spiders, harvestmen and pseudoscorpions of *Nardus stricta* grassland in upland Scotland. *Agriculture Ecosystems & Environment*, 86, 39–57.

4.18   Use traditional breeds of livestock

• Two UK studies (one replicated) and a review reported differences in quantities of plant species grazed[1, 2], vegetation structure and invertebrate assemblages[3] between areas grazed with different breeds of sheep or cattle. A small, replicated study found that Hebridean sheep grazed more purple moor grass than Swaledale sheep[1, 2] but the resulting density of purple moor grass and heather did not differ[1]. A UK study found that at reduced grazing pressure, traditional and commercial cattle breeds created different sward structures and associated invertebrate assemblages[3].

• One replicated trial from France, Germany and the UK found grazing by traditional rather than commercial livestock breeds had no clear effect on the number of plant species[4] or the abundance of butterflies, grasshoppers, birds, hares, or ground-dwelling arthropods in general[5].

Background
This intervention involves stocking areas with traditional or rustic breeds of livestock. Traditional or rustic breeds of livestock are often recommended for nature conservation management.

A small, replicated study from 1992 to 1996 of four pasture plots in North Yorkshire, UK (1) found that Hebridean sheep (a minority breed) grazed more purple moor grass *Molinia caerulea* than Swaledale sheep (traditional upland breed) but the resulting density of purple moor grass and heather *Calluna vulgaris* did not differ. Hebridean sheep grazed significantly more purple moor grass leaves than Swaledales (61% vs 23%). But the density of purple moor grass leaves did not differ between Hebridean (2,389 m^2) and Swaledale plots (2,798 m^2). Overall, cover by heather did not change over time. In the Hebridean plots, heather cover doubled in the first 4 years (12% to 29%), then declined (22%); the overall increase was 3.7%/year. In Swaledale plots, cover increased over the first 2 years (3% to 7%) then declined dramatically to 1%, with an overall decline of 1%. Two areas of pasture dominated by purple moor grass, ungrazed for 2 years and unburnt for 10, were divided into 2 plots of 2 ha, 1 grazed by Swaledale sheep, 1 by Hebridean sheep (99 kg live weight/ha) between May and September. Numbers of grazed and ungrazed purple moor grass leaves were sampled within 2 quadrats (0.5 x 0.5 m) and lengths of grazed/ungrazed leaves measured at 100 points across plots following sheep removal in September each year. Plant species were sampled using a point sampling technique along a 15 m transect (45 cm intervals).

A 2004 literature review (2) found 11 studies that compared the feeding behaviour of different sheep or cattle breeds in fairly controlled conditions. Only one study monitored the effects on vegetation. This UK study (Newborn *et al.* 1993, later reported as (1)) found that Hebridean sheep caused more of a decline in purple moor grass *Molinia caerula* than Swaledale sheep, when this plant was growing with heather *Calluna vulgaris*. On the basis of findings from all 11 studies, the authors concluded that the different breeds of livestock have only minor differences in their feeding behaviour and these differences are mostly due to body size.

A small-scale study over three years on species-poor lowland grassland in the UK (3) found that at reduced grazing pressure, two cattle breeds created different sward structures and associated invertebrate assemblages (details were not provided). Three grazing treatments were studied: commercial breed Charolais × Holstein-Friesian steers at moderate (maintaining 3,000 kg herbage dry matter mass/ha) or lenient (4,500 kg herbage dry matter mass/ha) grazing pressures and North Devon steers (traditional breed) at lenient grazing pressure. Treatments were applied from May to September and grasslands received no fertilizer during the study.

A randomized, replicated trial from 2002 to 2004 in France, Germany and the UK, (4) (same study as (5)) found that grazing using traditional breeds of livestock made no difference to the number of plant species on agricultural grasslands compared to grazing with commercial breeds. There were on average 25, 17 and 10 plant species in plots grazed with commercial breeds (in France, Germany and the UK respectively) compared to 24, 17 and 11 plant species in plots grazed with traditional breeds. Productive, species-poor grasslands

(UK site), species-rich semi-natural grasslands (France) and moderately species-rich 'meso-trophic' grasslands (Germany) were used in the experiment. Sites were grazed continously with cows. Treatments were replicated three times at each site. Paddocks 0.4 to 3.6 ha in size were leniently grazed with either a traditional or a commercial breed. Plants were monitored in 10 fixed 1 m^2 quadrats in each paddock in April–May, June–July and August–September from 2002 to 2004. Other plant species seen within 5 m of the quadrats were also recorded. An additional study site grazed by sheep in Italy was also included in the analysis but is not reported here because it falls outside the geographical range of this synopsis.

A randomized, replicated study from 2002 to 2004 in France, Germany and the UK (5) (same study as (4)) found using traditional breeds of livestock to graze grasslands had no effect on the numbers of butterflies (Lepidoptera), grasshoppers (Orthoptera), birds, European hares *Lepus europaeus* or ground-dwelling arthropods in general, relative to commercial breeds. One insect group ('other' beetles (Coleoptera)) were more abundant in pitfall traps under grazing by traditional breeds, at two of the four sites: at the site in Germany this group mainly comprised sap beetles (Nitidulidae). The traditional cattle breeds were Devon, German Angus and Salers, compared with commercial Charolais x Fresian, Simmental and Charolais in the UK, Germany and France respectively. Each treatment (leniently grazed with traditional or commercial livestock) was replicated 3 times at each site, in 0.4 to 3.6 ha paddocks. Animals were monitored in 2002, 2003 and 2004. Butterflies were counted once a fortnight from May to September and grasshoppers sampled with a sweep net each month from June to October on three 50 m-long transects. Birds and hares were counted for seven minutes along a transect fortnightly from May to October. Ground arthropods were sampled in 12 pitfall traps in each paddock, in spring, summer and early autumn. An additional study site grazed by sheep in Italy was also included in the analysis but is not reported here because it falls outside the geographical range of this synopsis.

(1)   Newborn D. (2000) The value of Hebridean sheep in controlling invasive purple moor grass. *Aspects of Applied Biology*, 58, 191–196.
(2)   Rook A.J., Dumont B., Isselstein J., Osoro K., WallisDeVries M.F., Parente G. & Mills J. (2004) Matching type of livestock to desired biodiversity outcomes in pastures – a review. *Biological Conservation*, 119, 137–150.
(3)   Tallowin J.R.B., Rook. A.J. & Rutter. S.M. (2005) Impact of grazing management on biodiversity of grasslands. *Animal Science*, 81, 193–198.
(4)   Scimone M., Rook A.J., Garel J.P. & Sahin N. (2007) Effects of livestock breed and grazing intensity on grazing systems. 3. Effects on diversity of vegetation. *Grass and Forage Science*, 62, 172–184.
(5)   Wallis De Vries M.F., Parkinson A.E., Dulphy J.P., Sayer M. & Diana E. (2007) Effects of livestock breed and grazing intensity on biodiversity and production in grazing systems. 4. Effects on animal diversity. *Grass and Forage Science*, 62, 185–197.

Additional references
Newborn D., Wakeham A. & Booth F. (1993) *Grazing and the Control of Purple Moor Grass*. Game Conservancy Council Review.4.19

4.19   Employ areas of semi-natural habitat for rough grazing (includes salt marsh, lowland heath, bog, fen)

• A series of site comparison studies from the UK found that areas of heathland that had been re-seeded with grass to improve livestock grazing were avoided by nesting whimbrels[1] but were the main early spring feeding areas for them[2]. There was no difference in whimbrel chick survival between areas of heathland re-seeded with grass and those that had not[3].

- Two replicated studies from the UK found higher butterfly abundance and species richness[7] and a higher frequency of occurrence of songbirds and invertebrate-feeding birds[6] on areas of grazed semi-natural upland grassland than grazed improved pasture. However members of the crow family showed the opposite trend[6]. A review found that excluding cattle from fenland reduced the number of plant species and that low-medium grazing levels could have positive effects on fenland biodiversity but may need to be accompanied by additional management such as mowing[5].

- One study from the UK found northern lapwing nest survival and clutch size were greater on ungrazed than grazed marshes[4]. A replicated site comparison study from the UK found that the proportion of young grey partridges was negatively associated with rough grazing (in combination with several other interventions)[8].

Background

This intervention involves grazing semi-natural habitats that are not an integral part of modern farming practice, either enclosed or unenclosed. It includes the use of areas of peat bog, or other undrained wetland, lowland heath, saltmarsh and sand dunes. It does not include management of upland heath and moorland, species-rich grasslands or wood pasture. These are covered under habitat-specific interventions.

A site comparison study in the Shetland Islands, Scotland (1) found that areas of heath seeded with grass to improve them for livestock grazing were mostly avoided by nesting whimbrels *Numenius phaeopus* in favour of unimproved heathland. In 1986 and 1987, this study monitored whimbrels in five areas of heathland that had been partly seeded: four on the island of Fetlar and one on Unst. Eighty-nine percent (111 nests) of the nests were found in unseeded heathland. Most nests were on hummocks and amongst heather. Seeding with grass after ploughing or harrowing resulted in the loss of hummocks and most heather *Calluna vulgaris* and created a predominantly grassy habitat. Surface-seeding, without ploughing or harrowing, created less marked changes, with hummocks and heather retained, although hummock height was lowered and in some areas only dead or dying heather was present.

At the same study sites as (1), (2) found that areas of heath seeded with grass after ploughing or harrowing and older pastures were the main early spring feeding areas for at least 90% of whimbrel *Numenius phaeopus* pairs in the study. They monitored habitat use by individually marked whimbrels during the pre-laying period in spring 1987 and 1988 on five Shetland Island heathlands. The birds made little use of unimproved heathland (where most nest) or heathland areas seeded without ploughing/harrowing. The greatest quantities of prey species – earthworms (Lumbricidae) and larval crane flies (Tipulidae) – were found in the soil of ploughed/harrowed seeded areas of heath and older pastures, with more recently seeded areas holding the highest masses of crane fly larvae.

In a third study using the same Shetland Island heaths as (1), (3) found no significant difference in whimbrel *Numenius phaeopus* chick survival between chicks that used areas of heathland re-seeded with grass and those that did not. Individually marked chicks were monitored after hatching in 20, 23, and 26 broods in 1986, 1987 and 1988 respectively. In each year 35–65% of all chicks remained on heathland, while others (usually broods over 12 days old, from nests within 200 m of the alternative habitat) moved into other habitats.

A study of northern lapwing *Vanellus vanellus* nests on 12 grazed and 6 ungrazed marshes within an Environmentally Sensitive Area over 1 year in England (4) found that ungrazed marshes had greater clutch size and nest survival than those that were grazed. Clutch size was larger on ungrazed than grazed marshes and nest survival was significantly higher for 4-egg clutches (57%) than 3-egg clutches (21%). Overall, nest survival was 64% on ungrazed marshes compared with 34% on grazed marshes. On grazed marshes 58% of nests were lost to predation compared with 36% on ungrazed marshes and 20% of unpredated nests were trampled by livestock. Incubating birds also left their eggs significantly more frequently on grazed compared to ungrazed marshes because of disturbance. All marshes had been grazed the previous year. Sheep or cattle were introduced to 12 marshes at low stocking densities (i.e. 0.20–0.51 livestock units/ha). Lapwing nest data were collected between April and June 1997.

A 2006 review (5) describes the results of six studies that evaluated the impact of cattle-grazing on fens in Europe. Two studies found that excluding cattle from formerly grazed fens reduced the number of plant species (Bakker & Grootjans 1991, Matějková *et al.* 2003) with a shift to trees and shrubs sometimes occurring after 10–15 years of abandonment. Benefits of cattle-grazing to fen biodiversity are likely to depend on the level of grazing, although one freshwater wetland study found that grazing intensity hardly affected the number of invertebrate species (Steinman *et al.* 2003). The authors suggest that low to moderate levels of cattle grazing could maintain or increase the biodiversity of nutrient-rich fens that have not been overgrazed in the past. However, trampling associated with high grazing pressure can lead to soil degradation and two studies found, respectively, that the low stocking density required to avoid negative trampling impacts will often be too low to maintain biodiversity (Schrautzer *et al.* 2004) and additional measures to remove excess biomass (such as mowing) remain necessary (Kleyer 2004). One study found that mowing resulted in higher species richness than grazing in undrained chalk/limestone (calcareous) fens (Stammel *et al.* 2003).

A replicated trial in the UK (6) found that songbirds and invertebrate-feeding birds were recorded more often on semi-natural rough grazing than on upland improved pasture, but the opposite was true for crows (Corvidae). Bird numbers and species were recorded in plots of improved upland pasture grazed by cattle and sheep (with and without seasonal removal of grazing in summer; 10 replicates for each) and in plots of semi-natural rough grazing grazed by cattle from June to September (6 replicates). The proportion of surveys where songbirds and invertebrate feeders were recorded was greater on semi-natural rough grazing than on improved pasture. However, the effect on the number of individuals varied over the year. The number of birds of invertebrate-feeding species was greater on semi-natural grassland between May and July (338 birds, compared to 52 and 41 on improved treatments, with and without seasonal grazing removal) but greater on improved treatments between October and January (5,833 and 1,458 birds on improved treatments compared to 606 birds on semi-natural grassland). There were fewer birds of species in the crow family on semi-natural rough grazing plots at all times of year, but the difference was greatest during July to September (16 birds on rough grazing compared to 496 and 77 on improved plots). The location of the study was not given.

A replicated trial in 2005–2007 of cattle grazing on six experimental plots of semi-natural upland grassland dominated by purple moor grass *Molinia caerulea* in the UK (7) found more butterfly (Lepidoptera) species and significantly more individual butterflies than on permanently or partially grazed plots of improved pasture. Between 15 and 17 butterfly species (905–1,938 individual butterflies) were recorded on the semi-natural plots in each year compared to 7–11 species (42–156 butterflies) on 10 continually grazed and 5–10 species (21–67 butterflies) on 10 partially grazed plots of improved pasture. The semi-natural plots were grazed from June to September, while the partially grazed improved plots were grazed in spring and autumn but had livestock excluded from May to September and one silage cut taken. Butterfly transect counts were conducted weekly between mid-April and mid-September in 2005, 2006 and 2007.

A replicated site comparison study from 2004 to 2008 in England (8) investigated the impact of rough grazing on grey partridge *Perdix perdix* and found a negative relationship between a combined intervention (grazing, scrub control and the restoration of various semi-natural habitats) and the proportion of young partridges in the population in 2008. The study does not distinguish between the individual impacts of grazing, scrub control and the restoration of various semi-natural habitats. Spring and autumn counts of grey partridge were made at 1,031 sites across England as part of the Partridge Count Scheme.

(1)   Grant M.C. (1992) The effects of re-seeding heathland on breeding whimbrel *Numenius phaeopus* in Shetland. I. Nest distributions. *Journal of Applied Ecology*, 29, 501–508.
(2)   Grant M.C., Chambers R.E. & Evans P.R. (1992) The effects of re-seeding heathland on breeding whimbrel *Numenius phaeopus* in Shetland. II. Habitat use by adults during the pre-laying period. *Journal of Applied Ecology*, 29, 509–515.
(3)   Grant M.C., Chambers R.E. & Evans P.R. (1992) The effects of re-seeding heathland on breeding whimbrel *Numenius phaeopus* in Shetland. III. Habitat use by broods. *Journal of Applied Ecology*, 29, 516–523.
(4)   Hart J.D., Milsom T.P., Baxter A., Kelly P.F. & Parkin W.K. (2002) The impact of livestock on lapwing *Vanellus vanellus* breeding densities and performance on coastal grazing marsh. *Bird Study*, 49, 67–78.
(5)   Middleton B.A., Holsten B. & van Diggelen R. (2006) Biodiversity management of fens and fen meadows by grazing, cutting and burning *Applied Vegetation Science*, 9, 307–316.
(6)   Vale J.E. & Fraser M.D. (2007) Effect of sward type and management on diversity of upland birds. In J.J. Hopkins, A.J. Duncan, D.I. McCracken, S. Peel & J.R.B. Tallowin (eds.) *High Value Grassland: Providing Biodiversity, a Clean Environment and Premium Products. British Grassland Society Occasional Symposium No.38*, British Grassland Society, Reading. pp. 333–336.
(7)   Fraser M.D., Evans J.G., Davies D.W.R. & Vale J.E. (2008) Effect of sward type and management on butterfly numbers in the uplands. *Aspects of Applied Biology*, 85, 15–18.
(8)   Ewald J.A., Aebischer N.J., Richardson S.M., Grice P.V. & Cooke A.I. (2010) The effect of agri-environment schemes on grey partridges at the farm level in England. *Agriculture, Ecosystems & Environment*, 138, 55–63.

Additional references
Bakker J.P. & Grootjans A.P. (1991) Potential for vegetation regeneration in the middle course of the Drentsche Aa brook valley (The Netherlands). *Verhandlungen der Gesellschaft für Ökologie*, 20, 249–263.
Matějková I., van Diggelen R. & Prach K. (2003) An attempt to restore a central European species-rich mountain grassland through grazing. *Applied Vegetation Science*, 6, 161–168.
Stammel B., Kiehl K. & Pfadenhauer J. (2003) Alternative management on fens: response of vegetation to grazing and mowing. *Applied Vegetation Science*, 6, 245–254.
Steinman, A.D., Conklin, J., Bohlen, P.J. & Uzarski, D.G. (2003) Influence of cattle grazing and pasture land use on macroinvertebrate communities in freshwater wetlands. *Wetlands*, 23, 877–889.
Kleyer M. (2004) Freie Beweidung mit geringer Besatzdichte und Fräsen als alternative Verfahren zur Pflege von Magerrasen [Free grazing at low stocking rates and infrequent rototilling as alternative conservation management systems for dry nutrient-poor grasslands]. *Schriftenreihe für Landschaftspflege und Naturschutz*, 78, 161–182.
Schrautzer, J., Irmler, U. & Jensen, K. (2004) Auswirkungen großflächiger Beweidung auf die Lebensgemeinschaften eines nordwestdeutschen Flusstales [Effects of extensive grazing on species biotic communities of a northwest German river valley]. *Schriftenreihe für Landschaftspflege und Naturschutz*, 78, 39–62.

4.20   Maintain wood pasture and parkland

• A randomized, replicated, controlled trial in Sweden[1] found that annual mowing maintained the highest number of plant species on wood pasture.

A long-term randomized, replicated, controlled trial from 1972 to 1986 on a traditionally managed wood meadow in Sweden (1) found that mowing or grazing were needed to preserve the characteristic flora, with annual mowing maintaining the greatest number of plant species. Seven management regimes were compared, including mowing, grazing, burning, removing woody vegetation and abandonment, with two replicates for each treatment. The number of plant species increased significantly on plots mown every year, and remained stable on grazed plots. Reducing the frequency of mowing to every third year did not significantly reduce species richness. Periodic mowing may therefore be a way to preserve wood meadow flora when management resources are limited. Other treatments reduced species richness over time and plots became either dominated by trees (under abandonment) or dominated by a few tall herbs and/or grasses. Treatments were applied in 5 x 20 m plots from 1972 to 1986 with 2 replicates. Percentage cover of plant species was assessed in five or six 1 m^2 subplots/plot in 1972, 1980 and 1986.

(1) Hansson M. & Fogelfors H. (2000) Management of a semi-natural grassland; results from a 15-year-old experiment in southern Sweden. *Journal of Vegetation Science*, 11, 31–38.

4.21   Restore or create wood pasture

• One replicated controlled trial in Belgium[2] found that protection from grazing enhanced the survival and growth of tree seedlings planted in pasture.

• One replicated study in Switzerland[1] found that cattle browsing increased the mortality of tree saplings of four species and reduced average shoot production and total above-ground biomass. Browsing frequency decreased with increasing height of the surrounding vegetation.

A replicated study in May–October 2003 in pastures in the Swiss Jura Mountains (1) found that cattle browsing significantly reduced average shoot production and total above-ground biomass, and increased mortality of tree saplings of four species, whereas mowing had no impact on sapling growth. Browsing frequency decreased with increasing height of surrounding vegetation, and large saplings were browsed much more frequently than small saplings. Silver fir *Abies alba* was the most frequently browsed species, whilst beech *Fagus sylvatica* was least frequently browsed. Browsing frequency increased with grazing intensity. Sixteen blocks of eight saplings were planted in 2 paddocks, one of 3.3 ha and one of 4.5 ha. Each was grazed by twenty-two 18-month-old steers for 4 periods of 14–17 days. In the 4.5 ha paddock, four 15.5 x 18 m enclosures were set up and split into mown and control plots. Each was planted with four saplings of each species and size class. Vegetation in mown plots was cut to 3 cm 5 times during the experiment.

A replicated, controlled trial in southern Flanders, Belgium (2) found that protection from grazing for two years significantly enhanced the survival and growth of planted pedunculate oak *Quercus robur* and ash *Fraxinus excelsior* seedlings in wood pasture. Eighteen tree seedlings were planted in each of 56 plots (each 8 m²) across 4 nature reserves in April 2004. Seedlings were monitored until September 2006. Each plot was half grazed throughout and half ungrazed for two years (until April 2006). The plots either had grassland, rush *Juncus* spp., sedge *Carex* spp., bramble *Rubus fruticosus* or short ruderal vegetation. Seedling survival was higher in ungrazed than grazed plots in grassland and short ruderal vegetation for both tree species and also in rush-covered plots for oak only. In bramble-covered plots (and sedge and rush plots for ash) there was no difference in seedling survival between grazed and ungrazed plots. Seedling growth was significantly higher in ungrazed plots for both ash and oak, except in bramble plots. Bramble thus protected tree seedlings from grazing impacts, but suppressed growth.

(1) Vandenberghe C., Frelechoux F., Moravie M.A., Gadallah F. & Buttler A. (2007) Short-term effects of cattle browsing on tree sapling growth in mountain wooded pastures. *Plant Ecology*, 188, 253–264.
(2) Van Uytvanck J., Maes D., Vandenhaute D. & Hoffmann M. (2008) Restoration of woodpasture on former agricultural land: the importance of safe sites and time gaps before grazing for tree seedlings. *Biological Conservation*, 141, 78–88.

4.22   Exclude livestock from semi-natural habitat (including woodland)

- Seven studies (including four replicated controlled trials, of which one was also randomized and a review) from Ireland, Poland and the UK looked at the effects of excluding livestock from semi-natural habitats. Three studies (including one replicated controlled and randomized study) from Ireland and the UK found that excluding livestock benefited plants[2] and invertebrates[3, 4].

- Three studies (one replicated controlled and one replicated paired sites comparison) from Ireland and the UK found that excluding grazing did not benefit plants[4, 5, 7] or birds[6].

- Two studies (one replicated and controlled; one review)[1, 8] from Poland and the UK found that the impact of excluding grazing as a tool in habitat restoration was neutral or mixed.

Background

This intervention involves preventing livestock from grazing certain semi-natural habitats such as grasslands to benefit farmland wildlife such as rare plants.

See also 'Reduce grazing intensity on grassland (including seasonal removal of livestock)' for studies that excluded livestock from areas of permanent grassland.

A 1984 review of studies in the UK (1) concluded that natural restoration of target heather *Calluna vulgaris* moorland by removing grazing from upland grassland happens very slowly. The review describes two long-term studies (1950s–1970s) that monitored botanical changes following the exclusion of sheep from upland grassland plots in England (Welch & Rawes

1964, Rawes 1981, 1983) and Wales (Hughes *et al.* 1975, Hill 1983). By 1983 early vegetation changes were slow, mainly involving an altered balance of plant species already present on plots, and entry of heath species was limited. This may have been due to a lack of local seed sources or because seeds were unable to germinate in the close grass turf.

A randomized, replicated, controlled trial from 1990 to 1993 on grassland in Northern Ireland (2) found that protecting pasture field margins from grazing is likely to improve their wildlife value. When field margins were grazed by sheep (and fertilized), the plant and ground beetle (Carabidae) communities were more similar to those of the open field than to field margins. There were also greater soil temperature fluctuations in grazed margins and their hedges had more gaps. Three treatments and an unmanaged control were replicated three times in the margins of pasture fields on either side of mature hawthorn *Crataegus monogyna* hedges: fertilized and grazed; ploughed and sown with a game cover strip; and ploughed and left to recolonize naturally. Plants were sampled in July 1991 and August 1993 in quadrats positioned at intervals up to 9 m into each field. Ground beetles were sampled in March, May, July and September each year using 3 pitfall traps placed 1–2 m and 8–10 m on either side of the hedge.

A replicated, controlled trial in summer 1995 and 1996 in grassland in West Sussex, England (3) found significantly more invertebrates and invertebrate taxa in 1.5 x 0.5 m cages from which grazing sheep were excluded, than in uncaged areas of sheep-grazed grassland (28–38 individuals and 9–11 invertebrate taxa in cages vs 7–15 individuals and 3–5 invertebrate taxa outside). Grass within the grazing-exclusion cages was approximately 40 cm tall, whilst outside the cages it was 2 cm tall. Twelve 10 x 4 m plots, each with 1 grazing-exclusion cage were established in 1995, with an additional 2 plots used in 1996. Invertebrates were sampled within and outside the grazing-exclusion cage using a D-Vac suction sampler.

A replicated, paired sites comparison in 1999 of grassland habitats on 30 farms in Counties Laois and Offaly, Republic of Ireland (4) (same study as (5)) found that fencing of field and watercourse margins to exclude grazing did not benefit plant diversity. For watercourse margins (8 paired replicates) more plant species were found in unfenced than fenced margins (52 and 56 species on unfenced margins on Rural Environment Protection Scheme and non-Rural Environment Protection Scheme farms respectively; 50 and 48 species on fenced margins on Rural Environment Protection Scheme and non-Rural Environment Protection Scheme farms respectively). For field margins, fencing helped gappy hedges to re-establish and slightly increased the number of ground beetle species in margins. However, as the distance from the hedge to the fence increased (and the ungrazed margin became wider) the number of field margin plant species decreased. The authors recommend that fences to protect hedges should allow stock to graze underneath to maintain the field margin flora. Plants were surveyed in two hedgerows, their associated field margins and one watercourse margin on each farm. Ground beetles were surveyed in June and August.

A replicated, paired sites comparison study in 2000 in counties Laois and Offaly, Ireland (5) (same study as (4)) found that fenced watercourse margins on Rural Environment Protection Scheme farms did not have higher numbers of plant species than unfenced watercourse margins on non-Rural Environment Protection Scheme farms (14.7 and 16.1 plant species/margin respectively). Fifteen farms with Rural Environment Protection Scheme agreements at least 4 years old were paired with 15 similar farms without agreements. On each farm, a randomly selected watercourse margin was surveyed for plants: all plant species were recorded in two 5 x 3 m quadrats and percentage cover estimated in a 1 x 3 m quadrat within each margin. Eleven of the farm pairs enabled a fenced/unfenced comparison.

A replicated, controlled (paired) study of wet pasture in Leicestershire, UK (6) found that bird visit rates were significantly higher in areas with livestock (wet plots: 0.26; dry: 0.10) than in those where livestock had been excluded (wet: 0.17; dry: 0.06). Sampling involved 45 minute bird observations between April 2005–March 2007 (twice/month April–October, once/month November–March).

A series of trials at four upland grassland or heath sites in west Perthshire, Scotland (7) showed that excluding sheep caused changes in vegetation and the decline of scarce plant species at two of the sites. Livestock were excluded from three areas of upland grassland and one area of heath in 1998, 2000 or 2001. Vegetation was monitored in one to five years following exclusion. At one chalk grassland site, the species of conservation value creeping sibbaldia *Sibbaldia procumbens*, moss campion *Silene acaulis* and hair sedge *Carex capillaris* all declined. At the heath site, a population of the small white orchid *Pseudorchis albida* crashed from 50 spikes in 2001 to none in 2006 after grazing exclusion in 2000. At the two other grassland sites species composition was not monitored but the vegetation structure changed, with increased overall height and at one site, less vegetation below 8 cm.

A replicated, controlled study of a degraded species-poor meadow in Central Poland (8) found that livestock exclusion had less of an effect on the restoration of plant community composition than topsoil removal and hay transfer. Deep soil removal (40 cm) with hay addition resulted in a community closest to the donor meadows, particularly where grazing was excluded. Species richness in grazed plots was slightly higher after hay transfer (23 vs 18 without transfer); in non-grazed plots hay transfer had no effect. Two plots (35 x 35 m) were subdivided to test combinations of the following treatments: topsoil removal (to 20 or 40 cm), hay transfer from a nearby meadow (collected mid-July 2004–2005; partly dried; stored for 1.5 months; spread in 5–7 cm layer) and livestock/exclusion. Data were obtained from plots on plant species distribution and abundance (2004–2007) and biomass (2006–2007); species composition of degraded meadows and donor meadow were also collected (2004, 2006 and 2007). The soil seed bank (top 5 cm) at the 2 topsoil removal depths and seed content of hay were also sampled in 2004.

(1) Ball D.F. (1984) Studies by ITE on the impact of agriculture on wildlife and semi-natural habitats in the uplands. In D. Jenkins (ed.) *Agriculture and the Environment*, NERC/ITE, Cambridge. pp. 155–162.
(2) Bell A.C., Henry T. & McAdam J.H. (1994) Grassland management and its effect on the wildlife value of field margins. *Proceedings of the Joint Meeting Between the British Grassland Society and the British Ecological Society. Grassland management and nature conservation*. Leeds University, 27–29 September 1993, British Grassland Society Occasional Symposium 28. pp. 185–189.
(3) Wakeham-Dawson A., Szoszkiewicz K., Stern K. & Aebischer N.J. (1998) Breeding skylarks *Alauda arvensis* on Environmentally Sensitive Area arable reversion grass in southern England: survey-based and experimental determination of density. *Journal of Applied Ecology*, 35, 635–648.
(4) Feehan J., Gillmor D.A. & Culleton N.E. (2002) The impact of the Rural Environment Protection Scheme (REPS) on plant and insect diversity. *Tearmann*, 2, 15–28.
(5) Feehan J., Gillmor D. & Culleton N. (2005) Effects of an agri-environment scheme on farmland biodiversity in Ireland. *Agriculture, Ecosystems & Environment*, 107, 275–286.
(6) Defra (2007) *Wetting up Farmland for Birds and Other Biodiversity*. Defra BD1323.
(7) Holland J.P., Pollock M.L. & Waterhouse A. (2008) From over-grazing to under-grazing: are we going from one extreme to another? *Aspects of Applied Biology*, 85, 25–30.
(8) Klimkowska A., Kotowski W., Van Diggelen R., Grootjans A.P., Dzierża P. & Brzezińska K. (2010) Vegetation re-development after fen meadow restoration by topsoil removal and hay transfer. *Restoration Ecology*, 18, 924–933.

Additional references
Welch D. & Rawes M. (1964) The early effects of excluding sheep from high-level grasslands in the northern Pennines. *Journal of Applied Ecology*, 1, 281–300.
Hughes R.E., Dale J., Lutman J. & Thomson A.G. (1975) Effects of grazing on upland vegetation in Snowdonia. *Annual Report of Institute of Terrestrial Ecology*, 1974, 46–50.
Rawes M. (1981) Further results of excluding sheep from high level grasslands in the northern Pennines. *Journal of Ecology*, 69, 651–669.

Hill M.O. (1983) Effects of grazing in Snowdonia. *Annual Report of Institute of Terrestrial Ecology*, 1982, 31–32.

Rawes M. (1983) Changes in two high altitude blanket bogs after the cessation of sheep grazing. *Journal of Ecology*, 71, 219–235.

4.23   Mark fencing to avoid bird mortality

• We have captured no evidence for the effects of marking fencing to avoid bird mortality on farmland wildlife.

Background

Fences erected around young plantations (to exclude deer and other browsers) or to delineate property can be hard to see and low-flying birds such as grouse can be killed by flying into them.

4.24   Create open patches or strips in permanent grassland

• Two studies (both randomized, replicated and controlled) investigated the effects of creating open strips in permanent grassland. One trial from the UK[1] found that more Eurasian skylarks used fields containing open strips, but variations in skylark numbers were too great to draw conclusions from this finding. One trial from Scotland[2] found insect numbers in grassy headlands initially dropped when strips were cleared.

Background

Open patches and strips in permanent grassland can be used, in a similar way to skylark plots (see 'Arable farming: Create skylark plots') to provide short, open swards for ground-nesting birds.

A randomized, replicated, controlled trial in winter 1995–1996 in southern England (1) found more Eurasian skylarks *Alauda arvensis* on seven fields with open strips than in seven control fields without strips, but the variation in numbers was so great that these differences were not significant (2–55 skylarks/km^2 on treated fields vs 0 on controls). Open strips were created in a grid pattern 25 m apart using a tine-cultivator in November 1995. Experimental fields were still significantly more open in May 1996, but the swards had closed entirely by February 1997. The number of skylarks was recorded on 3 visits/month from December 1995 to February 1996 on 14 fields.

A randomized, replicated, controlled trial in 1997 and 1998 in Scotland (2) found that conservation management of permanent pasture field headlands (protection from summer grazing and clearing strips in the sward using herbicides) substantially increased the number of chick-food insects. The study measured the effect of different combinations of grazing and herbicide strip treatments on the numbers of true bugs (Heteroptera), sawfly larvae (Symphyta) and caterpillars (Lepidoptera). Treatments began in 1997 and insects were sampled in June and July in 1997 and 1998. By 1998, headlands protected from summer grazing had 19–32 times more chick-food insects (609 true bugs, 75 sawflies and 18 caterpillars, on average per 10 sample) than grazed headlands (19 true bugs, 2 sawflies and 1 caterpillar). Clearing strips in the sward (to allow birds easier access to insects) initially reduced the numbers of caterpillars and sawflies but they recovered by 1998 when vegetation in bare areas had regrown.

(1) Wakeham-Dawson A. & Aebischer N.J. (1998) Factors determining winter densities of birds on environmentally sensitive area arable reversion grassland in southern England, with special reference to skylarks (*Alauda arvensis*). *Agriculture, Ecosystems & Environment*, 70, 189–201.
(2) Haysom K.A., McCracken D.I., Roberts D.J. & Sotherton N.W. (2000) Grassland conservation head-lands: a new approach to enhancing biodiversity on grazing land. *Proceedings of the British Grassland Society Conference. Grazing Management: The principles and practice of grazing, for profit and environmental gain, within temperate grassland systems.* Harrogate, UK, 29 February–2 March. pp. 159–160.

4.25   Provide short grass for birds

- A replicated UK study[1] found that common starlings and northern lapwing spent more time foraging on short grass, compared to longer grass, and that starlings captured more prey in short grass.

Background
Vegetation height is important in determining the value of a grassland to wildlife, with short vegetation allowing birds access to the ground for foraging and potentially reducing predation risk. However, high vegetation can provide more complex environments and more habitats.
See also 'Raise mowing height on grasslands'.

A replicated study from January to May 2002 of 15 northern lapwing *Vanellus vanellus* chicks on one grassland site in the Isle of Islay, UK and 20 common starlings *Sturnus vulgaris* on 1 grassland site each in Oxfordshire, UK (1) found that both species experienced significantly greater foraging success in shorter grass. For lapwing chicks, foraging rate declined as grass height increased. Starlings spent 30% more time actively foraging and captured 33% more prey in short grass, although intake rate (captures per second of active foraging) did not differ between long and short grass. Invertebrate abundance did not differ between long and short grass. Fertilizer application and water level was manipulated to provide a range of grass heights on the lapwing site. Starlings were observed in enclosures placed within intensively managed permanent pasture that was mown to either 3 cm (short grass) or 13 cm (tall grass).

(1) Devereux C.L., McKeever C.U., Benton T.G. & Whittingham M.J. (2004) The effect of sward height and drainage on common starlings *Sturnus vulgaris* and northern lapwings *Vanellus vanellus* foraging in grassland habitats. *Ibis*, 146, 115–122.

5 Residential and commercial development

Key messages

Maintain traditional farm buildings
We have captured no evidence for the effects of maintaining traditional farm buildings on farmland wildlife.

Provide bat boxes, bat grilles, improvements to roosts
We have captured no evidence for the effects of providing bat boxes, bat grilles or improvements to roosts on farmland wildlife.

Provide owl nest boxes (tawny owl, barn owl)
Two studies (one before-and-after study) from the Netherlands and the UK found providing nest boxes increased barn owl populations. A replicated study from the UK found that a decrease in the proportion of breeding barn owls was not associated with the number of nest boxes.

5.1 Maintain traditional farm buildings

- We have captured no evidence for the effects of maintaining traditional farm buildings on farmland wildlife.

Background
Farmland wildlife such as birds and bats may use certain features of traditional farm buildings for nesting and roosting. These features may include ledges, crevices and roof spaces.
See also 'Provide owl nest boxes (Tawny owl, Barn owl)' for studies looking at the effects of installing nest boxes for owls in farm buildings.

5.2 Provide bat boxes; bat grilles; improvements to roosts

- We have captured no evidence for the effects of providing bat boxes, bat grilles or improvements to roosts on farmland wildlife.

Background
This intervention involves providing artificial nest sites for bats or improvements to bat roosts. Summer roost sites where bats rest during the day may be found in buildings, hollow trees, under bridges or in caves (Entwistle *et al.* 2001). Artificial roost sites in the form of bat boxes can provide additional roosting habitat; they are typically made of wood and may be placed on buildings or trees (Entwistle *et al.* 2001). Existing roosting sites may be protected from disturbance by installing bat grilles. Grilles allow bats to access their roost sites whilst excluding humans (Mitchell-Jones 2004).

Entwistle A.C., Harris S., Hutson A.M., Racey P.A., Walsh A., Gibson S.D., Hepburn I. & Johnston J. (2001) *Habitat Management for Bats – A Guide for Land Managers, Land Owners and their Advisors*. Joint Nature Conservation Committee, Peterborough.
Mitchell-Jones A.J. (2004) Conserving and creating bat roosts. In A.J. Mitchell-Jones & A.P. McLeish (eds.) *Bat Workers' Manual, 3rd Edition*. Joint Nature Conservation Committee, Peterborough. pp. 111–134.

5.3   Provide owl nest boxes (tawny owl; barn owl)

- Two studies from the UK (a before-and-after study and a controlled study) found that the provision of owl nest boxes in farm buildings maintained barn owl nesting and roosting activity[3] and resulted in an increase in population density[2]. A study from the Netherlands found that the barn owl population increased with increased availability of nest boxes[1].

- A replicated, controlled study in Hungary[4] found that juvenile barn owls fledged from nest boxes were significantly less likely to be recovered alive than those reared in church towers.

- A replicated study from the UK investigating barn owl nest site use, found that the number of occupied nest sites and the proportion breeding decreased from 2001 to 2009, but were unaffected by the number of boxes[5].

Background

This intervention may involve the installation of nest boxes suitable for owls such as barn owl *Tyto alba* in farm buildings. See also 'All farming systems: Provide nest boxes for birds'.

A study of 200 barn owl *Tyto alba* nest boxes over 10 years in the east of the Netherlands (1) found that with increasing availability of boxes over 90% of broods in the area became established in boxes and populations increased. Occupancy rates were 32% in Liemers and 36% in Achterhoek, although boxes were occupied significantly less frequently in Liemers (2.8 vs 4.5 years). Between 1967–1975 less than 25% of the barn owl broods in the area were in nest boxes but with increasing availability in 1976–1984 over 75% of broods and in 1985–1993 over 90% of broods were in boxes. The number of fledged young was significantly higher in nest boxes (3.4) compared to natural nest sites (2.5). In Leimers there was a significant decline in the breeding population from the early 1960s (7–20 breeding pairs/100 km^2) to the end of the 1970s (2–3); the population then stabilized in the 1980s and increased in the 1990s (11–13). In Achterhoek, the population also decreased in the 1960s (to 3–5) but then increased from the 1970s (1976–1982: 5–8) through to the 1990s with increasing availability of nest boxes. In 1967–1984 75 nest boxes were put up in suitable buildings in Liemers and 135 in Achterhoek. Boxes were surveyed in 1967–1993. Landscape structure and habitats differed in Liemers (350 km^2) and Achterhoek (550 km^2), which were either side of a river.

A before-and-after study at a 150 km^2 site in Norfolk, UK (2) found that barn owl *Tyto alba* population density increased from 15 pairs/100 km^2 in 1989 to 27 pairs/100 km^2 in 1993, following the provision of 60 nest boxes. Nest boxes were used at the same rate as natural nest sites and pairs using boxes in trees produced more eggs (but not significantly more fledglings) than other nest types. Nest boxes were located in buildings (43 boxes; a maximum of 11 used in a single year) and on trees (17 boxes; a maximum of 5 used in a single year).

A small controlled study in 1990–1993 in Devon and Cornwall, UK (3) found that activity in buildings used by barn owls *Tyto alba* as nesting and/or roosting sites dropped by 68% in

9 areas following the conversion or demolition of the building but was maintained in 3 other areas where a cavity and access hole were incorporated into the conversion or another nearby (<50 m away) building. There were no changes in eight control areas.

A replicated, controlled study in Hungary in 1995–2003 (4) using ring-recapture data found that juvenile barn owls *Tyto alba* fledged from nest boxes were significantly less likely to be recovered alive than those reared in church towers (25% of 75 nest box-reared birds recovered alive 1 year after fledging vs 40% of 116 church tower-raised birds). This difference in survival was only apparent in the first year after fledging, with similar proportions being recovered 6 years after fledging (28% of nest box-reared birds vs 41% of church tower-raised birds).

A replicated study of almost 800 barn owl *Tyto alba* nest sites (mainly boxes) over 10 years across the UK (5) found that owls were present at 70% of the 5,466 sites visited and bred at 54%. Occupancy and the proportion breeding decreased from 2001 to 2009 but were not affected by the number of boxes at each site. Occupancy was greater in rough grassland than arable land, and was lowest in pastoral areas. However the proportion breeding did not differ with habitat. Brood size was larger in rough grassland than arable sites. Female weight at laying did not differ over time whereas average laying date tended to be earlier and clutch and brood size tended to increase over the study, but not significantly. Occupancy rates were highest in pole boxes at northern sites for jackdaw *Coloeus monedula* and in the south and east for kestrel *Falco tinnunculus*. Jackdaw was positively correlated with the number of nest boxes present, kestrel and stock dove *Columba oenas* were not. Jackdaw was negatively correlated with occupancy of the nest box by barn owl. The Wildlife Conservation Partnership monitored 159–200 nest boxes ('pole box' or 'A-frame' in trees). From 2002 an additional 365–593 Barn Owl Monitoring Programme Network sites were also surveyed. Repeat visits were made to sites during the barn owl nesting season (April–October) to assess occupancy, gather breeding statistics and ring adults and chicks.

(1) de Bruin O. (1994) Population ecology and conservation of the barn owl *Tyto alba* in farmland habitats in Liemers and Achterhoed (The Netherlands). *Ardea*, 82, 1–109.
(2) Johnson P.N. (1994) Selection and use of nest sites by barn owls in Norfolk, England. *Journal of Raptor Research*, 28, 149–153.
(3) Ramsden D.J. (1998) Effect of barn conversions on local populations of Barn Owl *Tyto alba*. *Bird Study*, 45, 68–76.
(4) Klein Á., Nagy T., Csörgő T. & Mátics R. (2007) Exterior nest-boxes may negatively affect Barn Owl *Tyto alba* survival: an ecological trap. *Bird Conservation International*, 17, 273–281.
(5) Dadam D., Barimore C.J., Shawyer C.R. & Leech D.I. (2011) *The BTO Barn Owl Monitoring Programme: Final Report 2000–2009*.

6 Agri-chemicals

Leave headlands in fields unsprayed (conservation headlands)
Twenty-two studies from 14 experiments (including 2 randomized, replicated, controlled) from 5 countries found conservation headlands had higher invertebrate or plant diversity than other habitats; 12 studies from 10 experiments (3 randomized, replicated, controlled) did not. Twenty-seven studies from 15 experiments (of which 13 were replicated and controlled) from 5 countries found positive effects on abundance or behaviour of some wildlife groups. Nineteen studies from 13 experiments (12 replicated, controlled) from 4 countries found similar, or lower, numbers of birds, invertebrates or plants on conservation headlands than other habitats.

Buffer in-field ponds
We have captured no evidence for the effects of buffering in-field ponds on farmland wildlife.

Provide buffer strips alongside water courses (rivers and streams)
Three studies (including one replicated site comparison) from the Netherlands and the UK found riparian buffer strips increased diversity or abundance of plants, invertebrates or birds and supported vegetation associated with water vole habitats. Two replicated site comparisons from France and Ireland found that farms with buffer strips did not have more plant species than farms without strips.

Reduce chemical inputs in grassland management
Six studies (including a randomized, replicated, controlled before-and-after trial) from three countries found stopping fertilizer inputs on grassland improved plant or invertebrate species richness or abundance. Two reviews from the Netherlands and the UK found no fertilizer or low fertilizer input grasslands favour some birds and invertebrates. Five studies (including two replicated trials, of which one was randomized and one replicated) from three countries found no clear effects on invertebrates or plants.

Restrict certain pesticides
A small UK study found two fungicides that reduced insect abundance less than an alternative. A replicated, controlled trial in Switzerland found applying slug pellets in a band at the field edge was as effective as spreading the pellets across the field.

Make selective use of spring herbicides
A randomized, replicated, controlled study from the UK found that spring herbicides had some benefits for beneficial weeds and arthropods.

Use organic rather than mineral fertilizers
Fourteen studies (including four randomized, replicated, controlled trials) from six countries found areas treated with organic rather than mineral fertilizers had more plants or invertebrates or higher diversity. A randomized, replicated, controlled trial from the UK found

no effect on weed numbers. Two studies (including a small trial from Belgium) found organic fertilizers benefited invertebrates; a UK review found that in large quantities they did not.

Reduce fertilizer, pesticide or herbicide use generally

Thirty-four studies (including a systematic review) from ten countries found reducing fertilizer, pesticide or herbicide inputs benefited some invertebrates, plants or birds. Twenty-five studies (including seven randomized, replicated, controlled trials) from eight countries found negative or no clear effects on some invertebrates, plants or birds.

6.1   Leave headlands in fields unsprayed (conservation headlands)

- Twenty-two studies from 14 replicated, controlled experiments (of which 2 were randomized), including 2 reviews, from a total of 32 studies from 20 experiments (of which 17 were replicated and controlled), including 3 reviews, from Finland, Germany, the Netherlands, Sweden and the UK that investigated species richness and diversity of farmland wildlife found that conservation headlands contained higher species richness or diversity of invertebrates[6, 10, 11, 18–20, 22, 23, 29, 31, 44] or plants[9, 11, 16, 19, 21, 25, 27, 30, 32, 33, 36, 53] than other habitat types. Twelve studies (including a review) from ten replicated experiments (of which eight were controlled and three controlled and randomized) found that some or all invertebrates[35, 40, 45, 47, 56] or plants[15, 19, 25, 26, 43, 49, 52] investigated did not have higher species richness or diversity on conservation headlands compared to other habitat types. This included both replicated, controlled studies investigating bee diversity[45, 47]. Two replicated studies from the UK[43, 49] found that unfertilized conservation headlands had more plant species than fertilized conservation headlands.

- Positive effects of conservation headlands on abundances or behaviours of some or all species investigated were found by 27 studies from 15 replicated experiments (of which 13 were controlled) including 5 reviews out of a total of 36 studies from 20 experiments (17 replicated, controlled) including 5 reviews from Finland, Germany, the Netherlands, Sweden and the UK that investigated birds[9, 14, 31, 46, 56] (some studies looked at number of visits), mammals[12, 31, 50, 51] (some studies looked at number of visits), invertebrates[1, 2, 4, 6–11, 13, 15, 17, 18, 20, 29, 31, 32, 34, 36, 44, 48, 50, 51, 55, 56] and plant abundance/cover[1, 9, 11, 12, 15, 19, 21, 27, 32, 36, 50, 51, 56]. One review from the UK[37] found a positive effect on grey partridge populations but did not separate the effects of several other interventions including conservation headlands. Nineteen studies from 13 replicated (12 controlled) experiments and a review from Finland, Germany, the Netherlands and the UK found that some or all species of birds[14, 28, 46], invertebrates[2, 7, 10, 15, 20, 22, 23, 35, 39, 40, 45, 47, 48] or plants[1, 13, 26, 56] investigated were at similar, or lower, abundances on conservation headlands compared to other management. One review from the UK and a study in Germany found conservation headlands had a positive effect on plants and some but not all invertebrates[41] or rare arable weeds[5] but did not specify how.

- All eight studies from the UK and Sweden that investigated species' productivity, from three replicated (two controlled) experiments, including two reviews, found that grey partridge productivity or survival was higher in conservation headlands (or in sites with conservation headlands), compared to other management[1–3, 9, 17, 30, 55, 57]. One replicated study from the UK[57] found that conservation headlands did not increase the ratio of young to old partridges. A before-and-after study from the UK[38] found that some invertebrates in conservation headlands survived pesticide applications to neighbouring fields. A review found that crop margins reduce the effects of spray drift on butterflies[42].

- A replicated study from Germany[40] and a review[54] found that conservation headlands appeared to prevent or reduce the establishment and spread of pernicious weeds.

Background
Conservation headland management involves restricted fertilizer, herbicide and insecticide spraying in a 6 m margin of sown arable crop. The prescription allows selected herbicide applications to control injurious weeds or invasive alien species.

A replicated, controlled study of cereal headlands on an arable farm in northeast Hampshire, UK (1) found that grey partridge *Perdix perdix* brood size, abundance of invertebrates (chick food and pest predators) and weed density tended to be greater on unsprayed compared to sprayed headlands. Grey partridge brood size was significantly larger on plots with unsprayed (6.4) compared to sprayed headlands (2.2). Abundance of chick food species (true bugs (Heteroptera), caterpillars (Lepidoptera) and sawfly (Hymenoptera: Symphyta) larvae, leaf beetles (Chrysomelidae) and weevils (Curculionidae)) was significantly higher and aphid predators (spiders (Araneae); ground beetles (Carabidae); rove beetles (Staphylinidae)) tended to be greater in unsprayed (chick food: 180/50 sweeps; predators: 7.8) compared to sprayed headlands (chick food: 62/50 sweeps; predators: 4.6). Weed densities tended to be higher on unsprayed ($5/m^2$) compared to sprayed headlands ($3/m^2$), but only 1 of 21 species was significantly higher. Three areas were split into 2 treatment plots: sprayed with conventional pesticides or 6 m headlands left unsprayed. Grey partridge brood size was recorded from August–September 1983. Insects were sampled using a sweep net (50 sweeps in June) and weed species were recorded within 10 quadrats in each headland. This study was part of the same experimental set-up as (2–4, 6, 9, 24).

A replicated, controlled study in 1980–1983 in arable fields on a farm in Hampshire, UK (2) found that grey partridge *Perdix perdix* broods were significantly larger in 1983 on plots with conservation headlands compared to control headlands sprayed with fungicides and herbicides (averages of 5.1–10.3 chicks/brood for 29 broods in unsprayed areas vs 1.8–2.4 chicks/brood for 39 broods on controls). No differences were found in 1980–1981, before conservation headlands were implemented. However, more broods were found on conventional fields, reflecting more pairs (49 vs 37) in the spring. Areas with conservation headlands had significantly higher abundances of true bugs (Heteroptera), leaf beetles (Chrysomelidae) and weevils (Curculionidae) than sprayed headlands (1.9 individuals/50 net sweeps vs 1.4 individuals/50 sweeps for true bugs; 1.0 individuals/50 sweeps vs 0.7 individuals/50 sweeps for leaf beetles and weevils). Sawflies (Hymenoptera) and butterfly/moth (Lepidoptera) larvae abundance did not vary significantly. The author argued that larger broods were the result of higher chick survival due to more food insects being present. This study was part of the same experimental set-up as (1, 3, 4, 6, 9, 24).

A replicated, controlled study in 1984 on the same farm in Hampshire as in (2) and on eight sites in East Anglia, UK (3) found that grey partridge *Perdix perdix* broods had significantly higher survival and were significantly larger on plots with conservation headlands compared to control plots with conventionally-sprayed headlands (average of 75% survival and 7.8–10.0 chicks/brood for 5 broods on conservation headland plots vs 60% and 4.7–7.5 chicks/brood for 4 broods on conventional plots; 196 broods surveyed). This paper also describes similar, although less conclusive effects on two non-native gamebirds (red-legged partridge *Alectoris rufa* and ring-necked pheasant *Phasianus colchicus*). This study was part of the same experimental set-up as (1, 2, 4, 6, 9, 24).

A replicated, controlled, paired study in 1984 of headlands in 14 arable fields in Hampshire, UK (4) found that butterfly (Lepidoptera) abundance was greater on conservation (unsprayed) headlands than on sprayed headlands. Twenty-two species of butterfly were recorded, 21 of which were on conservation and 17 on sprayed headlands. Significantly more individuals were found on conservation (868) than on sprayed headlands (297). Of the seventeen species recorded on more than one transect section, 13 were significantly more abundant on the conservation (11–140) than sprayed headlands (0–59). For half of the 14 fields a 6 m strip around the edge (headland) was left unsprayed, the remainder received conventional pesticide applications. Butterflies were sampled along a transect at least once a week from 9 May to 15 August 1984. Sprayed and conservation headlands were paired with similar adjacent habitats. This study was part of the same experimental set-up as (1–3, 6, 9, 24).

A study of cereal fields on 20 farms in Germany between 1978 and 1981 (5) found that unsprayed field margins had a positive effect on rare and endangered arable weeds. Results are not provided, but authors note that the project was such a success, particularly on the calcareous soils of the Eifel Mountains and intensively farmed soils of the Lower Rhine, that various states in the Federal Republic began providing financial support to protect biodiversity. Farmers made a total of 15 km, later 20 km, of 2–3 m wide unsprayed field margins within their cereal fields to protect arable weeds.

A continuation of the same replicated, controlled, paired study as in (4) in Hampshire, UK (6) found that butterfly (Lepidoptera) abundance was greater on conservation headlands than on conventional headlands over a further three years (1985–1987). Between 1984 and 1987, 29 species of butterfly were recorded, of which 13–21 were on conservation and 13–17 on conventional headlands each year. Significantly more individuals were found on conservation headlands (222–472/km) than on conventional headlands (80–259/km) in all years. This study was part of the same experimental set-up as (1–4, 9, 24).

A replicated, controlled study of the headland of a wheat field in Hampshire, UK (7) found that plots not sprayed with herbicides had significantly higher densities of arthropods than sprayed plots. This was particularly the case for non-pest species which are important for feeding birds and predatory arthropod groups. No significant between-treatment differences were found in the total pitfall trap catch of the two most common ground beetles (Carabidae) *Pterostichus melanarius* and *Agonum dorsale*. However, a significantly greater proportion of female *A. dorsale* were caught in treated plots than in untreated plots. Unsprayed headland plots had greater weed species, densities, biomass and cover. Along one field boundary the headland crop was divided into eight 100 x 12 m plots, which were alternately sprayed and unsprayed with herbicides in April 1988. Five vacuum-suction samples were taken (0.5 m^2) before and 5 times (up to 90 days) after spraying. Five pitfall traps (7 cm diameter) were placed within gaps in enclosure boundaries (6 x 10 m) within plots and were emptied twice weekly from 20 June to 29 July. Weeds were assessed in 10 quadrats/plot (0.25 m^2) in June.

A replicated, controlled study of 2 headlands over 2 years 1989–1990 in England (8) found that in 1990 there were significantly higher proportions of some species of hoverfly (Syrphidae) adults (marmalade hoverfly *Episyrphus balteatus*, *Metasyrphus* spp.) in conservation headlands (20–24%) compared to fully sprayed headlands (9–12%). There were also higher proportions of *E. balteatus* adults feeding and lower proportions inactive in conservation (feeding: 15–90%; inactive: 0–85%) compared to fully sprayed headlands (feeding: 0–35%; inactive: 0–100%). Behaviour of *Metasyrphus corollae* did not differ with treatment.

There were no significant differences in 1989. Weed density and floral area tended to be higher in conservation compared to sprayed plots. Headlands were divided into 3 or 5 replicate plots of 75–100 m x 12 m wide, each containing the 2 pesticide treatments. A set route was walked to record hoverflies encountered during a fixed time period. Weekly counts of weeds and hoverfly eggs on wheat were made in 15–21 quadrats/plot from May–July. Aphids (Aphidoidea) were also recorded but results are not included here.

A paired, replicated, controlled study in the 1980s in cereal fields in southern and eastern England (9) found higher plant species richness (on average 7 vs 2 species/0.25 m^2), biomass (10 vs 1 g/0.25 m^2) and percentage weed cover (14% vs 3%) in conservation headland plots compared with fully sprayed headland plots on one Hampshire farm. Several species of rare arable weeds occurred more frequently and in higher abundance in conservation headlands. Total numbers of chick-food items, true bugs (Heteroptera), sawflies (Tenthredinidae) butterfly/moth (Lepidoptera) larvae and beetles (Coleoptera) were higher in conservation headlands on the Hampshire farm ((2), Sotherton 1989) and butterfly/moth abundance was higher in field margins adjacent to conservation headlands with 2–4 more species observed there ((4), (6), Dover 1991). Examination of the digestive tract of polyphagous beetles (beetles that feed on many types of food) revealed that a higher proportion of beetles were better fed in conservation headlands. In every year 1983–1986 in southern and eastern England, the brood size of grey partridge *Perdix perdix* was higher on blocks of cereal fields with conservation headlands (6–10 chicks respectively) compared with conventionally sprayed headlands (3–8 chicks) (Sotherton & Robertson 1990). Breeding density of grey partridges on the Hampshire farm increased from 4 to 12 pairs/km^2 between 1979 and 1986. No such increases were recorded on adjacent farms where pesticide regimes remained unchanged. The yield of grain from conservation headlands was 6–10% lower than that from fully sprayed headlands. Grain moisture levels were around 1% higher and weed seed contamination was also higher in conservation headlands. The Hampshire part of this study was part of the same experimental set-up as (1–4, 6, 24).

A replicated, controlled study of headlands (outer 6 m) of 8 barley fields over 1 year, 1988, at 3 locations within the Breckland Environmentally Sensitive Area in East Anglia, UK (10) (same study as (13)) found that ground beetles (Carabidae) and true bugs (Heteroptera), but not spiders (Araneae), were more abundant in conservation headlands (restricted pesticides) than sprayed headlands. Ground beetles and true bugs were significantly more abundant in conservation headlands (ground beetles: 160; true bugs: 30) than sprayed headlands (ground beetles: 70; true bugs: 25), spiders did not differ significantly (110 vs 100). Ground beetles were twice as abundant in crops adjacent to conservation headlands than adjacent to sprayed headlands. There were significantly more ground beetle and true bug species in conservation headlands (ground beetles: 18; true bugs: 3 species) than in sprayed headlands (ground beetles: 15; true bugs: 2); species diversity did not differ significantly (ground beetles: 5–8; true bugs: 2; spiders: 2–3). Spider diversity and biovolume increased with age of site. True bug nymphs (for example field damsel bug *Nabis ferus*) penetrated further into the crop adjacent to conservation headlands than sprayed headlands. Thirty-five pitfall traps were set up in 6 x 50 m grids at each site (in headlands and crops) and emptied after 14 days in June–July 1988. A Dietrick Vacuum sampler was used along 5 transect lines (0–15 m into the crop); 2 samples were taken each of 5 subsamples (each 0.4 m^2).

A replicated, controlled study in summer 1988 in four headlands, adjacent arable fields and field margins in western Germany (11) found that vascular plants, hoverflies (Syrphidae) and ground beetles (Carabidae) all benefited from extensive management in unsprayed headlands. Species richness of all three taxa, as well as abundance and diversity of hoverflies and ground beetles and vegetation cover were higher in the unsprayed margins than in the adjacent fields and the conventional control edge. Field margins adjacent to conservation headlands also held higher numbers of plant, hoverfly and ground beetle species than the margins next to the control edge. Three unsprayed headlands (up to seven years old) and one conventional field edge were compared to three adjacent conventional cereal fields and four field margins next to the headlands. Plants were surveyed in May to September. Ground beetles were sampled weekly from April to August in 6 pitfall traps (9.5 cm diameter) on each site. Hoverflies were monitored between May and August. Visual observations were made along a 100 m transect in each site. In addition, 6 yellow bowls (23 x 16 x 5 cm) were placed in the field margins and 1 in the field centre.

A replicated, controlled study in 1986–1988 of two wheat fields in Oxfordshire, UK (12) found that conservation and unsprayed headlands were used more frequently by wood mice *Apodemus sylvaticus* than sprayed headlands and mid-field. Preference indices for wood mice were six for conservation headlands, six to seven for unsprayed headlands, two to four for sprayed plots and three for mid-field. Mice showed a significant preference for the mid-field over sprayed headlands. Conservation and unsprayed headland plots contained significantly higher densities of black grass *Alopecurus myosuroides*, wild oats *Avena* spp., sterile brome *Bromus sterilis* and forget-me-not *Myosotis arvensis*. Abundance did not differ between the sprayed headland and mid-field for any weed species. In 1986, 4 of 15 orders of invertebrate were significantly more abundant in unsprayed than sprayed headlands (springtails (Collembola): 6 vs 1 m^2; true bugs (Hemiptera): 23 vs 8 m^2; flies (Diptera): 142 vs 24 m^2; parasitoid wasps (Parasitica): 18 vs 5 m^2). In 1987 there was no significant difference between invertebrate abundance in sprayed, unsprayed or conservation headland plots. In one field alternate plots (20 x 10 m) along the headland were either conventionally sprayed or unsprayed or in 1987, conservation headland plots. Vegetation was sampled in 5–10 quadrats (0.25 x 0.25 m) in 8–11 plots/treatment in July. In the conventionally sprayed control field, plants were sampled in quadrats of 0.06 m^2 at 1, 5 and 8 m from the hedge and in the centre of the field. Invertebrates were sampled at 3–5 random positions within 3–8 plots/treatment using a D-Vac sampler in July 1986–1987. Wood mice were radio-tracked at 10 min intervals at night in May–August.

Further results for ground beetles (Carabidae) from a replicated, controlled study of headlands of eight barley fields in 1988 in East Anglia, UK (10) are presented in a second paper (13). As shown by (10) ground beetles tended to be more abundant in conservation headlands (3–14/trap) than sprayed headlands (3–6/trap) and the main crop (3–9/trap). Species richness was greater on conservation headlands (32 species) than fully sprayed headlands (24) but similar to the main crop (31). Significantly higher numbers of ground beetles were found in headlands than field verges (0–4/trap). There was no significant difference between numbers in verges adjacent to different treatments. There was no significant difference between the vegetation cover under different treatments or in the crop. Plant cover was measured in five 25 x 25 cm quadrats in each grid.

A replicated, controlled study in 1992–1993 of arable fields on eight farms in the Netherlands (14) found that unsprayed field margins had a higher abundance of blue-headed wagtail *Motacilla flava flava* than sprayed edges. Blue-headed wagtails made 1.5–2.4 visits/km to unsprayed margins compared to just 0.5 visits/km for sprayed margins. Numbers of Eurasian skylark *Alauda arvensis* and meadow pipits *Anthus pratensis* did not differ significantly in sprayed and unsprayed margins (skylark: 0.4 vs 0.2–0.4; meadow pipit: 0.1 vs 0.1). Blue-headed wagtails and skylarks visited field margins more than field centres and sprayed edges bordering ditches more than sprayed edges adjacent to a second plot. Strips 6 m wide along field edges were left unsprayed by herbicides and insecticides (total length 2,560–3,790 m/year) and were compared to sprayed edges in the same field and to the sprayed field in 1992–1993. Farmland birds were sampled using a linear transect census with all birds visiting field margins recorded and a similar size strip in the centre of each field recorded. Birds were sampled 10–12 times between April and mid-July. This study was part of the same experimental set-up as (19–21, 29, 31).

A replicated, controlled study in 1992–1994 of headlands of spring cereal fields on four farms in central and southern Finland (15) found that arable weed density and the abundance of some insect groups were higher in unsprayed headlands compared to sprayed headlands; weed diversity did not differ. Weed density was significantly higher in unsprayed conservation headlands (275–420/m) than sprayed headlands (160–371/m). However, numbers of species were similar in both treatments (31–38 vs 31–36). The following insect groups were more abundant in conservation headlands than sprayed headlands: leafhoppers/planthoppers/aphids (Homoptera) 112–1,401 individuals vs 85–706; flies (Diptera) 77–80 vs 69–74; bees/wasps/ants (Hymenoptera) 34–58 vs 29–46; true bugs (Heteroptera) 9–109 vs 7–43; beetles (Coleoptera) 7–14 vs 5–7. In contrast, thrips (Thysanoptera) were more abundant in sprayed headlands (746–1,846 vs 591–960). Twelve conservation (no pesticides) and control headlands (herbicide and insecticides) 4–6 m wide and 100–200 m long were established. Weed abundance was sampled in 3 pairs (0.5 and 3 m from the crop edge) of 0.25 m^2 quadrats per headland in late July. Insects were sampled using a D-Vac (5 x 10 s per sample) at the same locations as plant quadrats in early July. A sweep net sample (2 x 15 sweeps) was also taken from each headland.

A replicated site comparison study of cereal fields in the Lower Rhine area of Germany (16) found that plant species diversity was higher in fields with unsprayed margins than sprayed margins. Plant diversity (Shannon index) was higher in fields with unsprayed margins (–2.1 to –2.4 vs -1). The average number of species was also higher in unsprayed margins compared to sprayed centres of fields in winter (4–10 species vs 0–2) and summer crops (3–7 species vs 1–3). By the end of the study there were 100 species recorded within the study site (mean 44 species/field), including 9 categorized as highly endangered by the Red Data Book, which had recovered in the local area. Fields were either managed according to the regulations of the government field margins programme (5 m margins, no pesticides, limited fertilizer) or were managed conventionally with intensive pesticide and fertilizer use. Plant species diversity and floristic richness were sampled along transects within cereal fields.

A replicated, controlled study in 1991–1993 of cereal fields on 10 pairs of farms in central and southern Sweden (17) found that grey partridge *Perdix perdix* brood size, chick survival and abundance of invertebrates tended to be higher on farms with unsprayed headlands (6 m wide) compared to those sprayed conventionally. Mean brood size tended to be higher on experimental farms (half headlands unsprayed: 7–9 chicks) than on control farms (sprayed: 3–8). Numbers of broods (10–19 vs 4–16), chick survival rate (26–54% vs 11–47%) and numbers of partridge pairs in the spring (20–30 vs 15–24) also tended to be higher on experimental farms. However, none of these differences were statistically significant. Mean density of chick food insect groups (true bugs (Heteroptera), aphids/leafhoppers/planthoppers (Homoptera), weevils (Curculionidae), leaf beetles (Chrysomelidae), larvae of butterflies/moths (Lepidoptera) and sawflies (Tenthredinidae)) was significantly higher on unsprayed (25–74) compared to sprayed headlands of wheat (5–32). Farm pairs (control and experimental) were within 5 km of each other and of similar size, cropping and agricultural practice. On the experimental farm, the headlands left unsprayed (50%) were swapped each year (1991–1993). Partridge counts were undertaken in spring and after harvest using dogs to flush birds. Ten invertebrate samples (0.5 m^2) were taken from each headland during the first week in July using vacuum-suction.

A small replicated, controlled study from 1990 to 1992 of the headland (outer 6 m) of a wheat field at Ixworth Thorpe on the southern edge of Breckland in East Anglia, UK (18) found that ground beetles (Carabidae) were more abundant in conservation headlands (no herbicides or insecticides) than sprayed headlands (as main field). Conservation headlands had a significantly greater abundance of ground beetles (1,474) than sprayed headlands (938). Species diversity was higher in conservation headlands (41) than sprayed headlands (35). Different species reacted differently to treatments. There were a number of species that were restricted to conservation headlands and one restricted to sprayed headlands. Numbers of species and overall abundance varied with season. Two 120 m strips of each treatment were established in a randomized block design along 1 headland of the 19 ha field. Ground beetles were sampled using 3–5 pitfall traps in the middle of each plot, 3 m from the field boundary. Catches were collected every 1–2 weeks from February–August. Aphid (Aphidoidea) numbers were also sampled but are not presented here.

A replicated, controlled, paired study of arable field edges from 1990 to 1992 in the Netherlands (19) found that unsprayed field margins had greater plant cover, broad-leaved species, butterfly (Lepidoptera) abundance and insect groups than sprayed margins. Plant cover was significantly higher in 6 m (outer 3 m: 35%; inner 3 m: 26%) and 3 m unsprayed margins (36%) than sprayed margins (outer 3 m: 6%; inner 3 m: 3%). Numbers of broad-leaved species were also higher in 6 m (outer 3 m: 13 species; inner 3 m: 11) and 3 m (12) unsprayed strips than sprayed edges (outer 3 m: 3; inner 3 m: 2). Grass species did not differ (2–3). Numbers of butterfly species were significantly higher in unsprayed margins (6–7/300 m^2) compared to sprayed margins (1–2/300 m^2). Density did not differ between 3 m (6/300 m^2) and 6 m (7/300 m^2) unsprayed margins. Numbers on adjacent ditch banks were also higher for unsprayed (18–20) than sprayed margins (9–11). The number of insect groups in the upper vegetation was higher in the unsprayed (12–14) than sprayed margins (8–11). The predominant groups were flower-visiting insects such as hoverflies (Syrphidae) and ladybirds (Coccinellidae). Insect density was also significantly higher in unsprayed (3 m: 53/100 m; 6 m: 31/100 m) compared to sprayed margins (3 m: 20/100 m; 6 m: 12/100 m). Margins 3 m x 100 m and 6 m x 400 m were left unsprayed by herbicides and insecticides and compared to sprayed edges in the same field. Plant species were sampled in 75 m^2 plots within margins in June. Butterflies were sampled on 3 m (8 farms) and 6 m (6 farms) margins 11 times between mid-May–July. Insects in the upper parts of plants were sampled twice/plot at the end of June with a sweep net. This study was part of the same experimental set-up as (14, 20, 21, 29, 31).

A replicated, controlled, paired study of wheat field edges on 10 farms from 1992 to 1993 in the Netherlands (20) found that unsprayed field margins had greater insect diversity and abundance in the upper parts of plants than sprayed margins. The average number of insect groups was higher in the unsprayed margins (12–14) than sprayed margins (8–11). Insect density was also significantly higher in the unsprayed winter wheat margins (31–41 vs 10/100 m). Of the 18 groups found on 50% of the sites, 11 in 1992 and 9 in 1993 were significantly more abundant on unsprayed edges. The greatest effect was on flower-visiting insects and aphid predators (unsprayed: 62–73% of all insects; sprayed: 24–32%). Only three groups (long-legged flies (Dolichopodidae), crane flies (Tipulidae) and moths (Heterocera)) were less abundant in the unsprayed edges. Strips 6 m x 450 m along field edges were left unsprayed by herbicides and insecticides and were compared to sprayed edges in the same field. Insects were sampled once or twice in June. Ten sub-samples, 1.5 m from the field edge, were taken using a sweep net (total area sampled 20 m^2/100 m). Aphid (Aphidoidea) abundance and dispersal was also recorded, but results are not presented here. This study was part of the same experimental set-up as (14, 19, 21, 29, 31).

A replicated, controlled study in 1990–1994 of arable fields on 15 farms in the Netherlands (21) found that unsprayed field margins contained higher plant diversity, abundance and more important/rare plant species than sprayed margins or fields. Species diversity was significantly higher in unsprayed edges than in sprayed edges (sugar beet: 24 vs 16 species/75 m^2; potatoes: 17 vs 8; winter wheat: 17 vs 6). Field centres had the lowest diversity (2–10 species/75 m^2). Thirteen, 9 and 30 species were found only in the unsprayed edges in sugar beet, potatoes and wheat respectively. Unsprayed edges had significantly higher floristic values (scoring system based on the importance of different plant species in terms of rarity) than sprayed edges: by a factor 5.2 in sugar beet, 2.8 in potatoes and 7.2 in winter wheat; values were lowest in field centres. The cover (biomass and height) of farmland plants was significantly higher in unsprayed compared to sprayed edges (8–52% vs 1–13%) and lowest in field centres (0–3%). A total of 5–20 fields were studied from 1990 to 1994. Strips 3–6 m x 100 m long along field edges were left unsprayed by herbicides and insecticides and were compared to sprayed edges in the same field and to the sprayed field. Vegetation was sampled in 75 m^2 plots in mid June to mid July. This study was part of the same experimental set-up as (14, 19, 20, 29, 31).

A replicated, controlled study in summer 1995 in five different field margin types and one control (winter wheat field) in an intensively farmed, homogenous landscape near Göttingen, Germany (22) (same study as (23, 40)) found higher arthropod species richness on potted mugwort *Artemisia vulgaris* plants placed in unsprayed cereal strips (headlands) compared to the cereal field, but not compared to other margin types. The predator-prey ratio in the headland did not differ from the control but was significantly lower than in a six-year-old uncultivated field margin. The effect of unsprayed headlands on individual arthropod numbers was species-dependent with some species (e.g. the aphid *Macrosiphoniella oblonga* and the fly *Oxyna parietina*), but not all, being found in higher individual numbers in the headlands than in the control. Investigated margin types apart from the unsprayed cereal headlands were wildflower strips (wildflower seed mixture or *Phacelia* spp. only) and uncultivated margins (one- and six-years-old). There were four replicates of each margin type. Potted mugwort plants were placed in all margin types and the control. All herbivores feeding on the plant and their predators were recorded during six visits in June and July. In September, all mugwort plants were dissected in the lab to assess numbers of arthropods feeding within the plants.

A replicated, controlled study in summer 1995 investigating five field margin types plus controls in intensively managed farmland near Göttingen, Germany (23) (same study as (22, 40)) found higher species richness of arthropods colonizing potted mugwort *Artemisia vulgaris* plants in fertilized but unsprayed cereal strips than in the unsprayed cereal control edges. However, arthropod species numbers on mugwort did not differ between cereal strips and any of the other established margin types. Besides the cereal strips, 1- and 6-year-old naturally regenerated margins, wildflower strips (19 species sown), phacelia strips (*Phacelia tanacetifolia* plus 3 species), and cereal control edges were investigated. Potted mugwort plants (four pots) were placed in all margin types and the controls. Mugwort plants were visited six times in June and July to count all herbivores feeding on the plants and their predators before being taken to the lab in September to assess all arthropods feeding within the plants. Vegetation of all margins was surveyed in June.

A replicated, controlled, paired study in 1985–1987 of butterfly (Lepidoptera) behaviour in headlands of 14 cereal fields in northeast Hampshire (24) found that flight speeds tended to be slower and more time was spent resting, interacting and foraging in conservation headlands (no broadleaved herbicides) than those with conventional herbicide applications. Flight and transit speeds of male Pieridae and transits of female green-veined white *Pieris napi* were significantly slower in conservation headlands. In contrast gatekeeper *Pyronia tithonus* males (in 1986) were significantly slower in the sprayed headlands; sample sizes were too small for other species. In fields with sprayed headlands, spring emerging large white *P. brassicae*, green-veined white and small white *P. rapae* were principally associated with the hedgerow, whilst in fields with conservation headlands they were associated with the headlands. In sprayed headlands, the principal activity was flight, whereas in conservation headlands there was an increase in time spent resting, interacting and particularly foraging. Butterflies that emerged in the summer tended to have less of an association with conservation headlands than spring-emerging butterflies. Limited data were available for meadow brown *Maniola jurtina* and gatekeeper. Half of the 14 fields were sprayed with conventional pesticides and the other half had conservation headlands. The behaviour and location (hedgerow or headland) of 5 species of butterfly were observed during the middle of the day along 4–8 headlands. This study was part of the same experimental set-up as (1–4, 6, 9).

A replicated, controlled study in 1992–1994 of cereal headlands in Sweden (25) found that numbers of plant species, including species of conservation interest, were higher in margins without herbicide and pesticide applications at one of two sites. On Öland, there were significantly more species in field margins with no herbicides (6 species) and with no herbicides or fertilizers and reduced sowing rates (5.0) than in conventional plots (3.0). The same pattern was seen for cover per species (no herbicides: 8%; no herbicides/pesticides: 6%; conventional: 4%). Numbers did not differ significantly in Uppland (species: 5, 4 and 3; cover: 4%, 5% and 4% in the different treatment plots respectively). On Öland frequencies of rare species increased, particularly in plots with no herbicides (1992: 8–10 rare species; 1994: 45–50) compared to conventional plots (1992: 2; 1994: 28). Red List species established in all Öland plots over the study (1992: 0; 1994: 9–12) with higher cover in plots with no herbicides (1994: 3–6% vs 1%). No rare or Red List species occurred at Uppland. On Öland 4 fields with 2 to 4 treatment blocks and in Uppland, 2 fields with 2 blocks were established with treatment plots of 6 x 20 m within margins. Plant species were recorded in each plot on two or three visits each year.

A replicated, controlled, randomized study of cereal headlands on 26 farms in East Anglia, UK (26) found no significant difference in plant species richness or density between conservation and sprayed headlands, but plant composition did differ. Although figures tended to be higher in conservation headlands (6–12 m wide; restricted pesticide applications; selected herbicides only) there was no significant difference between conservation and conventionally sprayed headlands in terms of species richness (10 vs 5), plant density (99 vs 47) or grass tiller count (73 vs 69). Conservation headlands had a significantly greater proportion of annuals and biennials to perennials (0.7 vs 0.5) and ratio of broadleaves (dicotyledons) to grasses (monocotyledons) (0.8 vs 0.6). On nine farms within the Breckland Environmentally Sensitive Area three conservation headlands and one sprayed headland were selected. One conservation headland was randomly selected from each of an additional 17 farms. Three transects 50 m apart were located within randomly selected 100 m sections of headlands. Along these, 3 quadrats (0.5 x 0.5 m) at 1, 3 and 5 m from the field boundary in 6 m headlands and 2, 6 and 10 m from the boundary in 12 m headlands were surveyed.

A replicated, controlled study of headlands of three winter rye fields in 1991 in the Netherlands (27) found that weed species richness, abundance and biomass was higher in unfertilized crop edges than those that had received fertilizer. Species richness (17 vs 14 species/m^2), abundance (276 vs 170 plants/m^2) and biomass (88 vs 35 g/m^2) were higher in unfertilized headlands. Species richness, plant numbers and total weed biomass decreased significantly with distance from the field boundary in fertilized plots. Some individual species followed the same trend; others were more abundant in the fertilized plots and some showed no overall effect of treatment. Thirty-six plots were established within headlands; half were fertilized and half unfertilized; no herbicides were applied. Above-ground weed biomass and the number of individuals of each species were sampled in quadrats (0.5 x 2 m) at distances of 0.25, 1.25 and 2.25 m from the boundary in August.

A study of habitat use by yellowhammers *Emberiza citrinella* in 1994 and 1995 on a mixed farm in Leicestershire, UK (28) found that conservation headlands were not used significantly more than adjacent crops (1.0 vs 0.7 yellowhammers flushed). The outer tramline of each field was walked three times in June 1994 and 1995. Yellowhammers flushed from the conservation headland and equivalent area in the adjacent crop were recorded.

A replicated, controlled, paired study from 1990 to 1992 of arable field edges on 12 farms in the Netherlands (29) found that unsprayed field margins had greater butterfly (Lepidoptera) abundance than sprayed margins. Butterfly numbers were significantly higher in the unsprayed edges of winter wheat in both years (10–12 butterflies/100 m^2) and potatoes in 1992 (5/100 m^2) compared to sprayed edges (wheat: 2–3; potato: 1). The same was true for numbers of species: unsprayed winter wheat (3–4 species/100 m^2) and potatoes in 1992 (3/100 m^2) compared to sprayed edges (wheat: 1–2; potato: 1). In all six individual species, abundance was greater in unsprayed compared to sprayed edges (in one or both years and crops). Strips 6 m x 100 m or 400 m long along field edges were left unsprayed by herbicides and insecticides and were compared to sprayed edges in the same field. Butterflies were sampled once a week on the crop edges and adjacent ditch banks nine times from mid-May to July in 1990 and 1992. This study was part of the same experimental set-up as (14, 19–21, 31).

A 1998 literature review (30) looked at the effect of agricultural intensification and the role of set-aside on the conservation of farmland wildlife, particularly gamebirds and endangered annual arable wildflowers. It found a replicated study on farms in three English counties showing that there were greater numbers of arable weed species in headland plots of winter cereals that received no herbicide and fertilizer (23 species) compared to those that received fertilizer (no herbicide; 20 species) and the controls (normal fertilizer and herbicide applications; 6–8 species) (Wilson 1994). A further three studies were found, two in the UK (2, 4) and one in Sweden (Chiverton 1999) showing that gamebird (grey partridge *Perdix perdix*) chick survival rates were significantly higher in conservation headlands with reduced pesticide inputs compared to controls receiving the usual pesticide application.

A replicated, controlled study from 1990 to 1993 of 16 arable fields on farms in the Netherlands (31) found that unsprayed field margins had greater plant and insect diversity and abundance and more visits by blue-headed wagtails *Motacilla flava flava* and field mice *Apodemus* spp. than sprayed margins. Results for vegetation, butterflies (Lepidoptera), insects in the upper parts of the vegetation and birds are presented in (14, 19–21, 29). Unsprayed margins contained greater diversity and abundance of ground-dwelling invertebrate species (14–17 vs 12–14) and ground beetle (Carabidae) activity density tended to be significantly higher than in sprayed margins. Orb-weaving spiders (Araneida) (in winter wheat) and beetles (Coleoptera) (in sugar beet) were more abundant in unsprayed edges than in sprayed edges (in one year); other ground-dwelling groups did not differ with treatment. More field mouse visits were recorded in unsprayed (38 visits) than in sprayed cereal edges (27). Strips 3–6 m x 100–450 m long along field edges were left unsprayed by herbicides and insecticides and were compared to sprayed edges in the same field. Floristic value was sampled in 75 m^2 plots (mid-June to mid-July). Insects were sampled by: sweep netting in wheat (June), butterfly transects along crop edges (mid-May to July) and pitfall traps (11cm diameter, May–July).

Blue-headed wagtail, Eurasian skylark *Alauda arvensis* and meadow pipit *Anthus pratensis* visits were recorded in 6 m winter wheat edges using a linear transect census and small mammals were live trapped in winter wheat.

A 1999 review of research on unsprayed field margins in northwest Europe (32) found that both plants and invertebrates were enhanced in areas with unsprayed margins. Three studies reported that numbers of plant species and abundance were higher in or adjacent to unsprayed margins (e.g. (25, 33)). One study also found that rare arable weeds returned (Schumacher 1984). Three studies reported that buffer strips of 8–23 m were necessary for caterpillars of the large white butterfly *Pieris brassicae* for more toxic insecticides, whereas strips of 1 m were sufficient for other insecticides (Sinha *et al.* 1990, Davis *et al.* 1991, de Jong & van der Nagel 1994). An additional study reported that a buffer strip 3 m wide strongly decreased the effects on aquatic species next to a sprayed field (de Snoo & de Wit 1998).

A replicated, controlled, paired study in 1991–1992 of ditch banks on arable farms in the Netherlands (33) found that plant diversity and the value of the vegetation in terms of species rarity was significantly higher on ditch banks along unsprayed edges of winter wheat (65 species; floristic value: 2,201) than those sprayed with pesticides (50 species; floristic value: 1,181). There was no significant difference on banks along unsprayed and sprayed edges of sugar beet (species: 48 and 41; floristic values: 3,616 and 3,029 respectively) and potato crops (species: 46 and 41; floristic values: 1,961 and 1,864 respectively). Frequency and cover of species and floristic value of vegetation (scoring system based on the importance of different plant species in terms of rarity) was recorded in two plots on each ditch, one along a sprayed and one an unsprayed edge of sugar beet (7), potato (8) and winter wheat (20) fields in June–July.

A replicated, randomized study in Oxfordshire, UK (34) found that from 1995 to 1996 spraying naturally generated margins with herbicides resulted in significantly lower numbers of invertebrates than leaving them unsprayed (650 vs 1,275 invertebrates respectively). The same was true for spiders (Araneae) in all seasons (56–138 vs 107–392), true bugs (Heteroptera) in September (8 vs 27) and leafhoppers (Auchenorrhyncha) in July and September (39–60 vs 112–171). Existing field margins (0.5 m wide) were extended by 1.5 m in October 1987. These were rotavated and left to naturally regenerate or sown with a wildflower seed mix. Six management treatments were applied with six replicates in a randomized block design. Fifty metre-long plots received one of six treatments: sprayed once a year in summer; uncut; cut once in summer; cut in spring and summer; cut in spring and autumn; cut in spring and summer (hay left lying). Invertebrates were sampled using a D-Vac suction sampler at 10 m intervals along each plot in May, July and September in 1995–1996.

A replicated, controlled, randomized study in 1996–1997 of a silage field in Scotland (35) found that ground beetle (Carabidae) abundance and diversity was not consistently higher in headlands (no fertilizers or pesticides) than in main fields (fertilizers, no pesticides). Total abundance of ground beetles was higher mid-field (128–278 individuals) than in the headland (43–201). The field boundary had an intermediate abundance in 1996 (77 individuals) and a higher abundance in 1997 (612). In 1996 species diversity was higher mid-field (13) than in the headland (9–12). This trend was reversed in 1997 (headland: 10–14; mid-field: 6); the field boundary had the highest diversity both years (16–32). The headland received three treatments: uncut, annual cut (August) and three annual cuts (May, June, August). These were assigned randomly within blocks to three 10 x 10 m plots. Cattle were excluded in April–October and plots were intermittently grazed by sheep in October–February. Ground beetles were sampled using pitfall traps in late May to mid-July and late August to early October. There was a line of 3 pitfall traps in the centre of each plot within the field boundary and 3 rows in the main field 80 m from the field edge. Results from 1997 are also presented in (44).

A replicated, controlled study from 1989 to 1991 of headlands in two arable fields in England (36) found that plant and invertebrate diversity was greater in unsprayed plots compared to those receiving autumn herbicide applications. Thirty-two of 34 broadleaved plant species (dicotyledons) were found in unsprayed plots compared to 19 in sprayed plots; total dicotyledon species cover and total cover were also significantly greater in unsprayed plots. Numbers of true bugs (Heteroptera) (unsprayed: 29; sprayed: 23 species), leafhoppers (Auchenorrhyncha) (unsprayed: 1–13 groups/0.5 m^2; sprayed: 0–9), total beetles (Coleoptera) (10–55 vs 2–40), weevils (Curculionidae) (0.3–5.0 vs 0–1.3), rove beetles (Staphylinidae) (0–30 vs 0–22), total flies (Diptera) (40–280 vs 20–240), chick-food insects (7–33 vs 2–25), total arthropods (80–1,175 vs 40–1,165) and spiders (Araneae) (2–40 vs 2–28) were significantly greater in the untreated plots compared to sprayed plots. Aphids (Aphididae), sawflies (Symphyta) and butterfly/moth (Lepidoptera) larvae did not differ between treatments. Field headlands were divided into (6–8) plots (100 x 6 m), half (and crops) were sprayed with herbicides in autumn and half were unsprayed. No insecticides or fungicides were applied. Plants were sampled in 10 quadrats (0.25 m^2) per plot in May and late June/early July (1989–1991). Arthropods were sampled on five occasions (May–July); five samples were collected per plot using a D-Vac insect sampler.

A 2000 literature review from the UK (37) found that populations of grey partridge *Perdix perdix* were 600% higher on farms with conservation measures aimed at partridges in place, compared to farms without these measures (Aebischer 1997). Measures included the provision of conservation headlands, planting cover crops, using set-aside and creating beetle banks.

A before-and-after study of two winter wheat fields over one year in Dorset, UK (38) found that some arthropod taxa survived within the unsprayed headland and appeared to re-colonize the mid-field surrounded by the headland more extensively compared to when there was no headland. Ground beetle (Carabidae) species (*Pterostichus melanarius, P. madidus*), spiders (Araneae: money spiders (Linyphiidae); wolf spiders (Lycosidae)), parasitic wasps (*Aphidius spp.*) and total predatory arthropods showed the greatest decline immediately after application of the insecticide dimethoate; rove beetles in the sub-family Aleocharinae did not decline. Numbers of the ground beetle *P. madidus*, money spiders and parasitic wasps *Aphidius spp.* decreased within the field and unsprayed headland. Numbers of *P. madidus* recovered faster within the field edge than mid-field and particularly within the unsprayed buffer zone and the mid-field area it enclosed. Money spiders were present across most areas of both fields 19 days after spraying, although in lower numbers than pre-spraying, parasitic wasps *Aphidius spp.* had not recovered 20 days after spraying. A grid of 75 and 29 pitfall traps were used in each field over 2 days on 5 occasions May–July 1997 and then 6, 20 and 34 days after spraying with the pesticide dimethoate (0.86 l/ha). A 6 m headland around half of one field was unsprayed. A D-Vac suction sampler was also used 3 times pre-spraying and at 6 and 20 days after treatment.

A replicated, controlled trial in 1999 on 5 arable farms in the West Midlands, UK (39) found fewer bees (Apidae) on conservation headlands compared to naturally regenerated margins (average less than 3 bees/transect in conservation headlands vs averages of 10–50 bees/transect in naturally regenerated margins). Bumblebees *Bombus* spp. and honey bees *Apis mellifera* were counted on 50 m transects in five 6 m-wide field margins managed as conservation headlands, and 10 naturally regenerated, uncropped field margins between 29 June and 9 August. Two unsown margins and one conservation headland were created on each farm.

A replicated study from April to September 1995 in five types of field margin around four cereal fields near Göttingen, Germany (40) (same study as (22, 23)) found that unsprayed margins sown with cereals (conservation headlands) suppressed the colonization of aggressive weeds. However, abundance of predators (mainly spiders (Araneae)) and predator-prey ratios in cereal sown margins were lower than in six-year-old naturally developed margins. Arthropod abundance, diversity and predator-prey ratios in the cereal margins did not differ from the rest of the studied margin types. The following margin types (3 m wide, 100–150 m long) were studied: 1-year old naturally developed, 6-year old naturally developed, sown with a mixture of 19 wildflower species, sown with phacelia *Phacelia* spp. mixture, and control strips sown with winter wheat or oats. Potted plants of mugwort *Artemisia vulgaris* (four pots per margin) and red clover *Trifolium pratense* (three pots per margin) were used to study plant-arthropod communities. Mugwort pots were set out in May and visited weekly to count all arthropods living on the plants, leaf miners and galls that had colonized the plants. In September, the plants were dissected and all larvae and pupae living inside the plants were individually reared in the lab to estimate parasitization rates. Red clover pots were set out in April. At five visits in June and July, flower heads were sampled, dissected and larvae and pupae were reared in the lab for species determination.

A 2002 review (41) of two reports (Wilson *et al.* 2000, ADAS 2001) evaluating the effects of the Pilot Arable Stewardship Scheme in two regions (East Anglia and the West Midlands) in the UK from 1998 to 2001 found that conservation headlands with restricted use of fertilizers, insecticides or both, benefited plants and true bugs (Hemiptera), but not bumblebees *Bombus* spp., ground beetles (Carabidae) or sawflies (Symphyta). There were total areas of 605 and 1,085 ha of conservation headlands in East Anglia and the West Midlands respectively. The effects of the pilot scheme on plants and invertebrates were monitored over three years, relative to control areas.

A 2003 literature review in Europe (42) found three studies that showed that unsprayed crop margins reduce the effects of spray drift on butterflies (Lepidoptera) (Longley *et al.* 1997, Longley & Sotherton 1997b, (31)). One study suggests that a 6 m buffer is not sufficient to eliminate spray drift (Longley & Sotherton 1997b), whereas other research suggests that a 6 m buffer provides no more protection than a 3 m buffer (31). One study reported that hoverflies (Syrphidae) were more abundant in unsprayed headlands (Cowgill *et al.* 1993).

A replicated study in the summers of 1999–2000 comparing ten different conservation measures on arable farms in the UK (43) found that conservation headlands without fertilizer appeared to be one of the three best options for the conservation of annual herbaceous plant communities. Wildlife seed mix (sown for birds and bees) and uncropped, cultivated margins were the other two options. Conservation headlands with fertilizer use had fewer plant species. The average numbers of plant species in the different conservation habitats were no-fertilizer conservation headlands 4.8; conservation headlands 3.5; wildlife seed mixtures 6.7; uncropped cultivated margins 6.3; undersown cereals 5.9; naturally regenerated grass margins 5.5; spring fallows 4.5; sown grass margins 4.4; overwinter stubbles 4.2; grass leys 3.1. Plants were surveyed on a total of 294 conservation measure sites (each a single field, block of field or field margin strip) on 37 farms in East Anglia (dominated by arable farming) and 38 farms in the West Midlands (dominated by more mixed farming). The ten habitats were created according to agri-environment scheme guidelines. Vegetation was surveyed once in each site in June–August in 1999 or 2000 in thirty 0.25 m^2 quadrats randomly placed in 50–100 m randomly located sampling zones in each habitat site. All vascular plant species rooted in each quadrat, bare ground or litter and plant cover were recorded.

A continuation of the same replicated, controlled, randomized study in Scotland as in (35), (44) found that ground beetle (Carabidae) species diversity and abundance was significantly higher in the headland (no fertilizers or pesticides) than in the main field (fertilizers, no pesticides) in 1997–1998. The total abundance of ground beetles was significantly higher in the headland (927–1,053 individuals) than main field (631–910). This was also the case for species diversity (headland: 38 species; main field: 23).

A replicated, controlled, paired-sites comparison in 2003 in East Anglia and the West Midlands, UK (45) found no significant difference in bumblebee *Bombus* spp. species richness and abundance when 16 conservation headlands were compared with paired conventional field margins. In both types of field margin, a few species of plant contributed to the vast majority of foraging visits by bumblebees, mainly creeping thistle *Cirsium arvense* and spear thistle *C. vulgare*. Nineteen farms were surveyed in East Anglia and 17 farms in the West Midlands. Three agri-environment scheme (Arable Stewardship Pilot Scheme (ASPS)) options were studied: field margins sown with a wildlife seed mixture (28 sites); conservation headlands with no fertilizer (16 sites); naturally regenerated field margins (18 sites). Fifty-eight conventional cereal field margins were used as a control and paired with ASPS sites. Bumblebees were surveyed along 100 x 6 m or 50 x 6 m transects twice, in July and August. Vegetation was surveyed in twenty 0.5 x 0.5 m quadrats.

A replicated study in 1999 and 2003 on farms in East Anglia and the West Midlands, UK (46) found that 5 of 12 farmland bird species analysed were positively associated with conservation headlands and a general reduction in herbicide use. These were corn bunting *Miliaria calandra* (a field-nesting species), chaffinch *Fringilla coelebs*, greenfinch *Carduelis chloris*, whitethroat *Sylvia communis* and yellowhammer *Emberiza citrinella* (all boundary-nesting species). The study did not distinguish between conservation headlands and a general reduction in herbicide use, classing both as interventions reducing pesticide use. A total of 256 arable and pastoral fields across 84 farms were surveyed.

A replicated, controlled trial from 2001 to 2004 across central and eastern England (47) found that unsprayed conservation headlands did not support more bumblebee *Bombus* spp. individuals or species than conventional cropped field margins. The number of flowers and flower species in conservation headlands was not significantly different from cropped field margins or margins sown with a tussocky grass mix. Six sites were studied and 2 experimental plots (50 m x 6 m) established in each cereal field along 2 margins. Six treatments were assigned to plots: wildflower mixture (21 native wildflower species and 4 fine grass species); pollen and nectar mixture (4 agricultural legume species: red clover, Alsike clover *T. hybridum*, bird's-foot trefoil and sainfoin *Onobrychis viciifolia* and 4 fine grass species); tussocky grass mixture; conservation headland; natural regeneration; crop (control treatment). Foraging bumblebees were counted from May to late August, on 6 m-wide transects between 6 and 11 times in each margin. Flower abundance was also estimated along the bumblebee transects in 2002, 2003 and 2004.

A replicated, controlled study in 2000–2003 of five grassland headlands on four intensively managed pastoral farms across Scotland (48) found that aphids/leafhoppers/planthoppers (Homoptera) and true bugs (Heteroptera) were more abundant in conservation headlands (no fertilizers, pesticides or grazing April–August) than conventional headlands and open fields. Homoptera had higher activity densities in conservation headlands (2.1) and field edges (conventional: 2.0; conservation: 1.9) than in conventional headlands (0.8) and open fields (0.6). Roundback slugs (Arionidae) showed the same pattern (2.3 conservation headlands; 2.1 conventional field edges; 2.4 conservation field edges; 0.7 conventional headlands; 0.3 open fields). True bugs were more abundant in conservation headlands (0.7) and field edges (1.1–1.2) than in open fields (0.2). Keelback slug (Limacidae) activity density was greater in both headlands (conventional: 1.9; conservation: 2.8) and field edges (2.3–2.7) than in open fields (1.1). Butterfly/moth (Lepidoptera) and sawfly (Symphyta) larvae showed a similar trend, whereas ground beetle (Carabidae) abundance did not differ with treatment

(3.5–3.6). Ground beetle activity density was highest in open fields (4.0). One headland in each field was divided into 2 areas of 6 x 100 m, a conventional and conservation headland. In each field, invertebrates were sampled with five pitfall transects of nine traps in the conservation headland, conservation field edge, conventional headland, conventional field edge and open field. Traps were set for 3–4 weeks in May–June and July–August 2000–2003.

A replicated site comparison study in 2004 and 2005 in the UK (49) found that conservation headlands without fertilizer had more plant species (17 species/margin) than standard conservation headlands (reduced pesticide only, 11 species/margin) and significantly more species than control margins in all plant groups except grasses. Standard conservation headlands did not have significantly more plant species than control margins (11 and 8 species/margin on average), but they did have a higher percentage of spring germinating plant species (63% of plant species were spring germinating, compared to 48% in control margins). Thirty-nine of each type of conservation headland managed under the Countryside Stewardship Scheme, were surveyed in 2005 and compared with 72 conventionally cropped field margins surveyed in 2004 or 2005. Margins were randomly selected from eight UK regions. Plants were surveyed in thirty 0.025 m^2 quadrats within a 100 m sampling zone of each margin and percentage cover across all quadrats estimated.

A 2007 review of published and unpublished literature (50) found experimental evidence of benefits of conservation headlands to plants (four studies: (15, 17, 27, 43)), rare arable plants (when fertilizer was also reduced, one study: (27)), invertebrates (some groups, five studies: (10, 15, 17, 24, 29) and mammals (three studies not summarized here, showed potential value through increased food resources).

A review of the effects of agri-environment scheme options on small mammals in the UK (51) found one study that reported that 12 radio-tracked wood mice *Apodemus sylvaticus* preferred unsprayed and conservation headlands (sprayed only with herbicides to control grasses) over sprayed headlands and mid-fields (12). Another study found that conservation headlands have higher abundances of insects and arable weeds, both of which are eaten by wood mice (9).

A replicated, controlled, randomized site comparison study in 2005 of field margins at 39 sites in England (52) (same study as (53)) found no significant difference in the number of rare arable plants in conservation headlands and the crop. No-fertilizer conservation headlands had higher numbers of rare arable plants (0.7/sample zone) than conservation headlands (reduced insecticides/pesticides) or the crop (0.1) but the difference was not significant. In total 18 species were found on no-fertilizer conservation headlands. There were no significant differences in diversity at 1, 3 or 5 m from the field edge within margins, although it tended to decline. There were significant regional differences in diversity. One of each margin type and an adjacent control was randomly selected in thirty-nine 20 x 20 km squares in England. Rare arable plants were sampled in 10 quadrats (0.5 x 0.5 m) at 3 distances (1, 3 and 5 m) from the field edge within a 100 x 6 m sample zone in June–July 2005.

A replicated, controlled, randomized site comparison study in 2005 of Countryside Stewardship scheme field margin options across England (53) (same study as (52)) found that arable plant species diversity was higher in no-fertilizer conservation headlands (4.1 species) and spring fallow (4.3) than in fertilized conservation headlands (2.4 species; reduced pesticide use) and cereal crop controls (1.4). A total of 39 randomly selected 20 x 20 km squares throughout England were visited to sample four Countryside Stewardship scheme options: uncropped margins, spring fallow and conservation headlands with and without fertilizer. A conventionally managed cereal crop (control) was also sampled at each of the farms visited.

A total of 195 field margin agreements were surveyed during June and July 2005. All plant species and 86 rare arable plants were investigated.

A 2008 review of control methods for competitive weeds in field margins managed to maintain uncommon arable plant populations in the UK (54) found that specific management regimes can reduce abundance of pernicious weeds in margins. One study found pernicious weeds were more likely in uncropped cultivated margins than in conservation or conventional headlands (Critchley *et al.* 2004); two studies found the latter two did not differ in weed abundance (Pinke 1995, Critchley *et al.* 2004). Three studies found lower weed abundance in fertilized conservation headlands (Pinke 1995, (25), Wilson 2000). In naturally regenerated margins, fertilizer increased 1 grass species and decreased 10 out of 14 rare plant species (Wilson 2000; Meek *et al.* 2007).

A 2009 literature review of agri-environment schemes in England (55) found evidence that grey partridge *Perdix perdix* broods were significantly larger in cereal fields with a 6 m unsprayed margin around them, compared to conventional fields. Two studies showed that more butterflies (Lepidoptera) were found in conservation headlands than in pesticide-sprayed areas ((24), Longley & Sotherton 1997a). One study (53) found that 264 plant species typically found in disturbed or arable habitats, including 34 rare and uncommon arable plants, were recorded in 3 agri-environment scheme options: uncropped cultivated margins (highest diversity); spring fallow; conservation headlands (lowest diversity).

A 2009 literature review of European farmland conservation practices (56) found that rare annual flowers were more abundant in conservation headlands than in adjacent crops, but less abundant than in uncropped field margins. Invertebrates were also more common in conservation headlands than in crops, but less diverse than in uncropped margins. Gamebirds made frequent use of conservation headlands, for shelter and foraging. The authors note that the effects on non-gamebirds are less certain.

A replicated site comparison study from 2004 to 2008 in England (57) found that grey partridge *Perdix perdix* overwinter survival was positively correlated with the proportion of a site under conservation headlands in 2007–2008 and with year-on-year density changes in 2006–2007. There was no relationship between the proportion of a site under conservation headlands and brood size or the ratio of young to old birds. Spring and autumn counts of grey partridge were made at 1,031 sites across England as part of the Partridge Count Scheme.

(1) Rands M.R.W., Sotherton N.W. & Moreby S.J. (1984) Some effects of cereal pesticides on gamebirds and other farmland fauna. *Proceedings of the Recent Developments in Cereal Production.* University of Nottingham, December. pp. 98–113.
(2) Rands M.R.W. (1985) Pesticide use on cereals and the survival of grey partridge chicks: a field experiment. *Journal of Applied Ecology,* 22, 49–54.
(3) Rands M.R.W. (1986) The survival of gamebird (galliformes) chicks in relation to pesticide use on cereals. *Ibis,* 128, 57–64.
(4) Rands M.R.W. & Sotherton N.W. (1986) Pesticide use on cereal crops and changes in the abundance of butterflies on arable farmland in England. *Biological Conservation,* 36, 71–82.
(5) Schumacher W. (1987) Measures taken to preserve arable weeds and their associated communities in central Europe. *British Crop Protection Council Monographs,* 35, 109–112.
(6) Dover J., Sotherton N. & Gobbett K.A.Y. (1990) Reduced pesticide inputs on cereal field margins: the effects on butterfly abundance. *Ecological Entomology,* 15, 17–24.
(7) Chiverton P.A. & Sotherton N.W. (1991) The effects on beneficial arthropods of exclusion of herbicides from cereal crop edges. *Journal of Applied Ecology,* 28, 1027–1039.
(8) Cowgill S.E. (1991) The foraging ecology of hoverflies and the potential for manipulating their distribution on farmland. PhD thesis. University of Southampton.
(9) Sotherton N.W. (1991) Conservation Headlands: a practical combination of intensive cereal farming and conservation. In L.G. Firbank, N. Carter, J.F. Derbyshire & G.R. Potts (eds.) *The Ecology of Temperate Cereal Fields,* Blackwell Scientific Publications, Oxford. pp. 373–397.
(10) Hassall M., Hawthorne A., Maudsley M., White P. & Cardwell C. (1992) Effects of headland management on invertebrate communities in cereal fields. *Agriculture Ecosystems & Environment,* 40, 155–178.

(11)   Raskin R., Glück E. & Pflug W. (1992) Floren- und Faunenentwicklung auf herbizidfrei gehaltenen Agrarflächen. Auswirkungen des Ackerrandstreifenprogramms [Development of flora and fauna on herbicide-free agricultural land]. *Natur and Landschaft*, 67, 7–14.

(12)   Tew T.E., Macdonald D.W. & Rands M.R.W. (1992) Herbicide application affects microhabitat use by arable wood mice (*Apodemus sylvaticus*). *Journal of Applied Ecology*, 29, 532–539.

(13)   Cardwell C., Hassall M. & White P. (1994) Effects of headland management on Carabid beetle communities in Breckland cereal fields. *Pedobiologia*, 38, 50–62.

(14)   de Snoo G.R., Dobbelstein R. & Koelewijn S. (1994) Effects of unsprayed crop edges on farmland birds. *British Crop Protection Council Monographs*, 58, 221–226.

(15)   Helenius J. (1994) Adoption of conservation headlands to Finnish farming. *British Crop Protection Council Monographs*, 58, 191–196.

(16)   Lösch R., Thomas D., Kaib U. & Peters F. (1994) Resource use of crops and weeds on extensively managed field margins. *Proceedings of the Field margins: integrating agriculture and conservation.* Coventry, UK, 18–20 April. pp. 203–208.

(17)   Chiverton P.A. (1994) Large-scale field trials with conservation headlands in Sweden. *British Crop Protection Council Monographs*, 58, 185–190.

(18)   Hawthorne A. & Hassall M. (1995) The effect of cereal headland treatments on carabid communities. *Arthropod Natural Enemies in Arable Land I – Density, Spatial Heterogeneity and Dispersal, Acta Jutlandica*, 70, 185–198.

(19)   de Snoo G.R. (1996) Enhancement of non-target insects: Indications about dimensions of unsprayed crop edges. In K. Booij & L. den Nijs (eds.) *Arthropod Natural Enemies in Arable Land II – Survival, Reproduction and Enhancement*, 71, Acta Jutlandica, Aarhus University Press. pp. 209–219.

(20)   de Snoo G.R. & de Leeuw J. (1996) Non-target insects in unsprayed cereal edges and aphid dispersal to the adjacent crop. *Journal of Applied Entomology-Zeitschrift Fur Angewandte Entomologie*, 120, 501–504.

(21)   de Snoo G.R. (1997) Arable flora in sprayed and unsprayed crop edges. *Agriculture, Ecosystems and Environment*, 66, 223–230.

(22)   Denys C. (1997) Do field margins contribute to enhancement of species diversity in a cleared arable landscape? Investigations on the insect community of mugwort (*Artemisia vulgaris* L). *Mitteilungen Der Deutschen Gesellschaft Fur Allgemeine Und Agewandte Entomologie, Band 11, Heft 1–6, Dezember 1997 – Entomologists Conference*, 11, 69–72.

(23)   Denys C., Tscharntke T. & Fischer R. (1997) Colonization of wild herbs by insects in sown and naturally developed field margin strips and in cereal fields. *Verhandlungen der Gesellschaft fur Okologie*, 27, 411–418.

(24)   Dover J.W. (1997) Conservation headlands: effects on butterfly distribution and behaviour. *Agriculture, Ecosystems & Environment*, 63, 31–49.

(25)   Fischer A. & Milberg P. (1997) Effects on the flora of extensified use of field margins. *Swedish Journal of Agricultural Research*, 27, 105–111.

(26)   Hodkinson D.J., Critchley C.N.R. & Sherwood A.J. (1997) A botanical survey of conservation headlands in Breckland Environmentally Sensitive Area, UK. *Proceedings of the 1997 Brighton Crop Protection Conference – Weeds, Conference Proceedings Vols 1–3.* Farnham. pp. 979–984.

(27)   Kleijn D. & van der Voort L.A.C. (1997) Conservation headlands for rare arable weeds: the effects of fertilizer application and light penetration on plant growth. *Biological Conservation*, 81, 57–67.

(28)   Stoate C. & Szczur J. (1997) Seasonal changes in habitat use by yellowhammers (*Emberiza citrinella*). *Proceedings of the 1997 Brighton Crop Protection Conference – Weeds.* Farnham. pp. 1167–1172.

(29)   de Snoo G.R., van der Poll R.J. & Bertels J. (1998) Butterflies in sprayed and unsprayed field margins. *Zeitschrift Fur Angewandte Entomologie*, 122, 157–161.

(30)   Sotherton N. (1998) Land use changes and the decline of farmland wildlife: an appraisal of the set-aside approach. *Biological Conservation*, 83, 259–268.

(31)   de Snoo G.R. (1999) Unsprayed field margins: effects on environment, biodiversity and agricultural practice. *Landscape and Urban Planning*, 46, 151–160.

(32)   de Snoo G.R. & Chaney K. (1999) Unsprayed field margins – what are we trying to achieve? *Aspects of Applied Biology*, 54, 1–12.

(33)   de Snoo G.R. & van der Poll R.J. (1999) Effect of herbicide drift on adjacent boundary vegetation. *Agriculture, Ecosystems and Environment*, 73, 1–6.

(34)   Haughton A.J., Bell J.R., Gates S., Johnson P.J., Macdonald D.W., Tattersall F.H. & Hart B.H. (1999) Methods of increasing invertebrate abundance within field margins. *Aspects of Applied Biology*, 54, 163–170.

(35)   Haysom K.A., McCracken D.I., Foster G.N. & Sotherton N.W. (1999) Grass conservation headlands – adapting an arable technique for the grassland farmer. *Aspects of Applied Biology*, 54, 171–178.

(36)   Moreby S.J. & Southway S.E. (1999) Influence of autumn applied herbicides on summer and autumn food available to birds in winter wheat fields in southern England. *Agriculture, Ecosystems & Environment*, 72, 285–297.

(37)   Aebischer N.J., Green R.E. & Evans A.D. (2000) From science to recovery: four case studies of how research has been translated into conservation action in the UK. In N.J. Aebischer, A.D. Evans, P.V.

Grice & J.A. Vickery (eds.) *Ecology and Conservation of Lowland Farmland Birds*, British Ornithologists' Union, Tring. pp. 43–54.

(38)   Holland J.M., Winder L. & Perry J.N. (2000) The impact of dimethoate on the spatial distribution of beneficial arthropods in winter wheat. *Annals of Applied Biology*, 136, 93–105.

(39)   Kells A.R., Holland J.M. & Goulson D. (2001) The value of uncropped field margins for foraging bumblebees. *Journal of Insect Conservation*, 5, 283–291.

(40)   Denys C. & Tscharntke T. (2002) Plant-insect communities and predator-prey ratios in field margin strips, adjacent crop fields, and fallows. *Oecologia*, 130, 315–324.

(41)   Evans A.D., Armstrong-Brown S. & Grice P.V. (2002) The role of research and development in the evolution of a 'smart' agri-environment scheme. *Aspects of Applied Biology*, 67, 253–264.

(42)   Bat Conservation Trust (2003) *Agricultural Practice and Bats: A review of current research literature and management recommendations*. Defra BD2005.

(43)   Critchley C., Allen D., Fowbert J., Mole A. & Gundrey A. (2004) Habitat establishment on arable land: assessment of an agri-environment scheme in England, UK. *Biological Conservation*, 119, 429–442.

(44)   Haysom K.A., McCracken D.I., Foster G.N. & Sotherton N.W. (2004) Developing grassland conservation headlands: response of carabid assemblage to different cutting regimes in a silage field edge. *Agriculture, Ecosystems & Environment*, 102, 263–277.

(45)   Pywell R.F., Warman E.A., Carvell C., Sparks T.H., Dicks L.V., Bennett D., Wright A., Critchley C.N.R. & Sherwodd A. (2005) Providing foraging resources for bumblebees in intensively farmed landscapes. *Biological Conservation*, 121, 479–494.

(46)   Stevens D.K. & Bradbury R.B. (2006) Effects of the Arable Stewardship Pilot Scheme on breeding birds at field and farm-scales. *Agriculture, Ecosystems & Environment*, 112, 283–290.

(47)   Carvell C., Meek W.R., Pywell R.F., Goulson D. & Nowakowski M. (2007) Comparing the efficacy of agri-environment schemes to enhance bumble bee abundance and diversity on arable field margins. *Journal of Applied Ecology*, 44, 29–40.

(48)   Cole L.J., McCracken D.I., Baker L. & Parish D. (2007) Grassland conservation headlands: their impact on invertebrate assemblages in intensively managed grassland. *Agriculture, Ecosystems & Environment*, 122, 252–258.

(49)   Critchley C.N.R., Walker K.J., Pywell R.F. & Stevenson M.J. (2007) The contribution of English agri-environment schemes to botanical diversity in arable field margins. *Aspects of Applied Biology*, 81, 293–300.

(50)   Fisher G.P., MacDonald M.A. & Anderson G.Q.A. (2007) Do agri-environment measures for birds on arable land deliver for other taxa? *Aspects of Applied Biology*, 81, 213–219.

(51)   Macdonald D.W., Tattersall F.H., Service K.M., Firbank L.G. & Feber R.E. (2007) Mammals, agri-environment schemes and set-aside - what are the putative benefits? *Mammal Review*, 37, 259–277.

(52)   Walker K.J., Critchley C.N.R. & Sherwood A.J. (2007) The effectiveness of new agri-environment scheme options in conserving rare arable plants. *Aspects of Applied Biology*, 81, 301–308.

(53)   Walker K.J., Critchley C.N.R., Sherwood A.J., Large R., Nuttall P., Hulmes S., Rose R. & Mountford J.O. (2007) The conservation of arable plants on cereal field margins: an assessment of new agri-environment scheme options in England, UK. *Biological Conservation*, 136, 260–270.

(54)   Critchley C.N.R. & Cook S.K. (2007) *Long-term Maintenance of Uncommon Plant Populations in Agri-environment Scheme in England. Phase 1 Scoping Study*. Defra/ADAS BD1630.

(55)   Natural England (2009) *Agri-environment Schemes in England 2009: A review of results and effectiveness*. Natural England, Peterborough.

(56)   Vickery J.A., Feber R.E. & Fuller R.J. (2009) Arable field margins managed for biodiversity conservation: a review of food resource provision for farmland birds. *Agriculture, Ecosystems & Environment*, 133, 1–13.

(57)   Ewald J.A., Aebischer N.J., Richardson S.M., Grice P.V. & Cooke A.I. (2010) The effect of agri-environment schemes on grey partridges at the farm level in England. *Agriculture, Ecosystems & Environment*, 138, 55–63.

Additional references

Schumacher W. (1984) Gefährdete Ackerwildkräuter können auf ungespritzten Feldrändern erhalten warden [Endangered wild herbs can be protected/conserved on unsprayed field margins]. *Mitteilungen der LÖLF*, 9, 14–20.

Sotherton N.W. (1989) Farming methods to reduce the exposure of non-target arthropods to pesticides. In P.C. Jepson (ed.) *Pesticides and Non-target Invertebrates*. Intercept Ltd., Wimborne. pp. 195–212.

Sinha S.N., Lakhani K.H. & Davis B.N.K. (1990) Studies of the toxicity of insecticidal drift to the first instar larvae of the large white butterfly *Pieris brassica* (Lepidoptera: Pieridae). *Annals of Applied Biology*, 116, 27–41.

Sotherton N.W. & Robertson P.A. (1990) Indirect impacts of pesticides on the production of wild gamebirds in Britain. In K.E. Church, R.E. Warner & S.J. Brady (eds.). *Perdix V, Gray Partridge and Ring-necked Pheasant Workshop*. Kansas Department of Wildlife and Parks, Emporia. pp. 84–102.

Davis B.N.K, Lakhani K.H., Yates T.J. & Frost A.J. (1991) Bioassays on insecticide spray drift: the effects of wind speed on the mortality of *Pieris brassica* larvae (Lepidoptera) caused by diflubenzuron. *Agriculture, Ecosystems and Environment*, 36, 141–149.

Dover J.W. (1991) The conservation of insects on arable farmland. In N.W. Collins & J. Thomas (eds.) *The Conservation of Insects and their Habitats*. Academic Press, New York. pp. 293–318.

Cowgill S.E., Wratten S.D. & Sotherton N.W. (1993) The effect of weeds on the numbers of hoverfly (Diptera: Syrphidae) adults and the distribution and composition of their eggs in winter wheat. *Annals of Applied Biology*, 123, 499–515.

de Jong F.M.W. & van der Nagel M.C. (1994) A field bioassay for side-effects of insecticides with larvae of the large white butterfly (*Pieris brassica* L.). *Medical Faculty Landbouww. University of Gent*, 59/2a, 347–355.

Wilson P.J. (1994) Botanical diversity in arable field margins. *British Crop Protection Council Monographs*, 58, 53–58.

Pinke G. (1995) The significance of unsprayed field edges as refugia for rare arable plants. *Acta Agronomica Ovariensis* 37, 1–11.

Aebischer N.J. (1997) Gamebirds: management of the grey partridge in Britain. In M. Bolton (ed.) *Conservation and the Use of Wildlife Resources*. Chapman & Hall, London. pp. 131–151.

Longley M., Cilgi T., Jepson P.C. & Sotherton N.W. (1997) Measurements of pesticide spray drift deposition into field boundaries and hedgerows: 1. Summer applications. *Environmental Toxicology and Chemistry*, 16, 165–172.

Longley M. & Sotherton N.W. (1997a) Factors determining the effects of pesticides upon butterflies inhabiting arable farmland. *Agriculture, Ecosystems and Environment*, 61, 1–12.

Longley M. & Sotherton N.W. (1997b) Measurements of pesticide spray drift deposition into field boundaries and hedgerows: 2. Autumn applications. *Environmental Toxicology and Chemistry*, 16, 173–178.

de Snoo G.R. & de Wit P.J. (1998) Buffer zones for reducing pesticide drift to ditches and risks to aquatic organisms. *Ecotoxology and Environmental Safety*, 41, 112–118.

Chiverton P.A. (1999) The benefits of unsprayed cereal crop margins to grey partridges *Perdix perdix* and pheasants *Phasianus colchicus* in Sweden. *Wildlife Biology*, 5, 83–92.

Wilson P.J. (2000) Management for the conservation of arable plant communities. In P. Wilson & M. King (eds.) *Fields of Vision. A Future for Britain's Arable Plants*. RSPB, Sandy. pp. 38–47.

Wilson S., Baylis M., Sherrott A. & Howe G. (2000) *Arable Stewardship Project Officer Review*. Farming and Rural Conservation Agency report.

ADAS (2001) *Ecological Evaluation of the Arable Stewardship Pilot Scheme, 1998–2000*. ADAS report.

Critchley C.N.R., Fowbert J.A. & Sherwood A.J. (2004) Botanical assessment of the Arable Stewardship Pilot Scheme, 2003. *ADAS report to the Department for Environment, Food and Rural Affairs April 2004.*

Meek W.R., Pywell R.F., Nowakowski M. & Sparks T.H. (2007) Arable field margin management techniques to enhance biodiversity and control barren brome, *Anisantha sterilis*. In C. Britt, A. Cherrill, M. le Duc, R. Marrs, R. Pywell, T. Sparks, I. Willoughby (eds.) *Vegetation Management*, Aspects of Applied Biology 82. pp. 133–141.

6.2   Buffer in-field ponds

- We have captured no evidence for the effects of buffering in-field ponds on farmland wildlife.

Background

This intervention involves providing buffer strips around in-field ponds. These buffer strips serve the same purpose as those along rivers or streams with the aim of reducing pollution from agricultural systems and providing habitat.

See also 'Provide buffer strips alongside water courses (rivers and streams)' for studies looking at the effects of providing buffer strips along streams or rivers on farmland.

6.3   Provide buffer strips alongside water courses (rivers and streams)

- Three studies (including one replicated site comparison) from the Netherlands and the UK reported that the provision of riparian buffer strips had a positive influence on plant, invertebrate and bird diversity or abundance[5, 6] and supported vegetation associated with habitats preferred by water voles[1].

- Two replicated site comparison studies from France and Ireland found that the provision of riparian buffer strips on farms did not result in an increase in the number of plant species when compared to farms without buffer strips[2–4].

- One replicated site comparison study found ground beetle diversity was higher in grazed riparian zones and narrow fenced strips than in wide riparian buffer strips[7]. However the ground beetle assemblages in wide riparian buffer strips were more distinct from the adjacent pasture field assemblages than either the grazed riparian zones or narrow fenced strips.

Background

Riparian buffer strips (uncultivated strips at the edge of waterways) are increasingly being used to help to reduce diffuse pollution from agricultural systems. Agricultural field margins can provide important habitat for declining farmland species (see 'All farming systems: Create uncultivated margins around intensive arable or pasture fields', 'All farming systems: Plant grass buffer strips/margins around arable or pasture fields'); it is therefore important to understand the influence of riparian buffer strip management, placement and structure on biodiversity.

A site comparison study in 1995 and 1997 of two areas under the Habitat Scheme Water Fringe Option in Wiltshire and Kent, UK (1) found that the scheme, which includes establishment of riparian buffer strips (10–30 m wide), resulted in river bank vegetation associated with habitats preferred by water voles *Arvicola terrestris* (in terms of plant species and vegetation heights). Wetland, grassland or ruderal plant species dominated river bank vegetation in the three rivers studied in Wiltshire (40% wetland, 27% grassland and 6% ruderal species) and comprised over half the species along the one river studied in Kent (21% wetland, 20% grassland and 15% ruderal species). The 3 sites sampled in Wiltshire in 1997 had a relatively high frequency of vegetation up to 60 cm tall, which has been shown to be important for water voles. In Kent, two of the three sites also tended to have a higher frequency of taller vegetation. The species composition of bankside vegetation on agreement land was sampled within 20 m long representative sections at 27 sites in the 2 areas in 1995. At six sites surveyed in 1997 plant species were allocated to different height classes and the number of species in each class summed for each of five 4 m sub-sections.

A replicated paired sites comparison in 1999 of grassland habitats on 30 farms in Counties Laois and Offaly, Ireland (2) (same study as (3)) found that fenced watercourse margins on Rural Environment Protection Scheme farms did not have higher numbers of plant species than unfenced watercourse margins on non-Rural Environment Protection Scheme farms. For watercourse margins (8 paired replicates) more plant species were found in unfenced than fenced margins (52 and 56 species on unfenced margins on Rural Environment Protection Scheme and non-Rural Environment Protection Scheme farms respectively; 50 and 48 species on fenced margins on Rural Environment Protection Scheme and non-Rural Environment Protection Scheme farms respectively). Watercourse margins were fenced to exclude grazing livestock. Plants were surveyed on one watercourse margin on each farm.

A replicated paired sites comparison study in 2000 in counties Laois and Offaly, Ireland (3) (same study as (2)) found that fenced watercourse margins on Rural Environment Protection Scheme farms did not have higher numbers of plant species than unfenced watercourse margins on non-Rural Environment Protection Scheme farms (14.7 and 16.1 plant species/margin respectively). Fifteen farms with Rural Environment Protection Scheme agreements at least 4 years old were paired with 15 similar farms without agreements. On each farm, a randomly selected watercourse margin was surveyed for plants: all plant species were recorded in two 5 x 3 m quadrats and percentage cover estimated in a 1 x 3 m quadrat within each margin. Eleven of the farm pairs enabled a fenced/unfenced comparison.

A replicated site comparison study in 2005 and 2006 in Seine-et-Marne, France (4) found that the number of plant species was higher on farms that did not have buffer strips (mostly along rivers to prevent water pollution) relative to farms that did include these measures. The numbers of plant species in this comparison are not given and the number of farms with and without these buffer strips not specified. Twenty-six fields from 17 farms were sampled 3 times in 2005 (April; June; September). Sixty-four fields from 31 farms (including all those surveyed in 2005) were sampled twice in 2006 (April and July). Plants were recorded in 10 permanent, regularly spaced, 1 m^2 (0.5 x 2 m) quadrats along the permanent margins of each field.

A replicated site comparison study from 1999 to 2004 in the Netherlands (5) found that ditch bank plant diversity was significantly higher on farms with ecologically-managed ditches with $\geq$ 3 m-wide field margin buffer strips (36–65 plant species/400 m^2) compared to conventionally managed farms without buffer margins or ecological management (26–34 plant species/400 m^2). The number of plant species on ecologically-managed ditch banks with buffer strips was also higher than ditch banks without buffer strips or ecological management on organic farms (32–52 plant species/400 m^2). Ecologically managed strips were cut once in September and the cuttings removed to reduce nutrient input. Cutting date varied on conventional and organic farms but cuttings were never removed. Four ecologically managed farms, 18 conventional and 20 organic arable farms were studied. On ecologically managed farms, plant species surveys of 100 m of ditch bank spread over the whole farm were undertaken once a year from 1999 to 2004. On organic (in 2001) and conventional (2003) farms plant species presence was recorded on 10 x 25 m of ditch bank along a transect (May–June).

A single-site study from 2004 to 2006 in Leicestershire, UK (6) found that a sequence of seven constructed pools and a riparian buffer strip provided habitat for plant, invertebrate and bird diversity including previously absent species. Pools supported 30 aquatic plant species (macrophytes) including 2 locally scarce species (9–18 species/pool). Six nationally scarce and 4 locally uncommon water beetles (Coleoptera) were found in the pools (total 84 invertebrate species; 24–52 species/pool). Five species of grasshopper and cricket (Orthoptera) previously absent from the site were recorded, in addition to 12 hoverfly (Syrphidae) species of which 2 were scarce or new county records. More whitethroat *Sylvia communis*, reed bunting *Emberiza schoeniclus* and moorhen *Gallinula chloropus* territories were found following establishment of the wetland (4, 3 and 1 territories with pools/buffer strip vs 1, 1 and 0 prior to pools/buffer strip). The buffer strip was also used by lapwing *Vanellus vanellus*, yellowhammer *Emberiza citrinella* (breeding species), common snipe *Gallinago gallinago* and jack snipe *Lymnocryptes minimus* (overwintering species). The field drain fed wetland was constructed in 1998. The pool sequence was a maximum of 20 m wide within a riparian buffer strip approximately 70 m wide by 100 m long. Aquatic plants were listed and aquatic macroinvertebrates sampled (three minutes/pool; June 2004–2005) in six of seven pools. Grasshoppers and crickets (June 2005–2006) were sampled and a ten-visit territory mapping bird survey (May–June 2006) undertaken within the buffer strip; birds had also been surveyed in 1992.

A replicated, site comparison study from 2004 to 2006 in Scotland (7) found that there were more plant species in riparian zones (grazed and ungrazed strips) compared to the adjacent intensively managed pasture fields. Ground beetle (Carabidae) diversity was greater in grazed riparian zones and in narrow ungrazed strips than in wide buffer strips or adjacent fields. However, ground beetle assemblages in wide buffer strips were more distinct from adjacent field assemblages than those in narrow strips or grazed riparian zones. There were no significant differences between the numbers of ground beetles or plant species in narrow or wide ungrazed buffer strips. Three types of riparian zone on 7 farms were studied: open sites (no fence between the field and the watercourse; grazed by livestock), narrow strips (strips less than 2 m-wide fenced off around watercourse; ungrazed) and wide buffer strips (strips more than 4 m-wide fenced off around watercourse; ungrazed). Two transects were sampled at 22 locations; 1 adjacent to the watercourse, the other 4–6 m into the field from the fenceline (dividing the riparian zone from the field) or for unfenced, open sites 4–6 m from the watercourse transect. For wide buffer strip sites an additional transect was sampled, halfway between the fenceline and the watercourse transect. Ground beetles were sampled along transects during two 4-week periods (June and July) using pitfall traps (75 mm diameter). Vegetation composition was sampled using a quadrat (1 x 1 m) survey.

(1) Critchley C.N.R., Hodkinson D.J. & McKenzie S.E. (1999) Potential benefits to water voles (*Arvicola terrestris*) of waterside buffer strips in an agri-environment scheme. *Aspects of Applied Biology*, 54, 179–184.
(2) Feehan J., Gillmor D.A. & Culleton N.E. (2002) The impact of the Rural Environment Protection Scheme (REPS) on plant and insect diversity. *Tearmann*, 2, 15–28.
(3) Feehan J., Gillmor D. & Culleton N. (2005) Effects of an agri-environment scheme on farmland biodiversity in Ireland. *Agriculture, Ecosystems & Environment*, 107, 275–286.
(4) Chateil C., Abadie J.C., Gachet S., Machon N. & Porcher E. (2007) Can agri-environmental measures benefit plant biodiversity? An experimental test of the effects of agri-environmental measures on weed diversity. *Proceedings of the Vingtième conférence du columa journées internationales sur la lutte contre les mauvaises herbes*. Dijon, 11–12 December. pp. 356–366.
(5) Manhoudt A.G.E., Visser A.J. & de Snoo G.R. (2007) Management regimes and farming practices enhancing plant species richness on ditch banks. *Agriculture, Ecosystems & Environment*, 119, 353–358.
(6) Stoate C., Whitfield M., Williams P., Szczur J. & Driver K. (2007) Multifunctional benefits of an agri-environment scheme option: riparian buffer strip pools within 'Arable Reversion'. *Aspects of Applied Biology*, 81, 221–226.
(7) Cole L.J., Morton R., Harrison W., McCracken D.I. & Robertson D. (2008) The influence of riparian buffer strips on carabid beetle (Coleoptera, Carabidae) assemblage structure and diversity in intensively managed grassland fields. *Biodiversity and Conservation*, 17, 2233–2245.

6.4   Reduce chemical inputs in grassland management

- A total of 16 studies (including 5 reviews) investigated the effects of reducing inputs in permanent grasslands. Six studies from the Netherlands, Switzerland and the UK (including one review and four replicated studies, of which one was also controlled and one a randomized and controlled before-and-after trial) found that stopping fertilizer inputs in permanent grassland resulted in an increase in plant species richness[1, 3, 6, 8], reduced the rate of plant species loss[2] and attracted a higher abundance or species richness of some or all invertebrates studied[12, 14, 17]. One review from the Netherlands found that low fertilizer input grasslands favour common meadow bird species[3]. One review[11] found a study showing that densities of some invertebrates were higher in unfertilized plots compared with those receiving nitrogen inputs.

- Two replicated, controlled trials from the Czech Republic and the UK (1 randomized) found that applying fertilizer to permanent grasslands reduced plant species richness or diversity[15] and that the effects on plant communities were still apparent 16 years after the cessation of fertilizer application[16].

- Four studies from Ireland, the Netherlands and the UK (including two replicated trials, of which one was randomized, one controlled and a review) found that reducing fertilizer inputs on grassland had no clear or rapid effect on plant species richness[4, 5, 10, 13]. A review found no clear effect of reducing fertilizer inputs on the density of soil-dwelling invertebrates[9]. One replicated study found that fertilizer treatment only affected seed production of a small number of meadow plants[7]. One replicated study from the UK[18] found lower invertebrate abundance on plots with reduced fertilizer inputs but the differences were not significant.

Background

This intervention may involve reducing the amount of chemical inputs applied to permanent grasslands or ceasing inputs altogether. Chemical inputs to permanent grasslands may include fertilizers such as nitrogen (N), phosphorous (P) or potassium (K). Reducing chemical inputs on permanent grasslands is often used in conjunction with other actions such as delaying the first mowing or grazing date on grasslands; see also 'Livestock farming: Reduce management intensity on permanent grasslands'.

A replicated trial from 1972 to 1988 in the Drentsche A nature reserve, Drenthe, the Netherlands (1) found that stopping fertilizer inputs on grassland mown annually for hay led to a gradual change in plant species composition and an increase in the number of species in two out of three experimental fields. All three fields had a maximum number of plant species recorded in the middle of the study (between 1980 and 1985) followed by a slight decrease in the number of species as species initially present were replaced. In all fields, previously dominant grass species Yorkshire fog Holcus lanatus and creeping bent Agrostis stolonifera were replaced by creeping buttercup Ranunculus repens and sweet vernal grass Anthoxanthum odoratum, amongst others. In 1988 there were 23 and 28 species on 2 peaty fields, which had risen from 19 and 20 species in 1974 and 30 plant species on a sandy field, the same number as in 1974. Plant species were monitored in six 4 m² quadrats each June from 1974 to 1988, on two 50 x 10 m fields on peaty soil and one 20 x 10 m field on sandy soil. Fertilizer application stopped in 1972 and all fields were cut for hay either once or twice, but no earlier than July.

Two long-term replicated trials near Wageningen in the Netherlands (2) found that ceasing fertilizer inputs reduced the rate of plant species loss over 30 years, relative to conventional fertilizer application rate, but did not affect plant species loss over 17 years relative to a reduced rate of fertilizer application. The first experiment compared no fertilizer with 160 kg N/ha/year from 1958 to 1988, with 2 replicate 16 x 2.5 m plots of each treatment. Fertilized plots changed from 39 plant species to 10, while unfertilized plots had a slight but not significant drop in the number of species, from 33 to around 25. Plants were monitored annually in fifty 0.25 m² quadrats in each plot. The second experiment compared no fertilizer with 50 kg N/ha/year, from 1972 to 1989, with 4 replicate 10 x 10 m plots for each treatment. The number of plant species steadily declined in both treatments from 20 in 1973 to 12 in 1989. All plots were mown twice a year, with hay removed.

A 1994 review of methods to restore grasslands in the Netherlands (3) found three experiments in which the plant community became more similar to species-rich, lower nutrient grassland over 20 years, after the cessation of chemical fertilizer input. The location of these experiments is not given; they may be those reported in (1). Plant species of fertilized, nutrient-rich grasslands (such as perennial rye grass *Lolium perenne* and broad-leaved dock *Rumex obtusifolius*) were gradually replaced by species of lower-nutrient grassland (such as red fescue *Festuca rubra* and sweet vernal grass *Anthoxanthum odoratum* on two sites). One site had dry sandy soil and two sites had wet peaty soils. Fertilizer applications were stopped in 1972 and hay was cut and removed in July each year. Plants were monitored from 1972 to 1992. The method was not considered as effective on the wet sites. The review briefly described the results of an extensive Dutch study on the effects of grassland management on birds (Dijkstra 1991). The study showed that reducing fertilizer to 200 kg N/ha favoured common meadow birds such as northern lapwing *Vanellus vanellus*, Eurasian oystercatcher *Haematopus ostralegus* and black-tailed godwit *Limosa limosa*. Reducing fertilizer to 50 kg N/ha favoured common redshank *Tringa totanus*, common snipe *Gallinago gallinago* and ruff *Philomachus pugnax*. Reducing to no fertilizer reduced meadow bird numbers. Data were not given.

A trial from 1972 to 1989 on a grassland in the Netherlands (4) found that the number of plant species hardly increased over a 12 year period of reduced fertilization and the abundance of herbaceous (non-grass) species decreased. From 1986 to 1988 no fertilizer was applied and there was no increase in the number of plant species. By 1990 only 2 herb species (cuckoo flower *Cardamine pratensis* and meadow buttercup *Ranunculus acer* now *R. acris*) were present in more than 5% of 100 vegetation samples. The authors suggest that plant diversity did not increase because of the dense growth of a small number of competitive grass species. Species that could have colonized were present on the field borders and ditches. The 6.6 ha grassland in the Netherlands (location unknown) had been fertilized at 200 kg N/ha, grazed and mown for silage for many years. From 1973–1985 fertilizer was reduced to 50 kg N/ha. The stocking rate was 5 steers/ha from April to July, then 3.5 steers/ha until October. From 1986–1988 no fertilizer was applied. Three paddocks of 2.2 ha had stocking rates of 2.3, 3.6 or 4.9 steers/ha from April to October.

A replicated, controlled trial from 1990 to 1995 on two sown agricultural grasslands in Scotland (5) found that ceasing fertilizer input had only small effects on the vegetation over six years, if grazing continued. On plots with a grass height of 4 cm in summer (75% of the number of ewes relative to fertilized grassland), white clover *Trifolium repens* increased from 10% to over 20% by 1994. White clover did not increase on plots with a grass height of 8 cm in summer (42–57% of the number of ewes relative to fertilized grassland). Perennial rye grass *Lolium perenne* was more sensitive to autumn grazing pressure and decreased on unfertilized plots grazed to 4 cm in the autumn (from around 60% to 33–35% by 1994). Both white clover and perennial rye grass were sown agricultural varieties. Unsown species only increased substantially in unfertilized plots left ungrazed. There were two replicates of each treatment at each of two upland sites: Hartwood Research Station in central Scotland and Sourhope Research Station in southeast Scotland. The percentage cover of different plant species in each plot was measured in 1990, 1992 and 1994 using a point quadrat.

A replicated, controlled, before-and-after trial from 1991 to 1993 on a species-rich hay meadow at Tadham Moor in Somerset, UK (6) found that ceasing fertilizer application for three years led to a gradual increase in the number of plant species. All plots where fertilizer was stopped in 1991, having been applied at four different levels from 1986 to 1989, had fewer plant species than unfertilized control plots throughout the experiment (they did not revert to the original condition in three years). However, in 1992 and 1993 Yorkshire fog *Holcus lanatus* and perennial rye grass *Lolium perenne* declined in the plots without fertilizer and were being replaced by common bent *Agrostis capillaris* and crested dog's-tail *Cynosurus cristatus*. In 1993 plots without fertilizer had more plant species than plots with continued fertilizer and species richness was increasing at an estimated 1 species/m²/year in all treatments. The average estimated time for the vegetation to revert to the original community following three years of fertilizer application was between 4 (at 25 kg N/ha/year) and 8 years (for 50–200 kg N/ha/year). From 1986–1989 experimental plots (1.5 x 5 m) were fertilized at 5 different levels: 0 (control plots), 25, 50, 100 and 200 kg N/ha/year. There were three replicates of each treatment. From 1990 to 1993 half of each plot continued with the same fertilizer treatment as before; the other half stopped receiving any fertilizer. Plants were monitored on sixteen 1 m² quadrats/plot, 15 times between 1991 and 1993.

A replicated study over one year of a neutral meadow grassland at a farm in England (7) found that fertilizer treatments significantly affected seed numbers for a small number of plant species. More downy oat-grass *Avenula pubescens* seed was recorded where fertilizer had been applied; conversely there was less ribwort plantain *Plantago lanceolata*, yellow rattle *Rhinanthus minor* and common daisy *Bellis perennis*. The meadow was divided into 9 contiguous 20 x 30 m plots in 1990. Treatments were mineral fertilizer (June cut), no fertilizer (July cut; autumn cattle grazing; spring sheep grazing) and a September cut. The long-term management of the meadow prior to the experiment involved manure spreading in April–May each year. Vegetation was cut at 3 cm in 3 randomly placed 0.06 m² quadrats and seed collected.

A review of experimental evidence (largely published in German language) in 1997 (8) described a 40-year experiment with different fertilizer application rates on the Eggenalp, in the Bernese Oberland, Switzerland (Baumberger *et al.* 1996) which found that the number of plant species was highest on unfertilized plots and decreased as fertilizer rates increased. There were fewer than 40 species in all fertilized plots (fertilizer application rates not given here).

A 1998 review of how soil animals, especially nematodes (Nematoda) and microarthropods, change according to management of agricultural grasslands (9) found two studies showing a higher density of nematodes or microarthropods in organically managed or low input grasslands, compared to intensively managed grasslands with conventional chemical inputs (Siepel 1996, Yeates *et al.* 1997). One also found higher numbers of microarthropod species on low input grasslands (Siepel 1996). This is in contrast to four studies that found that adding mineral fertilizer can increase numbers of nematodes in the short term (Edwards & Lofty 1969, Coulson & Butterfield 1978, King & Hutchinson 1980, Bardgett *et al.* 1993), although one study found reduced abundance and diversity of microarthropods after nitrogen fertilizer was added to grassland (Siepel & van de Bund 1988). The authors argue that soil communities are functionally different under low input systems.

A 1998 review of case studies in France, gathered from published and unpublished literature (10) found no monitoring results from France for the effects of reduced management intensity on agriculturally improved grasslands. Three Dutch studies were cited (including (2)) which showed that stopping fertilization does not cause a rapid increase in plant species richness.

A 2000 literature review of grassland management practices in the UK (11) found one study that reported that densities of invertebrates such as species of mites and ticks (Acari), springtails (Collembola), flies (Diptera), beetles (Coleoptera) and millipedes and centipedes (Myriapoda) were higher in unfertilized permanent pasture than pasture receiving over 140 kg nitrogen/ha/year (Curry 1994). One study found that although soil macro-invertebrate densities did not differ in fields with and without farmyard manure applications, bird usage was higher in those that had received moderate applications (Tucker 1992).

A randomized, replicated, controlled trial from 2003 to 2006 on 4 farms in southwest England (12) (same study as (14, 17)) found that 50 x 10 m plots of permanent pasture, cut just once in May or July or not at all during the summer and left unfertilized, supported greater numbers and more species of beetles (Coleoptera) in suction traps, true bugs (Hemiptera) and plant hoppers (Hemiptera: Fulgoroidea) as well as greater abundances of spiders (Araneae), crane and St Mark's flies (Diptera) and more species of woodlice (Isopoda) than control fertilized plots cut in May and July (managed for silage). Small insectivorous birds (dunnock *Prunella modularis*, wren *Troglodytes troglodytes* and European robin *Erithacus rubecula*) and seed-eating finches (Fringillidae) and buntings (Emberizidae) preferred the less intensively managed treatments (particularly the plots uncut in summer) to control plots for foraging. There were 12 replicates of each management type, monitored over 4 years.

A replicated, randomized study of three intensive cattle farms in Ireland over three to five years (13) found that the number of plant species per field was significantly higher with reduced fertilizer application on only one of three farms, although numbers per quadrat increased with decreased fertilizer application at all sites. At Johnstown Castle, the total number of species per field with no fertilizer application (10 species) was significantly higher than fields with 225 (5–7 species) or 390 kg N/ha (6–8 species). There was no significant difference at Solohead (80 kg N: 5–8 species; 175 kg N: 5–10 species; 225 kg N: 3–8 species; 350 kg N: 6–9 species) or Grange (88 kg N: 10–13 species; 225 kg N: 8–15 species). However, the average number of species per quadrat decreased with increasing levels of fertilizer (e.g. Johnstown Castle: 0 kg N: 4 species, 390 kg N: 3 species; Solohead: 80 kg N: 3 species, 350 kg N: 2 species; Grange: 88 kg N: 4 species, 225 kg N: 3 species). Fertilizer treatments were applied in a randomized block design with two to five replicates per treatment. Vegetation was sampled in 50 quadrats (3 dm^3) in each field 3 to 5 years after the treatments commenced.

A randomized, replicated, controlled trial in 2003–2005 on four farms in southwest England (14) (same study as (12, 17)) found that 50 x 10 m plots of permanent pasture cut just once in July or not at all during the summer and left unfertilized attracted a greater abundance and more species of beetle (Coleoptera) than control fertilized plots cut in May and July (managed for silage), in the third year of monitoring. Plots without fertilizer added also had higher proportions of seed- and flower-feeding beetle species in the community. There were 12 replicates of each management type monitored over 3 years.

A randomized, replicated, controlled study in 1999–2005 on four hay meadow sites in Cumbria and Monmouthshire, UK (15) found that applying fertilizer reduced the number of plant species. The number of species declined at all sites when 24 t/ha/year of farmyard manure was applied. The maximum level of manure that could be applied without reducing species richness depended on past site management. The study compared two pairs of unimproved and semi-improved meadows. On the semi-improved Cumbrian meadow, which had previously been fertilized, species richness was unaffected when manure was applied at 12 t/ha/year. However, on the Monmouthshire meadows, which had no recent history of fertilizer use, species richness was reduced by even low levels of manure (≤ 6 tonnes/ha/year). The effects of liming also depended on past site management. Treatments were applied in March/April in 7 x 5 m plots from 1999 to 2005, with plants surveyed annually in May in three 1 m^2 quadrats/plot. Treatments were replicated in three plots at each study site.

A replicated, controlled trial in the Czech Republic (16) found that 16 years after fertilizer applications were stopped, the effects of different rates of fertilizer applications on plant communities were still apparent. Cover of tall nitrogen-loving grasses (especially Yorkshire fog *Holcus lanatus*) in the treatment with 400 kg N/ha plus PK remained significantly higher than in control plots in 2007. Meadow buttercup *Ranunculus acris* was most negatively affected by former application rates. Although species richness was not statistically significantly affected by treatment the number of species decreased from controls to treatments receiving 400 kg N/ha plus PK after 16 years. The following treatments were applied to 5 x 6 m plots of alluvial meadow foxtail *Alopecurus pratensis* grassland: unfertilized, phosphorous and potassium (PK), 100 kg N/ha plus PK, 200 kg N/ha plus PK, 300 kg N/ha plus PK, 400 kg N/ha plus PK, from 1966 to 1990 or 1975 to 1990 for the final 2 treatments. Annual application rates of P and K were 40 and 100 kg/ha, respectively. There were four replicates of each treatment. Fertilizer application was stopped in 1991 in half of each plot and the responses of plant communities monitored until 2007.

A randomized, replicated, controlled trial from 2003 to 2006 in southwest England (17) (same study as (12, 14)) found plots of unfertilized permanent pasture cut just once in July or not cut at all during the summer attracted more adult butterflies (Lepidoptera) but not more butterfly species or common bumblebees *Bombus* spp. than control fertilized plots cut in May and July (managed for silage). Plots cut just once in May, plots cut twice either unfertilized or ungrazed and plots with a higher cutting height did not support more adult butterflies than control plots. Caterpillars were more abundant in unfertilized plots cut just once in May or July, or not at all in summer, than in other treatments. None of the grass treatments supported more common bumblebee species or individuals than control plots. Experimental plots 50 x 10 m were established on permanent pastures (more than 5-years-old) on 4 farms. There were nine different management types, with three replicates/farm, monitored over four years. Seven management types involved different management options for grass-only plots, including conventional silage practices; no cutting in summer; early summer cut (May); late summer cut (July); raised mowing height. Bumblebees and butterflies were surveyed along a 50 m transect line in the centre of each experimental plot once a month from June to September annually. Butterfly larvae were sampled on two 10 m transects using a sweep net in April and June–September annually.

A replicated, controlled study from 2006 to 2009 on four permanent improved grassland fields in Herefordshire and North Yorkshire, UK (18) found fertilized plots with moderate grazing and without early cattle exclusion had consistently higher invertebrate and bird food invertebrate abundance compared to plots with the same grazing treatments but with reduced fertilizer application (eg. 2009: approximately 45 average total invertebrate numbers in reduced fertilizer vs approximately 80 in normal fertilizer plots), however these differences were not significant. Plots without early cattle exclusion (grazing until October) received either reduced fertilizer input (a single 50 kg N/ha application) or normal fertilizer input (three 50 kg N/ha applications over the growing season). Grazing control was based upon weekly measures of grass height. Prior to the experiment, fields had received fertilizer inputs of approximately 150 kg N/ha/year. Vegetation composition was surveyed in four 1 x 1 m quadrats/plot. Invertebrates were surveyed using sweep nets, Vortis suction sampling and pitfall traps.

(1) Olff H. & Bakker J.P. (1991) Long-term dynamics of standing crop and species composition after the cessation of fertilizer application to mown grassland. *Journal of Applied Ecology*, 28, 1040–1052.
(2) Berendse F., Oomes M.J.M., Altena H.J. & Elberse W.T. (1992) Experiments on the restoration of species-rich meadows in the Netherlands. *Biological Conservation*, 62, 59–65.

(3)   Bakker J.P. (1994) Nature management in Dutch grasslands. *Proceedings of the British Grassland Society Occassional Symposium*. 28. pp. 115–124.
(4)   Neuteboom J.H., t'Mannetje L., Lantinga E.A. & Wind K. (1994) Botanical composition, yield and herbage quality of swards of different age on organic meadowlands. *Proceedings of the 15th Meeting of the European Grassland Federation*. pp. 320–323.
(5)   Marriott C.A., Bolton G.R., Common T.G., Small J.L. & Barthram G.T. (1996) Effects of extensification of sheep grazing systems on animal production and species composition of the sward. *Proceedings of the 16th Meeting of the European Grassland Federation*. Grado, Italy. pp. 505–509.
(6)   Mountford J.O., Lakhani K.H. & Holland R.J. (1996) Reversion of vegetation following the cessation of fertilizer application. *Journal of Vegetation Science*, 7, 219–228.
(7)   Smith R.S., Pullan S. & Shiel R.S. (1996) Seed shed in the making of hay from mesotrophic grassland in a field in northern England: Effects of hay cut date, grazing and fertilizer in a split-split-plot experiment. *Journal of Applied Ecology*, 33, 833–841.
(8)   Nösberger J. & Kessler W. (1997) Utilisation of grassland for biodiversity. *Proceedings of the International Occasional Symposium of the European Grassland Federation: Management for grassland biodiversity*. Warszawa-Lomza, Poland, 19–23 May. pp. 33–42.
(9)   Bardgett R.D. & Cook R. (1998) Functional aspects of soil animal diversity in agricultural grasslands. *Applied Soil Ecology*, 10, 263–276.
(10)   Muller S., Dutoit T., Allard D. & Grevilliot F. (1998) Restoration and rehabilitation of species-rich grassland ecosystems in France: a review. *Restoration Ecology*, 6, 94–101.
(11)   Wakeham-Dawson A. & Smith K.W. (2000) Birds and lowland grassland management practices in the UK: an overview. *Proceedings of the Spring Conference of the British Ornithologists' Union, March 27–28, 1999*. Southampton, England. pp. 77–88.
(12)   Defra (2007) *Potential for Enhancing Biodiversity on Intensive Livestock Farms (PEBIL)*. Defra BD1444.
(13)   Geijzendorffer I.R. (2007) Integrating botanical diversity and management of agricultural grassland. PhD thesis. University College, Dublin.
(14)   Woodcock B.A., Potts S.G., Pilgrim E., Ramsay A.J., Tscheulin T., Parkinson A., Smith R.E.N., Gundrey A.L., Brown V.K. & Tallowin J.R. (2007) The potential of grass field margin management for enhancing beetle diversity in intensive livestock farms. *Journal of Applied Ecology*, 44, 60–69.
(15)   Kirkham F.W., Tallowin J.R.B., Sanderson R.A., Bhogal A., Chambers B.J. & Stevens D.P. (2008) The impact of organic and inorganic fertilizers and lime on the species-richness and plant functional characteristics of hay meadow communities. *Biological Conservation*, 141, 1411–1427.
(16)   Hrevušová Z., Hejcman M., Pavlů V.V., Hakl J., Klass paudisová M. & Mrkvičk J. (2009) Long-term dynamics of biomass production, soil chemical properties and plant species composition of alluvial grassland after the cessation of fertilizer application in the Czech Republic. *Agriculture Ecosystems & Environment*, 130, 123–130.
(17)   Potts S.G., Woodcock B.A., Roberts S.P.M., Tscheulin T., Pilgrim E.S., Brown V.K. & Tallowin J.R. (2009) Enhancing pollinator biodiversity in intensive grasslands. *Journal of Applied Ecology*, 46, 369–379.
(18)   Defra (2010) *Modified Management of Agricultural Grassland to Promote In-field Structural Heterogeneity, Invertebrates and Bird Populations in Pastoral Landscapes*. Defra BD1454.

Additional references
Edwards C.A. & Lofty J.R. (1969) The influence of agricultural practices on soil micro-arthropod populations. In J.R. Sheal (ed.) *The Soil Ecosystem*. Systematics Association, London. pp. 237–246.
Coulson J.C. & Butterfield J.E.L. (1978) An investigation of the biotic factors determining the rates of plant decomposition on blanket bog. *Journal of Ecology*, 66, 631–650.
King L.K. & Hutchinson K.J. (1980) The effects of superphosphate and stocking intensity on grassland microarthropods. *Journal of Applied Ecology*, 17, 581–591.
Siepel H. & van de Bund C.F. (1988) The influence of management practices on the microarthropod community of grassland. *Pedobiologia*, 31, 339–354.
Dijkstra H. (1991) Natuur- en landschapsbeheer door landbouwbedrijven: eindverslag van het COAL-onderzoek [Nature and landscape management by agricultural enterprises: final report of the COAL study]. *University of Wageningen Monograph. COAL-publikatie report*. 60.
Tucker G.M. (1992). The effects of agricultural practice on field use by invertebrate-feeding birds in winter. *Journal of Applied Ecology*, 29, 779–790.
Bardgett R.D., Frankland J.C. & Whittaker J.B. (1993) The effects of agricultural practices on the soil biota of some upland grasslands. *Agriculture, Ecosystems and Environment*, 45, 25–45.
Curry J.P. (1994) *Grassland Invertebrates*. London, Chapman & Hall.
Baumberger N., Koch B., Thomet P., Christ H. & Gex P. (1996) Entwicklung der artenvielfalt im lang-zeitversuch Eggenalp [Development of species diversity in the Eggenalp long-term experiment]. *Agrarforschung*, 3, 275–278.
Siepel H. (1996) Biodiversity of soil microarthropods: the filtering of species. *Biodiversity and Conservation*, 5, 251–260.

Yeates G.W., Bardgett R.D., Cook R., Hobbs P.J., Bowling P.J., & Potter J.F. (1997) Faunal and microbial diversity in three Welsh grassland soils under conventional and organic management regimes. *Journal of Applied Ecology*, 34, 453–471.

6.5   Restrict certain pesticides

- A small-scale study in the UK[1] found that using the fungicides Propiconazole and Triadimefon reduced chick food insect abundance less than using Pyrazophos. A replicated, controlled trial in Switzerland[2] found that applying metaldehyde slug pellets in a 50 cm band along the field edge adjacent to wildflower strips provided equivalent crop protection to broadcasting the pellets across the whole field.

Background

Certain agricultural pesticides and chemicals may have detrimental effects on farmland wildlife. This intervention involves restricting the use of certain chemicals either by using less harmful alternatives or limiting the extent of their use.

See also 'Reduce fertilizer, pesticide or herbicide use generally'.

A small-scale study in 1984–1985 of cereal fields treated with foliar fungicides (fungicides for fungal leaf diseases) in the UK (1) found that chick food insect abundance was reduced to a greater extent following applications of Pyrazophos compared to other fungicides. Compared to untreated crops, chick food insects were reduced by 31–70% in crops treated with Pyrazophos, 10% with Propiconazole and 3% with Triadimefon applications. The effect of Pyrazophos was greater when applied to crops at an earlier growth stage (GS) (GS37: 70% reduction in chick food insects; GS50: 45%; GS60: 31%). Following Pyrazophos applications, total predatory arthropods were reduced by 25–48%, aphid-specific predators 35–84% (17% with Triadimefon) and parasitoids 34–55%. Fungicides were sprayed at GS50 in winter wheat in 1984. Pyrazophos was also sprayed at GS60 in spring barley (1984) and GS37 in winter barley (1985). Chick food insects were sampled by sweep nets or suction sampling.

A replicated, controlled trial of slug (class: Gastropoda) control techniques in autumn 1996 in two oilseed rape *Brassica napus* arable fields near Bern, Switzerland (2) found that metaldehyde pellets applied in a 50 cm band alongside wildflower strips provided equivalent crop protection to broadcasting pellets across the entire field. Untreated control plots suffered severe crop damage. Slug sampling showed that *Arion lusitanicus* caused the majority of damage in plots 1 m from wildflower strips. The treatments trialled were broadcasting pellets across the entire plot at a density of 10 kg/ha, 50 cm of pellets alongside wildflower strips at 20 kg/ha and at 40 kg/ha and control plots with no metaldehyde applied. Slugs were sampled on eight nights using bait stations.

(1) Sotherton N.W. & Rands M.R.W. (1987) *Predicting, Measuring and Minimizing the Effects of Pesticides on Farmland Wildlife on Intensively Managed Arable Land in Britain.* Proceedings of the 6th International Congress of Pesticide Chemistry. Pesticide science and biotechnology. Ottawa, Canada, 10–15 August 1986, 433–436.
(2) Friedli J. & Frank T. (1998) Reduced applications of metaldehyde pellets for reliable control of the slug pests *Arion lusitanicus* and *Deroceras reticulatum* in oilseed rape adjacent to sown wildflower strips. *Journal of Applied Ecology*, 35, 504–513.

6.6 Make selective use of spring herbicides

- A replicated, controlled, randomized study in the UK[1] found that spring herbicides had some benefits for beneficial weeds and arthropods.

> **Background**
> This intervention aims to reduce herbicide inputs in fields which do not contain undesirable weed species. It involves not applying an autumn herbicide and reducing spring herbicide applications to a single treatment of a selective herbicide. Reducing spring herbicide treatment to a single application may allow a diverse weed community to develop.

A replicated, controlled, randomized study from 2003 to 2005 of arable fields at three sites in the UK (1) found that spring herbicides had some benefits for beneficial weeds and arthropods. Species richness and cover of beneficial weeds tended to be higher with single spring or post-emergence herbicide applications than pre-emergence or combinations of applications; figures were lowest in plots with three annual applications. Cover of undesirable weeds was higher in single spring or pre-emergence applications than combined treatments. Single applications tended to reduce arthropod abundance less than sequences of herbicides, although post-emergence and pre-emergence applications were detrimental to some taxa. There were 3 or 5 replicate plots (3 or 4 x 24 m) of each treatment per site: untreated, pre-emergence, post-emergence or spring applications or combinations of each 2/all herbicide applications. Vegetation was sampled in 5 quadrats (0.25 m^2) in each plot (June 2003–2005). Arthropods were sampled using a D-Vac suction sampler (5 sub-samples of 10s/plot) in a sub-set of treatments (June).

(1) Jones N.E. & Smith B. (2007) Effects of selective herbicide treatment, row width and spring cultivation on weed and arthropod communities in winter wheat. *Aspects of Applied Biology*, 81, 39–46.

6.7 Use organic rather than mineral fertilizers

- Seventeen studies (including three reviews) from Austria, Belgium, Germany, Ireland, Switzerland and the UK looked at the effects of using organic rather than mineral fertilizers. Fourteen studies (including two reviews and seven replicated and controlled studies, of which four were also randomized) from Austria, Belgium, Germany, Ireland, Switzerland and the UK found that areas treated with organic rather than mineral fertilizers supported higher plant diversity and cover[4] or species richness[14], increased earthworm abundance[16] or diversity, biomass and density[12] and increased abundance[2, 3, 5, 6, 8, 9, 11, 13, 17–19] and/or species richness[3, 7] of some or all invertebrates investigated.

- A literature review[15] found organic fertilizers without pesticides produced the highest earthworm biomass. A small trial in Belgium found more predatory beetles on an arable field two years after organic fertilizer application than on a control plot[1].

- One randomized, replicated, controlled trial in the UK found that using organic rather than mineral fertilizers did not affect the abundance of three weed species[10]. A replicated study from Ireland found that the application of farmyard manure had no long-term effect on invertebrates[2], whilst two studies from the UK found the increase in arthropod predators and springtails was only seen at a local not a field scale[17]. A review found one study from the UK reporting that heavy applications of slurry can be toxic to common earthworms[16].

Background

The use of chemicals in agricultural management, such as synthetic fertilizers, may have a detrimental effect on farmland biodiversity. Organic fertilizers such as farmyard manure (including green manure or crop residues), slurry and other composts provide an alternative to synthetic or mineral fertilizers. Mineral fertilizers are manufactured preparations including nitrogen N, phosphorous P or potassium K, or 'NPK'.

A small unreplicated trial at Huele, Belgium, from 1978 to 1980 (1) found more predatory beetles (Coleoptera) on an arable field two years after manure application than on a control plot or a plot manured the year before monitoring. A single two hectare field was split into three plots: one control plot and two plots with manure applied at 40,000 kg/ha. One plot had manure in October 1978, the other in October 1979. The field was planted with potatoes, then wheat. More ground beetles (Carabidae) and rove beetles (Staphylinidae) were found on the plot manured in 1978. There were 2,197 ground beetles and 1,456 rove beetles in total, compared to fewer than 1,800 ground beetles and fewer than 1,300 rove beetles on the other plots. There was no difference between plots in the total number of male spiders (Araneae) but there were significantly fewer female spiders on the plot manured in 1978 than on the other 2 plots (379 female spiders in total, compared to over 430 on the other plots). In all three arthropod groups, individual species responded differently, although most species were caught more often on a manured plot.

A replicated study in an arable field in Ireland (2) found that application of farmyard manure resulted in an initial, temporary increase in invertebrate taxa, including beneficial arthropods, but overall catch diversity did not differ significantly with organic fertilizer application. Inorganic fertilizers were applied in typical applications to sown sugar beet *Beta vulgaris*. Three treatments were applied, each replicated in two 10 x 25 m plots: application of pre- and post-emergence herbicides (control: Lenacil and Phenmedipham), application of pre- and post-emergence herbicides plus farmyard manure and no herbicide application. Percentage weed cover was estimated in 5 quadrats (0.09 m^2) in each plot in June 1979. Nine pitfall traps/plot (5.6 cm diameter) were set for four 7-day trapping periods (May–September).

A small study of seven cereal fields over one year in Belgium (3) found that organic manure increased the abundance and number of species of ground beetles (Carabidae). A field with a 60 t/ha application of organic manure (and organophosphorus insecticide) had significantly higher abundance (1,128 individuals) and number of species (20 species) than a field with no organic manure (and organochloride insecticide; 14 species). Ground beetle abundance was highest (1,690 individuals) when, as well as applying organic manure (30 t/ha), green manure was applied in late summer and turned under the soil in early spring. Species diversity was highest with the highest concentration of organic manure (60t/ha; 20 species). An application of aldicarb insecticide with organic manure did not affect the number of individuals but slightly reduced the number of species. However, without manure, the insecticide resulted in a three-fold reduction in the number of individuals. Fields differed in organic manure (none; 30 t/ha; 60 t/ha) and insecticide (aldicarb; lindane; E-605). Ten pitfall traps were placed in a row in each field, 4 m apart and were sampled from April to September.

A replicated, controlled, randomized study from 1988 to 1991 of an upland permanent pasture at Bronydd Mawr Research Centre, Powys, UK (4) found that plant diversity and herb cover was significantly higher in grassland with organic fertilizer than mineral fertilizer applications. Plots with farmyard manure and slurry had significantly higher species diversity (both 28% of species) than high (300 kg N/ha; 13% herbs) and low (100 kg N/ha; 18%) mineral applications and similar to unfertilized plots (31%). Nitrogen fertilizers resulted in a significant decrease in species diversity in the hedge bottom; in 1991 only 11 hedgerow species were present on mineral N treatments, 50% less than organic and control treatments. Herb cover was also lower in high (16–21%) and low (18–23%) N applications compared to farmyard manure (28–34%), slurry (27–28%) and the control (33–34%). Vegetation production was significantly higher with high N applications (1,697 g dry matter/m^2) than other treatments (low N: 1,413; farmyard manure: 1,343; slurry: 1,175; control 973 g dry matter/m^2). Sheep grazed grassland plots (7 x 4 m) extending into the hedge bottom were established. A randomized block design with three replicates was set up with the five treatments. Vegetation was sampled monthly within plots between April–November 1988–1991 and hedge bottoms were sampled in spring, summer and autumn each year.

A replicated, randomized, controlled trial in 1986–1987 in Northern Ireland (5) found that ground beetle (Carabidae) abundance was higher in plots of Brussels sprouts *Brassica oleracea* that received mineral fertilizer followed by organic (manure or slurry) fertilizer applications compared to control plots receiving mineral fertilizers only. Over the three year period more ground beetles were caught in no-barrier pitfall traps in the manure or slurry-treated plots than control (inorganic fertilizer-only) plots (average number of total ground beetles/trap/day: 0.46 manure plots; 0.39 slurry; 0.36 control; 0.26 straw). Ground beetles were more abundant in manure plots than controls for both barrier and no-barrier traps. In 1985 and 1986 within the planted area of plots, total catches of ground beetles were 13% and 5% higher in manure and slurry plots and 26% lower in straw plots compared with controls. The most common ground beetle *Bembidion lampros* was also more abundant in manure plots than controls. In 1985 and 1986 the largest number of springtails (Collembola) was found in manure plots; control plots had the lowest number of springtails. In 1985 and 1986 fly larvae (Diptera) and earthworms (Lumbricidae) were more abundant in manure and straw plots but the differences were not significant (no numbers given). The largest number of cabbage root fly *Delia radicum* eggs were found in slurry followed by control plots when ground beetles were excluded. There were 5 replicates of four 10 x 10 m plots. In 1985 and 1986 all plots were treated with 100 kg N, 50 kg P and 100 kg K/ha, followed by 4 different treatments: 0.5 t cattle manure/plot; 455 l cattle slurry/plot; 3 bales winter barley/plot; control (no additional treatment except herbicide). Plots were then treated with herbicide. Brussels sprouts were planted on 27 May in 1985 and 21 May 1986. In 1985 there were three pitfall traps/plot, recording from May–December. In each plot, 5 soil samples were taken from around unprotected plants and 5 soil samples from plants protected with a plastic barrier (6 cm high, 38 cm diameter; internal soil level raised to allow beetles to escape but not enter). Similar sampling was carried out in 1986, with beetles recorded weekly from 6 January to 2 December from three pitfall traps, and collected weekly April–December from 10 pitfall traps surrounded by plastic barriers (barriers used to stop egg predation). Cabbage root fly eggs were counted 10 June–21 October and pupae were collected from soil from 4 plants/plot on 10 January. No organic or inorganic fertilizers were applied to the plots in 1987 but cauliflower plants were planted in the plots and beetles and cabbage root fly egg-laying surveyed.

A controlled study in 1991–1992 on two arable fields northeast of Wien, Austria (6) (same study as (7, 8)) found that the number of emerging parasitic wasps (Hymenoptera) and flies

(Diptera) was generally higher in a plot fertilized with compost than in the mineral fertilized control field. However, the effect of fertilizer on the number of emerging arthropods varied strongly between arthropod families. The parasitic wasp families (Ichneumonidae, Braconidae and Proctotrupoidea) all emerged in significantly higher numbers on the compost fertilized plot. The same was true for two of the more common fly families, gall midges (Cecidomyiidae) and non-biting midges (Chironomidae), and partly dark-winged fungus gnats (Sciaridae), whereas two families, balloon flies (Empididae) and humpbacked flies (Phoridae), were found more often in the control field. None of the presented families or species emerged in highest numbers in an unfertilized plot. Two plots (185 x 10 m each) were established in a 4 ha organic winter rye field. One was left unfertilized; the other was fertilized with compost. A nearby conventional winter cereal field served as a control. The control was fertilized with mineral fertilizer and treated with herbicides in 1991 only. Emerging arthropods were sampled from May–November 1991 and May–August 1992 in 5 photo-eclectors placed along a line (20 m apart) in each habitat. The eclectors were emptied every second week and moved every month. Data from six sampling dates were used in this paper.

A small controlled study in three fields on an organic farm at Obere Lobau, Austria (7) (same study as (6, 8)) found that numbers of species, but not abundance of spiders (Araneae) and ground beetles (Carabidae) were higher in arable fields with compost rather than inorganic fertilizer applications. Numbers of ground beetle species were higher in compost and unfertilized plots (18 species) than inorganic plots (12), as was species diversity (Shannon's H: unfertilized 2.1; compost 1.8; inorganic 1.2). Ground beetle abundance did not vary with treatment (4–5 individuals/trap). There were variations in the responses of different species with treatment. Numbers of spider species were higher in compost and unfertilized plots (30) compared to inorganic plots (21); species diversity did not differ (Shannon's H: 2.2–2.3). Seven money spider (Linyphiidae) species made up approximately 85% of spiders in all treatments, thus numbers of additional species varied. Spider abundance did not vary with treatment (6–7/trap). Two plots (185 x 10 m) in a 4 ha wheat/rye field were either unfertilized since 1989 or fertilized with compost (80 t/ha in 1989 and 1991). A 7.6 ha field (potatoes/bean/cereal) received inorganic fertilizer (1990: 30 N, 75 P, 120 K kg/ha; 1991: 112 N, 104 Ca kg/ha) and herbicides. Five ground photoeclectors (0.25 m²) were placed 20 m apart in the centre of plots to sample arthropods. Traps were moved each month and emptied every 2 weeks, 5–6 times between May–November 1991–1992.

In the same small controlled study in three fields on an organic farm at Obere Lobau, Austria as (6, 7), (8), dominant arthropod groups tended to have a higher abundance in arable fields with compost rather than inorganic fertilizer applications. In the second year, the majority of dominant arthropod groups (15 of 24) had significantly higher abundance in compost plots; these included springtails (Isotomidae, Entomobryidae), rove beetles (Staphylinidae), long-horned flies (Nematocera) and dark-winged fungus gnats (Sciaridae). Three groups were more abundant in the unfertilized plots and six in inorganic plots. Differences between treatments were only consistent over two years for a few groups. Non-biting midges (Chironomidae) and long-legged flies (Dolichopodidae) were significantly more abundant in compost plots, spiders (Araneae) and hypogastrurid springtails (Hypogastruridae) in unfertilized plots and humpbacked flies (Phoridae) and slender springtails (Entomobryidae) in inorganic fertilizer plots.

A study of arable fields in 1989–1992 and 1994 in Germany (9) found that soil microbial activity, feeding activity of soil fauna and the abundance of springtails (Collembola) and mites (Acari) were higher in plots with organic rather than mineral fertilizers. Soil microbial activity did not differ between treatments in April (17–20 micro fluorescein g/dry matter/h) but it was significantly higher with compost treatments (farmyard manure 25–32; farmyard manure plus hornmeal 24–31; composted organic household waste 25–34) compared to mineral fertilization (20–27). Compost application significantly increased feeding activity compared to mineral fertilization (farmyard manure 1–5 perforated baits/d; farmyard manure plus hornmeal 1–5; household waste 1–6; mineral fertilization 1–2). The abundance of springtails and mites showed the same pattern. Composts were applied at 60 Mg fresh matter/ha and hornmeal at 0.6 Mg/ha. Fields were on rotation from winter wheat to oil radish, potatoes and winter barley. Soil biological activity was measured with the bait-lamina test (April–August 1994) and rate of fluorescein diacetate hydrolysis (topsoil 0–10 cm samples). Springtails and mites were sampled 13 times between 1989 and 1992 using a modified Kempson extractor.

A randomized, replicated, controlled trial from 1990 to 1992 in Suffolk, UK (10) found that the use of organic rather than mineral fertilizers did not affect the abundance of three weed species, sterile brome *Bromus sterilis*, common poppy *Papaver rhoeas* and cleavers *Galium aparine*. Abundance of the three species did not differ between plots treated with organic poultry manure and those treated with conventional NPK fertilizer. From October 1989 winter wheat plots were treated with either composted poultry manure or conventional NPK fertilizer, applied at 240 kg N/ha/year. The weed species were sown either singly or together, or left to grow naturally in control plots. There were three 9 m^2 replicate plots for each combination of treatments. Weed growth was monitored from 1990 to 1992.

A 1999 literature review (11) found six studies testing the effects of using organic rather than mineral fertilizers. Five studies found more ground beetles (Carabidae) with organic manure (in one case green manure) than mineral fertilizer; these included (1, 3, 5). One study (7) found no difference in the total numbers of ground beetles between compost and mineral fertilizer.

A replicated, controlled, randomized study between 1978–1998 of arable farming in Switzerland (12) investigated the effect of organic and conventional systems (including use of only mineral fertilizers) on arthropod, earthworm (Lumbricidae), weed and microorganism abundance and diversity. Organic systems had greater earthworm diversity (7–8 species vs 6), density (365–450 individuals/m^2 vs 247) and biomass (242–261 g/m^2 vs 183) compared to conventional systems. Earthworm density and biomass was lowest in conventional systems with mineral fertilizers (143 individuals/m^2; 117 g/m^2) and unfertilized plots (217 individuals/m^2; 137 g/m^2). Microorganism biomass was higher in organic systems (312–406 mg microbial C/kg) than conventional systems with manure and mineral fertilizer (271–285 mg microbial C/kg), conventional systems with mineral fertilizer only (171–244) and unfertilized plots (177–208). Ground beetle (Carabidae) diversity was higher in organic (35–38 species) than conventional systems (32), as was the density of ground beetles (99–113 vs 55), rove beetles (Staphylinidae) (37–40 vs 23) and spiders (Araneae) (58–76 vs 33). Organic systems received approximately 50% less fertilizer (farmyard manure only) than conventional systems. The study was a randomized block design with treatments replicated in 3–6 plots in each of 4 blocks (96 plots of 100 m^2).

A 2000 literature review (13) looked at which agricultural practices can be altered to benefit ground beetles (Carabidae). It found four European studies (2, 3, 5, 6) showing that adding organic manure or compost to agricultural soil increased the numbers of ground beetles relative to sites treated with mineral fertilizer.

A long-term replicated, controlled trial from 1956 to 1995 on alpine pasture in the Bernese Oberland region of Switzerland (14) found that the type of fertilizer used (slurry, PK or NPK) affected the number of plant species, with a significantly greater number of species found on plots fertilized with slurry (on average 36 species) than with NPK fertilizer (on average 29 species) and an intermediate number of species found on plots fertilized with PK. The type of fertilizer did not affect species diversity. Fertilization over a 40 year period reduced the number and diversity of plant species. Plant abundance and diversity (Shannon's H) were greatest in unfertilized plots, where over 60 species were recorded. N was applied at 83 kg/ha, P as 90 kg/ha phosphate (P_2O_5) and K as 180 kg/ha potash (K_2O). There were four replicates.

A small 2000 literature review on aspects of organic farming (15) found that organic fertilizers can enhance ground-dwelling arthropods through a richer supply of detritus-eating soil invertebrates (saprophagous mesofauna) (2). Organic fertilizers without the use of pesticides produced the highest earthworm biomass (Bauchhenss 1991).

A 2000 literature review of grassland management practices in the UK (16) found one study that reported that although the abundance of common earthworms *Lumbricus terrestris* tended to increase with the addition of farmyard manure, heavy applications (more than 500 m^3/ha/ year) of slurry can be toxic (Curry 1994). Another study found that although densities of larger soil invertebrates did not differ in fields with and without farmyard manure applications, bird usage was higher in those that had received moderate applications (Tucker 1992).

Two replicated, controlled trials in Warwickshire, UK (17) found that adding compost or green manure to wheat fields increased numbers of arthropod predators and springtails (Collembola) in the soil at or close to where the compost was added. The effect was local and did not translate to a field-scale increase in numbers of ground active arthropod predators when 1.5 to 3 m-wide strips of compost were added to fields. There were also fewer cereal aphids (pests) (Aphidoidea) in plots with compost applied, but in field-scale experiments this difference was not statistically significant. In the small scale experiment, half of 160 plots 30 x 35 cm in size (2000) or 20 plots of 4 m^2 (2001 and 2002) were treated with mushroom compost in April; half were not. Arthropod predators in the soil were sampled within sunken bowls between April and May each year. In the large-scale experiment 20 x 10 m plots were treated with compost in one 3 m-wide strip, two 1.5 m-wide strips, or not given compost. There were six replicates of each treatment. Arthropod predators were sampled in a large pitfall trap and two 0.5 m^2 quadrats 1–6 m away from the experimental treatments. Springtails were counted from soil cores in the compost strips and 1 m and 6 m away.

A replicated, controlled study in 2004 of grass/clover *Trifolium* spp. fields in Switzerland (18) found that spider (Araneae) but not ground beetle (Carabidae) or rove beetle (Staphylinidae) abundance was significantly greater in plots with organic fertilizers compared to those with synthetic fertilizers. Spider activity density was 80% greater in organic plots in April and October (1.1/m^2) than conventional plots (0.9/m^2). Spider diversity did not differ significantly with treatment. Ground-running spider abundance (78% *Pardosa* spp.) was significantly greater in organic compared to conventional plots in April (1.0 vs 0.6/m^2), August (1.2 vs 1.0/m^2) and October (1.2 vs 0.9/m^2). In contrast, foliage-running spiders were more abundant in conventional plots in May (0.5 vs 0/m^2). There was no effect of treatment on potential prey (aphids (Aphididae), leafhoppers (Auchenorrhyncha), and globular (Sminthuridae) and slender (Entomobryidae) springtails (Collembola)). Treatments were replicated in a 12 x 4 Latin square design of 10 x 20 m plots. Half of each plot received high levels of manure or synthetic fertilizers. No pesticides were applied. Five pitfall traps were placed within a fenced enclosure (1.8 m^2) established within each plot before each of the 5 annual cuts (April–November 2004). After 14 days invertebrates were identified to group and activity densities calculated. Potential prey taxa were sampled by suction sampling (2 m^2 area) and a soil core (20 cm diameter) in each sub-plot in October.

A replicated, controlled, randomized study of arable fields over two years in England (19) found that wolf spider (Lycosidae), ground beetle (Carabidae) and true bug (Hemiptera) abundance tended to be higher in plots with organic fertilizers. In contrast, rove beetles (Staphylinidae), money spiders (Linyphiidae), hoverflies (Syrphidae) and parasitoid wasps (Braconidae) tended to be more abundant in plots with conventional fertilizer applications. Effects depended on year and crop type (grass/clover *Trifolium* spp., cereals, vegetables). There was no effect of treatments on net-winged flies (Neuroptera) and parasitic wasps (Proctotrupoidea). A field was divided into 4 blocks (122 x 122 m), each with 32 plots (24 x 12 m). Treatments were conventional or organic (no) pesticide applications, and conventional (inorganic) or organic (none or compost) fertilizers. Invertebrates were sampled using 5 monthly samples from 5 pitfall traps/plot from May–September and three 1 minute suction samples/plot in the first week of July, August and September 2005 and 2006.

(1) Pietraszko R. & De Clercq R. (1982) Influence of organic matter on epigeic arthropods. *Mededelingen van de Faculteit Landbouwwetenschappen Universiteit Gent*, 47, 721–728.
(2) Purvis G. & Curry J.P. (1984) The influence of weeds and farmyard manure on the activity of carabidae and other ground-dwelling arthropods in a sugar-beet crop. *Journal of Applied Ecology*, 21, 271–283.
(3) Hance T. & Gregoirewibo C. (1987) Effect of agricultural practices on carabid populations. *Acta Phytopathologica Et Entomologica Hungarica*, 22, 147–160.
(4) Jones D. & Haggar R.J. (1993) Impact on nitrogen and organic manures on yield and botanical diversity of a grassland field margin. *Proceedings of the British Grassland Society Winter Meeting: Forward with grass into Europe*. Great Malvern, Worcestershire, 16–17 November 1992. pp. 135–138.
(5) Humphreys I.C. & Mowat D.J. (1994) Effects of some organic treatments on predators (Coleoptera, Carabidae) of cabbage root fly, *Delia-radicum* (l) (Diptera, Anthomyiidae), and on alternative prey species. *Pedobiologia*, 38, 513–518.
(6) Idinger J. (1995) Ground photoelector evaluation of Diptera and parasitoid Hymenoptera in unfertilized, mineral nitrogen and compost fertilized grain fields *Mitteilungen Der Deutschen Gesellschaft Fur Allgemeine Und Angewandte Entomologie Band 10, Heft 1–6, Dezember 1995*, 10, 553–556.
(7) Idinger J., Kromp B. & Steinberger K.H. (1996) Ground photoelector evaluation of the numbers of carabid beetles and spiders found in and around cereal fields treated with either inorganic or compost fertilizers. *Acta Jutlandica*, 71, 255–267.
(8) Idinger J. & Kromp B. (1997) Ground photoelector evaluation of different arthropod groups in unfertilized, inorganic and compost-fertilized cereal fields in eastern Austria. *Biological Agriculture & Horticulture*, 15, 171–176.
(9) Pfotzer G.H. & Schuler C. (1997) Effects of different compost amendments on soil biotic and faunal feeding activity in an organic farming system. *Biological Agriculture & Horticulture*, 15, 177–183
(10) McCloskey M.C., Firbank L.G., Watkinson A.R. & Webb D.J. (1998) Interactions between weeds of winter wheat under different fertilizer, cultivation and weed management treatments. *Weed Research*, 38, 11–24.
(11) Kromp B. (1999) Carabid beetles in sustainable agriculture: a review on pest control efficacy, cultivation impacts and enhancement. *Agriculture, Ecosystems & Environment*, 74, 187–228.
(12) Fliessbach A., Mäder P., Dubois D. & Gunst L. (2000) Results from a 21 year old field trial. Organic farming enhances soil fertility and biodiversity. *FiBL Dossier*, 15.
(13) Holland J.M. & Luff M.L. (2000) The effects of agricultural practices on Carabidae in temperate agroecosystems. *Integrated Pest Management Reviews*, 5, 109–129.
(14) Koch B. & Meister E. (2000) Graded management intensity of grassland systems for enhancing floristic diversity. In D. Gagnaux & J.R. Poffet (eds.) *Livestock Farming Systems: Integrating animal science advances in the search of sustainability*, EAAP Publication No. 97, Wageningen Pers, Wageningen. pp. 176–178.
(15) Pfiffner L. (2000) Significance of organic farming for invertebrate diversity-enhancing beneficial organisms with field margins in combination with organic farming. *Proceedings of the International Workshop: The relationship between nature conservation, biodiversity and organic agriculture*. Vignola, Italy 1999. pp. 52–66.
(16) Wakeham-Dawson A. & Smith K.W. (2000) Birds and lowland grassland management practices in the UK: an overview. *Proceedings of the Spring Conference of the British Ornithologists' Union March 27–28, 1999*. Southampton, England. pp. 77–88.
(17) Bell J.R., Traugott M., Sunderland K.D., Skirvin D.J., Mead A., Kravar-Garde L., Reynolds K., Fenlon J.S. & Symondson W.O.C. (2008) Beneficial links for the control of aphids: the effects of compost applications on predators and prey. *Journal of Applied Ecology*, 45, 1266–1273.

(18) Birkhofer K., Fliessbach A., Wise D.H. & Scheu S. (2008) Generalist predators in organically and conventionally managed grass-clover fields: implications for conservation biological control. *Annals of Applied Biology*, 153, 271–280.
(19) Eyre M.D., Sanderson R.A., Shotton P.N. & Leifert C. (2009) Investigating the effects of crop type, fertility management and crop protection on the activity of beneficial invertebrates in an extensive farm management comparison trial. *Annals of Applied Biology*, 155, 267–276.

Additional references
Bauchhenss J. (1991) Regenwurmtaxozönosen auf Ackerflächen unterschiedlicher Düngungs- und Pflanzenschutzintensitäten [Earthworm taxonomic communities on arable land with different fertilization and plant protection intensities]. *Bayerisches Landwirtschaftliches Jahrbuch*, 68, 335–354.
Tucker G.M. (1992) The effects of agricultural practice on field use by invertebrate-feeding birds in winter. *Journal of Applied Ecology*, 29, 779–790.
Curry J.P. (1994) *Grassland Invertebrates*. London, Chapman & Hall.

6.8   Reduce fertilizer, pesticide or herbicide use generally

- Of 38 individual studies from Austria, the Czech Republic, Denmark, Finland, France, Germany, Ireland, the Netherlands, Sweden and the UK investigating the effects of reducing fertilizers, pesticides or herbicides, 34 studies (23 replicated, of which 6 were also controlled and randomized, 1 review and 1 systematic review) found benefits to some invertebrates[1, 3–5, 7–11, 14–16, 23–26, 29, 31, 36, 37, 40, 41, 43–45], plants[13, 17–19, 22, 34, 37, 41, 42, 47], or farmland birds[21, 30, 33, 39, 44, 46]. Twenty-five studies (16 replicated, of which 7 were also randomized and controlled and 1 review) found negative, mixed, minimal or no effects on some invertebrates[1–4, 7, 9, 12, 14, 20, 23, 24, 27, 28, 31, 32, 37, 38, 40, 41, 43, 47], farmland birds[6, 33, 35, 38, 44, 46, 47] or plants[19, 32, 38, 42].

- Ten studies (six replicated, controlled studies of which two were randomized) from three countries found positive effects of reducing or stopping pesticide applications on invertebrates[4, 7, 8, 14, 23, 41, 43, 44], plants[17, 41] or birds[21, 30, 44]. Eight studies (two replicated, controlled and randomized, one paired before-and-after trial) from four countries found inconsistent or no effects on some invertebrates[7, 14, 20, 23, 27, 28, 41, 43] and birds[6, 44].

- Ten studies (nine replicated, five also controlled and a European systematic review) from four countries found positive effects of reducing or stopping herbicide use on plants[18, 19, 34, 37, 47], invertebrates[3, 10, 36, 37, 45] or birds[33]. Five replicated studies (two also controlled and randomized) from three countries found no or mixed effects on birds[33, 47], invertebrates[3, 37, 47] and plants[19].

- Five studies (three replicated, controlled, of which two were randomized) from four countries found positive effects of reducing or stopping fertilizer applications on invertebrates[1, 40], Eurasian skylark[44] or plants[19, 22]. Four studies (three replicated, controlled and randomized) from two countries found reducing or stopping fertilizer inputs had no, or no consistent effects on some invertebrates[4, 40] and farmland birds[35, 44]. Two studies from the UK (one replicated) found plots where fertilizer inputs were not reduced tended to have higher earthworm biomass[1, 2] or abundance[2].

- Fifteen studies (three replicated, controlled, of which one was also randomized, five site comparisons and one review) from seven countries looked at the effects of reducing or stopping applications of two or more inputs: pesticides, herbicides or fertilizers. Thirteen studies found positive effects of reducing two or more inputs on some or all invertebrates[5, 11, 15, 16, 24–26, 29, 31], plants[13, 19, 42], soil organisms[9] and birds[39, 46] studied. Seven studies found negative or no effects of reducing combinations of inputs on some invertebrates[9, 12, 24, 31, 32, 38], plants[32, 38, 42] or birds[38, 46].

Background

Pesticide, herbicide and fertilizer applications may have a negative impact on farmland wildlife. This intervention may involve reducing or ceasing applications of pesticides (such as insecticides; fungicides), herbicides and fertilizers.

Several European countries (Denmark, the Netherlands and Sweden) introduced initiatives in the 1990s to reduce pesticide applications (Pretty 2005). In Denmark, a Pesticide Action Plan was introduced in 1986 with the aim of reducing total pesticide applications by 50% in 10 years, in order to reduce impacts on biodiversity and groundwater resources (Jørgensen & Kudsk 2006). Reductions continued and by 2004 Danish farmers had reduced inputs of pesticides by 56% (kg of active ingredient) and 20% (treatment frequency index) (Jørgensen & Kudsk 2006).

See also 'Leave headlands in fields unsprayed (conservation headlands)', which also has monitoring of biodiversity in response to reduced fertilizer, pesticide and herbicide applications.

Pretty J. (2005) Sustainability in agriculture: recent progress and emergent challenges. *Issues in Environmental Science and Technology*, 21, 1–15.
Jørgensen L.N. & Kudsk P. (2006) *Twenty Years' Experience with Reduced Agrochemical Inputs: Effects on farm economics, water quality, biodiversity and environment*. Proceedings of the HGCA conference – Arable crop protection in the balance: Profit and the environment. 25–26 January, 16, 1–10.

A study of three-and-a-half year grass/clover *Trifolium* spp. leys from 1950 to 1956 in the UK (1) found that plots that were grazed rather than cut and with nitrogen applications had higher earthworm (Lumbricidae) mass in some years. Overall abundance of worms was 17–39/cubic foot. Plots with nitrogen applications had a significantly greater weight of earthworms than those without nitrogen applications in 1953 (9–10 vs 5–6 g/cubic feet) but not overall (8 vs 7 g/cubic feet). However, there tended to be lower numbers of worms in plots with nitrogen (17–30) than those without nitrogen (19–39/cubic foot). One plot of each treatment combination was established in each of the six years (the study also compared cutting and grazing management). Fertilizer was applied at 280 lb N/acre/year. Earthworms were sampled when plots were ploughed out of leys in the autumn (1953–1956). Four samples of two cubic feet of soil were sampled in each plot per year.

A replicated trial on an experimental farm in eastern Scotland (2) found fewer earthworms (Lumbricidae) at lower nitrogen application rates. There were 78 earthworms/m^2 and 0.42 tonnes earthworm/ha in plots with no nitrogen, compared to 106 earthworms/m^2 and 0.53 tonnes/ha in plots with 100 kg N/ha. The highest nitrogen treatment (150 kg N/ha) had fewer earthworms but higher biomass because there were more large-bodied species like Lumbricus terrestris (93 earthworms/m^2; 0.59 tonnes/ha). Earthworm biomass decreased with decreasing nitrogen application at a rate of 0.06 t/ha for every 50 kg N/ha. Only one of the eight species recorded, Allolobophora rosea, was more abundant in plots with lower fertilizer inputs (9 earthworms/m^2 at 0 kg N/ha, compared to 3.7 earthworms/m^2 at the highest rate of 150 kg/ha). The experiment was replicated eight times. Spring barley crops were managed from 1967 until 1973 with either 0, 50, 100 or 150 kg N/ha added annually.

A replicated study of an arable field in Ireland (3) found that invertebrate abundance tended to be greater where no herbicides were applied compared to sprayed areas. A greater number of detritus feeders (2,136 vs 637–674), particularly beetles (Coleoptera) and larval and adult flies (Diptera) and herbivores (2,061 vs 174–333), were found in the unsprayed plots compared to sprayed plots, once weed populations were established. Overall predator numbers differed little between treatments (unsprayed: 2,422; sprayed: 2,142–2,356), although more predatory rove beetles (Staphylinidae) (324 vs 78–149) and parasitic wasps (Hymenoptera) (376 vs 72–87) were found in unsprayed plots towards the end of the sampling period. Ground beetles (Carabidae), which were the most numerous predators, showed no difference between treatments (unsprayed: 1,312; sprayed: 1,543–1,606). Inorganic fertilizers were applied in typical applications to sown sugar beet Beta vulgaris. Three treatments were then applied, each replicated in two plots (10 x 25 m): application of pre- and post-emergence herbicides (control: Lenacil and Phenmedipham), application of pre- and post-emergence herbicides plus farmyard manure and no herbicide application. Percentage weed cover was estimated in five 0.09 m^2 quadrats/plot in June 1979. Nine pitfall traps/plot (5.6 cm diameter) were set for four 7-day trapping periods (May–September).

A replicated, randomized, controlled study of arable fields between 1982 and 1984 in England (4) found that the abundance of soil nematodes (Nematoda), slugs (Gastropoda) and fly (Diptera) larvae was greater in plots without pesticide (insecticide) applications. In spring 1983 numbers of nematodes were significantly higher in the plots without pesticide applications (5.5–6.5/50 g) compared to plots sprayed 2 weeks previously (2.5–3.5). Numbers of slugs did not differ between treatments in the first year but were significantly lower in sprayed plots in 1983–1984 (1–3 vs 10–26/tile). Overall, numbers of fly larvae were higher in plots without pesticide applications (25–75 individuals/replicate cores vs 7–65). Fertilizer did not tend to have a significant effect on soil invertebrate numbers. Four replicated, randomized blocks each comprising 10 plots (6 x 3 m) were established. Treatments were three different grass (Italian rye grass Lolium multiflorum; perennial rye grass L. perenne; existing mixed ley) and two wheat regimes ('Norman' and 'Armada' varieties) with (phorate and aldicarb; three applications) and without pesticide treatments. Fertilizer was applied to all plots (except wheat 'Norman' in 1982) and fungicides applied to wheat when required. Invertebrates were sampled in the spring and autumn after pesticide applications. Free-living soil nematodes and fly larvae were sampled by taking two or four randomly located soil samples (soil corer: 2.5 x 15 cm and 6.5 x 8 cm respectively) from each plot. Slugs were sampled using two wooden tiles/plot with slug pellets underneath, which were collected after 4–7 days.

A replicated study on five arable fields in Austria (5) found that fields with no pesticides (fungicides or herbicides) or fertilizers had a greater diversity of ground beetles (Carabidae) than those with conventional chemical applications. Wheat fields with no spraying had greater numbers of ground beetle species (43–50 species) and individuals (5–6 individuals/trap/day) than those that received conventional pesticide and fertilizer applications (species: 38–40; individuals: 2–3/trap/day). Conventionally farmed sugar beet Beta vulgaris fields had similar numbers to conventional winter wheat (1/trap/day). Fields differed in terms of weed control (mechanical or herbicides), disease control (none or fungicides) and manuring (green/compost/stone meal or mineral). One or two wheat and/or sugar beet fields were under each treatment in 1982 and 1983. Invertebrates were sampled using a line of 6–10 pitfall traps in the centre of each field from May–July.

A site comparison study of arable farmland over 7 years as part of the Boxworth Project in Cambridgeshire, UK (6) (same study as (7)) found that only 1 of 11 bird species declined in numbers with high pesticide inputs and none of 4 species had reduced breeding performance. The percentage of the total territories in the high input area for the 11 species remained fairly constant during the project (pre-treatment: 44–46%; treatment: 40–50%). Common starling Sturnus vulgaris showed a significant decline in percentage of breeding territories in the high input area relative to the low input area from 1984 to 1987 (45% to 28%); numbers recovered in 1988 (41%). There were no significant differences in breeding performance between treatment areas for tree sparrow Passer montanus or starling in terms of numbers of breeding pairs, numbers of young fledged, numbers of young fledged/pair or causes of nest losses. The only difference was that the percentage of first broods that failed to produce at least one fledgling tended to be lower in the high input area in baseline years, but increased more than the low input areas in treatment years. However, overall production of young was not reduced in the high input area. Sample sizes were small for blue tit Parus caeruleus and great tit P. major nests. Pre-treatment years were 1982–1983 and treatment years 1984–1988. The 'Common Bird Census' method was used to monitor birds with 10 visits from spring to early summer. A total of 220–244 nest boxes were put up in each area, which were checked weekly during the breeding season.

A controlled study as part of the Boxworth project comparing arable farmland with high and reduced pesticide inputs over five years in Cambridgeshire, UK (7) (same study as (6)) found that the abundance of invertebrate herbivores, carnivores and parasitoids tended to be higher in areas with reduced pesticide applications, whereas detritus-feeding invertebrates did not differ with treatment. On average, total numbers of herbivores were 50% lower, predators 53% (39–70%) lower and parasitoids 39–79% lower in the conventional area compared to the reduced pesticide areas. Numbers varied with year, and numbers of some taxa were higher in conventional areas in some years. Numbers of detritus-feeders did not differ significantly between treatments. There were two treatment areas, one with conventional and the other reduced pesticide applications (selective insecticides and slug/snail pesticides). Invertebrate density was sampled in the middle of each field using a Dietrick vacuum sampler at intervals of 7–10 days between mid-April and harvest. Each sample comprised 5 sub-samples (each 0.09 m^2) taken 10 m apart.

A replicated, controlled study in summer 1989–1990 in eight sites on one arable farm near Bonn, Germany (8) found that a 50% reduction in pesticide application could control an aphid (Aphidoidea) outbreak as efficiently as the normal application. In farming systems with no insecticide use at all natural predators reduced aphid populations to the same low levels (<5 aphids/plant), but the population decline occurred one week later than in the systems with pesticide use. Predatory arthropod populations also declined after pesticide treatment. Predator levels remained rather low in the normal pesticide system, however in the 50% reduced pesticide system they recovered in 3 weeks after pesticide application. Four farming systems were compared with two replicates each: conventional farming (normal pesticide use), integrated farming (50% reduction in pesticide use), 'minimum' farming (no insecticides, strongly reduced herbicide use) and 'no pesticide' farming (no pesticide use). Aphids and their predators were counted visually and with sweep nets once a week from April.

A site comparison study at the Lovinkhoeve Experimental Farm, Noordoostpolder, the Netherlands (9) found a higher biomass of microbes, protozoa, nematodes (Nematoda) and earthworms (Lumbricidae), but not of mites (Acari) and springtails (Collembola), in the upper 10 cm of an arable soil with reduced fertilizer and pesticide inputs than in a conventionally managed soil. At lower depth (10–25 cm) there were no consistent differences in soil animals. The reduced input plot had 8.9 kg C/ha of earthworms in the top 10 cm, and 4.7 kg C/ha at 10–25 cm depth. No earthworms were recorded in conventional plots. Total biomass of nematodes in the upper layer was 0.79 kg C/ha in reduced input plots and 0.3 kg C/ha in the conventional plots. Reduced input plots had applications of 65–170 kg nitrogen fertilizer/ha/year compared to 130–285 kg N/ha on conventional plots. They also had reduced tillage. The experiment began in 1985. Soil samples were taken from three areas of each plot under winter wheat in 1986.

A randomized, replicated trial from 1987 to 1991 on the Oxford University Farm, Wytham, Oxfordshire, UK (10) found fewer adult meadow brown butterflies Maniola jurtina on 2 m-wide naturally regenerated field margins that were sprayed with herbicide once in summer, compared to any margins that were not sprayed. There were 3–4 meadow browns/50 m sprayed plot on average compared to 4–13 meadow browns/50 m on unsown, uncut margins that were not sprayed. There was no difference between treatments in abundance of meadow brown larvae (3 larvae/plot on average). Two metre-wide field margins were established around arable fields in October 1987, rotavated and left to naturally regenerate from March 1988. Fifty metre-long plots were either uncut and unsprayed, subject to one of four different cutting regimes but unsprayed, uncut but sprayed once a year with herbicide (glyphosate) in late June or July. There were six replicates of each treatment. Adult meadow brown butterflies *Maniola jurtina* were monitored weekly along walked transects in the experimental plots from June to September 1989 and from April to September 1990 and 1991. Meadow brown larvae were sampled in spring 1991, by sweep netting and visual searching.

A replicated, controlled study of seven arable fields on three farms in England (11) found that reduced pesticide inputs tended to result in higher numbers of arthropods. Applications of chlorpyrifos spray in the conventional plots resulted in decreased numbers of ground beetle species (Carabidae: Bembidion aeneum; B. lunulatum; B. obtusum), one water scavenger beetle species (Hydrophilidae: Helophorus aquaticus), springtails (Collembola: Entomobrya multifasciata; Isotoma viridis; Sminthurides signatus; S. viridis) and money spiders (Linyphiidae: particularly Erigone spp.). Some of these species disappeared from sprayed plots and did not recover for a year. Spraying with deltamethrin also resulted in a decrease in water scavenger beetles Helophorus spp., several money spider species and one ground beetle species B. lunulatum; the latter did not recover for 15 months. Fields were divided in half with one receiving conventional pesticide applications, and the other reduced pesticides, i.e. lower herbicide and fungicide and no insecticides (1991–1996). All other practices were the same. Arthropods were monitored using a D-Vac suction sampler and pitfall traps. In each plot, 4 samples were taken, each comprising 5 sub-samples (total area 0.46 m²) between 25 and 125 m from the shared field margin. Four pitfall traps (9 cm diameter) were also located in each field half (12 m apart) and were operated for 7-day periods. This study was part of the same project (SCARAB – Seeking Confirmation About Results At Boxworth) as (16, 20).

A small replicated, controlled trial at two sites in Lower Saxony, Germany (12) found that aphids (Aphidoidea) and their insect predators were less abundant in wheat fields not treated with fertilizers, insecticides or herbicides in 1992 compared to conventionally farmed fields or fields with low fertilizer use and no insecticide. A maximum of 80 aphids/wheat stem were recorded on plots with no chemicals compared to 300 aphids/stem in the conventional farm system and close to 300 aphids/stem in fields with a 50% reduction in nitrogen fertilizer application (105 kg N/ha compared to the conventional 210 kg N/ha) and no insecticide (herbicides were used). Fields with no chemical use had no more than 20 aphid predator larvae/m²; hoverflies (Syrphidae), ladybirds (Coccinellidae) and lacewings (Chrysopidae); compared with up to 60–70 larvae/m² under conventional farming and up to 40 larvae/m² with 50% fertilizer reduction and no insecticide. Under conventional farming, ladybirds were only recorded on plots not treated with insecticide. In farming systems with reduced or no chemical use, ladybirds were the dominant aphid predator in most months. This study was carried out on areas of 35 to 45 ha at 2 sites (2 replicates of each farming system). Aphids and their predators were counted on 150 wheat stems twice a week and suction trapped every 2 weeks during the 1992 growing season. This study was part of the same project (INTEX – Integrated Farming and Extensification of Agriculture) and was carried out in partly the same research site as (13, 24).

A replicated, controlled study in 1990–1994 on three arable farms in Lower Saxony, Germany (13) found significantly higher plant species diversity, weed cover and seed numbers in the seed bank in an 'integrated' farming system with a 50% reduction in chemical inputs (fertilizer; pesticides/herbicides) than in a conventional farming system. Species richness, weed cover and seeds in the soil were also higher in the extensive (no input) farming system than in the conventional system, but did not differ from the integrated farming. Crop cover, however, was significantly reduced only in the extensive farming system. Thus, a 50% reduction in herbicide use was the most efficient way of combining the economic interests of agriculture (crop yield) with weed protection. On three farms, field trials with three different farming systems were compared: conventional farming (normal pesticide/herbicide use and fertilization), integrated farming (50% reduction in pesticide/ herbicide use; 25–40% reduction in mineral fertilization) and extensive farming (no pesticide/herbicide use; no mineral fertilization). Plants were monitored several times a year in 4 permanent plots (10 x 10 m) on 2 of the farms. Soil samples (0–5 cm and 5–30 cm depth) were taken in March 1990 and 1993 on all three farms. Seeds were germinated in the laboratory for 20 months after different growth stimulations. This study was part of the same project (INTEX – Integrated Farming and Extensification of Agriculture) and was carried out in partly the same research site as (12, 24).

A replicated, controlled, randomized study in arable fields in Finland (14) (same study as (23)) found that ground beetle (Carabidae) abundance was higher in reduced pesticide compared to conventional pesticide plots. This was true in 1993 and 1994; the opposite trend was seen in 1992. Spring species tended to be more affected by pesticides than autumn species. Overall there was no significant difference in beetle abundance between cultivation treatments: customary (deep ploughing; conventional fertilizer use; no undergrowth) vs integrated (soil treatment with cultivator only; reduced fertilizer use; undersown grass/ clover Trifolium spp.). There were 6 replicate blocks and treatments (in 0.7 ha plots) which were fully randomized within blocks (1 treatment combination/plot). Treatments were conventional pesticide applications, reduced pesticides or no pesticides (control) and customary or integrated cultivation. Beetles were sampled with pitfall traps at 12, 66 and 120 m into each crop 8–10 times (1 week/sample) between sowing and harvest.

A small, controlled study of three arable fields on two farms near Braunschweig, Germany (15) found that arthropod numbers and species richness tended to increase with a reduction in management intensity, largely a reduction in fertilizer and pesticide inputs. Arthropod abundance, number of spider (Araneae) species, numbers of juvenile spiders, abundance and number of ladybird (Coccinellidae) species increased with a reduction in inputs. Abundance of beneficial species and length of their activity period also tended to increase with decreased fertilizer and pesticides. Specific species differed in their response to treatment and the intensity of effects depended on type of lifecycle. Reduced inputs increased the activity density of wolf spiders (Lycosidae) and decreased the proportion of pioneer species. Spider species with a wide range of ecological living conditions tended to increase with an increase in pesticides. In 1989–1992 four plots within an arable field received different management intensities: no fertilizers/pesticides, extensive or integrated cultivation with medium fertilizer/ pesticide inputs and intensive cultivation with high fertilizer/pesticide inputs. In 1992–1995 four plots within an arable field received farming regimes that differed in the input of fertilizers and pesticides (high input; 30–50% reduction; none), crop rotation (three/four course), tillage, weed control (mechanical/chemical), cultivars, drilling technique and catch crops. A long-term set-aside was also sampled. Six to eight emergence traps and pitfall traps sampled arthropods within each treatment. Traps were collected every 2–4 weeks throughout the year. Results for pest species are not included here.

A replicated, controlled study of three arable rotation fields on three farms in England (16) found that reduced pesticide inputs tended to result in higher numbers of springtails (Collembola). Numbers of the springtails *Entomobrya multifasciata* and *Lepidocyrtus* spp. were significantly greater in reduced pesticide plots compared to conventional plots. In the conventional plots, these species tended to disappear following chlorpyrifos applications in particular and *Lepidocyrtus* spp. numbers then remained low for five years. *Sminthurinus elegans* also declined after chlorpyrifos applications, but tended to recover by the following year and have greater numbers in conventional plots. Fields were divided in half with one receiving conventional pesticide applications and the other reduced pesticides, i.e. lower herbicide and fungicide where possible and no insecticides (1991–1996). All other practices were the same. Springtails were monitored using a D-Vac suction sampler. In each plot, 5 samples were taken, each comprising 5 sub-samples (total area 0.46 m^2) between 25–50, 50–75, 75–100 and 100–125 m from the shared field margin. This study was part of the same project (SCARAB – Seeking Confirmation About Results At Boxworth) as (11, 20).

A replicated, controlled study in summer 1991–1994 on up to 13 farms and 2 experimental sites in the Province of Bayern, Germany (17) found that the 2 management types with restricted pesticide use (organic farming and controlled contract production, 'KVA') had a more positive effect on plant species richness (average ranges for the sites: organic: 18.4–22.6 species; KVA: 16.9–19.0 spp.; controls: 12.4 to 14.6 spp.) than the Bavarian culture landscape programme or control farms (15.6 and 13.8 spp. respectively). Farms under organic or controlled contract production both had restrictions concerning pesticide use. In the Bavarian culture landscape programme, no such restrictions existed but some less common crops (e.g. flax and grass seeds) can be included in the crop rotation. Vegetation was surveyed between June and September on total areas between 100 and 400 m^2. Cereal crops were surveyed yearly, cut set-asides several times a year. Note that no statistical analyses were performed on these data.

A replicated, controlled study of arable fields at three sites within the TALISMAN MAFF-funded experiment in England (18) found that seed bank density and weed density were higher with reduced (50%) herbicide applications. At High Mowthorpe, plots with reduced (50%) herbicide had significantly higher seed densities (3,181–16,231/m^2) than those with conventional applications (1,764–11,300/m^2). At Boxworth, the same was true for spring-cropped plots (25,824 vs 8,780/m^2). At Boxworth, broadleaved plant seed weights were significantly higher with reduced compared to conventional herbicides (35–151 vs 24–91 mg/m^2); treatments did not differ at High Mowthorpe. Plant density tended to be higher on plots with reduced herbicides (4–18/m^2) compared to conventional herbicide applications (3–16/m^2). At Boxworth, only broadleaved plant species/groups differed between treatments whereas at Drayton higher weed numbers were consistently found on reduced herbicide plots. At Boxworth there were two replicates in two blocks; at the other two sites, there was one replicate in three blocks. Conventional fertilizer, fungicide and insecticide levels were applied. Seed banks were sampled at Boxworth and High Mowthorpe after harvest from three sub-samples (60 combined soil cores) in each plot. Weed density was sampled in 15 quadrats/plot at the 3 sites after harvest (August–September) and in October–November.

A replicated, controlled, randomized study on a low productive grassland and high productive fallow arable field in the Netherlands (19) found that decreased fertilizer applications resulted in an increase in the number of plant species; minimal effects of herbicide applications were found on fallow land. There were significantly more plant species in the plots receiving no fertilizer (grassland: 16 species/m^2; fallow: 23–27/m^2) compared to those with 25% (grass: 14/m^2; fallow: 19–23/m^2) and 50% (grass: 13/m^2; fallow: 19–23/m^2) of conventional fertilizer applications. In the grassland there was no significant effect of herbicide whereas in the fallow land there was an effect in the final assessment when 0 and 5% of conventional herbicide application plots had significantly greater species diversity than 50% of conventional herbicide application plots (24–25 vs 22/m^2). The most species-rich plots were the 0% herbicide–0% fertilizer plots (grass: 15/m^2; fallow: 31/m^2) and the 10%–0% plots (grass: 15/m^2); 50–50% fertilizer plots had the least species (grass: 10/m^2; fallow: 20/m^2). Forty-eight plots (2 x 2 m) were established on a low productivity grassland and an adjacent fallow field sown with 30 broadleaved grassland species. Fertilizer treatments were: 0, 25 and 50% of the conventional application (110 kg N/ha/year). Herbicide (fluroxypyr) treatments were: 0, 5, 10 and 50% of the standard agricultural dose (200 g/ha). Vegetation composition was assessed in April–May (grass and fallow) and September (fallow only) 1993–1996.

A replicated study of arable fields on three farms in England (20) found that overall earthworm (Lumbricidae) populations did not differ significantly under conventional and reduced pesticide inputs. The only significant difference between treatments was found in autumn 1993 when earthworm density was higher in reduced pesticide treatments (35–50% of normal application) than controls at 2 of the farms (Warwickshire: 1,529 vs 1,149/m^2; North Yorkshire: 409 vs 346); the reverse was true at the third farm (Nottinghamshire: 35 vs 45). Differences in earthworm densities were much greater between farms than between fields within farms. Species and age composition differed between farms but the treatment effect was not consistent between fields, even within the same farm. Seven fields over three arable farms were split in two: one half received a conventional pesticide regime and the other a re-duced (35–50%) input and no insecticides (1991–1993). Earthworms were sampled in spring and autumn (1993–1994) from three 50 x 50 cm quadrats/plot by hand-digging and using 0.2% formalin solution (20 min period). This study was part of the same project (SCARAB – Seeking Confirmation About Results At Boxworth) as (11,16).

A site comparison before-and-after study from 1989 to 1994 in Sussex, England (21) found that survival rates of grey partridge *Perdix perdix* chicks were significantly higher on 21 km^2 of arable farmland that received irregular insecticide applications compared to a 7 km^2 farm with insecticide applications 4 times a year (average 34% survival on low application farms vs 22% on the high application farm). Before the start of intensive insecticide application (1970–1988) survival on the farm had been similar to, or higher than, that on the surround-ing farms (27% survival on low application farms vs 30% on the intensive application farm). Chick survival rates (up to the age of approximately six weeks) were calculated each year and compared between areas with intensive and irregular insecticide applications. A long-term data set (1970–1988) collected prior to this study was used to investigate chick survival prior to insecticide application on the intensive application farm.

A replicated, controlled study of former arable fields at 6 sites in Sweden (22) found that after 10 years there were twice as many plant species in unfertilized compared to fertilized set-aside (30 species in the least fertile site; 10 in the most fertile). Cutting and planting a cover crop also had a positive effect on the number of plant species. At each site 2 plots (10 x 20 m) were sown with a grass cover crop and 2 were left bare. Each year, one of each pair had fertil-izer added (equivalent to 150 kg N/ha) and half of every plot was cut and cuttings removed (late July). Vegetation cover was assessed in the centre of each plot (8 x 1 m^2) in 1975–1986.

A replicated, controlled, randomized study in arable fields in Finland (23) (same study as (14)) found that spider (Araneae) abundance was greater in reduced pesticide compared to conventional plots. This was the case in 1992 and 1994 (reduced fertilizer: peak 17–31/3 traps/week; conventional: 12–23); there was no significant difference in 1993. Conventional pesticide use decreased money spider (Linyphiidae) numbers in all years (peak: 9–12 vs 10–20/3 traps/week) but wolf spider (Lycosidae) catches only in 1994. Only one of the species tested (*Erigone atra*, money spider family) differed significantly between pesticide regimes. There was no significant difference in spider abundance between cultivation treatments: customary (deep ploughing; conventional fertilizer use; no undergrowth) vs integrated (soil treatment with cultivator only; reduced fertilizer use; undersowing with grass/clover *Trifolium* spp.). There were 6 replicate blocks and the treatments (in 0.7 ha plots) were fully randomized within blocks (one treatment combination/plot). Treatments were conventional pesticide applications or reduced pesticides and customary or integrated cultivation. Spiders were sampled with pitfall traps at 12, 66 and 120 m into each crop 8–10 times (1 week/ sample) between sowing and harvest.

A controlled trial of different farming systems at Reinshof experimental farm, Lower Saxony, Germany (24) found that the highest number of rove beetles (Staphylinidae) was caught under conventional farming with reduced inputs (50% reduction in nitrogen fertilizer, no insecticide, although herbicides were used) (7,897 beetles in total, compared to 6,581 in the control plot with conventional farming). The reduced input field did not have more rove beetle species than conventional farming (39 and 42 species respectively). Extensive farming with no nitrogen fertilizer, herbicides or insecticides had the lowest number of rove beetles (5,038 beetles, from 40 species). Rove beetles were monitored with pitfall and/or emergence traps throughout the year. The experiment was run from 1990 to 1994. There were three or four fields under each treatment, representing the full crop rotation. Monitoring was only in the wheat field from each system each year. The study also included integrated farming (30% of the nitrogen fertilizer used in the conventional system and 50% of the pesticides/herbicides, along with other measures) and extensive farming. The authors suggest that integrated farming without fertilizer does not create a favourable environment for beetles because plant growth is sparse. This study was part of the same project (INTEX – Integrated Farming and Extensification of Agriculture) and was carried out in partly the same research site as (12, 13).

A review of literature (25) found evidence that decreases in ground beetle fauna (numbers of species and individuals) caused by intensive agriculture can be reversed by reducing pesticide and fertilizer use (three European studies, including (15)). Different species responded differently.

A study of spiders (Araneae) in an apple orchard in the Czech Republic (26) found that an integrated pest management strategy resulted in higher spider diversity than conventional pesticide applications. The number of spider species was highest on the plot with reduced fungicide and no herbicide applications and mixed planting (49 species, 1,212 spiders), followed by reduced fungicide and no herbicide applications and sown grass (45 species, 1,497 spiders), conventional spraying resulted in the lowest number of species (39 species, 1,252 spiders). Conventional applications caused much greater fluctuations in late summer spider populations and had lower numbers of spiders after winter (4/plot) than plots under integrated pest management (9–10/plot). Half of the 2 ha orchard received normal applications of fungicides and herbicides; the other half received less frequent applications of fungicides and no herbicides (integrated pest management). Half of the latter was sown with buckwheat *Fagopyrum esculentum*, common millet *Panicum miliaceum*, dill *Anethum graveolens* and horse bean *Faba vulgaris* in 1992–1993 and coriander *Coriandrum sativum* in 1994–1995. The other half was sown with red fescue *Festuca rubra*. Spiders were sampled by tapping single branches (25 trees) over a 0.25 m^2 cloth and sweeping ground cover with a 0.25 m^2 net at weekly intervals (April–October 1992–1995). Cardboard traps (30 x 100 cm^2) were also attached to 10 tree trunks in each plot overwinter at a height of 50 cm.

A 2000 literature review (27) looked at which agricultural practices can be altered to benefit ground beetles (Carabidae). It found four European studies that examined the effect of reduced pesticide use on ground beetles. One, the UK SCARAB project (11), found no long-term effect. The other three ((14, 15), Holland *et al.* 1998) found mixed effects.

A small replicated trial in 1997 at an experimental farm in Normandy, France (28) (same study as (31)) found that the biodiversity of small arthropods (mites (Acari), springtails (Collembola) and others) was not consistently higher on arable land that had reduced insecticide and fungicide use compared to conventionally managed arable land. Half of each field was managed under integrated farming techniques, with reduced pesticide use on average over five cropping years in the previous eight. The comparison was replicated on three fields. In two, the integrated management also involved no deep ploughing. Here, the difference was more consistent (significantly higher biodiversity under integrated management in five out of six monitoring months). Monitoring was between January and June 1997. The authors concluded that tillage had more influence on small soil arthropods than reduced pesticide use.

A before-and-after study in an arable field in England (29) found that abundance and diversity of springtails (Collembola) was significantly lower under conventional pesticide applications than reduced applications (no insecticides, minimal herbicides and fungicides). The springtail *Entomobrya nicoleti* disappeared from the plot with conventional pesticide application during the first year and did not recover during the three year study. There was no evidence of an effect on populations of the springtail *E. nicoleti* at the field edge. *Lepidocyrtus* spp. also declined with the conventional spraying regime in the field but not at the field edge. *Orchesella cincta* and *Tomocerus* spp. were found only in field edge samples. Half of a field (grass and winter wheat rotation) received conventional pesticide applications and the other half received reduced input; insecticides were excluded from a 6 m headland around the crop (1991–1996). Treatments were reversed 1996–1999. Arthropods were monitored on 3 occasions/year using suction sampling (25–125 m each side of a hedgerow) and pitfall traps 75 m from hedgerow and at the field edge adjacent to a ditch beside the hedgerow.

A small replicated, controlled study from May–June 1992–1998 in Leicestershire, UK (30) found that the abundance of nationally declining songbirds and species of conservation concern significantly increased on a 3 km^2 site where pesticide use was restricted (alongside several other interventions) although there was no overall difference in bird abundance, species richness or diversity between the experimental and three control sites. Numbers of nationally declining species rose by 102% (except for Eurasian skylark Alauda arvensis and yellowhammer Emberiza citrinella). Nationally stable species rose (insignificantly) by 47% (8 species increased; 4 decreased).

A small replicated trial at an experimental farm in Normandy, France (31) (same study as (28)) found more spiders (Araneae) and ground beetles (Carabidae) but fewer rove beetles (Staphylinidae) in arable fields managed with limited use of herbicides and fungicides, and no insecticides, than in control conventionally managed fields. The experimental plots were also managed without deep ploughing, so it is difficult to separate the effects of ploughing from the effects of reducing pesticide use. However, both ground beetles and spiders were also more abundant in sub-plots that restricted pesticide use entirely (no fungicides) and restricted herbicide use even more, whereas rove beetles were not. The authors suggested that spiders and ground beetles were sensitive to both pesticide application and ploughing, with spiders being the most sensitive while rove beetles are less sensitive to pesticide application and prefer deep-ploughed fields. Management was over 11 years from 1990 to 2001. Insects and spiders were monitored in May and June from 1999 to 2001.

A replicated, paired sites comparison study in 2000 on 28 arable farms in County Wexford, Ireland (32) found that wider uncultivated margins (average 181 cm wide) with reduced agro-chemical inputs (fertilizer, herbicide and pesticide) on Rural Environment Protection Scheme farms did not have higher plant or ground beetle (Carabidae) diversity or abundance than margins on non-Rural Environment Protection Scheme farms (average 145 cm). There were around 11 plant species and 21–22 ground beetle species/margin on both types of farm. Four-teen farms with Rural Environment Protection Scheme agreements at least 4-years-old were paired with 14 similar farms without agreements. On each farm, two randomly selected field margins were surveyed for plants and ground beetles. In each margin all plant species were recorded in two 5 x 3 m quadrats and percentage cover estimated in a 1 x 3 m quadrat. Ground beetles were sampled in four pitfall traps/field margin (8 traps/farm) set at 10 m intervals in early June and late August.

A replicated study in 1999 and 2003 on farms in East Anglia and the West Midlands, UK (33) found that 5 of 12 farmland bird species analysed were positively associated with a general reduction in herbicide use and conservation headlands. The study did not distin-guish between conservation headlands and a general reduction in herbicide use, classing both as interventions reducing pesticide use. The five species positively associated with reducing pesticide use were corn bunting Miliaria calandra (a field-nesting species), chaffinch Fringilla coelebs, greenfinch Carduelis chloris, whitethroat Sylvia communis, and yellow-hammer Emberiza citrinella (all boundary-nesting species). A total of 256 arable and pastoral fields across 84 farms were surveyed.

A replicated, controlled study in 2000–2001 on cereal fields of three different farms in western Germany (34) found that both plant species richness and vegetation cover was higher in plots not sprayed with herbicide (spray windows) than in the sprayed part of the field centre. The increase in species richness in spray windows was similar for all five differ-ent plant categories considered. Whereas vegetation cover of herbs increased from 10% (field centre) to 50% (spray windows), no such increase was observed for grass cover. Note that no statistical analyses were performed on these data. Spray windows were created as unsprayed plots in the centre of arable fields on one integrated and two conventionally managed farms. Plant species richness and vegetation cover were recorded in both 'spray windows' and the sprayed part of the field. Plants were categorized as belonging to five different groups: Red-Listed species, declining species, unthreatened arable weeds, arable ruderal species and non-arable ruderal species. Information about crop rotation and herbicide application was obtained directly from the farmers.

A randomized, replicated, controlled trial from 2003 to 2006 on four farms in southwest England (35) found that no more foraging birds were attracted to twelve 50 x 10 m plots of permanent pasture with no fertilizer compared to 12 control (conventionally managed) plots. Experimental plots were managed in the same way as control plots except for the lack of fertilizer and all plots were cut twice in May and July and grazed in autumn/winter. There were 12 replicates of each management type, monitored over 4 years.

A systematic review of 23 studies (36) found that restricting herbicide inputs to crop edges tended to increase arthropod abundance. Studies mainly excluded or selectively used herb-icides; studies excluding fungicides or insecticides separately were not available. Studies focused on ground beetles (Carabidae), true bugs (Heteroptera), rove beetles (Staphylinidae), butterflies (Lepidoptera) and grouped bird 'chick-food' insects. Abundance of true bugs was up to almost 13 times higher where herbicide use was restricted or where herbicides and fungicides or insecticides were restricted. For other invertebrates restricted use generally increased abundance or had no impact. Only two species exhibited a significant decrease in abundance. In most (20 out of 23) studies, the possibility of confounding outcomes due to pesticide and fertilizer inputs could not be discounted.

A replicated, controlled, randomized study from 2003 to 2005 of arable fields at three sites in the UK (37) found that reduced frequency applications of herbicide resulted in higher species richness and abundance of beneficial weeds and tended to increase arthropod abundance (but not always). Plant species richness and cover of beneficial weeds tended to be highest in untreated and single spring or post-emergence application plots and lowest in those with three applications. In 2004 the inclusion of a pre-emergence herbicide reduced cover of beneficial weeds compared to other treatments. Cover of undesirable weeds was higher in single pre-emergence or spring applications than combined treatments. A post-emergence application was as effective at controlling undesirable weeds as sequences of herbicides. Untreated plots tended to support more arthropods than those with herbicides, but not always. Single applications tended to reduce arthropod abundance less than sequences of herbicides, although post-emergence and pre-emergence applications were detrimental to some taxa. There were 3 or 5 replicate plots (3 or 4 x 24 m) of each treatment per site: untreated, pre-emergence, post-emergence or March applications or combinations of 2/all herbicide applications. Vegetation was sampled in 5 quadrats (0.25 m^2) in each plot (June 2003–2005). Arthropods were sampled using a D-Vac suction sampler (5 sub-samples of 10s/plot) in a sub-set of treatments (June).

A replicated trial in 2004–2006 in Cheshire, Staffordshire and north Shropshire, England (38) (same study as (47)) found no differences in plant, insect or bird numbers between conventional and minimum input barley fields. Sixteen trial fields were sown with spring barley each on a separate dairy or mixed farm. One half of each barley field was managed conventionally, the other half managed with minimum pesticide inputs (no insecticide after 15 March; no broadleaved herbicide after 31 March; limited grass-specific herbicide). Plants, invertebrates and birds were monitored on each field in summer 2005 and winter 2005–2006.

A replicated site comparison on 186 overwinter stubble fields in Devon, England (39) found that cirl bunting Emberiza cirlus foraged at significantly higher densities on stubble fields under a 'Special Project' agri-environment option compared to stubbles under standard agri-environment schemes with a conventional pesticide regime (approximately 0.45 birds/ha for 102 special project stubble fields vs 0.05 birds/ha for 52 conventional wheat stubbles and 0.15 birds/ha for 32 conventional barley stubbles). The special project stubbles were also preferentially selected to some extent by four other species of songbird. The special project was designed to encourage cirl buntings and allowed the use of fungicides, growth regulators and specified herbicides to control grass weeds but prohibited the use of insecticides and herbicides to control broadleaved weeds.

A replicated, controlled, randomized study of undersown and conventional cereal systems in Denmark (40) found that money spider (Linyphiidae) web density increased with reduction in fertilizer; the same was true for springtail (Collembola) density in conventional but not undersown crops. Money spider web density tended to be higher in undersown crops with no fertilizer (peak 250–300/m^2) than low fertilizer input (200–250/m^2 and in conventional crops with low fertilizer input (150–200/m^2) than high-input (100–150/m^2). Springtail density was significantly higher in the fertilized (2,350/m^2) than unfertilized undersown crops (1,600/m^2) but higher in the low-input (1,250/m^2) compared to high-input conventional crops (300/m^2). Sixteen experimental plots (12 x 50 m) were established in a randomized block design. Treatments were wheat with clover Trifolium spp. undersown, with or without nitrogen fertilization

(50 kg/ha), or conventional wheat with low (50 kg/ha) or high nitrogen fertilization (160 kg/ha), only the latter received pesticide applications. Money spider web densities, vegetation density (lower layer only, i.e. clover and weed layer) were sampled between May–October 1995–1997. Money spiders and springtails were sampled in 1996.

A paired before-and-after trial in summer 2004–2006 in one arable field in central Germany (41) found higher numbers of aphids (Aphidoidea) and their arthropod predators ('predator units') in the half field with reduced pesticide treatment than in the control (normal pesticide application) part of the same field after insecticide treatment. No clear effect of reduced pesticide use could be found on ground beetles (Carabidae) as contradicting results were found in all three years. Weed cover was very low in all years and sites (often <1% after herbicide treatment) but significantly more plants were found in the low intensity part of the field in the third study year. Pesticide use on one half of a conventionally managed arable field was reduced to 50%, whereas the other half with 100% pesticide input was used as a control. Aphids, their predators and arable weeds were monitored before and after each pesticide treatment at five points along a line perpendicular to the field edge. Ground beetles were caught weekly in six pitfall traps in each site in June and July. Plants were recorded as plants present/m^2 before treatments and as plant cover after treatments. This study is also described in an additional publication (Schumacher & Freier 2006).

A site comparison study in 1998 and 2003 of ten 1.1 km^2 plots in Austria (42) showed that grasslands managed for extensive mixed agriculture or intensive livestock farming contained a greater number of plant species when the use of pesticides and fertilizers was reduced. On arable farmland reducing pesticide use had no effect on the number of plant species present, except for on mixed extensive arable land where fields with no agro-chemicals applied during critical periods had significantly more plant species than traditionally managed fields. For areas of mixed arable farmland in mountainous areas, fields without any agro-chemicals had a greater number of plant species than fields where the use of agro-chemicals was merely reduced. The number of broadleaved plant species in each plot was determined according to the relevés method of sampling vegetation during field surveys in April–September of 1998 and 2003.

A replicated, controlled, randomized study of arable fields over two years in England (43) found that crop protection measures (normal or no pesticide applications) had less impact on insect and spider (Araneae) abundance than type of fertilizer. Wolf spider (Lycosidae), ground beetle (Carabidae), ladybird (Coccinellidae) and true bug (Hemiptera) abundance tended to be higher in plots with organic (compost) compared to inorganic fertilizers and those with no pesticides. In contrast, rove beetles (Staphylinidae), money spiders (Linyphiidae), hoverflies (Syrphidae) and parasitoid wasps (Braconidae) tended to be more abundant in plots with conventional fertilizer and/or pesticide applications. Ground beetles were more abundant in no pesticide vegetable plots in both years, but more abundant in conventionally sprayed bean plots in one year. Effects depended on crop type (grass/clover Trifolium spp., cereals, vegetables) and year. There was no effect of treatments on net-winged flies (Neuroptera) and Proctotrupoidea (parasitoids). In both years the organic fertilizer and conventional pesticide combination had the greatest effect on invertebrates; the organic fertilizers and no pesticide combination also had a significant effect. A field was divided into 4 blocks (122 x 122 m), each with 32 plots (24 x 12 m). Treatments were: conventional or organic (no) pesticide applications and conventional (inorganic) or organic (none or compost) fertilizers. Invertebrates were sampled using five monthly samples from five pitfall traps/plot from May–September and three one-minute suction samples/plot in the first week of July, August and September 2005 and 2006.

A controlled study in 2000–2005 on 61 ha of farmland in Bedfordshire, England (44) found that both winter and summer densities of most farmland bird species and ground beetles (Carabidae) were higher on areas with no pesticide input compared to areas with conventional levels of pesticides (higher summer densities with no pesticides for 10 of 14 species, although only Eurasian skylark Alauda arvensis, yellow wagtail Motacilla flava and linnet Carduelis cannabina showed a significant increase; all songbirds and 16 of 19 species recorded in winter were at higher densities on zero-pesticide fields). Skylarks were also significantly higher on areas with no fertilizer inputs, but no other species were affected by fertilizer reduction.

A replicated study of autumn-sown and spring-sown barley on four farms in Scotland (45) found that arthropod abundance was higher with fewer herbicide applications. Peak season (July) counts of total arthropod abundance in autumn and spring-sown barley were significantly higher in fields that received 1 herbicide application (28/sample) than fields receiving 2 applications (18–21/sample). This was also the case for many individual orders, particularly beetles (Coleoptera) (spring barley 1 application: 14/sample, 2 applications: 12; autumn barley 1: 12, 2: 4) and spiders (Araneae) (spring barley 1 application: 1.5, 2: 0.75; autumn barley 1: 2.5, 2: 1.75). A total of five spring and five autumn barley fields were selected from four farms (two of each crop type). No insecticides were applied but fields received one or two herbicide applications. Arthropods were sampled on 5 occasions in each field (April–August 2004) using a leaf vacuum (15 cm diameter). Sampling was undertaken at intervals (5 or 30 m) along 2–5 parallel transects (100 m apart) across the width of each field.

A replicated site comparison study from 2004 to 2008 in England (46) found that reduced chemical inputs in combination with overwinter stubbles were associated with smaller grey partridge Perdix perdix brood sizes. However, year-on-year partridge density was positively associated with this combination of interventions. There was no relationship between reduced chemical inputs in combination with overwinter stubbles and grey partridge overwinter survival or the ratio of young to old birds. Spring and autumn counts of grey partridge were made at 1,031 sites across England as part of the Partridge Count Scheme.

A replicated, controlled study from April–July and November–February in 2004–2006 on 16 livestock farms in the West Midlands, England (47) (same study as (38)) found that there were no differences in bird usage of barley fields between fields sprayed with only a narrow-spectrum herbicide (amidosulfuron at 25–40 g/ha) and those sprayed with both a narrow- and a broad-spectrum herbicide. Broadleaved plant cover was higher on plots treated with only a narrow-spectrum herbicide but only in the first year of barley production. Invertebrate biomass did not differ between treatments. Insect-eating songbirds and crows (Corvidae) showed a reduced use of broad-spectrum-sprayed fields in summer and late summer respectively but all other groups used fields at equal rates. Barley fields on the farms were split with half being used for each treatment. Narrow-spectrum herbicide was applied in April–May and broad-spectrum in July. All plots were treated with mineral fertilizer; many received fungicide applications but very few received insecticides.

(1) Heath G.W. (1962) The influence of ley management on earthworm populations. *Grass and Forage Science*, 17, 237–244.
(2) Gerard B.M. & Hay R.K.M. (1979) The effect on earthworms of ploughing, tined cultivation, direct drilling and nitrogen in a barley monoculture system. *Journal of Agricultural Science*, 93, 147–155.
(3) Purvis G. & Curry J.P. (1984) The influence of weeds and farmyard manure on the activity of carabidae and other ground-dwelling arthropods in a sugar-beet crop. *Journal of Applied Ecology*, 21, 271–283.
(4) Linzell B.S. & Madge D.S. (1986) Effects of pesticides and fertilizer on invertebrate populations of grass and wheat plots in Kent in relation to productivity and yield. *Grass and Forage Science*, 41, 159–174.

(5)   Kromp B. (1989) Carabid beetle communities (Carabidae, Coleoptera) in biologically and conventionally farmed agroecosystems. *Agriculture, Ecosystems & Environment*, 27, 241–251.

(6)   Fletcher M.R., Jones S.A., Greig-Smith P.W., Hardy A.R. & Hart A.D.M. (1992) Population density and breeding success of birds. In P. Greig-Smith, G. Frampton & T. Hardy (eds.) *Pesticides, Cereal Farming and the Environment: The Boxworth project*. HMSO, London. pp. 160–174.

(7)   Vickerman G. (1992) The effects of different pesticide regimes on the invertebrate fauna of winter wheat. In P. Greig-Smith, G. Frampton & T. Hardy (eds.) *Pesticides, Cereal Farming and the Environment: The Boxworth project*, HMSO, London. pp. 82–109.

(8)   Henze M. & Şengonca Ç. (1993) The influence of different farming systems on the population dynamics of aphids and their predators in winter wheat. *Communications of the Deutschen Gesellschaft fur Allgemeine und Angewandte Entomologie*, 8, 615–622.

(9)   Ruiter P.C.D., Moore J.C., Zwart K.B., Bouwman L.A., Hassink J., Bloem J., Vos J.A.D., Marinissen J.C.Y., Didden W.A.M., Lebrink G. & Brussaard L. (1993) Simulation of nitrogen mineralization in the below-ground food webs of two winter wheat fields. *Journal of Applied Ecology*, 30, 95–106.

(10)   Feber R.E., Smith H. & MacDonald D.W. (1994) The effects of field margin restoration on the meadow brown butterfly (*Maniola jurtina*). *British Crop Protection Council Monographs*, 58, 295–300.

(11)   Frampton G.K., Cilgi T. & Wratten S.D. (1994) The MAFF 'SCARAB' project: long-term consequences for farmland arthropods of pesticide use in the UK. *Bulletin OILB SROP*, 17, 245–257.

(12)   Hasken K.H. & Poehling H.M. (1995) Effects of different intensities of fertilizers and pesticides on aphids and aphid predators in winter wheat. *Agriculture Ecosystems & Environment*, 52, 45–50.

(13)   Schmidt W., Waldhart R. & Mrotzek R. (1995) Extensification in arable systems: effects on flora, vegetation and soil seed bank – results of the INTEX-project (University of Göttingen). *Tuexenia*, 415–435.

(14)   Huusela-Veistola E. (1996) Effects of pesticide use and cultivation techniques on ground beetles (Col, Carabidae) in cereal fields. *Annales Zoologici Fennici*, 33, 197–205.

(15)   Büchs W., Harenberg A. & Zimmermann J. (1997) The invertebrate ecology of farmland as a mirror of the intensity of the impact of man? An approach to interpreting results of field experiments carried out in different crop management intensities of a sugar beet and an oil seed rape rotation including set-aside. *Biological Agriculture & Horticulture*, 15, 83–107.

(16)   Frampton G.K. (1997) The potential of Collembola as indicators of pesticide usage: Evidence and methods from the UK arable ecosystem. *Pedobiologia*, 41, 179–184.

(17)   Hilbig W. (1997) Effects of extensification programmes in agriculture on segetal vegetation. *Tuexenia*, 295–325.

(18)   Jones N.E., Burn A.J. & Clarke J.H. (1997) The effects of herbicide input level and rotation on winter seed availability for birds. *Proceedings of the 1997 Brighton Crop Protection Conference – Weeds, 3*. pp. 1161–1166.

(19)   Kleijn D. & Snoeijing G.I.J. (1997) Field boundary vegetation and the effects of agrochemical drift: botanical change caused by low levels of herbicide and fertilizer. *Journal of Applied Ecology*, 34, 1413–1425.

(20)   Tarrant K.A., Field S.A., Langton S.D. & Hart A.D.M. (1997) Effects on earthworm populations of reducing pesticide use in arable crop rotations. *Soil Biology & Biochemistry*, 29, 657–661.

(21)   Aebischer N.J. & Potts G.R. (1998) Spatial changes in grey partridge (*Perdix perdix*) distribution in relation to 25 years of changing agriculture in Sussex, UK. *Gibier faune sauvage, Game Wildlife*, 15, 293–308.

(22)   Hansson M. & Fogelfors H. (1998) Management of permanent set-aside on arable land in Sweden. *Journal of Applied Ecology*, 35, 758–771.

(23)   Huusela-Veistola E. (1998) Effects of perennial grass strips on spiders (Araneae) in cereal fields and impact on pesticide side-effects. *Journal of Applied Entomology*, 122, 575–583.

(24)   Krooss S. & Schaefer M. (1998) The effect of different farming systems on epigeic arthropods: a five-year study on the rove beetle fauna (Coleoptera: Staphylinidae) of winter wheat. *Agriculture, Ecosystems & Environment*, 69, 121–133.

(25)   Kromp B. (1999) Carabid beetles in sustainable agriculture: a review on pest control efficacy, cultivation impacts and enhancement. *Agriculture, Ecosystems & Environment*, 74, 187–228.

(26)   Pekar S. (1999) Effect of IPM practices and conventional spraying on spider population dynamics in an apple orchard. *Agriculture, Ecosystems and Environment*, 73, 155–166.

(27)   Holland J.M. & Luff M.L. (2000) The effects of agricultural practices on Carabidae in temperate agroecosystems. *Integrated Pest Management Reviews*, 5, 109–129.

(28)   Cortet J., Ronce D., Poinsot-Balaguer N., Beaufreton C., Chabert A., Viaux P. & de Fonseca J.P.C. (2002) Impacts of different agricultural practices on the biodiversity of microarthropod communities in arable crop systems. *European Journal of Soil Biology*, 38, 239–244.

(29)   Frampton G.K. (2002) Long-term impacts of an organophosphate-based regime of pesticides on field and field-edge Collembola communities. *Pest Management Science*, 58, 991–1001.

(30)   Stoate C. (2002) Multifunctional use of a natural resource on farmland: wild pheasant (*Phasianus colchicus*) management and the conservation of farmland passerines. *Biodiversity and Conservation*, 11, 561–573.

(31)   Chabert A. & Beaufreton C. (2005) Impact of some agricultural practices on carabidae beetles. *IOBC/ WPRS Bulletin*, 28, 101–109.
(32)   Feehan J., Gillmor D. & Culleton N. (2005) Effects of an agri-environment scheme on farmland biodiversity in Ireland. *Agriculture, Ecosystems & Environment*, 107, 275–286.
(33)   Stevens D.K. & Bradbury R.B. (2006) Effects of the Arable Stewardship Pilot Scheme on breeding birds at field and farm-scales. *Agriculture, Ecosystems & Environment*, 112, 283–290.
(34)   Wehke S., Zoldan J.W., Frankenberg T. & Ruthsatz B. (2006) Possibilities of farming to promote weed flora in the western Hunsrück (Germany). In W. Büchs (ed.) *Möglichkeiten und Grenzen der Ökologisierung der Landwirtschaft – Wissenschaftliche Grundlagen und praktische Erfahrungen – Beiträge aus dem Arbeitskreis 'Agrarökologie'*, Mitteilungen der Biologischen Bundesanstalt für Land- und Forstwirtschaft, Berlin. pp. 57–65.
(35)   Defra (2007) *Potential for Enhancing Biodiversity on Intensive Livestock farms (PEBIL)*. Defra BD1444.
(36)   Frampton G.K. & Dorne J. (2007) The effects on terrestrial invertebrates of reducing pesticide inputs in arable crop edges: a meta-analysis. *Journal of Applied Ecology*, 44, 362–373.
(37)   Jones N.E. & Smith B. (2007) Effects of selective herbicide treatment, row width and spring cultivation on weed and arthropod communities in winter wheat. *Aspects of Applied Biology*, 81, 39–46
(38)   Mortimer S.R., Westbury D.B., Dodd S., Brook A.J., Harris S.J., Kessock-Philip R., Chaney K., Lewis P., Buckingham D.L. & Peach W.J. (2007) Cereal-based whole crop silages: potential biodiversity benefits of cereal production in pastoral landscapes. *Aspects of Applied Biology*, 81, 77–86.
(39)   Bradbury R.B., Bailey C.M., Wright D. & Evans A.D. (2008) Wintering cirl buntings *Emberiza cirlus* in southwest England select cereal stubbles that follow a low-input herbicide regime. *Bird Study*, 55, 23–31.
(40)   Gravesen E. (2008) Linyphiid spider populations in sustainable wheat-clover bi-cropping compared to conventional wheat-growing practice. *Journal of Applied Entomology*, 132, 545–556.
(41)   Schumacher K. & Freier B. (2008) Effects of a low-input pesticide use on tritrophic systems in winter wheat and pea. *Mitteilungen Der Deutschen Gesellschaft Fur Allgemeine Und Angewandte Entomologie*, 16, 343–346.
(42)   Wrbka T., Schindler S., Pollheimer M., Schmitzberger I. & Peterseil J. (2008) Impact of the Austrian agri-environmental scheme on diversity of landscapes, plants and birds. *Community Ecology*, 9, 217–227.
(43)   Eyre M.D., Sanderson R.A., Shotton P.N. & Leifert C. (2009) Investigating the effects of crop type, fertility management and crop protection on the activity of beneficial invertebrates in an extensive farm management comparison trial. *Annals of Applied Biology*, 155, 267–276.
(44)   Henderson I.G., Ravenscroft N., Smith G. & Holloway S. (2009) Effects of crop diversification and low pesticide inputs on bird populations on arable land. *Agriculture, Ecosystems & Environment*, 129, 149–156.
(45)   Douglas D.J.T., Vickery J.A. & Benton T.G. (2010) Variation in arthropod abundance in barley under varying sowing regimes. *Agriculture, Ecosystems & Environment*, 135, 127–131.
(46)   Ewald J.A., Aebischer N.J., Richardson S.M., Grice P.V. & Cooke A.I. (2010) The effect of agri-environment schemes on grey partridges at the farm level in England. *Agriculture, Ecosystems & Environment*, 138, 55–63.
(47)   Peach W.J., Dodd S., Westbury D.B., Mortimer S.R., Lewis P., Brook A.J., Harris S.J., Kessock-Philip R., Buckingham D.L. & Chaney K. (2011) Cereal-based wholecrop silages: a potential convservation measure for farmland birds in pastoral landscapes. *Biological Conservation*, 144, 836–850.

Additional references
Holland J.M., Cook S.K., Drysdale A., Hewitt M.V., Spink J. & Turley D. (1998) *The Impact on Non-target Arthropods of Integrated Compared to Conventional Farming: Results from the LINK Integrated Farming Systems project*. 1998 Brighton Crop Protection Conference – Pests and Disease 2, 625–630.
Schumacher K. & Freier B. (2006) Impact of low-input plant protection on functional biodiversity in wheat and pea. *Bulletin OILB SROP*, 29, 121–124.

7 Transport and service corridors

Key messages

Manage land under power lines to benefit wildlife
We have captured no evidence for the effects of managing land under power lines to benefit wildlife.

7.1   Manage land under power lines to benefit wildlife

• We have captured no evidence for the effects of managing land under power lines to benefit wildlife.

Background
This intervention may involve managing the unfarmed land under power lines to provide benefits to wildlife. For instance, one study in the USA found more bee species under power lines managed for wildlife (areas of dense scrub) than in areas of annually mown grassland (Russell *et al.* 2005).

Russell K.N., Ikerd H. & Droege S. (2005) The potential conservation value of unmowed powerline strips for native bees. *Biological Conservation*, 124, 133–148.

8 Hunting and trapping (for pest control, food or sport)

Key messages

Avoid use of lead shot
We have captured no evidence for the effects of avoiding the use of lead shot on farmland wildlife.

Provide 'sacrificial' grasslands to reduce the impact of wild geese on crops
All six studies from the UK (including four replicated, controlled trials) found that managing grasslands for geese increased the number of geese using these areas. Four of these studies found geese were moving within the study sites.

Use scaring devices (e.g. gas guns) and other deterrents to reduce persecution of native species
A replicated, controlled trial in Germany found phosphorescent tape was more effective than normal yellow tape at deterring one of three mammal species.

Enforce legislation to protect birds against persecution
Two before-and-after studies from Denmark and the UK found increased numbers or survival of raptors under legislative protection.

Use alerts to reduce grey partridge by-catch during shoots
We have captured no evidence for the effects of using alerts to reduce grey partridge by-catch during shoots on farmland wildlife.

8.1   Avoid use of lead shot

- We have captured no evidence for the effects of avoiding the use of lead shot on farmland wildlife.

Background
Spent lead shot may be ingested by wildlife such as waterfowl, scavenging birds and mammals which may result in lead poisoning. This intervention involves avoiding the use of lead shot, for instance by using alternative shot types such as steel or copper (Knott *et al.* 2009).

Knott J., Gilbert J., Green R.E. & Hoccom D.G. (2009) Comparison of the lethality of lead and copper bullets in deer control operations to reduce incidental lead poisoning; field trials in England and Scotland. *Conservation Evidence*, 6, 71–78.

8.2 Provide 'sacrificial' grasslands to reduce the impact of wild geese on crops

- All six studies from the UK (including four replicated, controlled trials) found that managing grasslands for geese increased the number grazing there. Two replicated, controlled studies found that fertilized and cut areas were grazed by more white-fronted geese[1] or brent geese[4] than control areas. A replicated, controlled trial found that re-seeded and fertilized wet pasture fields were used by more barnacle geese than control fields and that fertilized areas were used less than re-seeded ones[3]. A replicated, controlled study found that spring fertilizer application increased the use of grassland fields by pink-footed geese[6]. A replicated study found that plots sown with white clover were preferred by dark-bellied brent geese compared to plots sown with grasses[5].

- However, four of the studies[1, 3, 5, 6] found that the birds were moving within a relatively small area (i.e. within the study site) and therefore the grasslands may not reduce conflict with farmers.

Background
There have been dramatic increases in many species of goose in recent decades (Madsen *et al.* 1999) and this has led to increasing conflict with farmers, as many species graze on arable land, potentially ruining crops. One potential solution, to reduce conflict whilst maintaining the populations of geese is to provide 'sacrificial grasslands' – areas set aside for geese to feed on, which keeps them away from agricultural fields.

To be useful, such areas need to be more attractive than neighbouring fields. Management to attract geese to these areas can include re-seeding grasslands with grass species or legumes such as clover *Trifolium* spp., or fertilizing grasslands to increase productivity so that these areas can support more birds.

Madsen J., Cracknell G. & Fox A.D. (1999) *Goose Populations of the Western Palearctic: A review of status and distribution*. Wetlands International Publication No. 48. Wetlands International, Wageningen, The Netherlands.

A replicated, controlled trial in the winter of 1972–1973 at a 6 ha pasture (periodically flooded by saltwater) in Gloucestershire, UK (1) found that significantly more greater white-fronted geese *Anser albifrons* fed on fertilized and cut areas compared to control areas (overall average of 30–35% of geese on cut, fertilized areas vs 17–20% on control areas; maximum of 65% use of cut, fertilized areas vs 20% for controls). Preferences decreased over time as preferred areas lost vegetation and became more crowded. Vegetation from experimental areas had a higher nitrogen content than that from control areas. Fertilization consisted of 125 kg/ha of 'nitro-chalk' – 25% nitrogen – applied in mid-October. In mid-October the grass was also cut to approximately 8 cm.

A before-and-after study in Gloucestershire, England (2) found that up to 87% of geese on a grassland site used a 130 ha area managed for them in 1975–1976. The main management practice was to change the stocking regime of the site.

A replicated, controlled trial in 1984–1987 on a reserve on the island of Islay, west Scotland (3) found that more barnacle geese *Branta leucopsis* used wet pasture fields if they were re-seeded or fertilized than if they were unmanaged. However, fewer geese used fertilized fields than re-seeded ones. Fertilizers were either 34.5% nitrogen in pellet form (at 125 kg/ha) or 'nitrochalk' – 25% nitrogen in granular form – (at 175 kg/ha) and spread in October (wet and dry fields) and March (dry fields only). However, increases in barnacle geese were due to a redistribution of local birds, rather than new birds visiting the reserve. The author therefore suggests that improving the reserve grasslands will only minimally reduce conflict with farmers elsewhere on the island.

A series of replicated, controlled trials on grassland sites at two reserves in Essex, England between 1990 and 1992 (4) found that brent geese *Branta bernicla* grazed at significantly higher densities on fertilized and cut areas, compared to unfertilized areas, but only at high levels of fertilizer application (50 kg N/ha used: 28–30 droppings/m^2 for fertilized areas vs 23–28 droppings/m^2 for controls; 18 kg N/ha used: 30–35 droppings/m^2 for fertilized areas vs 25–35 droppings/m^2 for control areas). There were no differences between trials using organic and inorganic fertilizer.

A replicated study in the winters of 1992–1993 and 1993–1994 on an arable field on Thorney Island, West Sussex, England (5) found that dark-bellied brent geese *Branta bernicla bernicla* preferentially foraged on plots sown with white clover *Trifolium repens*, compared to 3 grass species (10–13 droppings/m^2 for 12 clover plots vs 0–5 droppings/m^2 for 36 grass plots). There were no differences between grass species (perennial rye grass *Lolium perenne*, red fescue *Festuca rubra* or timothy *Phleum pratense*). Plots were established in spring 1991 and preferences were found in both years, although more geese used grass plots in 1993–1994.

A replicated, controlled study in 1990–1993 at a reserve in Aberdeenshire, Scotland (6) found that spring fertilizer application in 1990–1991 significantly increased the use of grassland fields by pink-footed geese *Anser brachyrynchus*, until applications of approximately 80 kg N/ha (1990: average of 13–14 goose droppings/m^2 with no application vs 18–22 droppings/m^2 with 40 kg N/ha, 28 droppings/m^2 with 80 kg/m^2 and 27–31 droppings/m^2 with 120–160 kg N/ha; patterns in 1991 were similar but with fewer droppings). However, 2 slow-release fertilizers did not affect foraging densities in winter 1990–1992 (average of 24.5–26.7 droppings/m^2 for fertilized vs 24 droppings/m^2 for control grasslands). Split fertilizer application did not increase field use, compared to a single application (average of 11 droppings/m^2 for fields with split applications vs 10 droppings/m^2 for single applications), although the authors note it may reduce nitrogen leaching.

(1) Owen M. (1975) Cutting and fertilizing grassland for winter goose management. *The Journal of Wildlife Management*, 39, 163–167.
(2) Owen M. (1977) The role of wildfowl refuges on agricultural land in lessening the conflict between farmers and geese in Britain. *Biological Conservation*, 11, 209–222.
(3) Percival S.M. (1993) The effects of reseeding, fertilizer application and disturbance on the use of grasslands by barnacle geese, and the implications for refuge management. *Journal of Applied Ecology*, 30, 437–443.
(4) Vickery J.A., Sutherland W.J. & Lane S.J. (1994) The management of grass pastures for brent geese. *Journal of Applied Ecology*, 31, 282–290.
(5) McKay H.V., Milsom T.P., Feare C.J., Ennis D.C., O'Connell D.P. & D.J. H. (2001) Selection of forage species and the creation of alternative feeding areas for dark-bellied brent geese *Branta bernicla bernicla* in southern UK coastal areas. *Agriculture, Ecosystems & Environment*, 84, 99–113.
(6) Patterson I.J. & Fuchs R.M.E. (2001) The use of nitrogen fertilizer on alternative grassland feeding refuges for pink-footed geese in spring. *Journal of Applied Ecology*, 38, 637–646.

8.3   Use scaring devices (e.g. gas guns) and other deterrents to reduce persecution of native species

- One replicated, controlled trial in Germany found phosphorescent tape was more effective than normal yellow tape at deterring deer from an area but had no effect on wild boar or European hare[1].

Background
Native wildlife can have a significant impact on agricultural crops or features of the farmed landscape through grazing, browsing and uprooting. This intervention involves using scaring devices to discourage native wildlife from areas susceptible to damage.

A replicated, controlled study from May to November 1997 of four grassland fields and one cultivated field with willow *Salix* spp. stools (coppiced willow stumps) in central Germany (1) found that phosphorescent tape was more effective than normal yellow tape in deterring deer (Cervidae) but had no effect on wild boar *Sus scrofa* or brown hare *Lepus europaeus*. At the four grazing sites, areas surrounded by phosphorescent tape were avoided by red deer *Cervus elaphus* for four months and roe deer *Capreolus capreolus* for three weeks. Red deer entered areas fenced with yellow non-phosphorescent tape after one week and roe deer after just one day. All deer species kept out of an area of willow fenced with phosphorescent strips for three weeks; after that roe deer (but no red deer) tracks were found within the area. Soft PVC tape (40 cm wide) was attached to 1.3 m iron posts at a height of 1 m. Four game grazing fields each had two 300 m^2 areas fenced off using phosphorescent strips and 2 with non-phosphorescent tape. After two months all four areas were mowed and control and experimental fields swapped. Mammal presence was assessed using droppings and tracks.

(1) Wölfel H. (1981) Testreihen zur Wirksamkeit von Leuchtbandfolien mit phosphoreszierenden Pigmenten bei der Wildschadensverhütung [Test trials on the effectiveness of strips of film with phosphorescent pigments in the prevention of damage by game]. *Zeitschrift für Jagdwissenschaft*, 27, 168–174.

8.4   Enforce legislation to protect birds against persecution

*   Two before-and-after studies[1, 2] evaluated effects of legislative protection on bird species in Europe. Both found that legislation protects bird populations. One[2] found increased population levels of raptors in Scotland, following protective legislation. One found increased survival of kestrels in Denmark[1] with stricter protection, but not necessarily population-level responses.

Background
Perhaps the most commonly used intervention in response to declining species is to provide legal protection for the species.

A before-and-after study examining 524 common kestrels *Falco tinnunculus* recovered during 1917–1980 in Denmark (1) found that estimated survival rates of birds ringed as chicks increased during 1967–1972 (66% annual survival) compared to 1945–1966 (50%), following the introduction of legal protection for all birds of prey in 1967. However, the increase in survival rate following kestrel-specific legislation in 1926 was insignificant (45% for 1917–1925 vs 55% for 1926–1939) and there was a significant fall in 1973–1980 (to 53%). There were similar (although insignificant) patterns for birds ringed as juveniles or adults. There were significant decreases in the proportion of recoveries that were shot following each piece of legislation, from 1917–1925 (59% of 29) to 1926–1939 (14% of 35) and again from 1945–1966 (17% of 76) to 1976–1980 (2% of 192).

A before-and-after study on a grouse moor in Dumfries and Galloway, south Scotland (2) found that the numbers of hen harriers *Circus cyaneus* and peregrine falcons *Falco peregrinus* increased after birds were given full protection from persecution in 1990 (harriers increased from 2 pairs in 1992 to 20 pairs in 1997, whilst peregrines increased from 2 to 6 pairs). However, following the discontinuation of moor management in 2000, harriers declined again to two to four pairs in 2003–2006. Both species were legally protected since 1961 but until 1990 many were still killed illegally on the moor. Three wader species and red grouse *Lagopus lagopus* all declined following harrier protection and the cessation of management. Meadow pipits *Anthus pratensis* and stonechats *Saxicola rubicola* both declined as harriers increased but increased again after 2000. Carrion crows *Corvus corone* increased from 2000 after they were no longer shot by gamekeepers.

(1) Noer H. & Secher H. (1983) Survival of Danish kestrels *Falco tinnunculus* in relation to protection of birds of prey. *Ornis Scandinavica*, 14, 104–114.
(2) Baines D., Redpath S., Richardson M. & Thirgood S. (2008) The direct and indirect effects of predation by hen harriers *Circus cyaneus* on trends in breeding birds on a Scottish grouse moor. *Ibis*, 150, 27–36.

8.5   Use alerts to reduce grey partridge by-catch during shoots

- We have captured no evidence for the effects of using alerts to reduce grey partridge by-catch during shoots on farmland wildlife.

Background

This intervention involves using alerts to avoid shooting grey partridges *Perdix perdix*. Alerts may include using whistles to indicate when a covey of grey partridge has been flushed (Buner & Aebischer 2008).

Buner F. & Aebischer N.J. (2008) *Guidelines for Re-establishing Grey Partridges through Releasing*. Game & Wildlife Conservation Trust, Fordingbridge.

9 Natural system modification

Key messages

Manage heather by swiping to simulate burning
A replicated, controlled trial from the UK found heather moorland subject to flailing had fewer plant species than burned plots but more species than unflailed plots.

Manage heather, gorse or grass by burning
A long-term replicated, controlled trial in Switzerland found burning of chalk grassland did not increase the number of plant species. A replicated, controlled trial in the UK found more plant species on burned than unburned heather moorland.

Raise water levels in ditches or grassland
Eight studies (including two replicated, controlled trials) from Denmark, the Netherlands and the UK found raising water levels increased numbers of birds, invertebrates or plants or allowed wet grassland plant species to establish more rapidly. Three studies (two replicated) from the Netherlands and the UK found raising water levels had negative, limited or no effects on plants or birds. A replicated study from the UK found unflooded pastures had a greater weight of soil invertebrates than flooded pastures.

Remove flood defence banks to allow inundation
A controlled before-and-after study from the UK found a stretch of river that was allowed to flood had more bird species and territories than a channelized section. A study from Belgium found flooding and mowing increased plant species richness in meadow plots.

Re-wet moorland
We have captured no evidence for the effects of re-wetting moorland on farmland wildlife.

Create scrapes and pools
Five studies (including a replicated, controlled, paired trial) from Sweden and the UK found creating scrapes and pools provided habitat for birds, invertebrates or plants or increased invertebrate diversity. Two replicated studies (one controlled, paired) from Ireland and the UK found mixed or no differences in invertebrate numbers between created ponds and controls or natural ponds. A study in Sweden found fewer fish species in constructed than natural wetlands.

9.1   Manage heather by swiping to simulate burning

- A replicated, controlled trial in Northern Ireland found that heather moorland subject to flailing to simulate burning had more plant species eight years after the management than control unflailed plots but fewer plant species than burned plots[1].

> **Background**
> This intervention may involve cutting or swiping heather to simulate the effects of burning.
> Cutting/swiping can be carried out using a flail mounted on a tractor (Defra 2007).

Defra (2007) *The Heather and Grass Burning Code – 2007 Version*. Defra Publications, London.

A replicated, controlled trial in Northern Ireland (1) found that heather moorland plots subject to flailing to simulate burning in 1996 had more plant species in 2004, eight years after management, than control unmanaged plots but fewer species than burned plots. Flailed plots had 26 species/site on average (average of 13 moss and liverwort species (bryophytes)) compared to 28 species/site on average on burned plots (average of 15 moss and liverwort species) and 20 species/site on control plots (10 moss and liverwort species). One year after the management, in 1997, both flailed and control plots had 23 plant species (11 moss and liverwort species) on average. The cover by mosses and liverworts increased significantly between 1997 and 2004 on flailed sites and on burned sites (numbers not given). Flailed sites had lower cover of heather *Calluna vulgaris* in 2004 than 8 burned sites in the same study (about 30% compared to 35% heather cover). Six sites managed by flailing in 1996 to stimulate heather regeneration were surveyed in 1997 and again in 2004. Plants were surveyed in four 4 m^2 quadrats per site. Adjacent unmanaged control areas were surveyed at each site.

(1) McEvoy P.M., Flexen M. & McAdam J.H. (2006) The Environmentally Sensitive Area (ESA) scheme in Northern Ireland: ten years of agri-environment monitoring. *Biology and Environment: Proceedings of the Royal Irish Academy, Section B*, 106B, 413–423.

9.2   Manage heather, gorse or grass by burning

- A long-term replicated controlled trial in Switzerland found that annual spring burning of calcareous grassland did not increase plant species richness relative to abandoned plots, after 15 years[1].

- A replicated controlled trial in Northern Ireland found that heather moorland subject to a single burn had more plant species eight years after the management than control unburned plots[2].

> **Background**
> This intervention may involve controlled burns to stimulate heather re-growth, modify the heathland/grassland plant communities or create a patchwork of heather of different ages.
> See also 'Livestock farming: Maintain upland heath/moorland' for studies that used burning as a management tool to maintain upland heath/moorland and 'Manage heather by swiping to simulate burning' for studies that use swiping as an alternative to burning.

A long-term replicated, controlled trial from 1978 to 1993 in the Jura Mountains, Switzerland (1) found calcareous grassland plots that were burned annually in February or March had fewer plant species after 13–15 years than plots cut annually or every second year (in October or July). Burned plots had 50 plant species/40 m^2 and 31 species/m^2 on average compared to 53 species/40 m^2 and 37 species/m^2 on average in plots cut every year or 2 years. Burned plots did not have significantly more plant species than abandoned control plots with no management. Burning also changed the species composition, reducing the cover of one of the three most abundant species, meadow brome *Bromus erectus*, from around 40% in July-mown plots to around 10%. There were three replicate 50 m^2 plots for each treatment and the experimental management regimes were carried out from 1978 to 1993. The percentage cover of plant species was estimated in 40 m^2 and 1 m^2 sample areas in each plot in the last week of June 1991 and 1993.

A replicated, controlled trial in Northern Ireland (2) found that heather moorland plots subject to burning in 1996 had more plant species in 2004, eight years after management, than control unmanaged plots. Burned plots had 28 species/site on average (average of 15 moss and liverwort species) and control plots had 24 species/site (10 moss and liverwort species). One year after the management in 1997, both burned and control plots had had 22 plant species (8–10 moss and liverwort species) on average. The cover by mosses and liverworts increased significantly between 1997 and 2004 on burned sites (numbers not given). Burned sites had higher cover of heather in 2004 than 6 flailed sites in the same study (about 35% compared to 30% heather *Calluna vulgaris* cover). Eight sites burned in 1996 to stimulate heather regeneration were surveyed in 1997 and again in 2004. Plants were surveyed in four 4 m^2 quadrats per site. Adjacent unburned control areas were surveyed at each site.

(1) Ryser P., Lagenhauer R. & Gigon A. (1995) Species richness and vegetation structure in a limestone grassland after 15 years management with six biomass removal regimes. *Folia Geobotanica and Phytotaxonomica*, 30, 157–167.
(2) McEvoy P.M., Flexen M. & McAdam J.H. (2006) The Environmentally Sensitive Area (ESA) scheme in Northern Ireland: ten years of agri-environment monitoring. *Biology and Environment: Proceedings of the Royal Irish Academy, Section B*, 106B, 413–423.

9.3   Raise water levels in ditches or grassland

- Seven studies from Denmark, the Netherlands and the UK (two replicated, controlled studies and two before-and-after studies) found that raising water levels in ditches or grassland was associated with increased bird numbers[3, 9, 11], breeding bird numbers[3, 6, 7, 12], plant species that favour wet conditions[1] and invertebrate numbers or biomass[8, 9] in agricultural landscapes.

- Two replicated studies from the Netherlands and the UK found that raising water levels resulted in a net loss of plant species[1] and did not affect lapwing foraging rate[5]. A review found three studies reporting that re-wetting soils on old arable fields is not an effective method of reducing nutrient levels and restoring species-rich grassland[10]. A replicated study from the UK found that unflooded pastures contained a high biomass of soil macroinvertebrates of importance to breeding wading birds[4].

- A controlled, randomized study from the Netherlands found that raising the water level resulted in a more rapid establishment of species typical of wet grassland, than vegetation management[2]. A review of agri-environment schemes from the UK found studies that suggested more expensive agri-environment scheme options for wetland habitats, such as controlling water levels, were more effective at providing good habitat for wading birds than easier-to-implement options[13].

Background
Wet habitats have been lost from agricultural systems as a result of filling ditches, loss of ponds and removal of water from fields by surface run-off and extensive under-

field drainage. These habitats provided important resources for a variety of species and re-wetting may offer an important mechanism for facilitating a reverse in the decline of birds and other farmland biodiversity.

Studies which raise water levels along with a combination of other measures, such as vegetation management, to restore wet grassland are included in the intervention 'Livestock farming: Restore or create traditional water meadows'.

A replicated trial from 1987 to 1989 on what was once a species-rich wet meadow at the Veen-kampen, near Wageningen in the Netherlands (1) found that areas with raised water levels lost plant species overall, but species favouring wet conditions increased. Cuckoo flower *Cardamine pratensis*, creeping buttercup *Ranunculus repens*, water foxtail *Alopecurus geniculatus* and creeping bent grass *Agrostis stolonifera* increased on wet and intermediately wet plots but the total number of plant species fell from around 20 to around 15. Dry plots had 14–18 plant species throughout. The area was divided into three compartments. One had water levels 30–70 cm below the soil surface in summer and 5–40 cm below in winter, like the surrounding farmland. A wet and an intermediate compartment had embankments built and water added. In the wet compartment, summer water levels were 10–50 cm below the surface and winter levels 0–20 cm below the surface. The intermediate compartment was in-between wet and dry levels. Other experimental treatments were tested in these compartments. No fertilizer was applied during the experiment and plots were mown for hay once or twice each year. Plants were monitored annually in fifty 0.25 m^2 quadrats/plot.

A controlled, randomized study of a former improved pasture in the Netherlands (2) found that raising the water level resulted in a more rapid establishment of species typical of wet grassland than vegetation management (cutting and removing hay; cutting, mulching and leaving hay; topsoil removal to 5 cm followed by cutting and removing hay). In 1985 the water level was raised to its former level in 1 area (1.5–2 ha); the other area was left dry. Plant species composition was recorded annually (20–50 samples/plot).

A 2000 literature review of grassland management practices in the UK (3) reported that there were numerous studies detailing the success of providing high ditch-water or water table levels in restoring breeding and wintering bird numbers (e.g. Andrews & Rebane 1994; Evans *et al.* 1995).

A replicated study in 1993–1995 at 17 UK lowland grassland sites (12 with winter flooding introduced in the previous 1–14 years) (4) found that unflooded pastures contained high biomass of soil macroinvertebrates (mainly crane fly (Tipulidae) larvae and earthworms (Lumbricidae)) of importance to breeding wading birds. Conversely, grasslands with a long history of winter flooding had a much lower soil macroinvertebrate biomass, comprising mainly a few semi-aquatic earthworm species.

A replicated study from January–March 2002 of 15 northern lapwing *Vanellus vanellus* chicks on one grassland site in the Isle of Islay, UK (5) found that raising water levels in the grassland did not affect lapwing foraging rate. Foraging rate increased with decreasing vegetation height and was greater in ditches than on rigs. Soil moisture, however, did not significantly affect foraging rate after sward height and rig versus ditch effects were factored out. The timing of fertilizer application (to promote grass growth) and water level in ditches was manipulated at the field scale, which resulted in a range of soil moisture levels and vegetation heights. Water level was controlled through sluiced canals that ran along field boundaries and in-field ditches. The authors point out that spring 2002 was particularly wet and may have confounded any effect of added soil moisture.

A before-and-after study of grazing marshes in eastern England (6) found that opening up existing footdrains, creating new ones and reconnecting drains to ditches resulted in an increase in breeding wading bird numbers. Northern lapwing *Vanellus vanellus* numbers increased from 19 pairs in 1993 to 85 pairs in 2003 and common redshank *Tringa totanus* rose from 4 to 63 pairs. Numbers of winter wildfowl also increased over the period and changes in vegetation communities to those more tolerant of inundation occurred. In 1993 water levels were raised by 45 cm. From 1995 approximately 600 m of footdrains were opened/year; from 2000 onwards approximately 2,000 m of footdrains were opened or added. Grazing intensity was also reduced from 1.5–2 head of cattle to 0.7 head/ha and fertilizer inputs were stopped.

A before-and-after study at Campfield Marsh RSPB Reserve, Cumbria, England (7) found that 5 years after water levels were raised in August 1995 breeding common snipe *Gallinago gallinago* and northern lapwing *Vanellus vanellus* recolonized the site and that, over the reserve as a whole, breeding Eurasian curlew *Numenius arquata* densities were 5.5 pairs/km^2 (one of the highest UK breeding densities). Five fields comprising 23 ha of former cattle-grazed, species-poor perennial rye grass *Lolium perenne* dominated grassland and arable cropland were restored. Over the five years vegetation also shifted towards target plant communities characteristic of wet grassland.

A replicated, controlled study of 32 ditches in arable and pastoral land in 2005 in Leicestershire, UK (8) (same study as (9)) found that bunded ditches, which dammed water, had significantly greater invertebrate biomass than controls (dry weight: 10 g/m^2 vs 4 g/m^2). Invertebrate families other than flies (Diptera) showed a more mixed response to bunding. Ditches were bunded (small dams placed across ditches) and slightly widened in 5–20 m lengths, with equal length control sections approximately 50 m upstream. Five insect emergence traps (0.5 mm mesh, surface area 0.1 m^2) were spaced along each section. Samples were collected every two weeks (April–August 2005) and invertebrates identified to family and recorded as biomass estimates.

A replicated, controlled (paired) study of wet pasture and drainage ditches in arable and pastoral areas in Leicestershire, UK (9) (same study as (8)) found that wetting-up resulted in higher invertebrate and bird numbers. The following were significantly higher in bunded (dammed ditches) compared to non-bunded ditches: bird visit rates (1.0 vs 0.5 visits/month), emergent aquatic insect biomass (1,400 vs 900 individuals/m^2), surface-active flies (Diptera) adults (in arable ditches in 2005; 85–100 vs 60–65/sample) and fly larvae and butterfly/moth (Lepidoptera) larvae (in pastoral ditches in 2006). There was no difference for invertebrates active in the grass layer. Vascular plant species richness was lower and bare ground cover higher in bunded ditches than controls in 2005 due to disturbance. In wet pasture bird visit rates were significantly higher (livestock: 0.26; livestock-excluded: 0.17 visits) than in control dry plots (livestock: 0.10; livestock-excluded: 0.06). Sampling involved bird observations (45 minutes, 1–2/month; both features), fixed/floating traps for emergence of aquatic insects (ditches), pitfall traps and sweep-netting for terrestrial invertebrates (ditches) and a botanical quadrat survey (0.25–0.5 m^2; ditches). Data was obtained between April 2005–March 2007; birds all year, other groups spring-summer.

A 2007 review of experimental evidence on how to restore species-rich grassland on old arable fields (10) found three studies showing that re-wetting soils rich in organic matter works only a little (around 20% less available nitrogen – Oomes 1991, Berendse *et al.* 1994) or increases nutrients (20% increase in available nitrogen – Eschner & Liste 1995).

A replicated, controlled study in 1999–2001 and 2004–2005 in Jutland, Denmark (11) found that permanent grassland fields under an agri-environment scheme designed to increase water levels had significantly higher numbers of three species of wading bird (northern lapwing

Vanellus vanellus; black-tailed godwit *Limosa limosa;* common redshank *Tringa totanus*) after the scheme was implemented (2004–2005) compared to numbers before the scheme (1999–2001). However, this was only the case for fields that successfully retained water (40 breeding pairs of northern lapwing before the scheme and 90 after for wet fields vs approximately 2 pairs before and 5 after for dry fields). In addition, fields that were dry before the scheme and wet after showed a greater increase in lapwing numbers (280–290% increase) than fields that were wet beforehand (130–170% increase). There were no increases in lapwing numbers on restored grasslands (formerly cropland), whether or not they were under the scheme, or on control fields (i.e. not under the scheme) that failed to retain water. Numbers increased on control fields that retained water, but the numbers found on them were no different from those expected if increases were uniformly distributed across the landscape (i.e. birds did not appear to be selecting the fields preferentially). Eurasian oystercatchers *Haematopus ostrolagus* did not increase on any field types and the authors note that regional wader numbers were still far lower than in 1978–1988. The scheme involved blocking drainage pipes and 'rills' (drainage channels) as well as reducing fertilizer inputs, grazing intensity and restricting when mowing could take place. A total of 615 fields were studied. The four species were surveyed twice during the breeding season (April–May) and the number of each species and their location recorded.

A replicated study in 2005–2006 of 70 fields with wet features at 9 lowland pastoral sites in eastern England (12) found that the probability of a field being used by nesting northern lapwing *Vanellus vanellus* was significantly greater with an increase of foot drain floods. Foot drains are shallow channels used historically for drainage. Foot drain floods are areas where water overtops the foot drain. Fields with foot drain floods held the highest densities of nesting pairs. Nests were more likely to be located within 50 m of foot drain floods and chicks more likely to forage near foot drain floods (in wet mud patches created by receding water). Fields with foot drains, foot drain floods and isolated pools were visited at least once a week (March–July 2005–2006) and the number of lapwing pairs displaying parental behaviour within a 10-min sampling period were used as a measure of brood density. Habitat variables and percentage of wet ground were collected around each nest site and the distance measured to the nearest foot drain, pool and flood.

A 2009 literature review of agri-environment schemes in England (13) found studies that suggested more expensive agri-environment scheme options for wetland habitats (such as controlling water levels) were more effective at providing good habitat for wading birds than easier-to-implement options.

(1) Berendse F., Oomes M.J.M., Altena H.J. & Elberse W.T. (1992) Experiments on the restoration of species-rich meadows in the Netherlands. *Biological Conservation*, 62, 59–65.

(2) Oomes M.J.M., Olff H. & Altena H.J. (1996) Effects of vegetation management and raising the water table on nutrient dynamics and vegetation change in a wet grassland. *Journal of Applied Ecology*, 33, 576–588.

(3) Wakeham-Dawson A. & Smith K.W. (2000) Birds and lowland grassland management practices in the UK: an overview. *Proceedings of the Spring Conference of the British Ornithologists' Union March 27–28, 1999.* Southampton, England. pp. 77–88.

(4) Ausden M., Sutherland W.J. & James R. (2001) The effects of flooding lowland wet grassland on soil macroinvertebrate prey of breeding wading birds. *Journal of Applied Ecology*, 38, 320–338.

(5) Devereux C.L., McKeever C.U., Benton T.G. & Whittingham M.J. (2004) The effect of sward height and drainage on common starlings *Sturnus vulgaris* and northern lapwings *Vanellus vanellus* foraging in grassland habitats. *Ibis*, 146, 115–122.

(6) Smart M. & Coutts K. (2004) Footdrain management to enhance habitat for breeding waders on lowland wet grassland at Buckenham and Cantley Marshes, Mid-Yare RSPB Reserve, Norfolk, England. *Conservation Evidence*, 1, 16–19.

(7) Lyons G. (2005) Botanical monitoring of restored lowland wet grassland at Campfield Marsh RSPB Reserve, Cumbria, England. *Conservation Evidence*, 2, 43–46.

(8)   Aquilina R., Williams P., Nicolet P., Stoate C. & Bradbury R. (2007) Effect of wetting-up ditches on emergent insect numbers. *Aspects of Applied Biology*, 81, 261–262.

(9)   Defra (2007) *Wetting up Farmland for Birds and Other Biodiversity*. Defra BD1323.

(10)   Diggelen R.v. (2007) Habitat creation: nature conservation of the future? *Aspects of Applied Biology*, 82, 1–11.

(11)   Kahlert J., Clausen P., Hounisen J. & Petersen I. (2007) Response of breeding waders to agri-environmental schemes may be obscured by effects of existing hydrology and farming history. *Journal of Ornithology*, 148, 287–293.

(12)   Eglington S.M., Gill J.A., Bolton M., Smart M.A., Sutherland W.J. & Watkinson A.R. (2008) Restoration of wet features for breeding waders on lowland grassland. *Journal of Applied Ecology*, 45, 305–314.

(13)   Natural England (2009) *Agri-environment Schemes in England 2009: A review of results and effectiveness*. Natural England, Peterborough.

Additional refererences

Oomes M.J.M. (1991) Effects of groundwater level and the removal of nutrients on the yield of non-fertilized grassland. *Acta Oecologia*, 12, 461–470.

Andrews J. & Rebane M. (1994) *Farming & Wildlife: A practical management handbook*. Royal Society for the Protection of Birds, Sandy.

Berendse F., Oomes M.J.M., Altena H.J. & De Visser W. (1994) A comparative study of nitrogen flows in two similar meadows affected by different groundwater levels. *Journal of Applied Ecology*, 31, 40–48.

Eschner D. & Liste H.H. (1995) Stoffdynamik wieder zu vernassender Niedermoore [Substance dynamics in the fens after rewetting]. *Zeitschrift fur Kurlturtechnik und Landentwickung*, 36, 113–116.

Evans C., Street S., Benstead P., Cadbury J., Hirons G., Self M. & Wallace H. (1995) Water and sward management for conservation: a case study of the RSPB's West Sedgemoor Reserve. *RSPB Conservation Review*, 9, 60–72.

9.4   Remove flood defence banks to allow inundation

- One controlled, before-and-after study from the UK found more bird territories and species on a stretch of river modified to allow inundation of river edges compared to a channelized section of river[1].

- One study from Belgium found that a combination of mowing and flooding resulted in increased plant species richness in meadow plots, but infrequently flooded, mown plots had more plant species than frequently flooded, non-mown plots[2].

Background

Recent major flooding events have resulted in a change in European water management policies, from flood exclusion strategies to the use of former floodplains through reconnection with main rivers. This change will lead to an increase in flood frequency and may provide opportunities for the restoration of floodplain ecosystems.

A controlled, before-and-after study on the river Roding in Essex, England (1) found that in 1982 there were more territories and more species of bird on a 3 km stretch of the river that was modified in 1979 to reduce flooding in the area compared to an adjacent 500 m stretch of river that was channelized in 1974 (52 territories of 9 species on the modified stretch vs 3 territories of 2 species on the channelized stretch). The experimental stretch had one bank excavated to create a 0.3 m high shelf (a 'flood beam') just above the level of the main channel. This meant that the main channel continued to carry water during dry periods (at a rate of 2 m^3/s) but during heavy rains the beam would carry water as well (at up to 40 m^3/s), increasing the width and the flow capacity of the river.

A study of two meadows in adjacent nature reserves on former natural floodplains of the River Demer, Belgium (2) found that mowing and flooding meadows resulted in increased plant diversity. Mown, frequently flooded plots had higher plant species richness (average 16 species/plot) than non-mown, frequently flooded (10 species) or mown, infrequently flooded (12 species) plots. Overall, there was a significant negative correlation between species richness and standing crop. Data were obtained from 2 reserves: 1 frequently flooded (150 ha; flooded at least once a year for 5–50 days) and 1 (600 ha) in which part was infrequently inundated (about once every 5 years). In each reserve 10 plots (2 x 2 m) were randomly selected in annually June-mown fields and 5 in non-mown fields. Plant species composition was recorded in each in early July 2005 and standing biomass mid–end of July.

(1)   Raven P. (1986) Changes in the breeding bird population of a small clay river following flood alleviation works. *Bird Study*, 33, 24–35.
(2)   Gerard M., El Kahloun M., Rymen J., Beauchard O. & Meire P. (2008) Importance of mowing and flood frequency in promoting species richness in restored floodplains. *Journal of Applied Ecology*, 45, 1780–1789.

9.5   Re-wet moorland

• We have captured no evidence for the effects of re-wetting moorland on farmland wildlife.

Background
In the past many moorland habitats were subject to draining through the creation of grips or drainage ditches. This intervention involves re-wetting moorland through blocking ditches and raising water levels to benefit characteristic moorland vegetation and wildlife.

9.6   Create scrapes and pools

• Three studies from Sweden and the UK (including two site comparisons, one of which was replicated) found that the creation of scrapes and pools provided habitat for a range of plant, invertebrate or bird species[3, 5] and resulted in increased aquatic macroinvertebrate diversity[4]. One of these studies found constructed pools supported locally or nationally scarce species of plant and water beetle[3].

• A study in Sweden found that a combination of large surface area, high shoreline complexity and shallow depth resulted in increased bird, bottom-dwelling invertebrate and aquatic plant diversity[1]. However there were fewer fish species than in natural wetlands[1]. Two replicated studies from Ireland and the UK (one controlled, paired study and a site comparison) found that bird visit rates were higher but invertebrate numbers varied in ditch-fed paired ponds compared with dry controls[2] and total macroinvertebrate and beetle richness did not differ between artificial and natural ponds, although communities did differ[6].

Background
Creating scrapes and pools in wetlands and wet grasslands can help create a heterogenous habitat with varying vegetation types and water levels. Scrapes consist of pools or strips of shallow water which dry during the summer. The muddy margins may provide suitable feeding conditions for invertebrates and wading birds (Natural England 2010).

Natural England (2010) *Illustrated Guide to Ponds and Scrapes.* Natural England Technical Information Note TIN079.

A study of 32 recently constructed (1–7 years) wetlands in an intensive agricultural area in southern Sweden (1) found that a combination of large surface area, high shoreline complexity and shallow depth increased bird, aquatic plant and bottom-dwelling invertebrate diversity. Fish species richness was lower than in natural wetlands. There were 15–54 species of bottom-dwelling invertebrates per wetland, increasing with wetland age up to approximately 5 years, when numbers levelled off. Wetland bird species richness increased with wetland area up to about 4 ha (12 species). There were 0–2 species of amphibian and 18–51 aquatic plant species per wetland. Wetland plant species richness increased with shoreline complexity, but aquatic plant richness decreased with increasing depth. Fish species richness was lower in constructed wetlands (0–5 species) than natural wetlands (more than 100 years old) in the same region (0–9 species). Sampling was undertaken in 2000 (aquatic plants in 2001). Aquatic plant cover was visually estimated (July and November), vegetation was sampled 0–5 m above shore (September) and submersed vegetation was sampled by throwing an anchor 15 m out into the water (5–15 times). Birds were sampled by walking around each wetland twice during the breeding season (mid-May to early June) and invertebrates at the bottom of the wetland were surveyed by kick-sampling along four 1 m lengths/wetland. Electro-fishing was undertaken in a 50 m stretch and amphibian larvae sampling in a 100 m length of the shallow, littoral zone.

A replicated, controlled paired study of 8 created ditch-fed paired ponds in field corners and 10 surface scrapes in arable field margins in Leicestershire, UK (2) found that bird visit rates were significantly higher in ditch-fed paired ponds (1 visit/month) than dry controls (0.5 visits/month), particularly in the summer months; sample sizes were too small to analyse visits to scrapes. Paired ponds in field corners are fed with water from a nearby ditch. Surface-active adult flies (Diptera) were more abundant and fly larvae and butterfly/moth (Lepidoptera) larvae (in 2005) less abundant in the scrapes than the controls. Numbers of invertebrates active in the grass layer were lower in scrapes than nearby unmanipulated plots. Vegetation was more heterogenous (diversity and height), grass cover lower and bare ground more extensive in the scrapes than the control areas. Sampling involved bird observations (45 minutes; 1–2/month), pitfall traps and sweep-netting for terrestrial invertebrates (scrapes) and botanical quadrat (0.25–0.5 m^2) survey (scrapes). Data was obtained between April 2005–March 2007; birds all year, other groups spring–summer.

A single-site study from 2004 to 2006 in Leicestershire, UK (3) found that a sequence of seven constructed pools within a riparian buffer strip provided habitat for a range of plant and invertebrate species. Pools each supported 9–18 species of aquatic plant (macrophytes) (30 overall) and 24–52 species of aquatic invertebrates (84 overall); these included the locally scarce marsh dock *Rumex palustris* and 6 nationally scarce and 4 locally uncommon water beetles (Coleoptera). The field drain-fed wetland was constructed in 1998. The pool sequence was a maximum of 20 m wide, within a riparian buffer strip approximately 70 m wide by 100 m long. Aquatic plants were listed and aquatic macroinvertebrates sampled (3 minutes/pool, June 2004–2005) in 6 of 7 pools.

A site comparison study of 36 newly created dual-purpose wetlands on agricultural land in Sweden (4) found that wetland creation increased aquatic macroinvertebrate diversity in agricultural landscapes. Wetlands had between 6–51 aquatic macroinvertebrates (total 176). Flight-dispersed insects dominated macroinvertebrate species richness: beetles (Coleoptera): 6; dragonflies and damselflies (Odonata): 4; caddisflies (Trichoptera): 5; true bugs (Heteroptera): 5; flies (Diptera): 4; mayflies (Ephemeroptera): 3; slugs and snails (Gastropoda): 2;

butterflies and moths (Lepidoptera): 1; leeches (Hirudinea): 1; others: 3. The estimated gain per created wetland ranged from 1 to 33 species. Sub-regions with high wetland density had higher species diversity and accumulation, but not different macroinvertebrate assemblage composition compared to sub-regions with low or moderate wetland densities. Species richness increased with wetland age and assemblage similarity increased with plant cover. Species richness in existing mature ponds (more than 50-years-old) was approximately 10% higher than created wetlands. Composition showed overall similarity, diversity was similar, but the rate of species accumulation differed between new and mature water bodies. The 300 ha area of wetlands was created from 1996 to 2004 in natural depressions of former pasture, crop or fallow land by soil excavations and damming existing waterways or drainage systems. Wetlands were largely permanent, flow-through water bodies (<2 ha). Fifteen percent of wetlands in sub-regions with low, moderate and high densities of created wetlands (i.e. 13, 8 and 15) were sampled. A D-shaped hand net was swept twice at 15 points along each wetland margin in May 2004. Twenty-five mature ponds in the region had been sampled in April each year 1996–2003.

A replicated site comparison study in March–July 2005 to 2007 within nine grazed wet grassland sites in Broadland, eastern England (5) found that installation of shallow wet features provided valuable foraging areas for northern lapwing *Vanellus vanellus* chicks. The wet features also supported more than twice the biomass of surface-active invertebrates and a greater abundance of aerial invertebrates than the grazing marsh. Chick foraging rates and estimated biomass intake (monitored May–July 2006) were 2–3 times higher in wet features. Later in the breeding season when water levels were low, chick body condition was significantly higher in fields with footdrain densities of more than 150 m/ha. Invertebrate abundance was estimated in wet footdrain, dry footdrain, wet pool, dry pool and vegetated grazing marsh habitats. Each year, chicks (<100 g) were weighed and bill length measured to determine growth rates.

A replicated site comparison study in 2006 of five recently developed Integrated Constructed Wetlands in a catchment in Ireland (6) found that the total number of macroinvertebrate taxa and beetle (Coleoptera) taxa did not differ between Integrated Constructed Wetlands and natural ponds, although communities did differ. A total of 134 taxa were found in Integrated Constructed Wetland ponds, 116 of which were in the last pond, compared to 129 taxa in natural ponds. Although taxon richness and beetle richness did not differ between natural and Integrated Constructed Wetland ponds, overall communities and beetle communities differed significantly. There were 151 taxa with the 2 pond types, of which 92 taxa (61%) were common to both types of ponds, 35 (23%) were found only in natural ponds and 24 (16%) only in Integrated Constructed Wetland ponds. There was no significant difference between the numbers of taxa in Integrated Constructed Wetlands and the river sites. Of 169 total taxa, 53 (31%) were found in both sites, 64 (38%) in only Integrated Constructed Wetlands and 52 (31%) only at river sites. Five Integrated Constructed Wetlands (consisting of interconnected ponds) and five natural ponds within pasture were sampled in March–April and July–August 2006. Sampling involved three 3-minute multi-habitat net samples (mesh: 1 mm) and 10 horizontal activity traps in each of the different pond habitats (ponds >10 cm deep). Nine sites on Annestown River, upstream and/or downstream of discharges from the final Integrated Constructed Wetlands were also sampled. Two 3-minute multi-habitat kick samples were collected (mesh: 0.5 mm) in each of the different river habitats (mesohabitats).

(1) Hansson L.-A., Brönmark C., Nilsson P.A. & Åbjörnsson K. (2005) Conflicting demands on wetland ecosystem services: nutrient retention, biodiversity or both? *Freshwater Biology*, 50, 705–714.
(2) Defra (2007) *Wetting up Farmland for Birds and Other Biodiversity*. Defra BD1323.

(3)   Stoate C., Whitfield M., Williams P., Szczur J. & Driver K. (2007) Multifunctional benefits of an agri-environment scheme option: riparian buffer strip pools within 'Arable Reversion'. *Aspects of Applied Biology*, 81, 221–226.

(4)   Thiere G., Milenkowski S., Lindgren P.E., Sahlen G., Berglund O. & Weisner S.E.B. (2009) Wetland creation in agricultural landscapes: biodiversity benefits on local and regional scales. *Biological Conservation*, 142, 964–973

(5)   Eglington S.M., Bolton M., Smart M.A., Sutherland W.J., Watkinson A.R. & Gill J.A. (2010) Managing water levels on wet grasslands to improve foraging conditions for breeding northern lapwing *Vanellus vanellus*. *Journal of Applied Ecology*, 47, 451–458.

(6)   Jurado G.B., Johnson J., Feeley H., Harrington R. & Kelly-Quinn M. (2010) The potential of Integrated Constructed Wetlands (ICWs) to enhance macroinvertebrate diversity in agricultural landscapes. *Wetlands*, 30, 393–404.

10 Invasive and other problematic species

Control invasive non-native plants on farmland (such as Himalayan balsam; Japanese knotweed)
Two randomized, replicated, controlled trials in the Czech Republic found that removing all giant hogweed flower heads at peak flowering time reduced seed production.

Control bracken
A systematic review found repeated herbicide applications reduced bracken abundance but that cutting may have been equally effective. A laboratory trial found the same herbicide could inhibit the growth of mosses under certain conditions.

Control scrub
A replicated site comparison from the UK found the number of young grey partridge per adult was negatively associated with management that included scrub control.

Control weeds without damaging other plants in conservation areas
Two studies (one randomized, replicated, controlled) from the UK found that after specific plants were controlled, new plants established or diversity increased. A replicated, controlled laboratory and grassland study found a specific herbicide had negative impacts on one beetle species. Eleven studies investigated different methods of controlling plants.

Control grey squirrels
We have captured no evidence for the effects of controlling grey squirrels on farmland wildlife.

Control mink
A systematic review found that trapping may be an effective method of reducing American mink populations. A study in the UK found mink were successfully eradicated from a large area by systematic trapping.

Control predatory mammals and birds (foxes, crows, stoats and weasels)
Eight studies (including a systematic review) from France and the UK found that predator control (sometimes alongside other interventions) increased the abundance, population size or productivity of some birds. A randomized, replicated, controlled study from the UK did not.

Protect individual nests of ground-nesting birds
Two randomized, replicated, controlled studies from Sweden found nest exclosures increased measures of ground-nesting bird productivity, however both found bird numbers or adult predation rates were unaffected or negatively affected by exclosures.

Erect predator-proof fencing around important breeding sites for waders
We have captured no evidence for the effects of erecting predator-proof fencing around important sites for waders on farmland wildlife.

Remove coarse fish

We have captured no evidence for the effects of removing coarse fish on farmland wildlife.

Manage wild deer numbers

We have captured no evidence for the effects of managing wild deer numbers on farmland wildlife.

Provide medicated grit for grouse

A controlled study from the UK found higher red grouse productivity where medicated grit was provided.

10.1   Control invasive non-native plants on farmland (such as Himalayan balsam; Japanese knotweed)

• Two randomized, replicated, controlled trials in the Czech Republic[1, 2] found that removing all flower heads of giant hogweed plants at peak flowering time dramatically reduced seed production in giant hogweed.

Background

This intervention involves controlling invasive non-native plant species such as giant hogweed *Heracleum mantegazzianum*, Himalayan balsam *Impatiens glandulifera* and Japanese knotweed *Fallopia japonica*. Invasive non-native plants may have negative effects on both native plants and animals (Natural England 2011). Techniques to control these species can involve a range of methods including mechanical control such as cutting or cultivation, or herbicide application (Environment Agency 2010).

Environment Agency (2010) *Managing Invasive Non-native Plants in or Near Fresh Water*. Environment Agency report.
Natural England (2011) *Horizon-scanning for Invasive Non-native Plants in Great Britain*. Natural England Commissioned Report NECR053.

A randomized, replicated, controlled study in summer 1993 in a pastureland in the Krivoklat Protected Landscape Area, Czech Republic (1) found that removing all flower heads (umbels) of giant hogweed *Heracleum mantegazzianum* reduced seed production by 95% in that year. The effect on seed production of removing all flower heads and leaves was not significantly different from removing flower heads alone. Removing only the terminal flower head showed no significant difference in seed production compared to the control. The timing of flower head removal was planned to coincide with the peak flowering period. The study site was divided into eight blocks of four plants, with one plant in each block receiving each of four tissue removal treatments: removal of all flower heads and leaves, removal of all flower heads, removal of the terminal flower head and a control.

A randomized, replicated, controlled study in July 2002 and June 2003 in ten pastureland sites in Slavkovský les Protected Landscape Area, Czech Republic (2) found that giant hogweed *Heracleum mantegazzianum* seed production was dramatically reduced by removal of flower heads (umbels) and less heavily reduced by removal of leaves. Timing of tissue removal also significantly affected seed production, with 80% of plants that had flower heads removed on 2 July 2003 regenerating and producing some seed compared to 30–60% of plants that were treated on 9–10 July 2002. For treatments that removed the whole plant, cutting the stem above the basal rosette was equally as effective as removing the basal rosette, and cutting 15 cm below the ground was the only treatment that killed the plants. Additionally, they found that 84% of cut flower heads could still produce seed if left on the ground, and 24% of these seeds germinated. The 2002 experiment used 10 different tissue removal treatments, each applied to one randomly selected plant at each of ten sites. The 2003 experiment used 70 plants randomly selected at 1 site. Two different treatments, cutting 5 cm above the ground and cutting just above the leaf rosette, were applied to 10 plants each on 7 June, 20 June and 2 July. Ten plants were also used as a control.

(1) Pysek P., Kucera T., Puntieri J. & Mandak B. (1995) Regeneration in *Heracleum mantegazzianum* – response to removal of vegetative and generative parts. *Preslia*, 67, 161–171.
(2) Pysek P., Krinke L., Jarosik V., Perglova I., Pergl J. & Moravcova L. (2007) Timing and extent of tissue removal affect reproduction characteristics of an invasive species *Heracleum mantegazzianum*. *Biological Invasions*, 9, 335–351.

10.2   Control bracken

- One systematic review[2] found that the herbicide asulam reduced bracken abundance if applied repeatedly, but that cutting may be equally effective.

- A replicated laboratory trial in the UK[1] found that the herbicide asulam inhibited the growth of 3 common moss species that commonly grow in association with bracken, when exposed over 3 weeks, but not if only exposed for 24 hours.

Background

Bracken *Pteridium aquilinum* is native to the UK, but in many areas it is considered invasive. It has commonly been controlled by application of the herbicide asulam, or by cutting, although other techniques are sometimes used.

A laboratory study in the UK (1) found that growth and development of three moss species were significantly inhibited by continuous exposure to the herbicide asulam over three weeks, but not by 24-hour exposure. The three moss species are widely distributed in the UK and frequently grow in association with bracken *Pteridium aquilinum*, so they are likely to be exposed when bracken is controlled using asulam. *Campylopus introflexus* was the least sensitive species tested and *Polytrichum formosum* the most sensitive, with a 10-fold difference in sensitivity between the two. The sensitivity of *Bryum rubens* lay between the two but was closer to that of *C. introflexus* than *P. formosum*. Mosses were exposed in sterile cultures to low concentrations (0.001–1 g/l) of the herbicide asulam for 24 hours or continuously and their growth measured over 3 weeks.

A systematic review of methods to control bracken (2) found that the herbicide asulam reduces bracken abundance but regeneration can be rapid and multiple applications are necessary. Complete eradication has not been demonstrated. Available evidence suggests that cutting may be as effective as asulam application, particularly if there are two cuts in the same growing season. Further research is needed to compare the effectiveness of different ways of applying asulam and to compare cutting and asulam. There was no robust experimental evidence on the effectiveness of rolling, burning or grazing to control bracken. The review examined the effectiveness of using the herbicide asulam, cutting, hand-pulling, rolling, livestock grazing and burning to control bracken.

(1) Rowntree J.K., Sheffield E. & Burch J. (2005) Growth and development of mosses are inhibited by the common herbicide asulam. *Bryologist*, 108, 287–294.
(2) Stewart G.B., Tyler C. & Pullin A.S. (2005) Effectiveness of current methods for the control of bracken (*Pteridium aquilinum*). Systematic Review No. 3. *Centre for Evidence Based Conservation*.

10.3   Control scrub

• A replicated site comparison study from the UK[1] found a negative relationship between the number of young grey partridge per adult and a combined intervention of scrub control, rough grazing and the restoration of various semi-natural habitats.

Background
Scrub may consist of vegetation dominated by bushes, shrubs and tree saplings (Mortimer *et al*. 2000). Scrub on farmland can add habitat complexity and heterogeneity. However, if scrub dominates non-productive land on farms it may lead to declines in species that require grassland and other farmland habitats. Scrub control may include cutting, grazing or herbicide application (Mortimer *et al*. 2000).

Mortimer S.R., Turner A.J., Brown V.K., Fuller R.J., Good J.E.G., Bell S.A. Stevens P.A., Norris D., Bayfield N. & Ward L.K. (2000) *The Nature Conservation Value of Scrub in Britain*. Joint Nature Conservation Committee Report 308.

A replicated site comparison study from 2004 to 2008 on agricultural sites across England (1) investigated the impact of scrub control on grey partridge *Perdix perdix*. There was a negative relationship between a combined intervention (scrub control, rough grazing and the restoration of various semi-natural habitats) and the ratio of young to old partridges in 2008. The study did not distinguish between the individual impacts of scrub control, rough grazing and the restoration of various semi-natural habitats. Spring and autumn counts of grey partridge were made at 1,031 sites across England as part of the Partridge Count Scheme.

(1) Ewald J.A., Aebischer N.J., Richardson S.M., Grice P.V. & Cooke A.I. (2010) The effect of agri-environment schemes on grey partridges at the farm level in England. *Agriculture, Ecosystems & Environment*, 138, 55–63.

10.4   Control weeds without damaging other plants in conservation areas

• Two studies looked at the effects of controlling weeds on the surrounding vegetation. One study from the UK found that new populations of rare arable plants established following the control of perennial weeds in a nature conservation area[9]. A replicated, controlled and randomized study in the UK found that using grass-specific herbicide reduced grass diversity and resulted in increases in broadleaved plants[3].

• Eleven studies investigated different methods of controlling plants. A review found that specific management regimes can reduce the abundance of pernicious weeds in nature conservation areas[10]. Four replicated controlled studies (one also randomized) from Denmark and Germany found cutting and infection with fungal pathogens were effective methods for controlling creeping thistle[5–8] and one replicated, randomized, controlled trial from the UK found long-term control was achieved by lenient grazing[12]. A replicated, controlled and randomized study in Germany found that weevils could be used to infect creeping thistle with systemic rust[11]. One study found a non-native beetle was unsuitable for controlling creeping thistle because it had poor survival in the UK climate[1]. A replicated controlled study found that spraying a high concentration of herbicide killed less than half of broad-leaved dock plants[2]. A replicated, controlled, randomized study found that black grass was eliminated with a December treatment of grass-specific herbicide[3]. A small replicated study found that Hebridean sheep grazed more purple moor grass than Swaledale sheep[4].

- Two replicated, controlled laboratory and grassland studies found negative impacts of the herbicide asulam on green dock beetles[2].

Background

This intervention involves controlling perennial weeds in nature conservation areas including field margins or fallow land. Invasive weed species can have an impact on the conservation and restoration of biodiversity, but control with herbicides can also be detrimental to non-target species (English Nature 2003). The use of mechanical methods and biological control, including the use of pathogens, are important alternatives to chemical control strategies, particularly where adverse effects of agrochemicals on the environment must be minimized or avoided (English Nature 2003).

The majority of evidence on the effectiveness of these interventions is currently available for creeping thistle *Cirsium arvense*, which is a persistent perennial weed that causes problems in arable fields, grasslands, pastures and within nature conservation areas. Limited evidence is available for purple moor grass *Molinia caerulea*, which can threaten the conservation of heather *Calluna vulgaris* dominated moorland or heathland.

See also 'Control invasive non-native plants on farmland' for studies which looked at methods of controlling non-native plants.

English Nature (2003) *The Herbicide Handbook: Guidance on the use of herbicides on nature conservation sites.* English Nature/FACT, West Yorkshire.

A series of replicated experiments in the UK from 1969 to 1971 (1) tested the suitability of the non-native herbivorous beetle *Haltica carduorum* as a biocontrol agent for the injurious weed creeping thistle *Cirsium arvense* and found it unsuitable due to low survival in the UK climate.

A replicated, controlled laboratory study (2) found that the systemic herbicide asulam had little influence on the survival or rate of development of the green dock beetle *Gastrophysa viridula* when used as a contact agent on eggs, first-instar larvae or adults. However, the ingestion of asulam-contaminated broad-leaved dock *Rumex obtusifolius* leaves reduced female fertility by 64% (a reduction in number of egg batches rather than in numbers of eggs per batch) and extended the time of reaching the adult stage by 3–4 days. There was no effect on female longevity. Presence of asulam in dock leaves had no significant influence on egg-laying or feeding site selection. Fifteen replicates of each of the following treatments were sprayed on batches of 30 eggs, 20 first-instar larvae or one adult: high (5 ml/l) or low (1.25 ml/l) asulam concentration and a water control. In addition, 45 first-instar larvae were fed dock sprayed with one of the 3 treatment solutions. Female beetles that were carrying eggs were also put in 20 'choice chambers' with leaves sprayed with each treatment; location, area of leaf consumed, position and numbers of egg batches were recorded after 6 hours.

The same authors (2) also undertook a replicated, controlled study of asulam use in unmanaged grassland in the UK and found that broad-leaved dock *Rumex obtusifolius* survival was unaffected by beetle grazing or spraying with a low concentration of asulam (1.25 ml/l) and at high asulam concentrations (5 ml/l) only 40% of plants were killed. Numbers of first generation green dock beetle *Gastrophysa viridula* were similar on treated and untreated plants, but numbers of the second generation were significantly higher on untreated than treated plants. Shoot and root dry-weights of asulam-treated plants were significantly lower than untreated ones. Beetle-grazing did not further reduce the dry weight of asulam-treated plants, but did reduce the dry weight of those treated with water. Second generation beetles laid 4 times as many eggs on untreated plots and survival from eggs to larvae was 25% compared to just 4–12% on treated plots. Four blocks of 6 plots (4 m²) received the 3 treatments, each in the presence or absence of beetle-grazing. One-month-old docks were planted (16/plot) in April 1981. In May 144 adult beetles were released in randomly allocated 'grazing plots' (12/plot); herbicide treatments were applied in June. Numbers of beetle eggs, larvae and adults were counted on four plants on each grazed plot from May–October. Dock plant material was harvested in August and October 1981 and February and April 1982 and dry weights and leaf areas (untreated plots) obtained.

A replicated, controlled, randomized study of a wildflower margin at a farm in Oxfordshire, UK (3) found that applying grass-specific herbicide did not affect overall plant species diversity, however grass diversity was reduced and broadleaved plants increased in sprayed plots. Applying a grass-specific herbicide in December eliminated black grass *Alopecurus myosuroides*; plots not treated with herbicide were dominated by black grass. Sown crested dog's-tail *Cynosurus cristatus* was eliminated by a second treatment of herbicide in April; late mowing in June also decreased this species. The wildflower/grass seed was sown on 21 contiguous margin plots (3 x 12 m). Plots were grouped into three blocks, within which they randomly received one of seven treatments: unmanaged, cut April, cut April and May, cut May, cut in May and June, cut in June or grass-specific herbicide (fluazifop-P-butyl) application in April. Cuttings were removed. Half of each plot received grass-specific herbicide in December. Plant composition of sub-plots was sampled in five 0.1 m² quadrats in July 1995.

A small, replicated study from 1992 to 1996 of four pasture plots in North Yorkshire, UK (4) found that Hebridean sheep grazed more purple moor grass *Molinia caerulea* than Swaledale sheep, but the resulting density of purple moor grass and heather *Calluna vulgaris* did not differ. Hebridean sheep grazed significantly more purple moor grass leaves than Swaledales (61% vs 23%). However, the density of purple moor grass leaves did not differ between Hebridean (2,389 m²) and Swaledale plots (2,798 m²). Overall, cover by heather did not change over time: in the Hebridean plots, heather cover doubled in the first 4 years (12% to 29%) then declined (22%); the overall increase was 3.7%/year. In Swaledale plots cover increased over the first 2 years (3% to 7%) then declined dramatically (to 1%) with an overall decline of 1%. Two areas of pasture dominated by purple moor grass, ungrazed for 2 years and unburnt for 10, were divided into 2 plots of 2 ha, 1 grazed by Swaledales, 1 by Hebridean sheep (99 kg live weight/ha) between May and September. Numbers of grazed and ungrazed purple moor grass leaves were sampled within 2 quadrats (0.5 x 0.5 m) and lengths of grazed/ungrazed leaves measured at 100 points across plots following sheep removal in September 1992–1996. Plant species were sampled using a point sampling technique along a 15 m transect (45 cm intervals) using an 8 pinned frame with the tallest vegetation touching each pin identified.

A replicated, controlled study in 1998–1999 on fallow land at the University of Göttingen, Germany (5) found that cutting reduced creeping thistle *Cirsium arvense* reproductive success, but combining cutting with infection with rust fungus *Puccinia punctiformis* further restricted sexual reproduction and was therefore a more effective control strategy. Potted thistle plants

(15 cm tall) were transferred outdoors and 1 of 4 treatments (10 replicates of each) were applied in June 1998 and 1999: cutting at 30 cm, application of a spore suspension of the rust fungus, cutting and rust application and controls. Thistle size, number of flower buds and fertile flower heads and above and below ground (1999) dry matter were measured.

A replicated, controlled study in 1998–1999 in Germany (6) found that pathogens reduced the reproductive capacity of potted creeping thistle *Cirsium arvense*. This was particularly the case with the fungus *Phoma hedericola* and a combination of *P. hedericola* with *Mycelia sterila*, *Phoma destructiva* and *Phoma nebulosa*. Infections with pathogens, both singly and in combination, tended to increase disease severity compared to control plants; this was not the case for *P. punctiformis* or combined *P. punctiformis* and *P. hedericola* in the first year. Experiments were conducted under semi-field conditions at Braunschweig. Target plants (7–20) were inoculated with fungal isolates: *Phoma hedericola*, *Phoma destructiva*, *Mycelia sterila* and *Puccinia puncti-formis* (*Phoma nebulosa* in 1999) applied individually, or a combination of *P. punctiformis* and *P. hedericola* (in 1998) or a combination of all 4 (1999). Twenty plants were controls. Plants were evaluated each week (May–August) for disease symptoms (tissue breakdown; tissue death; pale/yellow leaves) and plant development.

A replicated, randomized, controlled study in 1998–2000 on fallow land at the University of Göttingen, Germany (7) found that fungal pathogens resulted in a decline in cover of creeping thistle *Cirsium arvense*. Creeping thistle cover decreased significantly (60 to 5%) with an associated increase in co-occurring species in the experimental area over 3 years (1998–2000) following inoculations with the fungal pathogens *Puccinia punctiformis* and *Phoma destructiva*. There was no significant difference between disease incidence of *P. punctiformis* on creeping thistle in plots following single and triple inoculations, or in control plots (99%). Combined treatment with *P. punctiformis* and *P. destructiva* increased the disease incidence with *P. de-structiva* compared to control, fungicide and *P. punctiformis* treatments in the third year (2–7%). Ten treatments (6 replicates) and a control (12 replicates) were randomly assigned to 72 plots. *P. punctiformis* and *P. destructiva* were applied once in June, July or August or in all three months/ year (1998–2000) or both were applied together in June. These were compared with a fungi-cide treatment (Opus Top) twice/year and untreated controls. Plots were monitored monthly (May–September) to determine the percentage of creeping thistle infected, disease severity (*P. punctiformis*) and the percentage cover of creeping thistle and other species.

A replicated, controlled study in an arable field in Denmark (8) found that increased mowing and hoeing frequency tended to reduce the amount of above ground creeping thistle *Cirsium arvense* biomass in the subsequent year (up to 73% compared to control). The presence of a suppressive crop (grass/white clover *Trifolium repens* mixture or red clover *T. pratense*) tended to further reduce creeping thistle. Differences in barley yield were only explained by the amount of creeping thistle biomass in one experiment, where the weed was most abundant. Four adjacent sub-fields, divided into four blocks, were subject to a combination of cropping (grass/clover in half of the plots; spring barley with grass/clover undersown in half of the plots; spring barley with red clover undersown in half of the plots) and mechanical (0, 2, 4, 6 passes with a mower; 0, 3, 6 or 0, 1, 3, 5 passes with a hoe) treatments over a 3-year period (between 2000–2004). Heights and dry weights of above-ground shoots of creeping thistle were obtained before harvest in the third year, as was barley yield.

At Ranscombe Farm, a nature reserve managed for arable plants in the north Kent Downs, UK (9), a four hectare field sprayed with glyphosate in September 2005 produced new populations of the rare arable plants blue pimpernel *Anagallis arvensis* subspecies *foemina* and ground pine *Ajuga chamaepitys* the following year. The field had been managed for some years with shallow cultivation in autumn only and had increasing abundances of perennial weeds such as docks *Rumex* spp. and couch grass *Elytrigia repens*. The entire field was sprayed with glyphosate in September 2005, then deep ploughed or 'disced' in February 2006.

A 2008 review of control methods for competitive weeds in uncropped cultivated margins managed to maintain uncommon arable plant populations in the UK (10) found that specific management regimes can reduce abundance of pernicious weeds in margins. Abundance of pernicious weeds tended to increase if uncropped cultivated margins were not cultivated annually in two studies (Critchley 1996b, 2000a). However five studies found weeds also build up on margins cultivated annually, particularly with the same annual cultivation regime (Critchley 1996a,b, Critchley *et al.* 2004, Critchley *et al.* 2006, Still & Byfield 2007). Eight studies found that abundance of specific weed species depended on timing or method of cultivation (Marshall 1998, Critchley 2000b, Ford 2000, Critchley *et al.* 2004, Critchley *et al.* 2005, Critchley *et al.* 2006, Corsie 2007, Still & Byfield 2007). Rotating margin sites was found to reduce weed abundance in two studies (Davies *et al.* 1994, Wilson 2000) and four studies found specific timing, frequency and height of cutting decreased certain species (Marshall 1998, Carvell *et al.* 2004, Corsie 2007, Westbury *et al.* 2008). Twelve studies reported that particular weed species could be targeted with specific timing of herbicide applications in different margin types (Boatman 1991, Varney *et al.* 1995, Wilson 1995, Marshall 1998, Boatman *et al.* 1999, Ford 2000, Marshall 2002, Boatman 2007, Corsie 2007, Meek *et al.* 2007, (9), Still & Byfield 2007). However, rare arable plant species can also be susceptible to specific management regimes (Wilson *et al.* 1990, Wilson 1995, Wilson & King 2003).

A replicated, controlled, randomized study in 2004–2005 in a former agricultural field near Bern, Germany (11) found that creeping thistle *Cirsium arvense* could be infected with a systemic rust fungus *Puccinia punctiformis* using the weevil *Ceratapion onopordi* as a disease carrier. There was a significantly higher rust incidence within 1 m of weevil-treated thistle shoots (34 shoots infected) compared to controls (1 infected). Overall, within a radius of 1 m 27% of weevil-treated shoots had rust infections compared to 3% of control shoots. There was no significant effect of the treatment within radii of 0.3 m or above 1 m. Therefore, rust infections could be induced between 0.3–1 m from weevil-treated thistles. The field had been sown with a mixture of wildflower seeds, grass and clover *Trifolium* spp. In April 2004 60 thistle shoots (≥ 1 m apart) in the wildflower strip were randomly assigned as either infected (with 1 female weevil powdered with rust spores (1,000 spores/female)) or controls. Weevils were confined to shoots for 72 hours using a cylinder sealed at the top; controls received only the cylinder. Systemically infected thistles were located and assigned to the nearest experimental shoot, within radii of 0.3, 1, 2 or 3 m in April–July 2005.

A replicated, randomized, controlled study in 2000–2005 on the effectiveness of control strategies on creeping thistle *Cirsium arvense* numbers at two pastoral farms in England and Wales (12) found that lenient grazing was most effective for long-term control. In a lowland cattle and sheep system and an upland sheep system, thistle numbers decreased under lenient grazing (cattle grazed to 8–10 cm; sheep 6–8 cm) whereas they remained constant or increased under heavy grazing regimes (cattle grazed to 5–7 cm; sheep 3–5 cm). Herbicide wiping gave the most rapid and effective control and cutting was one of the least effective measures, however, the effects of all weed control sub-treatments were short-lived at both sites. Six grazing treatments (a combination of lenient and heavy grazing) were applied to plots in a randomized block design with three replicates. Five weed control sub-treatments (thistle cutting, herbicide wiping, cutting and wiping, controls) were undertaken within grazing treatments using a split-plot design with replication at the block level; hay cutting on a three year rotation was also undertaken at one site. Data on thistle shoot density and effects on non-target broadleaved plants (rooted frequency) were obtained within each sub-treatment plot (2000–2005). This study is also described in a 2004 Defra report (Pywell *et al.* 2004).

(1)   Baker C.R.B., Blackman R.L. & Claridge M.F. (1972) Studies on *Haltica carduorum* Guerin (Coleoptera: Chrysomelidae) an alien beetle released in Britain as a contribution to the biological control of creeping thistle, *Cirsium arvense* (L.) Scop. *Journal of Applied Ecology*, 9, 819–830.
(2)   Speight R.I. & Whittaker J.B. (1987) Interactions between the chrysomelid beetle *Gastrophysa viridula*, the weed *Rumex obtusifolius* and the herbicide Asulam *Journal of Applied Ecology*, 24, 119–129.
(3)   Marshall E.J.P. & Nowakowski M. (1996) Interactions between cutting and a graminicide on a newly-sown grass and wild flower field margin strip. *Aspects of Applied Biology*, 44, 307–312.
(4)   Newborn D. (2000) The value of Hebridean sheep in controlling invasive purple moor grass. *Aspects of Applied Biology*, 58, 191–196.
(5)   Kluth S., Kruess A. & Tscharntke T. (2003) Influence of mechanical cutting and pathogen application on the performance and nutrient storage of *Cirsium arvense*. *Journal of Applied Ecology*, 40, 334–343.
(6)   Guske S., Schulz B. & Boyle C. (2004) Biocontrol options for *Cirsium arvense* with indigenous fungal pathogens. *Weed Research*, 44, 107–116.
(7)   Kluth S., Kruess A. & Tscharntke T. (2005) Effects of two pathogens on the performance of *Cirsium arvense* in a successional fallow. *Weed Research*, 45, 261–269.
(8)   Graglia E., Melander B. & Jensen R.K. (2006) Mechanical and cultural strategies to control *Cirsium arvense* in organic arable cropping systems. *Weed Research*, 46, 304–312.
(9)   Still K.S. (2007) A future for rare arable plants. *Aspects of Applied Biology*, 81, 175–182.
(10)   Critchley C.N.R. & Cook S.K. (2008) *Long-term Maintenance of Uncommon Plant Populations in Agri-environment Schemes in England. Phase 1 Scoping Study*. Defra/ADAS BD1630.
(11)   Wandeler H., Nentwig W. & Bacher S. (2008) Establishing systemic rust infections in *Cirsium arvense* in the field. *Biocontrol Science and Technology*, 18, 209–214.
(12)   Pywell R.F., Hayes M.J., Tallowin J.B., Walker K.J., Meek W.R., Carvell C., Warman L.A. & Bullock J.M. (2010) Minimizing environmental impacts of grassland weed management: can *Cirsium arvense* be controlled without herbicides? *Grass and Forage Science*, 65, 159–174.

Additional references
Wilson P.J., Boatman N.D. & Edwards P.J. (1990) Strategies for the conservation of endangered arable weeds in Great Britain. *Proceedings of the EWRS Symposium 1990 – Integrated Weed Management in Cereals*. European Weed Research Society, Wageningen, 93–101.
Boatman N.D. (1991) Selective control of cleavers (*Galium aparine*) in conservation headlands with quinmerac. *Brighton Crop Protection Conference – Weeds – 1991*, 669–676.
Davies D.H.K. & Carnegie H.M. (1994) Vegetation patterns and changes in field boundaries and conservation headlands in Scottish arable fields. *British Crop Protection Council Monographs*, 58, 173–178.
Varney P.L., Scott T.A.J., Cooke J.S. & Ryan P.J. (1995) Clodinafop-propargyl – a useful tool for management of conservation headlands. *Brighton Crop Protection Conference – Weeds – 1995*, 967–972.
Wilson P.J. (1995) The potential for herbicide use in the conservation of Britain's arable flora. *Brighton Crop Protection Conference – Weeds – 1995*, 961–966.
Critchley C.N.R. (1996a) Monitoring as a feedback mechanism for the conservation management of arable plant communities. *Aspects of Applied Biology*, 44, 239–244.
Critchley C.N.R. (1996b) Vegetation of arable field margins in Breckland. PhD thesis, University of East Anglia.
Marshall E.J.P. (1998) *Guidelines for the Siting, Establishment and Management of Arable Field Margins, Beetle Banks, Cereal Conservation Headlands and Wildlife Seed Mixtures*. Institute of Arable Crops Research report to MAFF.
Boatman N.D., Bence S. & Jarvis P. (1999) Management and costs of conservation headlands on heavy soil. In N.D. Boatman, D.H.K. Davies, K. Chaney, R. Feber, G.R. de Snoo, & T.H. Sparks (eds.) *Field Margins and Buffer Zones: Ecology, management and policy*. Aspects of Applied Biology, 54. pp. 147–154.
Critchley C.N.R. (2000a) Ecological assessment of plant communities by reference to species traits and habitat preferences. *Biodiversity and Conservation*, 9, 87–105.
Critchley C.N.R. (2000b) The conservation ecology of arable plants: what role for research? In P. Wilson & M. King (eds.) *Fields of Vision: A future for Britain's arable plants*. RSPB, Bedfordshire. pp. 80–87.
Ford S. (2000) Can arable fields be managed specifically for arable plant communities? In P. Wilson & M. King (eds.) *Fields of Vision: A future for Britain's arable plants*. RSPB, Bedfordshire. pp. 57–60.
Wilson P.J. (2000) Management for the conservation of arable plant communities. In P. Wilson & M. King (eds.) *Fields of Vision: A future for Britain's arable plants*. RSPB, Bedfordshire. pp. 38–47.
Marshall E.J.P. (2002) Weeds and biodiversity. In R.E.L. Naylor (ed.) *Weed Management Handbook, 9th Edition*. Blackwell Publishing, Oxford. pp. 75–92.
Wilson P.J. & King M. (2003). *Arable Plants – A Field Guide*. English Nature/Wildguides Ltd.

Carvell C., Meek W.R., Pywell R.F., & Nowakowski M. (2004) The response of foraging bumblebees to successional change in newly created arable field margins. *Biological Conservation*, 118, 327–339.

Critchley C.N.R., Fowbert J.A. & Sherwood A.J. (2004) *Botanical Assessment of the Arable Stewardship Pilot Scheme, 2003*. ADAS report to Defra.

Pywell R., Tallowin J. & Masters G. (2004) *Effects of Grazing Management on Creeping Thistle and Other Injurious Weeds and Integration of Grazing with Weed Control*. Defra BD1437.

Critchley C.N.R., Fowbert J.A. & Sherwood A.J. (2005) *Re-assessment of Uncropped Wildlife Strips in Breckland Environmentally Sensitive Area*. ADAS report to Defra.

Critchley C.N.R., Fowbert J.A. & Sherwood A.J. (2006) The effects of annual cultivation on plant community composition of uncropped arable field boundary strips. *Agriculture, Ecosystems and Environment*, 113, 196–205.

Boatman N.B. (2007) Potential effects of Environmental Stewardship on arable weeds: uptake of options relevant to conservation of the arable flora and weed control issues. *44th British Crop Protection Council Annual Weed Review, 2007*. British Crop Protection Council, Reading.

Corsie C. (2007) *Worcestershire Wildlife Trust Important Arable Areas Project*. Report to Plantlife International. Worcestershire Wildlife Trust, Worcester.

Meek W.R., Pywell R.F., Nowakowski M. & Sparks T.H. (2007) Arable field margin management techniques to enhance biodiversity and control barren brome, *Anisantha sterilis*. In C. Britt, A. Cherrill, M. le Duc, R. Marrs, R. Pywell, T. Sparks, I. Willoughby (eds.) *Vegetation Management*, Aspects of Applied Biology, 82. pp. 133–141.

Still K. & Byfield A. (2007) *New Priorities for Arable Plant Conservation*. Plantlife, Salisbury.

Westbury D.B., Woodcock B.A., Harris S.J., Brown V.K. & Potts S.G. (2008) The effects of seed mix and management on the abundance of desirable and pernicious unsown species in arable buffer strip communities. *Weed Research*, 48, 113–123.

10.5   Control grey squirrels

- We have captured no evidence for the effects of controlling grey squirrels on farmland wildlife.

Background

This intervention involves controlling grey squirrels *Sciurus carolinensis* to benefit farmland wildlife. Grey squirrels are naturally found in North America but have been introduced to the UK and northern Italy (Linzey *et al.* 2008). Grey squirrels may cause damage to woodlands through bark stripping and in the UK they compete with the native red squirrel *S. vulgaris* for food and habitat and also carry the squirrel pox virus (Natural England 2011).

Linzey A.V., Koprowski J. & NatureServe (Hammerson G.) (2008) *Sciurus carolinensis*. IUCN Red List of Threatened Species. Version 2012.2. Available at www.iucnredlist.org/details/42462/0. Accessed 31 January 2013.

Natural England (2011) *Urban Grey Squirrels*. Natural England Technical Information Note TIN056.

10.6   Control mink

- A systematic review[1] found seven studies demonstrating that trapping appears to be an effective method of reducing American mink populations, but firm conclusions could not be made due to limitations in experimental design.

- A large-scale trapping programme in the UK[2] demonstrated that American mink have been successfully eradicated over a large area and this may have been associated with some localized water vole expansions.

Background

The UK water vole *Arvicola amphibius* population has declined by 96% since 1950, which is largely thought to be due to predation by invasive non-native American mink *Mustela vison*. Mink also prey on fish, birds and other mammals and can have a significant effect on local wildlife (Defra 2005). This intervention may involve controlling mink populations through trapping.

Defra (2005) *Mink*. Defra Rural Development Service. Rural Development Service Technical Advice Note 02.

A 2005 systematic review investigating the effectiveness of trapping in reducing or eradicating American mink *Mustela vison* populations in all habitats (1) found evidence from seven studies demonstrating that mink populations decreased but that firm conclusions could not be made because of experimental design limitations. Due to the lack of controls for comparison, decreases could not be attributed solely to trapping. There was no robust investigation into other factors that could also be acting upon the populations at the same time. A lack of available data meant that statistical analyses could not be performed.

A large-scale systematic trapping programme in an area of moorland, livestock farms and forestry centred on the Cairngorms National Park, UK (2) found that American mink *Mustela vison* were eradicated over a large area, conserving upland populations of water vole *Arvicola amphibius*. No mink were captured in 2006 when most traps were in catchment headwaters in close proximity to water vole colonies. However, capture rate increased rapidly as traps were added downstream (below 300 m). By December 2009 376 mink had been caught (47% female) and an area of 10,000 km^2 appeared to be free of breeding mink. There was some evidence of localized water vole expansions, but recolonization of the lowlands was expected to be slow. Capture rate increased with connectivity to mink in other sub-catchments and was highest from July–December. Mink rafts were used at 2 km intervals in each sub-catchment and were systematically moved downstream from the headwaters of the 5 main river catchments. Rafts were also retained upstream to remove immigrants. Once mink footprints were recorded on a raft, a trap was set. The project involved 186 local volunteers, including gamekeepers, conservation professionals, residents and land managers.

(1) Tyler C., Clark E. & Pullin A.S. (2005) Do management interventions effectively reduce or eradicate populations of the American Mink, *Mustela vison*? Systematic Review No. 7. *Centre for Evidence-Based Conservation*.
(2) Bryce R., Oliver M.K., Davies L., Gray H., Urquhart J. & Lambin X. (2011) Turning back the tide of American mink invasion at an unprecedented scale through community participation and adaptive management. *Biological Conservation*, 144, 575–583.

10.7 Control predatory mammals and birds (foxes, crows, stoats and weasels)

- A total of nine individual studies from France and the UK (including five replicated controlled studies and a systematic review) looked at the effects of removing predators on birds. Three studies found that controlling predatory mammals or birds (sometimes alongside other interventions) increased the abundance or population size of some birds[1, 6, 9, 10]. One of these studies from the UK found numbers of nationally declining songbirds increased on a site where predators were controlled, but there was no overall difference in bird abundance, species richness or diversity between predator control and no-control sites[6].

- Five studies (including two replicated and controlled, and two before-and-after trials) from the UK[2, 3, 5, 7, 8] found some evidence for increased productivity, nest or reproductive success or survival of birds following bird or mammal predator control (sometimes alongside other interventions). A randomized, replicated, controlled study found hen harrier breeding success was no different between areas with and without hooded crow removal[4].

- A global systematic review[11] including evidence from European farmland found that reproductive success of birds increased with predator removal.

Background

This intervention involves controlling predatory mammals and birds, which may provide benefits to birds nesting on farmland. Predators commonly controlled include carrion crow *Corvus corone* and red fox *Vulpes vulpes*.
See also 'Control mink' for studies looking at the results of controlling non-native American mink *Mustela vison*.

A replicated, controlled study in the spring of 1974 on a cereal farm in France (1) found that grey partridge *Perdix perdix* were significantly more abundant in areas provided with 'partridge cafeterias', which included stoat *Mustela erminea* box-traps, than in control areas. A total of 48 pairs (1 pair/4.7 ha) and 4 single birds were recorded in the southern section of the farm (224 ha), where 27 partridge cafeterias had been constructed. The northern section (200 ha) with no cafeterias had 24 pairs (1 pair/8.3 ha). As well as the stoat traps, cafeterias comprised a barrel with a feed mixture (grain and weed seeds), mouse *Mus* spp. traps, a mini-midden to provide maggots and insects and a sand bath, sheltered by a leaning roof that collected rainwater in a drinking trough.

A replicated, controlled study at two farmland and woodland sites in southern England between 1985 and 1990 (2) found that grey partridge *Perdix perdix* breeding success and brood sizes were significantly higher when predators were controlled compared to years without removal. This led to August partridge numbers being 75% higher and breeding numbers the next year being 36% higher. Over 3 years this led to breeding densities being 2.6 times greater when predators were removed. Predators removed through trapping and shooting were predominantly red foxes *Vulpes vulpes*, carrion crows *Corvus corone* and black-billed magpies *Pica pica*.

A study at three farmland sites in central England in 1992–1998 (3) (partly the same study as (6) and extended in (9)) found that nest survival rates of four songbird species were negatively related to the breeding density of carrion crows *Corvus corone* following the control of nest predators. These species were Eurasian blackbird *Turdus merula*, song thrush *T. philomelos*, dunnock *Prunella modularis* and yellowhammer *Emberiza citrinella*. Non-significant negative relationships were also found for whitethroat *Sylvia communis* and chaffinch *Fringilla coelebs* nesting success and predator densities. Brown rats *Rattus norvegicus*, red foxes *Vulpes vulpes*, stoats *Mustela erminea*, weasels *M. nivalis*, carrion crows and black-billed magpies *Pica pica* were controlled through trapping and shooting. Between 151 and 951 nests of each species were studied.

A randomized, replicated, controlled study in 1999–2000 on Orkney Mainland, Scotland (4) found that the breeding success of hen harriers *Circus cyaneus* was no different in nine territories where hooded crows *Corvus (Corone) cornix* were removed compared to territories without crow removal. The number of clutches/male, clutch size, hatching success and laying date were not affected, although experiments with artificial nests containing chicken eggs showed that predation had been reduced by crow removal (12 of 18 clutches surviving vs 2 of 18). A total of 113 crows were removed from the 9 territories.

A before-and-after study between 1996 and 1998 at a farmland site in eastern England (5) found that daily survival rates of Eurasian skylark *Alauda arvensis* nests in non-rotational set-aside were significantly higher (96% daily survival for 168 nests) following the introduction of intensive control of mammalian predators than when predator control was either 'light' (95.6% survival for 51 nests) or absent (91% survival for 192 nests). There was no significant difference between light control and no control. These differences resulted in average overall survival rates of 41%, 23% and 12% for heavy, light and no control, respectively. The main species targeted were mustelids (Mustelidae), hedgehogs *Erinaceus europaeus* and red foxes *Vulpes vulpes*.

A small replicated, controlled study from May–June 1992–1998 in Leicestershire, England (6) (partly the same study as (3) and extended in (9)) found that the abundance of nationally declining songbird species and species of conservation concern significantly increased over time on a 3 km^2 site where predators were controlled. However there was no overall difference in bird abundance, species richness or diversity between the experimental and control sites. Numbers of nationally declining species rose by 102% (except for Eurasian skylark *Alauda arvensis* and yellowhammer *Emberiza citrinella*). Nationally stable species rose (insignificantly) by 47% (8 species increased; 4 decreased). The other interventions employed at the same site were: managing hedges, wild bird cover strips, beetle banks, supplementary feeding and reducing chemical inputs generally.

A before-and-after study on a mixed farmland-woodland site in central England (7) found that the fledging success of spotted flycatcher *Muscicapa striata* nests was significantly higher when predators (grey squirrels *Sciurus carolinensis*, brown rats *Rattus norvegicus*, red foxes *Vulpes vulpes*, black-billed magpies *Pica pica* and carrion crows *Corvus corone*) were controlled (77% for 11 nests in 1997–2001) than when there was no control (16% for 28 nests in 2002–2004).

A replicated, controlled trial at 13 lowland wet grassland sites in England and Wales between 1996 and 2003 (8) found no overall increase in the success of 3,139 northern lapwing *Vanellus vanellus* nests during 4 years with predator control compared to 4 years without. However, when differences in initial predator densities were accounted for, control did improve survival, having a greater impact at sites with higher predator densities. At 2 sites where predators were controlled for all 8 years nesting success was not significantly different from the 11 other sites. Predators were red fox *Vulpes vulpes* and carrion crow *Corvus corone*, with average declines of 40% for foxes and 56% for crows.

A before-and-after study on a mixed farm in central England (9) between 1992 and 2007 (a continuation of the data series used in (3, 6)), found that controlling predator populations (carrion crow *Corvus corone*, black-billed magpie *Pica pica*, red fox *Vulpes vulpes* and other mammals) appeared to increase blackbird *Turdus merula* breeding populations. However, the authors caution that the study was not experimental and that other explanations for the trends seen could not be eliminated.

A controlled study in 2002–2009 on mixed farmland in Hertfordshire, England (10) found that the number of grey partridge *Perdix perdix* increased significantly on an experimental site, where predators were controlled (along with several other interventions), but only slightly on a control site without predator control. This increase was apparent in spring (from fewer than 3 pairs/km^2 in 2002 to 12 in 2009, with a high of 18 pairs/km^2 on the experimental site, compared to approximately 1 pair/km^2 on the control site in 2002, increasing to approximately 4 in 2009) and autumn (from fewer than 10 birds/km^2 in 2002 to approximately 65 in 2009, with a high of 85 birds/km^2, compared to approximately 4 birds/km^2 on the control site in 2002, increasing to approximately 15 birds/km² in 2009). Predators controlled were red fox *Vulpes vulpes*, stoat *Mustela erminea*, brown rat *Rattus norvegicus*, carrion crow *Corvus corone* and black-billed magpie *Pica pica*. The experimental site also had supplementary food provided and habitat creation.

A 2010 global systematic review covering habitats including European farmland (11) found that removing predators tended to lead to increased reproductive (hatching and fledging) success and breeding populations in birds. On mainlands, but not islands, predator removal also tended to increase post-breeding population size. Whether predators were native or not, the population trend of the bird population and whether the species was migratory or a game species did not affect responses to predator removal.

(1) Westerskov K.E. (1977) Covey-oriented partridge management in France. *Biological Conservation*, 11, 185–191.
(2) Tapper S.C., Potts G.R. & Brockless M.H. (1996) The effect of an experimental reduction in predation pressure on the breeding success and population density of grey partridges *Perdix perdix*. *Journal of Applied Ecology*, 33, 965–978.
(3) Stoate C. & Szczur J. (2001) Could game management have a role in the conservation of farmland passerines? A case study from a Leicestershire farm. *Bird Study*, 48, 279–292.
(4) Amar A. & Redpath S.M. (2002) Determining the cause of the hen harrier decline on the Orkney Islands: an experimental test of two hypotheses. *Animal Conservation*, 5, 21–28.
(5) Donald P.F., Evans A.D., Muirhead L.B., Buckingham D.L., Kirby W.B. & Schmitt S.I.A. (2002) Survival rates, causes of failure and productivity of skylark *Alauda arvensis* nests on lowland farmland. *Ibis*, 144, 652–664.
(6) Stoate C. (2002) Multifunctional use of a natural resource on farmland: wild pheasant (*Phasianus colchicus*) management and the conservation of farmland passerines. *Biodiversity and Conservation*, 11, 561–573.
(7) Stoate C. & Szczur J. (2006) Potential influence of habitat and predation on local breeding success and population in spotted flycatchers *Muscicapa striata*. *Bird Study*, 53, 328–330.
(8) Bolton M., Tyler G., Smith K. & Bamford R. (2007) The impact of predator control on lapwing *Vanellus vanellus* breeding success on wet grassland nature reserves. *Journal of Applied Ecology*, 44, 534–544.
(9) White P.J.C., Stoate C., Szczur J. & Norris K. (2008) Investigating the effects of predator removal and habitat management on nest success and breeding population size of a farmland passerine: a case study. *Ibis*, 150, 178–190.
(10) Aebischer N.J. & Ewald J.A. (2010) Grey Partridge *Perdix perdix* in the UK: recovery status, set-aside and shooting. *Ibis*, 152, 530–542.
(11) Smith R.K., Pullin A.S., Stewart G.B. & Sutherland W.J. (2010) Effectiveness of predator removal for enhancing bird populations. *Conservation Biology*, 24, 820–829.

10.8   Protect individual nests of ground-nesting birds

• Two replicated, randomized, controlled studies from Sweden found that providing nest exclosures offered some benefits to ground-nesting birds. One study found that protected nests had higher average daily survival rates than unprotected nests for both common redshank and northern lapwing, however, this study also reported higher predation of adult redshank on protected nests[1]. One study found that the average hatching rate for southern dunlin was higher for protected rather than unprotected nests[2]. This study also found no difference in the number of fledglings, breeding adults or new recruits during two periods with and without nest protection[2].

Background

If fencing does not work to exclude predators (for example, predatory birds) or is not a viable option, it may be possible to protect individual nests using a variety of cages and exclosures. These must be able to allow chicks and adults to get in and out, but not predators, and should be quick to install to minimize the chances of parents abandoning nests.

A replicated, randomized, controlled trial in 2002 and 2004 at 3 grazed pasture sites in south-west Sweden (1) found that nests protected with cages (truncated cone steel cages with 6.5–8.5 cm spacings between vertical bars and 4 x 4 cm steel netting on top) had significantly higher average daily survival rates than unprotected nests for both common redshank *Tringa totanus* (99.7% for 34 protected nests vs 96% for 32 unprotected nests in 2002) and northern lapwing *Vanellus vanellus* (99% for 37 protected nests vs 97% for 153 unprotected nests in 2002 and 2004). However, there was higher predation of adult redshank on protected nests and possibly higher abandonment by lapwings (9 redshank adults from 8 protected nests were predated vs a single bird from 31 unprotected nests).

A replicated, controlled, before-and-after study from 1999 to 2004 on pastures in southwest Sweden (2) found that the average hatching rate of southern dunlin *Calidris alpina schinzii* nests was significantly higher for nests protected by steel cages (20 cm high truncated cones with 7.5 cm gaps between vertical bars and 4 x 4 cm steel mesh covering the top) than for unprotected nests (67% of 25 protected nests survived to hatching vs 41% of 61 unprotected nests). Protected nests were also more likely to hatch more than 1 chick (80% of 25 protected nests vs 57% of 60 unprotected nests). Predation rates on brooding adults were unaffected (7% of 57 adults at protected nests predated vs 13% of 16 adults at unprotected nests). However, comparing 1993–1998 (when no nests were protected) with 1999–2004 (when some nests were protected) revealed that there was no significant change in either the number of fledglings/breeding adults or the number of new recruits/breeding adults produced by the study sites.

(1) Isaksson D., Wallander J. & Larsson M. (2007) Managing predation on ground-nesting birds: the effectiveness of nest exclosures. *Biological Conservation*, 136, 136–142.
(2) Pauliny A., Larsson M. & Bloqvist D. (2008) Nest predation management: effects on reproductive success in endangered shorebirds. *Journal of Wildlife Management*, 72, 1579–1583.

10.9   Erect predator-proof fencing around important breeding sites for waders

• We have captured no evidence for the effects of erecting predator-proof fencing around important sites for waders on farmland wildlife.

Background
As well as direct predation on adults, predators can have a devastating impact on bird populations through predating eggs and chicks too young to defend themselves or run away. Species such as hedgehogs *Erinaceus europaeus* or other birds can all affect bird populations in this way and in many cases it is not desirable or practical to remove these species.
This intervention involves erecting predator-proof fencing around important breeding sites for waders to exclude predators.

10.10   Remove coarse fish

• We have captured no evidence for the effects of removing coarse fish on farmland wildlife.

Background

This intervention involves removing coarse fish from waterbodies on farmland. Coarse fish include species such as common carp *Cyprinus carpio* which are typically stocked at high levels for recreational fishing. These species may have a negative impact on aquatic wildlife through competition for resources, predation and damage to habitats. Removal of coarse fish may benefit wildlife including waterbirds by reducing competition (Phillips 1992).

Phillips V.E. (1992) Variation in winter wildfowl numbers on gravel pit lakes at Great Linford, Buckinghamshire, 1974–79 and 1984–91, with particular reference to the effects of fish removal. *Bird Study*, 39, 177–185.

10.11   Manage wild deer numbers

- We have captured no evidence for the effects of managing wild deer numbers on farmland wildlife.

Background

This intervention involves controlling numbers of wild deer on farmland. Wild deer populations can cause damage to woodlands, crops and other wildlife habitats through browsing and bark stripping. The introduction of non-native deer species to the UK and a lack of predators to naturally control deer populations means that management of deer numbers may be required to prevent damage to farmland habitats (Mayle 1999).

Mayle B. (1999) *Managing Deer in the Countryside – Practice Note July 1999*. Forestry Commission, Edinburgh, UK.

10.12   Provide medicated grit for grouse

- A controlled study in England found that red grouse had higher productivity in areas where medicated grit was provided[1].

Background

Grouse commonly eat quartz grit to aid digestion of heather (Newborn & Foster 2002). This intervention involves using grit treated with medication to administer drugs, such as those used to treat parasitic worm infections, to grouse populations.

Newborn D. & Foster R. (2002) Control of parasite burdens in red grouse *Lagopus lagopus scoticus* through the indirect application of anthelmintics. *Journal of Applied Ecology*, 39, 909–914.

A controlled experiment during 1996–2000 on two moors in northern England (1) found that red grouse *Lagopus lagopus scoticus* in an area provided with quartz grit treated with anthelmintic drugs (drugs to treat parasitic worm infections) raised between 38% and 100% more chicks than grouse in a control area (treatment areas: 4.9–7.1 chicks/hen estimated from 36 radio-tagged birds and 4.9–6.7 chicks/hen estimated from 125 birds seen on counts using pointing dogs vs control areas: 1.9–4.8 chick/hen from 36 tagged birds and 2.8–4.5 chicks/hen from 117 on dog counts) and had significantly lower levels of infection of the parasitic nematode *Trichostrongylus tenuis* (34% fewer worms over 5 years). This was despite the fact that the medicated areas did not have larger broods or higher hatching success (medicated areas: 9.6 eggs/clutch and 90% hatching success for 161 clutches; control areas: 9.4 eggs/clutch and 94% hatching rate for 153 clutches). Survival rates of adults did not vary between medicated and control areas.

(1) Newborn D. & Foster R. (2002) Control of parasite burdens in red grouse *Lagopus lagopus scoticus* through the indirect application of anthelmintics. *Journal of Applied Ecology*, 39, 909–914.

11 Education and awareness

Key messages

Provide training for land managers, farmers and farm advisers
A study from the UK found farmers who were trained in how to implement agri-environment schemes created better quality wildlife habitat over five years.

Provide specialist advice and/or assistance preparing conservation plans
We have captured no evidence for the effects of providing specialist advice and/or assistance in preparing conservation plans on farmland wildlife.

11.1 Provide training for land managers, farmers and farm advisers

- One study from the UK found farmers who were trained in how to implement agri-environment schemes created better quality wildlife habitat over five years[1].

Background
This refers to training events on general or specialized aspects of wildlife conservation on farms. It is a way of building capacity and knowledge amongst farmers and land managers.

A recently completed project under the UK's Rural Economy and Land Use (RELU) programme showed that farmers who were trained in how to implement agri-environment schemes created better quality wildlife habitat, in terms of flower resources for bees (Apidae) and seeds for birds, over five years. This was reflected in local increases in target species of bird and bee. The results from this project are not yet published but briefly described in RELU Policy and Practice note number 37 (1). Details of the experimental design are not given.

(1) Rural Economy and Land Use Programme (2012) *Improving the Success of Agri-environment Initiatives.* RELU Policy and Practice Note, number 37.

11.2 Provide specialist advice and/or assistance preparing conservation plans

- We have captured no evidence for the effects of providing specialist advice and/or assistance in preparing conservation plans on farmland wildlife.

Background
This is when external consultants or organizations provide services to help farmers and land managers implement conservation measures.

Index

uncultivated margins 118, 119, 122, 123, 124

undersowing spring cereals 204, 205

Bombus terrestris (buff-tailed bumblebee) 10, 85, 93, 94, 261

Bombus terrestris/lucorum 85, 261

Borago officinalis
flower strips 79, 85, 88, 93, 94, 100, 105
planting wild bird cover 65, 67, 70

bow-winged grasshopper *see Chorthippus biguttulus*

Brachycera 330

bracken 419, 421

Braconidae
flower strips 78, 80
organic fertilizers 382, 385
reducing agri-chemical use 398

bramble *see Rubus fruticosus*

Branta bernicla bernicla (dark-bellied brent geese) 405

Branta bernicla (brent geese) 405

Branta leucopsis (barnacle geese) 202, 404

Brassica carinata 79

Brassica juncea (mustard) 67, 71, 105

Brassica napa (winter rape) 75

Brassica napus (oilseed rape) 38, 62, 64, 111, 223, 378

Brassica oleracae viridus 64, 65, 71

Brassica oleracea (cabbage) 103, 106, 236

Brassica spp.
flower strips 106
planting wild bird cover 60, 62, 63, 64, 67, 70, 71

brent geese *see Branta bernicla*

Bristol, United Kingdom 86, 117, 131, 264

Briza media (quaking grass) 261

broad-fruited cornsalad *see Valerianella rimosa*

Broadland, United Kingdom 417

broad-leaved cudweed *see Filago pyramidata*

broad-leaved dock *see Rumex obtusifolius*

Bromus erectus (meadow brome) 295, 410

Bromus hordeaceus 257

Bromus inermis 273

Bromus interruptus 234, 295

Bromus spp. 258

Bromus sterilis (sterile brome) 216, 354, 383

brown argus *see Aricia agestis*

brown hare *see Lepus europaeus*

brown rat 430

bryophytes 409

Bryum rubens 421

buckwheat *see Fagopyron esculentum*

buff-tailed bumblebee *see Bombus terrestris*

bulbous buttercup *see Ranunculus bulbosus*

bumblebees *see Bombus* spp.

buntings *see* Emberizidae

Burhinus oedicnemus (Eurasian thick-knee)
agri-environment schemes 9, 12, 13, 19
cultivated, uncropped margins 195, 197
permanent grassland 305
set-aside areas 32
uncultivated margins 117

burning 409–410

bush-crickets *see* Tettigoniidae

butterflies *see* Lepidoptera

cabbage moth *see Mamestra brassicae*

cabbage root fly *see Delia radicum*

cabbage weevils *see Ceutorhynchus* spp.

caddisflies 416

Calathus fuscipes 236

Calendula officinalis (dwarf marigold) 79, 85, 88

Calidris alpina schinzii (southern dunlin) 433

Calluna spp. 42

Calluna vulgaris (ling heather)
burning 410
excluding livestock from semi-natural habitat 341
hedge planting 56
semi-natural habitat for rough grazing 337
species-rich, semi-natural grassland 253
swiping 409
traditional breeds of livestock 335
upland heath/moorland 329, 332
weed control in conservation areas 423

Calluno-Genistion pilosae communities 322

Caltha palustris (marsh marigold) 321

Cambridgeshire, United Kingdom
cultivated, uncropped margins 196
flower strips 82
grass buffer strips 128–129, 143
hedgerow management 45, 46

Lightning Source UK Ltd.
Milton Keynes UK
UKHW020035220121
377440UK00002B/40